Lecture Notes in Physics

Edited by J. Ehlers, München, K. Hepp, Zürich
...enhahn, München, H. A. Weidenmüller, Heidelberg
...Zittartz, Köln
...ing Editor: W. Beiglböck, Heidelberg

126

Complex Analysis, Microlocal Calculus and Relativistic Quantum Theory

Proceedings of the Colloquium
Held at Les Houches, Centre de Physique
September 1979

Edited by D. Iagolnitzer

Springer-Verlag
Berlin Heidelberg New York 1980

Editor

Daniel Iagolnitzer
Service de Physique Théorique, CEN Saclay
BP 2, 91190 Gif-sur-Yvette/Frace

ISBN 3-540-09996-4 Springer-Verlag Berlin Heidelberg New York
ISBN 0-387-09996-4 Springer-Verlag New York Heidelberg Berlin

Printed in Germany

Printing and binding: Beltz Offsetdruck, Hemsbach/Bergstr.
2153/3140-543210

INTERNATIONAL COLLOQUIUM ON COMPLEX ANALYSIS, MICROLOCAL CALCULUS AND RELATIVISTIC QUANTUM THEORY

Les Houches, Sept.3-13, 1979

Organization Committee : J.M. BONY (Maths, Orsay), J. BROS (Phys. Théor., Saclay) D. IAGOLNITZER* (Phys. Théor., Saclay), J. LASCOUX (Phys. Théor., Polytechnique), B. MALGRANGE (Maths, Grenoble), F. PHAM (Maths, Nice), P. SCHAPIRA (Maths, Paris Nord)

*Main organizer

Participants

ABRAHAM, D., Oxford, England
ANDRONIKOV, E., Paris-Nord
BALABANE, M., Paris-Nord
BARLET, D., Nancy
BENGEL, G., Münster, Germany
BERG, B., Berlin, Germany
BERGERE, M., Saclay
BILLIONNET, C., Polytechnique, Paris
BONY, J.M., Orsay
BOUTET DE MONVEL, L., ENS-Paris
BROS, J., Saclay
CANDELPERGHER, A., Nice
CHAZARAIN, J., Nice
CHUDNOVSKY, D., Saclay
CHUDNOVSKY, G., Columbia Univ.USA
CLERC, J.L., Nancy
COLIN DE VERDIERE, Y., Grenoble
DOUADY, A., Orsay
EPSTEIN, M., I.H.E.S. Paris
ERMINE, J.L., Bordeaux
FAVILLI, F., Pisa, Italy
GAY, R., Bordeaux
GERARD, R., Strasbourg
GLASER, V., CERN, Geneva, Switzerland
GRIGIS, L., Orsay
GRUBB, G., Copenhagen, Denmark
HATORI, T., Tokyo, Japan
HONERKAMP, J. Freiburg, Germany
HOUZEL, C., Paris-Nord
IAGOLNITZER, D., Saclay
ION, P.D.F., Heidelberg, Germany
INTISSAR, R., Maroc
JAEKEL, M.T., ENS-Paris
JIMBO, M., Kyoto, Japan
KANTOR, J.M., Paris VII
KAROWSKI, M., Berlin, Germany
KASHIWARA, M., Kyoto, Japan
KATZ, A., Polytechnique, Paris
KAWAÏ, T., Kyoto, Japan
LABESSE, J.P., Dijon
LASCOUX, J.L., Polytechnique, Paris
LOKAMCHAN, L., Nice
LODAY, M., Strasbourg
LAURENT, Y., Orsay
LIEUTENANT, J.L., Liège
MAILLARD, J.M., ENS-Paris
MAISONOBE, L., Nice
MALGRANGE, B., Grenoble
MARES RUBEN, G., Polytechnico, Mexico
MEBKHOUT, Z., Orléans
MIWA, T., Kyoto, Japan
MONTEIRO, T., Paris-Nord
MORIMOTO, M., Tokyo, Japan
MORI, Y., Okinawa, Japan
MULASE, M., Kyoto, Japan
PESANTI, D., Orsay
PHAM, F., Nice
RADULESCU, N., Nice
RAMIS, J.P., Strasbourg
ROMBALDI, J., Nice
ROOS, A., France
SABBAH, C., Paris
SATO, M., Kyoto, Japan
SCARPALEZOS, D., Paris
SCHAPIRA, P., Paris-Nord
SEBBAR, A., Bordeaux
SJOSTRAND, J., Orsay
SOMMER, G., Bielefeld, Germany
STORA, R., CERN, Geneva, Switzerland
TALEB, S., Paris
TATARU-MIHAI, P., Max-Planck, Germany
TREPEAU, J.M., Reims
UNTERBERGER, A., Reims
UNTERGERGER, J., Reims
VAN DEN ESSEN, A., Nijmegen, Netherlands
VERDIER, J.L., ENS-Paris
VOROS, A., Saclay
XIA DAOXING, Shanghai, China

Organization Committee of Centre de Physique des Houches :

M.T. BEAL-MONOD (Phys. Solides, Orsay) D. THOULOUZE (CNRS, Grenoble)

FOREWORD

This volume presents the Proceedings of the Colloquium on "Complex Analysis, Microlocal Calculus and Relativistic Quantum Theory", held at the Centre de Physique des Houches in September 1979.

This Colloquium originated in the contacts developed during the seventies between two groups of (French and Japanese) mathematicians and a group of theoretical physicists who, with different motivations and approaches, were led to related, or even common, problems in the study of the singularity structure of functions (or distributions, hyperfunctions, microfunctions,...) of interest either in mathematics, in complex analysis and differential and microdifferential calculus, or in physics, in relativistic quantum theory. These contacts, including in particular previous meetings organized by F. Pham in Nice in 1973 and by M. Sato and collaborators in Kyoto in 1976, had proved to be useful. The present Colloquium, which was extended to related domains of common interest, has allowed the presentation of the important new developments and the exchanges that had appeared to be desirable or needed. Let us note, in this connection, that the separation between mathematicians and physicists is not always very neat: as will appear in these Proceedings, several of them in both groups have been led recently to contributions in either domain, the differences appearing mainly in the emphasis and in the main motivations and character of the various contributions.

The topics treated have been classified in four parts. The two first parts are mainly mathematical, while the last two are more oriented towards physics. Some indications on the connections between various parts are given later.

Part I presents the recent developments of microfunction theory and of the microlocal, or microdifferential, calculus (holonomic systems, second microlocalization) and related topics (essential support theory, ...). Other miscellaneous mathematical developments, on singularities of solutions of partial differential equations, pseudodifferential operators and generalizations, spectrum of operators, asymptotic expansions, monodromy, ... will be found in Part II.

Part III is mainly devoted to the rigorous study of the general analytic and microanalytic structure of Green functions and of the S-matrix (i.e., of collision amplitudes), in axiomatic quantum field theory and in S-matrix theory, a domain in which appreciable progress has been made recently for multiparticle processes. Recent developments in the related study of Feynman integrals, and a few other topics, are also included.

Finally, an important part of these Proceedings (Part IV) is devoted to the explicit determination of the S-matrix and, in some cases, of Green functions for various models of field theory in two space-time dimensions, and to related physical and mathematical developments with emphasis on those aspects that have, or should prove to have, a general character and give hopes of further developments. Approaches based on general principles and applying to theories "with soliton behaviour" are first presented (Sect.A). The more direct approach developed recently for quantization and solution of completely integrable systems is then introduced in Sect.B, where a general analysis of such systems, in connection with the isospectral deformation, is presented on the other hand. Finally, the recent developments on holonomic quantum fields, in connection with the isomonodromic deformation, and the corresponding solution of the Ising and other models by Sato-Miwa-Jimbo is presented in Sect.C. Complements on the Ising model are also included.

While each part has its own unity, we emphasize however that the above division is to some extent arbitrary, both because the distinction between physics and mathematics is not always very neat and in view of the connections that will appear between the various parts. For instance, some texts of Part I (Iagolnitzer, Kashiwara-Kawaï, Van Den Essen) are directly relevant to the study of the S-matrix and of Feynman integrals. As another example, the study of the general structure of the S-matrix in Part III gives insight into some aspects of the two-dimensional models considered in Part IV. (Note, however, that possible extensions of specific aspects of these models to more dimensions seem to lead to structures that differ from the usual ones). Finally, a number of connections, not described here in detail, will appear between various texts of Parts I,II and IV.

The Proceedings include short contributions, which are a summary or an introduction to more complete works published elsewhere, and longer contributions, which either present original work or are review works (with some original aspects) in domains in which there was a need for an up-to-date and clear presentation of recent developments.

On behalf of the Organization Committee, I would like to thank the directors of the Centre de Physique des Houches, and in particular Mrs. M.T. Beal-Monod, for their efficient help. I also wish to thank all participants and lecturers who contributed to the success of this Colloquium, and more particularly here all lecturers who, by their efforts in the preparation of their manuscripts, will contribute to the usefulness of these Proceedings. I am finally pleased to thank Mrs. E. Cotteverte for her efficient help in the final preparation of the manuscript.

D. IAGOLNITZER

CONTENTS

The list of invited lectures given during the Colloquium whose manuscripts have not been received is given at the end.

PART I - MICROFUNCTIONS, MICROLOCAL CALCULUS AND RELATED TOPICS

PART II - MISCELLANEOUS MATHEMATICAL DEVELOPMENTS

PART III - GENERAL STRUCTURE OF GREEN FUNCTIONS AND COLLISION AMPLITUDES

PART IV - TWO-DIMENSIONAL MODELS AND RELATED DEVELOPMENTS

Other Lectures Delivered During the Colloquium

L. BOUTET DE MONVEL - Quantized Contact Manifolds

Y. COLIN DE VERDIERE - Spectre Joint d'OPD qui Commutent

H. EPSTEIN, V. GLASER - Unitarity and Analyticity in Axiomatic Quantum Field Theory

R. GERARD - Théorie des Résidus Associés à une Connexion avec Singularités Régulières

G. GRUBB - On the Spectrum of Pseudo-Differential Elliptic Boundary Value Problems

C. HOUZEL - Théorème d'Image Directe pour les Systèmes Micro-Différentiels

F. PHAM - Singularities of Phase Integrals and Picard-Lefschetz Formula

J.L. VERDIER - Equations Différentielles Géométriques

A. VOROS - The Zeta Function of the Quartic Oscillator

PART I

MICROFUNCTIONS, MICROLOCAL CALCULUS AND RELATED TOPICS

ESSENTIAL SUPPORT THEORY AND u=0 THEOREMS

D. IAGOLNITZER

DPh-T, CEN Saclay, BP n°2, 91190 Gif-sur-Yvette, France

The standard results of essential support theory[1], or of hyperfunction theory[2], give no information on the essential support (= singular spectrum)of a product of distributions f',f" at u=0 points X (i.e. points where the essential supports of f' and f" contain opposite directions), even when the product itself is well defined in the neighborhood of X. The same remark applies also to products of bounded operators, such as the products of collision operators encountered in some applications in S-matrix theory (see the lecture of the author in Part III). The u=0 problem encountered there is crucial and has been at the origin of several works. The approach to the u=0 problem developed in the framework of the theory of holonomic microfunctions by Kashiwara-Kawai is presented in [3] and references therein. We present here u=0 results[4] obtained in the framework of essential support theory, on products of square integrable functions that satisfy a general regularity property R at X with respect to the relevant directions of their essential support. Analogous results[4a] also hold for products of bounded operators. Details in both cases will be given in [4c].

The basic facts on the essential support are recalled in Sect.1. Property R is probably linked with the second microlocalization introduced recently in microfunction theory. It is introduced in Sect.2 at the end of which a simple example is given in terms of analyticity properties. The announced u=0 theorem is then presented in Sect.3. As in [3], the result is similar to the standard u≠0 one, except that limiting procedures that may enlarge the essential support have to be introduced. The latter are however different, the respective conditions of application of the results being themselves different in general.

1. ESSENTIAL SUPPORT

Being given a tempered distribution f defined on the space $\mathbb{R}^n$ of n real variables $x=x_1,\ldots,x_n$, the essential support of f is defined, at each point X, as a cone with apex at the origin in the space $\mathbb{R}^n$ of the dual variables $u=u_1,\ldots,u_n$ composed of the "singular directions" along which the generalized, or localized,

Fourier transform of f at X does *not* fall off exponentially in a well specified sense. Namaly, let F_γ be defined, for every $\gamma > 0$, by

$$F_\gamma(u;X) = \int f(x)\, e^{-iu.x-\gamma|u|(x-X)^2}\, dx \qquad (1)$$

or, with an auxiliary real variable v, let :

$$F(u;v,X) = \int f(x)\, e^{-iu.x-v(x-X)^2}\, dx \quad . \qquad (2)$$

Then a direction $\hat{u}_o$ is by definition outside $ES_X(f)$, or $(X,\hat{u}_o) \notin ES(f)$ if there exist a neighbouring cone $V(\hat{u}_o)$ of $\hat{u}_o$ with apex at the origin, $\alpha > 0, \gamma_o > 0$, as also a polynomial P and $q \geq 0$ such that :

$$|F(u;v,X)| < [P(|u|)\, v^{-q}]\, e^{-\alpha v} \qquad (3)$$

in the region $u \in V(\hat{u}_o)$, $0 < v < \gamma_o|u|$.

The important factor in these bounds is the factor $e^{-\alpha v} \equiv e^{-\alpha\gamma|u|}$, which expresses exponential fall off in the direction $\hat{u}_o$ for all $\gamma > 0$ sufficiently small with a rate of fall off at least proportional to γ . (Bounds of the form (3) without this factor are always satisfied). Whereas the exponential fall-off at $\gamma=0$, i.e. of the usual Fourier transform $\tilde{f}$ of f, corresponds to analyticity properties independent of the real point X (see e.g. [1]), the above notion of essential support characterizes by duality the real points where f is analytic or is the boundary value of an analytic function, and more generally possible decompositions of f into sums of boundary values of analytic functions from specified directions (which may depend on X) : see details in [1], where the notion of essential support and the results above are extended to general distributions defined in $\mathbb{R}^n$ or on a real analytic manifold.

The characterization of analyticity properties obtained[2] in hyperfunction theory in terms of the notion of *singular spectrum* is similar to above, except that the boundary values involved in the decompositions of f may à priori be hyperfunctions, even when f itself is a distribution. It is, however, proved in [5] that the two notions do coïncide for distributions (and coïncide with Hormander's "analytic wave front set").

2. REGULARITY PROPERTY R (for square integrable functions)

The regularity property R is a condition on the way rates of exponential fall off of the generalized Fourier transform F_γ tend to zero in certain situations when directions of the essential support are approached. It asserts more precisely that certain uniform bounds are then satisfied (see below).

Being given X and $\hat{u}_o \notin ES(f)$, it follows from results of [1] that α in the bounds (3) can always be chosen arbitrarily close to

$$\mathrm{Max}_{\alpha'>0}\ \{\alpha' ; \hat{u}_o \notin ES_x(f)\ ,\ \forall x \text{ s.t. } (x-X)^2 < \alpha'\}\quad ,$$

(with appropriate choices of $V(\hat{u}_o)$ and $\gamma_o \geq 0$).

Let us then consider a direction $\hat{u}_o$ that now belongs to the boundary of $ES_X(f)$, and moreover to the boundary of $\bigcup_{x \in N} ES_x(f)$, where N is some real neighborhood of X. The above result entails that there exists $\alpha > 0$ (depending only on N) such that bounds of the form (3) be satisfied at X with this common *uniform* α for all directions $\hat{u}$ outside $\bigcup_{x \in N} ES_x(f)$. The choice of $V(\hat{u})$, $\gamma_o > 0$, ... depends on the other hand on $\hat{u}$, and as a matter of fact $\gamma_o(\hat{u})$ necessarily tends to zero when the direction $\hat{u}_o$ is approached, since $\hat{u}_o \in ES_X(f)$. The main content of property R at $(X,\hat{u}_o)$, when it holds, is the condition that $\gamma_o(\hat{u})$ should not tend to zero faster than linearly with respect to the angle of $\hat{u}$ with the boundary of $\bigcup_{x \in N} ES_x(f)$.

In order to introduce the precise statement of property R for square integrable functions, we mention the following results that always hold in this case. First[4a], the bracket $[P(|u|)v^{-q}]$ can always be replaced in the bounds (3) by a square integrable function d_v of u whose norm is independent of v. This function may à priori depend on the direction considered outside $ES_X(f)$. On the other hand, there always exists, as easily seen, a uniform square integrable function d of u such that the bounds $|F(u;v,X)| < d(u)$, without the exponential fall-off factor $e^{-\alpha v}$, be satisfied everywhere, in the whole region $v \geq 0$.

<u>Property R</u> (for square integrable functions)

"Being given X and a direction $\hat{u}_o$ of $\partial(ES_X(f))$, property R is by definition satisfied by f at $(X,\hat{u}_o)$ if, being given any real neighborhood N of X, there exist a neighboring cone $V(\hat{u}_o)$ of $\hat{u}_o$ with apex at the origin, $\alpha > 0$, $\chi > 0$ and a square integrable function d_v of u, whose norm is independent of v, such that :

$$|F(u;v,X)| < d_v(u)\, e^{-\alpha v} \tag{4}$$

in the region $u \in V(\hat{u}_o)$, $0 \leq v < \chi$ dist. $\{u, \bigcup_{x \in N} ES_x(f)\}$" .

<u>Example</u>.

Let f be, in the neighborhood of X, the boundary value of an analytic function $\underline{f}$ from the directions of an open (convex) cone Γ in Imz-space (where z=x+iy is the complexified variable of x), in which case $ES_X(f) \subset C$, where C is the closed dual cone of Γ. The above analyticity property means that there exist a neighborhood N of X and an open set B with profile Γ at the origin in y-space, such that $\underline{f}$ is analytic in $\{z=x+iy : x \in N, y \in B\}$. B has Γ as its profile at the origin if, being given any open cone Γ' with apex at the origin whose closure is contained (apart from the origin) in Γ, there exists $\rho > 0$ such that $\Gamma' \cap \{|y| < \rho\} \subset B$; ρ may, however, shrink to zero in general when Γ' expands to Γ.

Let $\hat{e}$ be a direction of the boundary $\partial\Gamma$ of Γ , if it exists, for which this is not the case, more precisely for which there exist $\rho(\hat{e}) > 0$ and a neighbouring cone $V(\hat{e})$ of $\hat{e}$ with apex at the origin such that the set $\{y; |y|<\rho(\hat{e}), y \in V(\hat{e}) \cap \partial\Gamma\}$ belongs to the closure of B. If, moreover, $\underline{f}$ is bounded in its analyticity domain when this set is approached, then property R is satisfied at $(X,\hat{u}_o)$ for any $\hat{u}_o$ of ∂C such that $\hat{u}_o.\hat{e}=0$. The converse is also essentially true.

3. U=0 THEOREM

Theorem - "Let f', f" be square integrable functions such that property R is satisfied by f' and f" at $(X,\hat{u}_o)$ and $(X,-\hat{u}_o)$ respectively, for any direction $\hat{u}_o$ such that $\hat{u}_o \in ES_X(f')$, $-\hat{u}_o \in ES_X(f'')$. Then :

$$ES_X(f'f'') \subset \{u;\exists\, u'_n,u''_n,\ X'_n,\ X''_n,\ u'_n \in ES_{X'_n}(f'),\ u''_n \in ES_{X''_n}(f''),$$
$$X'_n \to X,\ X''_n \to X,\ u'_n + u''_n \to u \text{ when } n\to\infty\}\ . \qquad (5)''$$

The (easy) proof is based on the formula :

$$F(u;v,X) = \int F'(u';\tfrac{v}{2},X)\, F''(u-u';\tfrac{v}{2},X)\, du' \qquad (6)$$

where f=f'f". The result is a generalization of the particular case obtained when f',f" satisfy the properties of the example of Sect.2 : if $\Gamma' \cap \Gamma''$ is empty (in which case X is a u=0 point), but if $\partial\Gamma' \cap \partial\Gamma''$ contains a direction $\hat{e}$ (or a set of such directions) such that the properties described at the end are satisfied by f' and f", then $ES_X(f)$ is contained in its dual half-space (or in their intersection).

REFERENCES

[1] D. Iagolnitzer, in *Structural Analysis of Collision Amplitudes*, ed. by R. Balian and D. Iagolnitzer, North-Holland, Amsterdam (1976), p.295, and references.

[2] M. Sato, T. Kawai, M. Kashiwara, in *Microfunctions and Pseudo-differential Equations*, Lecture Notes in Mathematics, Springer-Verlag, Heidelberg (1973), p.265.

[3] M. Kashiwara, T. Kawai, first text in this volume.

[4] D. Iagolnitzer, a) Comm. Math. Phys. 63, 49 (1978), b) Exposé 1, Séminaire Goulaouic-Schwartz 1978-79, c) in preparation.

[5] J.M. Bony, in Séminaire Goulaouic-Schwartz 1976-77.

THE THEORY OF HOLONOMIC SYSTEMS WITH REGULAR SINGULARITIES AND ITS RELEVANCE TO PHYSICAL PROBLEMS

Masaki KASHIWARA
Département de Mathématiques, Université de Paris-Sud
91405, Orsay, France
and
Research Institute for Mathematical Sciences,
Kyoto University, Kyoto 606, Japan

AND

Takahiro KAWAI
Research Institute for Mathematical Sciences
Kyoto University, Kyoto 606, Japan

The purpose of this report is to give a partial resume of our articles [11], [12] and indicate their relevance to physical problems. So we have tried to make this report easy to understand, especially for physicists, by sometimes sacrificing the generality of the statement. We have also concentrated our attention on topics which, we hope, relate to physical problems. In this report we use the same notations as in [12].

In the theory of linear ordinary differential equations and its applications, equations with regular singularities play a central role. Hence one might naturally wish to extend the theory to several variables case. Then one natural choice would be to concentrate our attention on holonomic systems, partly because they share the basic finiteness theorem with ordinary differential equations ([4], [8]) and partly because, not general systems, but holonomic systems are particularly important in applications (e.g. [9], [13]. [15], [20],...).

Now, let $\mathcal{M}$ be a holonomic $\mathcal{E}_X$-Module on $\Omega \subset T^*X - T^*_X X$. Let Λ be its characteristic variety. (Hence Λ is a Lagrangian variety possibly with singularities.) For each point p in the non-singular locus Λ_{reg} of Λ, we can find a suitable contact transformation φ which is defined in a neighborhood U of p and brings $\Lambda_{reg} \cap U$ to the form $T^*_Y X \cap \varphi(U)$ for a non-singular hypersurface Y in X. We choose a coordinate system $x = (x_1,\dots,x_n)$ near $\pi(\varphi(p))$ (*) so that $Y = \{x_1 = 0\}$ and $\varphi(p) = (0;dx_1\infty)$. Then by definition, $\mathcal{M}$ is with regular singularities along $T^*_Y X$ at $\varphi(p)$ if and only if $\mathcal{M}$ is isomorphic to the following $\mathcal{E}_X$-Module $\mathcal{N}$ in a neighborhood of $\varphi(p)$:

$$\mathcal{N} \ : \ (p_j I - A_j)U = 0 \qquad (j=1,\dots,n),$$

where U is a column vector with N unknown functions, I is the identity matrix of size $N \times N$, $p_1 = x_1 D_1$, $p_j = D_j$ $(j=2,\dots,n)$ and A_j $(j=1,\dots,n)$ is an $N \times N$ matrix of micro-differential operators which satisfies the following condition:

(1) There exists an integer m such that the order of each component of each power A_j^{ℓ} $(\ell \geqq 0)$ is less than m.

(When a matrix A of micro-differential operators satisfies the condition (1), A is said to be of order at most 0.)

We define the notion of a holonomic $\mathcal{E}_X$-Module with R.S. (= the abbreviation of regular singularities) by requiring that it is with regular singularities along Λ at each point in Λ_{reg}. A holonomic $\mathcal{D}_X$-Module $\mathcal{L}$ is said to be with R.S. if and only if so is $(\mathcal{E}_X \underset{\mathcal{D}_X}{\otimes} \mathcal{L})|_{T^*X - T^*_X X}$.

See [12] Chap. I and Chap. V for more intrinsic definition of systems with R.S. and results which show the naturality of the above definition.

An important class of holonomic $\mathcal{D}_X$-Modules is that of the D-type equations. A D-type equation corresponds to the sheaf of meromorphic sections of a regular integrable connection in the sense of Deligne [2], namely, the sheaf of Nilsson class functions. In terms of $\mathcal{D}_X$-Modules, it is characterized as follows:

Definition 1. A holonomic $\mathcal{D}_X$-Module $\mathcal{L}$ is said to be of D-type (=Deligne type) with singularities along a hypersurface $Y = f^{-1}(0)$, if and only if it satisfies the following three conditions:

(2) $SS(\mathcal{L})$ (**) $\subset \pi^{-1}(Y) \cup T^*_X X$

(3) $\mathcal{L}$ is with regular singularities along $T^*_{Y_{reg}} X$.

(4) $f: \mathcal{L} \longrightarrow \mathcal{L}$ is bijective.

(*) Here and in what follows, π denotes the canonical projection from T^*X to X.

(**) $SS(\mathcal{L})$ denotes the characteristic variety of the $\mathcal{D}_X$-Module $\mathcal{L}$.

Remark 1. Since there may be some irreducible components other than $T^*_Y X$ in $SS(\mathcal{L})$, the condition (3) does not imply immediately that $\mathcal{L}$ is with R.S. The proof for this fact requires (at least, at present) the use of $\mathcal{D}^\infty$, the sheaf of linear differential operators of infinite order. ([12] Chap. V, Theorem 5.2.3.)

Example 1. Let f be a holomorphic function defined on X. Suppose that f is not identically zero on any connected components of X. Then, for generic α, $\mathcal{D}_X f^\alpha$ is a holonomic system of D-type. ([11] Remark 1.2. The precise condition which guarantees α to be generic is also given there.)

Example 2. Let f be the same as in Example 1. Denote $f^{-1}(0)$ by Y. Let $\mathcal{M}$ be a holonomic $\mathcal{D}_X$-Module which is with regular singularities along $T^*_{Y_{reg}} X$ at each point x in Y_{reg}. Then the localization $\mathcal{M}_f^{(*)}$ of $\mathcal{M}$ along Y is a holonomic system of D-type along Y. ([12] Chap. II, §3, Proposition 2.3.4.)

The importance of the D-type equation lies in the fact that it is "generic" in the sense of Theorem 1 below.

Definition 2. A Lagrangian variety Λ in $T^*X - T^*_X X$ is said to be in a generic position at p in Λ if and only if $\Lambda \cap \pi^{-1}(\pi(p)) = \mathbb{C}^\times p$ holds in a neighborhood of p.

Theorem 1. Let $\mathcal{M}$ be a holonomic $\mathcal{E}_X$-Module defined in a neighborhood of p in $T^*X - T^*_X X$. Suppose that $\Lambda \underset{def}{=} \operatorname{Supp} \mathcal{M}$ is in a generic position at p in Λ. Then there exist a holonomic system $\mathcal{L}$ of D-type along $\pi(\Lambda)$, a holonomic $\mathcal{D}$-sub-Module $\mathcal{P}$ of $\mathcal{L}$ with $SS(\mathcal{P}) \subset T^*_X X$ (**) and an injective $\mathcal{D}_{X,\pi(p)}$-linear homomorphism $\phi: \mathcal{M}_p \to (\mathcal{L}/\mathcal{P})_{\pi(p)}$.

If, in addition, any order of any sections of $\mathcal{M}$ is not an half integer, then we do not need to divide $\mathcal{L}$ by a sub-Module $\mathcal{P}$, namely, $\mathcal{M}_p$ can be imbedded into $\mathcal{L}_{\pi(p)}$.

(*) Here $\mathcal{M}_f$ is equal to $\bigcup_{j=0}^{\infty} f^{-j}\mathcal{M}$, and it is known to be holonomic. ([5], Theorem 3.1.)

(**) Hence $\mathcal{P}$ is isomorphic to $\mathcal{O}_X^r$ for some integer r.

Remark 2. The assumption that Λ is in a generic position at p is not too restrictive, because a suitable contact transformation brings Λ to this form.

In the course of the proof of Theorem 1, we also obtain the following

Theorem 2. Assume the same conditions on $\mathcal{M}$ and Λ as in Theorem 1. Then $\mathcal{M}_p$ is a finitely generated $\mathcal{D}_{X,\pi(p)}$-Module. Furthermore we have

$$\mathcal{E}_{X,p'} \underset{\pi^{-1}\mathcal{D}_{X,\pi(p)}}{\otimes} \mathcal{M}_p = \begin{cases} \mathcal{M}_p & \text{if } p' = p \\ 0 & \text{if } p' \text{ is in } \pi^{-1}\pi(p) - T^*_X X - \mathbb{C}^\times p. \end{cases}$$

This theorem is useful in investigating the analytic structure of Feynman amplitudes at cuspidal points of Landau-Nakanishi surfaces, which are far from the physical region. (Cf. [13] §3.)

Remark 3. As a matter of fact, a much more striking result is proved in [12] (Chap. IV, Theorem 4.1.1): Theorem 1 holds for an arbitrary holonomic $\mathcal{E}_X$-Module if we use micro-differential operators of infinite order, namely, there exist a holonomic system $\mathcal{L}$ of D-type, its $\mathcal{D}$-sub-Module $\mathcal{P}$ with $SS(\mathcal{P}) \subset T^*_X X$ and a $\mathcal{D}^\infty_{X.\pi(p)}$-linear homomorphism

$$\phi\colon (\mathcal{E}^\infty_X \underset{\mathcal{E}_X}{\otimes} \mathcal{M})_p \longrightarrow (\mathcal{D}^\infty_X \underset{\mathcal{D}_X}{\otimes} (\mathcal{L}/\mathcal{P}))_{\pi(p)}$$

such that the associated homomorphism

$$\tilde{\phi} \underset{\text{def}}{=} 1 \otimes \phi\colon (\mathcal{E}^\infty_X \underset{\mathcal{E}_X}{\otimes} \mathcal{M})_p \rightarrow \mathcal{E}^\infty_{X,p} \underset{\mathcal{D}^\infty_{X,\pi(p)}}{\otimes} (\mathcal{L}/\mathcal{P})_{\pi(p)}$$

is an injective $\mathcal{E}^\infty_p$-linear homomorphism.

In fact, this imbedding theorem is the most important result in [12]. Most of the main results of [12] follow from this. So, here we will sketch some basic ideas of the proof. We like to mention that the method used in [9] to discuss the hierarchical principle for Feynman amplitudes is one of the essential ingredients of our proof.

We first introduce a vector space C, which is an analogue of the space of germs of microfunctions. It is the space of holomorphic functions defined on a cone with its apex at the origin, modulo the space of holomorphic functions defined on a neighborhood of the origin. (See (4.5.1) of [12] Chap. IV, §5 for the rigorous definition.) This space is introduced to make clear the action of micro-differential operators on holomorphic functions. (Cf. [6], [1] and [9].) We can then prove that $V = \mathrm{Hom}_{\mathcal{E}_p^\infty}(\mathcal{M}_p^\infty, C)$ is finite-dimensional and that $\mathcal{M}_p^\infty$ is imbedded into $\mathrm{Hom}_{\mathbb{C}}(V,C)$. Let $\{s_j\}_{j=1}^N$ be a system of generators of $\mathcal{M}$, let $\{\phi_\nu\}_{\nu=1}^\ell$ be a base of V and let $\varphi_{j,\nu}$ be a holomorphic function whose modulo class in C is $\phi_\nu(s_j)$. Then we can prove that $\varphi_{j,\nu}$ can be extended to a multi-valued holomorphic functions on $X-\pi(\Lambda)$. Furthermore, thus extended functions are with finite determination. Since $\pi(\Lambda)$ is a hypersurface, we can choose a holomorphic function f on X such that $\pi(\Lambda) = f^{-1}(0)$. In this case, the following result is proved by Deligne [2].

Let φ denote one of $\varphi_{j,\nu}$. Then there exists Nilsson class functions ψ_k and single-valued holomorphic fucntion a_k defined on $X-f^{-1}(0)$ such that $\varphi = \sum_{k=1}^m a_k\psi_k$ holds.

Using this result we can obtain a much sharper result to the effect that φ has the form $\sum_{k=1}^m Q_k\psi_k$ with linear differential operators Q_k (of course, possibly with infinite order). See [12] Chap. II, §2 Theorem 2.2.4 for the precise statement and the proof. Here, instead of repeating the proof, we present a heuristic, but instructive argument which, we hope, convince the reader that there must exist such an operator Q_k.

We know ([5], Theorem 2.7) that there exists a non-zero polynomial $b(s)$ which satisfies

$$(5)\qquad b(s)f^s\varphi_1 = P(s,x,D_x)(f^{s+1}\varphi_1)$$

for some $P(s) \underset{\text{def}}{=} P(s,x,D_x)$ in $\mathcal{D}_X[s]$, i.e., for some linear differential operator which is a polynomial in s. Then, a superficial application of (5) would entail

$$(6)\qquad f^{-p}\varphi_1 = \frac{P(-p)\cdots P(-1)}{\prod_{s=1}^{p} b(-s)}\varphi_1.$$

Since a_1 is single-valued on $X-f^{-1}(0)$, it has the form

$$(7) \qquad \sum_{p=-\infty}^{\infty} a_{1,p} f^p ,$$

for holomorphic functions $a_{1,p} = a_{1,p}(x)$. Hence, combining (6) and (7), we might be finished by choosing

$$\sum_{p=1}^{\infty} a_{1,-p} \frac{P(-p)\cdots P(-1)}{\prod_{s=1}^{p} b(-s)} + \sum_{p=0}^{\infty} a_{1,p} f^p$$

as Q_1. Of course, this is a very formal reasoning, because $b(-p)$ may be zero and because the estimation of the growth order of the coefficients of Q_1 is not done. Probably, it will not be a very good strategy to try to modify this formal reasoning so that it may become rigorous in the general case, even though it might be possible for rather simple f's. In fact, the proof given in [12] Chap. II is completely different from this type of the reasoning. (Actually, it is much more transcendental.) Still we hope it will convince the reader of the existence of the required Q_k.

Once we get Theorem 1, it is fairly easy to prove the following

Theorem 3. Let Y be a submanifold of X. Let $\mathcal{M}$ be a holonomic $\mathcal{E}_X$-Module with respect to which Y is non-characteristic. Then the tangential system (i.e., the system restricted to Y) $\mathcal{M}_Y$ of $\mathcal{M}$ is also with R.S.

Intuitively speaking, this is an immediate consequence of Theorem 1, because the restriction of a D-type equation must be with R.S. See [12] Chap. V, §3 for a more precise reasoning. As a matter of fact, one should be a little careful in the reasoning, for Y may not be non-characteristic with respect to the D-type equation $\mathcal{L}$ used in Theorem 1.

By a quantized contact transformation, we can obtain a corresponding result to the effect that, if we integrate a holonomic $\mathcal{E}_X$-Module $\mathcal{M}$ with R.S. along finite fiber(*) $\varphi : X \to Y$, then we get a

(*) Here "finite fiber" means that $\rho_\varphi^{-1}(\mathrm{Supp}\,\mathcal{M}) \cap \varpi_\varphi^{-1}(U) \to U$ is a finite map for $\rho_\varphi : X \underset{Y}{\times} T^*Y \to T^*Y$ and $\varpi_\varphi : X \underset{Y}{\times} T^*Y \to T^*X$.

holonomic $\mathcal{E}_Y$-Module $\varphi_* \mathcal{M}$ with R.S.

For the application to physical problems, the following results (Theorems 4 and 5) are more useful, since the conditions on Y is much less restrictive. The additional condition needed there, namely, the condition that $\mathcal{M}$ is a $\mathcal{D}_X$-Module, is usually not too difficult to verify in investigating the problems which appear in physics, especially in connection with Feynman amplitudes.

Theorem 4. Let $\mathcal{M}$ be a holonomic $\mathcal{D}_X$-Module with R.S. Let Y be a sumbmanifold of X. Then the tangential system $\mathcal{M}_Y$ of $\mathcal{M}$ is a holonomic $\mathcal{D}_Y$-Module with R.S.

Theorem 5. Let $\mathcal{M}$ be a holonomic $\mathcal{D}_X$-Module with R.S. Assume that the map $\varphi : X \longrightarrow Y$ is projective, i.e., φ is imbedded into $Y \times P^N(\mathbb{C}) \longrightarrow Y$ for some N. Then the integral of $\mathcal{M}$ along fiber $\varphi : X \longrightarrow Y$ is a holonomic $\mathcal{D}_X$-Module with R.S.

Since the proof of these two theorems are complicated, we content ourselves with only referring the reader to [12] Chap. V, §3 and Chap. VI, §2.

Using Theorem 5 we can improve several results of ours ([7],[10], ...) in that the holonomic $\mathcal{D}$-Module whose existence is guaranteed in the papers quoted above is actually with R.S. In particular, we have the following

Theorem 6. Analytically renormalized Feynman amplitude $f_D(p)$ satisfies a holonomic system of linear differential equations *with R.S.* whose characteristic variety is contained in the extended Landau variety $\tilde{\mathcal{L}}(D)$. (Cf. [10], Theorem 7.)

On the other hand, in applying Theorem 4 to physical problems, we encounter a serious trouble, for it does not say anything about the characteristic variety of $\mathcal{M}_Y$. However, obtaining some (desirably, complete, of course) information on $SS(\mathcal{M}_Y)$ is crucially important in analyzing the singularity structure of the S-matrix and related functions. In fact, one unpleasant feature of analytic S-matrix theory (the so-called u = 0 problem) is closely tied with the lack of such an information. (See [3],[15] and [14] for this point.) We bypassed this problem in several cases by finding some information on $SS(\mathcal{M})$

for very special $\mathcal{M}$, i.e., $\mathcal{M} = \mathcal{D}_X(\prod_{\ell=1}^{N} f_\ell(x)_+^{\lambda_\ell})$. The recent development in mathematics attained in [12] has enabled us to obtain the following satisfactory results. (Theorems 7 and 8 below.)

Definition 3. Let f be a holomorphic function on X.

(i) $\tilde{W}_f$ is, by definition, the closure of $\{(\sigma;x,\xi) \in \mathbb{C}\times T^*X;\ x \in X,\ f(x) \neq 0,\ \xi = \sigma d \log f(x)\}$ in $\mathbb{C} \times T^*X$

(ii) $W_0 \underset{\text{def}}{=} \{(x,\xi) \in T^*X;\ (0;x,\xi) \in \tilde{W}_f\}$. When it is necessary to make the dependence of W_0 on f explicit, we sometimes use the notation $W_0(f)$ instead of W_0.

Remark 4. It is known that W_0 is a Lagrangian variety.

Example 3. Let X be $\mathbb{C}^2$ and let f be $x_1(x_1 - x_2^2)$. Denote by S_1 (resp., S_2) the hypersurface $\{x \in \mathbb{C}^2 ;\ x_1 = 0\}$ (resp., $\{x \in \mathbb{C}^2;\ x_1 - x_2^2 = 0\}$). Then $W_0(f) = T^*_{\mathbb{C}^2}\mathbb{C}^2 \cup T^*_{S_1}\mathbb{C}^2 \cup T^*_{S_2}\mathbb{C}^2 \cup T^*_{\{0\}}\mathbb{C}^2$.

Theorem 7. Let f be a holomorphic function on X and denote $f^{-1}(0)$ by Y. Let $\mathcal{L}$ be a holonomic system of D-type along Y. Then its characteristic variety $SS(\mathcal{L})$ is contained in $W_0(f)$.

The proof of this theorem essentially uses the desingularization theorem of Hironaka. The key-point in the proof is that a D-type equation remains to be a D-type equation after the desingularization procedure. ([11], §1.)

Next, by the aid of the result mentioned in Example 2, we can deduce the following result from Theorem 7.

Theorem 8. Let M be a real analytic manifold and X its complexification. Let φ be a locally integrable function on M. Assume that there exist a holomorphic function f on X and a left Ideal $\mathcal{J}$ of $\mathcal{D}_X$ which satisfy the following conditions (8)∿(12).

(8) f is not identically zero on any connected component of X and $f|_M$ is real-valued.

(9) $\mathcal{J}\varphi = 0$ holds on $M - f^{-1}(0)$.

(10) $\mathcal{M} \underset{\text{def}}{=} \mathcal{D}_X/\mathcal{J}$ is a holonomic $\mathcal{D}_X$-Module.

(11) $SS(\mathcal{M}) \subset T^*_X X \cup \pi^{-1}(f^{-1}(0))$.

(12) $\mathcal{M}$ is with regular singularities along $T^*_{Y_{reg}} X$, where $Y=f^{-1}(0)$.

Then there exists a left Ideal $\mathcal{J}'$ of $\mathcal{D}_X$ which satisfies the following conditions:

(13) $\mathcal{J}'\varphi = 0$ holds on M.

(14) $\mathcal{J}' = \mathcal{J}$ on $X-f^{-1}(0)$.

(15) $\mathcal{M}' \underset{\text{def}}{=} \mathcal{D}_X/\mathcal{J}'$ is a holonomic $\mathcal{D}_X$-Module with R.S.

(16) $SS(\mathcal{M}')$ is contained in $W_0(f)$.

<u>Remark 5.</u> As the following example shows, $\mathcal{J}' \neq \mathcal{J}$ in general.

<u>Example 4.</u> Let $\mathbb{R}$ be M, let x be f and choose the Heaviside function $Y(x)$ as φ. If we choose $\mathcal{D}\frac{d}{dx}$ as $\mathcal{J}$, then conditions (8)∿(12) are all satisfied. In this case we must choose $\mathcal{D}(x\frac{d}{dx})$ as $\mathcal{J}'$.

The reader will see clearly that Theorem 8 provides us with a powerful tool in dealing with the restriction and the multiplication of holonomic hyperfunctions, if they are "tame" enough. Here the tameness means the quantitative aspect of φ, i.e., its local integrability or something like that. Such a tameness condition is used to prove that $f_+^\lambda\varphi$ can be holomorphically extended to $\{\lambda \in \mathbb{C};\ \mathrm{Re}\ \lambda > -\varepsilon\}$ for some $\varepsilon > 0$. See [11], §2 for details.

Before ending this report, we mention some comparison theorems which justify our terminology "with regular singularities."

<u>Theorem 9.</u> Let $\mathcal{M}$ and $\mathcal{N}$ be holonomic $\mathcal{E}_X$-Modules with R.S. Then

$$\mathbb{R}\,\mathcal{H}om_{\mathcal{E}_X}(\mathcal{M}, \mathcal{N}) = \mathbb{R}\,\mathcal{H}om_{\mathcal{E}_X}(\mathcal{M}, \mathcal{E}_X^\infty \underset{\mathcal{E}_X}{\otimes} \mathcal{N}).$$

As an easy corollary of this theorem, we also have the following

<u>Theorem 10.</u> Let $\mathcal{M}$ be a holonomic $\mathcal{D}_X$-Module with R.S. Then, for each point x in X and each integer j, the natural homomorphism

$$\mathcal{E}xt^j_{\mathcal{D}_X}(\mathcal{M}, \mathcal{O}_{X,x}) \longrightarrow \mathcal{E}xt^j_{\mathcal{D}_X}(\mathcal{M}, \hat{\mathcal{O}}_{X,x})$$

is an isomorphism. Here $\hat{\mathcal{O}}_{X,x}$ denotes the ring of formal power series around x, i.e., $\varprojlim_k \mathcal{O}_{X,x}/\mathfrak{m}^k$ for the maximal ideal $\mathfrak{m}$ of $\mathcal{O}_{X,x}$.

As a matter of fact, we have the following more general result.

Theorem 11. Let $\mathcal{M}$ and $\mathcal{N}$ be holonomic $\mathcal{E}_X$-Modules with R.S. Then

$$\mathbb{R}\,\mathcal{H}om_{\mathcal{E}_X}(\mathcal{M},\mathcal{N}) = \mathbb{R}\,\mathcal{H}om_{\mathcal{E}_X}(\mathcal{M},\hat{\mathcal{E}}_X \underset{\mathcal{E}_X}{\otimes} \mathcal{N}).$$

Here $\hat{\mathcal{E}}_X$ denotes the sheaf of formal micro-differential operators, i.e., $\varprojlim_k \mathcal{E}_X/\mathcal{E}_X(k)$, where $\mathcal{E}_X(k)$ is the sheaf of micro-differential operators of order at most k.

Remark 6. We know that the converse statement of Theorem 10 is also true for a holonomic $\mathcal{D}_X$-Module $\mathcal{M}$. ([12] Chap. VI, §4, Theorem 6.4.1.) Hence the validity of the comparison theorem characterizes the holonomic $\mathcal{D}_X$-Module with R.S. (Cf. Ramis [19], where the validity of the comparison theorem is used as the definition of Fuchsian systems. See also the remark at the end of §4 of [12] Chap. VI.) Note, however, that the validity of the comparison theorem at the 0-th cohomology group does *not* imply that $\mathcal{M}$ is with R.S. (even in the one-dimensional case). (Cf. Malgrange [16].)

We do not give even the sketch of the proof of Theorem 9 here, because probably it is not very interesting to physicists. The reader who is interested in it is referred to [12] Chap. VI. However, we mention one thing which may be helpful for the understanding of the meaning of Theorem 9.

Let f be a holomorphic function on X. Denote $\mathcal{D}_X f^\lambda$ ($\lambda \in \mathbb{C}$) by $\mathcal{L}$. Define $\mathcal{N}$ by $\mathcal{L}_f$, the localization of $\mathcal{L}$ along $f^{-1}(0)$. Then $\mathcal{N}$ is a holonomic $\mathcal{D}_X$-Module with R.S. and it consists of multi-valued holomorphic functions of the form φf^λ for a meromorphic function φ having poles along $f^{-1}(0)$. Then $\mathcal{D}_X^\infty \underset{\mathcal{D}_X}{\otimes} \mathcal{N}$ consists of φf^λ with φ having essential singularities along $f^{-1}(0)$. (Cf. Mebkhout [17].) Hence in this case Theorem 9 can be expressed in a rather symbolical and loose language as follows:

Let $\mathcal{M}$ be a holonomic $\mathcal{D}_X$-Module with R.S. Then the solution spaces of $\mathcal{M}$ are identical whether we admit only pole-type singularities or essential singularities as its solutions.

Appendix

Utilizing this opportunity, we report that the recent progress in the theory of holonomic systems has enabled us to prove the conjecture in [13] (p.123) with a trivial modification. Hence the results stated in [13] (pp.123∿129) with prefix C- such as C-Theorem 4.6, i.e., the results whose proof rely on the conjecture hold as they stand. (The modification, i.e., the inclusion of Case I in the statement does not require any essential change of the argument given in [13].) The result which replaces the conjecture is the following

Theorem A.1. Let X be a complex manifold and p a point in $T^*X - T^*_X X$. Let $\mathcal{M} = \mathcal{E}_X/\mathcal{J}$ be a holonomic $\mathcal{E}_X$-Module defined on a neighborhood of p, where $\mathcal{J}$ is a left Ideal of $\mathcal{E}_X$. Suppose that $V \underset{\text{def}}{=} \operatorname{Supp} \mathcal{M}$ has the form $V_1 \cup V_2$, where V_1 and V_2 are Lagrangian submanifolds of T^*X and they intersect along a submanifold V_{12} of T^*X with $\operatorname{codim}_{V_j} V_{12} = 1$ $(j=1,2)$. Suppose further that $T_p V_{12} = T_p V_1 \cap T_p V_2$ holds. Assume that $\mathcal{M}$ is simple both on $V_1 - V_{12}$ and on $V_2 - V_{12}$. Then either one of the following two cases happens:

Case I. There exist simple holonomic $\mathcal{E}_X$-Modules $\mathcal{M}_j$ $(j=1,2)$ supported respectively by V_j $(j=1,2)$ such that $\mathcal{M} = \mathcal{M}_1 \oplus \mathcal{M}_2$ holds on a neighborhood of p.

Case II. In a neighborhood of p, $\mathcal{M}$ is isomorphic to a holonomic $\mathcal{E}_X$-Module $\mathcal{E}_X/\mathcal{J}'$ with the symbol Ideal $\sigma(\mathcal{J}')$ being reduced.

Proof. By Theorem 5.3.1 of [21], Chap. II, we can reduce the problem to the case where X is an open subset of $\mathbb{C}^2$. Hence, by a quantized contact transformation, we may assume that $V_1 = \{(x,y;\xi,\eta) \in T^*\mathbb{C}^2;\ \xi=y=0\}$ and $V_2 = \{(x,y;\xi,\eta) \in T^*\mathbb{C}^2;\ x=y=0\}$.

Now, we know ([12] Chap. V, §1, Theorem 5.1.6) that there exists a coherent $\mathcal{E}(0)$-sub-Module $\mathcal{L}$ of $\mathcal{M}$ such that $\mathcal{E}\mathcal{L} = \mathcal{M}$ holds and that $\overline{\mathcal{L}} \underset{\text{def}}{=} \mathcal{L}/\mathcal{L}(-1)$ satisfies the following conditions:

(A.1) $$J_V \overline{\mathcal{L}} = 0.$$

(A.2) The support of any section of $\overline{\mathcal{L}}$ has codimension 0 in V.

Here and in what follows, J_V denotes $\{f \in \mathcal{O}_{T^*X}(0);\ f|_V = 0\}$, where $\mathcal{O}_{T^*X}(k)$ denotes the sheaf of holomorphic functions on T^*X which are homogeneous of degree k with respect to the fiber coordinate. In the sequel we sometimes abbreviate $\mathcal{O}_{T^*X}(k)$ to $\mathcal{O}(k)$. It follows from the assumption of the simplicity of $\mathcal{M}$ that $\bar{\mathcal{L}}$ is isomorphic to $\mathcal{O}_{T^*X}(0)/J_{V_1} \oplus \mathcal{O}_{T^*X}(0)/J_{V_2}$ on $V - V_{12}$, where $J_{V_j} = \{f \in \mathcal{O}_{T^*X}(0);\ f|_{V_j}=0\}$ $(j=1,2)$. Let us now recall the following lemma in algebraic geometry. We omit its proof here.

Lemma A.2. Let C_0, C_1 and C_2 be respectively the curves $\{(z_1,z_2) \in \mathbb{C}^2;\ z_1z_2=0\}$, $\{(z_1,z_2) \in \mathbb{C}^2;\ z_2=0\}$ and $\{(z_1,z_2) \in \mathbb{C}^2;\ z_1=0\}$. Let $\mathcal{F}$ be an $\mathcal{O}_{C_0}$-Module such that $\mathcal{H}^0_{\{0\}}(\mathcal{F}) = 0$ and $\mathcal{F}|_{C_0-\{0\}} = \mathcal{O}_{C_1} \oplus \mathcal{O}_{C_2}$, where $\mathcal{O}_{C_j} = \mathcal{O}_{\mathbb{C}^2}/J_j$ with $J_j = \{f \in \mathcal{O}_{\mathbb{C}^2};\ f|_{C_j}=0\}$ $(j=1,2)$. Then $\mathcal{F}$ is isomorphic to either $\mathcal{O}_{C_0}$ or $\mathcal{O}_{C_1} \oplus \mathcal{O}_{C_2}$.

We find by this lemma that either $\bar{\mathcal{L}} = \mathcal{O}(0)/J_V$ or $\bar{\mathcal{L}} = (\mathcal{O}(0)/J_{V_1}) \oplus (\mathcal{O}(0)/J_{V_2})$.

We first consider the case where $\bar{\mathcal{L}} = (\mathcal{O}(0)/J_{V_1}) \oplus (\mathcal{O}(0)/J_{V_2})$. Let u and v be sections of $\mathcal{L}$ which give $1 \bmod J_{V_1}$ and $1 \bmod J_{V_2}$, respectively. Then we have

(A.3) $\quad \mathcal{L} = \mathcal{E}(0)u + \mathcal{E}(0)v$

(A.4) $\quad (xD_x + 2yD_y)u,\ (xD_x + 2yD_y)v \in \mathcal{L}.$

(A.5) $\quad D_xu \in \mathcal{L}$ and $xD_yv \in \mathcal{L}.$

Now the following result is proved in [18] (Theorem 3.2):

Lemma A.3. Let P be a square matrix of micro-differential operators which are of order at most zero and defined in a neighborhood of $p = (0;dy) \in T^*\mathbb{C}^2$. Then there exists an invertible matrix U of micro-differential operators of order at most 0 which satisfies the following:

(A.6) $\quad \sigma_0(U) = I$

(A.7) $\quad (xD_x+2yD_y-P) = U(xD_x+2yD_y-A)U^{-1}$, where $A = \sigma_0(P)(p)$.

By virtue of this result we may assume from the first that u and v satisfy

(A.8) $\quad (xD_x + 2yD_y - A)\binom{u}{v} = 0$

for a constant matrix A.

Now we prove the following

Lemma A.4. Let u be a column vector of N unknown functions. Assume that u satisfies

$$(xD_x + 2yD_y - A)u = 0$$

for some constant matrix A. Let $P(x,y,D_x,D_y)$ be a row vector of length N defined in a neighborhood of $p = (0;\, dy)$ which satisfies $Pu = 0$. Then, writing $P = \sum_{j\in\mathbb{Z}} P_j$ with $[\vartheta, P] = jP_j$, where $\vartheta = xD_x + 2yD_y$, we find $P_j u = 0$ for every j.

Proof. Let $\mathcal{J}$ be the sub-Module of $\mathcal{E}(0)^N$ which consists of Q such that $Qu = 0$. Let $Q_1, \cdots, Q_r$ be a system of generators of $\mathcal{J}$ and let Q denote the $r \times N$ matrix with $Q_1, \cdots, Q_r$ as its row components. Then $Qu = 0$ holds by the definition. On the other hand, we have

(A.9) $\quad ([\vartheta, Q] + QA)u = \vartheta Qu - Q(\vartheta - A)u = 0.$

Hence there exists $R \in M(r,r;\, \mathcal{E}(0))$ such that $[\vartheta, Q] + QA = RQ$, i.e., $Q(\vartheta - A) = (\vartheta - R)Q$. Again by Lemma A.3 there exists $U \in GL(r, \mathcal{E}(0))$ and $B \in M(r,r;\, \mathbb{C})$ such that $\vartheta - R = U(\vartheta - B)U^{-1}$. Hence, by replacing Q with $U^{-1}Q$, we may assume from the first that $Q(\vartheta - A) = (\vartheta - B)Q$ holds, i.e., $[\vartheta - Q] = BQ - QA$ holds. Hence, writing $Q = \sum_{j\in\mathbb{Z}} Q_j$ with $[\vartheta, Q_j] = jQ_j$, we obtain

$$jQ_j = BQ_j - Q_j A$$

for every j. This implies that $Q_j = 0$ holds except for finitely many j. Furthermore, by the repeated application of (A.9), we find

$$(\sum_j Q_j f(j+A))u = 0$$

holds for each polynomial $f(\lambda)$. On the other hand, for each integer j, there exists a polynomial $f_j(\lambda)$ such that

$f_j(j+A) = I$ and $f_j(k+A) = 0$ for $k \neq j$ with $Q_k \neq 0$. Therefore $Q_j u = 0$ holds for every j.

Now let P be the row vector in the statement of Lemma A.4. Then there exists $T \in M(1,r;\mathcal{E})$ such that $P = RQ$ holds. Writing $R = \sum_{j \in \mathbb{Z}} R_j$ with $[\vartheta, R_j] = jR_j$, we have $P_j = \sum_k R_{j-k}Q_k$. Since $Q_k u = 0$ holds, we obtain $P_j u = 0$. Q.E.D.

Now we resume the proof of the theorem. It follows from (A.5) that there exists $P \in M(2.2;\mathcal{E}(0))$ such that

$$\begin{pmatrix} D_x & \\ & xD_y \end{pmatrix}\begin{pmatrix} u \\ v \end{pmatrix} = P\begin{pmatrix} u \\ v \end{pmatrix}.$$

Dividing P by $xD_x + 2yD_y - A$, we may assume that P does not contain y. By Lemma A.4 we may further assume that $[\vartheta, P] = -P$. On the other hand, since the operator P is of order at most 0 and does not contain y, it is a linear combination of $x^i D_x^j D_y^{-j-k}$ with $i,j,k \geqq 0$.

Since $[\vartheta, x^i D_x^j D_y^{-j-k}] = (i-j+2j+2k)x^i D_x^j D_y^{-j-k}$ and since $i-j+2j+2k = i+j+2k$ cannot be -1, we must have $P = 0$. Hence we obtain the following equations with a,b,c and d in $\mathbb{C}$:

$$\text{(A.10)} \quad \begin{cases} (xD_x+2yD_y)u = au + bv \\ (xD_x+2yD_y)v = cu + dv \\ D_x u = 0 \\ xv = 0. \end{cases}$$

Hence we have

$$\text{(A.11)} \quad (yD_y - a/2)u = bv/2$$

and

$$\text{(A.12)} \quad (yD_y - 1/2 - d/2)v = cu/2.$$

Therefore we have $D_x bv = (2yD_y - a)D_x u = 0$ and $xbv = 0$. This implies $bv = 0$. Similarly we obtain $cu = 0$. Thus we finally obtain

$$(A.13)\quad \begin{cases} (yD_y - a/2)u = D_x u = 0 \\ (yD_y - 1/2 - d/2)v = xv = 0. \end{cases}$$

This implies that $\mathcal{M}$ is the direct sum of $\mathcal{E}/(\mathcal{E}(yD_y - a/2) + \mathcal{E}D_x)$ and $\mathcal{E}/(\mathcal{E}(yD_y - 1/2 - d/2) + \mathcal{E}x)$, which are respectively simple holonomic systems supported by V_1 and V_2.

In the case where $\mathcal{L} = \mathcal{O}(0)/J_V$, we take a section u of $\mathcal{L}$ which gives $1 \bmod J_V$. Then we have $\mathcal{L} = \mathcal{E}(0)u$ and $\mathcal{M} = \mathcal{E}u$. This implies that $\mathcal{M}$ satisfies the property stated in Case II of the theorem. Q.E.D.

References

[1] Bony, J. M. and P. Schapira: Propagation des singularité analytiques pour les solutions des équations aux dérivées partielles. Ann. Inst. Fourier, Grenoble, 26, 81-140 (1976).

[2] Deligne, P.: Equations differentielles à points singuliers réguliers. Lecture Notes in Math. No.163, Berlin-Heidelberg-New York: Springer, 1970.

[3] Iagolnitzer, D.: The u=0 structure theorem. Commun. math. Phys., 63, 49-96 (1978).

[4] Kashiwara, M.: On the maximally overdetermined system of linear differential equations, I. Publ. RIMS, Kyoto Univ., 10, 563-579 (1975).

[5] ————: On the holonomic systems of linear differential equations. II. Inventiones math., 49, 121-135 (1978).

[6] Kashiwara, M. and T. Kawai: Micro-hyperbolic pseudo-differential operators I. J. Math. Soc. Japan, 27, 359-404 (1975).

[7] ————: Micro-local properties of $\prod_{j=1}^{n} f_{j+}^{s_j}$. Proc. Japan Acad., 51, 270-272 (1975).

[8] ————: Finiteness theorem for holonomic systems of micro-differential equations. Proc. Japan Acad., 52, 341-343 (1976).

[9] ————: Holonomic character and local monodromy structure of Feynman integrals. Commun. math. Phys., 54, 121-134 (1977).

[10] ————: Holonomic systems of linear differential equations and Feynman integrals. Publ. RIMS, Kyoto Univ., 12 Suppl., 131-140 (1977).

[11] ————: On the characteristic variety of a holonomic system with regular singularities. Adv. in Math, 34, 163-184 (1979).

[12] ————: On holonomic systems of micro-differential equations. III —Systems with regular singularities—. To appear. (RIMS Preprint No.293.)

[13] Kashiwara, M., T. Kawai and T. Oshima: A study of Feynman integrals by micro-differential equations. Commun. math. Phys., 60, 97-130 (1978).

[14] Kashiwara, M., T. Kawai and H. P. Stapp: Micro-analyticity of the S-matrix and related functions. Commun. math. Phys., 66, 95-130 (1979).

[15] Kawai, T. and H. P. Stapp: Discountinuity formula and Sato's conjecture. Publ. RIMS, Kyoto Univ., 12 Suppl., 155-232 (1977).

[16] Malgrange, B.: Sur les points singuliers des équations differentielles. Enseignement Math. 20, 147-176 (1974).

[17] Mebkhout, Z.: Local cohomology of analytic spaces. Publ. RIMS, Kyoto Univ., 12 Suppl., 247-256 (1977).

[18] Oshima, T.: Singularities in contact geometry and degenerate pseudo-differential equations. J. Fac. Sci. Univ. Tokyo, IA, 21, 43-83 (1974).

[19] Ramis, J. P.: Variations sur le theme "GAGA". Lecture Notes in Math. No.694, pp.228-289, Berlin-Heidelberg-New York: Springer, 1978.

[20] Sato, M.: Recent development in hyperfunction theory and its application to physics. Lecture Notes in Phys. No.39, pp. 36-48, Berlin-Heidelberg-New York: Springer, 1975.

[21] Sato, M., T. Kawai and M. Kashiwara: Microfunctions and pseudo-differential equations. Lecture Notes in Math. No.287, pp. 265-529. Berlin-Heidelberg-New York: Springer, 1973.

SECOND-MICROLOCALIZATION AND ASYMPTOTIC EXPANSIONS

Masaki KASHIWARA
Département de Mathématiques,
Université de Paris-Sud
91405, Orsay, France

and

Research Institute for Mathematical Sciences,
Kyoto University, Kyoto 606, Japan

AND

Takahiro KAWAI
Research Institute for Mathematical Sciences,
Kyoto University, Kyoto 606, Japan

Abstract

The asymptotic expansions of (holonomic) microfunctions is neatly dealt with by the second micro-localization.

§1. Introduction.

Several years ago, Jeanquartier [5] proved the following:

For a real-valued real analytic function $f(x)$ defined on a real analytic manifold of dimensions n, $\delta(t-f(x))$ has an asymptotic expansion of the form

$$(1.1) \qquad \sum_{\mu=0}^{n-1} \sum_{\nu=1}^{N} \sum_{j=0}^{\infty} a_{\nu,\mu,j}(x) t^{\lambda_\nu + j} (\log t)^{\mu}.$$

for distributions $a_{\nu,\mu,j}(x)$, as t tends to zero. Here the meaning of the asymptotic expansion is given through the pairing with a C^∞-function $\varphi(x)$ with compact support.

Recently, by using a group-theoretic technique, Barbasch-Vogan [1] proved that any eigendistribution on a real semi-simple group G has an asymtpotic expansion and they studied some properties of the initial term of the expansion. Here we note that an eigendistribution satisfies, by definition, the following system of linear differential equations :

$$(1.2)\quad \begin{cases} Pu = \chi(P)u & \text{for each } P \text{ in } \mathfrak{z} \\ L_A u = 0 & \text{for each } A \text{ in the Lie algebra of } G, \end{cases}$$

where $\mathfrak{z}$ is the ring of G-biinvariant differential operators on G, χ is the character of $\mathfrak{z}$ and L_A is the vector field on G corresponding to the inner automorphism of G on G.

We can show that $\delta(t-f(x))$ satisfies a holonomic system with R.S.(*) and that (1.2) is also a holonomic system with R.S. As we show later (§4.2) each microfunction solution of a holonomic system with R.S. has an asymptotic expansion, whose meaning will be clarified in subsequent sections. Hence the result of Jeanquartier and that of Barbasch-Vogan are explained in a neat and unified manner by employing the theory of holonomic systems with R.S.

We now introduce the following space $M^{(r)}$ so that we may consider such asymptotic expansions of distributions mentioned above in a more general situation .

Definition 1.1. A distribution u on $\mathbb{R}^n$ belongs to $M^{(r)}$ $(r\in\mathbb{R})$ if and only if there exist positive constants ε, m and C such that

$$(1.3)\qquad \left|\int u(tx)\varphi(x)\,dx\right| \leqq Ct^{-\frac{n}{2}-r}\sum_{|\alpha|\leqq m}\sup|D^\alpha\varphi|$$

holds for every φ in $C_0^\infty(\{x\in\mathbb{R}^n; |x|<\varepsilon\})$.

These spaces $M^{(r)}$ can be used as a scale in considering the asymptotic expansion of a distribution u as follows:

We say that u has an asymptotic expansion $\sum_{j=0}^{\infty} u_j$ and we write $u \sim \sum_{j=0}^{\infty} u_j$ if there exist a sequence $\{u_j\}_{j\geqq 0}$ of distributions which satisfy the following condition:

(1.4) For any $r\in\mathbb{R}$, there exists $N_0\in\mathbb{Z}$ such that $u-\sum_{j=0}^{N_0} u_j \in M^{(r)}$.

In particular, this condition implies the following:

(1.5) For any $r\in\mathbb{R}$, there exists $N_0\in\mathbb{Z}$ such that $u_j\in M^{(r)}$ for any $j \geqq N_0$.

(*) See Kashiwara-Kawai [8], [9] for the theory of holonomic systems with regular singularities.

In what follows we exclusively consider the case where u_j is homogeneous of degree $\lambda+j$ for some $\lambda \in \mathbb{C}$, or, a little more generally, the case where $(\sum_{\nu=1}^{n} x_\nu D_\nu - \lambda - j)^p u_j = 0$ holds for some $p \in \mathbb{N}$. In the latter case, u_j belongs to $M^{(-\mathrm{Re}\lambda - j - \frac{n}{2} + \varepsilon)}$ for any $\varepsilon > 0$.

In order to exemplify the notion introduced here, we give the following example, where the j-th terms in the right hand side are homogeneous of degree $2\lambda+j$ and $3\lambda+\frac{1}{2}+j$, respectively.

$$(1.6)\quad (y^2-x^3)_+^\lambda \sim \sum_{j=0}^{\infty} \frac{\Gamma(-\lambda+\frac{1}{2})}{4^j j!\,\Gamma(-\lambda+\frac{1}{2}+j)} x^{3j} D_y^{2j} |y|^{2\lambda}$$

$$+ \sum_{j=0}^{\infty} \frac{\sqrt{\pi}\,\Gamma(\lambda+1)}{4^j j!\,\Gamma(\lambda+j+\frac{3}{2})\cos\pi\lambda} x^{3j} D_y^{2j} (-x_+^{3\lambda+\frac{3}{2}} + \sin\pi\lambda\, x_-^{3\lambda+\frac{3}{2}})\,\delta(y).$$

The asymptotic expansion done by using $M^{(r)}$ as the scale can be regarded as an asymptotic expansion with respect to the origin $\{x=0\}$. Hence it is natural to try to micro-localize the notion as follows:

Let Λ be the conormal bundle of the origin, i.e., $\sqrt{-1}\,T^*_{\{0\}}\mathbb{R}^n = \{(x,\sqrt{-1}\xi) \in \mathbb{R}^n \times \sqrt{-1}\,\mathbb{R}^n;\ x=0\}$. Let $p = (0,\sqrt{-1}\xi_0)$ be a point in Λ. In this situation, we introduce the following spaces $M^{(r)}_{\Lambda,p}$ $(r \in \mathbb{R})$.

<u>Definition 1.2.</u> A distribution u belongs to $M^{(r)}_{\Lambda,p}$ if and only if the following condition is satisfied:

There exist $\chi(x) \in C_0^\infty(\mathbb{R}^n)$ with $\chi(0) \neq 0$, an open neighborhood U of ξ_0 and constants C and m such that

$$(1.7)\quad \left|\iint \exp(-\sqrt{-1}\langle x,\tau\xi\rangle)(\chi u)(x)\,dx\,\psi(\xi)\,d\xi\right| \leqq C\tau^{r-\frac{n}{2}} \sum_{|\alpha|\leqq m} \sup |D_\xi^\alpha \psi(\xi)|$$

holds for $\tau \geqq 1$ for every $\psi \in C_0^\infty(U)$.

The following lemma easily follows from the definition.

<u>Lemma 1.3.</u> (i) If (1.7) holds for some $\chi \in C_0^\infty(\mathbb{R}^n)$ with $\chi(0) \neq 0$, then (1.7) holds for any $\tilde{\chi} \in C^\infty(\mathbb{R}^n)$ such that $\tilde{\chi}(0) \neq 0$ and that supp $\tilde{\chi}$ is contained in the domain of definition of u.
(ii) If p is not in WF(u),(*) then u belongs to $M^{(r)}_{\Lambda,p}$ for every r.

(*) See Hörmander [3] for the definition of the wave front set.

(iii) If u belongs to $M^{(r)}_{\Lambda,p}$ and $a(x)$ is a C^∞-function defined in a neighborhood of the origin, then au belongs to $M^{(r)}_{\Lambda,p}$. If we further suppose that $a(0) = 0$, then au belongs to $M^{(r-1)}_{\Lambda,p}$.

(iv) Let u be a tempered distribution and denote by v its Fourier transform. If there exist an open neighborhood U of ξ_0 and constants C, m such that

$$(1.8)\quad \left|\int v(\tau\xi)\psi(\xi)d\xi\right| \leqq C\tau^{r-\frac{n}{2}} \sum_{|\alpha|\leqq m} \sup|D^\alpha\psi|$$

holds for $\tau \geqq 1$ and for every $\psi \in C_0^\infty(U)$, then u belongs to $M^{(r)}_{\Lambda,p}$.

(v) If u belongs to $M^{(r)}_{\Lambda,p}$, then $\partial u/\partial x_k$ belongs to $M^{(r+1)}_{\Lambda,p}$.

(vi) Let P be a classical pseudo-differential operator of order at most 0. If u belongs to $M^{(r)}_{\Lambda,p}$, then Pu belongs to $M^{(r)}_{\Lambda,p}$.

(vii) The following three conditions are equivalent:

(a) $u \in M^{(r)}$

(b) $u \in M^{(r)}_{\Lambda,p}$ for any $p \in \Lambda$

(c) $u \in M^{(r)}_{\Lambda,p}$ for $p = (0;0)$

(viii) For each distribution u there exists r such that u belongs to $M^{(r)}_{\Lambda,p}$.

Remark. The converse of (ii) is not true, namely, the fact that u belongs to $M^{(r)}_{\Lambda,p}$ for every r does not imply that $WF(u)$ does not contain p; For example, let n be 1 and consider the C^∞-function $f(x) \underset{\text{def}}{=} \exp(-1/x^2)$ and a continuous function $g(x)$ which is not C^∞ at x_j, where $\{x_j\}$ is a sequence tending to 0. Then (iii) and (viii) combined implies that $u(x) \underset{\text{def}}{=} f(x)g(x)$ is in $M^{(r)}_{\Lambda,p}$ for any p and r. However $WF(u)$ contains some p in Λ, because x_j is contained in sing supp u.

We can also prove the following result.

Theorem 1.4. Let $a(x,\theta)$ be a C^∞-function defined in a neighborhood of $(x,\theta) = (0,\theta_0) \in \mathbb{R}^n \times \mathbb{R}^N$ $(N \geqq n)$. Suppose that $a(x,\theta)$ satisfies the following conditions:

$$(1.9)\quad a(0,\theta) = 0$$

$$(1.10)\quad \left(\frac{\partial^2 a(x,\theta)}{\partial x_i \partial \xi_j}\right)_{i,j} \text{ has rank } n \text{ at } (0,\theta_0).$$

Denote by p the point $(0, \sqrt{-1}d_x a(0,\theta_0)) \in \sqrt{-1}T^*_{\{0\}}\mathbb{R}^n$. Let u be a compactly supported distribution which belongs to $M^{(r)}_{\Lambda,p}$. Define a function $w(\theta)$ by

$$\int \exp(-\sqrt{-1}a(x,\theta))u(x)dx - \int \exp(-\sqrt{-1} \langle \mathrm{grad}_x a(0,\theta),x\rangle)u(x)dx.$$

Then there exist constants C, m and $\varepsilon(>0)$ such that

$$(1.11) \quad \left|\int w(\tau\theta)\varphi(\theta)d\theta\right| \leqq C\tau^{r-\frac{n}{2}-2} \sum_{|\alpha|\leqq m} \sup|D^\alpha_\theta \varphi|$$

holds for $\tau \geqq 1$ and for every φ in $C^\infty_0(\{\theta; |\theta-\theta_0| < \varepsilon\})$.

It immediately follows from this result that a Fourier integral operator(*) A of order at most 0 such that the associated canonical transformation Φ preserves Λ sends $M^{(r)}_{\Lambda,p}$ to $M^{(r)}_{\Lambda,\Phi(p)}$. In view of this fact we can define $M^{(r)}_{\Lambda,p}$ for every Lagrangian submanifold Λ of $\sqrt{-1}T^*\mathbb{R}^n$ by using a suitable phase function $a(x,\theta)$, even though we exclusively consider the case where $\Lambda = \sqrt{-1}T^*_{\{0\}}\mathbb{R}^n$ in this section. For example, if $\Lambda = \{(x',x''; \sqrt{-1}(\xi',\xi'') \in \mathbb{R}^\ell \times \mathbb{R}^{n-\ell} \times \sqrt{-1}(\mathbb{R}^\ell \times \mathbb{R}^{n-\ell}); x'=0, \xi''=0\}$, then a distribution u belongs to $M^{(r)}_{\Lambda,p}$ $(p=(0;\sqrt{-1}(\xi'_0,0))$ if and only if there exist $\chi \in C^\infty_0(\mathbb{R}^n)$ with $\chi(0)\neq 0$ and constants C, m and $\varepsilon > 0$ such that

$$(1.12) \quad \left|\int\left(\int \exp(-\sqrt{-1}\langle\tau\xi',x'\rangle)(\chi u)(x',x'')dx'\right)\psi(\xi',x'')d\xi'dx''\right| \leqq C\tau^{r-\frac{\ell}{2}} \sum_{|\alpha|+|\beta|\leqq m} \sup|D^\alpha_{\xi'}D^\beta_{x''}\psi|$$

holds for $\tau \geqq 1$ and for every $\psi(\xi',x'')$ in $C^\infty_0(\{(\xi',x'')\in \mathbb{R}^\ell\times\mathbb{R}^{n-\ell}; |\xi'-\xi'_0|+|x''| < \varepsilon\}$.

Now we want to investigate how these scaling spaces are related to micro-differential equations with regular singularities. Let us denote by $\Lambda^{\mathbb{C}}$ the complexification of $\Lambda = \sqrt{-1}T^*_{\{0\}}\mathbb{R}^n$, i.e., $\Lambda^{\mathbb{C}} = T^*_{\{0\}}\mathbb{C}^n$. Then $\mathcal{E}_{\Lambda^{\mathbb{C}}}(m)$ is, by definition, $\{P = \sum_j p_j(x,D) \in \mathcal{E}; p_j(x,\xi)$ has the zero of degree $(j-m)$ along $\Lambda^{\mathbb{C}}\}$. We abbreviate $\mathcal{E}_{\Lambda^{\mathbb{C}}}(0)$ to $\mathcal{E}_{\Lambda^{\mathbb{C}}}$. It then follows from Lemma 1.3 (iii) that

(*) See Hörmander [3] for the definition.

(1.13) $\qquad \mathcal{E}_{\Lambda^{\mathbb{C}}}(-1)M^{(r)}_{\Lambda,p} \subset M^{(r-1)}_{\Lambda,p}.$

Note also that, if a micro-differential operator P belongs to $\mathcal{E}_{\Lambda^{\mathbb{C}}}$, then it has the form

(1.14) $\qquad \sum_{0\leq|\alpha|\leq m} A_\alpha(D)x^\alpha + Q,$

where $A_\alpha(D)$ is homogeneous of degree $|\alpha|$ and Q belongs to $\mathcal{E}_{\Lambda^{\mathbb{C}}}(-1)$. In order to see how $\mathcal{E}_{\Lambda^{\mathbb{C}}}(m)$ is related to the asymptotic expansions, we consider a simple case where a distribution solution $u(x)$ of the equation $Pu = 0$ has an asymptotic expansion of the form $\sum_{j=0}^{\infty} u_{\lambda+j}(x)$ where $u_{\lambda+j}$ is homogeneous of degree $\lambda+j$. It is clear that $u_{\lambda+j}$ belongs to $M^{(-\mathrm{Re}\lambda-j-\frac{n}{2})}$. In this case, we find the following relation:

(1.15) $\qquad Pu \equiv \sum_{0\leq|\alpha|\leq m} A_\alpha(D)x^\alpha u_\lambda(x) \quad \mathrm{mod}\ M^{(-\mathrm{Re}\lambda-1-\frac{n}{2})}.$

If we define a homogeneous differential operator $L(P)(\xi,D_\xi)$ by

$$\sum_{0\leq|\alpha|\leq m} A_\alpha(\sqrt{-1}\xi)(\sqrt{-1}\frac{\partial}{\partial\xi})^\alpha,$$

then we obtain another equation

(1.16) $\qquad L(P)\hat{u}_\lambda(\xi) = 0,$

where $\hat{u}_\lambda(\xi) = \int \exp(-\sqrt{-1}\langle x,\xi\rangle)u_\lambda(x)dx$. If we denote the image of L by $\mathcal{A}_{\Lambda^{\mathbb{C}}}$, $\mathcal{A}_{\Lambda^{\mathbb{C}}}$ is isomorphic to $\mathcal{E}_{\Lambda^{\mathbb{C}}}/\mathcal{E}_{\Lambda^{\mathbb{C}}}(-1)$. (Cf. [8], Chap. I, §5.) It follows from the definition that $\mathcal{A}_{\Lambda^{\mathbb{C}}}$ consists of homogeneous linear differential operators of degree 0 defined on $\Lambda^{\mathbb{C}}$. The following Theorem 1.5 shows the importance of the associated equation (1.16), especially because we will later prove that, if u is a solution of holonomic $\mathcal{E}$-Module with R.S., then its asymptotic expansion is determined by its initial term. (See §4.2, Theorem 4.2.12 for the precise statement.)

<u>Theorem 1.5</u>. Let X be a complex manifold and Λ a Lagrangian submanifold of T^*X. Let $\mathcal{J}$ be a left Ideal of $\mathcal{E}_X$ such that $\mathcal{M} \underset{\mathrm{def}}{=} \mathcal{E}_X/\mathcal{J}$ is a holonomic $\mathcal{E}_X$-Module with R.S. Then

$$\mathcal{L}: \quad L(P)v = 0 \qquad (P \in \mathcal{J} \cap \mathcal{E}_\Lambda)$$

is a holonomic system with R.S. on $T^*\Lambda$.

Proof. By a quantized contact transformation, we may assume the following:

(1.17) X is an open subset of $\mathbb{C}^{1+n} = \{(t,x);\ t \in \mathbb{C},\ x \in \mathbb{C}^n\}$.

(1.18) Λ is the conormal bundle of $Y \underset{\text{def}}{=} \{(t,x) \in X;\ t=0\}$.

(1.19) Supp $\mathcal{M}$ is in a generic position (in the sense of [8] Chap. I, §6).

Then we know ([8] Chap. V, §1) that there exists a holonomic $\mathcal{D}_X$-Module $\mathcal{F}$ with R.S. and a section $\tilde{u}$ of $\mathcal{F}$ such that

(1.20) $\mathcal{E}_X \underset{\pi^{-1}\mathcal{D}_X}{\otimes} \mathcal{F} \cong \mathcal{M}, \quad \mathcal{H}^0_Y(\mathcal{F}) = 0$

and

(1.21) $1 \otimes \tilde{u}$ corresponds to the section $u = (1 \bmod \mathcal{J})$ of $\mathcal{M}$ by the above isomorphism.

We also know ([6]) the following:

(1.22) $\mathcal{D}[s](t^s u)$ is a coherent subholonomic $\mathcal{D}_X$-Module.

Here and in what follows, $\mathcal{D}[s]$ denotes $\mathbb{C}[s] \underset{\mathbb{C}}{\otimes} \mathcal{D}_X$. By using the same reasoning as in [6] we can also show that

(1.23) $\mathcal{D}[s](t^s u)/\mathcal{D}[s](t^{s+1}u)$ is a holonomic $\mathcal{D}_Y$-Module with R.S.

Let us now denote by $\mathcal{D}'$ the sub-Algebra of $\mathcal{D}_X$ generated by $\mathcal{O}_X$, D_{x_j} and tD_t. Then, for any P in $\mathcal{D}'$, $t^s P t^{-s}$ belongs to $\mathcal{D}[s]$. Moreover, any section of $\mathcal{D}_X$ can be written in the form $\sum_{j \geq 0} D_t^j t^s P_j t^{-s}$ with $P_j \in \mathcal{D}'$. Here we note the following two lemmas.

Lemma 1.6. Let $P(s)$ be a section of $\mathcal{D}[s]$ of the form

$\sum_{j\geq 0} D_t^j t^s P_j t^{-s}$. If $P(s)t^s u = 0$ holds, then $P_j u = 0$ holds for any j.

Proof. It follows from the assumption that $\sum_{j\geq 0} D_t^j t^s P_j u = 0$ holds. It suffices to show that $P_j u = 0$ $(j>r)$ entails $P_r u = 0$ in this case for any integer r. If $P_j u = 0$ $(j>r)$, then we have

$$0 = \sum_{j=0}^{r} D_t^j t^s P_j u = \sum_{0\leq k\leq j\leq r} s(s-1)\cdots(s-k+1)t^{s-k}D_t^{j-k}P_j u.$$

Looking at the coefficients of s^r in the right hand side, we find $t^{s-r}P_r u = 0$. Hence we have $P_r u = 0$. Q.E.D.

Lemma 1.7. We have an injective $\mathcal{D}_Y$-linear homomorphism

$$\mathcal{D}'u/t\,\mathcal{D}'u \longrightarrow \mathcal{D}[s](t^s u)/\mathcal{D}[s](t^{s+1}u)$$

by assigning $t^s Pu$ to Pu $(P \in \mathcal{D}')$.

Proof. Let P be an element of $\mathcal{D}'$. If $t^s Pu$ belongs to $\mathcal{D}[s](t^{s+1}u)$, then we can find $P(s)=\sum_j D_t^j t^{-s} P_j t^s$ $(P_j \in \mathcal{D}')$ such that

$$t^s Pu = P(s)(t^{s+1}u) = \sum_j D_t^j t^s P_j tu.$$

Hence it follows from the preceding lemma that

$$Pu = P_0(tu) \in \mathcal{D}'(tu) = t\,\mathcal{D}'u.$$

Q.E.D.

Now we resume the proof of Theorem 1.5. Let $\mathcal{J}'$ denote the Ideal of $\mathcal{D}_Y$ which annihilates $t^s u$ mod $\mathcal{D}[s](t^{s+1}u)$. Then it follows from (1.23) that $\mathcal{D}_Y/\mathcal{J}'$ is a holonomic $\mathcal{D}_Y$-Module with R.S. On the other hand, Lemma 1.7 implies

$$\mathcal{J}'u \subset t\,\mathcal{D}'u.$$

Hence for any P in $\mathcal{J}'$, we can find Q in $t\,\mathcal{D}'$ such that $(P+Q)u = 0$. Since $\mathcal{D}' \subset \mathcal{E}_\Lambda$ and $L(P+Q) = L(P)$, the system $\mathcal{L}$ in Theorem 1.5 is stronger than the system $\mathcal{D}_Y/\mathcal{J}'$. This means that $\mathcal{L}$ is a holonomic $\mathcal{D}_Y$-Module with R.S. Q.E.D.

So far we have developed the theory of asymptotic expansions by the aid of scaling spaces $M^{(r)}$ (or its micro-localization $M_\Lambda^{(r)}$). However, it would be much more desirable if we could find a suitable sheaf on which another suitable sheaf of rings of operators acts and in which the asymptotic expansion of (a class of) microfunctions can be considered. In the example (1.6), we can rewrite the right hand side

$$(1.24)\quad \left(\sum_{j=0}^{\infty} c_j x^{3j} D_y^{2j}\right)|y|^{2\lambda} + \left(\sum_{j=0}^{\infty} c_j' x^{3j} D_y^{2j}\right) \times \left\{\frac{\sqrt{\pi}\,\Gamma(\lambda+1)}{\Gamma(\lambda+\frac{3}{2})\cos\pi\lambda}\left(-x_+^{3\lambda+\frac{3}{2}} + \sin\pi\lambda\; x_-^{3\lambda+\frac{3}{2}}\right)\right\}\delta(y),$$

where $c_j = \Gamma(-\lambda+\frac{1}{2})/4^j j!\,\Gamma(-\lambda+\frac{1}{2}+j)$ and $c_j' = \Gamma(\lambda+\frac{3}{2})/4^j j!\,\Gamma(\lambda+\frac{3}{2}+j)$.

In view of the growth order of c_j and c_j', it is easy to see that the infinite series of operators appearing in (1.24) do not preserve the local character but that it has "propagation velocity" of order $|x|^{\frac{3}{2}}$. (Cf. Kashiwara-Kawai [7]). Having this in mind, we will introduce the sheaves $\tilde{\mathcal{D}}_\Lambda^\infty$ and $\tilde{\mathcal{E}}_\Lambda^\infty$ in the next section. We note that it has turned out that $\tilde{\mathcal{E}}_\Lambda^\infty$ coincides with the sheaf $\mathcal{E}_\Lambda^{2\infty}$ introduced by Laurent [14], [15] in different context.

§2. The second-microlocalization of operators.

As we mentioned at the end of the introduction, we want to find a sheaf of operators which is suitable for the manipulation of the asymptotic expansion of microfunctions. For this purpose we introduce the sheaves $\tilde{\mathcal{D}}_\Lambda^\infty$ and $\tilde{\mathcal{E}}_\Lambda^\infty$ starting from the sheaf $\mathcal{M}$ of simple holonomic $\mathcal{E}_X$-Module supported by a complex Lagrangian submanifold Λ of T^*X. Our procedure for finding the desired sheaves is similar to the way of constructing the sheaves of hyperfunctions and (micro-) differential operators starting from the sheaf of holomorphic functions. For example, the sheaf $\mathcal{D}_{\mathbb{C}^n}^\infty$ of linear differential operators (of infinite order) on $\mathbb{C}^n$ is, by definition $\mathcal{H}_\Delta^n(\mathcal{O}_{\mathbb{C}^n} \hat{\otimes} \Omega_{\mathbb{C}^n}^n)$, where $\Delta = \{(x,y)\in\mathbb{C}^n\times\mathbb{C}^n;\ x=y\}$ and $\Omega_{\mathbb{C}^n}^n$ denotes the sheaf of holomorphic n-forms. Note that $\Omega^n_{\mathbb{C}^n}$ is a dual sheaf of $\mathcal{O}_{\mathbb{C}^n}$ and locally isomorphic to $\mathcal{O}_{\mathbb{C}^n}$. We will follow this procedure in defining the sheaf $\tilde{\mathcal{D}}^\infty_\Lambda$ (Definition 2.2). We first prepare several notations.

Let X be a complex manifold of dimension n and Λ a non-singular Lagrangian submanifold of T^*X. We denote by γ the projection from $T^*X - T^*_XX$ to P^*X. In what follows, we will consider the problem locally in Λ, and hence we may assume that $\Lambda = \gamma^{-1}\tilde{\Lambda}$ with $\tilde{\Lambda} = \gamma(\Lambda)$.

Let $\mathcal{M}$ be a simple holonomic $\mathcal{E}_X$-Module with support Λ and let $\mathcal{M}^*$ denote $\mathcal{E}xt^n_{\mathcal{E}_X}(\mathcal{M}, \mathcal{E}_X)$. Let $\mathcal{N}$ be $\mathcal{M} \hat{\otimes} \mathcal{M}^*$ and $\Delta^a \underset{\text{def}}{=} T^*_\Delta(X\times X) = \{(x,y;\xi,\eta) \in T^*(X\times X);\ x=y,\ \xi+\eta=0\}$. Here we note the following

Lemma 2.1. The system $\mathcal{N}$ is independent of the choice of $\mathcal{M}$.

Proof. Let $\mathcal{M}'$ be another simple holonomic system with support Λ. Then there exists a constant λ such that $\mathcal{M}'$ is isomorphic to $\mathcal{M}'' \underset{\text{def}}{=} \mathcal{E}_{(\lambda)} \otimes_{\mathcal{E}} \mathcal{M}$, where $\mathcal{E}_{(\lambda)}$ denotes the sheaf of micro-differential operators of fractional order $\lambda+j$ $(j\in\mathbb{Z})$. Furthermore the isomorphism φ from $\mathcal{M}'$ to $\mathcal{M}''$ is unique up to constant factor. On the other hand, $\mathcal{M}''^* = \mathcal{E}_{(-\lambda)} \otimes_{\mathcal{E}} \mathcal{M}^*$ and hence $\mathcal{M}'' \hat{\otimes} \mathcal{M}''^* = (\mathcal{E}_{(\lambda)} \otimes_{\mathcal{E}} \mathcal{M}) \hat{\otimes} (\mathcal{E}_{(-\lambda)} \otimes_{\mathcal{E}} \mathcal{M}^*) = \mathcal{M} \hat{\otimes} \mathcal{M}^* = \mathcal{N}$. Therefore φ and $\varphi^{*-1}: \mathcal{M}''^* \to \mathcal{M}'^*$ give rise to an isomorphism ψ from $\mathcal{M}' \hat{\otimes} \mathcal{M}'^*$ onto $\mathcal{M}'' \hat{\otimes} \mathcal{M}''^* = \mathcal{M} \hat{\otimes} \mathcal{M}^*$. Since $((c\varphi)^*)^{-1} = c^{-1}\varphi^{*-1}$ for $c \in \mathbb{C}^\times$, ψ does not depend on φ. Q.E.D.

After this observation we introduce the following

Definition 2.2. $\tilde{\mathcal{D}}^\infty_\Lambda \underset{\text{def}}{=} \mathcal{H}^n_{\Delta^a}(\mathcal{E}^\infty_{X\times X} \otimes_{\mathcal{E}_{X\times X}} \mathcal{N})$.

We can also consider the micro-localization of $\tilde{\mathcal{D}}^\infty_\Lambda$ by the same procedure used in S-K-K [17] Chap. II to define the sheaf of micro (=pseudo)-differential operators starting from $\mathcal{O}^{(0,n)}_{X\times X}$. In what follows, $\widetilde{\Lambda\times\Lambda^{a}}^{\Delta^a *}$ denotes the comonoidal transform of $\Lambda\times\Lambda^a$ with center Δ^a, i.e., $(\Lambda\times\Lambda^a - \Delta^a) \sqcup S^*_{\Delta^a}(\Lambda\times\Lambda^a)$. The projection from $\widetilde{\Lambda\times\Lambda^{a}}^{\Delta^a *}$ to $S^*_{\Delta^a}(\Lambda\times\Lambda^a)$ and the projection from $S^*_{\Delta^a}(\Lambda\times\Lambda^a)$ to $P^*_{\Delta^a}(\Lambda\times\Lambda^a)$ are denoted by π and γ respectively.

Definition 2.3.

(i) $\tilde{\mathcal{E}}^{\mathbb{R}}_{\Lambda} \underset{\text{def}}{=} \mathcal{H}^{n}_{S^*_{\Delta^a}(\Lambda\times\Lambda^a)}(\pi^{-1}\mathcal{N}^{\infty})$.

(ii) $\tilde{\mathcal{E}}^{\infty}_{\Lambda} \underset{\text{def}}{=} \gamma^{-1}\gamma_* \tilde{\mathcal{E}}^{\mathbb{R}}_{\Lambda}$.

Here and in what follows $\mathcal{N}^{\infty}$ denotes $\mathcal{E}^{\infty} \otimes_{\mathcal{E}} \mathcal{N}$. Since $\mathcal{N}$ is independent of the choice of $\mathcal{M}$, the definitions introduced above are independent of the choice of $\mathcal{M}$. Hence we usually choose $\mathcal{C}_{Y|X}$ as $\mathcal{M}$ when Λ has the form $T^*_Y X$ for a non-singular submanifold of X.

We now begin to explain how to obtain the concrete expression of the germ of $\tilde{\mathcal{D}}^{\infty}_{\Lambda}$. The corresponding result for $\tilde{\mathcal{E}}^{\infty}_{\Lambda}$ is obtained by combining the results given below and the reasoning of S-K-K [17] Chap. II, §1.4, so we will state the result (Theorem 2.12), leaving the detailed arguments to the reader. See also Laurent [14].

We begin our discussion with proving several vanishing theorems needed for the concrete calculation of the relative cohomology groups introduced above.

Proposition 2.4. Let Y be a non-singular hypersurface of $X = \mathbb{C}^n$ and let Λ denote $T^*_Y X$. Let U be an open subset of Λ such that each fiber of $U \to \gamma(U)$ is contractible and that $\gamma(U)$ is a Stein manifold. Then we have

(2.1) $H^j(U; \mathcal{C}^{\infty}_{Y|X}) = 0 \quad (j\neq 0)$

Proof. Since $V \underset{\text{def}}{=} \gamma(U)$ is Stein, there exists a fundamental system of Stein open neighborhoods $W_k \subset X$ of V. Then $W_k \cap (X-Y)$ is also Stein. Denote by j the inclusion map from X-Y into X and let $\mathcal{P}$ denote the sheaf $j_* j^{-1} \mathcal{O}_X$. Denote $\mathcal{P}/\mathcal{O}_X$ by $\mathcal{F}$. Then it follows from the definition that

(2.2) $\mathcal{C}^{\infty}_{Y|X} \cong \gamma^{-1}\mathcal{F} \oplus \gamma^{-1}\mathcal{O}_X$.

On the other hand, we have

(2.3) $H^j(W_k; \mathcal{P}) = H^j(W_k - Y; \mathcal{O}_X) = 0$

for $j \geqq 1$. Hence we have

$$(2.4) \quad H^j(W_k; \mathcal{F}) = 0 \quad (j \geq 1).$$

Therefore we find

$$(2.5) \quad H^j(U; \mathcal{C}^{\infty}_{Y|X}) = \varinjlim_k H^j(W_k; \mathcal{F} \otimes \mathcal{O}_X) = 0$$

for $j \geqq 1$. Q.E.D.

By a quantized contact transformation we can easily deduce the following corollary from Proposition 2.4.

Corollary 2.5. Let Λ be a Lagrangian submanifold of T^*X and $\mathcal{M}$ a simple holonomic $\mathcal{E}_X$-Module with support Λ. Then for any p in $\Lambda - T^*_X X$, we can find a neighborhood W of p which satisfies the following:

For each open subset U of W satisfying the conditions

$$(2.6) \quad \gamma(U) \text{ is a Stein manifold}$$

and

$$(2.7) \quad \text{each fiber of } U \to \gamma(U) \text{ is contractible,}$$

we find

$$H^j(U; \mathcal{M}^{\infty}) = 0 \quad \text{for } j \neq 0.$$

Next we deduce the following vanishing theorem (Theorem 2.6) from Corollary 2.5. In what follows, X denotes $\mathbb{C}^n$, V denotes $T^*_{\{0\}}\mathbb{C}^n \times (T^*_{\{0\}}\mathbb{C}^n)^a \cong T^*_{\{0\}}\mathbb{C}^{2n} \cong \mathbb{C}^{2n}$ and $\mathcal{N}$ denotes $\mathcal{C}_{\{0\}|\mathbb{C}^n} \hat{\otimes} \mathcal{C}_{\{0\}|\mathbb{C}^n} \cong \mathcal{C}_{\{0\}|\mathbb{C}^{2n}}$. We denote a point in $\mathbb{C}^{2n}$ by (x,y) and a point in $T^*_{\{0\}}\mathbb{C}^{2n}$ by $(0,0;\xi,\eta)$. Let us take a point p in $\Delta^a \cap V - T^*_{X\times X}(X\times X) = \{(0,0;\ \xi,\eta);\ \xi=-\eta\neq 0\}$. By a linear transformation we may assume that p is $(0,0;\ \xi^0,-\xi^0)$ with $\xi^0 = (0,\cdots,0,1)$. Define V_ℓ by $\{(0,0;\ \xi,\eta)\in V;\ \xi_\ell+\eta_\ell \neq 0\}$ for $\ell=1,\ldots,n$ and let U be a sufficiently small neighborhood of p which satisfies conditions (2.6) and (2.7). Then it follows from Corollary 2.5 that

$$H^j(U\cap V_{\nu_1}\cap\cdots\cap V_{\nu_k};\mathcal{N}^\infty)=0$$

holds for $j\neq 0$ and $1\leq\nu_1<\cdots<\nu_k\leq n$ $(0\leq k\leq n)$. Since $\{U\cap V_\ell\}$ is an open covering of $U-(\Delta^a\cap V)$, we obtain by a theorem of Leray

$$H^j_{\Delta^a}(U;\mathcal{N}^\infty)=0 \quad \text{for } j>n \tag{2.8}$$

and

$$H^n_{\Delta^a}(U;\mathcal{N}^\infty)=\mathcal{N}^\infty(U\cap\bigcap_{\ell=1}^{n}V_\ell)/(\sum_{j=1}^{n}\mathcal{N}^\infty(U\cap(\bigcap_{\ell\neq j}V_\ell)). \tag{2.9}$$

On the other hand, Theorem 1.2.2 of [8] Chap. I, §2 asserts that

$$\mathcal{E}xt^j_{\mathcal{E}_{X\times X},\Delta^a\cap V}(\mathcal{E}_{X\times X},\mathcal{N}^\infty)=0$$

holds for $j<\operatorname{codim}_{T^*(X\times X)}(\Delta^a\cap V)-\operatorname{proj\,dim}\mathcal{N}$. It is easy to see that

$$\operatorname{codim}_{T^*(X\times X)}(\Delta^a\cap V)=3n,\ \operatorname{proj\,dim}\mathcal{N}=2n$$

and

$$\mathcal{E}xt^j_{\mathcal{E}_{X\times X},\Delta^a\cap V}(\mathcal{E}_{X\times X},\mathcal{N}^\infty)=\mathcal{H}^j_{\Delta^a}(\mathcal{N}^\infty).$$

Thus we obtain the following

<u>Theorem 2.6</u>. $\mathcal{H}^j_{\Delta^a}(\mathcal{N}^\infty)=0 \quad (j\neq n)$.

We next try to find a symbol sequence corresponding to an element in $\tilde{\mathcal{D}}^\infty_{\Lambda,p}$ $(p\in\Lambda)$.

First we note that

$$\mathcal{H}^n_{\Delta^a}(\mathcal{N}^\infty)_p=\varinjlim_{U\ni p}\mathcal{N}^\infty(U\cap\bigcap_{\ell=1}^{n}V_\ell)/(\sum_{j=1}^{n}\mathcal{N}^\infty(U\cap(\bigcap_{\ell\neq j}V_\ell)))$$

holds by (2.9). Let P be in $\tilde{\mathcal{D}}^\infty_{\Lambda,p}$. Let $(\sum_j p_j(D_x,D_y))\delta(x,y)$ be a corresponding element in $\mathcal{N}^\infty(U\cap\bigcap_{\ell=1}^{n}V_\ell)$ for sufficiently small U which satisfies conditions (2.6) and (2.7). Here and in what follows we denote by $\delta(x,y)$ the generator of $\mathcal{N}$. It follows from the definition that $p_j(\xi,\eta)$ is homogeneous of degree j and satisfies the following conditions (2.10) and (2.11):

(2.10) For each $\varepsilon > 0$ and each compact subset K of $U \cap (\bigcap_{\ell=1}^{n} V_\ell)$ there exists a constant $C_{\varepsilon,K}$ such that

$$\sup_K |p_j| \leqq C_{\varepsilon,K}\varepsilon^j/j! \qquad (j \geqq 0)$$

holds.

(2.11) For each compact subset K of $U \cap (\bigcap_{\ell=1}^{n} V_\ell)$, there exists a constant R_K such that

$$\sup_K |p_j| \leqq R_K^{-j}(-j)! \qquad (j < 0)$$

holds.

By the Laurent expansion, we have

$$(2.12) \quad p_j(\xi,\eta) = \sum_{\alpha\in\mathbb{Z}^n} p_{j,\alpha}(\xi) \prod_{\ell=1}^{n} (\xi_\ell+\eta_\ell)^{-\alpha_\ell-1}.$$

It is then easy to see that, if we define $q_j^{(\ell)}(\xi,\eta)$ by

$$\sum_{\substack{\alpha_1,\ldots,\alpha_{\ell-1}\geqq 0\\ \alpha_\ell<0}} p_{j,\alpha}(\xi) \prod_{\ell=1}^{n} (\xi_\ell+\eta_\ell)^{-\alpha_\ell-1},$$

then $\sum_j q_j^{(\ell)}(D)\delta$ belongs to $\mathcal{N}^\infty(U \cap (\bigcap_{k\neq\ell} V_k))$. Hence we may assume from the first that $p_j(\xi,\eta)$ has the form

$$(2.13) \quad \sum_{\alpha\in\mathbb{Z}_+^n} p_{j,\alpha}(\xi) \prod_{\ell=1}^{n} (\xi_\ell+\eta_\ell)^{-\alpha_\ell-1}.$$

Here we note that $p_{j,\alpha}(\xi)$ is homogeneous of degree $j+|\alpha|+n$ and defined on a neighborhood of ξ^0. Furthermore Cauchy's integral formula combined with (2.10) and (2.11) entails the following:

There exists a neighborhood W of $\xi^0 = (0,\ldots,0,1)$ on which p_j's are defined and satisfy the following conditions:

(2.14) For each $\varepsilon > 0$ and $\kappa > 0$ there exists a constant $C_{\varepsilon,\kappa}$ such that

$$\sup_W |p_{j,\alpha}| \leqq C_{\varepsilon,\kappa}\, \varepsilon^j \kappa^{|\alpha|}/j! \qquad (j \geqq 0)$$

(2.15) For each $\kappa > 0$ there exists a constant R_κ such that

$$\sup_W |p_{j,\alpha}| \leqq (-j)!\,\kappa^{|\alpha|} R_\kappa^{-j} \qquad (j < 0).$$

On the other hand, we have the following formula:

$$(2.16)\quad \prod_{\ell=1}^{n} (D_{x_\ell} + D_{y_\ell})^{-1-\alpha_\ell} \delta(x,y) = \frac{x^\alpha}{\alpha!} \prod_{\ell=1}^{n} (D_{x_\ell} + D_{y_\ell})^{-1} \delta(x,y).$$

Hence, as a section of $\mathcal{N}^\infty(U \cap (\bigcap_{\ell=1}^{n} V_\ell))$,

$$\sum_{\substack{j\in\mathbb{Z} \\ \alpha\in\mathbb{Z}_+^n}} p_{j,\alpha}(D_x) \prod_{\ell=1}^{n} (D_{x_\ell} + D_{y_\ell})^{-\alpha_\ell - 1} \delta(x,y)$$

is equal to

$$\sum_{\substack{j\in\mathbb{Z} \\ \alpha\in\mathbb{Z}_+^n}} \frac{1}{\alpha!} p_{j,\alpha}(D_x) x^\alpha \prod_{\ell=1}^{n} (D_{x_\ell} + D_{y_\ell})^{-1} \delta(x,y).$$

Furthermore we have

$$(2.17)\quad (D_{x_\ell} + D_{y_\ell}) \prod_{k=1}^{n} (D_{x_k} + D_{y_k})^{-1} \delta(x,y) = \prod_{k\neq\ell} (D_{x_k} + D_{y_k})^{-1} \delta(x,y) \in \mathcal{N}^\infty(U \cap (\bigcap_{k\neq\ell} V_k))$$

and we find

$$(2.18)\quad (x_\ell - y_\ell) \prod_{k=1}^{n} (D_{x_k} + D_{y_k})^{-1} \delta(x,y) = 0.$$

Hence, as an element of $H^0(U; \widetilde{\mathscr{D}}_\Lambda^\infty)$,

$$w \underset{\text{def}}{=} \prod_{k=1}^{n} (D_{x_k} + D_{y_k})^{-1} \delta(x,y)$$

satisfies the system of differential equations

$$\begin{cases} (D_{x_\ell} + D_{y_\ell}) w = 0 & (\ell = 1, \dots, n) \\ (x_\ell - y_\ell) w = 0 & (\ell = 1, \dots, n). \end{cases}$$

Hence w can be identified with $\delta(x-y)$. Therefore w can be identified with the identity operator, and hence, $\sum_j p_j(D_x, D_y)\delta(x,y)$

can be identified with $\left(\sum_{\substack{j\in\mathbb{Z}\\ \alpha\in\mathbb{Z}_+^n}} \frac{1}{\alpha!} p_{j,\alpha}(D_x)x^\alpha\right)$.

Now, if we define $q_{j,i}(\xi,x)$ by $\sum_{|\alpha|=i} \frac{1}{\alpha!} p_{j-i-n,\alpha}(\xi)x^\alpha$, then $q_{j,i}(\xi,x)$ is a holomorphic function defined on a neighborhood Ω of ξ^0 and a polynomial in x, where Ω is independent of i and j. It is homogeneous of degree i in x and of degree j in ξ. Further, they satisfy the following growth conditions:

(2.19) For each $\varepsilon > 0$ and each compact set K of $\Omega \times \mathbb{C}_x^n$, there exists a constant $C_{\varepsilon,K}$ such that

$$\sup_K |q_{j,i}| \leqq \frac{C_{\varepsilon,K}\varepsilon^{i+j}}{i!(j-i)!}$$

holds if $j \geqq i$.

(2.20) For each $\varepsilon > 0$ and each compact set K in $\Omega \times \mathbb{C}_x^n$, there exists a constant $R_{\varepsilon,K}$ such that

$$\sup_K |q_{j,i}| \leqq \frac{(i-j)!}{i!} \varepsilon^i R_{\varepsilon,K}^{i-j}$$

holds if $i > j$.

Now we introduce the following

Definition 2.7. The symbol sequence of $P \in \tilde{\mathcal{D}}_\Lambda^\infty(\Omega)$ $(\Omega \subset \Lambda = T^*_{\{0\}}\mathbb{C}^n)$ is, by definition, the doubly indexed sequence $\{q_{j,i}(\xi,x)\}_{\substack{i\in\mathbb{Z}_+\\ j\in\mathbb{Z}}}$ which satisfies the following conditions:

(2.21) $q_{j,i}(\xi,x)$ is a holomorphic function defined on $\Omega \times \mathbb{C}_x^n$ which is homogeneous of degree j in ξ and a polynomial of degree i in x and which satisfies the growth conditions (2.19) and (2.20).

Using this definition, we can summalize our result in the following form:

Theorem 2.8. An operator P in $\widetilde{\mathcal{D}}^{\infty}_{\Lambda}(\Omega)$ determines a symbol sequence $\{q_{j,i}(\xi,x)\}_{\substack{i\in\mathbb{Z}_+\\ j\in\mathbb{Z}}}$ which satisfies conditions (2.19) and (2.20) in a unique manner and each symbol sequence $\{q_{j,i}(\xi,x)\}_{\substack{i\in\mathbb{Z}_+\\ j\in\mathbb{Z}}}$ satisfying the the above conditions determine an operator P belonging to $\widetilde{\mathcal{D}}^{\infty}_{\Lambda}(\Omega)$ by defining it by $\sum_{j,i} q_{j,i}(D_x,x)$. Here $q_{j,i}(D_x,x)$ means that the multiplication by x should be applied first and next the action by D_x is considered.

Remark 2.9. We often express the symbol sequence of $P\in\widetilde{\mathcal{D}}^{\infty}_{\Lambda}(\Omega)$ as $\{p_{i,j}(x,\xi)\}_{\substack{i\in\mathbb{Z}_+,\\ j\in\mathbb{Z}}}$ where $p_{i,j}$ is homogeneous of degree j in ξ and a homogeneous polynomial of degree i in x. When the symbol sequence is given in this form, we assign $P\in\widetilde{\mathcal{D}}^{\infty}_{\Lambda}(\Omega)$ to it by setting $P = \sum_{i,j} p_{i,j}(x,D_x)$, namely, the (micro-)differentiation is applied first and next comes the multiplicaiton by x. In this case $\{p_{i,j}(x,\xi)\}$ is related to $\{q_{i,j}(\xi,x)\}$ in Theorem 2.8 by

$$p_{i,j}(x,\xi) = \sum_{\alpha} \frac{1}{\alpha!} D^{\alpha}_{\xi} D^{\alpha}_{x} q_{j+|\alpha|,i+|\alpha|}(\xi,x)$$

and

$$q_{j,i}(\xi,x) = \sum_{\alpha} \frac{(-1)^{|\alpha|}}{\alpha!} D^{\alpha}_{\xi} D^{\alpha}_{x} p_{i+|\alpha|,j+|\alpha|}(x,\xi).$$

Furthermore $\{p_{i,j}(x,\xi)\}$ satisfies the same growth conditions that $\{q_{j,i}(\xi.x)\}$ satisfies. Hence $\widetilde{\mathcal{D}}^{\infty}_{\Lambda}$ is closed under the operation of taking the formal adjoint.

Remark 2.10. Let $P = \sum_j p_j(x,D)$ be in $\mathcal{E}^{\infty}_X|_{\Lambda}$. Then we can assign $\sum_{i,j} p_{i,j}(x,D)$ in $\widetilde{\mathcal{D}}^{\infty}_{\Lambda}$ to P, by defining $p_{i,j}(x,\xi)$ to be the part of $p_j(x,\xi)$ that is homogeneous of degree i in x. This assignment is consistent with the embedding $\mathcal{E}^{\infty}_X|_{\Lambda} \rightarrowtail \widetilde{\mathcal{D}}^{\infty}_{\Lambda}$. Note also that the Ring structure of $\widetilde{\mathcal{D}}^{\infty}_{\Lambda}$ coincides with that of $\mathcal{E}^{\infty}_X|_{\Lambda}$ when it is restricted to this subsheaf, namely, the composition $\sum_{m,n} r_{m,n}(x,D_x)$

of two operators $\sum_{i,j} p_{i,j}(x,D_x)$ and $\sum_{k,\ell} q_{k,\ell}(x,D_x)$ in $\tilde{\mathcal{D}}^\infty_\Lambda$ is given by the following:

(2.22) $$r_{m,n}(x,\xi) = \sum_{\substack{m=i+k-|\alpha| \\ n=j+\ell-|\alpha|}} \frac{1}{\alpha!} D^\alpha_\xi p_{i,j}(x,\xi) D^\alpha_x q_{k,\ell}(x,\xi)$$

<u>Remark 2.11</u>. Theorem 2.8 establishes the correspondence between an operator in $\tilde{\mathcal{D}}^\infty_\Lambda$ and a suitable symbol sequence, when $\Lambda = T^*_{\{0\}}\mathbb{C}^n$. An analogous result can be obtained in the case where $\Lambda = T^*_Y\mathbb{C}^n$ with an affine variety Y of $\mathbb{C}^n$ in the following manner:

Let Y be given by $\{x \in \mathbb{C}^n;\ x' \underset{\text{def}}{=} (x_1,\dots,x_d) = 0\}$ and $\Lambda = T^*_Y\mathbb{C}^n$. Denote $(\xi_{d+1},\dots,\xi_n)$ by ξ''. Then there exists a one-to-one correspondence between $\tilde{\mathcal{D}}^\infty_\Lambda(\Omega)$ $(\Omega \subset \Lambda)$ and the set of doubly-indexed sequences $\{q_{i,j}(x,\xi)\}_{\substack{i\in\mathbb{Z}_+ \\ j\in\mathbb{Z}}}$ which satisfy the following condition (2.23) and growth conditions (2.19) and (2.20).

(2.23) $q_{i,j}(x,\xi)$ is a holomorphic function defined on $\Omega \times \mathbb{C}^n_{(x',\xi'')}$ which is homogeneous of degree j in ξ and a homogeneous polynomial of degree i in (x',ξ'').

In a similar way we can find the characterization of operators in $\tilde{\mathcal{E}}^\infty_\Lambda$. (Cf. S-K-K [17] Chap. II, §1.4.)

<u>Theorem 2.12</u>. Let Λ be $T^*_{\{0\}}\mathbb{C}^n$ and let P be an operator in $\tilde{\mathcal{E}}^\infty_\Lambda(U)$, where U is an open subset of $T^*\Lambda \cong \mathbb{C}^n_x \times \mathbb{C}^n_\xi$. Then there exists a symbol sequence $\{p_{i,j}(x,\xi)\}_{i,j\in\mathbb{Z}}$ which satisfies the following conditions (2.24) $\sim$ (2.28) so that P can be expressed as $\sum_{i,j} p_{i,j}(x,D_x)$. Conversely, if $\{p_{i,j}(x,\xi)\}_{i,j\in\mathbb{Z}}$ satisfies the conditions (2.24) $\sim$ (2.28), then $\sum_{i,j} p_{i,j}(x,D_x)$ defines an operator in $\tilde{\mathcal{E}}^\infty_\Lambda(\Omega)$.

(2.24) $p_{i,j}(x,\xi)$ is a holomorphic function defined on U and it is homogeneous of degree i in x and of degree j in ξ.

(2.25) For each $\varepsilon > 0$ and each compact subset K of U, there exists a constant $C_{\varepsilon,K}$ such that

$$\sup_K |p_{i,j}| \leqq \frac{C_{\varepsilon,K}}{j!} \varepsilon^j$$

holds for $j \geqq i \geqq 0$.

(2.26) For each $\varepsilon > 0$ and each compact subset K of U, there exists a constant $R_{\varepsilon,K}$ such that

$$\sup_K |p_{i,j}| \leqq \frac{(i-j)!}{i!} R_{\varepsilon,K}^{i-j} \varepsilon^i$$

holds for $i \geqq 0$ and $j < i$.

(2.27) For each compact subset K of U, there exists a constant R_K such that

$$\sup_K |p_{i,j}| \leqq (-j)! R_K^{-j}$$

holds for $j < i < 0$.

(2.28) For each compact subset K of U, there exists a constant R_K such that for each $\varepsilon > 0$ we can find another constant $C_{\varepsilon,K}$ so that

$$\sup_K |p_{i,j}| \leqq \frac{(-i)!}{(j-i)!} C_{\varepsilon,K} \varepsilon^{j-i} R_K^{-i}$$

holds for $i < 0$ and $j \geqq i$.

<u>Remark 2.13</u>. If Λ is $T_Y^*\mathbb{C}^n$ with $Y = \{x \in \mathbb{C}^n;\ x_1 = \cdots = x_d = 0\}$, then by replacing the condition (2.24) with the following condition (2.29), we obtain the same result.

(2.29) $p_{i,j}(x,\xi)$ is a holomorphic function defined on U and it is homogeneous of degree i in $(x_1,\ldots,x_d,\xi_{d+1},\ldots,\xi_n)$ and of degree j in ξ.

§3. Construction of the sheaves on which $\tilde{\mathcal{D}}^\infty_\Lambda$ and $\tilde{\mathcal{E}}^\infty_\Lambda$ act.

In this section we construct sheaves $\tilde{\mathcal{B}}_\Lambda$ and $\tilde{\mathcal{C}}_\Lambda$ on which $\tilde{\mathcal{D}}^\infty_\Lambda$ and $\tilde{\mathcal{E}}^\infty_\Lambda$ act respectively. The first subsection is devoted to the proof of two theorems which relate the cohomology groups on tangential sphere bundles and those on cotangential sphere bundles. The results obtained there are effectively used in the second subsection to study basic properties of $\tilde{\mathcal{B}}_\Lambda$ and $\tilde{\mathcal{C}}_\Lambda$.

§3.1. Correspondence between sheaves on sphere bundles and co-sphere bundles.

Let M be a topological space and V a real vector bundle over M with fiber dimension d. Let V^* be the dual bundle of V. Denote by S (resp., S^*) the sphere bundle associated with V (resp., V^*), namely, $S = (V-M)/\mathbb{R}^+$ and $S^* = (V^*-M)/\mathbb{R}^+$. Denote by γ (resp., γ^*) the projection from $V-M$ (resp., V^*-M) to S (resp., S^*). Let D denote $\{(x,\bar{v}_x,\bar{\eta}_x) = (x,\bar{v}_x;x,\bar{\eta}_x) \in S \underset{M}{\times} S^* ;\ \langle v_x,\eta_x\rangle \geqq 0\}$, where $(\bar{v}_x,\bar{\eta}_x)$ is the equivalence class of $(x,v_x,\eta_x) \in (V-M) \underset{M}{\times} (V^*-M)$. We denote by τ (resp., π) the projection from S onto M (resp., from S^* onto M). The projection from D onto S^* (resp., onto S) is also denoted by τ (resp., π).

<u>Theorem 3.1.1</u>. Let $\mathcal{F}$ be a complex of sheaves on S and let $\mathcal{G}$ denote the complex of sheaves $\mathbb{R}\tau_*\pi^{-1}\mathcal{F}$. Let K be a closed and properly convex[(*)] set of S^*. Let $K°$ be the polar of K. Then we have

$$(3.1.1)\quad \mathbb{R}\Gamma_K(S^*;\mathcal{G}) = \mathbb{R}\Gamma_{\tau^{-1}\pi(K)}(K°;\mathcal{F})[1-d].$$

<u>Proof</u>. Set $M' = \pi(K)$. Then we have

$$\mathbb{R}\Gamma_K(S^*;\mathcal{G}) = \mathbb{R}\Gamma_K(S^*;\mathbb{R}\Gamma_{\pi^{-1}(M')}(\mathcal{G}))$$

(*) Convexity means that each fiber $K_x \underset{\text{def}}{=} K \cap \pi^{-1}(x)$ is convex, i.e., $\gamma^{*-1}(K_x)$ is convex. Proper convexity means that $\gamma^{*-1}(K_x)$ does not contain a line.

and

$$\mathbb{R}\Gamma_{\tau^{-1}(M')}(K^\circ;\mathcal{F}) = \mathbb{R}\Gamma(K^\circ;\ \mathbb{R}\Gamma_{\tau^{-1}(M')}(\mathcal{F})).$$

Since $\mathbb{R}\Gamma_{\pi^{-1}(M')}(\mathcal{G}) = \mathbb{R}\tau_*\pi^{-1}\mathbb{R}\Gamma_{\tau^{-1}(M')}(\mathcal{F})$ holds, by replacing M with M', we may assume from the first that

(3.1.2) $\quad \pi(K) = M.$

Furthermore, by considering a flabby resolution of $\mathcal{F}$, we can reduce the problem to the case where $\mathcal{F}$ is a flabby sheaf.

Let f denote $\pi|_{\tau^{-1}(S^*-K)}$. Then we have

(3.1.3) $\quad \mathbb{R}\Gamma(S^*-K;\mathcal{G})$

$$= \mathbb{R}\Gamma(S^*-K;\ \mathbb{R}\tau_*\pi^{-1}\mathcal{F})$$

$$= \mathbb{R}\Gamma(\tau^{-1}(S^*-K);\ \pi^{-1}\mathcal{F})$$

$$= \mathbb{R}\Gamma(S;\ \mathbb{R}f_*f^{-1}\mathcal{F}).$$

On the other hand, we also have

(3.1.4) $\quad \mathbb{R}\Gamma(S^*;\mathcal{G})$

$$= \mathbb{R}\Gamma(\tau^{-1}S^*;\ \pi^{-1}\mathcal{F})$$

$$= \mathbb{R}\Gamma(\pi^{-1}S;\ \pi^{-1}\mathcal{F})$$

$$= \mathbb{R}\Gamma(S;\mathcal{F}).$$

Hence, by combining (3.1.3) and (3.1.4) with the long exact sequence of relative cohomology groups, we obtain the following triangle:

(3.1.5)
$$\begin{array}{ccc} & \mathbb{R}\Gamma_K(S^*;\mathcal{G}) & \\ \swarrow & & \nwarrow {\scriptstyle +1} \\ \mathbb{R}\Gamma(S;\mathcal{F}) & \longrightarrow & \mathbb{R}\Gamma(S;\ \mathbb{R}f_*f^{-1}\mathcal{F}). \end{array}$$

On the other hand, the following triangle follows from the definition of $\mathbb{R}\mathcal{D}ist_f(\mathcal{F})$ (cf. S-K-K [17] p.270 Remark):

(3.1.6)
$$\begin{array}{ccc} & \mathbb{R}\Gamma(S;\ \mathbb{R}\mathcal{D}ist_f(\mathcal{F})) & \\ \swarrow & & \nwarrow{\scriptstyle +1} \\ \mathbb{R}\Gamma(S;\mathcal{F}) & \longrightarrow & \mathbb{R}\Gamma(S;\ \mathbb{R}f_*f^{-1}\mathcal{F}). \end{array}$$

Therefore we find

(3.1.7) $\quad \mathbb{R}\Gamma_K(S^*;\mathcal{G}) = \mathbb{R}\Gamma(S;\ \mathbb{R}\mathcal{D}ist_f(\mathcal{F})).$

Thus the problem is reduced to the calculation of $\mathbb{R}\mathcal{D}ist_f(\mathcal{F})$. In order to calculate it, we appeal to a limitting procedure. For this purpose we choose a decreasing family of open and properly convex subset U_n of S^* so that $K = \bigcap_{n\geqq 0} U_n$ holds. Let f_n denote $\pi|_{\tau^{-1}(S^*-U_n)}$. Note that

(3.1.8) $\quad f_n^{-1}(x,v_x) = \{(x,v_x,\eta_x) \in S \underset{M}{\times} S^*\ ;\ \ (x,\eta_x) \in U_n^c \text{ and } \langle v_x,\eta_x\rangle \geqq 0\},$

where $U_n^c = S^*-U_n$.

First suppose that (x,v_x) is not contained in U_n°, the polar set of U_n. Then $f_n^{-1}(x,v_x)$ is a non-empty contractible set. In fact, it follows from the assumption that there exists (x,η^0) in U_n such that $\langle v_x,\eta^0\rangle < 0$. Hence $f_n^{-1}(x,v_x)$ contains $(x,v_x,-\eta^0)$. Then, for any η in $f_n^{-1}(x,v_x)$ and any t such that $0 \leqq t \leqq 1$,

$$t\eta - (1-t)\eta^0 \in f_n^{-1}(x,v_x).$$

In fact, if $(x,\ t\eta-(1-t)\eta^0)$ were in U_n, then $(x,t\eta)$ $=(x,t\eta-(1-t)\eta^0+(1-t)\eta^0)$ should be in U_n because of its convexity.

Thus we have seen that $f_n^{-1}(x,v_x)$ is contractible to the point $(x,v_x,-\eta^0)$ in this case. Hence we have

(3.1.9) $\quad \mathbb{R}\mathcal{D}ist_{f_n}(\mathcal{F})_{(x,v_x)} = 0, \quad \text{if } (x,v_x) \notin U_n^\circ.$

Next we consider the case where (x,v_x) is contained in U_n°. In this case, U_n is contained in $\{(x,v_x)\}^\circ$. On the other hand, it follows from the definition that (x,v_x,η_x) is contained in $f_n^{-1}(x,v_x)$ if and only if (x,η_x) is contained in $\{(x,v_x)\}^\circ \cap U_n^c$. Since U_n is open, $f_n^{-1}(x,v_x)$ is a closed annulus, namely, homotopic to S^{d-2}. Therefore we have

$$(3.1.10)\quad \mathbb{R}\mathcal{D}ist_{f_n}(\mathcal{F})_{(x,v_x)} = \mathbb{R}\Gamma(S^{d-2} \to \{\text{point}\}; \mathcal{F}_{(x,v_x)})$$

$$= \mathcal{F}_{(x,v_x)}[1-d], \quad \text{if} \quad (x,v_x) \quad \text{in} \quad U_n^\circ.$$

Thus we find

$$(3.1.11)\quad \mathbb{R}\mathcal{D}ist_{f_n}(\mathcal{F}) = \mathcal{F}[1-d]\Big|_{U_n^\circ}$$

and hence

$$(3.1.12)\quad \mathbb{R}\Gamma(S; \mathbb{R}\mathcal{D}ist_{f_n}(\mathcal{F})) = \mathbb{R}\Gamma(U_n^\circ; \mathcal{F})[1-d].$$

Since $\{U_n\}_{n\geqq 0}$ is a decreasing family of proper convex sets whose limit set K is a closed and proper convex set each of whose fiber is not void, $\{U_n^\circ\}_{n\geqq 0}$ is an increasing family of properly convex sets whose limit set is K°. On the other hand, we have assumed that $\mathcal{F}$ is flably. Therefore $\{H^j(U_n^\circ;\mathcal{F})\}_{n\geqq 0}$ satisfies the (ML)-condition. Hence we have

$$(3.1.13)\quad \varprojlim_n H^{j+1-d}(U_n^\circ;\mathcal{F}) = H^{j+1-d}(K^\circ;\mathcal{F}).$$

Combining (3.1.7), (3.1.12) and (3.1.13), we finally obtain

$$\mathbb{R}\Gamma_K(S^*;\mathcal{G}) = \mathbb{R}\Gamma(S; \mathbb{R}\mathcal{D}ist_f(\mathcal{F}))$$

$$= \mathbb{R}\Gamma(K^\circ;\mathcal{F})[1-d]. \qquad \text{Q.E.D.}$$

In application, we usually use Theorem 3.1.1 when $\mathcal{F}$ has a special structure. It enables us to obtain more concrete expression of the cohomology groups in question, as we show below. In what follows, we consider the problem in the following geometric situation:

Let N be a submanifold of codimension d of a real analytic manifold M. Denote by $\widetilde{{}^N M}$ the (real) monoidal transform of M with center N, i.e., $\widetilde{{}^N M} = (M-N) \sqcup S_N M$. We denote by τ the projection from $\widetilde{{}^N M}$ to M. In what follows, the triplet $(S_N M, S_N^* M, N)$ corresponds to (S, S^*, M) in Theorem 3.1.1.

<u>Theorem 3.1.2</u>. Let $\mathcal{K}$ be a sheaf on M and define $\mathcal{F}$ by $\mathbb{R}\Gamma_{S_N M}(\tau^{-1}\mathcal{K})$. Let $\mathcal{G}$ be $\mathbb{R}\tau_* \pi^{-1}\mathcal{F}$ and let K be a closed and properly convex subset of $S^* = S_N^* M$. Then we have

$$(3.1.14) \qquad H^j_K(S^*; \mathcal{G}) = \varinjlim_{Z,U} H^{j+1-d}_{Z\cap U}(U; \mathcal{K}),$$

where U ranges over a system of open neighborhood of $\pi(K)$ and Z ranges over the set of closed subset of U such that $Z \cap N \subset \pi(K)$ and $C_N(Z) \cap K^\circ = \phi$. Here $C_N(Z)$ denotes the normal cone of Z along N.

<u>Proof</u>. Theorem 3.1.1 asserts

$$H^j_K(S^*; \mathcal{G}) = H^{j+1-d}_{\tau^{-1}\pi(K)}(K^\circ; \mathcal{F}).$$

Hence it suffices to show that

$$H^j_{\tau^{-1}\pi(K)}(K^\circ; \mathcal{F}) = \varinjlim_{Z,U} H^j_Z(U; \mathcal{K}).$$

On the other hand, it follows from the definition of $\mathcal{F}$ that

$$\mathbb{R}\Gamma_{\tau^{-1}\pi(K)}(K^\circ; \mathcal{F}) = \mathbb{R}\Gamma_{\tau^{-1}\pi(K)\cap K^\circ}(W; \tau^{-1}\mathcal{K}),$$

where W is an open subset of $\widetilde{{}^N M}$ such that $W \cap S_N M = K^\circ$. Hence we obtain the following triangle:

$$(3.1.15) \qquad \begin{array}{ccc} & \mathbb{R}\Gamma_{\tau^{-1}\pi(K)}(K^\circ; \mathcal{F}) & \\ \swarrow & & \nwarrow {\scriptstyle +1} \\ \varinjlim_{W} \mathbb{R}\Gamma(W; \tau^{-1}\mathcal{K}) & \longrightarrow & \varinjlim_{W} \mathbb{R}\Gamma(W-(K^\circ \cap \tau^{-1}\pi(K)); \tau^{-1}\mathcal{K}) \end{array},$$

where W ranges over the set of neighborhoods of K° in $\widetilde{{}^N M}$ such that $W \cap S_N M = K°$. Then we have

$$(3.1.16) \qquad \varinjlim_{W} \mathbb{R}\Gamma(W;\ \tau^{-1}\mathcal{K}) = \mathbb{R}\Gamma(K°;\ \tau^{-1}\mathcal{K}).$$

Now we note that a fiber of $K°$ over a point x in N is contractible or $(S_N M)_x$ according as x is in $\pi(K)$ or not. Hence we obtain the following triangle:

$$(3.1.17) \qquad \begin{array}{ccc} & \mathbb{R}\Gamma(N;\mathcal{K}) & \\ \swarrow & & \nwarrow {\scriptstyle +1} \\ \mathbb{R}\Gamma(K°;\tau^{-1}\mathcal{K}) & \longrightarrow & \mathbb{R}\Gamma(N-\pi(K);\mathcal{K})[1-d]. \end{array}$$

Next let us define a subset $\tilde{W}$ of $M-N$ by $\tau(W-S_N M)$. Then we have $W-(K° \cap \tau^{-1}\pi(K)) = \tau^{-1}(\tilde{W} \cup (N-\pi(K)))$, because $K°-\tau^{-1}\pi(K) = \tau^{-1}(N-\pi(K))$. On the other hand, we easily see

$$\mathbb{R}\mathcal{D}ist_{\tau}(\mathcal{K}) = \mathcal{K}|_N[-d].$$

Therefore we obtain the following triangle:

$$(3.1.18) \qquad \begin{array}{ccc} & \mathbb{R}\Gamma(\tilde{W} \cup (N-\pi(K));\mathcal{K}) & \\ \swarrow & & \nwarrow {\scriptstyle +1} \\ \mathbb{R}\Gamma(W-(K° \cap \tau^{-1}\pi(K));\mathcal{K}) & \longrightarrow & \mathbb{R}\Gamma(N-\pi(K);\mathcal{K})[1-d]. \end{array}$$

Comparing (3.1.15), (3.1.16), (3.1.17) and (3.1.18), we find the following triangle:

$$\begin{array}{ccc} & \mathbb{R}\Gamma_{\tau^{-1}\pi(K)}(K°;\mathcal{F}) & \\ \swarrow & & \nwarrow {\scriptstyle +1} \\ \mathbb{R}\Gamma(N;\mathcal{K}) & \longrightarrow & \varinjlim_{W} \mathbb{R}\Gamma((W-S_N M) \cup (N-\pi(K));\mathcal{K}). \end{array}$$

Here we note that $\mathbb{R}\Gamma(N;\mathcal{K}) = \varinjlim_{U} \mathbb{R}\Gamma(U;\mathcal{K})$ holds when U ranges over the set of neighborhoods of N. Hence, by setting $W' = (W-S_N M) \cup (N-\pi(K))$, we obtain

$$\begin{array}{ccc} & \varinjlim\limits_{U,W} \mathbb{R}\Gamma_{U-W'}(U;\mathcal{K}) & \\ \swarrow & & \nwarrow {\scriptstyle +1} \\ \varinjlim\limits_{U} \mathbb{R}\Gamma(U;\mathcal{K}) & \longrightarrow & \varinjlim\limits_{U,W} \mathbb{R}\Gamma(U\cap W';\mathcal{K}) \end{array}$$

Thus we finally obtain

$$\mathbb{R}\Gamma_{\tau^{-1}\pi(K)}(K^\circ;\mathcal{F}) = \varinjlim_{U,W} \mathbb{R}\Gamma_{U-W'}(U;\mathcal{K}). \tag{3.1.19}$$

Since $U-W'$ ranges over the set of closed subset Z of U such that $Z\cap N = K$ and that $C_N(Z) \subset K^\circ$, (3.1.19) is the required result.

Q.E.D.

§3.2. Definition and basic properties of sheaves $\widetilde{\mathscr{B}}_\Lambda$ and $\widetilde{\mathcal{C}}_\Lambda$.

In §2 we introduced sheaves $\widetilde{\mathscr{D}}^\infty_\Lambda$ and $\widetilde{\mathcal{E}}^\infty_\Lambda$ for a Lagrangian submanifold Λ of T^*X. As we observed there, $\widetilde{\mathcal{E}}^\infty_\Lambda$ deserves the name of second-microlocalization of operators. Hence it is natural to try to find sheaves $\widetilde{\mathscr{B}}_\Lambda$ and $\widetilde{\mathcal{C}}_\Lambda$ on which $\widetilde{\mathscr{D}}^\infty_\Lambda$ and $\widetilde{\mathcal{E}}^\infty_\Lambda$ act respectively, as in the case of (micro-)differential operators. In order to do this job, we hereafter assume that X is a complexification of a real analytic manifold M of dimension n and that the purely imaginary locus $\Lambda^{\mathbb{R}}$ of Λ, i.e., $\Lambda \cap \sqrt{-1}T^*M$, is also a (purely imaginary) Lagrangian submanifold of $\sqrt{-1}T^*M$. Throughout this subsection we choose and fix a simple holonomic $\mathcal{E}_X$-Module $\mathcal{M}$ with support Λ. We denote $\mathcal{E}^{\mathbb{R}}_X \otimes_{\mathcal{E}_X} \mathcal{M}$ by $\mathcal{M}^{\mathbb{R}}$. In parallel with the discussion in §2, we will introduce sheaves $\widetilde{\mathscr{B}}_\Lambda$ and $\widetilde{\mathcal{C}}_\Lambda$ by using the same procedure employed in defining the sheaf $\mathscr{B}_M$ of hyperfunctions and the sheaf $\mathcal{C}_M$ of microfunctions. In defining new sheaves $\widetilde{\mathscr{B}}_\Lambda$ and $\widetilde{\mathcal{C}}_\Lambda$, we start with $\mathcal{M}^{\mathbb{R}}$ instead of $\mathcal{O}_X$. In what follows, we denote by π the projection from $\widetilde{\Lambda^{\mathbb{R}}\Lambda^*}$ to Λ.

<u>Definition 3.2.1</u>. (i) $\widetilde{\mathscr{B}}_\Lambda \underset{\text{def}}{=} \mathscr{H}^n_{\Lambda^{\mathbb{R}}}(\mathcal{M}^{\mathbb{R}})$

$$\text{(ii)} \quad \tilde{\mathcal{C}}_\Lambda \underset{\text{def}}{=} \mathcal{H}^n_{T^*_{\Lambda^{\mathbb{R}}}\Lambda} (\pi^{-1}\mathcal{M}^{\mathbb{R}})^a .$$

Since we can easily prove that $\mathcal{M}^{\mathbb{R}}$ is a $\tilde{\mathcal{D}}^\infty_\Lambda$-Module, $\tilde{\mathcal{B}}_\Lambda$ is also a $\tilde{\mathcal{D}}^\infty_\Lambda$-Module. Using the same reasoning as in S-K-K [17] Chap. III, §1, we can also verify that $\tilde{\mathcal{C}}_\Lambda$ is an $\tilde{\mathcal{E}}^\infty_\Lambda$-Module.

In order to study basic properties of $\tilde{\mathcal{B}}_\Lambda$ and $\tilde{\mathcal{C}}_\Lambda$, we need several cohomology vanishing theorems.

We first show that $\Lambda^{\mathbb{R}}$ is purely n-codimensional with respect to $\mathcal{M}^{\mathbb{R}}$, namely, we prove the following

<u>Theorem 3.2.2</u>. $\quad \mathcal{H}^j_{\Lambda^{\mathbb{R}}}(\mathcal{M}^{\mathbb{R}}) = 0 \quad$ for $\; j \neq n$.

<u>Proof</u>. Let p be a point in $\Lambda^{\mathbb{R}}$. Since the problem is local in T^*X, we can find a real contact transformation which brings $(\Lambda^{\mathbb{R}}, p)$ to $(\sqrt{-1}T^*_NM, p^0)$ for a non-singular hypersurface N of M. We may choose a local coordinate system $z = (z_1,\ldots,z_n) = (x_1+\sqrt{-1}y_1,\ldots, x_n+\sqrt{-1}y_n)$ around $\pi(p^0) \in X$ so that $M = \{y_1=\cdots=y_n=0\}$, $N = \{x_1=y_1=\cdots=y_n=0\}$ and $p^0 = (0;\sqrt{-1}dx_1)$ hold. We denote by Y the complex hypersurface $\{z_1=0\}$. By a quantized contact transformation, we can bring $\mathcal{M}$ to a simple holonomic system whose support is T^*_YX. Hence we may assume without loss of generality that $\mathcal{M}^{\mathbb{R}}$ is $\mathcal{C}^{\mathbb{R}}_{Y|X}$. (S-K-K [17] Chap. II, Definition 1.1.4 and Theorem 4.2.5.) Then Theorem 3.1.2 implies the following:

$$(3.2.1.) \quad \mathcal{H}^j_{\sqrt{-1}T^*_NM}(\mathcal{C}^{\mathbb{R}}_{Y|X})_{p^0} = \varinjlim_{U,Z} H^j_{U\cap Z}(U;\mathcal{O}_X),$$

where U ranges over the set of open neighborhoods of 0 in $\mathbb{C}^n$ and Z ranges over the set of closed subsets of $\mathbb{C}^n$ which satisfy the following two conditions:

$$(3.2.2) \quad Z\cap Y \subset N$$

$$(3.2.3) \quad C_Y(Z) \cap \{z=x+\sqrt{-1}y \in \mathbb{C}^n;\ y_1 > 0\} = \phi.$$

Here we have identified T_YX with X by using the linear coordinate system on X. Furthermore, by shrinking U and replacing Z with $\overline{U\cap Z}$ if necessary, we may assume without loss of generality

(3.2.4) $\quad Z$ is compact

(3.2.5) $\quad Z \subset \{z \in \mathbb{C}^n;\ y_1 \leqq |x_1|\}.$

In what follows, for compact subset Z of $\mathbb{C}^n$, $\hat{Z}$ denotes $\{z \in \mathbb{C}^n;\ |f(z)| \leqq \max_Z |f|$ for each entire function $f\}$. We say that a compact set Z is holomorphically convex, if $\hat{Z} = Z$ holds.

We now recall the following result ([10]).

<u>Proposition 3.2.3</u>. Let Z_1 and Z_2 be compact subsets of $\mathbb{C}^n$. Assume that both Z_1 and Z_2 are holomorphically convex. Then we have

$$H^j_{Z_1 - Z_2}(\mathbb{C}^n - Z_2;\ \mathcal{O}_{\mathbb{C}^n}) = 0 \qquad \text{for} \quad j \neq n.$$

We will prove the vanishing of (3.2.1) by using this result.

Let $\mathfrak{Z}$ denote the family of compact subsets Z of $\mathbb{C}^n$ satisfying conditions (3.2.2), (3.2.3), (3.2.4) and (3.2.5). We shall show that $\hat{Z}$ belongs to $\mathfrak{Z}$ if Z belongs to $\mathfrak{Z}$. First we show that $\hat{Z}$ satisfies the condition (3.2.2). Note that the condition (3.2.2) is equivalent to

(3.2.6) $\quad (Z-M) \cap Y = \phi.$

This condition implies the following:

For each $\varepsilon > 0$, there exists $\delta_\varepsilon > 0$ such that

(3.2.7) $\quad Z \cap \{z \in \mathbb{C}^n;\ |y'| \geqq \varepsilon,\ |z_1| \leqq \delta_\varepsilon\} = \phi.$

Here and in what follows, z', x' and y' denote $(z_2, \ldots, z_n)$, $(x_2, \ldots, x_n)$ and $(y_2, \ldots, y_n)$, respectively.

Let z^0 be a point in $Y-M$. Define an open subset R_c $(c > 0)$ of $\mathbb{C}$ by $\mathbb{C} - \{w = u + \sqrt{-1}v \in \mathbb{C};\ v \geq c\}$. We will show that there exists an entire function $f(z)$ $(z \in \mathbb{C}^n)$ which satisfies

(3.2.8) $\quad f(z^0) \notin R_c$

and

(3.2.9) $\quad f(Z) \subset R_c.$

Since R_c is a simply connected open subset of $\mathbb{C}$, R_c is a Runge domain. Hence there exists an entire function $\varphi(w)$ $(w \in \mathbb{C})$ which satisfies

(3.2.10) $\quad \operatorname{Re} \varphi(f(z^0)) \geqq 0$

and

(3.2.11) $\quad \operatorname{Re} \varphi(f(z)) < 0 \qquad \text{if} \quad z \in Z.$

Therefore the existence of such $f(z)$ will entail that $\hat{Z}$ satisfies the condition (3.2.2).

Now we try to find such $f(z)$. We may assume without loss of generality that

$$\operatorname{Im} z^0 = (0,b,0,\ldots,0) \qquad (b > 0).$$

Define $f(z)$ by $z_1 - \sqrt{-1}a(z_2-x_2^0)^2$, where a is a positive constant which we shall specify later. Choose δ_ε as in (3.2.7) for $\varepsilon = b/2$. First let us consider the case where $|y'| \leqq \varepsilon = b/2$. Let g and h denote $\operatorname{Re}(-\sqrt{-1}a(z_2-x_2^0)^2)$ and $\operatorname{Im}(-\sqrt{-1}a(z_2-x_2^0)^2)$, respectively. Then we have

$$g^2 = 4a^2(x_2-x_2^0)^2 y_2^2 \leqq 4a^2\varepsilon^2(x_2-x_2^0)^2$$

and

$$h = a(y_2^2-(x_2-x_2^0)^2) \leqq a\varepsilon^2 - a(x_2-x_2^0)^2.$$

It then follows that

(3.2.12) $\quad h + \dfrac{g^2}{4a\varepsilon^2} \leqq a\varepsilon^2.$

If there were z in Z such that $f(z) \notin R_c$, then we should have

(3.2.13) $\quad \begin{cases} x_1 + g = 0 \\ y_1 + h = t \end{cases}$

with $t \geqq c$. In view of (3.2.5), we should then conclude

(3.2.14) $|g| + h \geqq c.$

Combining (3.2.12) and (3.2.14), we should obtain

$$|g| - \frac{g^2}{4a\varepsilon^2} \geqq c - a\varepsilon^2,$$

and hence

(3.2.15) $g^2 - 4a\varepsilon^2|g| + 4ac\varepsilon^2 - 4a^2\varepsilon^4 = (|g| - 2a\varepsilon^2)^2 + 4a\varepsilon^2(c - 2a\varepsilon^2) \leqq 0.$

If we choose $c = ab^2$, $f(z^0) \notin R_c$. On the other hand, we have

$$c - 2a\varepsilon^2 = ab^2 - 2a(\tfrac{b}{2})^2 = \frac{ab^2}{2} > 0 .$$

This implies that (3.2.15) is not satisfied. Hence, for this choice of c, $f(Z \cap \{|y'| < \varepsilon\})$ is contained in R_c. Note that we need not have fixed a so far. Next let us consider the case where $|y'| \geqq \varepsilon$ $(= b/2)$. Since Z is compact, there exists a constant κ such that

$$Z \subset \{z \in \mathbb{C}^n; |z| \leqq \kappa\}.$$

In this case it follows from (3.2.5) and (3.2.7) that

(3.2.16) $\left|\frac{z_1}{a} - \sqrt{-1}t\right| \geqq \frac{\delta_\varepsilon}{\sqrt{2}a}$

holds for every positive t. Hence if we choose a (>0) so that

$$-\frac{\delta_\varepsilon}{\sqrt{2}a} + (\kappa + |x_2^0|)^2 < b^2$$

and

$$\frac{\delta_\varepsilon}{\sqrt{2}a} - (\kappa + |x_2^0|)^2 > 0$$

hold, then $f(z)/a$ is contained in R_{b^2}. Then it is obvious that for sufficiently small $a > 0$, $f(z)/a$ is contained in R_{b^2}. Hence $f(z)$ is contained in R_c. Thus we have verified that $\hat{Z}$ satisfies the condition (3.2.2).

Next we show that $\hat{Z}$ satisfies the condition (3.2.3). The condition (3.2.3) is equivalent to the following:

For any $\varepsilon > 0$, there exists δ_ε such that

$$(3.2.17)\qquad Z \subset \{z \in X;\ y_1 \leqq \varepsilon |x_1|\} \cup \{z \in X;\ |z_1| \geqq \delta_\varepsilon\}.$$

Then it is easy to see that $\hat{Z}$ satisfies the condition (3.2.17), and hence (3.2.3). It is also clear that $\hat{Z}$ satisfies conditions (3.2.4) and (3.2.5), if Z satisfies them. Therefore we may replace $\mathfrak{Z}$ with $\mathfrak{Z}_0 \underset{\mathrm{def}}{=} \{Z \in \mathfrak{Z};\ \hat{Z} = Z\}$ when we consider the inductive limit in (3.2.1). Next we shall show that, for each $\varepsilon > 0$ and each Z_1 in $\mathfrak{Z}_0$, we can find a compact and holomorphically convex set Z_2 such that

$$(3.2.18)\qquad 0 \notin Z_2$$

and

$$(3.2.19)\qquad Z_1 - Z_2 \subset \{z \in \mathbb{C}^n;\ |z| < 4\varepsilon\} \underset{\mathrm{def}}{=} U_\varepsilon.$$

If this is the case, we have

$$(3.2.20)\qquad \varinjlim_{Z_1, U} H^j_{Z_1 \cap U}(U;\ \mathcal{O}_{\mathbb{C}^n})$$

$$- \varinjlim_{Z_1, U} H^j_{Z_1 \cap U \cap (Z_2^c)}(U \cap Z_2^c;\ \mathcal{O}_{\mathbb{C}^n}).$$

$$= \varinjlim_{Z_1, U} H^j_{Z_1 - Z_2}(U \cap Z_2^c;\ \mathcal{O}_{\mathbb{C}^n}).$$

Here Z_2^c denotes $\mathbb{C}^n - Z_2$. Then it follows from Proposition 3.2.3 and the excision theorem that

$$H^j_{Z_1 - Z_2}(U \cap Z_2^c;\ \mathcal{O}_{\mathbb{C}^n}) = H^j_{Z_1 - Z_2}(Z_2^c;\ \mathcal{O}_{\mathbb{C}^n}) = 0$$

holds for $j \neq n$.

Now we embark on the construction of required Z_2. Let K denote the set $\{w = u + \sqrt{-1}v \in \mathbb{C};\ v \leq |u| - \beta\}$ and let $\varphi(z)$ denote $z_1 - \sqrt{-1}\alpha(z_2^2 + \cdots + z_n^2)$. Here α and β are sufficiently small positive constants which we will specify later under the constraint

$$(3.2.21)\qquad \beta < 2\alpha\varepsilon^2.$$

We now define Z_2 by $\varphi^{-1}(K) \cap Z_1$. Since Z_1 is compact and since we may assume that Z_1 is contained in $\{z = x+\sqrt{-1}y \in \mathbb{C}^n;\ 2y_1 \leqq |x_1|\}$, $Z_1 \cap \{|z_1| \geqq \varepsilon\} - U_\varepsilon$ is contained in Z_2 if we choose α and β sufficiently small. In view of (3.2.7) we can also conclude that $Z_1 \cap \{|y'| \geqq \varepsilon\} - U_\varepsilon$ is contained in Z_2. Hence it suffices to show that $Z_1' \underset{\text{def}}{=} Z_1 \cap \{|x'| \geqq 3\varepsilon,\ |y'| \leqq \varepsilon,\ |z_1| \leqq \varepsilon\} - U_\varepsilon$ is contained in Z_2. In order to see this, we have only to verify that $\operatorname{Im} \varphi(z) \leqq |\operatorname{Re}\varphi(z)| - \beta$ holds on Z_1', namely,

$$y_1 + \alpha(|y'|^2 - |x'|^2) \leqq |x_1 + 2\alpha\langle x', y'\rangle| - \beta \tag{3.2.22}$$

holds on Z_2'. Since $y_1 \leqq |x_2|/2$ holds on Z_1, (3.2.22) is satisfied if

$$\alpha(|y'|^2 - |x'|^2 + 2|\langle x', y'\rangle|) \leqq -\beta \tag{3.2.23}$$

holds. Since $|x'| \geqq 3\varepsilon$ and $|y'| \leqq \varepsilon$ hold on Z_1, (3.2.23) follows from the constraint (3.2.21). Q.E.D.

<u>Corollary 3.2.4</u>. The sheaf $\widetilde{\mathcal{B}}_\Lambda$ is flabby.

<u>Proof</u>. Since the flabby dimension of $\mathcal{M}^{\mathbb{R}}$ is equal to or at most n, this immediately follows from the theorem.

By a reasoning similar to the proof of Theorem 3.2.2, we can prove

<u>Theorem 3.2.5</u>. There exists an injective $\mathcal{E}_X^\infty$-homomorphism from $\mathcal{C}_M|_\Lambda$ to $\widetilde{\mathcal{B}}_\Lambda$. More precisely, if we choose an $\mathcal{E}_X$-linear injective homomorphism φ from $\mathcal{M}$ into $\mathcal{C}_M$, then there exists ψ such that the composition $\psi\circ\varphi\colon \mathcal{M} \to \widetilde{\mathcal{B}}_\Lambda$ coincides with the composition of the canonical homomorphism from $\mathcal{M}$ to $\mathcal{M}^{\mathbb{R}}$ and that from $\mathcal{M}^{\mathbb{R}}$ to $\widetilde{\mathcal{B}}_\Lambda$.

Probably the same argument will work for the proof of the following conjecture. However, so far, we have been unable to prove the conjecture completely. (The point which we cannot prove at the moment is the vanishing of $\mathcal{H}^j_{T^*_\Lambda\mathbb{R}^\Lambda}(\pi^{-1}\mathcal{M}^{\mathbb{R}})_p$ $(j \neq n)$ at p where $\omega = \sum \xi_j dx_j$ vanishes.)

<u>Conjecture 3.2.6</u>. (i) $\mathcal{H}^{j}_{T^*_{\Lambda}\mathbb{R}^{\Lambda}}(\pi^{-1}\mathcal{M}^{\mathbb{R}}) = 0 \quad (j \neq n)$

(ii) $$0 \longrightarrow \mathcal{M}^{\mathbb{R}} \longrightarrow \widetilde{\mathcal{B}}_{\Lambda} \xrightarrow{\widetilde{\mathrm{sp}}} \pi_* \widetilde{\mathcal{C}}_{\Lambda} \longrightarrow 0$$

is exact.

In what follows, for a section f of the sheaf $\widetilde{\mathcal{B}}_{\Lambda}$, we denote supp $\widetilde{\mathrm{sp}}(f)$ by $\widetilde{SS}_{\Lambda} f$.

§4. Asymptotic expansion of solutions of holonomic systems with regular singularities.

In this section we clarify the meaning of the asymptotic expansion of a microfunction solution of a holonomic system with R.S.(*) by the aid of sheaves $\tilde{\mathcal{D}}^{\infty}_{\Lambda}$ and $\tilde{\mathcal{B}}_{\Lambda}$. Our procedure is as follows:

First, for a (microfunction) solution of a holonomic system with R.S., we find a micro-differential equation of a special type which it satisfies. (Theorem 4.1.1 and Proposition 4.1.6 in §4.1.) Next we employ $\tilde{\mathcal{D}}^{\infty}_{\Lambda}$ to bring each equation of that type to the simple and canonical equation whose $\tilde{\mathcal{B}}_{\Lambda}$-solution is easy to describe. Using this result we clarify the meaning of the asymptotic expansion of a microfunction solution of the equation in question with respect to a Lagrangian manifold Λ, namely, we grasp the notion of the asymptotic expansion of a microfunction by regarding it as a section of the sheaf $\tilde{\mathcal{B}}_{\Lambda}$.

§4.1. Holonomic systems with R.S. and indicial polynomials.

The purpose of this subsection is to prove Theorem 4.1.1 stated below. The structure of the equation $(b(\vartheta)-P)u = 0$ appearing there will be investigated in the next sybsection by the aid of the sheaf $\tilde{\mathcal{D}}^{\infty}_{\Lambda}$. Before stating the theorem, we prepare some notations.

Let Λ be a Lagrangian submanifold of T^*X. Denote by I_{Λ} the defining Ideal of Λ and by $\mathcal{J}_{\Lambda}$ the Ideal $\{P \in \mathcal{E}_X(1); \sigma_1(P)|_{\Lambda} = 0\}$. Let $\mathcal{E}_{\Lambda}$ denote the sub-Ring of $\mathcal{E}_X$ generated by $\mathcal{J}_{\Lambda}$ and define $\mathcal{E}_{\Lambda}(m)$ by $\mathcal{E}_{\Lambda}\mathcal{E}_X(m)$ $(= \mathcal{E}_X(m)\mathcal{E}_{\Lambda})$. As we noted in §1, $\mathcal{E}_{\Lambda}/\mathcal{E}_{\Lambda}(-1)$ is locally isomorphic to $\mathcal{A}_{\Lambda}$, which is a sub-Ring of the sheaf of linear differential operators on Λ.

Now let us consider a micro-differential operator $\vartheta = \vartheta_1 + \vartheta_0 + \cdots$ which satisfies the conditions below. Here ϑ_1 and ϑ_0 are respectively the first order and the zeroth order term of ϑ.

(4.1.1) $\vartheta \in \mathcal{J}_{\Lambda}$

(4.1.2) $d\vartheta_1 \equiv \omega_X \; (= \sum_j \xi_j dx_j) \mod I_{\Lambda}\Omega^1$

(*) R.S. is the abbreviation of "regular singularities". See Kashiwara-Kawai [8], [9] for the definition.

$$(4.1.3) \quad \vartheta_0 - \frac{1}{2}\sum_j \frac{\partial^2 \vartheta_1}{\partial x_j \partial \xi_j} = 0 \quad \text{on } \Lambda.$$

Here Ω^1 denotes the sheaf of holomorphic 1-forms on T^*X. It is then easy to verify that such ϑ always exists and it is unique modulo $\mathcal{J}_\Lambda^2 \mathcal{E}_X(-1)$. For example, if $\Lambda = T_Y^*\mathbb{C}^n$ with $Y = \{x \in \mathbb{C}^n; x_1 = \ldots = x_d = 0\}$, then we may choose $\frac{1}{2}(\sum_{j=1}^{d}(x_jD_j + D_jx_j))$ as ϑ. In what follows, we always use the letter ϑ to denote such an operator. Now we have the following

<u>Theorem 4.1.1</u>. Let u be a section of a holonomic $\mathcal{E}_X$-Module with R.S. and let Λ be an arbitrary Lagrangian submanifold of T^*X. Then there exist a non-zero polynomial $b(s)$ and a micro-differential operator $P \in \mathcal{E}_\Lambda(-1)$ which satisfy the following conditions:

$$(4.1.4) \quad \deg b \geqq \text{order } P$$

$$(4.1.5) \quad (b(\vartheta) - P)u = 0.$$

In order to prove Theorem 4.1.1, we prepare several lemmas. In what follows $\mathcal{E}(m)$ denotes the sheaf of micro-differential operators of order at most m. For an $\mathcal{E}(0)$-Module $\mathcal{L}$, $\mathcal{L}(m)$ denotes $\mathcal{E}(m)\mathcal{L}$. The symbols $\mathring{T}^*X$ and $\mathring{\pi}$ denote respectively $T^*X - T_X^*X$ and the projection from $\mathring{T}^*X$ onto X.

<u>Lemma 4.1.2</u>. Let X be $\mathbb{C}^{1+n} = \{(t,x); t \in \mathbb{C}, x \in \mathbb{C}^n\}$, $V = \{(t,x;\tau,\xi) \in T^*X;\ t=0\}$ and $\Lambda = \{(t,x;\tau,\xi) \in T^*X;\ t=\xi=0, \tau \neq 0\}$. Let Θ denote tD_t. Let $\mathcal{M}$ be a holonomic $\mathcal{E}_X$-Module with R.S. which is defined on a neighborhood of a point p of Λ, and let $\mathcal{L}$ be a coherent $\mathcal{E}(0)$-sub-Module of $\mathcal{M}$. If $\operatorname{Supp}\mathcal{M}$ is contained in V, then there exists a non-zero polynomial $b(s)$ such that $b(\Theta)\mathcal{L} \subset \mathcal{L}(-1)$.

<u>Proof</u>. First we note that it suffices to show the existence of a coherent $\mathcal{E}(0)$-sub-Module $\mathcal{L}'$ of $\mathcal{M}$ and a non-zero polynomial $b(\Theta)$ such that $\mathcal{E}\mathcal{L}' = \mathcal{M}$, $b(\Theta)\mathcal{L}' \subset \mathcal{L}'(-1)$ and $\Theta\mathcal{L}' \subset \mathcal{L}'$ hold. In fact, if it is the case, then there exists an integer r such that $\mathcal{L} \cap \mathcal{L}'(-r) \subset \mathcal{L}(-1)$ holds. On the other hand,

$$b(\Theta-j)\mathcal{L}'(-j) = b(\Theta-j)D_t^{-j}\mathcal{L}' = D_t^{-j}b(\Theta)\mathcal{L}' \subset \mathcal{L}'(-j-1)$$

holds. Hence, by setting $b(\Theta-r+1)\cdots b(\Theta)$ to be $\tilde{b}(\Theta)$, we obtain

$\tilde{b}(\Theta)\mathcal{L}' \subset \mathcal{L}'(-r)$. Therefore $\tilde{b}(\Theta)\mathcal{L} \subset \mathcal{L}'(-r) \cap \mathcal{L} \subset \mathcal{L}(-1)$ holds.

Now we shall prove the existence of such $\mathcal{L}'$ and b. Since $\mathcal{M}$ is with R.S. and since Supp $\mathcal{M}$ is contained in V, $\mathcal{M}$ has regular singularities along $V \cap \mathring{T}^*X$. ([8] Chap.V, §1, Corollary 5.1.7.) Let M be the monodromy of $\mathcal{M}$ in the sense of [11]. Then, by the definition, M is an endomorphism of $\mathcal{M}$. Since End($\mathcal{M}$) is finite-dimensional, $\mathcal{M}$ is a direct sum of sub-Modules $\mathcal{M}_j$ such that $(M-\lambda_j)^{N_j}|_{\mathcal{M}_j} = 0$ for some $\lambda_j \in \mathbb{C}^\times$ and some integer N_j. Hence we may assume from the first that there exist $\lambda \in \mathbb{C}$ and $N \in \mathbb{Z}_+$ such that $(M-\exp(2\pi\sqrt{-1}\lambda))^N\mathcal{M} = 0$ holds. Since $\mathcal{M}$ has regular singularities along V, there exists a coherent $\mathcal{E}(0)$-sub-Module $\mathcal{L}$ such that $\mathcal{M} = \mathcal{E}\mathcal{L}$ and that $\Theta\mathcal{L} \subset \mathcal{L}$. Let $u_1,\ldots,u_m$ be a system of generators of $\mathcal{L}$ as an $\mathcal{E}(0)$-Module. Then there exist $P_{ij} \in \mathcal{E}(0)$ $(1 \leqq i,j \leqq m)$ such that $\Theta u_i = \Sigma P_{ij}u_j$. If we denote by u the column vector ${}^t(u_1, \ldots, u_m)$ and by P the matrix (P_{ij}), we have $\Theta u = Pu$. We know ([11]) that there exists $U \in GL(m; \mathcal{E}(0))$ and a matrix $A(x)$ of functions in x such that $\Theta - P = U(\Theta - A(x))U^{-1}$. Hence, by replacing u with $U^{-1}u$, we may assume that $\Theta u = A(x)u$ holds. Then it follows from the definition of M that $Mu = \exp(2\pi\sqrt{-1}A(x))u$ holds. Since $(M-\exp(2\pi\sqrt{-1}\lambda))^r u = 0$ holds, we have $(\exp(2\pi\sqrt{-1}A(x)) - \exp(2\pi\sqrt{-1}\lambda))^r u = 0$. Considering the problem on a neighborhood of x_0, we can find an integer c so that all the eigenvalues of $A(x)$ are contained in $\{t \in \mathbb{C};\ |t-\lambda| < c\}$ for any x in the neighborhood. Let us now define $b_0(s)$ and $G(x)$ by $\prod_{\nu=-c}^{c}(s-\nu)$ and $\frac{1}{2\pi\sqrt{-1}}\int_{|t-\lambda|=c+\frac{1}{2}} \frac{b_0(t)(t-A(x))^{-1}}{(\exp(2\pi\sqrt{-1}t)-\exp(2\pi\sqrt{-1}\lambda)}dt$.

Then $G(x)(\exp(2\pi\sqrt{-1}A(x))-\exp(2\pi\sqrt{-1}\lambda)) = b_0(A(x))$ holds, and, moreover, $G(x)$ and $A(x)$ are commutative. Hence we obtain

$$b_0(\Theta)^r u = b_0(A(x))^r u = G(x)^r(\exp(2\pi\sqrt{-1}A(x))-\exp(2\pi\sqrt{-1}\lambda))^r u = 0.$$

This entails $b_0(\Theta)^r\mathcal{L} \subset \mathcal{L}(-1)$. Q.E.D.

Lemma 4.1.3. Let $\mathcal{M}$ be a coherent $\mathcal{D}_X$-Module and let $\mathcal{F}$ be a coherent $\mathcal{O}_X$-sub-Module of $\mathcal{M}$ such that $\mathcal{D}\mathcal{F} = \mathcal{M}$ holds. Let $\mathcal{L}$ be a coherent $\mathcal{E}(0)$-sub-Module of $\mathcal{E} \underset{\mathcal{D}}{\otimes} \mathcal{M}$ which is generated by $\mathcal{F}$. Then we have the following:

(i) $\mathring{\pi}_*(\mathcal{L}(k)/\mathcal{L}(k-1)) = \mathcal{D}_k\mathcal{F}/\mathcal{D}_{k-1}\mathcal{F}$ holds for $k >> 0$. Here and in what follows $\mathcal{D}_k$ denotes the sheaf of linear differential operators of order at most k.

(ii) $\mathcal{D}_k\mathcal{F} = \{u \in \mathcal{M};\ 1 \otimes u$ is contained in $\mathcal{L}(k)$ on $\mathring{T}^*X\}$ holds for $k >> 0$.

(iii) If a section u of $\mathcal{M}$ belongs to $\mathcal{L}(k)$ at a point p of $\mathring{T}^*X$, then there exists a linear differential operator P of order m such that $Pu \in \mathcal{D}_{m+k-1}$ and $\sigma_m(P)(p) \neq 0$.

Proof. There exist integers r_1, r_2, N_0, N_1, N_2 $(r_1 \leqq r_2)$ and an exact sequence

$$0 \leftarrow \mathcal{M} \overset{\varphi}{\leftarrow} \mathcal{D}^{N_0} \overset{P_0}{\leftarrow} \mathcal{D}^{N_1} \overset{P_1}{\leftarrow} \mathcal{D}^{N_2}$$

such that $\varphi(\mathcal{O}^{N_0}) = \mathcal{F}$, $P_0(\mathcal{D}_k^{N_1}) \subset (\mathcal{D}_{k-r_1})^{N_0}$, $P_1(\mathcal{D}^{N_2}_{k-r_2}) \subset \mathcal{D}^{N_1}_{k-r_1}$ for $k \geqq 0$ and that

$$(4.1.6)\quad 0 \leftarrow \frac{\mathcal{D}_k\mathcal{F}}{\mathcal{D}_{k-1}\mathcal{F}} \leftarrow \left(\frac{\mathcal{D}_k}{\mathcal{D}_{k-1}}\right)^{N_0} \leftarrow \left(\frac{\mathcal{D}_{k-r_1}}{\mathcal{D}_{k-r_1-1}}\right)^{N_1} \leftarrow \left(\frac{\mathcal{D}_{k-r_2}}{\mathcal{D}_{k-r_2-1}}\right)^{N_2}$$

is exact for $k \geqq r_2$. Then the following sequence

$$\mathcal{E}(0)^{N_0} \overset{P_0}{\leftarrow} \mathcal{E}(-r_1)^{N_1} \overset{P_1}{\leftarrow} \mathcal{E}(-r_2)^{N_2}$$

is also exact. Since $\mathcal{M}$ is the cokernel of P_0, we obtain the exact sequence

$$0 \leftarrow \mathcal{L}(k) \leftarrow \mathcal{E}(k)^{N_0} \leftarrow \mathcal{E}(k-r_1)^{N_0}$$

for each k. Hence its symbol sequence

$$0 \leftarrow \mathcal{L}(k)/\mathcal{L}(k-1) \leftarrow (\mathcal{E}(k)/\mathcal{E}(k-1))^{N_0} \leftarrow (\mathcal{E}(k-r_1)/\mathcal{E}(k-r_1-1))^{N_1}$$

is also exact. Here we recall that $\mathcal{E}(k)/\mathcal{E}(k-1) \cong \mathcal{O}(k)$, the sheaf of holomorphic functions on $\mathring{T}^*X$ which are homogeneous of degree k with respect to the fiber coordinate. Then the following sequence (4.1.7) is exact for $k >> 0$ by Serre's theorem on the vanishing of cohomology groups of coherent sheaves on a projective space.

$$(4.1.7)\quad 0 \leftarrow \mathring{\pi}_*(\mathcal{L}(k)/\mathcal{L}(k-1)) \leftarrow \mathring{\pi}_*(\mathcal{O}(k)^{N_0}) \leftarrow \mathring{\pi}_*(\mathcal{O}(k-r_1)^{N_1}).$$

Since $\mathring{\pi}_*(\mathcal{O}(k)) \cong \mathcal{D}_k/\mathcal{D}_{k-1}$ holds for $k \geqq 0$, by comparing (4.1.6) and (4.1.7), we obtain

$$(4.1.8)\quad \mathring{\pi}_*(\mathcal{L}(k)/\mathcal{L}(k-1)) = \mathcal{D}_k\mathcal{F}/\mathcal{D}_{k-1}\mathcal{F}$$

for $k \geqq k_0$ with a sufficiently large k_0. This proves (i). Next we shall prove (ii) for $k \geqq k_0$. Let u be a section of $\mathcal{M}$ such that $1 \otimes u$ belongs to $\mathcal{L}(k)|_{\overset{\circ}{T}{}^*X}$ for some $k \geqq k_0$. Then there exists j such that $\mathcal{D}_j\mathcal{F}$ contains u. If j is larger than k, $1 \otimes u$ is zero in $\overset{\circ}{\pi}_*(\mathcal{L}(j)/\mathcal{L}(j-1))$. Hence (4.1.8) implies that $\mathcal{D}_{j-1}\mathcal{F}$ contains u. Thus the induction on j porves that $\mathcal{D}_k\mathcal{F}$ contains u. This proves (ii).

Finally let us prove (iii). Let S denote the set of points where $1 \otimes u$ does not belong to $\mathcal{L}(k)$. Then there exists a homogeneous polynomial a on T^*X such that $a(p) \neq 0$ and $a|_S = 0$. Let r be the homogeneous degree of a and let P be a linear differential operator of order r whose principal symbol is a. Then it follows from Hilbert's Nullstellensatz that $P^\nu u$ belongs to $\mathcal{L}(k+r\nu-1)$ for $\nu>>0$. Hence (ii) implies $P^\nu u \in \mathcal{D}_{k+r\nu-1}\mathcal{F}$. This proves (iii). Q.E.D.

Lemma 4.1.4. Let X, V and Λ be the same as in Lemma 4.1.2 and let $\mathcal{M}$ be a holonomic $\mathcal{D}_X$-Module with R.S. Assume that $\mathrm{Supp}(\mathcal{E} \underset{\mathcal{D}}{\otimes} \mathcal{M})$ is contained in V on a neighborhood of Λ. Let u be a section of $\mathcal{M}$. Then there exist a non-zero polynomial $b(s)$ and a linear differential operator P which satisfy the following:

(4.1.9) $b(tD_t)u = tPu,$

where $\operatorname{order} P \leqq \deg b$ and $P \in \mathcal{E}_\Lambda$.

Proof. We may assume without loss of generality that $\mathcal{M} = \mathcal{D}u$. Let $\tilde{u}$ be the section $1 \otimes u$ of $\mathcal{E} \otimes \mathcal{M}|_{\overset{\circ}{T}{}^*X}$. Then Lemma 4.1.2 guarantees that there exists a non-zero polynomial $b(s)$ such that $b(\Theta)\tilde{u} \in \mathcal{E}(-1)\tilde{u}$ holds on a neighborhood of $p \in \Lambda$, where $\Theta = tD_t$. Hence it follows from Lemma 4.1.3 that there exists a linear differential operator P of order m such that

(4.1.10) $\sigma_m(P)(p) \neq 0$

(4.1.11) $Pb(\Theta)u = Qu$ with $Q \in \mathcal{D}_{m-1}$.

By the condition (4.1.10) we may assume that P has the form

$$D_t^m - \sum_{j=0}^{m-1} A_j(t,x,D_x)D_t^j,$$

where A_j is of order at most $m-j$. Let Q be written in the form

$$\sum_{j=0}^{m-1} Q_j(t,x,D_x)D_t^j,$$

where Q_j is of order at most $m-1-j$. Then, setting $\tilde{b}(s) = b(s)s(s+1)\cdots(s+m-1)$, we obtain from (4.1.11)

$$\tilde{b}(\Theta)u = t^m D_t^m b(\Theta)u = t(\sum_{j=0}^{m-1}(A_j b(\Theta-m+1)+Q_j)t^{m-1}D_t^j)u.$$

This proves the required result. Q.E.D.

Lemma 4.1.5. Let X and Λ be the same as in Lemma 4.1.2. Let $\mathcal{M}$ be an arbitrary holonomic $\mathcal{D}_X$-Module with R.S. and u a section of $\mathcal{M}$. Then there exists a non-zero polynomial $b(s)$ and a linear differential operator P which satisfy the following:

(4.1.12) $b(tD_t)u = tPu,$

where $\operatorname{order} P \leqq \deg b$ and $P \in \mathcal{E}_\Lambda$.

Proof. For each positive integer N we denote by f_N the map from X to X defined by $f_N(t,x) = (t^N,x)$. Then the associated map f_{N*} is an isomorphism on $T^*X - \{t=0\}$, because $f_{N*}(t,x;\tau,\xi) = (t^N,x;\tau/Nt^{N-1},\xi)$. Let W be $\operatorname{Supp}(\mathcal{E}\otimes\mathcal{M})$. Then W is a Lagrangian subvariety of T^*X, and hence $C_\Lambda(W)\subset\{t=0\}\subset T_\Lambda(T^*X)$. ([12], Chap.X.) Therefore there exist an integer N and a constant C such that

$$t^{N-1}\tau^N \leqq C(|t\tau|+|\xi|)^N,$$

or equivalently,

$$(t\tau)^N \leqq Ct(|t\tau|+|\xi|)^N.$$

Therefore we have

(4.1.13) $(t\tau/N)^N \leqq Ct^N(|t\tau/N|+|\xi|)^N$

on $W' \underset{\text{def}}{=} (f_{N*})^{-1}(W\cap\{t\neq 0\})$.

The inequality (4.1.13) implies that

$$\overline{W'}\cap\Lambda = \phi.$$

In the sequel, we fix such an N and set $f = f_N$. Let $\mathcal{M}'$ be $f^*(\mathcal{M})$

and let $\tilde{u}$ be the section $1 \otimes u$ of $\mathcal{M}'$. Then $\mathcal{M}'$ is also with R.S. Further $\operatorname{Supp}(\mathcal{E} \otimes \mathcal{M}')$ is contained in $V \cap W'$. Hence $\mathcal{M}'$ satisfies all the conditions in the preceding lemma. Therefore there exist a polynomial $b(s)$ and $P \in \mathcal{D} \cap \mathcal{E}_\Lambda$ such that $b(tD_t)\tilde{u} = tP(t,x,tD_t,D_x)\tilde{u}$ holds with $\deg b \geqq \operatorname{order} P$. Hence, if we denote by ε the primitive N-th root of 1, we obtain

$$(4.1.14) \quad b(tD_t)\tilde{u} = \frac{1}{N}\sum_{i=0}^{N-1} \varepsilon^i tP(\varepsilon t,x,tD_t,D_x)\tilde{u}.$$

It is easy to see that $\frac{1}{N}\sum_{i=0}^{N-1} \varepsilon^i tP(\varepsilon t,x,tD_t,D_x)$ can be written in the form $t^N Q(t^N,x,tD_t,D_x)$, because it is invariant under the transformation $t \mapsto \varepsilon t$. Therefore (4.1.14) implies that

$$b(NtD_t)u = tQ(t,x,NtD_t,D_x)u$$

holds outside $\{t = 0\}$. Here we note that $\operatorname{order} Q = \operatorname{order} P \leqq \deg b$. Hence it follows from Hilbert's Nullstellensatz that there exists r such that

$$(4.1.15) \quad t^r b(N\Theta)u = t^{r+1}Q(t,x,N\Theta,D_x)u.$$

By multiplying the both sides of (4.1.15) by D_t^r, we finally obtain

$$\tilde{b}(\Theta)u = t(\Theta+2)\cdots(\Theta+r+1)Q(t,x,N\Theta,D_x)u$$

with $\tilde{b}(s) = (s+1)\cdots(s+r)b(Ns)$. This completes the proof. Q.E.D.

Now let us prove Theorem 4.1.1. By a suitable contact transformation, we can bring $\operatorname{Supp}\mathcal{M}$ to a generic position and Λ to T_Y^*X for a hypersurface Y of X. ([8] Chap. I, §6, Corollary 1.6.4.) Then we may assume that $\mathcal{M}$ is a holonomic $\mathcal{D}_X$-Module with R.S., because $\operatorname{Supp}\mathcal{M}$ is in a generic position. ([8] Chap. V, §1, Theorem 5.1.4.) Therefore Theorem 4.1.1 follows from Lemma 4.1.5. Q.E.D.

The following proposition easily follows from Theorem 4.1.1.

<u>Proposition 4.1.6</u>. Let $\mathcal{M}$ be a holonomic $\mathcal{E}_X$-Module with R.S. and let Λ be an arbitrary Lagrangian submanifold of T^*X. Then there exist a system of generators $u_1,\ldots,u_m$ of $\mathcal{M}$, a non-zero polynomial $b(s)$ and micro-differential operators $P_{ij} \in \mathcal{E}_\Lambda(-1)$ $(1 \leqq i,j \leqq m)$ which satisfy the following three conditions:

(4.1.16) $b(\vartheta)u_i = \sum_j P_{ij}u_j$

(4.1.17) $\deg b \geqq \text{order}\, P_{ij}$

(4.1.18) The difference of any two different roots of $b(s) = 0$ is not an integer.

Proof. It follows from Theorem 4.1.1 that there exist a coherent $\mathcal{E}(0)$-sub-Module $\mathcal{L}$ of $\mathcal{M}$ and a non-zero polynomial $b(s)$ of degree, say, r which satisfy

(4.1.19) $b(\vartheta)\mathcal{L} \subset \mathcal{J}_\Lambda^{r+1}\mathcal{L}(-1)$

and

(4.1.20) $\mathcal{M} = \mathcal{E}\mathcal{L}$.

We now show that we can choose $\mathcal{L}$ and b so that they satisfy the additional condition (4.1.18). For this purpose, we prove the following:

If $b(s)$ has the form $\tilde{b}(s)(s-\lambda)$ and satisfies (4.1.19) and (4.1.20), then there exists a coherent $\mathcal{E}(0)$-sub-Module $\mathcal{L}'$ which satisfies

(4.1.21) $(\vartheta-\lambda)\tilde{b}(\vartheta-1)\mathcal{L}' \subset \mathcal{J}_\Lambda^{r+1}\mathcal{L}'(1)$.

We will prove this fact by showing that

$$\mathcal{L}' = \mathcal{J}_\Lambda\mathcal{L}+(\vartheta-\lambda-1)(\mathcal{L}(1))$$

is a required one. This $\mathcal{L}'$ is clearly coherent $\mathcal{E}(0)$-Module. Furthermore we see by an easy computation that $b(\vartheta-1)(\mathcal{L}(1))$ is contained in $\mathcal{J}_\Lambda^{r+1}\mathcal{L}$. Hence we have

$$(\vartheta-\lambda)\mathcal{L} \subset \mathcal{L}'(-1).$$

Therefore we obtain

$$\tilde{b}(\vartheta-1)\mathcal{L}' \subset \tilde{b}(\vartheta-1)\mathcal{J}_\Lambda\mathcal{L}+b(\vartheta-1)(\mathcal{L}(1)) \subset \mathcal{J}_\Lambda^{r+1}\mathcal{L}.$$

This implies

$$(\vartheta-\lambda)\tilde{b}(\vartheta-1)\mathcal{L}'\subset(\vartheta-\lambda)\mathcal{J}_\Lambda^{r+1}\mathcal{L}\subset\mathcal{J}_\Lambda^{r+2}(-1)\mathcal{L}+\mathcal{J}_\Lambda^{r+1}(\vartheta-\lambda)\mathcal{L}\subset\mathcal{J}_\Lambda^{r+1}\mathcal{L}'(-1).$$

Thus we obtain the required result. Q.E.D.

§4.2. Transforming equations to the canonical one.

The purpose of this section is to find the canonical form of the micro-differential equations of the form (4.1.5) by using $\tilde{\mathcal{D}}_\Lambda^\infty$. Needless to say, $\tilde{\mathcal{D}}_\Lambda^\infty$ does not act as a sheaf homomorphism on the sheaf of microfunctions, and hence the nature of the reduction to the canonical form discussed in this subsection is different from that of the classical one (e.g. S-K-K [17] Chap. II, §5). Actually the reduction discussed here changes the characteristic variety of the equation in question. Still $\tilde{\mathcal{D}}_\Lambda^\infty$ acts on $\tilde{\mathcal{B}}_\Lambda$ as a sheaf homomorphism. Therefore the structure of $\tilde{\mathcal{B}}_\Lambda$-solutions can be investigated by our reduction and it enables us to clarify the meaning of the asymptotic expansion of a microfunction solution. Before stating our main theorem, we re-examine the example (1.6) discussed in §1 from the view point of the transformation of the equations.

We first consider the following system of equations which admits $(y^2-x^3)_+^s$ as its solution:

$$(4.2.1)\qquad \begin{cases} (\frac{1}{3}xD_x+\frac{1}{2}yD_y-s)u = 0 \\ (2yD_x+3x^2D_y)u = 0 \end{cases}$$

Choosing xD_x+yD_y+1 as ϑ, we obtain the following equation from (4.2.1).

$$(4.2.2)\qquad (\vartheta-3s-3/2)(\vartheta-2s-1)u = x^3D_y^2u/4.$$

For simplicity we assume that $s+1/2$ is not an integer. Then u admits an asymptotic expansion of the following form:

$$(4.2.3)\qquad u \sim \sum_{j=0}^{\infty} u_j + \sum_{j=0}^{\infty} v_j,$$

where u_j and v_j are homogeneous of degree $2s+j$ and $3s+1/2+j$, respectively. By (1.6) in §1, we find

$$(4.2.4)\qquad u_j = \frac{\Gamma(-s+1/2)}{j!\Gamma(-s+1/2+j)} \times (\tfrac{1}{4}x^3D_y^2)^j u_0$$

$$(4.2.5)\quad v_j = \frac{\Gamma(s+3/2)}{j!\,\Gamma(s+3/2+j)}\times(\tfrac{1}{4}x^3D_y^2)^j v_0,$$

namely,

$$(4.2.6)\quad u \sim (x^{3/2}D_y/2)^{s+1/2}\Gamma(-s+1/2)I_{-s-1/2}(x^{3/2}D_y)u_0$$
$$+(x^{3/2}D_y/2)^{-s-1/2}\Gamma(s+3/2)I_{s+1/2}(x^{3/2}D_y)v_0,$$

where $I_\nu(z)$ is the ν-th modified Bessel function. Further, if we define v by $(\vartheta-5s/2-5/4)u$, then we obtain

$$(4.2.7)\quad \begin{pmatrix} u \\ v \end{pmatrix} = \begin{pmatrix} A & B \\ C & D \end{pmatrix}\begin{pmatrix} u_0 \\ v_0 \end{pmatrix}$$

with $A = (z/2)^{s+1/2}\Gamma(-s+1/2)I_{-s-1/2}(z)$,

$B = (z/2)^{-s-1/2}\Gamma(s+3/2)I_{s+1/2}(z)$,

$C = (z/2)^{s+3/2}\Gamma(-s+1/2)I'_{-s-1/2}(z)$

and $D = (z/2)^{-s+1/2}\Gamma(s+3/2)I'_{s+1/2}(z)$,

where $z = x^{3/2}D_y$. Then Lommel's formula for Bessel functions entails

$$(4.2.8)\quad \begin{pmatrix} u_0 \\ v_0 \end{pmatrix} = \begin{pmatrix} D & -B \\ -C & A \end{pmatrix}\begin{pmatrix} u \\ v \end{pmatrix}$$

Hence this transformation reduces the equation (4.2.2) to

$$(4.2.9)\quad \begin{cases} (\vartheta-2s-1)u_0 = 0 \\ (\vartheta-3s-3/2)v_0 = 0. \end{cases}$$

This example shows that, essentially speaking, the asymptotic expansion in our sense is nothing but the transformation of the equation by $\tilde{\mathscr{D}}^{\omega}_{\Lambda}$. Hence we want to have a general result on the transformation of euqations. The following theorem gives us a satisfactory result in this direction.

<u>Theorem 4.2.1.</u> Let Λ be $T^*_Y\mathbb{C}^{1+n}$, where $Y = \{(t,x)=(t,x_1,\dots,x_n)\in\mathbb{C}^{1+n};\ t=0\}$. Assume that $A = (a_{\mu\nu}(x))_{1\leq\mu,\nu\leq N}$ and $Q = (Q_{\mu\nu}(x,D_x,D_t))_{1\leq\mu,\nu\leq N}$

satisfy the follwoing conditions (4.2.10) ~ (4.2.14) for some integers m_μ $(1 \leqq \mu \leqq N)$:

(4.2.10) $a_{\mu\nu}(x)$ is holomorphic on a neighborhood of the origin.

(4.2.11) $a_{\mu\nu} = 0$ if $m_\mu - m_\nu + 1 < 0$.

(4.2.12) If we denote by $\lambda_\mu(x)$ $(1 \leqq \mu \leqq N)$ the eigenvalues of the matrix A, then $\lambda_\mu(x)$ is holomorphic on a neighborhood of zero and $\lambda_\mu(0) - \lambda_\nu(0) \notin \mathbb{Z} - \{0\}$ $(1 \leqq \mu, \nu \leqq N)$.

(4.2.13) $Q_{\mu\nu} \in \mathcal{J}_\Lambda^{m_\mu - m_\nu + 2}(-1)$.

Then on a neighborhood of $\{(t,x;\tau,\xi);\ t=x=0,\ \xi=0\}$ there exists an invertible matrix $R(x,D_x,D_t)$ each of whose entry belongs to $\widetilde{\mathcal{D}}_\Lambda^\infty$ and which satisfies

(4.2.14) $(tD_t - A(x) + Q(x,D_x,D_t))R(x,D_x,D_t) = R(x,D_x,D_t)(tD_t - A(x))$.

Furthermore $R(x,D_x,D_t)$ can be expressed in the form $\sum_p R^p(x,D_x)D_t^{-p}$, where $R^p(x,D_x) = 0$ for $p < 0$, $R^0 = I$, the identity matrix and $\operatorname{order} R^p \leqq ap$ holds for $p > 0$ with $a = 2 + \max_{\mu,\nu}(m_\mu - m_\nu)$.

Corollary 4.2.2. Let $b(s)$ be a polynomial of degree m and Λ a Lagrangian submanifold of T^*X. Assume further that the difference of any two different roots of $b(s) = 0$ is not an integer. Let P be a micro-differential operator of order at most m which belongs to $\mathcal{E}_\Lambda(-1)$. Then we have

(4.2.15) $\widetilde{\mathcal{D}}_\Lambda^\infty / \widetilde{\mathcal{D}}_\Lambda^\infty(b(\vartheta) - P) \cong \widetilde{\mathcal{D}}_\Lambda^\infty / \widetilde{\mathcal{D}}_\Lambda^\infty b(\vartheta)$.

Proof of Corollary 4.2.2. By a suitable quantized contact transformation we may assume that $\Lambda = T_Y^*\mathbb{C}^{1+n}$ with $Y = \{(t,x) \in \mathbb{C}^{1+n};\ t=0\}$ and $\vartheta = tD_t + 1/2$. Furthermore, by the division theorem for micro-differential operators (S-K-K [17] Chap. II, §2, Theorem 2.2.2), we may assume without loss of generality that P is a polynomial of degree $(m-1)$ in t, i.e., $P = \sum_{j=0}^{m-1} P_j(x,D_x,D_t)\vartheta^j$. Then rewriting the equation $(b(\vartheta) - P)u = 0$ in the matrix form, we can apply Theorem 4.2.1 to obtain the required result. Q.E.D.

Proof of Theorem 4.2.1. We try to find an invertible matrix R in the following form:

$$(4.2.16)\quad R = \sum_{p=0}^{\infty} R^p(x,D_x)D_t^{-p},$$

where $R^p(x,D_x)$ is a matrix of linear differential operators of finite order and $R^0 = I$, the identity matrix. If we can construct R in this form, then its inverse can be constructed in a similar manner. We expand Q also in this form, namely,

$$(4.2.17)\quad Q = \sum_{j\geqq 1} Q^j(x,D_x)D_t^{-j},$$

where $Q^j = (Q^j_{\mu\nu})$ is a matrix of linear differential operators. Note that (4.2.13) implies

$$(4.2.18)\quad \text{order}\, Q^j_{\mu\nu} \leqq m_\mu - m_\nu + j + 1.$$

In order to satisfy (4.2.14) it suffices to define R^p $(p \geqq 1)$ successively by the following recursion formula:

$$(4.2.19)\quad pR^p - [R^p, A] = \sum_{\substack{j+k=p\\ j\geqq 1}} Q^j R^k \qquad (p \geqq 1).$$

Here we note that order $R^p \leqq ap$ $(p > 0)$ holds with $a = 2 + \max_{\mu,\nu}(m_\nu - m_\mu)$ in view of (4.2.18) and (4.2.19), if R^p exists.

Let us now define an $N \times N$ matrix $A^*(y)$ and a column vector $\tilde{R}^p$ of length N^2 by $(a_{\nu\mu}(y))_{1\leq\nu,\mu\leq N}$ and ${}^t(\tilde{r}^p_{11}(x,y,D_x),\ldots,\tilde{r}^p_{1N}(x,y,D_x),\tilde{r}^p_{21}(x,y,D_x),\ldots,\tilde{r}^p_{N1}(x,y,D_x),\ldots,\tilde{r}^p_{NN}(x,y,D_x))$, respectively. Then, instead of solving (4.2.19), we solve the following equation.

$$(4.2.20)\quad (p-(I_N \otimes A^*(y) - A(x)\otimes I_N))\tilde{R}^p(x,y,D_x) = \sum_{\substack{j+k=p\\ j\geqq 1}} (Q^j(x,D_x)\otimes I_N)\tilde{R}^p(x,y,D_x),$$

where $A(x)\otimes I_N$ etc. denote the Kronecker product of the matrices. In fact, rewriting $\tilde{r}^p_{\mu\nu}(x,y,D_x)\delta(x-y)$ in the form $r^p_{\mu\nu}(x,D_x)\delta(x-y)$, we can find the required $R^p = (r^p_{\mu\nu})$ which satisfies (4.2.19) from the solution $\tilde{R}^p$ of (4.2.20). If we denote $I_N \otimes A^*(y) - A(x)\otimes I_N$ by $\mathfrak{A}$, then the assumption (4.2.12) guarantees that $(p-\mathfrak{A})^{-1}$ exists for $p = 1,2,\cdots$. In order to facilitate the argument, we introduce a dummy variable x_{n+1} and define an $N^2 \times N^2$ matrix C by

$$I_N \otimes \begin{pmatrix} (D_{x_{n+1}})^{m_1} & & 0 \\ & \ddots & \\ 0 & & (D_{x_{n+1}})^{m_N} \end{pmatrix}.$$

Then C is invertible on $\Omega = \{(t,x,y,x_{n+1};\tau,\xi,\eta,\xi_{n+1}) \in T^*\mathbb{C}^{1+2n+1};\ \xi_{n+1} \neq 0\}$. Hence $\mathcal{B}(p) \underset{\text{def}}{=} C^{-1}(p-\mathcal{O}\!\mathcal{l})^{-1}C$ is well-defined there. Furthermore, by using the assumption (4.2.11), we can prove that $\mathcal{B}(p)$ has the form

(4.2.21) $\displaystyle\sum_{\ell=0}^{b} C^{\ell}(p)$

where each entry $C^{\ell}_{\mu\nu}(p)$ of $C^{\ell}(p)$ is of order at most ℓ and it satisfies

$$N_{\ell}(C^{\ell}_{\mu\nu}(p);\delta) \leqq A/p^{\ell+1}.$$

Here N_{ℓ} denotes the formal norm introduced in Boutet de Monvel and Krée [2]. See Tahara [18] Lemma 1.2.8 for the proof of this assertion. Defining $\tilde{S}^p$ by $C^{-1}\tilde{R}^p$, we can rewrite (4.2.20) into the following form:

$$S^p = \mathcal{B}(p) \sum_{\substack{j+k=p \\ j \geqq 1}} (C^{-1}(Q_j \otimes I_N)C)S^k.$$

Thus we can write down S^p in terms of $\mathcal{B}(p)$ and $C^{-1}(Q_j \otimes I_N)C$ explicitly. Then, using the assumption (4.2.13) and the decomposition (4.2.21), we can verify that the i-th homogeneous part $s^p_i(x,y,\xi,\xi_{n+1})$ of the symbol sequence of S^p satisfies the following:

(4.2.22) For each $\varepsilon > 0$ and each compact subset K of Ω, there exist constants $C_{\varepsilon,K}$ and $M_{\varepsilon,K}$ such that

$$\sup_K |s^p_i| \leqq M_{\varepsilon,K} C^p_{\varepsilon,K} \varepsilon^i p!/i!$$

holds.

Note also that

(4.2.23) $s^p_i = 0$, if $i \leqq (b+2)p$.

From the estimate (4.2.22), we easily find that $R = \displaystyle\sum_{p=0}^{\infty} R^p(x,D_x)D_t^{-p}$

belongs to $M_N(\tilde{\mathscr{D}}^\infty_\Lambda)$. Q.E.D.

Remark 4.2.3. Let Λ be the same as in Theorem 4.2.1. Then in view of (4.2.23), the operator R (and R^{-1} also) acts on $\hat{\mathcal{C}}_{Y|X}$ (S-K-K [17] Chap. II, §3), where $X = \mathbb{C}^{1+n}$. On the other hand, it is easy to verify that

$$\mathcal{E}xt^j(\mathcal{M}_0, \mathcal{C}_{Y|X}) \cong \mathcal{E}xt^j(\mathcal{M}_0, \hat{\mathcal{C}}_{Y|X}) \qquad (j = 0,1)$$

holds for $\mathcal{M}_0 = \mathcal{E}^N_X / \mathcal{E}^N_X(tD_t - A(x))$. Hence the comparison theorem of this type also holds for $\mathcal{M} = \mathcal{E}^N_X / \mathcal{E}^N_X(tD_t - A(x) + Q(x, D_x, D_t))$.

Remark 4.2.4. Because of the form (4.2.16) of the intertwining operator R constructed above, we find that R and its inverse S actually belong to a smaller class of operators introduced by Laurent [14], namely, R and S belong to $M_N(\mathcal{E}^{2\infty}_V|_\Lambda)$, where $V = \{(t,x;\tau,\xi) \in T^*\mathbb{C}^{1+n};\ \xi = 0\}$.

In order to apply Theorem 4.2.1 to the study of microfunction solution of the equation in question, we prepare the following Lemma 4.2.5. In what follows, $\Lambda^{\mathbb{R}}$ denotes the purely imaginary locus of Λ, which is supposed to be Lagrangian.

Lemma 4.2.5. Let Λ be the same as in Theorem 4.2.1. Let u be a $\tilde{\mathscr{B}}_\Lambda$-solution of the following equation:

$$(4.2.24) \qquad (tD_t - \lambda(x))u = 0$$

Then we have

$$u = \Gamma(-\lambda(x))(t + \sqrt{-1}0)^{\lambda(x)}\varphi(x)$$

$$= 2\pi \exp(-\pi\sqrt{-1}\lambda(x)/2)(D_t/\sqrt{-1})^{-\lambda(x)-1}\delta(t)\varphi(x) \quad \text{with } \varphi(x) \in \mathscr{B}_{\mathbb{R}^n}.$$

Proof. By applying $(D_t/\sqrt{-1})^{-\lambda(x)-1} = \exp(-(\lambda(x)+1)\log(D_t/\sqrt{-1}))$ to (4.2.24), we may replace (4.2.24) by

$$(4.2.25) \qquad tu = 0$$

On the other hand, we have the following exact sequence:

$$(4.2.26) \qquad 0 \to \pi^{-1}\mathcal{O}_Y \to \mathcal{C}^{\mathbb{R}}_{Y|X} \overset{t}{\to} \mathcal{C}^{\mathbb{R}}_{Y|X} \to 0,$$

where π is the projection from T_Y^*X to Y. Since

$$\mathcal{H}^{j}_{\sqrt{-1}\,T^*_{\mathbb{R}^n}\mathbb{R}^{n+1}}(\pi^{-1}\mathcal{O}_{\mathbb{C}^n}) = \mathcal{H}^{j-1}_{\mathbb{R}^n}(\mathcal{O}_{\mathbb{C}^n}),$$

(4.2.26) implies

$$0 \to \mathcal{B}_{\mathbb{R}^n} \to \tilde{\mathcal{B}}_\Lambda \overset{t}{\to} \tilde{\mathcal{B}}_\Lambda \to 0.$$

This means that $u = \delta(t)\varphi(x)$ holds with $\varphi(x) \in \mathcal{B}_{\mathbb{R}^n}$. Q.E.D.

The following theorem is an immediate consequence of Theorem 4.2.1 and Lemma 4.2.5.

Theorem 4.2.6. Assume the same conditions as in Theorem 4.2.1. For simplicity, assume further that $A(x)$ is a constant matrix A. Let U be a microfunction solution of $(tD_t-A+Q(x,D_x,D_t))U = 0$. Then U has the following expression as a section of $\tilde{\mathcal{B}}_\Lambda$:

$$U = R\begin{pmatrix} \Gamma(-\lambda_1)(t+\sqrt{-1}0)^{\lambda_1}L_1 & & \\ & \ddots & \\ & & \Gamma(-\lambda_d)(t+\sqrt{-1}0)^{\lambda_d}L_d \end{pmatrix}\begin{pmatrix} \varphi_1(x) \\ \vdots \\ \varphi_N(x) \end{pmatrix},$$

where $R \in M_N(\tilde{\mathcal{D}}^\infty_\Lambda)$, $\lambda_1,\ldots,\lambda_d$ are the mutually different eigenvalues of A, L_j has the form

$$\begin{pmatrix} 1 & \log(t+\sqrt{-1}0) & (\log(t+\sqrt{-1}0))^2/2! & \cdots & (\log(t+\sqrt{-1}0))^{m_j-1}/(m_j-1)! \\ & 1 & \log(t+\sqrt{-1}0) & \cdots & (\log(t+\sqrt{-1}0))^{m_j-2}/(m_j-2)! \\ & & 1 & \ddots & \\ & & & \ddots & \vdots \\ & & & & 1 \end{pmatrix}$$

with m_j being the multiplicity of λ_j and $\varphi_\ell(x)$ is a hyperfunction in x.

Remark 4.2.7. If $A(x)$ is not supposed to be a constant matrix, U has the form

$$R\Gamma(-A(x))(t+\sqrt{-1}0)^{A(x)}\Phi(x)$$

with $\Phi(x) \in \mathcal{B}^N_{\mathbb{R}^n}$.

Corollary 4.2.8. Assume the same conditions as in Corollary 4.2.2. Then for each microfunction solution u of the equation $(b(\vartheta)-P)u = 0$, there exist $R_\mu \in \hat{\mathcal{D}}^\infty_\Lambda$ $(\mu = 1,\dots,N)$ and hyperfunctions ψ_μ $(\mu = 1,\dots,N)$ which satisfy $b(\vartheta)\psi_\mu = 0$ so that

$$u = \sum_{\mu=1}^{N} R_\mu \psi_\mu$$

holds as section of $\widetilde{\mathcal{B}}_\Lambda$. In particular, if $\Lambda = T^*_Y\mathbb{C}^{1+n}$ with $Y = \{(t,x) \in \mathbb{C}^{1+n}; t=0\}$, then u has the form

$$\sum_{\nu=1}^{d} \sum_{j=0}^{j_\nu-1} R_{\nu,j}(x,D_x,D_t)(t+\sqrt{-1}0)^{\lambda_\nu}(\log(t+\sqrt{-1}0))^j$$

where $R_{\nu,j} \in \widetilde{\mathcal{D}}^\infty_\Lambda$, $b(\lambda_\nu) = 0$ and j_ν is the multiplicity of the root λ_ν.

These results show that the sheaf $\widetilde{\mathcal{B}}_\Lambda$ is fitted for the study of the asymptotic expansion of a microfunction which satisfies a suitable micro-differential equation. Note that $R(x,D_x,D_t)(\varphi(x)(t+\sqrt{-1}0)^\lambda)$ can be formally rewritten in the form

$$\sum_{j=0}^{\infty} \varphi_j(x)(t+\sqrt{-1}0)^{\lambda+j},$$

if $\Lambda = T^*_Y\mathbb{C}^{1+n}$ and R is the operator constructed in Theorem 4.2.1.

In what follows, we call the expression $\sum_{\mu=1}^{N} R_\mu\psi_\mu$ $(R_\mu \in \widetilde{\mathcal{D}}^\infty_\Lambda)$ given in Corollary 4.2.8 to be the asymptotic expansion with respect to Λ of the microfunction solution u of the equation $(b(\vartheta)-P)u = 0$. Note that Proposition 4.1.6 guarantees that such an equation always exists if u satisfies a holonomic system with R.S.

Now, as far as we start from a single equation $(b(\vartheta)-P)u = 0$, Corollary 4.2.8 is the best possible one of the sort. However, in practical problems equations of this type usually appear in connection with holonomic systems. In such circumstances, as we mentioned in §1, the top term of the asymptotic expansion is controlled again by holonomic systems on Λ. For example, in the case of the example (4.2.1) discussed at the beginning of this subsection, u_0 and v_0 satisfy the following equations beside (4.2.9):

$$(xD_x/3+yD_y/2-s)w = yD_xw = 0.$$

This situation is best formulated by regarding the original holonomic system as $\tilde{\mathscr{D}}_\Lambda^\infty$-Module. In connection with such a formulation we first prepare some notions related to the Levi condition of the system. (Cf. Monteiro-Fernandes [16].) In what follows, Λ is a Lagrangian submanifold of T^*X and $\mathscr{J}_\Lambda$ and $J_\Lambda(1)$ denote $\{P\in \mathscr{E}_X(1);\sigma_1(P)|_\Lambda = 0\}$ and $\{f\in \mathscr{O}_{T^*X}(1);\ f|_\Lambda = 0\}$, respectively.

<u>Definition 4.2.9</u>. (i) $A \underset{\text{def}}{=} \bigoplus_k(\mathscr{J}_\Lambda^k/(\mathscr{J}_\Lambda^{k-1}+\mathscr{J}_\Lambda^{k+1}(-1))$

$$(= \bigoplus_k(J_\Lambda(1)^k/J_\Lambda(1)^{k-1}\mathscr{O}_{T^*X}(1))).$$

(ii) Let $\mathscr{M}$ be a coherent $\mathscr{E}_X$-Module and $\mathscr{L}$ its coherent $\mathscr{E}(0)$-sub-Module such that $\mathscr{M} = \mathscr{E}_X\mathscr{L}$ holds. Then $\mathrm{Ch}_\Lambda(\mathscr{M})$ is, by definition, a subvariety of $T_\Lambda(T^*X)$, the normal bundle along Λ, which is given by

$$\mathrm{supp}(\mathscr{O}_{T_\Lambda(T^*X)} \underset{A}{\otimes} (\bigoplus_k(\mathscr{J}_\Lambda^k\mathscr{L}/\mathscr{J}_\Lambda^{k-1}\mathscr{L}+\mathscr{J}_\Lambda^{k+1}(-1)\mathscr{L})).$$

<u>Remark 4.2.10</u>. The variety $\mathrm{Ch}_\Lambda(\mathscr{M})$ is independent of the choice of $\mathscr{L}$. We conjecture that it is an involutory subvariety of $T^*\Lambda \cong T_\Lambda(T^*X)$.

<u>Remark 4.2.11</u>. Denote by $\tilde{\sigma}_k$ the map from $\mathscr{J}_\Lambda^k$ to $J_\Lambda(1)^k/J_\Lambda(1)^{k-1}\mathscr{O}_{T^*X}(1)$, which is induced from the isomorphism between $\mathscr{J}_\Lambda^k/(\mathscr{J}_\Lambda^{k-1}+\mathscr{J}_\Lambda^{k+1}(-1))$ and $J_\Lambda(1)^k/J_\Lambda(1)^{k-1}\mathscr{O}_{T^*X}(1)$. Then, for a system $\mathscr{M}$ of the form $\mathscr{E}_X/\mathscr{J}$,

$$\mathrm{Ch}_\Lambda(\mathscr{M}) = \bigoplus_k A/(\tilde{\sigma}_k(\mathscr{J}_\Lambda^k\cap\mathscr{J})).$$

Now we begin the discussion on the structure of a holonomic system with R.S. regarded as a $\tilde{\mathscr{D}}_\Lambda^\infty$-Module. In order to simplify the description, we will consider the case where $X = \mathbb{C}^{1+n} = \{(t,x);t\in\mathbb{C},\ x\in\mathbb{C}^n\}$ and $\Lambda = \{(t,x;\tau,\xi)\in T^*X;\ t=0,\ \xi=0,\ \tau\neq 0\}$. Let $\mathscr{M}$ be a holonomic $\mathscr{E}_X$-Module with R.S. Take a coherent $\mathscr{E}_\Lambda$-sub-Module $\mathscr{N}$ of $\mathscr{M}$ such that $\mathscr{E}_X\mathscr{N} = \mathscr{M}$. If we set $\mathscr{N}(-1) = \mathscr{E}(-1)\mathscr{N}$, then $\mathscr{N}/\mathscr{N}(-1)$ is a coherent $\mathscr{A}_\Lambda$-Module. Here $\mathscr{A}_\Lambda = \mathscr{E}_\Lambda/\mathscr{E}_\Lambda(-1)$. By identifying $\mathscr{A}_\Lambda$ with the sheaf of homogeneous linear differential operators on Λ, we can regard $\mathscr{N}/\mathscr{N}(-1)$ as a system of differential equations on Λ, i.e., x-space. If $\mathscr{N} = \mathscr{E}_\Lambda u$ for a section u of $\mathscr{M}$, then $\mathscr{N}/\mathscr{N}(-1)$ is nothing but the following equations:

$$L(P)\bar{u} = 0 \quad \text{for} \quad P\in\mathscr{E}_\Lambda \quad \text{such that} \quad Pu = 0.$$

As we have shown in §1 (Theorem 1.5), $\mathcal{N}/\mathcal{N}(-1)$ is a holonomic system with R.S.

Now it follows from Propositin 4.1.6 that there exist a coherent $\mathcal{E}(0)$-sub-Module $\mathcal{L}$ of $\mathcal{M}$ and a non-zero polynomial $b(s)$ of degree, say, r which satisfy the following two conditions:

(4.2.27) $b(tD_t)\mathcal{L} \subset \mathcal{J}_\Lambda^{r+1}(-1)\mathcal{L}$

(4.2.28) The difference of any roots of $b(s)=0$ is not a non-zero integer.

Take a system of generators $u_1,\ldots,u_m$ of $\mathcal{L}$ and denote by u the column vector having $(tD_t)^{\alpha_0}D_{x_1}^{\alpha_1}\ldots D_{x_n}^{\alpha_n}u_j$ as components, where $\alpha_0,\ldots,\alpha_n$ range over all non-negative integers satisfying the constraint $\sum_{\nu=0}^{n}\alpha_\nu < r$ and $1 \leqq j \leqq m$. Let N be the length of the vector u. Then it is easy to see that u satisfies the equation

$$tD_t u = Au + Qu,$$

where A is a constant matrix and Q is a matrix of linear differential operators satisfying the conditions in Theorem 4.2.1. Here we note that the eigenvalues of A are the roots of $b(s) = 0$. Let us now consider the system $\tilde{\mathcal{M}} = \mathcal{E}^N/\mathcal{E}^N(tD_t-A-Q)$ and let $\tilde{u}$ be the corresponding vector whose components generate $\tilde{\mathcal{M}}$. Let f be the homomorphism from $\tilde{\mathcal{M}}$ onto $\mathcal{M}$ defined by assigning u to $\tilde{u}$. Let $\mathcal{M}'$ denote the kernel of f and let $\tilde{\mathcal{L}}$ be the $\mathcal{E}(0)$-Module generated by components of $\tilde{u}$. Then it is easy to see that $\tilde{b}(tD_t)\tilde{\mathcal{L}} \subset \mathcal{J}_\Lambda^{N+1}\tilde{\mathcal{L}}(-1)$ holds for the characteristic polynomial $\tilde{b}(s)$ of A. Note that deg $\tilde{b}$ is at most N. The Artin-Rees theorem can be modified to show that

$$(\mathcal{J}_\Lambda^k\tilde{\mathcal{L}})\cap\mathcal{M}' = \mathcal{J}_\Lambda^{k-k_0}(\mathcal{J}_\Lambda^{k_0}\tilde{\mathcal{L}}\cap\mathcal{M}')$$

holds for $k \geqq k_0$ if we take k_0 sufficiently large. We can easily prove that $\tilde{b}(tD_t)\mathcal{J}_\Lambda^k\tilde{\mathcal{L}} \subset \mathcal{J}_\Lambda^{k+N+1}\tilde{\mathcal{L}}(-1)$, and hence we obtain

$$\tilde{b}(tD_t)(\mathcal{J}_\Lambda^k\tilde{\mathcal{L}}\cap\mathcal{M}') \subset \mathcal{J}_\Lambda^{k+N+1}\tilde{\mathcal{L}}(-1)\cap\mathcal{M}' = \mathcal{J}_\Lambda^{N+1}(-1)(\mathcal{J}_\Lambda^k\tilde{\mathcal{L}}\cap\mathcal{M}').$$

Then, by the same reasoning used above, we can find a system of generators $v_1,\ldots,v_{\tilde{N}}$ of $\mathcal{J}_\Lambda^k\tilde{\mathcal{L}}\cap\mathcal{M}'$ such that the row vector $v = (v_1,\ldots,v_{\tilde{N}})$ satisfies

$$tD_t v = \tilde{A}v + \tilde{Q}v,$$

where $\tilde{A}$ is a constant matrix whose eigenvalues are the roots of $\tilde{b}(s) = 0$ (and hence those of $b(s) = 0$) and $\tilde{Q}$ satisfies the conditions in Theorem 4.2.1. Then, if we write $v = P(x,D_x,D_t)u$ with $P \in M(\tilde{N},N;\mathcal{E})$ which does not contain t, then we have $(tD_t-\tilde{A}-\tilde{Q})P = P'(tD_t-A-Q)$ with $P' \in M(\tilde{N},N;\mathcal{E})$. By comparing the coefficients of t, we can easily verify that $P' = P$. Hence we obtain

$$(tD_t-\tilde{A}-\tilde{Q})P = P(tD_t-A-Q)$$

It follows from Theorem 4.2.1 that there exist $U(x,D_x,D_t) \in GL(N;\tilde{\mathcal{D}}^{\infty}_{\Lambda})$ and $\tilde{U}(x,D_x,D_t) \in GL(\tilde{N};\tilde{\mathcal{D}}^{\infty}_{\Lambda})$ which satisfy

$$tD_t-A-Q = U(tD_t-A)U^{-1}$$

and

$$tD_t-\tilde{A}-\tilde{Q} = \tilde{U}(tD_t-\tilde{A})\tilde{U}^{-1}.$$

Hence we have

$$(tD_t-\tilde{A})(\tilde{U}^{-1}PU) = (\tilde{U}^{-1}PU)(tD_t-A).$$

Set $T = \tilde{U}^{-1}PU = \sum_j T^j(x,D_x)D_t^{-j}$. Then

(4.2.29) $\quad jT^j = \tilde{A}T^j - T^jA$

holds. Note that the difference of any eigenvalues of A and that of $\tilde{A}$ are not a non-zero integer by (4.2.28). Hence (4.2.29) implies $T^j = 0$ for $j \neq 0$. Therefore T is a linear differential operator in x. On the other hand, U^{-1} and $\tilde{U}$ have respectively the form $\sum_{j=0}^{\infty} G^j(x,D_x)D_t^{-j}$ and $\sum_{j=0}^{\infty} H^j(x,D_x)D_t^{-j}$ with $G^0 = H^0 = I$ and order G^j, order $H^j \leqq aj$ for some a. Hence, if we write $P = \sum_{j=0}^{\infty} P^j(x,D_x)D_t^{-j}$, then we have

(4.2.30) $\quad P^0 = T$ and order $P^j \leqq aj$.

Since $\mathcal{M}$ is an $\mathcal{E}_X$-Module generated by u with the relation $Pu = (tD_t-A-Q)u = 0$, $\tilde{\mathcal{D}}^{\infty}_{\Lambda} \underset{\mathcal{E}_X}{\otimes} \mathcal{M}$ is generated by $1 \otimes u$ with the same relation. Hence, if we take another generator $w \underset{\text{def}}{=} U^{-1}(1 \otimes u)$ of $\tilde{\mathcal{D}}^{\infty}_{\Lambda} \underset{\mathcal{E}_X}{\otimes} \mathcal{M}$, we find

(4.2.31) $(tD_t - A)w = Tw = 0$.

Hence $\tilde{\mathcal{D}}^{\infty}_{\Lambda} \underset{\mathcal{E}_X}{\otimes} \mathcal{M}$ has the form $\tilde{\mathcal{D}}^{\infty}_{\Lambda} \underset{\mathcal{A}_{\Lambda}}{\otimes} \mathcal{F}$, where $\mathcal{F}$ is the $\mathcal{A}_{\Lambda}$-Module generated by a generator w with the relation (4.2.31). Here we identify $\mathcal{A}_{\Lambda}$ with a sub-Module of $\tilde{\mathcal{D}}^{\infty}_{\Lambda}$. Let us take as $\mathcal{N}$ the $\mathcal{E}_{\Lambda}$-Module generated by u. Then by using the fact that $\mathcal{E}_{\Lambda}v \cap \mathcal{M}'$ $= \mathcal{E}^{\tilde{N}}_{\Lambda}Pu$ and (4.2.30) we can easily verify that $\mathcal{N}/\mathcal{N}(-1) = \mathcal{F}$ holds. Furthermore, by using Remark 4.2.11 and an invertibility theorem for operators in $\tilde{\mathcal{E}}^{\infty}_{\Lambda}$ (Laurent [14]), we can easily show the characteristic variety of $\mathcal{F}$ is contained in $Ch_{\Lambda}(\mathcal{M})$. Thus we obtain the following

Theorem 4.2.12. Let $\mathcal{M}$ be a holonomic $\mathcal{E}$-Module with R.S. and let $\mathcal{N}$ be a coherent $\mathcal{E}(0)$-sub-Module of $\mathcal{M}$ such that there exists a non-zero polynomial $b(s)$ which satisfies the condition (4.2.28) and $b(\vartheta)\mathcal{N} \subset \mathcal{N}(-1)$. Then we have

(i) $\mathcal{N}/\mathcal{N}(-1)$ is a holonomic $\mathcal{A}_{\Lambda}$-Module with R.S. whose characteristic variety is contained in $Ch_{\Lambda}(\mathcal{M})$.

(ii) $\tilde{\mathcal{D}}^{\infty}_{\Lambda} \underset{\mathcal{E}_{\Lambda}}{\otimes} \mathcal{M} = \tilde{\mathcal{D}}^{\infty}_{\Lambda} \underset{\mathcal{A}_{\Lambda}}{\otimes} (\mathcal{N}/\mathcal{N}(-1))$

This theorem implies that operators in $\tilde{\mathcal{D}}^{\infty}_{\Lambda}$ enable us to transform a holonomic system of micro-differential equations with R.S. to a differential equation in (t,x) variables. It also asserts that any microfunction solution of such a system is determined by its initial terms in its asymptotic expansions.

§5. Discussion on the analyticity of the S-matrix.

It is now commonly accepted that the analyticity properties of the S-matrix is most neatly expressed in terms of microfunctions and/or essential support theory. However, in order to study the structure of the S-matrix, we sometimes need more precise information than the microanalyticity of the S-matrix. For example, Iagolnitzer-Stapp [4] makes essential use of "no sprout assumption" in the study of pole-factorization theorem. This condition cannot be described in terms of the micro-analyticity.

The purpose of this section is to discuss how such a delicate analyticity problem is related to the theory expounded so far, and how it is related to the holonomic character of the S-matrix. (See Kawai-Stapp [13] for the discussion on the holonomic character of the S-matrix.) The theory of double-microlocalization with respect to an involutory manifold, which Laurent [14],[15] is now developing, is also very useful for this purpose.

Now let Λ be $T^*_Y\mathbb{C}^n$ with $Y = \{(z_1,z')\in\mathbb{C}^n;\ z_1=0\}$ and let V be $\{(z,\zeta)\in T^*\mathbb{C}^n;\ \zeta_2=\cdots=\zeta_n=0\}$. Let L and W denote their purely imaginary loci. For a microfunction f, $SS^2_W f$ (Kashiwara-Laurent [10]) and $\widetilde{SS}_L f$ are, by definition, subsets of $T_W(\sqrt{-1}T^*\mathbb{R}^n)$ and $T_L(\sqrt{-1}T^*\mathbb{R}^n)$, respectively. (Cf. the end of §3.2. Note that $\sqrt{-1}T^*_L\mathbb{R}^n$ and $T_L(\sqrt{-1}T^*\mathbb{R}^n)$ can be identified by the Hamiltonian map H, because L is Lagrangian. Actually $H(\sqrt{-1}dx_j) = -\sqrt{-1}\partial/\partial\xi_j$ and $H(\sqrt{-1}d\xi_j) = \sqrt{-1}\partial/\partial x_j$ hold.) Using the coordinate system we denote a point in $T_W(\sqrt{-1}T^*\mathbb{R}^n)$ and a point in $T_L(\sqrt{-1}T^*\mathbb{R}^n)$ respectively by $(x, \sqrt{-1}\xi_1;\ \sqrt{-1}\sum_{j=2}^n c_j\partial/\partial\xi_j)$ and $(x',\sqrt{-1}\xi_1;\sqrt{-1}(c_1\partial/\partial x_1+\sum_{j=2}^n c_j\partial/\partial\xi_j)$, where x,ξ_1 and c_j are real. Note that the subset of $T_L(\sqrt{-1}T^*\mathbb{R}^n)$ defined by $\omega = 0$ is identified with $T_W(\sqrt{-1}T^*\mathbb{R}^n)|_L$.

Let us now consider a microfunction f supported by $L^+ \overset{\text{def}}{=} \{(x,\sqrt{-1}\xi)\in\sqrt{-1}(T^*\mathbb{R}^n-T^*_{\mathbb{R}^n}\mathbb{R}^n);\ x_1=0,\ \xi_1>0,\ \xi' = (\xi_2,\ldots,\xi_n) = 0\}$. Then

$$(5.1)\quad SS^2_W(f)\subset\{(x,\sqrt{-1}\xi_1;\ c')\in T_W(\sqrt{-1}T^*\mathbb{R}^n);\ c'=0\}$$

is equivant to the assertion that the defining function $F(z)$ of the microfunction f is holomorphic on $\{z\in\mathbb{C}^n;\ |z|<\varepsilon,\ \mathrm{Im}\, z_1>0\}$, namely, f satisfies the no sprout assumption. (In the condition (5.1) $c' = (c_2,\ldots,c_n)$ is identified with $\Sigma c_j\partial/\partial\xi_j$.) Next consider the following condition

$$(5.2)\quad \widetilde{SS}_L(f)\subset\{(x',\sqrt{-1}\xi_1;\ c)\in T_L(\sqrt{-1}T^*\mathbb{R}^n);\ c=0\}.$$

Then, on the supposition that Conjecture 3.2.6 (ii) in §3.2 is valid, (5.2) is equivalent to the assertion that f belongs to $\mathcal{M}^{\mathbb{R}}(=\mathcal{E}^{\mathbb{R}}\otimes\mathcal{M})$ for a simple holonomic system $\mathcal{M}$ supported by Λ. In particular, the defining function $F(z)$ of f can be analytically continued to the universal covering space of Ω-Y, where Ω is a neighborhood of $0\in\mathbb{C}^n$. Hence the condition (5.2) is close to the holonomic character of f. However, if f is holonomic, we find another important property of f, namely, the finite determination property. This property is not implied by (5.2). The simplest example of f that satisfies condition (5.2) but that is not holonomic is given by $(x_1+\sqrt{-1}0)^{x_2}$. Thus the sheaf $\tilde{\mathcal{C}}_\Lambda$ supplies us with a link between the no sprout assumption and the holonomic character of the function in question. Note also that

(5.3) $SS^2_W(f) \cong \widetilde{SS}_L(f)$

holds for a microfunction f which satisfies an equation dealt with in Corollary 4.2.2. (See Remark 4.2.4.) It is also noteworthy that (5.3) is valid for an arbitrary holonomic microfunction f (i.e., not necessarily supported by L^+). In fact, we know that, for each holonomic $\mathcal{E}_X$-Module $\mathcal{M}$, we can find a holonomic $\mathcal{E}_X$-Module $\mathcal{M}_{reg}$ with R.S. so that

$$\mathcal{E}^\infty \otimes \mathcal{M} \cong \mathcal{E}^\infty \otimes \mathcal{M}_{reg}$$

holds. ([8] Chap. V, §2, Theorem 5.2.1. See also our report [9] in this colloquium.) Since $\mathcal{E}^\infty_X|_\Lambda$ is a subsheaf of $\tilde{\mathcal{D}}^\infty_\Lambda$ and $\mathcal{E}^{2\infty}_V|_\Lambda$, this fact implies (5.3).

We end this report by mentioning a fact which is a generalization of the no sprout assumption and will probably turn out to be useful in application.

Let f be a hyperfunction defined on $\mathbb{R}^n$ whose singularity spectrum is contained in a properly convex cone $C=\{\sqrt{-1}\xi\in\sqrt{-1}\,\mathbb{R}^n;\ \varphi(\xi)>0\}$ with φ being a real valued real analytic homogeneous function such that $d\varphi$ never vanishes on $V \underset{\mathrm{def}}{=} \{\xi\in\mathbb{R}^n-\{0\};\varphi(\xi)=0\}$. Let Γ be the dual cone of C, i.e., $\{y\in\mathbb{R}^n;\ \langle y,\xi\rangle>0\ \ (\xi\in C)\}$. Suppose that $SS^2_V f \subset \{(x,\sqrt{-1}\xi;\ c)\in T_V(\sqrt{-1}T^*\mathbb{R}^n);\ \varphi(\xi)=0,\ \sum_{j=1}^n c_j\partial\varphi/\partial\xi_j \geqq 0\}$. Then the defining function F(z) of f is holomorphic on $\{z\in\mathbb{C}^n;\ |\mathrm{Im}\,z|<\varepsilon(\mathrm{Re}\,z),\ \mathrm{Im}\,z\in\Gamma\}$, where ε is a positive-valued function on $\mathbb{R}^n$.

References

[1] Barbasch, D. and D. A. Vogan: The local structure of characters. To appear.

[2] Boutet de Monvel, L. and P. Krée: Pseudo-differential operators and Gevrey classes. Ann. Inst. Fourier, Grenoble, 17, 295-323 (1967).

[3] Hörmander, L. : Fourier integral operators. I. Acta Math., 127 79-183 (1971)

[4] Iagolnitzer, D. and H. P. Stapp: The pole-factorization theorem in S-matrix theory. Commun. math. Phys., 57, 1-30 (1977).

[5] Jeanquartier, P. : Développement asymptotique de la distribution de Dirac attachée à une fonction analytique. C. R. Acad. Sci. Paris, Série A, 271, 1159-1161 (1970).

[6] Kashiwara, M. : On the holonomic systems of linear differential equations. II. Inverntiones math., 49,121-135 (1978).

[7] Kashiwara, M. and T. Kawai: Micro-hyperbolic pseudo-differential operators I. J. Math. Soc. Japan, 27, 359-404 (1975).

[8] ——————: On holonomic systems of micro-differential equations. III —— Systems with regular singularities——. To appear. (RIMS preprint No.293.)

[9] ——————: The theory of holonomic systems with regular singularities and its relevance to physical problems. This proceedings.

[10] Kashiwara, M. and Y. Laurent: Théorèmes d'annulation et double microlocalisation. To appear.

[11] Kashiwara, M. and T. Oshima: Systems of differential equations with regular singularities and their boundary value problems. Ann. of Math., 106, 145-200 (1977).

[12] Kashiwara, M. and P. Schapira: Micro-hyperbolic systems. Acta Math., 142, 1-55 (1979).

[13] Kawai, T. and H. P. Stapp: Discontinuity formula and Sato's conjecture. Publ. RIMS,Kyoto Univ., 12 Suppl., 155-232 (1977).

[14] Laurent, Y.: Double microlocalisation et problème de Cauchy dans le domaine complex. Proc. of the conference at St. Cast, 1979.

[15] ——————: The report in this proceedings.

[16] Monteiro-Fernandes, T. : Thèse de 3ème cycle à l'Université, Paris-Nord (1978).

[17] Sato, M., T. Kawai and M. Kashiwara (referred to as S-K-K [17]): Microfunctions and pseudo-differential equations. Lecture Notes in Math. No.287, pp. 265-529, Berlin-Heidelberg-New York: Springer, 1973.

[18] Tahara, H. : Fuchsian type equations and Fuchsian hyperbolic equations. To appear in Japanese J. Math.

DEUXIEME MICROLOCALISATION

YVES LAURENT

Université Paris-Sud, Mathématiques
91405 ORSAY

Introduction :

Dans [13] , Sato, Kawaï et Kashiwara ont construit sur le fibré cotangent $T^* X$ à une variété analytique complexe X , le faisceau $\mathscr{E}_X$ des opérateurs microdifférentiels (qu'ils notent $\mathscr{P}_X^f$). La première propriété de ce faisceau est que les opérateurs différentiels sur X non caractéristiques dans une direction de $T^* X$, sont inversibles au voisinage de cette direction dans $\mathscr{E}_X$.

Nous construisons sur $T^*(T^* X)$ (ou plutôt sur un quotient de cet espace) le faisceau $\mathscr{E}^2$ des opérateurs "2-microdifférentiels". Nous dirons qu'un vecteur x^* est non microcaractéristique pour un système d'équations microdifférentielles si ce système est inversible au voisinage de x^* comme système d'équations 2-microdifférentielles. On retrouve ainsi la notion de vecteur microcaractéristique définie par Bony [1] pour un opérateur puis par Kashiwara-Schapira [8] pour un système.

En fait, nous construirons toute une famille de faisceaux semblables à $\mathscr{E}^2$ qui permettrons de définir pour un système d'équations microdifférentielles la notion de "vecteur microcaractéristique de type (r,s)" où (r,s) est un couple de rationnels. En particulier, pour $s = 1$ et $r = \infty$ on retrouve la définition de la Condition de Lévi pour un système de T . Monteiro [9] .

Dans le cas général nous démontrons un théorème de Cauchy pour les systèmes que nous relions d'une part au Problème de Cauchy pour les fonctions holomorphes ramifiées (généralisant Hamada -Leray-Wagschal [4]) et d'autre part au polygône de Newton d'un opérateur ou d'un système (généralisant au cas de plusieurs variables un théorème de Ramis [11]).

§1 . Opérateurs microdifférentiels de classe (r,s) :

Soient r et s deux nombres rationnels tels que $1 \leq s \leq r \leq +\infty$.

Si U est un ouvert cônique de $T^* \mathbb{C}^n = \mathbb{C}^n \times \mathbb{C}^n$, par définition un symbole de classe (r,s) sur U est une série formelle $P(x,\xi) = \sum_{j\in\mathbb{Z}} P_j(x,\xi)$, où $P_j(x,\xi)$ est une fonction holomorphe sur U, homogène de degré j en ξ, telle que :

A/ $\underline{r = s < +\infty}$

$\forall K$ compact de U, $\exists C > 0$, $\forall \varepsilon > 0$ $\exists C_\varepsilon > 0$ t.q.

(i) $\forall j \geq 0$ $\sup_K |P_j(x,\xi)| \leq C_\varepsilon \, \varepsilon^j \dfrac{1}{(j!)^r}$

(ii) $\forall j < 0$ $\sup_K |P_j(x,\xi)| \leq C_K^{-j}[(-j)!]^r$

B/ $\underline{1 \leq s < r < +\infty}$

$\forall K$ compact de U $\exists C > 0$ t.q.

(i) $\forall j \geq 0$ $\sup_K |P_j(x,\xi)| \leq C^j \dfrac{1}{(j!)^r}$

(ii) $\forall j < 0$ $\sup_K |P_j(x,\xi)| \leq C^{-j}[(-j)!]^s$

C/ Si $r = +\infty$ on suppose que $P_j(x,\xi) \equiv 0$ pour $j \geq j_0$ et si $s = +\infty$ on ne suppose aucune condition pour $j < 0$.

Si $y = \chi(x)$ est un difféomorphisme de $\mathbb{C}^n$ on pose :

$$(*) \quad \overline{P}(y,\eta) = e^{d_{x'} \cdot d_{\xi'}} P(x,\xi' + {}^tM(x,x')\eta) \Big|_{\substack{x = x' \\ \xi' = 0}}$$

avec $y = \chi(x)$, $d_{x'} \cdot d_{\xi'} = \sum \dfrac{\partial}{x'_i} \cdot \dfrac{\partial}{\partial \xi'_i}$ et $M(x,x')$ est la matrice définie par

$\chi(x) - \chi(x') = M(x,x').(x-x')$.

On voit facilement que si $P(x,\xi)$ est un symbole de classe (r,s) sur U, $\overline{P}(y,\eta)$ est un symbole de classe (r,s) sur l'image de U par $(x,\xi) \to \langle y = \chi(x), \eta = {}^tM(x,x)^{-1} \xi \rangle$.

Si X est une variété analytique complexe et U un ouvert de $T^* X$, un opérateur microdifférentiel de classe (r,s) est la donnée dans chaque carte locale de X d'un symbole de classe (r,s), tous ces symboles se transformant suivant la formule $(*)$ dans les changements de cartes.

Les opérateurs microdifférentiels de classe (r,s) sur $T^* X$ forment un faisceau que nous noterons $\mathcal{E}_X(r,s)$.

Remarque : $\mathcal{E}_X(\infty,\infty)$ est le faisceau $\hat{\mathcal{E}}_X$ des opérateurs microdifférentiels formels, $\mathcal{E}_X(\infty,1)$ est le faisceau $\mathcal{E}_X$ des opérateurs microdifférentiels d'ordre fini et enfin $\mathcal{E}_X(1,1)$ est le faisceau $\mathcal{E}_X^\infty$ des opérateurs microdifférentiels d'ordre infini (suivant les définitions de [13], [5], [7]).

La formule $P.Q(x,\xi) = \sum_{\alpha\in\mathbb{N}^n} \frac{1}{\alpha!} D_\xi^\alpha P(x,\xi) D_x^\alpha Q(x,\xi)$ permet de munir $\mathscr{E}_X(r,s)$ d'une structure de faisceau d'anneaux unitaires.

Pour $r = \infty$ et P opérateur de classe (∞,s) , le symbole principal de P , noté $\sigma(P)$, est le terme non nul de plus haut degré de P , c'est une fonction holomorphe homogène sur $T^* X$.

On peut réécrire pour ces faisceaux le chapitre II de [13] , $\mathscr{E}_X(\infty,s)$ jouant le rôle des opérateurs d'ordre fini et $\mathscr{E}_X(r,s)$ pour $1 \leq r \leq s$ celui des opérateurs d'ordre infini.

En particulier, si $P \in \mathscr{E}_X(\infty,s)$, il est inversible dans $\mathscr{E}_X(\infty,s)$ si et seulement si $\sigma(P)$ est inversible et pour tout transformation canonique homogène $\Phi : T^* X \to T^* Y$, il existe localement au-dessus de Φ un isomorphisme de faisceaux d'anneaux $\tilde{\Phi} : \mathscr{E}_X(r,s) \to \mathscr{E}_Y(r,s)$.

§2 . Opérateurs 2-microdifférentiels :

Soit X une variété analytique complexe et Λ une sous-variété (lisse) involutive homogène de $T^* X$. Dans la suite nous supposerons que Λ est régulière (i.e. la restriction à Λ de la 1-forme canonique de $T^* X$ ne s'annule pas) ou que Λ est maximalement dégénérée (i.e. l'ensemble des points où la 1-forme s'annule est une sous-variété lagrangienne lisse de $T^* X$).

Si Λ^a est l'antipodale de Λ , $\Lambda \times \Lambda^a$ est une sous-variété involutive de $T^* X \times X$ donc est munie d'un feuilletage canonique. Soit $\tilde{\Lambda}$ la réunion des feuilles de $\Lambda \times \Lambda^a$ qui passent par $\Lambda \times \Lambda^a \approx \Lambda$ $\tilde{\Lambda}$ est une sous-variété lagrangienne de $T^*(X \times X)$ au voisinage de $T^*_X X \times$ .. et $T^*_\Lambda \tilde{\Lambda}$ est un quotient de $T^*(T^*X)\times_{T^*X}\Lambda$.

Soit N un $\mathscr{E}_X$-module ($\mathscr{E}_X = \mathscr{E}_X(\infty,1)$) à gauche cohérent à un générateur et à caractéristiques simples sur Λ .($N \approx \mathscr{E}_X/I$ où I est un idéal de $\mathscr{E}_X$ dont l'idéal des symboles est réduit et définit Λ).

Soit $N^* = \mathcal{E}xt^{\operatorname{codim}\Lambda}_{\mathscr{E}_X}(N,\mathscr{E}_X) \otimes_{\mathcal{O}_X} \Omega_X^{\otimes -1}$ le dual de N (Ω_X est le faisceau des formes différentielles holomorphes de degré maximum sur X) et $N \hat{\otimes} N^*$ le produit de N et N^* (cf [13] ch II). Soit L le sous-module de $N \hat{\otimes} N^*$ engendré par les éléments de la forme $f(u) \otimes v - u \otimes f(v)$ pour $u \in N$, $v \in N^*$ et $f \in \operatorname{End}(N) \approx \operatorname{End}(N^*)$.

Alors le quotient $\tilde{N}$ de $N \hat{\otimes} N^*$ par L est un $\mathscr{E}_{X\times X}$-module holonôme de support $\tilde{\Lambda}$. (Si Λ est lagrangien $\tilde{\Lambda} = \Lambda \times \Lambda^a$ et $\tilde{N} = N \hat{\otimes} N^*$). On pose $\tilde{N}(r,s) = \mathscr{E}_X(r,s) \otimes_{\mathscr{E}_X} N$.

Soient $\pi : \widetilde{{}^\wedge\tilde{\Lambda}} = (\tilde{\Lambda}-\Lambda) \cup T^*_\Lambda\tilde{\Lambda} \to T^*_\Lambda\tilde{\Lambda}$ et $\gamma : T^*_\Lambda\tilde{\Lambda} - \Lambda$ $\underline{\underline{P}}^*_\Lambda\tilde{\Lambda}$ les projections canoniques ($\widetilde{{}^\wedge\tilde{\Lambda}}$ est muni de la topologie de coéclaté ([13] et [7]) et $\underline{\underline{P}}^*_\Lambda\tilde{\Lambda}$

est le fibré projectif), soit $\Omega_{X\times X}^{(0,\dim X)}$ le faisceau des formes différentielles holomorphes de type (0,dim X) sur X × X .

Définition 2.1. $\mathscr{C}_N^{2\mathbb{R}}(r,s) = \mathscr{H}^p_{T^*_{\tilde{\Lambda}}\tilde{\Lambda}}(\pi^{-1}\tilde{N}(r,s))^a \otimes_{\mathcal{O}_{X\times X}} \Omega_{X\times X}^{(0,\dim X)}$

(a application antipodale de $T^*_\Lambda \tilde{\Lambda}$ et $p = \operatorname{codim}_{T^*X}\Lambda$).

$\mathscr{C}_N^{2\infty}(r,s)$ est le faisceau défini par :

$$\mathscr{C}_N^{2\infty}(r,s)\,|_{T^*_\Lambda\Lambda-\Lambda} = \gamma^{-1}\gamma_*\,(\mathscr{C}_N^{2\mathbb{R}}(r,s)\,|_{T^*_\Lambda\Lambda-\Lambda})$$

et $\mathscr{C}_N^{2\infty}(r,s)|_\Lambda = \mathscr{C}_N^{2\mathbb{R}}(r,s)|_\Lambda \;(= \mathscr{H}^p_\Lambda(\tilde{N}(r,s)))$.

(ces définitions sont analogues à celles de $\mathscr{C}_X^{\mathbb{R}}$ et $\mathscr{C}_X^\infty$ de [13] et [7] dans lesquelles on a remplacé $\mathcal{O}_{X\times X}$ par N(r,s) et X par Λ).

Remarque : On montre que si $j \neq p$ $\mathscr{H}^j_{T^*_\Lambda\tilde{\Lambda}}(\pi^{-1}N(r,s)) = 0$.

Proposition 2.2. : $\mathscr{C}_N^{2\mathbb{R}}(r,s)$ et $\mathscr{C}_N^{2\infty}(r,s)$ sont munis de structures canoniques de faisceaux d'anneaux unitaires et on a un morphisme injectif de faisceau d'anneaux :

$$\pi^{-1}(\mathscr{C}_X(r,s)|_\Lambda) \to \mathscr{C}_N^{2\infty}(r,s)$$

Si Λ est fixée, les modules N à un générateur et à caractéristiques simples sur Λ sont localement isomorphes (par les transformations canoniques quantifiées associées à l'identité) donc, à isomorphisme local près, $\mathscr{C}_N^{2\infty}(r,s)$ ne dépend que de la variété Λ.

§3 . Symbole des opérateurs 2-microdifférentiels :

Dans ce paragraphe, nous supposons que X est muni des coordonnées locales $(x_1,\ldots,x_n,\, y_1,\ldots,y_p,\, z_1,\ldots,z_q)$ dans lesquelles $\Lambda = \{(x,y,z,\xi,\eta,\zeta) \in T^*X/\xi=0, z=0\}$

Si Λ est une sous-variété involutive régulière ou maximalement dégénérée de T^*X , alors d'après Oshima [12] , on peut toujours se ramener à ce cas par une transformation canonique homogène.

Si $Z = \{(x,y,z,x',y',z') \in X\times X \;/\; y = y',\; z = z' = 0\}$ alors $\Lambda = T^*_Z X\times X$.

Nous prendrons $N = \mathscr{C}_X / \mathscr{C}_X D_{x_1} + \ldots + \mathscr{C}_X D_{x_n} + \mathscr{C}_X z_1 + \ldots + \mathscr{C}_\Lambda z_q$ et alors $N = C_{Z|X\times X} \cdot T^*_\Lambda\Lambda$ est muni des coordonnées $(x,y,\eta,\zeta,x^*,\zeta^*)$.

Théorème 3.1. : Soit U un ouvert de $T^*_\Lambda\tilde{\Lambda}$, il y a bijection entre les sections de $\mathscr{C}_N^{2\infty}(r,s)$ sur U et les séries formelles $P = \sum_{(i,j)\in\mathbb{Z}^2} P_{ij}(x,y,\eta,\zeta,x^*,\zeta^*)$ où P_{ij} est une fonction holomorphe sur U , homogène de degré i en (x,ζ^*) et j en (η,ζ,x^*) avec les majorations suivantes :

$\forall K$ compact de U $\exists C > 0$ $\forall \varepsilon > 0$ $\exists C_\varepsilon > 0$ tels que (en posant $k = j - i$)

(i) $\forall k < 0$ $\forall i < 0$ $\sup_K |P_{i,k+i}| \leq C^{-k-i}((-k)!)^s(-i)!$

(ii) $\forall k < 0$ $\forall i \geq 0$ $\sup_K |P_{i,k+i}| \leq C_\varepsilon^{-k} \varepsilon^i((-k)!)^s/i!$

(iii) $(r = s)$ $\forall k \geq 0$ $\forall i < 0$ $\sup_K |P_{i,k+i}| \leq C_\varepsilon \varepsilon^k C^{-i}(-i)!/(k!)^r$

(iii)' $(r > s)$ $\forall k \geq 0$ $\forall i < 0$ $\sup_K |P_{i,k+i}| \leq C^{k-i}(-i)!/(k!)^r$

(iV) $(r = s)$ $\forall k \geq 0$ $\forall i \geq 0$ $\sup_K |P_{i,k+i}| \leq C_\varepsilon \varepsilon^{k+i} 1/(k!)^r i!$

(iV)' $(r > s)$ $\forall k \geq 0$ $\forall i \geq 0$ $\sup_K |P_{i,k+i}| \leq C_\varepsilon^k \varepsilon^i 1/(k!)^r i!$

Proposition 3.2. Si P et Q sont deux opérateurs de $\mathscr{C}_N^{2\infty}(r,s)$ de symboles $P = \Sigma P_{ij}$ et $Q = \Sigma Q_{ij}$, si $R = PQ$ est le produit de P et Q dans l'anneau $\mathscr{C}_N^{2\infty}(r,s)$, le symbole $R = \Sigma R_{ij}$ de R est donné par

$$R_{\lambda,\mu}(x,y,\eta,\zeta,x^*,\zeta^*) = \sum_{\substack{\lambda = i+k-(|\alpha|+|\gamma|) \\ \mu = j+l-(|\alpha|+|\beta|+|\gamma|)}} \frac{1}{\alpha!\beta!\gamma!} D_{x^*}^\alpha D_\eta^\beta D_\zeta^\gamma P_{ij} D_x^\alpha D_y^\beta D_{\zeta^*}^\gamma Q_{kl}$$

L'injection canonique de $\pi^{-1}(\mathscr{C}_X(r,s)|\,\Lambda)$ dans $\mathscr{C}_N^{2\infty}(r,s)$ se traduit de la manière suivante pour les symboles :

Soit P un opérateur de $\mathscr{C}_X(r,s)|\,\Lambda$, P a un symbole $P - \sum_{j\in\mathbb{Z}} P_j(x,y,z,\xi,\eta,\zeta)$ où P_j est une fonction définie au voisinage de Λ, homogène de degré j en (ξ,η,ζ).

P_j se développe de manière unique en série de Taylor sous la forme

$$P_j(x,y,z,\xi,\eta,\zeta) = \sum_{\alpha\in\mathbb{N}^n,\beta\in\mathbb{N}^q} P_j^{\alpha\beta}(x,y,\eta,\zeta) z^\beta \xi^\alpha .$$

Alors le symbole de l'image de P dans $\mathscr{C}_N^{2\infty}(r,s)$ est défini par

$$P = \sum_{i\in\mathbb{N},j\in\mathbb{Z}} P_{ij} \text{ avec } P_{ij}(x,y,\eta,\zeta,x^*,\zeta^*) = \sum_{|\alpha|+|\beta|=i} P_j^{\alpha\beta}(x,y,\eta,\zeta) x^{*\alpha} \zeta^{*\beta} .$$

§4 . Opérateurs d'ordre fini - Polygône de Newton d'un opérateur :

Plaçons nous tout d'abord dans le cadre du §3. i.e.

$\Lambda = \{(x,y,z,\xi,\eta,\zeta)/\xi = 0 \quad z = 0\}$.

Soit (r,s) un couple de rationnels tels que $1 \leq s \leq r \leq +\infty$ avec $s < +\infty$.

Définition 4.1. : Soit $(i_o, j_o) \in \mathbb{Z}^2$, $\mathscr{E}_N^2(r,s)[i_o,j_o]$ est le sous-faisceau de $\mathscr{E}_N^{2\infty}(r,s)$ des opérateurs $P = \sum_{(i,j)\in\mathbb{Z}^2} P_{ij}$ tels que :

A/ Si $1 \leq s < r \leq +\infty$

$P_{ij} \equiv 0$ si $\frac{1}{r} i + (j-1) > \frac{1}{r} i_o + (j_o-i_o)$

ou $\frac{1}{s} i + (j-i) > \frac{1}{s} i_o + (j_o-i_o)$.

B/ Si $1 \leq s = r < +\infty$

a) $P_{ij} \equiv 0$ si $\frac{1}{s} i + (j-i) > \frac{1}{s} i_o + (j_o-i_o)$

b) $\forall k$ $\lambda(k)$ t.q. $P_{ij} \equiv 0$ si $\frac{1}{s} i + (j-i) = k$ et $i < \lambda(k)$

et pour $k_o = \frac{1}{s} i_o + (j_o-i_o)$ on a $\lambda(k_o) = i_o$.

$\mathscr{E}_N^2(r,s) = \bigcup_{(i,j)\in\mathbb{Z}^2} \mathscr{E}_N^2(r,s)[i,j]$ est le faisceau des opérateurs 2-microdifférentiels de classe (r,s) (d'ordre fini).

Si P est une section de $\mathscr{E}_N^2(r,s)[i_o,j_o]$ et si P n'est section d'aucun sous-faisceau $\mathscr{E}_N^2(r,s)[i,j]$ strictement plus petit que $\mathscr{E}_N^2(r,s)[i_o,j_o]$ ((r,s)fixé) on pose $\sigma_\Lambda^{(r,s)}(P) = P_{i_o,j_o}$.

Polygône de Newton :

Si P est un opérateur 2-microdifférentiel de symbole $\sum P_{ij}$, le polygône de Newton de P , $N_\Lambda(P)$, est l'enveloppe convexe de la réunion des ensembles $\{(i,k) \in \mathbb{Z}^2/k \leq k_o \text{ , } i + k \leq i_o + k_o\}$ pour les couples (i_o,k_o) tels que P_{i_o,i_o+k_o} ne soit pas identiquement nul.

Si P est un opérateur microdifférentiel défini au voisinage de Λ , le polygône de Newton de P le long de Λ est le polygône de Newton del'image de P dans $\mathscr{E}_N^{2\infty}(1,1)$.

Remarque 4.2. Les faisceaux $\mathscr{E}_N^2(r,s)$, les symboles principaux et le polygône de Newton sont définis dans le cas particulier ou Λ à la forme du §3. mais ces définitions sont invariantes par transformation canonique de $T^* X$ et on peut donc considérer $\mathscr{E}_N^2(r,s)$, $\sigma_\Lambda^{(r,s)}$ et le polygône de Newton le long de Λ pour toute sous-variété involutive Λ régulière ou maximalement dégénérée de $T^* X$.

Remarque 4.3. Si $X = \underline{\underline{C}}$ et $\Lambda = T^*_{\{0\}}\underline{\underline{C}}$, le polygône de Newton le long de Λ défini ci-dessus n'est autre que le polygône de Newton classique d'un opérateur différentiel (Ramis [11]) (au changement de variable $i' = i+k$, $j' = -k$ près).

Remarque 4.4. Nous noterons $\mathcal{E}^2_N = \mathcal{E}^2_N(1,1)$ et $\sigma_\Lambda = \sigma_\Lambda^{(1,1)}$. Si P est un opérateur microdifférentiel $\sigma_\Lambda(P)$ est le symbole qui est noté $\sigma_\Lambda(\sigma(P))$ dans [8] .

Les zéros de $\sigma_\Lambda(P)$ sont les directions microcaractéristiques pour P au sens de [1] et [8] tandis que les zéros de $\sigma_\Lambda^{(\infty,1)}(P)$ sont les directions non fortement non microcaractéristiques pour P au sens de [9] .

Etant donné un opérateur microdifférentiel P défini au voisinage de Λ , les fonctions $P_{i,j}$ ou (i,j) est un sommet du polygône de Newton sont des symboles $\sigma_\Lambda^{(r,s)}(P)$ pour r et s convenables (r et s étant les pentes des côtés du polygône qui passent par (i,j) et donc sont invariants par transformation canonique de $T^* X$ (les autres symboles $\sigma_\Lambda^{(r,s)}(P)$ sont identiquement nuls).

Le théorème fondamental qui relie les faisceaux $\mathcal{E}^2_N(r,s)$ aux directions microcaractéristiques est le suivant :

Théorème 4.5. : Si U est un ouvert de $T^*_\Lambda \tilde{\Lambda}$, P une section de $\mathcal{E}^2_N(r,s)$ sur U ($1 \le s \le r \le +\infty$ et $s < +\infty$), et si $\sigma_\Lambda^{(r,s)}(P)$ ne s'annule pas sur U il existe $Q \in \Gamma(U,\mathcal{E}^2_N(r,s))$ tel que $P.Q = Q.P = 1$.

(Pour montrer ce théorème on utilise une norme formelle semblable à celle de Boutet de Monvel-Kree [3]).

On peut montrer pour $\mathcal{E}^2_N(r,s)$ un théorème de finitude analogue au théorème 3.1. de Boutet de Monvel [2] , de ce théorème on déduit d'une part un théorème de préparation et un théorème de division (cf [13] th. 2.2.1. et 2.2.2. ch II) et d'autre part le théorème suivant :

Théorème 4.6. :

1. Si r et s sont rationnels, $1 \le s \le r \le +\infty$ et $s < +\infty$, $\mathcal{E}^2_N(r,s)$ est un faisceau d'anneaux cohérent et noethérien, à fibres noethériennes.

2. $\mathcal{E}^2_N(r,s)$ est plat sur $\pi^{-1}(\mathcal{E}_X|_\Lambda)$ $(\pi : T^*_\Lambda \tilde{\Lambda} \to \Lambda)$ (et sur $\pi^{-1}(\mathcal{E}_X(\infty,s)|_\Lambda)$.

3. $\mathcal{E}^{2\infty}_N(r,s)$ est fidèlement plat sur $\mathcal{E}^2_N(r,s)$.

§5 . Vecteurs microcaractéristiques et Condition de Lévi pour un système - Problème de Cauchy dans le complexe .

Définition 5.1. : Soient Λ une sous-variété involutive homogène (régulière ou maximalement dégénérée) de $T^* X$ et N un $\mathcal{E}_X$-module cohérent à un générateur à caractéristiques simples sur Λ .

Soit M un $\mathcal{E}_X$-module cohérent défini au voisinage de Λ ; pour r et s rationnels, $1 \leq s \leq r \leq +\infty$, $s < +\infty$, on pose :

$$Ch^2_\Lambda(r,s)(M) = \text{support } (\mathcal{E}^2_N(r,s) \otimes_{\pi^{-1}(\mathcal{E}_X|_\Lambda)} \pi^{-1}M)$$

$$\text{avec } \pi : T^*_\Lambda \tilde{\Lambda} \to \Lambda \ .$$

(Variété microcaractéristique de type (r,s) de M le long de Λ).

Définition 5.2. : Soient M et N deux $\mathcal{E}_X$ modules cohérents , $N^* = \underline{\underline{R}}\, \mathrm{Hom}_{\mathcal{E}_X}(N,\mathcal{E}_X) \otimes_{\mathcal{O}_X} \Omega_X^{\otimes -1}$ [d] , (d = codim support N) , $M \hat{\otimes} N^*$ est un $\mathcal{E}_{X\times X}$-module cohérent et $T^*_X X \times X$ est une sous-variété lagrangienne de $T^* X \times X$, on pose :

$$Ch^2(r,s)(M,N) = Ch^2_{T^*_X X \times X}(r,s)(M \hat{\otimes} N^*) \ .$$

$T^*_X X \times X \approx T^* X$ et on considèrera $Ch^2(r,s)(M,N)$ comme un sous-ensemble de $T^* (T^*X)$.

Si Λ est une sous-variété lisse involutive régulière ou maximalement dégénérée de $T^* X$ et si N est un système à 1 générateur à caractéristiques simples sur Λ , on a une projection canonique $\varpi : T^*(T^*X)\times_{T^* X}\Lambda \to T^*_\Lambda \Lambda$ et on a pour tout $\mathcal{E}_X$-module cohérent M : $Ch^2(r,s)(M,N) = \omega^{-1}\, Ch^2_\Lambda(r,s)(M)$ (cf [9] th. 4.18).

On en déduit facilement du théorème 4.5. la proposition suivante :

Proposition 5.3. : Si on identifie $T^*_\Lambda \tilde{\Lambda}$ et $T_\Lambda (T^*X)$ à l'aide du hamiltonien $H : T^*(T^*X) \to T(T^*X)$, $Ch^2_\Lambda (1,1)(M)$ est l'ensemble des vecteurs microcaractéristiques pour M le long de Λ au sens de [8] , tandis que $Ch^2_\Lambda(\infty,1)(M)$ est l'ensemble des vecteurs non fortement non microcaractéristiques pour M le long de Λ au sens de [9] . La condition de Lévi ([9]) pour M s'écrit donc

$$Ch^2_\Lambda(\infty,1)(M) = Ch^2_\Lambda(1,1)(M).$$

Soit $x^* \in T^*_\Lambda \tilde{\Lambda}$, on définit la suite des "pentes critiques" de M le long de Λ au point x^* comme suit

$$s_o = \sup \{s/ x^* \notin Ch^2_\Lambda(s,1)(M)\} \text{ et par récurrence sur } k :$$

$$s_{k+1} = \sup\{s/x^* \notin CH^2_\Lambda(s,s_k)(M)\} \ .$$

On définit ainsi une suite croissante finie de rationnels :

$$1 \leq s_o < s_1 < \dots < s_p < s_{p+1} = +\infty \quad .$$

Si $M = \mathscr{E}/\mathscr{E}P$ pour un opérateur P , les pentes critiques de M le long de Λ sont les valeurs $\frac{-1}{\lambda}$ ou λ parcourt les pentes du polygône de Newton de P le long Λ .

Si $X = \underline{\underline{C}}$, $\Lambda = T^*_{\{0\}}\underline{\underline{C}}$ la suite $s_o,\dots,s_p$ est la suite des "valeurs exceptionnelles de s" de Ramis [11] .

On peut de même définir la suite des pentes critiques d'un couple de systèmes (M,N).

Remarquons enfin que si Λ est une variété lagrangienne lisse, si M est un système holonôme de variété caractéristique Λ , alors $Ch^2_\Lambda(1,1)(M) \subset \Lambda$ (Λ identifié à la section nulle de $T^*_\Lambda \tilde{\Lambda} = T^* \Lambda$) et M est à points singuliers réguliers au sens de Kashiwara-Kawaï [6] si et seulement si $Ch^2_\Lambda(\infty,1)(M) \subset \Lambda$.

Nous allons maintenant étudier le problème de Cauchy dans le domaine complexe :

Définition 5.4. : Une sous-variété Y de X sera dite non microcaractéristique de type (r,s) pour un couple de $\mathscr{E}_X$-modules cohérent (M,N) si :

$T^*_{Y \times_X T^*X}(T^*X) \cap Ch^2(r,s)(M,N)$ est contenu dans la section nulle de $T^*(T^*X)$.

Soient $\rho : (T^*X)\times_X Y \to T^*Y$ et $\omega: (T^*X)\times_X Y \to T^*X$ les projections canoniques et $d = \dim X - \dim Y$.

Théorème 5.5. : Soient M et N deux $\mathscr{E}_X$-modules cohérents définis au voisinage de $(T^*X)\times_X Y$ dans $T^* X$ et (r,s) deux rationnels tels que $1 \leq s \leq r \leq +\infty$, $s < +\infty$. Si Y est non microcaractéristique de type (r,s) pour (M,N) on a des isomorphismes :

$$\omega^{-1}\underline{\underline{R}}\, \mathrm{Hom}_{\mathscr{E}_X}(M,N(r,s)) \rightleftarrows \underline{\underline{R}}\, \mathrm{Hom}_{\mathscr{E}_X}(M,\mathscr{E}_{X\leftarrow Y}) \otimes_{\rho^{-1}\mathscr{E}_Y} (\mathscr{E}(r,s)_{Y\to X} \otimes_{\mathscr{E}_X} N)[d]$$

avec $N(r,s) = \mathscr{E}_X^{(r,s)} \otimes_{\mathscr{E}_X} N$.

Si $x^* \in (T^*X) \times_X Y$ est tel que $Ch(M) \cap \rho^{-1}(\rho(X^*)) \subset \{x^*\}$ et $Ch(N) \cap \rho^{-1}(\rho(x^*)) \subset \{x^*\}$ l'isomorphisme précédent se réduit à :

$$\forall j \geq 0 \quad \mathrm{Ext}^j_{\mathscr{E}_X}(M,N^{(r,s)})_{x^*} \tilde{\to} \mathrm{Ext}^j_{\mathscr{E}_Y}(M_Y,N_Y^{(r,s)})_{\rho(x^*)}$$

où M_Y et N_Y désignent les modules induits par M et N sur Y [13] .

Le théorème 5.5. signifie que le problème de Cauchy est bien posé sur Y pour le couple $(M, N^{(r,s)})$. Il a été démontré dans le cas $r = s = 1$ (faisceau $\mathcal{E}_X^\infty$ des opérateurs d'ordre infini) par Kashiwara-Schapira [8], dans le cas $r = \infty$, $s = 1$ (faisceau $\mathcal{E}_X$ des opérateurs d'ordre fini) T. Monteiro [10] a adapté la démonstration de Kashiwara-Schapira. Notre méthode de démonstration est différente :

Rappelons ([13]) que si $Z = \{(x,z) \in X / x = 0\}$, $C_{Z|X}$ est défini par $C_{Z|X} = \mathcal{E}_X / \mathcal{E}_X x + \mathcal{E}_X D_z$.

On a $\underline{\underline{R}}\,\mathrm{Hom}_{\mathcal{E}_X}(M, N^{(r,s)}) \approx \underline{\underline{R}}\,\mathrm{Hom}_{\mathcal{E}_{X\times X}}(M \hat{\otimes} N^*, C^{(r,s)}_{X|X\times X})$

ce qui permet ([8]) de ramener le théorème au cas particulier ou N est de la forme $C_{Z|X}$ pour Z hypersurface de X transverse à Y.

D'après [13], si $\Sigma = (T^*_Z X) \times_Z (Y \cap Z)$ et $d = \mathrm{codim}_X Y$ on a un triangle :

$$C^{(r,s)}_{Z|X}|_\Sigma$$

$$\underline{\underline{R}}\,\Gamma_\Sigma(C^{(r,s)}_{Z|X})[2d] \to \underline{\underline{R}}\,\pi'_* \; C^{2\underline{\underline{R}}}_{Z\cap Y|Z|X}(r,s)[d]$$

avec $\Lambda = T^*_Z X$, $\pi : \widetilde{\Sigma}_{\Lambda^*} = (\Lambda - \Sigma) \cup T^*_\Sigma \Lambda \to \Lambda$, $\pi' \; T^*_\Sigma \Lambda - \Sigma \to \Sigma$ et

$$C^{2\underline{\underline{R}}}_{Z\cap Y|Z|X}(r,s) = \underline{\underline{R}}\Gamma_{T^*_\Sigma \Lambda}(\pi^{-1} C_{Z|X}(r,s))[d] .$$

d'où le triangle :

$$\underline{\underline{R}}\,\mathrm{Hom}_{\mathcal{E}_X}(M, \mathcal{C}_{Z|X}(r,s))|_\Sigma$$

$$\underline{\underline{R}}\,\mathrm{Hom}_{\mathcal{E}_X}(M, \underline{\underline{R}}\Gamma_\Sigma(C_{Z|X}(r,s))[2d] \to \underline{\underline{R}}\,\mathrm{Hom}_{\mathcal{E}_X}(M, \underline{\underline{R}}\pi'_* C^{2\underline{\underline{R}}}_{Z\cap Y|Z|X}(r,s))[d]$$

$\mathcal{E}^2_N(r,s)$ est plat sur $\pi^{-1}(\mathcal{E}_X|_\Lambda)$ $(N = C_{Z|X})$ donc

$$\underline{\underline{R}}\,\mathrm{Hom}_{\mathcal{E}_X}(M, \underline{\underline{R}}\,\pi'_* \; C^{2\underline{\underline{R}}}_{Z\cap Y|Z|X}(r,s)) = \underline{\underline{R}}\,\pi'_* \, (\underline{\underline{R}}\,\mathrm{Hom}_{\pi^{-1}\mathcal{E}_X}(\pi^{-1}M, C^{2\underline{\underline{R}}}_{Z\cap Y|Z|X}(r,s)|_{T^*_\Sigma \Lambda - \Lambda})$$

$$= \underline{\underline{R}}\,\pi'_* \, (\underline{\underline{R}}\,\mathrm{Hom}_{\mathcal{E}^2_N(r,s)}(\mathcal{E}^2_N(r,s) \underset{\pi^{-1}\mathcal{E}_X}{\otimes} \pi^{-1}M, C^{2\underline{\underline{R}}}_{Z\cap Y|Z|X}(r,s)|_{T^*_\Sigma \Lambda - \Sigma}) .$$

D'après l'hypothèse du théorème, les supports $(T^*_{Z\cap Y} Z) \times_Z (T^*_Z X)$ de $C^{2\underline{\underline{R}}}_{Z\cap Y|Z|X}(r,s)$ et $Ch^2_\Lambda(r,s)(M)$ de $\mathcal{E}^2_N(r,s) \underset{\mathcal{E}_X}{\otimes} \pi^{-1}M$ sont disjoints en dehors de la section nulle donc :

$$\underline{\underline{R}}\,\mathrm{Hom}_{\mathcal{E}_X}(M, \underline{\underline{R}}\pi'_* \; C^{2\underline{\underline{R}}}_{Z\cap Y|Z|X}(r,s)) = 0$$

donc $\underline{\underline{R}}\,\mathrm{Hom}_{\mathcal{E}_X}(M, C_{Z|X}(r,s))|_\Sigma \approx \underline{\underline{R}}\,\mathrm{Hom}_{\mathcal{E}_X}(M, \underline{\underline{R}}\Gamma_\Sigma(C_{Z|X}(r,s)))[2d]$.

Pour montrer le théorème 5.5., il reste à montrer que

$$\underline{\underline{R}}\,\mathrm{Hom}_{\mathcal{E}_X}(M, \underline{\underline{R}}\,\Gamma_\Sigma(C_{Z|X}(r,s)))_{x^*}[2d] = \underline{\underline{R}}\,\mathrm{Hom}_{\mathcal{E}_Y}(M_Y, C_{Z\cap Y}(r,s)\ \rho(x^*))$$

si x^* est un point tel que $Ch(M) \cap \rho^{-1}(\rho(x^*)) = \{x^*\}$ ce qui se fait par un théorème de dualité pour les $\mathcal{E}^2_N(r,s)$-modules analogue au théorème 3.5.6. ch. II de [13].

Suivant [8] , on peut déduire du théorème 5.5. un théorème sur le problème de Cauchy dans les fonctions holomorphes ramifiées :

Si Z est une hypersurface lisse de X et φ une équation de Z on définit : $\mathcal{O}^1_{[Z|X]} = \mathcal{O}_X \operatorname{Log}\varphi$.

Si $\lambda \in [1, +\infty]$, λ rationnel on pose $\mathcal{O}_X(\lambda) = \mathcal{E}_X(\lambda,1)|_{T^*_X X}$ et $\mathcal{O}^1_{[Z|X]}(\lambda) = \mathcal{O}_X(\lambda) \operatorname{Log}\varphi$.

$\mathcal{O}^1_{[Z|X]}(\lambda)$ est le faisceau des fonctions holomorphes ramifiées autour de Z qui s'écrivent $\sum_{j\geq 0} a_j(x)(\varphi(x))^{-j} + b(x) \operatorname{Log}(x)$ avec localement

$$|a_j(x)| \leq C^j \frac{1}{(j!)^{\lambda-1}} \text{ pour } \lambda < +\infty \quad \text{et} \quad a_j(x) \equiv 0 \text{ si } j \geq j_0 \text{ pour } \lambda = +\infty .$$

Si $(Z_i)_{i=1,\dots,r}$ sont des hypersurfaces de X on définit $\sum_{i=1}^{r} \mathcal{O}^1_{Z_i|X}(\lambda)$ comme le conoyau de l'application :

$$\mathcal{O}_X^{r\,1} \to \bigotimes_{i=1}^{r} \mathcal{O}^1_{Z_i|X} \qquad (f_1,\dots,f_{r-1}) \mapsto (f_1, f_2-f_1, \dots, f_{r-1})$$

Théorème 5.6. Soit Y une sous-variété de X, Z une hypersurface de Y , $(Z_i)_{i=1,\dots,r}$ des hypersurfaces de X transverses 2 à 2 et transverses à Y telles que $Z_i \cap Y = Z$ pour tout i.

Soit M un $\mathcal{O}_X$-module cohérent tel que :

a) $Ch\,M \cap \rho^{-1}(T^*_Z Y) \subset \bigcup_i T^*_{Z_i} X \quad (\rho : (T^*X)\times_X Y \to T^*Y)$

b) Y est non microcaractéristique de type $(\lambda,1)$ pour M sur $T^*_{Z_i} X \setminus T^*_X X$ pour tout i .

Alors pour tout j on a des isomorphismes :

$$\mathcal{E}xt^j_{\mathcal{O}_X}(M, \sum_{i=1}^{r} \mathcal{O}^1_{Z_i|X}(\lambda))|_Z \to \mathcal{E}xt^j_{\mathcal{O}_Y}(M_Y, \mathcal{O}^1_{Z|Y}(\lambda))|_Z .$$

Ce théorème a été démontré (sous des hypothèses plus restrictives)par Hamada-Leray-Wagschal [4] .

Le théorème suivant donne une interprétation analytique des pentes critiques d'un système :

Théorème 5.7. : Soit Λ une sous-variété involutive régulière ou maximalement dégénérée de T^*X et N un $\mathscr{E}_X$-module cohérent à un générateur, à caractéristiques simples sur Λ .

Soit M un $\mathscr{E}_X$-module cohérent et x un point de Λ tel que les pentes critiques de M le long de Λ en tous les points x^* de $(T^*_\Lambda \widetilde{\Lambda} \setminus \Lambda)$tels que $\pi(x^*) = x$ soient les mêmes.

Soient $1 \leq s_o < s_1 < \ldots < s_p < s_{p+1} = +\infty$ ces pentes critiques . pour tout (r,s) il existe $j \in [0,\ldots,p+1]$ tel que (posant $s_{-1}=1$)
$\forall k \geq 0$ $\mathrm{Ex}\,t^k_{\mathscr{E}_X}(M,N(r,s)) = \mathrm{E}\,t^k_{\mathscr{E}_X}(M,N(s_j,s_{j-1}))$.

Corollaire 5.8. : Soit $x \in X$ et M un $\mathscr{D}_X$-module holonôme de variété caractéristique $T^*_{\{x\}}X \cup T^*_X X$.

Soit $\Lambda = T^*_{\{x\}}X - \{x\}$. On suppose que pour tous les points de $T^*\Lambda - \Lambda$ les pentes critiques de M le long de Λ sont les mêmes.

Alors pour toute solution u de M dans les hyperfonctions à support dans $\{x\}$, il existe une et une seule de ces pentes critiques s_j telle que u soit une ultradistribution de classe (s_j) et ne soit pas de classe (s) pour $s > s_j$.

(Les ultradistributions de classe (s) a support dans $\{x\}$ sont les séries $\sum_{\alpha\in\underline{\underline{N}}^n} a_\alpha\, \delta^{(\alpha)}(y-x)$ avec $|a_\alpha| \leq C^{|\alpha|}\,(|\alpha|!)^{-s}$) et si $s = +\infty$ ce sont les distributions).

Dans le cas où X est de dimension 1, la condition sur M est toujours remplie et on retrouve un théorème analogue au théorème 1 de Ramis [11] .

Pour montrer le théorème 5.7. , on se ramène au cas de la dimension 1, grâce au théorème 5.5.

§6 . Théorème de prolongement :

Théorème 6.1. : Soient M et N deux $\mathscr{E}_X$-modules cohérents et r et s deux rationnels avec $1 \leq s \leq r \leq +\infty$, $s < +\infty$.

Soit Ω un ouvert de T^*X à bord de classe C^1 au voisinage d'un point x^* de T^*X . On suppose que la normale à Ω en x^* n'appartient pas à $Ch^2(r,s)(M,N)$, alors $\underline{\underline{R}}\,\Gamma_{\overline{\Omega}}(\underline{\underline{R}}\,\mathrm{Hom}_{\mathscr{E}_X}(M,N(r,s)))_{x^*} = 0$.

(Pour $r = s = 1$ ce théorème est démontré dans [7]).

Démonstration :

$$\underline{\underline{R}}\,\mathrm{Hom}_{\mathscr{E}_X}(M,N(r,s))_{x^*} \approx \underline{\underline{R}}\,\mathrm{Hom}_{\mathscr{E}_{X\times X}}(M \hat{\otimes} N^*, C_{X|X\times X}(r,s)_{(x^*,x^*)})$$

ce qui ramène au cas où N est un module à un générateur à caractéristiques simples sur une variété lisse Λ .

Il suffit alors de montrer que $\underline{\underline{R}}\,\Gamma_{\bar{\Omega}}(N(r,s))_{x^*}$ est un $\mathscr{E}^2_N(r,s)_{(x^*,\xi^*)}$-module si ξ^* est la normale à Ω en x^* .

En effet on a alors :

$$\underline{\underline{R}}\,\Gamma_{\bar{\Omega}}\,\underline{\underline{R}}\,\mathrm{Hom}_{\mathscr{E}_X}(M,N(r,s))x^* = \underline{\underline{R}}\,\mathrm{Hom}_{\mathscr{E}_{X,x^*}}(M_{x^*}, \underline{\underline{R}}\,\Gamma_{\bar{\Omega}}(N)_{x^*})$$

$$= \underline{\underline{R}}\,\mathrm{Hom}_{\mathscr{E}^2_{N(x^*,\xi^*)}(r,s)}((\mathscr{E}^2_N(r,s) \underset{\pi^{-1}\mathscr{E}_X}{\otimes} \pi^{-1}M)_{(x^*,\xi^*)}, \underline{\underline{R}}\Gamma_{\bar{\Omega}}(N)_{x^*}) = 0.$$

d'après l'hypothèse.

[1] J.M. BONY : Extension du théorème de Holmgren, Sem. Goulaouic-Schwarz 1975-76, exposé 17.

[2] L. BOUTET de MONVEL : Opérateurs pseudo-différentiels analytiques. Univ. Sci et Méd. Grenoble (75-76).

[3] L. BOUTET de MONVEL, P. KREE : Ann Inst. Fourier, Grenoble, 17, p. 295-323 (1967).

[4] T. HAMADA, J. LERAY, C. WAGSCHAL : Problème de Cauchy ramifié, J. Math. Pures et Appl. 55, 297-352 (1976).

[5] M. KASHIWARA : Cours sur les systèmes d'équations microdifférentielles. Prépubl. de l'Univ. Paris-Nord, C.S.P. 1978.

[6] M. KASHIWARA, T. KAWAI : Systems with regular singularities (Preprint).

[7] M. KASHIWARA, P. SCHAPIRA : Microhyperbolic systems, Acta Math 142, 1-55, 1979.

[8] M. KASHIWARA, P. SCHAPIRA : Problème de Cauchy dans le domaine complexe Inv. Math 46, 17-38 (1978).

[9] T. MONTEIRO : Thèse de 3è cycle à l'Université Paris-Nord (1978).

[10] T. MONTEIRO : article à paraître.

[11] J.P. RAMIS : Dévissage Gevrey Astérisque 59-60 (1978) p. 173-204.

[12] T. OSHIMA : Singularities in contact geometry. J. Fac Sci. Univ. of Tokyo, Sec IA, 21, (1974), 43-83.

[13] M. SATO, T. KAWAI, M. KASHIWARA : Hyperfunctions and pseudo-differential equations, Lect. Notes in Math. 287, Springer, 265-529.

SUR LE PROBLÈME DE HILBERT-RIEMANN

par Zoghman MEBKHOUT

SOMMAIRE

----

§.1 - INTRODUCTION

Nous présentons dans cet exposé le théorème d'existence de ([19]; chap.V) sur le problème de Hilbert-Riemann en dimension supérieure pour les faisceaux analytiquement constructibles. Alors qu'en dimension un le problème de Hilbert-Riemann (= 21ème problème de Hilbert) a connu et connaît encore une grande activité (voir par exemple les exposés de Sato-Jimbo-Miwa dans ce même colloque), le premier résultat en dimension supérieure est dû à P. Deligne [3] qui donne une description purement algébrique de toutes les représentations de dimension finie du groupe fondamental d'une variété algébrique lisse sur le corps des nombres complexes. Pour ce faire, P. Deligne utilise une compactification d'Hironaka comme A. Grothendieck [4] qui avait déjà obtenu le cas de la représentation triviale comme cas particulier du théorème de comparaison de Grothendieck [4]. Le modèle de démonstration de Grothendieck a été fondamental pour ce genre de questions, parce qu'on utilise de façon systématique la résolution des singularités. (voir [5],[9]). Ce bref article [4] contient en germe plusieurs théories dont la théorie des systèmes à points singuliers réguliers à plusieurs variables.

Il est bien connu que la catégorie de représentations de dimension finie du groupe fondamental d'une variété analytique complexe lisse connexe est équivalente à la catégorie des systèmes locaux, c'est-à-dire des faisceaux localement constants de dimension finie, elle-même équivalente à la catégorie des connexions analytiques intégrables. C'est le théorème d'existence de Frobenius. Cependant, pour l'étude des singularités, cette catégorie a des inconvénients facheux : par exemple, l'image directe d'un système local par une application propre cesse en général d'être un système local pour devenir un faisceau constructible. C'est la catégorie des faisceaux constructibles qui a les bonnes propriétés de stabilité par images directes et réciproques. De ce point de vue, il devient naturel de reformuler le problème de Hilbert-Riemann pour les faisceaux constructibles qui auront pour analogues analytiques les systèmes holonomes.

On démontre [19] que la catégorie dérivée des faisceaux constructibles est équivalente à la catégorie dérivée des systèmes différentiels linéaires holonomes d'ordre infini. C'est là une généralisation du théorème d'existence de Frobenius dont la démonstration nous fournira une réponse positive au problème de Hilbert-Riemann (local pour l'instant) pour les faisceaux constructibles. On doit donc s'attendre à un dictionnaire entre énoncés géométriques concernant les faisceaux constructibles et énoncées analytiques concernant les systèmes différentiels linéaires. On en examine, dans le dernier paragraphe, quelques exemples dont la dualité de Poincaré-Verdier qui a été le point de départ ([18] ; remarque 5.1).

Les démonstrations, qui ont des intermédiaires intéressants tels que la conservation par morphisme projectif de la régularité, ne sont qu'esquissées mais se trouvent en détails dans [19]. A ce titre, cet exposé est introductif au problème de Hilbert-Riemann tout comme l'étaient les exposés [20],[21] qui étaient, eux, concentrés sur le théorème de dualité pour les $\mathcal{D}_X$-modules.

L'auteur expose ses résultats et laisse le soin au lecteur de faire le lien ou de comparer s'il y a lieu avec d'autres travaux. Nous essayons de présenter les problèmes de la façon la plus naturelle possible en nous appuyant sur la bibliographie.

§.2 - POSITION DU PROBLEME

Soit $(X, \mathcal{O}_X)$ une variété analytique complexe lisse connexe de dimension n et Y une hypersurface de X. Notons U le complémentaire de Y et $j : U \to X$ l'inclusion canonique. Soit, d'autre part, $\mathcal{L}$ une monodromie sur U, c'est-à-dire soit une représentation du groupe fondamental $\pi_1(U)$, soit un système local. Classiquement, U est une variété algébrique et $X = \bar{U}$ un de ses compartifiés. On se pose alors deux questions de façon un peu vague :

Question $(HR)_1$: est-ce que $\mathcal{L}$ est la monodromie d'une équation différentielle sur U ?

Question $(HR)_2$: est-ce que $\mathcal{L}$ est la monodromie d'une équation différentielle "régulière" le long de Y ?

Par équation différentielle, on entend connexion intégrable dont on admettra la définition pour l'instant. Pour répondre à $(HR)_1$, il suffit de prendre la connexion $\mathcal{O}_U \otimes_{\mathbb{C}} \mathcal{L}$. Mais cette connexion est "irrégulière" en général. Dans [3], P. Deligne construit une connexion "<u>régulière</u>" pour répondre à $(HR)_2$. Si U est algébrique, alors cette connexion régulière est algébrisable.

Comme nous l'avons vu dans l'introduction, il faut reformuler les questions précédentes pour les faisceaux <u>constructibles</u>. Soit $\mathcal{F}$ un faisceau constructible sur X, c'est-à-dire tel qu'il existe une stratification (<u>de Whitney</u>) $X = \cup X_i$ pour laquelle la restriction $\mathcal{F}_{|X_i}$ à chaque strate est un système local. On se pose alors de façon aussi vague deux questions :

<u>Question(HR)'₁</u> : est-ce que $\mathcal{F}$ est solution d'un système holonome ?

<u>Question (HR)'₂</u> : est-ce que $\mathcal{F}$ est solution d'un système holonome "régulier" ?

Il convient de préciser les rapports entre les questions (HR) et les questions (HR)' qui ne sont semblables qu'en apparence. Alors que la question $(HR)_1$ est facile, il n'en est plus de même de la question $(HR)'_1$. Une réponse <u>locale</u> à la question $(HR)'_2$ est nécessaire pour répondre à la question $(HR)'_1$. Les questions (HR)' sont étroitement liées. Alors que le système local $\mathcal{L}$ est donné sur U dans les questions (HR), le faisceau constructible $\mathcal{F}$ est donné sur X. Pour récupérer la réponse des questions (HR) à partir des questions (HR)', il faut prendre un "prolongement" de $\mathcal{L}$ à X. Il existe bien sûr un prolongement canonique, à savoir $j_*\mathcal{L}$, mais il en existe d'autres, ce qui introduit un certain arbitraire. En fait, même en dimension un, il faut considérer les solutions supérieures pour une bonne compréhension du problème. Par exemple, <u>un</u> système différentiel peut donner naissance à un <u>complexe</u> de solution tout comme on verra qu'<u>un</u> faisceau $\mathcal{F}$ peut donner naissance à un <u>complexe</u> de systèmes différentiels. Pour toutes ces raisons, il apparaît <u>clairement</u> que le cadre <u>naturel</u> est les <u>catégories dérivées</u> qui sont désormais <u>fondamentales</u> même pour passer d'une solution locale à une solution globale. Nous renvoyons le lecteur non averti aux notes originales de <u>Verdier</u> (Lectures Notes in Math. n°569) qui restent la meilleure référence pour la question. Pour la commodité du lecteur, nous allons reprendre quelques notions sur les systèmes différentiels maintenant bien connues.

§.3 - SYSTEMES HOLONOMES (RAPPELS)

Soit toujours $(X, \mathcal{O}_X)$ une variété analytique complexe lisse de dimension n. Le faisceau des opérateurs différentiels $\mathcal{D}_X$ d'ordre fini est un faisceau cohérent d'anneaux unitaires non commutatifs. Il contient comme sous-faisceau d'anneaux le faisceau $\mathcal{O}_X$. En coordonnées locales $(x_1,\ldots,x_n)$, une section $P(x,\frac{d}{dx})$ de $\mathcal{D}_X$ est un <u>polynôme</u> de la forme

$$P(x,\frac{d}{dx}) = \sum_{|\alpha| \leqslant m} a_\alpha(x) \frac{d^\alpha}{dx^\alpha}$$

où $a_\alpha(x)$ sont des fonctions holomorphes ; $\alpha = (\alpha_1,\ldots,\alpha_n) \in N^n$; $|\alpha| = \alpha_1 + \ldots + \alpha_n$. Le faisceau $\mathcal{D}_X$ opère à gauche sur le faisceau $\mathcal{O}_X$ qui devient un système différentiel, à savoir le système de "De Rham". Plus généralement, un système différentiel d'ordre fini est un $\mathcal{D}_X$-module à gauche <u>cohérent</u>, c'est-à-dire admettant localement une présentation finie. Une connexion intégrable est simplement un système différentiel qui soit libre de type fini en tant que $\mathcal{O}_X$-module. Il suffit d'ailleurs qu'il soit de type fini. Si on a un vrai système différentiel, c'est-à-dire une matrice $P = P(x,\frac{d}{dx})$ de p lignes et q colonnes d'opérateurs différentiels, on lui associe le conoyau $\mathcal{M}$ de l'application linéaire $\mathcal{D}^q \overset{P}{\rightarrow} \mathcal{D}^p$ définie par P. C'est un système différentiel par construction.

Pour notre but, le faisceau $\mathcal{D}_X$ est insuffisant. Il faut introduire le $\mathcal{D}_X^\infty$ des opérateurs différentiels d'ordre <u>infini</u>. C'est un faisceau <u>non</u> cohérent d'anneaux unitaires non commutatifs. Il contient comme sous-faisceau d'anneaux $\mathcal{D}_X$ sur lequel il est <u>fidèlement plat</u> [22]. En coordonnées locales $x = (x_1,\ldots,x_n)$, une section locale $P^\infty(x,\frac{d}{dx})$ est une <u>série</u> de la forme

$$P^\infty(x,\frac{d}{dx}) = \sum_{|\alpha|=0}^{\infty} \frac{a_\alpha(x)}{\alpha!} \frac{d^\alpha}{dx^\alpha}$$

avec la condition

$$\lim_{|\alpha|\to 0} |a_\alpha(x)|^{\frac{1}{|\alpha|}} = 0$$

uniformément sur tout compact.

Le faisceau $\mathcal{D}_X^\infty$ opère à gauche sur $\mathcal{O}_X$. On définit un système différentiel d'ordre infini comme un $\mathcal{D}_X^\infty$-module à gauche $\mathcal{M}^\infty$ admettant une présentation locale de la forme :

$$(\mathcal{D}_X^\infty)^q \xrightarrow{P} (\mathcal{D}_X^\infty)^p \longrightarrow \mathcal{M}^\infty \longrightarrow 0 \quad ,$$

mais où la matrice P est d'ordre fini. On dit admissible dans [22]. Ceci ramène l'étude des $\mathcal{D}_X^\infty$-modules admissibles à l'étude des $\mathcal{D}_X$-modules cohérents. Mais un problème fondamental et difficile est de décider qu'étant donné un $\mathcal{D}_X^\infty$-module, s'il est admissible ou pas (voir le théorème 4.1).

La variété caractéristique d'un opérateur différentiel d'ordre fini est l'hypersurface du fibré cotangent T^*X défini par son symbole principal. Pour définir la variété caractéristique d'un système différentiel, une méthode consiste à introduire le faisceau des opérateurs micro-différentiels $\check{\mathcal{E}}_X$ d'ordre fini. C'est un faisceau cohérent d'anneaux unitaires non commutatifs sur T^*X. Si $\mathcal{M}$ est un système différentiel, la variété caractéristique $\check{SS}(\mathcal{M})$ est le sous-ensemble de T^*X défini par

$$\check{SS}(\mathcal{M}) = \text{support} \; (\check{\mathcal{E}}_X \underset{\pi^{-1}\mathcal{D}_X}{\otimes} \pi^{-1}\mathcal{M})$$

où π désigne la projection de T^*X sur X. Le faisceau $\check{\mathcal{E}}_X$ contient comme sous-faisceau d'anneaux $\pi^{-1}\mathcal{D}_X$ sur lequel il est fidèlement plat [22]. Le théorème fondamental ([22] théorème 5.3.2) affirme que la variété caractéristique $\check{SS}(\mathcal{M})$ d'un système différentiel non nul $\mathcal{M}$ est un sous-espace analytique involutif de T^*X. En particulier, on a l'inégalité fondamentale

$$\dim X = n \leq \dim \check{SS}(\mathcal{M}) \leq 2n = \dim T^*X \; .$$

Par définition, un système holonome $\mathcal{M}$ est un système tel que $\dim \check{SS}(\mathcal{M}) = n$. Par exemple, une connexion intégrable est un système holonome dont la variété caractéristique est la section nulle de T^*X. De ce point de vue, la notion de système holonome assouplit considérablement la notion de connexion intégrable en admettant des singularités dans la variété caractéristique en toute codimension. La variété caractéristique d'un système d'ordre infini est définie de la même façon grâce à l'admissibilité. On a la notion de système holonome d'ordre infini.

Soit un $\mathcal{D}_X$-module à gauche $\mathcal{M}$. On construit son complexe de De Rham DR($\mathcal{M}$) (voir par exemple [16] page 103) :

$$0 \longrightarrow \mathcal{M} \xrightarrow{\nabla} \mathcal{M} \underset{\mathcal{O}_X}{\otimes} \Omega_X^1 \longrightarrow \cdots\cdots\cdots\cdots \longrightarrow \mathcal{M} \underset{\mathcal{O}_X}{\otimes} \Omega_X^n \longrightarrow 0 \quad .$$

Si $\mathcal{M}$ est une connexion intégrable, le lemme de Poincaré affirme que le complexe DR($\mathcal{M}$) est cohomologiquement trivial en degrés positifs.

Le théorème d'existence de Cauchy affirme que les sections horizontales de $\mathcal{M}$, Ker(∇) qui est la cohomologie en degrés zéro sont un système local de vectoriels complexes de dimension finie. Si la variété caractéristique d'un système holonome a des singularités, le lemme de Poincaré et le théorème d'existence de Cauchy cessent d'avoir lieu. On les remplace par le théorème de Kashiwara [8] qui affirme que les solutions holomorphes d'un système holonome forment des faisceaux constructibles. Pour formuler de façon précise le théorème de Kashiwara, il y a intérêt tout de suite à travailler dans les catégories dérivées. On démontre (voir par exemple [16] page 104) que le complexe $\underline{\mathrm{DR}}(\mathcal{M})$ est isomorphe dans $D(\mathbb{C}_X)$, catégories dérivées de la catégorie des faisceaux d'espaces vectoriels complexes, à $\mathbb{R}\,\underline{\mathrm{hom}}_{\mathcal{D}_X}(\mathcal{O}_X, \mathcal{M})$.

D'autre part, le complexe $\mathbb{R}\,\underline{\mathrm{hom}}_{\mathcal{D}_X}(\mathcal{M}, \mathcal{O}_X)$ dont les faisceaux de cohomologie ne sont autres que les $\mathcal{E}xt^i_{\mathcal{D}_X}(\mathcal{M}, \mathcal{O}_X)$ est par définition le complexe des solutions holomorphes. La raison est simple. On peut voir facilement que le premier faisceau s'identifie au faisceau des solutions holomorphes du système dans une présentation locale. Alors le théorème de Kashiwara [8] s'énonce comme suit :

<u>Théorème 3.1.- Le complexe $\mathbb{R}\underline{hom}_{\mathcal{D}_X}(\mathcal{M}, \mathcal{O}_X)$ est (analytiquement) constructible si est un système holonome.</u>

Si $\mathcal{M}$ est une connexion intégrable, le théorème 3.1 se réduit au théorème d'existence de <u>Cauchy</u> conjugué au lemme de <u>Poincaré</u>. Nous noterons $D(\mathbb{C}_X)_c$ la catégorie dérivée des complexes <u>bornés</u> de $\mathbb{C}_X$-modules à cohomologie constructible. De même nous noterons $D(\mathcal{D}_X)_h$ (resp $D(\mathcal{D}_X^\infty)_h$) la catégorie des complexes de $\mathcal{D}_X$-modules (resp de $\mathcal{D}_X^\infty$-modules) à gauche <u>bornés</u> à cohomologie $\mathcal{D}_X$-holonome (resp $\mathcal{D}_X^\infty$-holonome). Il résulte facilement du théorème 3.1 que le foncteur $\mathbb{R}\underline{hom}_{\mathcal{D}_X}(*\ ,\ \mathcal{O}_X)$ (resp $\mathbb{R}\,hom_{\mathcal{D}_X^\infty}(*\ ,\ \mathcal{O}_X)$) envoie $D(\mathcal{D}_X)_h$ (resp $D(\mathcal{D}_X^\infty)_h$) dans $D(\mathbb{C}_X)_c$.

§.4 - ENONCE DU THEOREME D'EXISTENCE

Soit $\mathcal{F}$ un complexe de $D(\mathbb{C}_X)$. La structure de $\mathcal{D}^\infty$-module à gauche de $\mathcal{O}_X$ induit une structure de complexe de $D(\mathcal{D}_X^\infty)$ sur $\mathbb{R}\underline{hom}_{\mathbb{C}_X}(\mathcal{F}, \mathcal{O}_X)$.

<u>Théorème 4.1.- Le complexe $\mathbb{R}\underline{hom}_{\mathbb{C}_X}(\mathcal{F}, \mathcal{O}_X)$ appartient à $D(\mathcal{D}_X^\infty)_h$ si $\mathcal{F}$ appartient à $D(\mathbb{C}_X)_c$.</u>

De façon un peu plus parlante, le théorème 4.1 affirme que les <u>faisceaux</u> $\mathcal{E}xt^i_{\mathbb{C}_X}(\mathcal{F}, \mathcal{O}_X)$ sont des systèmes différentiels holonomes d'ordre <u>infini</u>. Des exemples simples (par exemple pour $\mathcal{F}$ égal au faisceau caractéristique d'une sous-variété Y lisse de X) montrent que $\mathbb{R}\underline{hom}_{\mathbb{C}_X}(\mathcal{F}, \mathcal{O}_X)$ n'appartient pas à $D(\mathcal{D}_X)_h$ à cause de la présence des <u>singularités essentielles</u>. D'où la nécessité du faisceau $\mathcal{D}_X^\infty$.

Par le théorème 3.1 et le théorème 4.1, on dispose de deux foncteurs

$$\mathbb{F}\ :\ D(\mathcal{D}_X^\infty)_h \longrightarrow D(\mathbb{C}_X)$$
$$\mathbb{F}(\mathcal{M}^\infty) = \mathbb{R}\underline{hom}_{\mathcal{D}^\infty}(\mathcal{M}^\infty, \mathcal{O}_X)$$

$$\mathbb{C} : D(\mathbb{C}_X)_c \longrightarrow D(\mathcal{D}_X^\infty)_h$$

$$\mathbb{C}(\mathcal{F}) = \mathbb{R}\underline{\mathrm{hom}}_{\mathbb{C}_X}(\mathcal{F}, \mathcal{O}_X) .$$

Proposition 4.2.- Les foncteurs $\mathbb{F}$ et $\mathbb{C}$ sont inverses l'un de l'autre et établissent donc une équivalence de catégorie entre $D(\mathcal{D}_X^\infty)_h$ et $D(\mathbb{C}_X)_c$.

La démonstration de cette proposition utilise d'une part le théorème de bidualité locale de nature analytique (théorème 2.1, chap.III ; [19]) qu'on peut résumer en disant que le système de De Rham " $\mathcal{O}_X$ " est dualisant dans $D(\mathcal{D}_X^\infty)_h$, d'autre part le théorème de bidualité locale de nature géométrique dû à Verdier [23] que l'on peut résumer en disant que le faisceau constant "$\mathbb{C}_X$" est dualisant dans $D(\mathbb{C}_X)_c$. Le théorème 4.1 et la proposition 4.2 donnent une réponse globale très précise globale à la question $(HR)_1'$ du paragraphe 2. La démonstration du théorème 4.1 apporte une réponse locale mais précise à la question $(HR)_2'$ comme nous allons le voir. L'auteur espère utiliser la méthode de la descente cohomologique comme pour le théorème de dualité ([19] ; chap.IV) pour répondre globalement à $(HR)_2'$. On peut donc s'attendre, en vertu de l'équivalence des catégories $D(\mathbb{C}_X)_c$ et $D(\mathcal{D}_X^\infty)_h$, à un dictionnaire entre énoncés géométriques et énoncés analytiques. Nous verrons de tels exemples dans le paragraphe 7. Il n'est pas étonnant que les démonstrations combinent les méthodes analytiques introduites dans [16],[17] pour traiter le cas du faisceau caractéristique d'un sous-espace analytique Y de X ayant des singularités arbitraires et les méthodes géométriques utilisant les dévissages de faisceaux constructibles et la résolution des singularités. Nous allons rappeler rapidement dans les paragraphes 5 et 6 ces différentes méthodes.

§.5 - REGULARITE ET DUALITE

Soit $\mathcal{M}$ un complexe de $D(\mathcal{D}_X)_h$ et Y un sous-espace analytique complexe de X défini par un Idéal $\mathcal{I}_Y$. Considérons le complexe de cohomologie locale algébrique introduit par Grothendieck dans [4] :

$$\underline{\mathbb{R}\Gamma}_{[Y]}(\mathcal{M}) \overset{\text{Déf}}{=} \mathbb{R} \varinjlim_{k} \underline{hom}_{\mathcal{O}_X}(\mathcal{O}_X/\mathcal{I}_Y^k, \mathcal{M}) \quad .$$

En fait $\underline{\mathbb{R}\Gamma}_{[Y]}(\mathcal{M})$ est un objet de $D(\mathcal{D}_X)$. Il faut préciser la structure de $\mathcal{D}_X$-modules. Le complexe $\underline{\mathbb{R}\Gamma}_{[Y]}(\mathcal{M})$ est un sous-complexe de $\underline{\mathbb{R}\Gamma}_Y(\mathcal{M})$, cohomologie locale ordinaire, où opère le faisceau $\mathcal{D}_X$ parce que les opérateurs différentiels sont locaux et diminuent le support. Pour voir que $\mathcal{D}_X$ opère sur $\underline{\mathbb{R}\Gamma}_{[Y]}(\mathcal{M})$, il suffit de vérifier qu'un champ de vecteurs le conserve. Ce qui est fait dans ([16] ; page 98).

On rappelle alors ([19], proposition 3.1, chap.III) que les conditions a), b) et c) suivantes sont équivalentes :

a) L'homomorphisme canonique dans $D(\mathbb{C}_X)$ est un isomorphisme

$$\underline{DR}(\underline{\mathbb{R}\Gamma}_{[Y]}(\mathcal{M})) \longrightarrow \underline{DR}(\underline{\mathbb{R}\Gamma}_Y(\mathcal{M})) \quad .$$

b) L'homomorphisme canonique dans $D(\mathbb{C}_X)$ est un isomorphisme

$$\mathbb{R}\,\underline{hom}_{\mathcal{D}_X}(\mathcal{M}, \mathcal{O}_X) \underset{\mathbb{C}}{\otimes} \mathbb{C}_Y \longrightarrow \mathbb{R}\,\underline{hom}_{\mathcal{D}_X}(\underline{\mathbb{R}\Gamma}_{[Y]}(\mathcal{M}), \mathcal{O}_X)$$

c) L'homomorphisme canonique dans $D(\mathcal{D}_X^\infty)$ est un isomorphisme :

$$\mathcal{D}_X^\infty \underset{\mathcal{D}_X}{\otimes} \underline{\mathbb{R}\Gamma}_{[Y]}(\mathcal{M}) \longrightarrow \underline{\mathbb{R}\Gamma}_Y(\mathcal{D}_X^\infty \underset{\mathcal{D}_X}{\otimes} \mathcal{M}) \quad .$$

Dans le cadre des $\mathcal{D}_X$-modules, les conditions a), b) et c) sont établies pour la première fois dans [17] pour le système de De Rham [la condition a) est le théorème 3.1, la condition b) est le théorème 1.1 et la condition c) est le théorème 1.2, [17]]. Pour un système général, la démonstration de l'équivalence des conditions a), b) et c) est toute pareille. Naturellement, les conditions a), b) et c) ont une longue histoire dans d'autres situations. En dimension un, la condition a) est due à Picard (les formes de deuxième espèce calculent l'homologie d'une surface de Riemann).

En dimension supérieure et Y un diviseur à croisements normaux, elle est due à Atiyah-Hodge (Integrals of the second kind on an algebraic variety. Ann. of Math. 62, 56-91, 1955). Pour Y, hypersurface singulière quelconque, la condition a) est le théorème de comparaison de Grothendieck [4]. Dans [3], P. Deligne caractérise par la condition a) une connexion régulière ([3], proposition 6.8).

Si on remarque que l'on a un isomorphe canonique dans $D(C_X)$ ([19], proposition 6.1, chap.II),

$$\mathbb{R}\underline{hom}_{\mathcal{D}_X}(\mathbb{R}\underline{\Gamma}_{[Y]}(\mathcal{M}),\ \mathcal{O}_X) \xrightarrow{\sim} \mathbb{R}\underline{hom}_{\mathcal{D}_X}(\mathcal{M},\ \mathcal{O}_{\widehat{X|Y}})$$

où $\mathcal{O}_{\widehat{X|Y}} = \varprojlim_k \mathcal{O}_X/\mathcal{J}_Y^k$ est le complété formel de X le long de Y qui est un $\mathcal{D}_X$-module à gauche, on constate que la condition b) est due à P. Deligne (pour $\mathcal{M} = \mathcal{O}_X$) (1969, non publié) à Herrera-Libermann (X compact, Duality and the De Rham Cohomology of Infinitesimal Neighbourhoods. Int. Math. 13, 97-126, 1971). Une autre démonstration est due à Hartshorne (On the De Rham Cohomology of Algebraic Varieties. Publ. I.H.E.S. n°45, 6-99, 1975). Cependant, ces auteurs ne considéraient que la cohomologie de Betti des variétés algébriques et ne considéraient pas l'aspect systèmes différentiels. En dimension un, Malgrange caractérise une équation différentielle régulière par la condition b) : "toute solution formelle est convergente" ([12], théorème 1.4).

La condition c) est en rapport à la théorie des résidus et le complexe dualisant en géométrique analytique. Pour un diviseur à croisements normaux, elle a été établie par des calculs directs [14],[15]. Le cas général a demandé plusieurs années d'efforts. Dans [16],[17], on a utilisé les cinq ingrédients suivants.

1°) Le complexe $\mathbb{R}\underline{\Gamma}_{[Y]}(\mathcal{M})$ appartient à $D(\mathcal{D}_X)_h$ si $\mathcal{M}$ appartient à $D(\mathcal{D}_X)_h$. C'est là un des résultats les plus importants de la théorie du polynôme de Bernstein-Sato ([1],[2],[9]) sous la forme que lui a donnée Kashiwara [9] pour une hypersurface. Le cas de la codimension quelconque se ramène par une suite de Mayer-Vietoris algébrique sous la forme de [17].

2°) Le complexe $\mathbb{R}\underline{hom}_{\mathcal{D}}(\mathcal{M}, \mathcal{O}_X)$ est constructible si $\mathcal{M}$ appartient à $D(\mathcal{D}_X)_h$ (Kashiwara [8]).

3°) Le faisceau constant $\mathbb{C}_X$ est dualisant dans $D(\mathbb{C}_X)_c$ (Verdier [23]).

4°) Le théorème de dualité locale dans $D(\mathbb{C}_X)_c$ qui affirme que $\underline{DR}(\mathcal{M})$ et $\mathbb{R}\underline{hom}_{\mathcal{D}}(\mathcal{M}, \mathcal{O}_X)$ sont duals l'un de l'autre dans $D(\mathbb{C}_X)_c$ si $\mathcal{M}$ appartient à $D(\mathcal{D}_X)_h$ ([19], chap. III, théorème 1.1).

5°) Le théorème de bidualité locale dans $D(\mathcal{D}_X^\infty)_h$ qui affirme que "$\mathcal{O}_X$" est dualisant dans $D(\mathcal{D}_X^\infty)_h$ ([19], chap.III, théorème 2.1).

Les ingrédients 4°) et 5°) sont dûs à l'auteur ([19], théorèmes 1.1 et 2.1, chap.III). Ils sont à l'origine du théorème de dualité globale [18]. Nous les avons communiqués à Kashiwara en janvier 1978. Les cinq ingrédients ont un intérêt en eux-mêmes. Une fois qu'on les a tous, il est facile de voir que les conditions a), b) et c) sont équivalentes ([19], proposition 3.1, chap.III). Mais un problème fondamental est de décider qu'étant donné un système holonome $\mathcal{M}$, s'il satisfait à l'une des conditions a), b) et c) le long d'un sous-espace convenable Y. A cause des singularités de Y, c'est là un problème difficile. Pour l'illustrer, il suffit de voir les efforts considérables qu'il a fallu pour démontrer que le système de De Rham $\mathcal{O}_X$ est régulier le long de tout sous-espace analytique Y de X [17].

Ce résultat qui peut être regardé comme théorème de comparaison algébrique-analytique locale a de nombreuses conséquences [16]. Soit $\mathcal{M}$ un complexe de $D(\mathcal{D}_X)_h$. Supp($\mathcal{M}$) est un sous-ensemble analytique de X. Il existe un sous-espace analytique fermé Sing($\mathcal{M}$) de Supp($\mathcal{M}$) en dehors duquel la cohomologie de son complexe des solutions holomorphes est formée par des systèmes locaux. On dira qu'un complexe de $D(\mathcal{D}_X)_h$ est régulier si $\mathcal{M}$ est régulier le long de Sing($\mathcal{M}$). Les complexes réguliers forment une sous-catégorie pleine de $D(\mathcal{D}_X)_h$ que nous noterons $D(\mathcal{D}_X)_{hr}$.

§.6 - DEMONSTRATIONS (INDICATIONS)

Pour démontrer le théorème 4.1, on dévisse le complexe $\mathcal{F}$ pour construire localement un complexe $\mathcal{M}$ de $D(\mathcal{D}_X)_h$ qui satisfait à la condition a) et donc à la condition c) qui nous fournira le théorème. Pour cela, nous suivons la méthode de J.L. Verdier [23] qui démontre que les faisceaux $\mathcal{E}xt^i_{\mathbb{C}_X}(\mathcal{F}, G)$ sont constructibles si $\mathcal{F}$ et G sont constructibles sur un espace analytique. Le cas des schémas est traité par Grothendieck dans (S.G.A.5, exposé I). Les autres ingrédients sont la résolution des singularités, le théorème de comparaison de ([17], théorème 1.2), l'extension canonique de P. Deligne [3], la régularité des systèmes holonomes, le théorème d'images directes pour un morphisme projectif sous l'hypothèse d'existence de bonnes filtrations globales. On invoque enfin le théorème de Grauert-Remmert comme Grothendieck [4] pour conserver la régularité par un morphisme projectif. On se bornera ici à indiquer les trois étapes.

6.1 - 1ère étape. Réduction de $D(\mathcal{D}^\infty_X)_h$ à $D(\mathcal{D}_X)_{hr}$.

La catégorie $D(\mathbb{C}_X)_c$ est triangulée. On peut supposer que $\mathcal{F}$ est un faisceau constructible. Le problème est local sur X. On peut filtrer le faisceau $\mathcal{F}$ par des sous-faisceaux tels que les quotients successifs soient localement constants sur un ouvert lisse d'un fermé analytique de X, complémentaire d'un fermé analytique. On peut donc supposer alors que $\mathcal{F}$ est localement constant sur un ouvert W lisse d'un fermé analytique Y de X dont le complémentaire Z dans Y est un fermé analytique de X et nul en dehors. Notons $j : W \to Y$, $\alpha : Y \to X$, $\beta = \alpha \circ j$ $W \to X$ et $\gamma : V \to X$ les inclusions canoniques si V désigne le complémentaire de Z dans X. En utilisant le théorème de comparaison ([17], théorème 1.2), on se ramène par des formules de dualité à :

$$\mathbb{R}\underline{hom}_{\mathbb{C}_X}(\mathcal{F}, \mathcal{O}_X) = \mathbb{R}\gamma_*\gamma^{-1}(\mathcal{D}^\infty_X \underset{\mathcal{D}_X}{\otimes} \mathbb{R}\Gamma_{[Y]}(\mathcal{O}_X) \underset{\mathbb{C}}{\otimes} \beta ! L)$$

où L désigne le faisceau localement constant constructible, sur W, $\mathbb{R}\underline{hom}_{\mathbb{C}_W}(j^{-1}\alpha^{-1}\mathcal{F}, \mathbb{C}_W)$.

Le point clef est la remarque suivante : si le complexe $\gamma^{-1}(\mathbb{R}\Gamma_{[Y]}(\mathcal{O}_X) \underset{\mathbb{C}}{\otimes} \beta!L)$ de $D(\mathcal{D}_W)_h$ se prolonge en un complexe "$\mathcal{M}$" de $D(\mathcal{D}_X)_h$ régulier le long de Z, alors on obtient le théorème 4.1. En effet, la condition c) nous donne :

$$\mathbb{R}\underline{hom}_{\mathbb{C}_X}(\mathcal{F}, \mathcal{O}_X) \xrightarrow{\sim} \mathbb{R}\gamma_*\gamma^{-1}(\mathcal{D}_X^\infty \underset{\mathcal{D}_X}{\otimes} \mathcal{M}) \xrightarrow{\sim}$$

$$\mathcal{D}_X^\infty \underset{\mathcal{D}_X}{\otimes} \mathbb{R} \varinjlim_k hom_{\mathcal{O}_X}(\mathcal{J}_Z^k, \mathcal{M}) \quad .$$

Mais $\mathbb{R} \varinjlim_k \underline{hom}_{\mathcal{O}_X}(\mathcal{J}_Z^k, \mathcal{M})$ appartient à $D(\mathcal{D}_X)_h$ puisque $\mathcal{M}$ appartient à $D(\mathcal{D}_X)_h$ en vertu de l'ingrédient 1 du paragraphe 5. Par la même occasion, on aura répondu localement à la question $(HR)_2'$ puisque $\mathbb{R} \varinjlim_k \underline{hom}_{\mathcal{O}_X}(\mathcal{J}_Z^k, \mathcal{M})$ est régulier.

6.2 - Résolution des singularités.

On peut trouver [7] un morphisme π projectif de variétés analytiques complexes lisses :

$$\begin{array}{ccccc} \tilde{X} & \supset & \tilde{Y} & \supset & \tilde{Z} \\ \pi \downarrow & & \downarrow & & \downarrow \\ X & \supset & Y & \supset & Z \end{array}$$

tel que π soit un isomorphisme hors de Z et que $\tilde{Y} = \pi^{-1}(Y)$ est lisse, $\tilde{Z} = \pi^{-1}(Z)$ est un diviseur à croisements normaux dans $\tilde{Y}$. C'est de cette façon qu'on résout localement les singularités d'un sous-espace Y de X qui joue le rôle d'espace ambiant lisse. Notons : $\tilde{j} = \tilde{W} = \pi^{-1}(W) \to \tilde{Y}$, $\tilde{\alpha} : \tilde{Y} \to \tilde{X}$, $\tilde{\beta} = \tilde{Z} \to \tilde{X}$ et $\tilde{\gamma} : \tilde{V} \to \tilde{X}$ les inclusions canoniques. La restriction $\tilde{L}$ de $\pi^{-1}\ \beta!L$ à $\tilde{W} = \tilde{Y}\backslash\tilde{Z}$ est un système local. On a donc une variété lisse $\tilde{Y}$ et une connexion intégrable $\mathcal{O}_{\tilde{W}} \underset{\mathbb{C}}{\otimes} \tilde{L}$ sur $\tilde{W}$ complémentaire d'un diviseur $\tilde{Z}$ à croisements normaux. C'est la situation favorite de P. Deligne [3]. En vertu du théorème d'existence [3], cette connexion se prolonge en une connexion $\tilde{\mathcal{N}}$ régulière méromorphe sur $\tilde{Y}$. Nous avons rappeler dans ([19], chap.V) la construction de cette extension canonique en l'adaptant au langage des $\mathcal{D}_{\tilde{Y}}$-modules. En tant que $\mathcal{O}_{\tilde{Y}}$-module $\tilde{\mathcal{N}}$ est égale à

$$\varinjlim_{k} \underline{\mathrm{hom}}_{\mathcal{O}_{\tilde{Y}}}(\mathcal{J}^{k}_{\tilde{Z}}, \mathcal{O}_{\tilde{Y}}) \otimes_{\mathbb{C}} E$$

où E est la fibre de $\tilde{L}$ en un point de $\tilde{W}$ qu'on peut supposer connexe. La considération de la monodromie locale permet de munir $\tilde{\mathcal{N}}$ d'une structure de $\mathcal{D}_{\tilde{Y}}$-module qui prolonge celle de $\mathcal{O}_{\tilde{W}} \otimes_{\mathbb{C}} \tilde{L}$. On peut observer que le $\mathcal{D}_{\tilde{Y}}$-module $\tilde{\mathcal{N}}$ est globalement filtré. D'autre part, il est régulier en ce sens que l'on a l'isomorphisme

$$\underline{\mathrm{DR}}(\tilde{\mathcal{N}}) \xrightarrow{\sim} \underline{\mathrm{DR}}(\tilde{j}_* \tilde{j}^{-1} \tilde{\mathcal{N}})$$

qui correspond à la condition a) du paragraphe 5.

Nous cherchons un prolongement $\tilde{\mathcal{M}}$ en tant que $\mathcal{D}_{\tilde{X}}$-module pour éviter les singularités de Y. On pose

$$\tilde{\mathcal{M}} = \mathcal{D}_{\tilde{X} \leftarrow \tilde{Y}} \otimes_{\mathcal{D}_{\tilde{Y}}} \tilde{\mathcal{N}}[-\operatorname{codim} \tilde{Y}] \quad .$$

Pour la définition de $\mathcal{D}_{\tilde{X} \leftarrow \tilde{Y}}$, voir [9]. On démontre par un lemme facile [19] que $\tilde{\mathcal{M}}$ est un prolongement régulier de $\pi^{-1}\gamma^{-1}(\mathbb{R}\Gamma_{[Y]}(\mathcal{O}_X) \otimes_{\mathbb{C}} \beta ! L)$.

6.3 - Images directes.

Le prolongement cherché sera

$$\mathcal{M} = \mathbb{R}\pi_* \, \mathcal{D}_{X \leftarrow \tilde{X}} \overset{L}{\otimes}_{\mathcal{D}_{\tilde{X}}} \tilde{\mathcal{M}} = \int_\pi \tilde{\mathcal{M}}.$$

Il faut vérifier deux choses. D'une part, que $\mathcal{M}$ appartient à $D(\mathcal{D}_X)_h$. Mais $\tilde{\mathcal{M}}$ est globalement filtré et π est projectif. Ce sont les conditions suffisantes pour lui appliquer le théorème d'images directes de [9] et $\mathcal{M}$ appartient à $D(\mathcal{D}_X)_h$. D'autre part, que $\mathcal{M}$ est régulier le long de Z. Ceci résulte du théorème suivant qui a un intérêt en soi.

Théorème 6.3.1.- L'homomorphisme canonique de $D(C_X)$ est un isomorphisme :

$$\underline{\mathrm{DR}}(\mathbb{R}\Gamma_{[Z]}(\mathcal{M})) \to \underline{\mathrm{DR}}(\mathbb{R}\Gamma_Z(\mathcal{M})).$$

Le théorème 6.3.1 achève la construction de "l'extension $\mathcal{M}$" qui est régulière puisqu'elle satisfait à la condition a). Pour démontrer le théorème 6.3.1, on construit le diagramme

$$\begin{array}{ccc} R\pi_* \, \underline{DR}(\underline{R\Gamma}_{[\tilde{Z}]}(\tilde{\mathcal{M}})) & \longrightarrow & R\pi_* \, \underline{DR}(\underline{R\Gamma}_{\tilde{Z}}(\tilde{\mathcal{M}})) \\ \downarrow & & \downarrow \\ \underline{DR}(\underline{R\Gamma}_{[Z]}(\mathcal{M})) & \longrightarrow & \underline{DR}(\underline{R\Gamma}_{Z}(\mathcal{M})) \end{array}$$

La formule de <u>projection</u> appliquée à π nous permet de montrer que la flèche verticale de droite est un isomorphisme. C'est là un isomorphisme purement <u>topologique</u>. Maintenant, pour montrer que la flèche verticale de gauche est un isomorphisme, le théorème de <u>Grauert-Remmert</u> (trivialité cohomologique des morphismes projectifs) nous réduits <u>ipso-facto</u> à le vérifier fibre par fibre (voir <u>Grothendieck</u> [4], note n°7, page 99).

La fibre de π étant <u>algébrique</u>, le système $\tilde{\mathcal{M}}$ <u>régulier</u> est donc <u>algébrisable</u>. On est réduit comme précédemment à un raisonnement purement <u>topologique</u> pour la topologie de <u>Zariski</u> via la formule de projection. La flèche verticale de gauche est un isomorphisme. Comme la première flèche horizontale est un isomorphisme correspondant à la <u>régularité</u> de $\tilde{\mathcal{M}}$ le long de $\tilde{Z}$, on obtient que la deuxième flèche horizontale est un isomorphisme, d'où la régularité de $\mathcal{M}$ le long de Z et le théorème 6.3.1 donc le théorème 4.1 et en même temps une réponse locale du problème de <u>Hilbert-Riemann</u> pour les faisceaux constructibles. Pour tous les détails, nous renvoyons à ([19], chap.V).

<u>Remarque 6.3.2</u>.- On peut voir que l'extension régulière "$\mathcal{M}$" comme objet de $D(\mathcal{D}_X)_h$ ne dépend pas de la résolution π choisie à cause du fait qu'on peut coiffer deux résolutions par une troisième. Les opérations mises en jeu commutent.

<u>Remarque 6.3.3</u>.- L'extension locale "$\mathcal{M}$" peut être globale dans certains cas. Par

exemple, si X est algébrique et $\mathcal{F}$ est algébriquement constructible, alors il sera globalement solution de "$\mathcal{M}$" objet de $D(\mathcal{D}_X)_h$ régulier. Ceci résulte du fait que $\mathcal{F}$• peut se dévisser globalement et qu'on dispose pour les variétés algébriques de résolutions globales des singularités.

Remarque 6.2.4.- On espère, en utilisant la méthode de la descente cohomologique comme pour le théorème de dualité ([19], chap.IV), "globaliser" l'extension $\mathcal{M}$ dans $D(\mathcal{D}_X)_{hr}$.

Résumons nos résultats par le diagramme suivant :

$$\begin{array}{ccc} D(\mathcal{D}_X^\infty)_h & \underset{\mathbb{G}}{\overset{\mathbb{F}}{\rightleftarrows}} & D(\mathbb{C}_X)_c \\ {\scriptstyle T}\nwarrow & & \nearrow{\scriptstyle S} \\ & D(\mathcal{D}_X)_{hr} & \end{array}$$

$$\mathbb{F}(\mathcal{M}^\infty) = \mathbb{R}\underline{hom}_{\mathcal{D}_X^\infty}(\mathcal{M}^\infty, \mathcal{O}_X)$$

$$\mathbb{G}(\mathcal{F}) = \mathbb{R}\underline{hom}_{\mathbb{C}_X}(\mathcal{F}, \mathcal{O}_X)$$

$$S(\mathcal{M}) = \mathbb{R}\underline{hom}_{\mathcal{D}_X}(\mathcal{M}, \mathcal{O}_X)$$

$$T(\mathcal{M}) = \mathcal{D}_X^\infty \underset{\mathcal{D}_X}{\otimes} \mathcal{M}$$

On a obtenu que $\mathbb{F}$ et $\mathbb{G}$ sont inverses l'un de l'autre (= $(HR)_1'$), que S, donc T, sont localement sur X surjectifs(= $(HR)_2'$ local).

§.7 - APPLICATIONS

Nous allons voir quelques exemples du dictionnaire géométrique-analytique ainsi obtenu.

7.1 - Dualité de Poincaré-Verdier.

Soit $\mathcal{F}$ un complexe de $D(\mathbb{C}_X)_c$. Il est donc solution d'un complexe de $D(\mathcal{D}_X^\infty)_h$, $\mathcal{M}^\infty$,

$$\mathcal{F} = \mathbb{R}\underline{hom}_{\mathcal{D}_X^\infty}(\mathcal{M}^\infty, \mathcal{O}_X) \quad .$$

En prenant les solutions globales, on trouve

$$\mathbb{H}^i(X ; \mathcal{F}) = \mathbb{E}xt^i_{\mathcal{D}^\infty_X}(X ; \mathcal{M}^\infty, \mathcal{O}_X) \quad .$$

De même, en prenant les De Rham globaux, on trouve ([19], chap.IV)

$$\mathbb{E}xt^{2n-i}_{\mathbb{C}_X, c}(X ; \mathcal{F}, \mathbb{C}_X) = \mathbb{E}xt^{2n-i}_{\mathcal{D}^\infty_X, c}(X ; \mathcal{O}_X, \mathcal{M}^\infty) \quad .$$

Le premier espace de gauche est le dual algébrique du deuxième espace de gauche. C'est la dualité de Poincaré si $\mathcal{F}$ est constant et de Verdier si $\mathcal{F}$ est quelconque. Le premier espace de droite est le dual algébrique du deuxième espace de droite. C'est la dualité pour les $\mathcal{D}^\infty_X$-modules [18]. Ainsi on obtient une description analytique de la dualité de Poincaré-Verdier qui a été à l'origine du problème de Hilbert-Riemann pour les faisceaux constructibles ([18], remarque 5.1). Comme on le voit, il a fallu résoudre le problème de Hilbert-Riemann pour traiter le cas d'un faisceau constructible ([19], chap.IV).

7.2 - Bidualité locale.

On sait que le faisceau "$\mathbb{C}_X$" est dualisant dans $D(\mathbb{C}_X)_c$ [23] ; c'est-à-dire que l'homomorphisme de $D(\mathbb{C}_X)$ canonique est un isomorphisme :

$$\mathcal{F} \longrightarrow \mathbb{R}\,\underline{hom}_{\mathbb{C}_X}(\mathbb{R}\,\underline{hom}_{\mathbb{C}_X}(\mathcal{F}, \mathbb{C}_X), \mathbb{C}_X) \quad .$$

C'est là un résultat géométrique. Son analogue analytique affirme que le système "$\mathcal{O}_X$" est dualisant dans $D(\mathcal{D}^\infty_X)_h$; c'est-à-dire que l'homomorphisme canonique de $D(\mathcal{D}^\infty_X)$ est un isomorphisme :

$$\mathcal{M}^\infty \longrightarrow \mathbb{R}\,\underline{hom}_{\mathbb{C}_X}(\mathbb{R}\,\underline{hom}_{\mathcal{D}^\infty_X}(\mathcal{M}^\infty, \mathcal{O}_X), \mathcal{O}_X) \; .$$

Sa démonstration utilise en plus de l'analyse : Dualité de Serre, espaces vectoriels topologiques nucléaires et définition cohomologique de $\mathcal{D}^\infty_X$ de M. Sato ([19], chap.III).

7.3 - Régularité et irrégularité.

Soit $\mathcal{M}$ un complexe de $D(\mathcal{D}_X)_h$ non nécessairement régulier et $\mathcal{F} = \mathbb{R}\,\underline{hom}_{\mathcal{D}_X}(\mathcal{M}, \mathcal{O}_X)$

son complexe des solutions holomorphes. Par la réponse <u>locale</u> de $(HR)_2'$, $\mathcal{F}$ est aussi solution d'un complexe de $D(\mathcal{D}_X)_{hr}$ $\mathcal{M}_R$ <u>régulier</u> :

$$\mathcal{F} \xrightarrow{\sim} \mathbb{R}\underline{hom}_{\mathcal{D}_X}(\mathcal{M}_R, \mathcal{O}_X) .$$

Mais, en vertu du théorème de bidualité précédent ("$\mathcal{O}_X$" est dualisant dans $D(\mathcal{D}_X^\infty)_h$) , on a :

$$\mathcal{M}^\infty = \mathcal{D}_X^\infty \underset{\mathcal{D}_X}{\otimes} \mathcal{M} \xrightarrow{\sim} \mathbb{R}\underline{hom}_{\mathbb{C}_X}(\mathcal{F}, \mathcal{O}_X) \xrightarrow{\sim} \mathbb{R}\underline{hom}_{\mathbb{C}_X}(\mathbb{R}\underline{hom}_{\mathcal{D}_X}(\mathcal{M}_R, \mathcal{O}_X), \mathcal{O}_X)$$
$$\xrightarrow{\sim} \mathcal{D}_X^\infty \underset{\mathcal{D}_X}{\otimes} \mathcal{M}_R .$$

On obtient <u>localement</u>

$$\mathcal{D}_X^\infty \underset{\mathcal{D}_X}{\otimes} \mathcal{M} \xrightarrow{\sim} \mathcal{D}_X^\infty \underset{\mathcal{D}_X}{\otimes} \mathcal{M}_R .$$

C'est-à-dire que, par une transformation d'ordre <u>infini</u>, on peut ramener un système <u>irrégulier</u> à un système <u>régulier</u>. C'est, semble-t-il, un résultat <u>fondamental</u> pour <u>Kawai-Kashiwara</u> qui ont annoncé le cas micro-local dans ce même colloque. C'est une simple conséquence de $(HR)_2'$ et du théorème de bidualité locale. De même, une fois qu'on aura une réponse globale à $(HR)_2'$, on déduira l'analogue global. De toute façon, les résultats de ces auteurs n'englobent pas le théorème d'existence ni l'équivalence de catégories entre $D(\mathbb{C}_X)_c$ et $D(\mathcal{D}_X^\infty)_h$.

7.4 - <u>Système de Gauss-Manin</u>.

Soit $\pi : \tilde{X} \to X$ un morphisme d'espace analytique où X est lisse mais où on ne fait pas d'hypothèse sur le couple $(\tilde{X}, \pi)$. Soit $\tilde{\mathcal{F}}$ un complexe de $D(\mathbb{C}_{\tilde{X}})_c$ et faisant l'hypothèse que $R\pi_*\tilde{\mathcal{F}}$ appartient à $D(\mathbb{C}_X)_c$. C'est là une hypothèse pas trop <u>restrictive</u> et qui englobe des situations <u>non</u> propres.

Par la réponse à $(HR)_2'$, il existe <u>localement</u> sur X un complexe de $D(\mathcal{D}_X^\infty)_{hr}$, soit $\mathcal{M}$, dont le complexe solution est $\mathbb{R}\pi_*\tilde{\mathcal{F}}$:

$$\mathbb{R}\pi_*\tilde{\mathcal{F}} \xrightarrow{\sim} \mathbb{R}\underline{hom}_{\mathcal{D}_X}(\mathcal{M}, \mathcal{O}_X) .$$

Nous appellerons le complexe $\mathcal{M}$, système de <u>Gauss-Manin</u> associé au couple $(\tilde{X}, \pi)$.

Sans les hypothèses de non singularité de $\tilde{X}$ et pour $\tilde{\mathcal{F}} = \mathbb{C}_{\tilde{X}}$, c'est le système de <u>Gauss-Manin</u> de la singularité π (dim $X = 1$). Pour le voir, on utilise le théorème de <u>dualité relative</u> pour les $\mathcal{D}_X$-modules holonomes. Soit $\tilde{\mathcal{M}}$ un complexe de $D(\mathcal{D}_{\tilde{X}})_h$, où on suppose que $\tilde{X}$ est lisse, tel que son intégrale le long des fibres $\int_\pi \tilde{\mathcal{M}} = \mathcal{M}$ appartient à $D(\mathcal{D}_X)_h$. C'est le cas si π est projectif et $\tilde{\mathcal{M}}$ admettant des filtrations globales [9], mais pas seulement.

<u>Théorème 7.4.1.</u>- <u>On a un isomorphisme canonique dans $D(\mathbb{C}_X)_c$:</u>

$$\mathbb{R}\pi_* \mathbb{R}\,\underline{hom}_{\mathcal{D}_{\tilde{X}}}(\tilde{\mathcal{M}}, \mathcal{O}_{\tilde{X}})[\dim \tilde{X}] \xrightarrow{\sim} \mathbb{R}\,\underline{hom}_{\mathcal{D}_X}(\mathcal{M}, \mathcal{O}_X)[\dim X].$$

Si $\tilde{\mathcal{M}} = \mathcal{O}_{\tilde{X}}$, alors $\mathcal{M} = \int_\pi \mathcal{O}_{\tilde{X}}$ est le système de Gauss-Manin dont le complexe solution est $\mathbb{R}\pi_*\mathbb{C}_{\tilde{X}}$ par le lemme de Poincaré. Le théorème 7.4.1 peut se démontrer à partir de sa version absolue [18].

Donnons un exemple de système de <u>Gauss-Manin</u> où $\tilde{X}$ n'est pas lisse. Soit $(\tilde{X},x)$ un germe d'espace analytique réduit. Soit $\pi : (\tilde{X},x) \to (\mathbb{C},0)$ un germe de fonction analytique. On suppose que $(\tilde{X},x)$ est plongée dans $(\mathbb{C}^N,0)$ et π est la restriction de $\tilde{\pi} : (\mathbb{C}^N,0) \to (\mathbb{C},0)$ à $(X,0)$.

Fixons ε, $0 < \varepsilon \leqslant \varepsilon_0$ et $0 < \eta \leqslant \varepsilon_0$. Posons $\tilde{X} = \overset{0}{B}_\varepsilon \cap \pi^{-1}(D_\eta)$ où $D = D_\eta$ est le disque de rayon η centré à l'origine dans $\mathbb{C}$, et $\overset{0}{B}_\varepsilon$ est la boule ouverte de rayon ε centrée à l'origine dans $\mathbb{C}^N$. Notons encore π le morphisme induit $\tilde{X} \to D$. <u>D.T. LÊ</u> nous a communiqué le résultat suivant [11]:

<u>Théorème 7.4.2.</u>- <u>Le complexe $\mathbb{R}\pi_*\mathbb{C}_{\tilde{X}}$ appartient à $D(\mathbb{C}_D)_c$.</u>

L'intérêt de ce résultat est qu'on ne fait aucune hypothèse de <u>singularité</u> sur $\tilde{X}$ ni de <u>propriété</u> sur π. Alors, par $(HR)'_2$, il lui correspond un complexe de $D(\mathcal{D}_D)_{hr}$, soit $\mathcal{M}$.

$$\mathbb{R}\pi_* \mathbb{C}_{\tilde{X}} = \mathbb{R}\,\underline{\mathrm{hom}}_{\mathcal{D}_D}(\mathcal{M}, \mathcal{O}_D) \quad .$$

Le complexe $\mathcal{M}$ doit rendre des services pour l'étude de la singularité $(\tilde{X},x) \to (\mathbb{C},0)$

tout comme la connexion de Gauss-Manin usuelle.

BIBLIOGRAPHIE

[1] BERNSTEIN I.N. - The Analytic Continuations of Generalized Functions with Respect to a Parameter. *Functional Anal. Appl.* 6, 26-40 (1972).

[2] BJÖRK J.E. - Rings of Differential Operators. North-Holland, Amsterdam (1979).

[3] DELIGNE P. - Equations Différentielles à Points Singuliers Réguliers. Lecture Notes in Math. n°163, Berlin, Heidelberg, New York (1970).

[4] GROTHENDIECK A. - On the De Rham Cohomology of Algebraic Varieties. *Publ. Math I.H.E.S.* n°29, 93-103 (1966).

[5] KATZ N. - The Regularity Theorem in Algebraic Geometry. Actes Congrès International des Mathématiciens de Nice, 1970, vol.I, 436-449. Gauthier-Villards, Paris

[6] KATZ N. - An Overview of Deligne's Work in Hilbert's Twenty-first Problem. *Proceedings of Symposia in Pure Mathematics A.M.S.*, vol.28, 537-557 (1976).

[7] HIRONAKA I. - Resolution of Singularities on an Algebraic Variety over a Field of Characteristic Zero. I-II *Ann. of Math* n°1 and n°2 (1964).

[8] KASHIWARA M. - On the Maximally Overdetermined Systems of Linear Differential Equations I. *Publ. R.I.M.S. Kyoto University* 10, 563-579 (1975).

[9] KASHIWARA M. - B-Functions and Holonomic Systems. *Invent. Math.* 38, 33-53 (1976).

[10] KASHIWARA M., OSHIWA I. - System of Differential Equations with Regular Singularities and their Boundary Value Problem. *Ann. of Math.* 106, 145-200 (1977).

[11] LÊ D.T. - Some Remarks on Relative Monodromy. Nordic Summer School I.N.A.V.F. Symposium in Mathematics. Oslo, August 5-25, (1976). Sijthoff Noordhoof (1977).

[12] MAKGRANGE B. - Sur les Points Singuliers Réguliers des Equations Différentielles. *Enseignement Math.* 20, 147-176 (1974).

[13] MALGRANGE B. - Le Polynôme de I.N. Bernstein d'une Singularité Isolée. Lecture Notes in Math. n°459, 99-119. Berlin, Heidelberg, New York, Springer (1976).

[14] MEBKHOUT Z. - La Valeur Principale des Fonctions à Singularités Essentielles. *C.R. Acad. Sc. Paris* 280, 205-207 (1975).

[15] MEBKHOUT Z. - La valeur Principale et le Résidu Simple des Formes à Singularités Essentielles. In Fonctions de Plusieurs Variables Complexes II. Lecture Notes in Math. n°482, 190-215, Berlin, Heidelberg, New York, Springer (1975).

[16] MEBKHOUT Z. - La Cohomologie Locale d'une Hypersurface. In Fonctions de Plusieurs Variables Complexes III. Lecture Notes in Math. n°670, 89-119, Berlin, Heidelberg, New York, Springer (1977).

[17] MEBKHOUT Z. - Local Cohomology of Analytic Spaces. *Publ. R.I.M.S. University* 12, 247-256, suppl. (1977).

[18] MEBKHOUT Z. - Théorèmes de Dualité pour les $\mathcal{D}_X$-Modules Cohérents. *C.R. Acad. Sc. Paris* 285, 785-787 (1977).

[19] MEBKHOUT Z. - Cohomologie Locale des Espaces Analytiques Complexes. Thèse de Doctorat d'Etat, University de Paris VII (1978), 126 pages. Soutenue le 15.2.1979.

[20] MEBKHOUT Z. - Dualité de Poincaré. Séminaire sur "Singularités" de Paris VII. 1977-1979. Publ. de l'Université de Paris VII (à paraître).

[21] MEBKHOUT Z. - The Poincaré-Serre-Verdier Duality. Proceeding of the Conf. of Algebraic Geometry, Copenhague (1978). Lecture Notes in Math. n°732, Berlin, Heidelberg, New York, Springer (1979).

[22] SATO M., KAWAI T., KASHIWARA M. - Micro-Functions and Pseudo-Differential Equations. Lecture Notes in Math. n°287, 265-529. Berlin, Heidelberg, New York, Springer (1973).

[23] VERDIER J.L. - Classe d'Homologie Associée à un cycle. Séminaire Douady-Verdier Astérisque 36-37, 101,151 (1976).

Conditions de positivité dans une variété symplectique complexe . Applications à l'étude des microfonctions (*)

Pierre SCHAPIRA

Département de Mathématiques
C.S.P. Université Paris-Nord
Avenue J.B. Clément,
93430 VILLETANEUSE

Let Ω be a strictly pseudo-convex (s.p.c.) open set with real analytic boundary $\partial\Omega$ in a complex analytic manifold X , $(T^*_{\partial\Omega}X)^+$ the exterior conormal bundle of $\partial\Omega$ in T^*X . We prove that if a conic set Λ of T^*X is positive with respect to $(T^*_{\partial\Omega}X)^+$ in the sense of Melin and Sjöstrand, Λ projects on X outside of Ω , and we prove the converse when Λ is the conormal bundle of a complex hypersurface. We also give a condition for two manifolds Λ_0 and Λ_1 to be locally of the type $(T^*_{\partial\Omega_0}X)^+$ and $(T^*_{\partial\Omega_1}X)^+$ for two s.p.c. open sets, in a common choice of symplectic coordonates. That allows us to construct injective morphisms of sheaves of microfunctions associated to Λ_0 or Λ_1.

(*) Ce texte est le même qu'une Note aux Comptes Rendus de l'Académie des Sciences, à paraître en 1980.

§1. Soit X une variété analytique complexe, $\mathcal{O}_X$ le faisceau des fonctions holomorphes sur X. Soit T^*X le fibré cotangent à X, ω la 1-forme canonique de T^*X, $\mathcal{E}_X$ le faisceau sur T^*X des opérateurs micro-différentiels d'ordre fini [6]. Nous notons π la projection de T^*X sur X et nous désignons par $\dot{T}^*X$ l'espace T^*X privé de la section nulle. Nous dirons qu'une partie Λ de $\dot{T}^*X$ est $\mathbb{R}^+$-conique (resp. $\mathbb{C}^\times$-conique) si Λ est invariante par l'action du groupe $\mathbb{R}^+$ (resp. $\mathbb{C}^*$). On note $\mathbb{C}^\times\Lambda$ l'enveloppe $\mathbb{C}^\times$-conique d'une partie Λ.

<u>Définition 1</u> : Nous appellerons variété $\mathbb{C}$-Lagrangienne une variété analytique complexe, $\mathbb{C}^\times$-conique, lagrangienne dans T^*X.
Nous appellerons variété $\mathbb{R}$-lagrangienne une variété analytique réelle, $\mathbb{R}^+$-conique, lagrangienne dans la variété symplectique réelle $((T^*X)^{\mathbb{R}}, \mathrm{Re}\, d\omega)$.
Nous dirons d'une variété $\mathbb{R}$-lagrangienne Λ qu'elle est I-symplectique si $\mathrm{Im}\, d\omega$ est symplectique sur Λ, ou ce qui est équivalent, si T^*X définit un complexifié de Λ.

<u>Exemple 1</u> : Soit Y une sous-variété analytique complexe de X. La variété $\mathbb{C}$-lagrangienne T^*_YX (le fibré conormal à Y dans X) est muni du faisceau de microfonctions $\mathcal{C}^{\mathbb{R}}_{Y|X}$ défini dans [6]. C'est un faisceau de $\mathcal{E}_X$-modules.

<u>Exemple 2</u> : Soit M une variété analytique réelle de complexifié X. La variété $\mathbb{R}$-lagrangienne et I-symplectique T^*_MX est muni du faisceau $\mathcal{C}_M$ des microfonction de M. Sato [6]. C'est un faisceau de $\mathcal{E}_X$-modules.

<u>Exemple 3</u> : Soit Ω un ouvert strictement pseudo-convexe (s.p.c.) de X défini par l'équation $\rho < 0$, où ρ est une fonction analytique réelle, avec $d\rho \neq 0$ pour $\rho = 0$. Soit $(T^*_{\partial\Omega}X)^+$ le fibré conormal extérieur à $\partial\,\Omega$:

$$(T^*_{\partial\Omega}X)^+ = \{(z,\zeta) \in T^*X \; ; \; f(z) = 0, \; \zeta = k\, d'f(z), \; k > 0\}$$

C'est une variété $\mathbb{R}$-lagrangienne et I-symplectique. On munit $(T^*_{\partial\Omega}X)^+$ du faisceau $\mathcal{C}^+_{\partial\Omega}$, image réciproque par π du faisceau

$$\mathcal{O}^+_{\partial\Omega} = \mathcal{H}^1_{X-\Omega}(\mathcal{O}_X)|\partial\Omega$$

des valeurs au bord sur $\partial\Omega$ des fonctions holomorphes. C'est un faisceau de $\mathcal{E}_X$-modules [4].

Soit ϕ une transformation canonique complexe homogène qui échange localement T^*_MX et $(T^*_{\partial\Omega}X)^+$. On peut alors, localement, "quantifier" ϕ en un isomorphisme $\hat{\phi}$ de $\mathcal{E}_X$-modules de $\mathcal{C}_M$ sur $\mathcal{C}^+_{\partial\Omega}$ (cf. [4], [5], ou [1] dans le cadre des distributions).

§2. Soient Λ_o et Λ_1 deux variétés $\mathbb{R}$-lagrangiennes et I-symplectiques de T^*X, et Λ une variété $\mathbb{C}$-lagrangienne, ces variétés étant définies au voisinage d'un point $p \in \dot{T}^*X$.

<u>Théorème 1</u> : a) On peut trouver une transformation canonique complexe homogène qui échange Λ_o en $(T^*_{\partial\Omega_o}X)^+$ et Λ en T^*_YX, où Ω_o est un ouvert s.p.c. de X et Y une hypersurface analytique complexe de X.

b) On suppose la codimension de $T_p\Lambda_o \cap T_p\Lambda_1$ dans $T_p\Lambda_o$ inférieure ou égale à un. On peut alors trouver une transformation canonique complexe homogène qui échange Λ_o en $(T^*_{\partial\Omega_o}X)^+$ et Λ_1 en $(T^*_{\partial\Omega_1}X)^+$, où Ω_o et Ω_1 sont deux ouverts s.p.c. de X.

§3. Rappelons la définition de A. Melin et J. Sjöstrand de la positivité [2], [3]. Soit d'abord V une variété analytique réelle, W un complexifié de V. On note ν l'"application" de W dans TV qui dans une carte locale, réelle sur V, à $z \in W$ associe $(\mathrm{Re}z, \mathrm{Im}z) \in TV$. Si α est une 1-forme sur V, la fonction $z \to <\alpha,\nu(z)>$ est bien définie modulo $I_3(V)$, l'espace des fonctions nulles à l'ordre 3 sur V.

Définition 2 [3] : Soit Λ_o une variété $\mathbb{R}$-lagrangienne et I-symplectique de T^*X , Λ une partie $\mathbb{R}^+$-conique. On dit que (Λ_o,Λ) est positif en $p \in \Lambda_o$ si l'on a sur Λ , au voisinage de p :

$$-\frac{1}{i} < \omega , \nu > \geqslant 0 \mod I_3(\Lambda_o)$$

Si $\Lambda_o = T^*_{\mathbb{R}^n}\mathbb{C}^n$, $T^*\mathbb{C}^n$ étant muni des coordonnées (z,ζ), cette condition s'exprime par :

$$- <\mathrm{Im}\ z, \mathrm{Im}\ \zeta> \geqslant - c(|\mathrm{Im}\ z| + |\mathrm{Re}\ \zeta|)^3$$

pour $(z,\zeta) \in \Lambda$ dans un voisinage de p , et pour une constante $c > 0$.

Théorème 2 : Soit Ω un ouvert s.p.c. de bord analytique réel et soit Λ une partie $\mathbb{C}^\times$-conique de T^*X . On suppose $((T^*_{\partial\Omega}X)^+,\Lambda)$ positif en un point p. Il existe alors un voisinage $\mathcal{U}$ de p tel que $\pi(\mathcal{U} \cap \Lambda) \cap \Omega = \emptyset$.

§4. On peut construire des variétés lagrangiennes positives de différentes manières.

Théorème 3 : Soit Ω un ouvert s.p.c. de bord analytique réel, Y une hypersurface analytique complexe ne rencontrant pas Ω . Alors $((T^*_{\partial\Omega}X)^+,T^*_YX)$ est positif en tous points de $(T^*_{\partial\Omega}X)^+$.

Soit M une variété analytique réelle de complexifié X , f une fonction holomorphe au voisinage de $x^\circ \in M$, dont la restriction à M est une fonction de type positif [6, ch. 1, déf. 3.1.4].

Soit Y l'hypersurface complexe des zéros de f. Alors $(T^*_M X, T^*_Y X)$ est positif en $(x°, idf(x°))$.

Soit $\phi(z,\theta)$ une fonction holomorphe au voisinage du point $p \in \mathbb{R}^n \times (i\mathbb{R}^n)$ dans $\mathbb{C}^n \times \mathbb{C}^n$, homogène de degré 1 en θ , et tangente à l'application $<z,\theta>$ en p . Soit T_ϕ la transformation canonique qui a $(z, \zeta) \in T^*\mathbb{C}^n$ associe (w,θ) par la relation

$$w = \phi'_\theta(z,\theta) \quad , \quad \zeta = \phi'_z(z,\theta) .$$

On déduit immédiatement de [3] le résultat suivant, utile pour les applications :

<u>Théorème 4</u> : On suppose $\mathrm{Re}\, \phi \geqslant 0$ pour $(z,\theta) \in \mathbb{R}^n \times (i\, \mathbb{R}^n)$. Alors $(T^*_{\mathbb{R}^n} \mathbb{C}^n , \mathbb{C}^\times(T_\phi(T^*_{\mathbb{R}^n} \mathbb{C}^n)))$ est positif en p .

§5. Soit Λ une variété $\mathbb{R}$-lagrangienne et I-symplectique (resp. $\mathbb{C}$-lagrangienne). Nous appellerons faisceau de microfonctions sur Λ un faisceau de $\mathcal{E}_X$-modules localement isomorphe, par transformation canonique qunatifiée au faisceau $\mathcal{C}_M$ de l'exemple 2 (resp. $\mathcal{C}^{\mathbb{R}}_{Y|X}$ de l'exemple 1). On peut alors énoncer :

<u>Théorème 5</u> : Soit Λ_o une variété $\mathbb{R}$-lagrangienne et I-symplectique, Λ une variété a) : $\mathbb{R}$-lagrangienne et I-symplectique, ou b) : $\mathbb{C}$-lagrangienne. On suppose $(\Lambda_o, \dot{\mathbb{C}}^\times\Lambda)$ positif en un point $p \in \dot{T}^*X$, et on suppose aussi, dans le cas de l'hypothèse a) , $T_p\Lambda_o \cap T_p\Lambda_1$ de codimension $\leqslant 1$ dans $T_p\Lambda_o$. Soit $\mathcal{C}_{\Lambda_o}$ et $\mathcal{C}_\Lambda$ des faisceaux de microfonctions sur Λ_o et sur Λ . Alors on peut trouver dans un voisinage de p , un morphisme injectif de $\mathcal{E}_X$-modules du faisceau $\mathcal{C}_\Lambda|\Lambda_o \cap \Lambda_1$ dans le faisceau $\Gamma_{\Lambda_o \cap \Lambda_1}(\mathcal{C}_{\Lambda_o})$.

Ce théorème permet de déduire des résultats d'hypo-ellipticité analytique sur Λ pour des systèmes micro-différentiels de résultats de propagation sur Λ_o [7].

Nous sommes heureux de remercier J. Sjöstrand avec qui nous avons eu plusieurs discussions fructueuses.

BIBLIOGRAPHIE

[1] L. BOUTET de MONVEL : Convergence dans le domaine complexe des séries de fonctions propres. C.R. Acad. Sc. Paris, t. 287 (1978) 855-856.

[2] A. MELIN, J. SJOSTRAND : Fourier integral operators with complex valued phase functions. Lecture Notes in Math. 459 Springer (1975) 120-223.

[3] A. MELIN, J. SJOSTRAND : Fourier integral operator with complex phase functions and parametrix for an interior boundary value problem. Comm. in Partial Diff. Eq. 1 (1976), 313-400.

[4] M. KASHIWARA : Systèmes micro-différentiels. Publ. Université PARIS-NORD (1976-77) et séminaires U.P.N. (76-77) non publiés.

[5] M. KASHIWARA, T. KAWAI : Some applications of boundary value problems for elliptic systems of linear differential equations. Ann. of Math. Studies, n° 93, Princeton. Princeton Univ. Press. A paraître.

[6] M. SATO, T. KAWAI, M. KASHIWARA : Hyperfunctions and pseudodifferential equations. Lecture Notes in Mat. 287, Springer (1973), 265-529.

[7] P. SCHAPIRA : Conditions de positivité dans une variété symplectique complexe. Applications à l'étude de l'hypo-ellipticité analytique. (Résumé). Rencontre sur les E.D.P. linéaires Saint-Cast, Mai 1979 et article à paraître.

FUCHSIAN SYSTEMS OF LINEAR DIFFERENTIAL EQUATIONS ASSOCIATED TO NILSSON CLASS FUNCTIONS AND AN APPLICATION TO FEYNMAN INTEGRALS

by Arno van den Essen(*)

§0. INTRODUCTION

In 1963 Regge proposed to construct partial differential equations for Feynman integrals, that generalize the standard hypergeometric equation. Ten years later, in 1973, V.A. Golubeva remarked that most probably a Feynman integral will appear as solution of a system of linear partial differential equations, which forms a generalization to several variables of the theory of ordinary linear differential equations of Fuchsian type (see [4]) . In 1975 Malgrange pointed out the great interest of having a theory of partial linear differential equations having regular singularities in the framework of $\mathcal{D}$-modules; the so-called Fuchsian modules (see [7]).

In the meantime several authors have been working on this subject: Kashiwara, Kawai, Oshima, Mebkhout, Ramis.

In my thesis, [2], I have introduced in a pure algebraic way a very general framework in which Fuchsian $\mathcal{D}$-modules are defined. For example, we study simultaneously modules over the rings of differential operators over the convergent and formal power series, the Weylalgebra $A_n(k)$ etc. etc.

In this note I will try to give an idea of some of the results obtained in my thesis and indicate that Nilsson class functions (and a fortiori Feynman integrals) are solutions of the Fuchsian $\mathcal{D}$-modules introduced there. At the end of this note I will discuss the connection between the theory of regular singularities introduced by the authors mentioned before and my theory of Fuchsian $\mathcal{D}$-modules.

We now pick up again the questions of Regge and Golubeva on the construction of differential equations for Feynman integrals.

Let F be a Feynman integral, or more general, a Nilsson class function (i.e. F is a multi-valued analytic function on $\mathbb{C}^n \setminus V(q)$ of so-called finite dimensional determination and satisfying a temperated growth condition; here q denotes a non-zero polynomial with complex coefficients, i.e. $q \in \mathbb{C}[z_1,\ldots,z_n]$, such that q has no multiple factor. (See [1] and [8] for more details.)

To such a Nilsson class function F we can associate the system of all linear differential equations (with polynomial coefficients) which F satisfies i.e.

(*) Supported in part by the Netherlands Organisation for the Advancement of Pure Research (Z.W.O.).

$$L = \{Q \in A_n(\mathbb{C}) \mid QF = o\}$$

$(A_n(\mathbb{C}) = \mathbb{C}[z_1,\ldots,z_n][\partial_1,\ldots,\partial_n]$; $\partial_i = \frac{\partial}{\partial z_i}$, see [1], Chap. I.)

So the A_n-module A_nF is isomorphic to A_n/L .

Our main result is:

THEOREM: A_nF is a Fuchsian A_n-module, Fuchsian along the principal ideal $(q) = \mathbb{C}[z_1,\ldots,z_n]q$ i.e. for every $\tau \in \mathrm{Der}_{\mathbb{C}}\mathbb{C}[z_1,\ldots,z_n]$ such that $\tau q \in (q)$ there exist $m \in \mathbb{N}$ and $a_o,\ldots,a_{m-1} \in \mathbb{C}[z_1,\ldots,z_n]$ such that

$$\tau^m F + a_{m-1}\tau^{m-1}F + \ldots + a_o F = o .$$

In particular, there exist $p \in \mathbb{N}$ and $a_{ij} \in \mathbb{C}[z_1,\ldots,z_n]$ such that

$$\begin{aligned}(q\partial_1)^p F + a_{p-1,1}(q\partial_1)^{p-1}F + \ldots + a_{o1}F &= o\\ &\vdots\\ (q\partial_n)^p F + a_{p-1,n}(q\partial_n)^{p-1}F + \ldots + a_{on}F &= o\end{aligned}$$

To prove this result we also introduce on $\mathbb{C}^n$ a sheaf $\underline{L}$ of left ideals in the sheaf $\underline{\mathcal{D}}$ of differential operators:

$$L_z = \{Q \in \mathcal{D}_z \mid QF = o\}$$

A non-trivial result, proved by Kashiwara and Kawai, asserts that $\underline{\mathcal{D}}/\underline{L}$ is a coherent sheaf of left $\underline{\mathcal{D}}$-modules and we prove:

Proposition: $\underline{\mathcal{D}}/\underline{L}$ is a Fuchsian $\underline{\mathcal{D}}$-module, i.e. each stalk $\mathcal{D}_z/L_z$ is a Fuchsian $\mathcal{D}_z$-module.

Our results on Fuchsian modules in [2] imply that these systems are holonomic.

§1. THE ALGEBRAIC THEORY OF FUCHSIAN MODULES

We now start introducing Fuchsian modules in a pure algebraic way.
Let A denote either $\mathbb{C}[z_1,\ldots,z_n]$ or $\mathcal{O}_n$, the ring of convergent power series in $z_1,\ldots,z_n$ over $\mathbb{C}$, or $\hat{\mathcal{O}}_n$, the ring of formal power series in $z_1,\ldots,z_n$ over $\mathbb{C}$. By D we denote the ring of $\mathbb{C}$-linear differential operators over A i.e. $D = A[\partial_1,\ldots,\partial_n]$. If $A = \mathbb{C}[z_1,\ldots,z_n]$ or $\mathcal{O}_n$ or $\hat{\mathcal{O}}_n$, then D is usually denoted by A_n, $\mathcal{D}_n$ and $\hat{\mathcal{D}}_n$ respectively. Finally, if R is a ring, then $\underline{\underline{M}}(R)$ denotes the set of all left R-modules which are finitely generated over R .

To avoid technical complications we shall restrict our attention to the case of D-modules without A-torsion.

1.1 D-modules and the ideal of singularities

Let M be a left D-module without A-torsion. Then we define its ideal of singularities, denoted $S(M)$, as follows

$1^{\underline{o}}$ If $M \in \underline{\underline{M}}(D)$, then

$$S(M) = \{g \in A \mid M_g \in \underline{\underline{M}}(A_g)\} .$$

Obviously $S(M)$ is a radical ideal in A.

$2^{\underline{o}}$ If M is arbitrary, we put

$$S(M) = \bigcap_{m \in M} S(Dm)$$

1.2 Examples: If $M = A$, then $S(M) = A$.
If $M = A_f f^{\alpha} (\alpha \in \mathbb{C})$, then $S(M) = Af$.
If $M = D$, then $S(M) = (o)$.

Observe that $S(M)$ is a principal ideal in these cases. Indeed by [2], Chap. II, Th. 1.24 we get

Proposition 1.3: $S(M)$ is a principal (radical) ideal, i.e. $S(M) = Af$ (for some $f \in A$).

An important property of the ideal $S(M)$ is, that it serves to describe holonomic D-modules as follows (see [2], Chap. III, Th. 3.1).

Proposition 1.4: Let $\underline{\underline{M}} \in M(D)$. Then:
M is a holonomic D-module iff $S(M) \neq (o)$.

1.5. Fuchsian D modules

Now we introduce Fuchsian D-modules without A-torsion and give some properties. To make the definition more understandable we recall first Yu. Manin's definition of a Fuchsian $\hat{\mathcal{D}}_1$-module. Put $\hat{\mathcal{O}} = \mathbb{C}[[t]]$, $\hat{\mathcal{D}}_1 = \hat{\mathcal{O}}[\frac{d}{dt}]$.

Definition 1.6: A left $\hat{\mathcal{D}}_1$-module M is called Fuchsian iff

$$\Sigma_{p=o}^{\infty} \hat{\mathcal{O}}(t\frac{d}{dt})^p m \in \underline{\underline{M}}(\hat{\mathcal{O}}), \quad \text{for all } m \in M . \tag{1.7}$$

In more concrete terms (1.7) amounts to say that for every $m \in M$ there exist an integer p, $p \geq o$ and elements $a_o, \ldots, a_{p-1} \in \hat{\mathcal{O}}$ such that

$$(t\frac{d}{dt})^p m = a_o m + a_1(t\frac{d}{dt})m + \ldots + a_{p-1}(t\frac{d}{dt})^{p-1} m .$$

Now back to the general situation. Let I be an ideal in A. Define

$$\mathcal{R}(I) = \{\tau \in \mathrm{Der}(A) \mid \tau a \in I, \text{ for all } a \in I\}$$

For $\tau \in R(I)$ and $m \in M$ put

$$E_\tau(m) = \Sigma_{p=o}^{\infty} A\tau^p m .$$

Definition 1.8: M is Fuchsian along I iff $E_\tau(m) \in \underline{\underline{M}}(A)$, for all $m \in M$ and all $\tau \in R(I)$. M is Fuchsian iff M is Fuchsian along $S(M)$.

1.9 Examples: $M = A$ and $M = A_f f^\alpha (\alpha \in \mathbb{C})$ are Fuchsian.

Proposition 1.10: M is Fuchsian along I iff both M is Fuchsian and $I \subset S(M)$ (see [2], Chap. II, Cor. 1.28 and Cor. 1.31) .

Corollary 1.11: M is Fuchsian iff M is Fuchsian along some ideal I.

Corollary 1.12: If M is Fuchsian, then $S(M) \neq (o)$.

Proof: If $S(M) = (o)$, then $E_\tau(m) \in \underline{\underline{M}}(A)$ for all $m \in M$ and all $\tau \in R((o)) = \mathrm{Der}(A) = R(A)$. So M is Fuchsian along A, implying $A \subset S(M) = (o)$, a contradiction.

From Prop. 1.4 and Cor. 1.12 we deduce

Proposition 1.14: If $M \in \underline{\underline{M}}(D)$ and Fuchsian, then M is holonomic.

1.15 A criterion for Fuchsian modules

Let $M \in \underline{\underline{M}}(D)$ (without A-torsion). Suppose $S(M) \neq o$ and let $f \neq o$, $f \in S(M)$ without multiple factor. Put $V = k \otimes_A M$, where k is the quotient field of A, say $d = \dim_k V$. By [2], Chap. II, Th. 4.4 and Chap. I, Th. 1.3 we get

THEOREM 1.16: There is equivalence between

1) M is a Fuchsian D-module.
2) $E_{f\partial_i}(m) \in \underline{\underline{M}}(A)$, for all $m \in M$ and all i.
3) For every irreducible component p of f and every $\partial \in \mathrm{Der}(A)$ such that $\partial p \notin Ap$, the differential operator $T = p(\partial p)^{-1}\partial$ on V has the following property:

If $e \in V$ is a cyclic vector of T in V (i.e. $(e, Te, \ldots, T^{d-1}e)$ is a k-basis of V) and $T^d e = -a_o e + \ldots + a_{d-1} T^{d-1} e$, then for all i $a_i \in A_{\mathfrak{p}}$ $(= \{a/b \in k \mid a \in A, b \notin \mathfrak{p}\})$, where $\mathfrak{p} = Ap$.

§2. ANALYTIC CONSEQUENCES

We shall now try to illuminate the above results by describing their analytic meaning.

Let X be a complex analytic manifold of dimension n and let $\underline{M} = \underline{\mathcal{D}}/\underline{L}$ be a coherent sheaf of left $\underline{\mathcal{D}}$-modules on X without $\underline{\mathcal{O}}$-torsion. By S_M we denote the set

$$S_M = \{z \in X \mid (z,\xi) \in V_M, \text{ for some } \xi \neq o\}$$

(here V_M denotes the characteristic variety of $\underline{M}$, see [1].)
Let $J(S_M,X)$ be the sheaf of holomorphic functions on X vanishing on S_M. Then we get (see [2], Chap. IV, Prop. 2.1)

Proposition 2.1: $S(M_x) = J(S_M,X)_x$, for all $x \in X$.

Let Y be an analytic subvariety of X with $\dim(Y) < n$. Suppose $S_M \subset Y$. By Th. 1.16 the following result can be proved ([2], Chap. IV, Th. 2.3)

THEOREM 2.2: There is equivalence between:
1) Every component of Y of codimension one contains a point y such that M_y is a Fuchsian $\mathcal{D}_y$-module.
2) M_x is a Fuchsian $\mathcal{D}_x$ module for all $x \in X$.

2.3 Fuchsian modules and Nilsson class functions

Let F and L_z be as before. We already remarked that $\underline{M} = \underline{\mathcal{D}}/\underline{L}$ is a coherent sheaf of left $\underline{\mathcal{D}}$-modules. It is not difficult to prove that $\underline{M}$ has no $\underline{\mathcal{O}}$-torsion and that $S_M \subset V(q)$. So to prove that M_z is a Fuchsian $\mathcal{D}_z$-module for all $z \in \mathbb{C}^n$ it suffices to prove this in the regular points of $V(q)$ by Th. 2.2. Hence we can assume $q = z_n$. But then, locally on some polydisc Δ around $z \in V(q)$ we have

$$F = \Sigma \varphi_{\alpha,h}\, z_n^{\alpha}(\log z_n)^h \quad (\alpha \in \mathbb{C},\ h \in \mathbb{Z},\ h \geq o)$$

where $\varphi_{\alpha,h}$ is holomorphic on Δ. It is easy to verify that $\mathcal{D}_z F$ is Fuchsian along $\mathcal{O}_z z_n$. Finally, to prove that $A_n F$ is a Fuchsian A_n-module, we use this local result for $\mathcal{D}_z$-modules and apply Th. 1.16. For more details we refer to [3].

§3. FINAL REMARKS

Remark 3.1. In the theory above we introduced Fuchsian $A_n, \mathcal{D}_n, \hat{\mathcal{D}}_n$-modules without $\mathbb{C}[z_1,\dots,z_n]$, $\mathcal{O}_n$, $\hat{\mathcal{O}}_n$-torsion respectively. In my thesis also Fuchsian modules with torsion are introduced. This is done by studying modules over quite general rings $\mathcal{D}$ of so-called universal differential operators. It is proved that Fuchsian modules (of finite type) are holonomic and that they are stable under localization i.e. iff $f \in A$ and M is a Fuchsian D-module, then M_f is also a Fuchsian D-module.

Remark 3.2: Let M be a $\mathcal{D}_n$-module of finite type without $\mathcal{O}_n$-torsion. Suppose M has regular singularities in the sense of [5] and [6]. Then, using Th. 2.2, it can be proved that M is a Fuchsian $\mathcal{D}_n$-module. Conversely, if M is a Fuchsian $\mathcal{D}_n$-module without $\mathcal{O}_n$-torsion, then it can be shown that M is a finite sum of modules of the form $\mathcal{D}_n F$, where F is some (local) Nilsson class function. Each such a module has regular singularities, so M also. Summarizing. If M is a finitely generated $\mathcal{D}_n$-module without $\mathcal{O}_n$-torsion then

M is Fuchsian iff M has regular singularities.

For a discussion about Fuchsian $\mathcal{D}_n$-modules with torsion and their relation with $\mathcal{D}_n$-modules having regular singularities we refer to [1], Chap. IV

REFERENCES

[1] J.-E. Björk, Rings of Differential Operators, North-Holland Mathematical Library Series (1979).

[2] A.R.P. van den Essen, Fuchsian Modules (Thesis, 1979, Katholieke Universiteit Nymegen, The Netherlands).

[3] A.R.P. van den Essen, C.R. Acad.Sc.Paris, t. 289, Ser. A, 103-105, (1979).

[4] V.A. Golubeva, Russian Math. Surveys, 31, no 2, 1976, 139-207.

[5] M. Kashiwara and T. Oshima, Ann. of Math. 106, 1977, 145-200.

[6] M. Kashiwara and T. Kawai, On holonomic systems of micro-differential equations III. Systems with regular singularities R.I.M.S-293, (1979).

[7] B. Malgrange, Springer Lecture Notes Vol. 459, 1976, 98-119.

[8] N. Nilsson, Arkiv för Matematik, 5, 1965, 463-475.

Mathematisch Instituut der Katholieke Universiteit, Toernooiveld, 6525 ED Nymegen, The Netherlands

Till 1 oktober 1980: Matematiska Institutionen, Stockholms Universitet, Stockholm, Sweden

PART II

MISCELLANEOUS MATHEMATICAL DEVELOPMENTS

SINGULARITES DES SOLUTIONS DES EQUATIONS AUX DERIVEES PARTIELLES NON LINEAIRES

Jean Michel BONY
Mathématiques - Université Paris-Sud
91405 Orsay, France

1. Notations et résultats

Nous utiliserons les espaces de Sobolev H^s dont nous ne rappelons pas la définition et les espaces de Hölder C^ρ, définis pour $\rho \in \mathbb{R} \setminus \mathbb{Z}$ qui jouissent de propriétés suivantes : pour $0 < \rho < 1$, on a $u \in C^\rho \iff |u(y) - u(x)| \leqslant C^{te} |y-x|^\rho$; les opérateurs pseudo-différentiels d'ordre k appliquent C^ρ dans $C^{\rho-k}$, pour ρ et $\rho - k \notin \mathbb{Z}$.

Nous nous intéressons aux régularités "microlocales" des distributions définies comme suit. Soit $(x_0, \xi_0) \in \mathbb{R}^n \times (\mathbb{R}^n \setminus \{0\})$, et soit u une distribution définie au voisinage de x_0. On dit que u appartient microlocalement à C^ρ[resp. H^s] en (x_0, ξ_0), si pour tout opérateur pseudo-différentiel K d'ordre 0, à symbole nul hors d'un petit voisinage "conique en ξ" de (x_0, ξ_0), on a $Ku \in C^\rho$ [resp. H^s].

Soit $\mathcal{F}[u] = F(x,u(x), \ldots, \partial^\beta u(x), \ldots) = 0$ une équation aux dérivées partielles non linéaire d'ordre m, où $F(x,u_0,\ldots,u_\beta,\ldots)$ est une fonction C^∞ des ses arguments. On supposera éventuellement que F est quasi-linéaire, et plus précisément qu'elle peut se mettre sous la forme :

$$F(x,u,\ldots) = \sum_{k_0<k\leqslant m} \sum_{|\alpha|=k} A_\alpha(x,u,\ldots,\partial^\beta u,\ldots)_{\beta\leqslant p(k)} \cdot \partial^\alpha u + A_{k_0} (x,u\ldots,\partial^\beta u)_{\beta\leqslant k_0}$$

Dans ce cas, on note $d = \mathrm{Max}(k_0, \frac{k+p(k)}{2})$ en faisant la convention $p(k) = -\infty$ lorsque les A_α correspondants ne dépendent que de x. On a $d = m$ pour une équation non quasi-linéaire, $0 \leqslant d \leqslant m - \frac{1}{2}$ pour une équation quasi-linéaire, et $d = -\infty$ pour une équation linéaire.

Si u est une fonction réelle de classe C^ρ avec $\rho > d$, les fonctions non linéaires et les produits apparaissant dans l'expression de F sont bien définis, et $F(x,u,\ldots, \quad ,\ldots) \in C^{\rho-d}$.

De même, si $u \in H^s$, l'expression $\mathcal{F}[u]$ est bien définie si on a la double condition : $s > n/2 + \mathrm{Max}(k_0,p(k))$ et $s > n/4 + d$.

Si u est une solution de $\mathcal{F}[u] = 0$, on définit l'équation caractéristique

$$p_m(x,\xi) = \sum_{|\alpha|=m} \frac{\partial F}{\partial u_\alpha} (x,u(x),\ldots,\partial^\beta u(x),\ldots)\,(i\xi)^\alpha = 0.$$

Un point (x_0,ξ_0) est dit caractéristique si $p_m(x_0,\xi_0) = 0$. Les bicaractéristiques sont les courbes intégrales du champ hamiltonien de p_m (ou de $i\,p_m$ pour m impair), c'est-à-dire les application $(x(t), \xi(t))$ vérifiant $\dot{x}(t) = \frac{\partial p_m}{\partial \xi}$, $\dot{\xi}(t) = -\frac{\partial p_m}{\partial x}$.

Théorème 1 : *Soit u une solution de classe C^ρ de $\mathcal{F}[u] = 0$, $\rho > d$. Alors en tout point (x_0,ξ_0) non caractéristique, u est microlocalement de classe $C^{2\rho + m - 2d}$ si $\rho + m - 2d \notin \mathbb{Z}$. Si $u \in H^s$, $s > n/2 + \mathrm{Max}(k_0,p(k))$, $s > n/4 + d$, on a microlocalement $u \in H^t$ pour $t < 2s - n/2 + m - 2d$.*

Théorème 2 : *Soit u une solution de classe H^s, s vérifiant les conditions ci-dessus. Supposons de plus que $p_m(x,\xi)$ soit de classe C^2. Soit (x_0,ξ_0) appartenant à la variété caractéristique, tel que u appartienne microlocalement à H^r en (x_0,ξ_0), avec $r < 2s - n/2 + m - 2d - 1$. Alors u appartient à H^r le long de la bicaractéristique issue de (x_0,ξ_0).*

Sous l'hypothèse $u \in H^s$, on choisit $\rho < s - n/2$, voisin de $s - n/2$. On a alors $u \in C^\rho$. Le gain de régularité est $\rho + m - 2d$ dans le théorème 1, quantité positive d'après les hypothèses. Le théorème 2 n'est intéressant que si $\rho + m - 2d > 1$, ce qui implique seulement $p_m(x,\xi) \subset C^1$. On n'a plus alors nécessairement unicité des courbes intégrales du champ hamiltonien. Toutefois, un raffinement du théorème 2 assure qu'une singularité en (x_0,ξ_0) se prolonge le long de l'une au moins des demi-bicaractéristiques issues de (x_0,ξ_0).

Remarques : Les théorèmes précédents ne s'appliquent qu'à des solutions ayant déjà une certaine régularité. Pour des solutions moins régulières, il peut se produire des phénomènes du type "ondes de choc", où le comportement des singularités est très différent.

D'autre part, les théorèmes 1 et 2 ne contrôlent la régularité microlocale que jusqu'à un certain ordre de régularité. Au delà, comme le montrent les exemples de B. Lascar [5] et J. Rauch [6], des phénomènes typiquement non linéaires apparaissent.

Enfin, les caractéristiques dépendent en général de la solution u elle-même. Les théorèmes 1 et 2 ne donnent une description géométrique exacte de la propagation des singularités que pour les équations quasi-linéaires dont la partie principale est linéaire. Toutefois, dans le cas général, des estimations portant sur u fournissent des estimations sur les domaines où u est régulière.

Nous ne donnons ici qu'un bref aperçu des démonstration des théorèmes 1 et 2. Les démonstrations detaillées figurent dans [1]. Certains cas de ces théorèmes ont également été démontrés par B. Lascar [5] et J. Rauch [6].

2. Un nouveau calcul symbolique

Soit $a(x,\xi)$ une fonction homogène de degré m en ξ, de classe C^∞ en ξ pour $\xi \neq 0$, à support compact en x et de classe C^ρ en x ($\rho > 0$ non entier).

Posons :

$$T_a u(x) = (2\pi)^{-n} \int e^{ix.\xi} \chi\left(\frac{\xi}{|\eta|},\eta\right) \hat{a}(\xi-\eta,\eta)\hat{u}(\eta)d\eta$$

où $\hat{a}$ est la transformée de Fourier de a par rapport à la première variable, et où χ est égale à 1 pour $|\xi-\eta| \leqslant \varepsilon_1|\eta|$ et égale à 0 pour $|\xi-\eta| \geqslant \varepsilon_2\eta$, avec $0 < \varepsilon_1 < \varepsilon_2 < 1$.

L'opérateur T_a applique C^σ dans $C^{\sigma-m}$, H^s dans H^{s-m} quels que soient σ (non entier) et s. Si χ_1 et χ_2 vérifient les conditions ci-dessus, la différence des deux opérateurs associés applique $C^\sigma(H^s)$ dans $C^{\sigma-m+\rho}$ $(H^{s-m+\rho})$. On se ramène en fait, à l'aide de la décomposition en harmoniques sphériques, à l'étude des "paramultiplications" : opérateur T_a associés à une fonction a(x) de classe C^ρ. On étudie ceux-ci notamment à l'aide du découpage "en couronnes dyadiques" de R. Coifman et Y. Meyer [2].

Dans un ouvert Ω de $\mathbb{R}^n$, on note Σ_ρ^m la classe des fonctions du type $a_m(x,\xi) + a_{m-1}(x,\xi) +\ldots+ a_{m-[\rho]}(x,\xi)$, définies dans $\Omega \times (\mathbb{R}^n\backslash\{0\})$, où a_{m-k} est homogène de degré m-k en ξ, C^∞ en ξ, $C^{\rho-k}$ en x.

On dit que A est un opérateur paradifférentiel (proprement supporté) dans Ω, d'ordre m et de classe C^ρ [$A \in Op(\Sigma_\rho^m)$] si A est un opérateur proprement supporté, et s'il existe $a(x,\xi) \in \Sigma_\rho^m(\Omega)$ tel que, pour tout compact K de Ω, quelles que soient φ_1 et φ_2 dans $C_0^\infty(\Omega)$, égales à 1 au voisinage de K, l'opérateur $A - \varphi_2 T_{\varphi_1 a}$, restreint aux distributions à support dans K, applique C^σ (resp. H^s) dans $C^{\sigma-m+\rho}$ (resp. $H^{s-m+\rho}$)

L'application qui à un opérateur paradifférentiel A associe son symbole $a = \sigma(A)$ est bien définie et surjective.

<u>Proposition 4</u> : Soient $A \in Op(\Sigma_\rho^{m_1})$ et $B \in Op(\Sigma_\rho^{m_2})$. Alors $A.B \in Op(\Sigma_\rho^{m_1+m_2})$ et on a :

$$\sigma(AB) = \sigma(A) \# \sigma(B) = \sum_{|\alpha|+\ell+k\leqslant[\rho]} (1/\alpha!)\partial_\xi^\alpha a_{m_1-k} D_x^\alpha b_{m_2-\ell}$$

L'adjoint A^* appartient à $Op(\Sigma_\rho^{m_1})$, et on a

$$\sigma(A^*) = [\sigma(A)]^* = \sum_{|\alpha|+k\leqslant[\rho]} (i/\alpha!)\ \partial_\xi^\alpha D_x^\alpha \bar{a}_{m_1-k}$$

3. Fonctions non linéaires

<u>Proposition 5</u> : Soient $a \in C^\rho$ et $u \in C^\sigma$

a) si $\rho > 0$, $\sigma > 0$, on a $au = T_a.u + T_u.a + r$ avec $r \in C^{\rho+\sigma}$.

b) si $\rho > 0$, $\sigma < 0$, $\rho + \sigma > 0$, on a $au = T_a.u + r$ avec $r \in C^{\rho+\sigma}$.

<u>Théorème 6</u> : *<u>Soit</u> $u \in C^\rho$, $\rho > 0$ et F de <u>classe</u> C^∞. <u>Alors</u>*

$$F(u(x)) = T_{F'(u(x))}.u(x) + r(x) \ \underline{\text{avec}}\ r \in C^{2\rho}$$

<u>Plus généralement</u>, <u>on a</u> :

$$F(u_1(x),\ldots,u_N(x)) = \Sigma T_{\partial F/\partial u_i}.u_i + r \ \underline{\text{avec}}\ r \in C^{2\rho}$$

si $u_j \in C^\rho$. <u>Si de plus</u> $u \in H^s$, $s > n/2$, <u>on a</u> $r \in H^{2s-n/2-\varepsilon}$.

D'après la proposition 5, on a $u^2 \equiv 2T_u.u \equiv T_{2u}.u(\text{mod } C^{2\rho})$. Une application répétée de cette proposition et du fait que (Prop. 4), $T_a . T_b \equiv T_{ab}$ (mod. ρ-régularisants) permet de démontrer le théorème lorsque F est un polynôme. Une estimation soigneuse des restes, et les théorèmes de Bernstein sur l'approximation des fonctions C^∞ par des polynômes permettent de conclure.

Corollaire 7 : Soit $u \in C^\rho$, $\rho > Max(k_o, p(k))$, *une solution de* $\mathcal{F}[u] = 0$. *Soit* P *un opérateur paradifférentiel d'ordre* m *et de classe* $C^{\rho+m-2d}$ *de symbole*

$$\sigma(P) = \sum_{\beta>2d-\rho} \partial F/\partial u_\beta \, (x,u(x),\ldots)(is)^\beta.$$

a) *Si* $\rho > d$, *on a* $Pu \in C^{2\rho-2d}$

b) *Si* $u \in H^s$, *avec* $s > n/2 + \rho$ *et* $s > n/4 + d$, *on a* $Pu \in H^{s+\rho-2d}$.

Dans le cas non quasi-linéaire, on a immédiatement, pour $\rho > m$

$$\mathcal{F}[u] \equiv \Sigma T_{\partial F/\partial u_\beta} \cdot \partial^\beta u \equiv T_{\Sigma \partial F/\partial u_\beta}(i\xi)^\beta \cdot u \ (\mathrm{mod.}\ C^{2\rho-2m}).$$

Dans le cas quasi-linéaire, on applique d'abord la proposition 5 aux produits $A_\alpha(x,u,\ldots)\,\partial^\alpha u$, puis le théorème 6 aux termes $A_\alpha(\ldots)$.

Indications sur les démonstrations des théorèmes 1 et 2

Soit donc, sous les hypothèses du théorème 1, (x_o,ξ_o) un point non caractéristique, et soit k(x,D) un opérateur pseudo-différentiel classique, d'ordre 0, dont le symbole s'annule au voisinage de la variété caractéristique de P, et dont le symbole principal est non nul en (x_o,ξ_o).

Soit $q \in \Sigma^{-m}_{\rho+m-2d}$ tel que $k = q \# p$, et $Q \in \mathrm{op}\,(\Sigma^{-m}_{\rho+m-2d})$ de symbole q. On a alors $k(x,D)\,u = QPu + Ru$ où R est $\rho + m - 2d$ régularisant. On a $Q[Pu] \in C^{2\rho+m-2d}$ (ou $H^{2s-n/2+m-2d-\varepsilon}$) et Ru appartient au même espace, ce qui prouve le théorème.

En ce qui concerne le théorème 2, c'est une conséquence immédiate du théorème linéaire suivant :

Théorème 8 : *Soit $P \in Op(\Sigma^m_\sigma)$, $\sigma > 1$, dont le symbole principal p_m est réel et de classe C^2. Soit $u \in H^s$ tel que $Pu \in H^t$ au voisinage d'un arc de bicaractéristique. Pour $r \leq Min(t+m-1, s+\sigma-1)$, si u est de classe C^r en un point de l'arc de bicaractéristique, u est de classe C^r au voisinage de l'arc entier.*

La démonstration de ce dernier théorème est assez technique, mais voisine dans son principe de la démonstration de Hörmander [4] dans le cas pseudo-différentiel, fondée sur une inégalité "d'énergie". Un point-clef de cette démonstration est le théorème suivant (inégalité de Garding précisée pour les opérateurs paradifférentiels), que nous démontrons par une méthode inspirée de [3].

Théorème 9 : *Soit $S \in Op(\Sigma^m_\tau)$, $\tau > 0$. On suppose le symbole principal $s_m(x,\xi) \geq 0$. Il existe alors $\mu > 0$ tel que*

$$Re\,(Su,u) \geq - C^{te}|u|_{m/2-\mu/2}\,.$$

BIBLIOGRAPHIE

[1] J. M. Bony : Calcul symbolique et propagation des singularités pour les équations aux dérivées partielles non linéaires (Prépublications Université Paris-Sud).

[2] R. Coifman, Y. Meyer : Au-delà des opérateurs pseudo-différentiels, Astérisque 57 (1978).

[3] A. Cordoba, C. Fefferman : Wave packets and Fourier integral operators. Comm. P.D.E. (1978).

[4] L. Hörmander : On the existence and the regularity of solutions of linear pseudo-differential equations. L'enseignement Math. 17 (1971) 99-163.

[5] B. Lascar : Singularités des solutions d'équations aux dérivées partielles non linéaires. C.R. Acad. Sc. Paris 287 A (1978) 527-529.

[6] J. Rauch : Singularities of solutions to semilinear wave equations. J. Math. Pures et Appli. (1979).

COMPORTEMENT SEMI CLASSIQUE DU SPECTRE D'UN HAMILTONIEN QUANTIQUE.

J. CHAZARAIN.

I - Introduction.

On considère dans $\mathbb{R}^n$ l'opérateur elliptique (hamiltonien quantique en physique) $Q = -\frac{h^2}{2}\Delta + V(x)$ dépendant du paramètre $h \in]0,1]$ (la constante de Planck en en physique). On suppose que la fonction potentielle V est dans $C^\infty(\mathbb{R}^n)$ à valeurs ≥ 0 et vérifie pour $|x| \to +\infty$ les conditions suivantes :

$$\partial_x^\alpha V(x) = \begin{cases} O(|x|^{2-|\alpha|}) & \text{pour } 0 \leq |\alpha| \leq 2 \\ O(1) & \text{pour } |\alpha| \geq 2 \end{cases}$$

et

$$V(x) \geq c|x|^2 \quad \text{avec } c > 0.$$

L'exemple le plus simple est l'oscillateur harmonique $-\frac{h^2}{2}\Delta + |x|^2$.

En utilisant les résultats de Beals [3] et Robert [19], on vérifie que l'on peut associer à Q un opérateur auto-adjoint positif dans $L^2(\mathbb{R}^n)$; son spectre est constitué de valeurs propres $(\lambda_j)_{j \geq 1}$ telles que :

$$0 < \lambda_1 \leq \lambda_2 \leq \dots \qquad \text{avec } \lambda_j \xrightarrow[j^\infty]{} +\infty.$$

Un problème constant en mécanique quantique est d'essayer d'exprimer le comportement asymptotique, quand $h \to 0$, des grandeurs liées à Q en fonction de grandeurs de la mécanique classique, c'est-à-dire définies à partir du champ hamiltonien $H_q(x,\xi) = \partial_\xi q(x,\xi)\partial_x - \partial_x q(x,\xi)\partial_\xi$ (ici $q(x,\xi) = \frac{|\xi|^2}{2} + V(x)$) (cf. Berry et Mount [4] et Voros [20]). Lorsque le champ H_q est complètement intégrable, on a de nombreux résultats sur l'expression asymptotique de familles de valeurs propres (Maslov [18] , Duistermaat [12] , Colin de Verdière [9] , Leray [17],...).

2. Résultats

Soit Φ^t le flot du champ hamiltonien H_q, on note $\mathscr{L}$ l'ensemble des périodes des solutions périodiques des équations de Hamilton Jacobi

$$(x'(t), \xi'(t)) = H_q(x(t),\xi(t))$$

La mesure spectrale de Q est définie par

$$\sigma(\lambda) = \sum_{j \geq 1} \delta(\lambda - \lambda_j),$$

on lui associe, par transformation de Fourier, la distribution

$$S(t) = \sum_{j \geq 1} \exp(-ih^{-1}t\ \lambda_j) \in \mathscr{S}'(\mathbb{R}).$$

On a le

Théorème 1 : Soit $\tau_o \in \mathbb{R}$ et $\rho \in C_o^\infty(\mathbb{R})$ tel que supp $\rho \cap \mathscr{L} = \emptyset$. Alors la fonction de h

$$I_h = < S_h(t),\ \rho(t)\ e^{-i\ \tau_o\ h^{-1}t} >$$

est $O(h^\infty)$, c'est-à-dire que pour tout $N \in \mathbb{N}$ il existe C_N tel que

$$|I_h| \leq C_N\ h^N,\ h \in]0,1] .$$

On peut donner une formulation plus parlante de ce théorème en utilisant la notion d'ensemble de fréquence d'une distribution asymptotique au sens de Guillemin et Sternberg [15] . Ici, $S_h(t)$ est une distribution de $\mathscr{S}'(\mathbb{R})$ qui dépend du paramètre h ; on dit que le point (t_o,τ_o) de $T^*\mathbb{R}$ n'est pas dans l'ensemble de fréquence de S (ensemble noté F[S]) s'il existe $\rho \in C_o^\infty(\mathbb{R})$ avec $\rho(t_o) \neq 0$ tel que

$$< S_h(t),\rho(t)e^{-i\tau h^{-1}} > = O(h^\infty) \text{ pour } \tau \text{ voisin de } \tau_o.$$

Alors le théorème 1 permet de montrer l'inclusion

$$\pi(F[S]) \subset \mathscr{L} \text{ , où } \pi : T^*\mathbb{R} \longrightarrow \mathbb{R}.$$

Ce type d'inclusion ressemble au résultat concernant la formule de Poisson pour un opérateur elliptique Q sur une variété compacte (cf. Chazarain [6] , Duistermaat et Guillemin [13]). Mais l'analogie est purement formelle, car ici la géométrie dépend essentiellement de V alors que dans le cas de la formule de Poisson, c'est la partie principale de Q, à savoir $\frac{|\xi|^2}{2}$, qui intervenait.

En fait, ce théorème est plutôt à rapprocher des travaux de Balian et Bloch [2] et Berry et Tabor [5].

Dans le cas où le flot hamiltonien est complètement périodique, on a un phénomène de concentration des valeurs propres au voisinage d'une progression arithmétique, c'est à rapprocher des résultats de Duistermaat et Guillemin [13], Weinstein [21], Colin de Verdière [11] avec la différence qu'il s'agit ici d'un comportement quand $h \to 0$.

Théorème 2 : On suppose Φ^t périodique et soit T la plus petite période positive. Alors, il existe $M \geq 0$ et $\alpha \in \mathbb{Z}$ tels que

$$\text{spectre } Q \subset \bigcup_{k \in N} [\tilde{\lambda}_k(h) - Mh^2, \tilde{\lambda}_k(h) + Mh^2]$$

avec $\tilde{\lambda}_k(h) = (\gamma + \frac{2\pi}{T} k + \alpha \frac{\pi}{2T})h$ et où $\gamma = \frac{1}{T} \int_0^T L(x(s), x'(s))ds$ désigne la moyenne du Lagrangien $L(x,x') = \frac{|x'|^2}{2} - V(x)$ sur une trajectoire de période T.

En plus de l'oscillateur harmonique, il y a de nombreux cas où les hypothèses du théorème 2 sont satisfaites. Voici une façon d'en construire quand n = 1. Soit $\theta(X)$ une fonction $C^\infty(\mathbb{R})$, paire, bornée, à dérivée bornée par une constante $k < 1$. Soit $X(x)$ la fonction réciproque de la fonction strictement croissante $x = X + \theta(X)$. On pose $V(x) = \frac{1}{2}(X(x))^2$, alors toutes les solutions de $x''(t) + V'(x(t)) = 0$ sont périodiques de période 2π. Si on suppose de plus que pour tout entier $j \geq 2$, on a $\theta^{(j)}(X) = O(\frac{1}{X})$, alors V(x) satisfait aux conditions de croissance à l'infini.

3. Esquisse de la démonstration du théorème 1.

Il ne peut être question de donner en une conférence la démonstration de ces deux théorèmes, aussi on va seulement expliquer celle du premier théorème.

Soit $U(t) = \exp(-ih^{-1}tQ)$ le groupe unitaire solution de l'équation de Schrödinger

$$(\star) \qquad ih\partial_t U - Q.U = 0, \quad U(0) = 1$$

Alors U est lié à la distribution S par la relation "$S(t) = \mathrm{tr}\, U(t)$ qui signifie que pour tout $\theta \in \mathcal{S}(\mathbb{R})$ l'opérateur $U_\theta = \int U(t)\, \theta(t)\, dt$ est à trace et vérifie

$$< S(t),\ \theta(t) = \mathrm{tr}\,(U_\theta).$$

Pour démontrer le théorème, il suffit de prouver que pour tout $N_o \in \mathbb{N}$ on a $I_h = O(h^{N_o})$; aussi on est conduit à construire une solution approchée $E(t)$ de $(\star)$ modulo h^N avec N assez grand.

On commence par construire $E(t)$ pour $|t| \leq T$ avec T assez petit, ce qui permettra de démontrer le théorème pour supp $\rho \subset]-T,T[$. On construit $E(t)$, au moins formellement, sous la forme

$$(E(t)u)(x) = (2\pi)^{-n} \iint_{\mathbb{R}^n \times \mathbb{R}^n} e^{+ih^{-1}(S(t,x,\eta)-y.\eta)} a(t,x,\eta;h) u(y)\, dy\, d\eta$$

où la phase $S(t,x,\eta)$ est solution de l'équation caractéristique

$$\partial_t S + q(x, \partial_x S) = 0 \qquad S|_{t=o} = x.\eta$$

et où l'amplitude $a(t,x,\eta;h) = h^{-n} \sum_{j=o}^{N} a_j(t,x,\eta) h^j$, avec a_j solution de l'équation de transport

$$\partial_t a_j + \partial_x S \partial_x a_j + \frac{1}{2} \Delta S . a_j = - \frac{1}{2} \Delta a_{j-1} \quad j \geq 0, \ldots, N$$

$$a_j|_{t=o} = \delta_{o,j} \quad \text{et } a_{-1} = 0.$$

Une construction analogue a été faite également par Fujiwara [14] et Kitada [16]. Pour dépasser le stade formel, il faut préciser le comportement en t, x, η des fonctions S et a_j. Posons $\lambda(x,\eta) = (1 + |x|^2 + |\eta|^2)^{1/2}$, la partie technique consiste à démontrer la

<u>Proposition</u> : Il existe $T > 0$ et $c > 0$ tels que $S(t,x,\eta)$ est défini et vérifie $|\partial_t S(t,x,\eta)| \geq c^2 \lambda^2(x,\eta)$, $\partial_t^p \partial_{x,\eta}^\alpha S(t,x,\eta) = O(\lambda^p)$ pour $|t| \leq T$, $(x,\eta) \in \mathbb{R}^{2n}$, $p \geq 2$, α multi-indice. De plus, on a $\partial_t^p \partial_{x,\eta}^\alpha a_j(t,x,\eta) = O(\lambda^p)$ pour tout $p \geq 0$ et α.

Cette proposition permet de montrer que, si on pose

$$J_h = \mathrm{trace}\ \left(\int E(t)\, \rho(t)\, e^{-i\tau h^{-1} t}\, dt \right),$$

on a l'expression

$$J_h = (2\pi)^{-n} \iiint e^{ih^{-1}(S(t,x,\eta)-x.\eta)}\ \rho(t)a(t,x,\eta;h)dt\ dx\ d\eta$$

et de plus, on peut appliquer le théorème de la phase non stationnaire, pour prouver que $J_h = O(h^\infty)$.

Ensuite, il faut vérifier que I_h a le même comportement asymptotique (modulo h^{N_o}) que J_h, c'est-à-dire que l'erreur $F(t) = U(t) - E(t)$ n'apporte qu'une contribution $O(h^{N_o})$ à la trace. Pour cela, on utilise un théorème de Asada et Fujiwara [1] qui permet de montrer que pour $|t| \leq T$ on a

$$||F(t)||_{\mathscr{L}(L^2,L^2)} = O(h^N).$$

Cette majoration, combinée avec le fait que l'opérateur Q^{-n} possède une trace telle que $|\text{trace } (Q^{-n})| = O(h^{-2n})$, entraine l'estimation :

$$\text{trace } \left(\int F(t)\ \rho(t)\ e^{-i\tau h^{-1}} dt \right) = O(h^{N_o})$$

pour N assez grand vis à vis de N_o.

Enfin, quand ρ n'est pas à support dans $]-T,T[$ on peut toujours supposer que son support est dans un intervalle du type $]kT, (k+1)T[$ et on utilise dans cet intervalle l'opérateur $E(t - kT).(E(T))^k$ comme approximation de $U(t)$.

Ces résultats ont été annoncés dans des notes aux C.R. Acad. Soc. [7] et [8].

REFERENCES :

[1] K. ASADA et D. FUJIWARA Jap. J. of Math. Vol. 4, n° 2, 1978, p. 299-361.

[2] R. BALIAN et C. BLOCH Ann. of Physics, vol. 85, 2, 1974, p. 514-545.

[3] R. BEALS Duke Math. J. 42, 1, 1975, p. 1-42.

[4] M.V. BERRY et K.E. MOUNT Rep. Prog. Phys., 35, 1972, p. 315-397.

[5] M.V. BERRY et M. TABOR J. Phys. A, vol. 10, n°3, 1977, p. 371-374.

[6] J. CHAZARAIN Inv. Math. 24, 1974, p. 75-82.

[7] J. CHAZARAIN C.R. Acad. Sc. 288, 1979, p. 725-728.

[8] J. CHAZARAIN — C.R. Acad. Sc. 288, 1979, p. 895-897.

[9] Y. COLIN DE VERDIERE — Inv. Math. 43, 1977, p. 15-52.

[10] Y. COLIN DE VERDIERE — Duke Math. J. 46, 1979, p. 169-182.

[11] Y. COLIN DE VERDIERE — C.R. Acad. Sc. 286, 1978, p. 1195-1197.

[12] J.J. DUISTERMAAT — Comm. Pure Appl. Math. 27, 1974, p. 207-281.

[13] J.J. DUISTERMAAT et V. GUILLEMIN — Inv. Math., 29, 1975, p. 39-79.

[14] D. FUJIWARA — Proc. Jap. Acad. 54, 1978, p. 62-66.

[15] V. GUILLEMIN et S. STERNBERG — Geometric Asymptotics A.M.S., 1977.

[16] H. KITADA — On the fundamental solution for schrödinger equations (à paraître).

[17] J. LERAY — Analyse lagrangienne et mécanique quantique, Collège de France 1976-77.

[18] V.P. MASLOV — Théorie des perturbations et méthodes asymptotiques, Dunod, 1972.

[19] D. ROBERT — Comm. Partial Diff. Equat. 3, 1978, p. 755-826.

[20] A. VOROS — Développements semi-classiques, thèse, Orsay, 1977.

[21] A. WEINSTEIN — Duke Math. J. 44, 1977, p. 883-892.

RATIONAL AND PADÉ APPROXIMATIONS TO SOLUTIONS OF LINEAR DIFFERENTIAL EQUATIONS AND THE MONODROMY THEORY

by G.V. Chudnovsky

(CNRS - Paris and Columbia University, New York)

Introduction. The purpose of this paper is to investigate rational and Padé approximations to solutions of linear differential equations from the perspective of the monodromy theory. The classical Padé approximation problem - the construction of a linear combination of a given function having zeroes of high order - is reduced to the analysis of local exponents of Fuchsian linear differential equations. The recurrent relations between Padé approximants are reduced to Riemann's contiguous relations [1], [2]. This reduction forces us to concentrate on monodromy properties of Fuchsian linear differential equations [2], [12]. We continue isomonodromy studies [2],[12],[13], and contiguous relations in this respect are nothing but special cases of the Bäcklund transformation [12],[17]. The present paper is important for applications; one possible application is the analytic continuation of solutions of differential equations represented by the asymptotic series (in the spirit of Stilties). Another application is to transcendental numbers, where the tool of Padé approximants [2],[15],[16] is most useful in proofs of irrationality and transcendence of values of analytic functions (e.g., e, π, log 2 or $\zeta(3)$[17]). In §1 we introduce our main object: the Riemannian module of multi-valued functions on $\mathbb{CP}^1$ which satisfy the monodromy conditions. In §2 we repeat known facts on Fuchsian equations in the dimension one. The relation between monodromy and exponential matrices and coefficients of linear differential equations is investigated in §§3,4 (in §3, equations with regular [10], in §4, with irregular singularities). Then §§5-7 are devoted to Fuchsian and Riemannian modules; we study the relations between scalar and matrix linear differential equations. We investigate local exponents of solutions of linear differential equations and apply this knowledge to rational approximations. In §8 proper definitions of p-point Padé approximations are introduced, and we prove the "perfectness" of the system $\{x^{\omega_1},\ldots,\ldots, x^{\omega_n}\}$ at a finite set in $\mathbb{C}^1$. Then in §9 a simple formula for rational approximations of vector solutions of matrix linear differential equations is proposed in terms of monodromy data. In §10 the recurrences between one-point Padé approximants are written down explicitly in terms of the coefficients of linear differential equations. At last, in §11 the conditions for perfectness and "almost perfectness" of the Padé approximations to solutions of linear differential equations are formulated.

§1. The Riemann modules of systems of functions satisfying monodromy conditions.

1.1. We repeat now the basic assumptions of Riemann's classic studies [1], following our paper [2]. We decided to follow our function - theoretic approach instead of the modern algebraic, methods of Fuchsian modules. Moreover, we prefer to call the corresponding modules not Fuchsian, but rather Riemannian; repeating the Lappo-Danilevski [3] notations of the "corps de Riemann".

1.2. We start from m finite singularities $a_1,\dots,a_m$ and we consider the Riemann surface $\Gamma(a_1,\dots a_m)$ which we get from the x-complex plane $\mathbb{CP}^1$ by cutting along m curves $(a_1,\infty),\dots,(a_m,\infty)$:

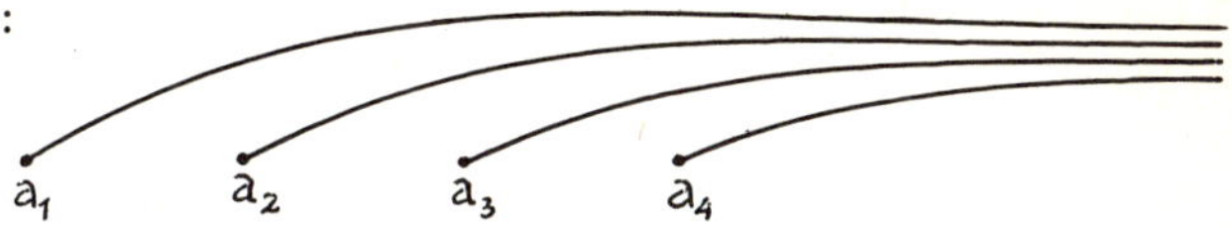

1.3. We pick up some fixed representation of the fundamental group $\pi^1(\mathbb{CP}^1\setminus\{a_1,\dots,a_m\})$ in $GL(n;\mathbb{C}^1)$. In other words we can take m non-singular matrices

$$V_1,\dots,V_m \; . \tag{1.3.1}$$

We assume of course that V_i is connected with the point a_i; we associate with the matrix V_i the linear transformation:

$$z\mapsto V_i z \; , \tag{1.3.2}$$

and so V_i is called an integral substitution associated with a_i: $i=1,\dots,m$. Our main object becomes the following ordered object:

$$\begin{pmatrix} V_1,\dots,V_m \\ a_1,\dots,a_m \end{pmatrix} . \tag{1.3.3}$$

1.4. Now we introduce the main analytic object connected with (1.3.3). This is the Fuchsian module of n-plets of the functions analytic in $\Gamma(a_1,\dots,a_m)$ defined over $\mathbb{C}[x]$.

Definition 1.4.1. Let $F(a_1,\dots,a_m)$ be the space of n-plets $(f_1(x),\dots,f_n(x))$ of analytic functions, each of them being uniform and holomorphic in $\Gamma(a_1,\dots,a_m)$ with the exception of the curves $(a_1,\infty),\dots,(a_m,\infty)$.

Definition 1.4.2. For any given sequence $\begin{pmatrix} V_1,\dots,V_m \\ a_1,\dots,a_m \end{pmatrix}$ we define the following module

$$S\begin{pmatrix} V_1,\dots,V_m \\ a_1,\dots,a_m \end{pmatrix} \tag{1.4.3}$$

over $\mathbb{C}[x]$. The module $S\begin{pmatrix} V_1,\dots,V_m \\ a_1,\dots,a_m \end{pmatrix}$ consists of n-plets $\vec{f}(x)=(f_1(x),\dots,f_n(x))$ from $F(a_1,\dots,a_m)$ such that for any $j=1,\dots,m$ and any point x_0 on the curve (a_j,∞), the limit values

$$\vec{f}^-(x_0)=(f_1^-(x_0),\dots,f_n^-(x_0)) \text{ and}$$

$$\vec{f}^+(x_0)=(f_1^+(x_0),\dots,f_n^+(x_0)),$$

when x tends to $x_0 \varepsilon (a_j,\infty)$ from $\Gamma(a_1,\dots,a_m)$ in negative (-) and positive (+) directions are linearly related via the transformation (1.3.2):

(1.4.4) $\qquad \vec{f}^-(x_0)^t = V_j \cdot \vec{f}^+(x_0)^t$ at $x_0 \varepsilon (a_j,\infty)$: $j=1,\dots,m$.

Instead of the functions $\vec{f}(x)=(f_1(x),\dots,f_n(x))$ defined on $\Gamma(a_1,\dots,a_m)$, we can consider the multi-valued functions $\vec{f}(x)$ on $\mathbb{CP}^1$ with singularities (i.e., branch points and logarithmic singularities) at $\{a_1,\dots,a_m\}$ and, possibly, at ∞. Instead of $V_1,\dots,V_m$ we speak about the representation of π^1 $(\mathbb{CP}^1 \setminus \{a_1,\dots,a_m,\infty\})$ in $GL(n;\mathbb{C})$, and for a fixed basis $\gamma_1,\dots,\gamma_m,\gamma_\infty$ of $\pi^1(\mathbb{CP}^1\setminus\{a_1,\dots,a_m,\infty\})$; $\gamma_1+\dots+\gamma_m+\gamma_\infty=0$; we take the matrices V_{γ_i} corresponding to γ_i : $i=1,\dots,m,\infty$. Then $V_{\gamma_1}\cdots V_{\gamma_m}\cdot V_{\gamma_\infty}=I$ and we replace (1.4.4) with

(1.4.5) $\vec{f}(x) \underset{\gamma}{\mapsto} V_\gamma \vec{f}(x)$

when x makes a circuit γ around the closed path on $\mathbb{CP}^1\setminus\{a_1,\dots,a_m,\infty\}$, and $V_\gamma \varepsilon GL(n;\mathbb{C})$. There for $\gamma=n_1\gamma_1+\dots+n_m\gamma_m+n_\infty\gamma_\infty$ we have $V_\gamma=\prod_{i=1}^m V_{\gamma_i}^{n_i}\cdot V_{\gamma_\infty}^{n_\infty}$. Naturally V_{γ_i} stands for V_i in (1.4.4): $i=1,\dots,m$.

1.5. Riemann's [1] main result (theorem 1.6.3 below) shows that the module $S\begin{pmatrix}V_1,\dots,V_m\\a_1,\dots,a_m\end{pmatrix}$ is at most n-dimensional over $\mathbb{C}[x]$. Moreover the construction of J. Plemelj [4] shows that the module $S\begin{pmatrix}V_1,\dots,V_m\\a_1,\dots,a_m\end{pmatrix}$ is indeed n-dimensional over $\mathbb{C}[x]$. Conditions (1.4.4) imply a certain Fuchsian linear differential equation of the order n satisfied by $\vec{f}(x)$.

One should remember though, that there is no single linear differential equation of the order n, satisfied by all elements $\vec{f}(x)$ from $S\begin{pmatrix}V_1,\dots,V_m\\a_1,\dots,a_m\end{pmatrix}$. Actually the sequence $\vec{f}(x)$ from (1.4.3), which have the same monodromy properties (1.4.4) may have local exponents at $x=a_k$ that differ by integers. We call n-plets $\vec{f}(x)$, from $S\begin{pmatrix}V_1,\dots,V_m\\a_1,\dots,a_m\end{pmatrix}$ contiguous to each other. Then the system of linear differential equations of order n satisfied by contiguous systems of functions is parametrized by $\mathbb{Z}^{n(m+1)-1}$.

1.6. Let's take some basis of $S\begin{pmatrix}V_1,\dots,V_m\\a_1,\dots,a_m\end{pmatrix}$:

(1.6.1) $\qquad \vec{\phi}_k(x)=(\phi_{k,l}(x))_{l=1,\dots,n}$: $k=1,\dots,n$.

Indeed it's enough to assume that $\vec{\phi}_1(x),\dots,\vec{\phi}_n(x)$ are linearly independent over $\mathbb{C}[x]$. Let's denote by $\Phi(x)$, the fundamental matrix of $S\begin{pmatrix}V_1,\dots,V_m\\a_1,\dots,a_m\end{pmatrix}$:

(1.6.2) $\qquad \Phi(x)=(\vec{\phi}_1^t,\dots,\vec{\phi}_n^t)=(\phi_{k,l}(x))^t_{k,l=1,\dots,n}$.

Now we can present the simple proof of the Riemann theorem about the dimension of $S\begin{pmatrix}V_1,\dots,V_m\\a_1,\dots,a_m\end{pmatrix}$:

Theorem 1.6.3 [1]. For any element $\vec{f}(x)=(f_1(x),\dots,f_n(x))$ of $S\begin{pmatrix}V_1,\dots,V_m\\a_1,\dots,a_m\end{pmatrix}$ we have the representation of $\vec{f}(x)$ as a linear combination of $\vec{\phi}_1(x),\dots,\vec{\phi}_n(x)$ with coefficients from $\mathbb{C}(x)$:

(1.6.4) $\qquad \vec{f}(x)=\sum_{j=1}^n q_j(x)\vec{\phi}_j(x)$, where uniform functions $q_j(x)$ in x (i.e. rational): j=1,...,n are defined in the following way:

$$(1.6.5)\qquad q_j(x)=\frac{\det(\vec{\phi}_1,\ldots,\vec{\phi}_{j-1},\vec{f},\vec{\phi}_{j+1},\ldots,\vec{\phi}_n)}{\det(\vec{\phi}_1,\ldots,\vec{\phi}_j,\ldots,\vec{\phi}_n)}\ :\ j=1,\ldots,n.$$

Proof. It is trivial that functions $q_j(x)$ defined by (1.6.5) indeed satisfy the equation (1.6.4). Now $q_j(x)$ is analytic in $\Gamma(a_1,\ldots,a_m)$. For x_0 from (a_i,∞): i=1,...,m we consider x_0 to be a limit point of x from $\Gamma(a_1,\ldots,a_m)$ in the negative (-) and positive (+) directions. Then we have:

$$(1.6.6)\qquad q_j(x_0^-)=\frac{\det(V_i)\cdot\det(\vec{\phi}_1(x_0^+),\ldots,\vec{f}(x_0^+),\ldots,\vec{\phi}_n(x_0^+))}{\det(V_i)\cdot\det(\vec{\phi}_1(x_0^+),\ldots,\vec{\phi}_j(x_0^+),\ldots,\vec{\phi}_n(x_0^+))}=q_j(x_0^+).$$

In other words, $q_j(x)$ are rational functions: j=1,...,n.

1.7. We can consider now any fundamental matrix

$$(1.7.1)\qquad \Phi\left(\begin{matrix}V_1,\ldots,V_m\\ a_1,\ldots,a_m\end{matrix}\middle|\ x\right)=(\vec{\phi}_1(x)^t,\ldots,\vec{\phi}_n(x)^t)$$

for linear independent elements $\vec{\phi}_1(x),\ldots,\vec{\phi}_n(x)$ from $S\left(\begin{matrix}V_1,\ldots,V_m\\ a_1,\ldots,a_m\end{matrix}\right)$ (i.e. linearly independent over $\mathbb{C}(x)$). Following Lappo-Danilevski [3], we call $\Phi\left(\begin{matrix}V_1,\ldots,V_m\\ a_1,\ldots,a_m\end{matrix}\middle|\ x\right)$ a basis of (1.4.3). As a consequence of Riemann's theorem 1.6.3 we get:

Theorem 1.7.2. For any two bases

$$\Phi\left(\begin{matrix}V_1,\ldots,V_m\\ a_1,\ldots,a_m\end{matrix}\middle|\ x\right)\ \text{and}\ \tilde{\Phi}\left(\begin{matrix}V_1,\ldots,V_m\\ a_1,\ldots,a_m\end{matrix}\middle|\ x\right)\ \text{of (1.4.2) we have:}$$

$$(1.7.3)\qquad \tilde{\Phi}\left(\begin{matrix}V_1,\ldots,V_m\\ a_1,\ldots,a_m\end{matrix}\middle|\ x\right)=\Phi\left(\begin{matrix}V_1,\ldots,V_m\\ a_1,\ldots,a_m\end{matrix}\middle|\ x\right)\cdot P(x)$$

for a rational matrix-function P(x) (from $GL(n;\mathbb{C}(x))$).

Proof. Indeed for

$$(1.7.4)\qquad P(x)=\Phi\left(\begin{matrix}V_1,\ldots,V_m\\ a_1,\ldots,a_m\end{matrix}\middle|\ x\right)^{-1}\cdot\tilde{\Phi}\left(\begin{matrix}V_1,\ldots,V_m\\ a_1,\ldots,a_m\end{matrix}\middle|\ x\right)$$

we have for $x\to x_0^{\pm}$ and $x_0\varepsilon(a_i,\infty)$; i=1,...,m: $P(x_0^-)=\Phi(x_0^-)^{-1}\tilde{\Phi}(x_0^-)=\Phi(x_0^+)V_i^{-1}V_i\tilde{\Phi}(x_0^+)$ $=P(x_0^+)$.

Thus (1.7.4) is a rational function in x. The formula (1.7.3) is proved.

1.8. We remark that the definition of the basis of $S\left(\begin{matrix}V_1,\ldots,V_m\\ a_1,\ldots,a_m\end{matrix}\right)$ is the following major property of the nonsingular matrix $\Phi\left(\begin{matrix}V_1,\ldots,V_m\\ a_1,\ldots,a_m\end{matrix}\middle|\ x\right)$ with elements being functions analytic in $\Gamma(a_1,\ldots,a_m)$:

$$(1.8.1)\qquad \Phi\left(\begin{matrix}V_1,\ldots,V_m\\ a_1,\ldots,a_m\end{matrix}\middle|\ x_0^-\right)=V_i\Phi\left(\begin{matrix}V_i,\ldots,V_m\\ a_1,\ldots,a_m\end{matrix}\middle|\ x_0^+\right)$$

for x_0 on (a_i,∞): i=1,...,m where $x_0^{\pm}$ means the limit $x\to x_0$ from inside of $\Gamma(a_1,\ldots,a_m)$ in a positive (+) or negative (-) direction.

1.9. The differential equations satisfied by an element of $S\left(\begin{matrix}V_1,\ldots,V_m\\ a_1,\ldots,a_m\end{matrix}\right)$ are again consequences of Riemann's theorem 1.6.3.

Indeed if $\vec{\phi}_j(x)$ is an element of $S\begin{pmatrix} V_1,\dots,V_m \\ a_1,\dots,a_m \end{pmatrix}$, then $\frac{d}{dx}\vec{\phi}_j(x)$ is also an element of $S\begin{pmatrix} V_1,\dots,V_m \\ a_1,\dots,a_m \end{pmatrix}$, since (1.4.3) can be differentiated. If $\vec{\phi}_1(x),\dots,\vec{\phi}_n(x)$ is the basis of $S\begin{pmatrix} V_1,\dots,V_m \\ a_1,\dots,a_m \end{pmatrix}$, then according to Theorem 1.6.3:

(1.9.1) $\frac{d}{dx}\vec{\phi}_j(x) = \vec{\phi}_1(x)P_{1j}(x)+\dots+\vec{\phi}_n(x)P_{nj}(x)$, $j=1,\dots,n$; where $P_{ij}(x)$ are rational functions in x: $i,j=1,\dots,n$.

In other words, for a basis $\Phi\left(\begin{matrix} V_1,\dots,V_m \\ a_1,\dots,a_m \end{matrix}\middle| x\right)$ of $S\begin{pmatrix} V_1,\dots,V_m \\ a_1,\dots,a_m \end{pmatrix}$ we have the following linear differential equation with rational coefficients:

$$(1.9.2)\qquad \frac{d}{dx}\Phi\left(\begin{matrix} V_1,\dots,V_m \\ a_1,\dots,a_m \end{matrix}\middle| x\right) = \Phi\left(\begin{matrix} V_1,\dots,V_m \\ a_1,\dots,a_m \end{matrix}\middle| x\right) R(x)$$

for $R(x)$ from $GL(n;\mathbb{C}(x))$.

<u>1.10.</u> The rational matrix-function $R(x)$ in (1.9.2) has poles in $\mathbb{CP}^1$ of the maximal order $k\geq 1$ (the case of $k=0$ is completely trivial). Then $k-1$ can be called the order of irregularity of (1.9.2). For $k=1$, the system (1.9.2) is called a regular one, and we write the canonical regular system of the equations (1.9.2) in the form [3]

$$(1.10.1)\qquad \frac{d}{dx}\Phi(x) = \Phi(x)\cdot\sum_{i=1}^{m}\frac{U_i}{x-a_i}$$

for $n\times n$ matrices U_j called differential substitutions. The generality of the equation (1.10.1) is explained by results of Plemelj [4] and Lappo-Danilevski [3]:

<u>Theorem 1.10.2.</u> [3], [4]. For any $\begin{pmatrix} V_1,\dots,V_m \\ a_1,\dots,a_m \end{pmatrix}$ there is a basis $Y\left(\begin{matrix} V_1,\dots,V_m \\ a_1,\dots,a_m \end{matrix}\middle| x\right)$ of the module (1.4.3) satisfying the canonical regular system of the equations.

$$(1.10.3)\qquad \frac{d}{dx}Y\left(\begin{matrix} V_1,\dots,V_m \\ a_1,\dots,a_m \end{matrix}\middle| x\right) = Y\left(\begin{matrix} V_1,\dots,V_m \\ a_1,\dots,a_m \end{matrix}\middle| x\right)\cdot\sum_{i=1}^{m}\frac{U_i}{x-a_i}.$$

As a consequence of 1.7.2 we obtain

<u>Corollary 1.10.4.</u> Two canonical regular systems of equations

$$\frac{d\Psi}{dx} = \Psi\cdot\sum_{i=1}^{m}\frac{U_i}{x-a_i} \quad\text{and}\quad \frac{d\Psi^*}{dx} = \Psi^*\cdot\sum_{i=1}^{m}\frac{U_i^*}{x-a_i}$$ belong to the same module $S\begin{pmatrix} V_1,\dots,V_m \\ a_1,\dots,a_m \end{pmatrix}$ if and only if there exists a rational matrix-function $G(x)$ such that

$$\frac{dG(x)}{dx} = \sum_{i=1}^{m}\frac{GU_i^*-U_iG}{x-a_i}.$$

<u>1.11.</u> We want to remark that if the module $S\begin{pmatrix} V_1,\dots,V_m \\ a_1,\dots,a_m \end{pmatrix}$ is considered over $\mathbb{C}(x)$, then $\vec{f}(x)$, in addition to branch points at $x=a_i$, may have poles in $\mathbb{CP}^1\setminus\{a_1,\dots,a_m,\infty\}$. Still we speak about $F(a_1,\dots,a_m)$; this ambiguity will cause serious troubles later.

§2. <u>The Classical theory of scalar Fuchsian linear differential equations</u>.

<u>2.1.</u> We repeat some known facts about Fuchsian scalar linear differential equations. Though the best way to consider these equations is to use matrix linear differential equations of the first order (cf. E. Hille [5]), we formulate the results of L. Fuchs'

theory independently. For the proofs see E. Hille [5], A. Forsyth [6], E. Ince [7], B. Golubev [8].

2.2. Let us consider a linear differential equation of the order n with rational function coefficients

(2.2.1) $\quad w^{(n)}(z)+P_1(z)w^{(n-1)}(z)+\ldots+P_n(z)w(z) - 0$

for $w^{(i)}(z) = d^i/dz^{i}\cdot w(z)$ and $P_j(z)\varepsilon\mathbb{C}(z)$: $j=1,\ldots,n$. We are interested in equations (2.2.1) with regular singularities only. First of all, all singularities of solutions are singularities of $P_j(z)$: $j=1,\ldots,n$. We remind that the singularity of the system (2.2.1) is called apparent [6] if it is a singularity of some of $P_j(z)$: $j=1,\ldots,n$ but not of the solutions of (2.2.1). The notion of a regular singularity for (2.2.1) is defined in many equivalent ways. The simples way is described by the roots of an indicial equation [6],[7]:

Definition 2.2.2. Let a be a singularity of (2.2.1), i.e., a singularity of one of the $P_j(z)$: $j=1,\ldots,n$. We look for formal solutions $w(z)$ of (2.2.1) of the form

(2.2.3) $\quad w(z) = (z-a)^r\cdot w_0(z) = (z-a)^r\cdot(a_0+a_1(z-a)+\ldots)$, $a_0\neq 0$, where $w_0(z)$ is an invertible element of $\mathbb{C}[[z-a]]$. Substituting $w(z)$ into (2.2.1) we obtain some algebraic equation $P_a(r)=0$ characterizing possible values of r. This equation is called an indicial equation of (2.2.1) at a and has degree $\leq n$. If the degree $P_a(r)$ is exactly n, then z=a is called a regular singularity of (2.2.1).

If z=a is a regular singularity of (2.2.1), then a fundamental system of solutions of the form (2.2.3) exists. However one must be aware of the case, when the roots of $P_a(r)$ differ by rational integers and logarithmic terms may appear in the expansion of $w(z)$ (see [5], [6], [7]). Using the method of G. Frobenius and L. Fuchs [9], we obtain the fundamental set of solutions at z=a of (2.2.1) as products of $(z-a)^r\cdot\log^k(z-a)$ by a convergent power series at z=a:

Theorem 2.2.4. Let z=a be a singularity of (2.2.1) and $r_1,\ldots,r_n$ be roots of an indicial equation $P_a(r)=0$ of (2.2.1) at z=a. Let $\{r_{1,1},\ldots,r_{1,k_1}\},\ldots,\{r_{l,1},\ldots r_{l,k_l}\}$ be l blocks of elements of $\{r_1,\ldots,r_n\}$, $n=k_1+\ldots+k_l$ such that in each block $\{r_{i,1},\ldots,r_{i,k_i}\}$ elements are equivalent (mod $\mathbb{Z}$), and elements of different blocks are nonequivalent (mod $\mathbb{Z}$), i.e. $r_{i,p}-r_{j,q}\varepsilon\mathbb{Z}$ if and only if i=j for $p=1,\ldots,k_i$; $q=1,\ldots,k_j$. Let the order of $r_1,\ldots,r_n$ be chosen in such a way that $r_{i,1}\geq,\ldots,\geq r_{i,k_i}$: $i=1,\ldots,l$. Then there exists a fundamental system of solutions of (2.2.1) having the form

(2.2.5) $\quad w_{1,p}(z) = (z-a)^{r_{i,p}}u^p_{i,p}(z)+(z-a)^{r_{i,p+1}}\cdot\log(z-a)\cdot u^p_{i,p+1}(z)+\ldots$

$$\ldots +(z-a)^{r_{i,k_i}}\log^{k_i-p}(z-a)\cdot u^p_{i,k_i}(z)$$

where $u^p_{i,q}(z)$ are analytic functions at z=a: $p\leq q\leq k_i$; $p=1,\ldots,k_i$; $i=1,\ldots,l$.

The origin of the partition of $\{r_1,\ldots,r_n\}$ into blocks is connected with the normal form of the monodromy matrix of the system of (2.2.1) at z=a, [10], [11]. Exactly, let $\vec{w}(z) = (w_1(z),\ldots,w_n(z))$ be the fundamental system of solutions of (2.2.1). We consider now any simple closed contour γ surrounding a, and not containing any singularity of (2.2.1) other than a. Let the function $\vec{W}(z) = (W_1(z),\ldots,W_n(z))$ be what $\vec{w}(z)$ becomes after z has described the circuit γ. Since the coefficients $P_j(z)$: j=1,...,n are unaltered by the description of the circuit γ, the equation as a whole is unchanged. Thus $\vec{W}(Z)$ is expressed in terms of $\vec{w}(z)$:

(2.2.6) $\quad \vec{W}(Z)^t = V_\gamma \cdot \vec{w}(z)^t$.

Let the canonical Jordan form of V_γ be the matrix $S_\gamma V_\gamma S_\gamma^{-1} = \begin{pmatrix} J_{\lambda_1} & & 0 \\ & \ddots & \\ 0 & & J_{\lambda_l} \end{pmatrix}$, where J_{λ_i} is a Jordan block corresponding to the number λ_i, and of the sizes k_i: i=1,...,l and $k_1+\ldots+k_l = n$. Now as is well-known [2], [7], [9], the roots r_j of the indicial equation are connected with eigenvalues μ_j of V_γ by a famous formula $\mu_j = e^{2\pi\sqrt{-1}r_j}$: j=1,...,n. Now $\{r_1,\ldots,r_n\}$ is divided into l blocks corresponding to the l Jordan blocks J_{λ_i} of $S_\gamma V_\gamma S_\gamma^{-1}$ according to a formula $e^{2\pi\sqrt{-1}r_{i,p}} = \lambda_i$: p=1,...,$k_i$; i=1,...,l.

2.3. There is another characterization of a regular singularity due to L. Fuchs [9]:

Lemma 2.3.1. The equation (2.2.1) has a regular singularity at its singularity z=a if and only if $P_i(z) = (z-a)^{-i}\cdot p_i(z)$ for $p_i(z)$ is analytic at z=a.

It is easy to get the most general form of the equations (2.2.1) having only regular singularities in the whole z-plane $\mathbb{CP}^1$ (including z=∞) [6], [8], [9]. These equations are called Fuchsian, and their form is as follows:

Theorem 2.3.2. Fuchsian linear differential equations of the order n have the following form:

$$(2.3.3) \qquad w^{(n)}(z) + \frac{Q_1(z)}{P(z)} w^{(n-1)}(z) + \frac{Q_2(z)}{P(z)^2} w^{(n-1)}(z) + \frac{Q_n(z)}{P(z)^n} w(z) = 0$$

for $P(z) = \prod_{j=1}^m (z-a_j)$, of the degree m and polynomials $Q_i(z)$ of degree $\leq i(m-1)$: i=1,...,n.

Now we make a few remarks about the local monodromy of Fuchsian linear differential equations. Due to the form (2.2.5) of the fundamental system of solutions of (2.2.1), we call the roots $\{r_1,\ldots,r_n\}$ of the indicial equation of (2.2.1) at z=a, local exponents of (2.2.1) corresponding at z=a. It's clear from the form (2.2.5), that the local exponents of (2.2.1) corresponding to z=a are independent of the choice of the fundamental solution of (2.2.1). Let's consider the local exponents of the equation (2.3.3).

We have the set of all singularities of (2.3.3) as the set $\{a_1,\ldots,a_m\}$, possibly including ∞.

If $$\frac{Q_i(z)}{P(z)^i} = \sum_{s=1}^m \frac{P_{m,s}}{(z-a_s)^i} + \frac{Q'_i(z)}{P(z)^{i-1}} : i=1,\ldots,n,$$

then the indicial equation corresponding to $z=a_s$ has the form

$$(r)_n + \sum_{i=1}^n P_{is}\,(r)_{n-i} = 0,$$

where $(r)_k = r\ldots(r-k+1)$. Let's denote by $\lambda_i^{(a_s)}$: i=1,...,n, local exponents of (2.3.3) at $z=a_s$: s=1,...,m, and by $\lambda_i^{(\infty)}$: i=1,...,n, the local exponents of (2.3.3) at z=∞ (with the local parameter 1/z for a large z). Then we have

$$P_{1s} - \frac{n(n-1)}{2} = -\sum_{i=1}^{n} \lambda_i^{(a_s)} : s=1,\dots,m$$

and

$$\sum_{s=1}^{m} P_{1s} - \frac{n(n-1)}{2} = \sum_{i=1}^{n} \lambda_i^{(\infty)} .$$

Corollary 2.3.4. The total sum of all local exponents of (2.3.3) corresponding to all singularities at $\mathbb{CP}^1$ is $(m-1)\cdot\frac{n(n-1)}{2}$:

$$\sum_{s=1}^{m} \sum_{i=1}^{n} \lambda_i^{(a_s)} + \sum_{i=1}^{n} \lambda_i^{(\infty)} = \frac{n(n-1)}{2}(m-1).$$

Another way to understand local multiplicities is to reduce the scalar Fuchsian equation (2.3.3) to a matrix linear differential equation of the first order with poles of the first order [5].

§3. The Global and local monodromy theory of matrix linear differential equations with regular singularities.

3.0. In this chapter we describe the monodromy theory of matrix linear differential equations with regular singularities following the algorithmic approach of Lappo-Danilevski [10]. We formulate Poincare-Lappo-Danilevski solutions of the direct and inverse monodromy problems in terms of expansions in polylogarithmic functions. In our presentation we follow papers [12] - [14], where some of the applications of the algorithmic approach to Feynman integrals are outlined.

3.1. We consider a canonical regular system (1.10.1) with differential substitutions U_j at regular singularities a_j: $j=1,\dots,m$:

(3.1.1) $$\frac{dY(x)}{dx} = \sum_{j=1}^{n} \frac{YU_j}{x-a_j} .$$

The system (3.1.1) with regular singularities at $x=a_j$ gives rise to a certain fixed representation $\begin{pmatrix} V_1,\dots,V_m \\ a_1,\dots,a_m \end{pmatrix}$ of $\pi^1(\mathbb{CP}^1\setminus\{a_1,\dots,a_m\})$, of 1.3 and 1.4. This means that $Y(x)$ as a matrix function on $\Gamma(a_1,\dots,a_m)$ satisfies the following monodromy property (cf. (1.4.4)):

(3.1.2) $$Y(x_0^-) = V_j\cdot Y(x_0^+)$$

for x_0 on the left (-) and right (+) sides of the cut (a_j,∞): $j=1,\dots,m$. In other words, let us consider $Y(x)$ to be a multi-valued function on $\mathbb{CP}^1$ with branch points at $a_1,\dots,a_m,\infty$, and let $\gamma_1,\dots,\gamma_m,\gamma_\infty$ be a fixed basis of $\pi^1(\mathbb{CP}^1\setminus\{a_1,\dots,a_m,\infty\})$, $\gamma_1+\dots+\gamma_m+\gamma_\infty = 0$. Then after enclosing a circuit along γ_i, the initial value $Y(x)$ changes via a linear transformation (1.3.2):

(3.1.3) $$Y(x) \underset{\gamma_j}{\mapsto} V_j Y(x) : j=1,\dots,m$$

and

(3.1.4) $$Y(x) \underset{\gamma_\infty}{\mapsto} V_\infty Y(x) \text{ with}$$

(3.1.5) $$V_i\dots V_m\cdot V_\infty = I.$$

3.2. The method of Lappo-Danilevsky enables us to find monodromy matrices V_j (so-called integral substitutions) through the coefficients U_j and vice versa. The natural tools for this are polylogarithmic functions [10], which were studied beginning with Euler and Abel.

For a fixed b different from $a_1,\dots,a_m,\infty$ we define as in [10] the system of polylogarithmic functions

(3.2.1) $\quad L_b(a_{j_1},\dots,a_{j_\nu}|x) \quad (j_1,\dots,j_\nu=1,\dots,m;\ \nu=1,2,3,\dots)$

by induction

$$L_b(a_{j_1}|x)=\int_b^x \frac{dx_1}{x_1-a_{j_1}} = \log\frac{x-a_{j_1}}{b-a_{j_1}};$$

(3.2.2)
$$L_b(a_{j_1},\dots,a_{j_\nu}|x)=\int_b^x \frac{L_b(a_{j_1},\dots,a_{j_{\nu-1}}|x_1)\,dx_1}{x_1-a_{j_\nu}}$$

Theorem 3.2.3. The fundamental solution $Y_b(x)$ of (3.1.1) satisfying

(3.2.4) $\quad Y_b(b) = I$

is represented as a series

(3.2.5) $$Y_b(x)=I+\sum_{\nu=1}^{\infty}\sum_{j_1,\dots,j_\nu}^{(1,\dots,m)} U_{j_1}\dots U_{j_\nu}\cdot L_b(a_{j_1},\dots,a_{j_\nu}|x),$$

that is entire in U_j and is uniformly convergent with respect to x in any finite domain D in $\Gamma(a_1,\dots,a_m)$.

Now we will express the monodromy matrices similarly. Let γ_i be closed contours from x_0 traveling around a_i and returning back to x_0 : i=1,...,m. Then one defines [10]:

$$P_j(a_{j_1}|b)=\int_{\gamma_j}\frac{dx}{x-a_{j_1}} = \begin{cases}2\pi\sqrt{-1} & : j=j_1\\ 0 & j\neq j_1\end{cases};$$

(3.2.6)
$$P_j(a_{j_1},\dots,a_{j_\nu}|b)=\int_{\gamma_j}\frac{L_b(a_{j_1},\dots,a_{j_{\nu-1}}|x)dx}{x-a_{j_\nu}}$$

Theorem 3.2.7 [10]. The monodromy matrices V_j in (3.1.3) which correspond to γ_j and to the fundamental solution $Y_b(x)$ of (3.1.1) satisfying (3.2.4) are defined by

(3.2.8) $$V_j = I+\sum_{\nu=1}^{\infty}\sum_{j_1,\dots,j_\nu}^{(1,\dots,m)} U_{j_1}\dots U_{j_\nu}\cdot P_j(a_{j_1},\dots,a_{j_\nu}|b) : j=1,\dots,m.$$

3.3. For applications it is necessary to consider together with monodromy matrices U_j, the so-called exponents W_j at $x=a_j$: j=1,...,m.

Definition 3.3.1. Let Y(x) be a (fundamental) solution of (3.1.1). We call a matrix W_j the exponent of Y(x) at $x=a_j$ if

(3.3.2) $\quad Y(x) = (x-a_j)^{W_j}H_j(x)$

where $H_j(x)$ and $H_j(x)^{-1}$ are holomorphic at $x=a_j$: j=1,...,m. At the infinite point the exponent W_∞ is defined by

(3.3.3) $\quad Y(x) = (x^{-1})^{W_\infty}H_\infty(x),$

where $H_\infty(x)$ and $H_\infty(x)^{-1}$ are holomorphic at $x=\infty$.

In the case when the exponents W_j of Y(x) exist, the relation with the corresponding monodromy matrices V_j of $Y(x)$ in (3.1.2), is very simple;

(3.3.4) $$e^{2\pi\sqrt{-1}W_j} = V_j \; : \; j=1,\ldots,m,\infty.$$

Theorem 3.3.5 [10]. Let differential substitution U_j in (3.1.1) be in the neighborhood of a zero matrix. For a fixed b we consider the monodromy matrices V_j of the solution $Y_b(x)$ of (3.1.1) defined in (3.2.8). Then V_j are in the neighborhood of a unit matrix and we can define the following matrices:

(3.3.6) $$W_j = \frac{1}{2\pi\sqrt{-1}} \log V_j = \frac{1}{2\pi i} \sum_{\nu=1}^{\infty} \frac{(-1)^{\nu-1}}{\nu} (V_j - I)^{\nu}$$

for $j=1,\ldots,m,\infty$. Then $W_j = W_j(b)$, defined by (3.3.6), is the unique exponent of the solution $Y_b(x)$ of (3.1.1), satisfying (3.2.4), at singularity $x=a_j$: $j=1,\ldots,m,\infty$.

Moreover, we can find exponents W_j of $Y(x)$ at $x=a_j$: $j=1,\ldots,m$ in the general case, when U_j are no longer close to a zero matrix. According to Theorem VIII, Ch. II [10] the exponents W_j are meromorphic functions of the differential substitutions $U_1,\ldots,U_m$. The only singularities of W_j as functions of $U_1,\ldots,U_m$, are those values of matrices U_j for which there are two eigenvalues of U_j which are different by a non-zero rational integer. In addition to a fundamental solution $Y_b(x)$ of (3.1.1) normed at x=b: (3.2.4), we can consider the so-called metacanonical solutions $\theta_j(x)$ of (3.1.1) normed at $x=a_j$ [3]. These solutions have the form

(3.3.7) $$\theta_j(x) = (x-a_j)^{K_j} \cdot \tilde{\theta}_j(x)$$

for $\tilde{\theta}_j(a_j) = I$.

Lemma 3.3.8. If U_j don't have eigenvalues that differ by non-zero integers, then there exists a unique metacanonical matrix

(3.3.9) $$\theta_j(x) - (x-a_j)^{U_j} \cdot \tilde{\theta}_j(x), \; \tilde{\theta}_j(a_j) = I \; : \; j=1,\ldots,m.$$

The same statement is true for $x=\infty$ if we put

(3.3.10) $$U_\infty = -\sum_{j=1}^{m} U_j,$$

and so

(3.3.11) $$\theta_\infty(x) = (x^{-1})^{U_\infty} \cdot \tilde{\theta}_j(x), \; \tilde{\theta}_j(\infty) = I.$$

3.4. We devote this part of the paper to establishing some simple relations between differential, integral and exponential matrices of the system (3.1.1). First of all, we define a differential substitution U_∞ at $x=a_\infty=\infty$ by (3.3.10), so

(3.4.1) $$\sum_{j=1}^{m,\infty} U_j = 0$$

Theorem 3.4.2 [10]. For any solution $Y(x)$ of (3.1.1) and monodromy problem (3.1.2): the eigenvalues of the monodromy matrix V_j are equal to the eigenvalues of $e^{2\pi\sqrt{-1}U_j}$; and eigenvalues of the exponent W_j are equal to eigenvalues of U_j: $j=1,\ldots,m,\infty$.

Other relations between U_j, V_j and W_j are contained in (3.1.5), (3.3.4) and (3.4.1). To simplify computation we refer the reader to [10], where the theory of functions of the matrix argument is presented. Here are some of the examples. If W=S diag $(\lambda_1,\ldots,$

$,\ldots,\lambda_n)\cdot S^{-1}$, then $x^W=S\cdot\mathrm{diag}(x^{\lambda_1},\ldots,x^{\lambda_n})\cdot S^{-1}$. We can determine all possible exponents W_j satisfying equations $e^{2\pi\sqrt{-1}W_j}=V_j$. Let $V_j=S\cdot[J_{\rho_i}(\omega_i),\ldots,J_{\rho_l}(\omega_l)]\cdot S^{-1}$ where $J_{\rho_i}(\omega_i)$ are Jordan blocks of size S_i corresponding to ω_i: $i=1,\ldots,l$; $\rho_i+\ldots+\rho_l=n$. Then for a fixed solution W_j of the equation (3.3.4), all other solutions W_j' of $e^{2\pi\sqrt{-1}W_j'}=V_j$ can be represented as

$$(3.4.3)\qquad W_j' = W_j + S\cdot[I_{\rho_i}\cdot r_1,\ldots,I_{\rho_l}\cdot r_l]\cdot S^{-1},$$

where I_{ρ_i} is a unit matrix of the size ρ_i and r_i is a rational integer: $i=1,\ldots,l$. Also matrices W_j and W_j' commute for all solutions W_j, W_j' of (3.3.4).

3.5. Let us make a few remarks about the solution of the inverse monodromy problem (namely the Riemann-Hilbert problem) of the reconstruction of the system of functions satisfying the given monodromy properties (1.4.4) or (3.1.3). The solution of this problem was proposed by Lappo-Danilevski [10] as the result of the inversion of the series (3.2.5), (3.2.8) and (3.3.6). First of all, for any object (1.3.3): $\begin{pmatrix}V_1,\ldots,V_m\\a_1,\ldots,a_m\end{pmatrix}$ and V_∞ satisfying (3.1.5), there exists a system of differential substitutions U_j: $j=1,\ldots,m$ such that the solution $Y(x)$ of the equation (3.1.1):

$$(3.5.1)\qquad Y_x = Y\sum_{j=1}^{m}\frac{U_j}{x-a_j}$$

satisfies the monodromy properties (3.1.2). Moreover, these differential substitutions U_j: $j=1,\ldots,n$ can be chosen in such a way that the fundamental solution $Y(x)$ of (3.5.1) has exponents W_j at $x=a_j$: $j=1,\ldots,m,\infty$ and $e^{2\pi\sqrt{-1}W_j}=V_j$: $j=1,\ldots,m$. Also in [10] the problem of the reconstruction of (3.1.1) having exponents W_j at $x=a_j$ is considered. First of all, the exponential matrices W_j at $x=a_j$: $j=1,\ldots,m,\infty$ uniquely determine the system (3.1.1). For given matrices $W_1,\ldots,W_m$ we can always find a system (3.1.1) having exponents W_j at $x=a_j$, having monodromy matrices $V_j=e^{2\pi\sqrt{-1}W_j}$ at $x=a_j$: $j=1,\ldots,m$, and having some exponent W_∞ at $x=\infty$. The exponent W_∞ is not arbitrary since

$$e^{2\pi\sqrt{-1}W_1}\cdots e^{2\pi\sqrt{-1}W_m}\cdot e^{2\pi i\sqrt{-1}W_\infty} = 1.$$

However, due to (3.4.3), the choice of W_∞ depends on an n-plet of rational integers. The variation of W_∞ with the fixed $W_1,\ldots,W_m$ is exactly equivalent to the construction of the one-point Padé approximation at $x=\infty$.

§4. The exponents of matrix linear differential equations in the irregular case.

4.0. We consider here the formal expansion of the solution of an arbitrary linear differential equation, with rational function coefficients, at the neighborhood of the irregular singularity [10].

4.1. Let's start with the most general matrix linear differential equation with the rational function coefficients:

$$(4.1.1)\qquad \frac{dY}{dx} = \sum_{j=1}^{m}\sum_{r=1}^{s}\frac{Y\cdot U_j^{(r)}}{(x-a_j)^r}\ ;$$

then matrices $U_j^{(1)},\dots,U_j^{(s)}$ are called differential substitutions of rank s-1 at $x=a_j$. Let Y(x) be a fundamental solution of (4.1.1) normed by (3.2.4) at x=b. Then making a circuit γ_i around the singularity a_i, we transform the fundamental solution Y(x) linearly as in (3.1.2) - (3.1.5):

$$Y_b(x) \underset{\gamma_i}{\longmapsto} V_i Y_b(x) \tag{4.1.2}$$

for the basis γ_i of $\pi^1(\mathbb{CP}^1 \setminus \{a_1,\dots,a_m,\infty\})$, $\gamma_1+\dots+\gamma_m+\gamma_\infty = 0$.

Then formally the solution $Y_b(x)$ can be represented in the neighborhood of $x=a_j$ as

$$Y_b(x) = G_b^{(j)}(x)\cdot\overline{Y}_b^{(j)}(x) \tag{4.1.3}$$

where $G_b^{(j)}(x)$ satisfies the system of equations

$$\frac{dG_b^{(j)}(x)}{dx} = \sum_{r=1}^{s} \frac{G_b^{(j)}(x)\cdot W_j^{(r)}}{(x-a_j)^r} \tag{4.1.4}$$

and

$$G_b^{(j)}(x) = (x-a_j)^{W_j^{(1)}} \cdot\tilde{G}_b^{(j)}(x); \tag{4.1.5}$$

where

$$\tilde{G}_b^{(j)}(x) = I + \sum_{p=1}^{\infty} C_j^{(p)}\cdot \frac{1}{(x-a_j)^p} \tag{4.1.6}$$

is an entire function of $(x-a_j)^{-1}$. In (4.1.3) both $\overline{Y}_b^{(j)}(x)$ and $\overline{Y}_b^{(j)}(x)^{-1}$ are holomorphic at $x=a_j$. The matrices $W_j^{(r)}$: r=1,...,s are called exponents (of the rank s-1) at $x=a_j$: $j=1,\dots,m,\infty$. Only the exponents $W_j^{(1)}$ are important for the monodromy matrices V_j:

$$e^{2\pi\sqrt{-1}W_j^{(1)}} = V_j \ :\ j=1,\dots,m,\infty. \tag{4.1.7}$$

<u>4.2.</u> Again, in the case when $U_j^{(r)}$ $(j=1,\dots,m;\ r=1,\dots,s)$ is in the neighborhood of the zero matrix, all the exponents $W_j^{(r)}$: $j=1,\dots,m;\ r=1,\dots,s$ exist, and representations (4.1.3) - (4.1.6) take place. However in this case the exponents are not unique and we have:

<u>Theorem 4.2.1.</u> Let's assume that $Y_b(x)$ satisfies (4.1.3) - (4.1.6) at $x=a_j$ and $Y_b(x)$ admits also representation

$$Y_b(x) = G'(x)\,\overline{Y}'(x) = (x-a_j)^{W_j^{(1)'}}\cdot\overline{G}'(x)\overline{Y}'(x), \tag{4.2.2}$$

$$\overline{G}'(x) = I + \sum_{p=1}^{\infty} C'_p\,(x-a_j)^{-p} \tag{4.2.3}$$

and $\overline{G}'(x)$ is an entire function in $(x-a_j)^{-1}$ and $\overline{Y}'(x)$ and $\overline{Y}'(x)^{-1}$ are holomorphic at $x=a_j$. Then $G'(x)$ satisfies an equation

$$\frac{dG'(x)}{dx} = \sum_{r=1}^{s} \frac{G'(x)W_j^{(r)'}}{(x-a_j)^r}\ . \tag{4.2.4}$$

with $e^{2\pi\sqrt{-1}W_j^{(1)}} = e^{2\pi\sqrt{-1}W_j^{(1)'}} = V_j$ and the equation

$$(4.2.5)\qquad \frac{dX}{dx} = \sum_{r=1}^{s} \frac{X\cdot W_j^{(r)'} - W_j^{(r)}X}{(x-a_j)^r}$$

admits a solution $X(x)$ such that X and X^{-1} are polynomial in x. Oppositely, let (4.2.5) admit a solution X such that X and X^{-1} are polynomials in x. Then $Y_b(x)$, together with representation (4.1.3) - (4.1.6) admits another representation of the form (4.2.2) - (4.2.4).

4.3. The investigation of the system (4.2.5) is closely connected with studies of contiguous relations and recurrent formulae for Padé approximations. In both cases we must consider the following matrix equation on an unknown matrix X[10]:

$$(4.3.1)\qquad [X,U] = \lambda X.$$

If $U = S\,\mathrm{diag}(\sigma_1,\ldots,\sigma_n)\cdot S^{-1}$, then a solution is possible only for $\lambda = \sigma_1-\sigma_k$. In this case let $(\tilde{C})_{ij} = 0$ if $i\neq j$ and $\{i,j\} \neq \{k,1\}$. If $C = S\tilde{C}S^{-1}$, then

$$(4.3.2)\qquad X = (\sigma_1-\sigma_k)\cdot[C,U] + [C,\ [C,U]]$$

satisfies

$$(4.3.3)\qquad [X,U] = (\sigma_1-\sigma_k)X.$$

§5. Scalar Fuchsian linear differential equations and their relation with matrix regular linear differential systems.

5.0. One of the most crucial problems in the study of n-point Padé approximations is the analysis of the local monodromies of solutions of Fuchsian equations. We will consider systems of the scalar Fuchsian linear differential equations satisfied by systems of contiguous functions (see §2). We are in the same situation as in §1 when the system of n scalar Fuchsian equations is equivalent to the matrix linear differential equation with regular singularities. However passing from the scalar to the matrix case and vice versa, we are losing local exponents and get apparent singularities. E.g., the Schlesinger equations [12] in the matrix case correspond to variations of the matrix coefficients with respect to real singularities; while in the scalar case PainlevéV[12] correspond to the variation of the positions of the apparent singularities as functions of real singularities.

5.1. First of all we want to imbed the scalar Fuchsian equation (2.3.3) into the canonical regular matrix linear differential equation (1.10.1). Let's consider the fundamental system $\vec{y}(x) = (y_1(x),\ldots,y_n(x))$ of the equation (2.3.3) of the order n:

$$(5.1.1)\qquad y^{(n)}(x) + \sum_{i=1}^{n} \frac{Q_i(x)}{P(x)^i}\, y^{(n-i)}(x) = 0,$$

where $P(x)=\prod_{j=1}^{m} (x-a_j)$ and $d(Q_i) \leq i(m-1)$: $i=1,\ldots,n$. Let $R_k(x)$: $k=1,\ldots,n$ be polynomials relatively prime with $P(x)$, whose degrees satisfy inequalities

$$(5.1.2)\qquad d(R_{n-i+1}) > d(Q_i) - d(P):\ i=1,\ldots,n;$$

$$(5.1.3)\qquad d(R_{k+1}) + d(P) > d(R_k):\ k=1,\ldots,n-1$$

and $R_n\equiv 1$. The system (5.1.2) - (5.1.3) is trivially solvable e.g., with $d(R_{n-i+1}) =$

$= (i-1)(m-1)$: $i=1,\dots,n$. We define

(5.1.4) $$\phi_{i,k}(x) = y_i^{(k-1)}(x)\cdot P(x)^{k-1}\cdot R_k(x): \ k=1,\dots,n;\ i=1,\dots,n.$$

Then we have

(5.1.5) $$\phi_{i,k,x} = \phi_{i,k+1}\frac{R_k}{P\cdot R_{k+1}} + \phi_{i,k}\left\{\frac{R_{kx}}{R_k} + k\frac{P_x}{P}\right\} : \ k=1,\dots,n-1$$

and

(5.1.6) $$\phi_{i,n,x} = \phi_{i,n}\cdot n\frac{P_x}{P} - \sum_{j=1}^{n} (-1)\frac{Q_j}{P\cdot R_{n-j+1}}\cdot\phi_{i,n-j+1}$$

The system (5.1.5) - (5.1.6) for the functions (5.1.4) can be represented in the canonical regular form (1.10.1). Indeed, let $\{c_1,\dots,c_d\}$ be the set of (simple) zeroes of the polynomials $R_k(x)=0$: $k=1,\dots,n-1$ and

(5.1.7) $$\Phi(x) = (\phi_{i,k}(x))^n_{i,k=1}.$$

Then the system (5.1.5) - (5.1.7) is equivalent to the canonical regular system for $\Phi(x)$

(5.1.8) $$\frac{d}{dx}\Phi(x) = \Phi(x)\left\{\sum_{j=1}^{m}\frac{U_j}{x-a_j} + \sum_{t=1}^{d}\frac{A_t}{x-c_t}\right\},$$

where the expressions of U_j, A_t follow from (5.1.5) - (5.1.6) and $x=c_t$ are, of course, apparent singularities.

5.2. Let us consider the situation when all R_i are relatively prime polynomials with simple zeroes, so that $d = \sum_{k=1}^{n} (n-k)(m-1) = (m-1)\frac{n(n-1)}{2}$. We can easily compute the eigenvalues of exponential matrices of the system (5.1.8) for (5.1.7) at $x=a_\nu$: $\nu=1,\dots,m,\infty$ and $x=c_t$: $t=1,\dots,d$. Let $\lambda_1^{(a_\nu)},\dots,\lambda_n^{(a_\nu)}$ be local exponents of (5.1.1) at $x=a_\nu$: $\nu=1,\dots,m,\infty$ (with $a_\infty=\infty$). Then the form of (5.1.4), (5.1.7) shows that the exponential matrix W_ν of (5.1.7) at $x=a_\nu$ has eigenvalues $\lambda_1^{(a_\nu)},\dots,\lambda_n^{(a_\nu)}$: $\nu-1,\dots,m$. The exponential matrix W_∞ of (5.1.7) at $x=\infty$ has eigenvalues $\lambda_1^{(\infty)}-(n-1)(m-1),\dots,\dots,\lambda_n^{(\infty)}-(n-1)(m-1)$, and the exponential matrix W_{c_t} of (5.1.8) at $x=c_t$ has eigenvalues: $0,0,\dots,0,1$ for $t=1,\dots,(m-1)\frac{n(n-1)}{2}$. By (3.4.1) we have $\sum_{j=1}^{m,\infty} U_j + \sum_{t=1}^{d} A_t = 0$ and according to (3.4.2), the eigenvalues of differential substitutions are the same as those of exponential matrices. Thus the sum of the eigenvalues of all exponential matrices of (5.1.8) is zero. This gives us

$$\sum_{\nu=1}^{m,\infty}\sum_{i=1}^{n}\lambda_i^{(a_\nu)} = \frac{n(n-1)}{2}(m-1),$$

and so we obtain another proof of (2.3.5).

In this way we transform a Fuchsian linear differential equation (5.1.1) of the order n with m finite regular singularities into a canonical regular matrix system (1.10.1) of the order n having m regular singularities and $(m-1)\frac{n(n-1)}{2}$ additional apparent singularities.

5.3. It is quite natural to transform any canonical regular system (1.10.1) into n scalar Fuchsian linear differential equations of the order n, satisfied by contiguous systems of functions. If we do this, we again add systems of apparent singularities to regular singularities. Such reduction is carried on in §6 (or §7) based on Riemann theorem 1.6.3. This reduction can be done straightforwardly taking up to n-1 derivatives of (1.10.1).

§6. Matrices of the local exponents of linear differential equations.

6.0. Now we want to define eigenvalues of the general matrix system of linear differential equations with regular singularities of the form (1.9.2). We start first of all with the most general matrix linear differential equation

$$\frac{dY(x)}{dx} = Y(x)\cdot A(x) \tag{6.0.1}$$

for the n x n matrix $A(x)$ with rational function coefficients. We assume that $\{a_1,\dots,a_m\}$ are the only singularities of the coefficients of $A(x)$. The concept of the monodromy group of the equation (6.0.1) is applied to an arbitrary equation with regular (§3) or irregular singularities (§4). If $\{\gamma_1,\dots,\gamma_m,\gamma_\infty\}$ is the basis of $\pi^1(\mathbb{CP}^1\setminus\{a_1,\dots,a_m,\infty\})$, $\gamma_1+\dots+\gamma_m+\gamma_\infty = 0$, then a fixed fundamental solution $Y(x)$, by analytic continuation along a closed path γ_ν undergoes a linear transformation (1.3.2):

$$Y(x) \underset{\gamma_\nu}{\longmapsto} V_\nu Y(x): \nu=1,\dots,m,\infty. \tag{6.0.2}$$

The formula (6.0.2) defines a representation of $\pi^1(\mathbb{CP}^1\setminus\{a_1,\dots,a_m,\infty\})$ in $GL(n;\mathbb{C}^1)$; and the corresponding subgroup $GL(n;\mathbb{C}^1)$ generated by $\{V_1,\dots,V_m,V_\infty\}$ is called the monodromy group of (6.0.1). The only general relation is

$$V_1\dots V_m V_\infty = I. \tag{6.0.3}$$

The existence of the monodromy matrices at $x=a_\nu$ makes it possible to define exponential matrices W_ν at $x=a_\nu$ as in 3.3.1 or as in (4.1.5) - (4.1.7).

Our main interest is concentrated on the eigenvalues of W_ν only. The exponential matrix W_ν doesn't necessarily exist nor is it unique, but the eigenvalues of W_ν do exist. E.g. in the case of the canonical regular system (1.10.1) the eigenvalues of W_ν are simply eigenvalues of U_ν by the Theorem 3.4.2 ($\nu=1,\dots,m,\infty$). Again by the Theorem 4.2.1, the eigenvalues of $W_\nu^{(1)}$ are determined by the system (6.0.1) up to the n-plet of rational integers (cf. (3.4.3)).

6.1. Let us proceed with the most general system (6.0.1) in the case of regular singularities only, and with the singular set $\{a_1,\dots,a_m,\infty\}$. For a given fundamental solution $Y(x)$ of (6.0.1) we have the monodromy matrices $V_1,\dots,V_m,V_\infty$ at $x=a_1,\dots,a_m,\infty$. If

$$Y(x) = (\vec{y}_1^{\,t},\dots,\vec{y}_n^{\,t}), \tag{6.1.1}$$

then each $\vec{y}_i(x)=(y_{i,1}(x),\dots,y_{i,n}(x))$ is an element of $S\begin{pmatrix}V_1,\dots,V_m\\ a_1,\dots,a_m\end{pmatrix}$ corresponding to $\begin{pmatrix}V_1,\dots,V_m\\ a_1,\dots,a_m\end{pmatrix}$: $i=1,\dots,n$. Now $\vec{y}_i(x)$ is, according to §1, 1.6.3 or [2], the system of solutions of the Fuchsian linear differential equations $L_i\{y\}=0$ of order n:

$$L_i\ \{y_{i,j}(x)\} = 0 : j=1,\dots,n;\ i=1,\dots,n. \tag{6.1.2}$$

We expand $\{a_1,\dots,a_m,\infty\}$ to include the set of all apparent singularities of all the equations (6.1.2). The monodromy substitutions corresponding to apparent singularities are trivial; however the addition of apparent singularities may change the local multiplicities by additive integers. For each system $\vec{y}_i(x)$, the monodromy matrices $V_1,\dots,V_m,V_\infty$ at $x=a_1,\dots,a_m,\infty$ are the same. Let $V_\nu=S_\nu^{-1}[J_{k_1}(\rho_1^\nu),\dots,J_{k_l}(\rho_l^\nu)]\cdot S_\nu$ and $\mu_1^{(a_\nu)},\dots,\mu_n^{(a_\nu)}$ be eigenvalues of V_ν arranged in l blocks of sizes $k_1,\dots,k_l$. The local exponents (roots of the indicial equation) $\lambda_{i,1}^{(a_\nu)},\dots,\lambda_{i,n}^{(a_\nu)}$ of the equation (6.1.2) at $x=a_\nu$ are related to the eigenvalues of V_ν by the famous formula

(6.1.3) $$e^{2\pi\sqrt{-1}\lambda_{i,j}^{(a_\nu)}} = \mu_j^{(a_\nu)}:\ i,j=1,\dots,n.$$

We define the vector $\vec{\lambda}_i^{(a_\nu)}$ of the local exponents of (6.1.2) at $x=a_\nu$ by:

(6.1.4) $$\vec{\lambda}_i^{(a_\nu)} = (\lambda_{i,1}^{(a_\nu)},\dots,\lambda_{i,n}^{(a_\nu)}),$$

and according to (6.1.3), one gets

(6.1.5) $$\vec{\lambda}_i^{(a_\nu)} \equiv \vec{\lambda}_{i'}^{(a_\nu)} (\mathrm{mod}\ \mathbb{Z}^n)\ :\ i,i'=1,\dots,n;\ \nu=1,\dots,m,\infty .$$

According to Corollary 2.3.4 we have

(6.1.6) $$\sum_{\nu=1,\dots,m,}\ \sigma(\vec{\lambda}_i^{(a_\nu)}) = \frac{n(n-1)}{2}(m-1)$$

for the trace $\sigma(\cdot)$ of $\mathbb{C}^n$-vector:

$$\sigma(\vec{\lambda}_i^{(a_\nu)}) = \sum_{j=1}^n \lambda_{i,j}^{(a_\nu)}.$$

The relation between $\vec{\lambda}_i^{(a_\nu)}$ and the eigenvalues $\vec{\mu}^{(a_\nu)} = (\mu_i^{(a_\nu)},\dots,\mu_n^{(a_\nu)})$ is contained in (6.1.3):

(6.1.7) $$\exp\{2\pi\sqrt{-1}\ \mathrm{diag}\ \vec{\lambda}_i^{(a_\nu)}\} = \mathrm{diag}\ \vec{\mu}^{(a_\nu)}:\ i=1,\dots,n;\ \nu=1,\dots,m,\infty.$$

The analytic structure of $Y(x)$ at $x=a_\nu$ is characterized by the normal form of the monodromy matrices V_ν. We have

(6.1.8) $$S_\nu Y(x) = Z_\nu(x)$$

together with following structure of $Z_\nu(x)$ based on the description of 2.2.4 and (2.2.5) of the system of solutions $\vec{y}_i S_\nu^t$ of (6.1.2). Following our agreement about the order of indexes in the local monodromy, we can write for $Z_\nu(x)$ in (6.1.8):

(6.1.9) $$Z_\nu(x)_{ij} = d_{ij}^{(\nu)}(x)(x-a_\nu)^{\lambda_{j,i}^{(a_\nu)}}\cdot \phi_{\nu,ij}(x):\ i,j=1,\dots,n$$

for $\phi_{\nu,ij}(x)$ being holomorphic and non-zero at $x=a_\nu$. The following is the exact structure of $d_{i,j}^\nu(x)$. Let $i=1,\dots,n$ be fixed, and let's consider $\lambda_{i,j}^{(a_\nu)}$ that correspond to the s-th block $J_{k_s}(\rho_s^\nu)$, i.e. $k_1+\dots+k_{s-1}+1 \le j \le k_1+\dots+k_s$ and $e^{2\pi\sqrt{-1}\lambda_{i,j}^{(a_\nu)}}=\rho_s^\nu$. Now if we arrange $\lambda_{i,j}^{(a_\nu)}$ in an order such that $\lambda_{i,j}^{(a_\nu)} \ge \lambda_{i,j+1}^{(a_\nu)}$: $j=k_1+\dots+k_{s-1}+1,\dots,k_1+\dots+k_s-1$ (this order may be different for a different i), then $d_{i,j}^\nu(x)$ has the form $(h_s=k_1+\dots+k_{s-1}+1)$:

$$d_{i,h_s}^\nu(x) = a_{i,h_s}^\nu;\ \text{and for } u = \frac{1}{2\pi\sqrt{-1}}\log(z-a_\nu),$$

(6.1.10) $$d^{\nu}_{i,h_s+1}(x) = a^{\nu}_{i,h_s+1}(x) + a^{\nu}_{i,h_s}\cdot u\cdot(x-a_{\nu})^{\lambda^{(a_\nu)}_{i,h_s}-\lambda^{(a_\nu)}_{i,h_s+1}};$$

...

$$d^{\nu}_{i,h_s+k_s-1}(x) = a^{\nu}_{i,h_s+k_s-1}(x)+\ldots+a^{\nu}_{i,h_s}\cdot\binom{u}{k_s-1}\cdot(x-a\ )^{\lambda^{(a_\nu)}_{i,h_s}-\lambda^{(a_\nu)}_{i,h_s+k_s-1}}$$

where $a^{\nu}_{i,j}(x)$ are regular at $x=a_{\nu}$. With this form of $d^{\nu}_{i,j}$ we have the complete description of the $Y(x)$ at $x=a_{\nu}$. We define the matrix of the local multiplicities of $Y(x)$ at $x=a_{\nu}$ by

$$Y(x)\sim\begin{pmatrix}\lambda^{(a_\nu)}_{1,1},\ldots,\lambda^{(a_\nu)}_{n,1}\\ \vdots \qquad\quad \vdots\\ \lambda^{(a_\nu)}_{1,n},\ldots,\lambda^{(a_\nu)}_{n,n}\end{pmatrix} \text{ at } x=a_{\nu}$$

or by

(6.1.11) $$Y(x)\sim(\vec{\lambda}_1^{(a_\nu)^t},\ldots,\vec{\lambda}_n^{(a_\nu)^t}) \text{ at } x=a_{\nu}.$$

In other words, we determine explicitly the analytic structure of $Y(x)$ at $x=a_{\nu}$, knowing only the matrix of the local multiplicities (also with the knowledge of the normal form of V_{ν}, if necessary).

6.2. We define now the matrix of the local exponents of the system (6.0.1) in the case, when (6.0.1) is of canonical form, e.g., of canonical regular form (1.10.1). In this case (6.0.1) may have an exponential matrix W_{ν} at $x=a_{\nu}$ as in 3.3.1. The existence of the exponential matrix implies particularly simple structure of the matrix of the local exponents of $Y(x)$ at $x=a_{\nu}$. In particular, right side in (6.1.5) changes to 0; this is explained by the presence of apparent singularities in (6.1.2) in addition to singularities of (6.0.1) (as we have seen in §5). Let's consider a fundamental solution $Y(x)$ of (6.0.1): $\frac{dY(x)}{dx}=Y(x)A(x)$ having at a regular singularity $x=a_{\nu}$ the monodromy matrix V_{ν}. Let

(6.2.1) $$V_{\nu} = S_{\nu}^{-1}\cdot[J_{k_1}(\rho_1^{\nu}),\ldots,J_{k_l}(\rho_l^{\nu})]\cdot S_{\nu}$$

for blocks $J_{k_i}(\rho_i^{\nu})$ of sizes k_i: $i=1,\ldots,l$; $k_1+\ldots+k_l=n$. For any given ρ_i^{ν} we have k_i eigenvalues $\mu_{i,1},\ldots,\mu_{i,k_i}$ of the exponential matrix W_{ν} at $x=a_{\nu}$ such that $e^{2\pi\sqrt{-1}\mu_{i,j}} = \rho_i^{\nu}$: $i=1,\ldots,l$. Here we assume that as in 3.3.1 the exponential matrix W_{ν} determine the structure of $Y(x)$ at $x=a_{\nu}$:

(6.2.2) $$Y(x) = (x-a_{\nu})^{W_{\nu}}H_{\nu}(x)$$

for $H_{\nu}(x)$ and $H_{\nu}(x)^{-1}$ being holomorphic at $x=a_{\nu}$. Let eigenvalues $\lambda_i^{(a_\nu)}$ of W_{ν} be or- according to $\mu_{1,1},\ldots,\mu_{1,k_1};\ldots;\mu_{l,1},\ldots,\mu_{l,k_l}$. Then looking at the normal form of W_{ν} we can write (§3 and [10]):

(6.2.3) $\quad S_\nu Y(x) = Z(x)$

with the matrix $Z(x)$ having the canonical form (cf. 6.1.):

(6.2.4) $$Z(x)_{ij} = d_{ij}^{(\nu)}(x)\cdot(x-a_\nu)^{\lambda_i^{(a_\nu)}} \phi_{\nu,ij}(x)$$

for $\phi_{\nu,ij}(x) = c_{i,j}^{(\nu)} + O(x-a_\nu)$ - being an analytic function at $x=a_\nu$ and the matrix $D_\nu = (d_{i,j}^{(\nu)}(x))_{i,j=1}^n$ has the block structure $D_\nu = \begin{pmatrix} D_{\nu,1} \\ \vdots \\ D_{\nu,1} \end{pmatrix}$ where $D_{\nu,s}$ are k_s x n matrices: $s=1,\ldots$ $\ldots,1$. In these notations we define:

$$(D_{\nu,s})_{1j} = a_{ij} + \binom{u}{1}a_{2j} + \ldots + \binom{u}{k_s-1}a_{k_s j},$$

(6.2.5) $$(D_{\nu,s})_{2j} = a_{2j} + \binom{u}{1}a_{3j} + \ldots + \binom{u}{k_s-2}a_{k_s j}, \ldots$$

$$(D_{\nu,s})_{k_s j} = a_{k_s j}$$

where $u = \frac{1}{2\pi\sqrt{-1}}\log(x-a_\nu)$. In fact the terms $d_{ij}^{(\nu)}(x)\cdot(x-a_\nu)^{\lambda_i}$ really determine the main contribution to $C_\nu Y(x)$:

(6.2.6) $$\det(\phi_{\nu,ij})_{i,j=1}^n\Big|_{x=a_\nu} = \det(c_{ij}^{(\nu)})_{i,j=1}^n \neq 0$$

Thus (6.2.6) shows that for $\vec{\lambda}^{(a_\nu)} = (\lambda_1^{(a_\nu)},\ldots,\lambda_n^{(a_\nu)})$ the matrix of the local exponents of $Y(x)$ at $x=a_\nu$ can be represented as

(6.2.7) $$Y(x) \sim (\vec{\lambda}^{(a_\nu)^t},\ldots,\vec{\lambda}^{(a_\nu)^t}) \text{ at } x=a_\nu.$$

If we assume that $Y(x)$ has exponential matrices W_ν for any $x=a_\nu$: $\nu=1,\ldots,m,\infty$ (see 3.3.5), then following 3.4.1 - 3.4.2 one gets that the sum of the eigenvalues of all W_ν: $\nu=1,\ldots,m,\infty$ is zero:

(6.2.8) $$\sum_{\nu=1,\ldots,m,\infty}\sum_{i=1}^n \lambda_i^{(a_\nu)} = 0$$

This "disagreement" with (6.1.6) is explained by the presence in (6.1.2) apparent singularities (i.e., in (6.1.2) the set of a_ν is larger than in (6.0.1)) and by the fact that in 6.1 the equation didn't have a canonical form or exponential matrices.

In the case when $Y(x)$ doesn't have any exponential matrix (or has several, like for the irregular singularity §4) the structure of $Y(x)$ is slightly more complicated and was discussed in 6.1. More precisely, instead of the canonical form (4.2.6) we had in (6.1.11), (6.1.4):

$$Y(x) \sim (\vec{\lambda}^{(a_\nu)^t},\ldots,\vec{\lambda}^{(a_\nu)^t}) (\text{mod } M_n(\mathbb{Z})) \text{ at } x=a_\nu.$$

6.3. Bases of Riemannian modules.

Now we can choose the basis of the Riemannian module $S\begin{pmatrix} V_1,\ldots,V_m \\ a_1,\ldots,a_m \end{pmatrix}$ in many ways. All of these bases of $\Phi(x)$ are characterized by the following data. For the module

$S\begin{pmatrix}V_1,\dots,V_m\\ a_1,\dots,a_m\end{pmatrix}$ we can choose according to the discussion in §1 and 6.2 above, a single vector $\vec{\lambda}^{(a_\nu)}=(\bar{\lambda}_1^{(a_\nu)},\dots,\bar{\lambda}_n^{(a_\nu)})\in\mathbb{C}^n$ corresponding to the local exponents of $S\begin{pmatrix}V_1,\dots,V_m\\ a_1,\dots,a_m\end{pmatrix}$ $\nu=1,\dots,m,\infty$. Here for the eigenvalues $\mu_1^\nu,\dots,\mu_n^\nu$ of V_ν we have $e^{2\pi\sqrt{-1}\bar{\lambda}_i^{(a_\nu)}}=\mu_i{}^\nu$: $i=1,\dots,n$; $\nu=1,\dots,m,\infty$ and

$$(6.3.1)\qquad \sum_{\nu=1,\dots,m,\infty}\sum_{i=1}^{n}\bar{\lambda}_i{}^{(a_\nu)}=0.$$

Now for any basis $\Phi(x)$ of $S\begin{pmatrix}V_1,\dots,V_m\\ a_1,\dots,a_m\end{pmatrix}$ we have the matrix of local multiplicities at $x=a_\nu$ of the form

$$(6.3.2)\qquad \Phi(x)\sim\begin{pmatrix}\lambda_{1,1}^{(a_\nu)} & \cdots & \lambda_{n,1}^{(a_\nu)}\\ \vdots & & \vdots\\ \lambda_{1,n}^{(a_\nu)} & & \lambda_{n,n}^{(a_\nu)}\end{pmatrix}\quad \text{at } x=a_\nu.$$

For any $\nu=1,\dots,m,\infty$ we have

$$(6.3.3)\qquad \Phi(x)\sim(\vec{\lambda}^{(a_\nu)^t},\dots,\vec{\lambda}^{(a_\nu)^t})(\mathrm{mod}\ M_n(\mathbb{Z}))\ \text{at } x=a_\nu.$$

It is obvious that for any matrices $\Lambda^{(a_\nu)}$ such that $\Lambda^{(a_\nu)}\equiv(\vec{\lambda}^{(a_\nu)^t},\dots,\vec{\lambda}^{(a_\nu)^t})$ $(\mathrm{mod}\ M_n(\mathbb{Z}))$, there exists a basis $\Phi(x)$ of $S\begin{pmatrix}V_1,\dots,V_m\\ a_1,\dots,a_m\end{pmatrix}$ having $\Lambda^{(a_\nu)}$ as its matrix of local multiplicities at $x=a_\nu$: $\nu=1,\dots,m$. However, according to 1.7.2, all of these bases are transformed into one another by right multiplication on a non-zero matrix from $GL(n,\mathbb{C}(x))$. These transformations: a)may change $\Lambda^{(a_\nu)}$ to $\Lambda^{(a_\nu)}+K_\nu$ with $K_\nu\in M_n(\mathbb{Z})$; b) may add new singularities to $\{a_1,\dots,a_m,\infty\}$ (either apparent or poles).

§7. Riemannian and Fuchsian modules.

7.0. We return now to the study of Riemannian modules started in §1. For this we use also the concept of the Fuchsian module of [19] (see lectures at this volume) when instead of the family of all Fuchsian equations with a given monodromy group, we consider a system of all solutions of a single Fuchsian equation, which is a module itself.

7.1. Let $\vec{f}(x)=(f_1(x),\dots,f_n(x))$ be any element of the space $S\begin{pmatrix}V_1,\dots,V_m\\ a_1,\dots,a_m\end{pmatrix}$ introduced in §1. The $\vec{f}(x)$ is a system of multivalued functions (or is uniform at $\Gamma(a_1,\dots,a_m)$) with branch points at $x=a_\nu$ and monodromy matrices V_ν for $\nu=1,\dots,m,\infty$ and $V_1\cdots V_mV_\infty=I$. Now we can consider a module $\mathcal{L}(\vec{f})$ generated by $\{f_1(x),\dots,f_n(x)\}$ over $\mathbb{C}$.

7.2. We already know that $\vec{f}(x)$ naturally generates a scalar Fuchsian linear differential equation (cf. §1):

Lemma 7.2.1. If $\vec{f}(x)$ belongs to $S\begin{pmatrix}V_1,\dots,V_m\\ a_1,\dots,a_m\end{pmatrix}$, then all elements of $\mathcal{L}(\vec{f})$ satisfy a scalar Fuchsian linear differential equation (with rational function coefficients):

$$(7.2.2)\qquad L\{y\}=0\quad\text{for } y\in\mathcal{L}(\vec{f}).$$

Here the order of $L\{\cdot\}$ is equal to $\dim_{\mathbb{C}}\mathcal{L}(\vec{f})$, and $\mathcal{L}(\vec{f})$ is the set of all solutions of (7.2.2).

Proof. Let $\dim_{\mathbb{C}}\mathcal{L}(\vec{f})=n$; then we consider $\vec{f}^{(i)}(x)=(f_1^{(i)}(x),\ldots,f_n^{(i)}(x))$: $i=0,1,\ldots,n$. Naturally, according to (1.4.4), all $\vec{f}^{(i)}(x)$ also belong to $S\begin{pmatrix}V_1,\ldots,V_m\\a_1,\ldots,a_m\end{pmatrix}$. The theorem 1.6.3 implies that the system $\vec{f}^{(i)}(x)$: $i=0,1,\ldots,n$ is linearly dependent over $\mathbb{C}[x]$: $\sum_{i=0}^{n} P_i(x)\vec{f}^{(i)}(x)=0$ for the polynomials $P_i(x)\varepsilon\mathbb{C}[x]$, which is equivalent to (7.2.2). If $\dim_{\mathbb{C}}\mathcal{L}(\vec{f})=r<n$, then we choose a $\mathbb{C}$-linearly independent basis $\vec{g}$ in $\mathcal{L}(\vec{f})$ and repeat the argument as before, changing the sizes of the monodromy matrices to $r \times r$.

Remark 7.2.2. As we have seen already, the construction of the scalar Fuchsian equations satisfied by elements of $\mathcal{L}(\vec{f})$ may introduce, in addition to real singularities $a_1,\ldots,a_m,a_\infty=\infty$, some apparent singularities, which are singularities of (7.2.2) but not of $\mathcal{L}(\vec{f})$.

7.3. A very important problem is to determine $\mathbb{C}$-linear independence of a given system of functions $\{f_1(x),\ldots,f_n(x)\}$. One of the best tests is based on the knowledge of the monodromy group of $\vec{f}(x)$.

Here is the simplest criterion showing when any non-zero n-plet of function $\vec{f}(x)$ on $\Gamma(a_1,\ldots,a_m)$ from $S\begin{pmatrix}V_1,\ldots,V_m\\a_1,\ldots,a_m\end{pmatrix}$ is a fundamental system of solutions of the equation (7.2.2), i.e., when $\dim_{\mathbb{C}}\mathcal{L}(\vec{f})=n$. We have:

Theorem 7.3.1. Let $\mathcal{M}$ be a monodromy group corresponding to a fixed representation $\pi^1(\mathbb{CP}^1\setminus\{a_1,\ldots,a_m,\infty\})\hookrightarrow GL(n,\mathbb{C}^1)$ and $V_1,\ldots,V_m$ are generators of $\mathcal{M}$, corresponding to a fixed basis $\gamma_1,\ldots,\gamma_m$ of $\pi^1(\mathbb{CP}^1\setminus\{a_1,\ldots,a_m,\infty\})$. Let's assume that among the elements of $\mathcal{M}$ there exists a matrix M with the following property: for any root λ_0 of the characteristic polynomial $\chi_M(\lambda)$ of M there is only one elementary divisor $(\lambda-\lambda_0)^{r_0}$ of $\chi_M(\lambda)$ corresponding to M. Then for any system $\vec{f}(x)$ from $S\begin{pmatrix}V_1,\ldots,V_m\\a_1,\ldots,a_m\end{pmatrix}$ elements of $\vec{f}(x)$ are $\mathbb{C}$-linearly independent, which means $\dim_{\mathbb{C}}\mathcal{L}(\vec{f})=n$.

The proof follows immediately from the description of the structure of the subgroups of the system of functions $\mathcal{L}(\vec{f})$ in the vicinity of singularities following Jurgens, Hamburger, and Fuchs studies as presented in [6].

7.4. By the definition of 7.1, $\vec{f}(x)$ is a system of generators of $\mathcal{L}(\vec{f})$, and we have

Lemma 7.4.1. For and $\vec{f}(x)$ from $S\begin{pmatrix}V_1,\ldots,V_m\\a_1,\ldots,a_m\end{pmatrix}$ and any matrix C from $GL(n,\mathbb{C}^1)$, the system $\vec{f}\cdot C^t$ is a system of generators of $\mathcal{L}(\vec{f})$ and belongs to the system

$$S\begin{pmatrix}CV_1C^{-1},\ldots,CV_mC^{-1}\\a_1,\ldots,a_m\end{pmatrix}.$$

This change of the variables doesn't affect the eigenvalues of the $CV_\nu C^{-1}$ and gives one the definition of local exponents of $\vec{f}(x)$ or of $\mathcal{L}(\vec{f})$ at any point $x=b$. First of all, at $x=a_\nu$ the local exponents $\lambda_1^{(a_\nu)},\ldots,\lambda_n^{(a_\nu)}$ of $\mathcal{L}(\vec{f})$ are defined explicitly in §2, Theorem 2.2.4 as roots of the indicial equation of the differential equation (7.2.2) satisfied by all elements of $\mathcal{L}(\vec{f})$. Moreover we can now describe local exponents corresponding to an arbitrary point $x=b$. We consider only the case when $\dim_{\mathbb{C}}\mathcal{L}(\vec{f})=n$, when we have n exponents. In the case when $\vec{f}(x)$ is a basis of $\mathcal{L}(\vec{f})$ and $x=b$ is different from singular points of the Fuchsian equation (7.2.2) satisfied by all elements of $\mathcal{L}(\vec{f})$, the existence and uniqueness of the solution of any initial value problem at x=b is guaranteed. This means that we can find a transformation of $\mathcal{L}(\vec{f})$: $\vec{f}\mapsto\vec{f}\cdot C_b$ reducing $\vec{f}$ to the form

$$\vec{f}\cdot C_b = (\phi_1^b(x),\ (x-b)\phi_2^b(x),\dots,(x-b)^{n-1}\phi_n^b(x))$$

with $\phi_i^b(x)$ regular and non-zero at x=b: i=1,...,n. Thus we define local exponents of $\mathcal{L}(\vec{f})$ at x=b as (0,1,2,...,n-1). If x=b is a singularity of the equation (7.2.2), but not the singularity of any of the functions from $\mathcal{L}(\vec{f})$, then the local exponents of $\mathcal{L}(\vec{f})$ are again rational integers $(r_1,\dots,r_n)$.

<u>Lemma 7.5.1.</u> If x=b is an apparent singularity of the system $\mathcal{L}(\vec{f})$ (i.e., a singularity of (7.2.2) but not of $\mathcal{L}(\vec{f})$), then local exponents $(r_1,\dots,r_n)$ of $\mathcal{L}(\vec{f})$ at x=b are n distinct nonnegative rational integers, and at least one of them is greater than or equal to n.

In other words, an apparent singularity x=b is characterized by the existence of an element $\phi(x)$ from $\mathcal{L}(\vec{f})$ such that $\phi(x)$ has a zero of an order $\geq$ n at x=b, but $\phi\not\equiv 0$.

<u>Corollary 7.5.2.</u> Let S be any finite subset of $\mathbb{CP}^1$ containing all the singularities (real and apparent) of the equation (7.2.2) satisfied by all the elements of $\mathcal{L}(\vec{f})$. If $\dim\mathcal{L}(\vec{f})=n$, and the local exponents of $\mathcal{L}(\vec{f})$ at x=c are $(\lambda_1^{(c)},\dots,\lambda_n^{(c)})$, then

$$(7.5.3)\qquad \sum_{c\in S}\sum_{i=1}^{n}\lambda_i^{(c)} = \frac{n(n-1)}{2}(|S|-2).$$

The formula (7.5.3) follows from (2.3.5) since for a regular point x=b, $\sum_{i=1}^{n}\lambda_i^{(b)} = \frac{n(n-1)}{2}$. Moreover we can easily generalize the Corollary 7.5.2 in the case of the Fuchsian linear differential equations with algebraic coefficients; or Fuchsian equations defined on a Riemann surface $\mathcal{T}$ of the genus $g\geq 0$ [19]. We have naturally by 7.5.2 and arguments of [19], and [6]:

<u>Corollary 7.5.4.</u> Let $\mathcal{T}$ be a Riemann surface of the genus $g\geq 0$, and let $\mathcal{L}(\vec{f})$ be the n-dimensional space of solutions of the system of Fuchsian linear differential equations of the order n with algebraic coefficients corresponding to $\mathcal{T}$. If S is any finite set in $\mathcal{T}$ containing all the singularities of the equation satisfied by elements of $\mathcal{L}(\vec{f})$ then for the local exponents $(\lambda_1^{(c)},\dots,\lambda_n^{(c)})$ of $\mathcal{L}(\vec{f})$ at x=c at $c\in\mathcal{T}$ we have

$$(7.5.5)\qquad \sum_{c\in S}\sum_{i=1}^{n}\lambda_i^{(c)} = \frac{n(n-1)}{2}(|S|+2g-2)$$

§8 <u>The definition of Padé approximations and perfectness properties.</u>

<u>8.0.</u> We start with the problem of a general p-point Padé approximation [2],[16],[20] to a system of functions $\vec{f}(x)=(f_1(x),\dots,f_n(x))$ defined in some domain of $\mathbb{C}$ [21],[20]:

<u>Definition 8.0.1.</u> Let $\vec{f}(x)=(f_1(x),\dots,f_n(x))$ be a system of n functions regular at complex points $x=\zeta_1,\dots,x=\zeta_p$, and let $m_1,\dots,m_n$ be fixed non-negative integers. Let the polynomials $P_1(x),\dots,P_n(x)$ of degrees $m_1,\dots,m_n$ be chosen such that the remain-

der function

(8.0.2) $R(x) = \sum_{i=1}^{n} P_i(x)f_i(x)$

satisfies the following conditions on the orders $\mathrm{ord}_{\zeta_k}R(x)$ of zeroes of $R(x)$ at $x=\zeta_k$:

(8.0.3) $\sum_{k=1}^{p} \mathrm{ord}_{\zeta_k}R(x) \geq \{\sum_{i=1}^{n} (m_i+1)\} - 1.$

In order to concentrate on nontrivial situations we can assume $\sum_{i=1}^{n} (m_i+1)\geq np+1$ and change (8.0.3) to

(8.0.4) $\sum_{k=1,\ \mathrm{ord}_{\zeta_k}R(x)\geq n}^{p} \mathrm{ord}_{\zeta_k}R(x) \geq \{\sum_{i=1}^{n} (m_i+1)\} - 1.$

Then the system of polynomials $\vec{P}(x)=(P_1(x),\dots,P_n(x))$ is called <u>an $(m_1,\dots,m_n)$-approximation to $\vec{f}(x)$</u> at p points $\zeta_1,\dots,\zeta_p$, and $R(x)$ is called a <u>remainder function.</u> We can write (8.0.2) shortly as $R(x)=\vec{P}(x)\cdot\vec{f}^t(x)$.

<u>Lemma 8.0.5.</u> For any n-plet $(m_1,\dots,m_n)\in\mathbb{Z}_+^n$ satisfying $\sum_{i=1}^{n}(m_i+1)\geq np+1$ there exists a non-zero system of polynomials $\vec{P}(x)=(P_1(x),\dots,P_n(x))\neq\vec{0}$ that is a system of $(m_1,\dots,\dots,m_n)$-approximations to $\vec{f}(x)$ at $\zeta_1,\dots,\zeta_p$. Moreover for any $h_1\geq n,\dots,h_p\geq n$ and any $(m_1,\dots,m_n)\in\mathbb{Z}_+^n$, there exists a system $\vec{P}(x)\neq\vec{0}$ of polynomials of degrees $(m_1,\dots,m_n)$ such that for $R(x)=\vec{P}(x)\cdot\vec{f}(x)^t$ we have

(8.0.6) $\mathrm{ord}_{\zeta_i}R(x) \geq h_k \ : k=1,\dots,p,$

provided only that

(8.0.7) $\sum_{i=1}^{n} (m_i+1) \geq \{\sum_{k=1}^{p} h_k\} + 1.$

<u>Remark 8.0.8.</u> It is reasonable to weaken the condition of analyticity of $\vec{f}(x)$ at $x=\zeta_k$. In the case of the system of functions $\vec{f}(x)$ from $S\begin{pmatrix}V_1,\dots,V_m\\a_1,\dots,a_m\end{pmatrix}$ we can look at the remainder $R(x)=\sum_{i=1}^{n} P_i(x)f_i(x)$ in the neighborhood of the branch points $x=a_\nu$, and we can formally define $\mathrm{ord}_{a_\nu}R(x)$ as the exponent with the smallest real part among all exponents in the development of $R(x)$ at $x=a_\nu$. This concept is different from the one traditionally considered. On the other hand, the $\mathrm{ord}_{\zeta_k}R(x)$ is defined in the normal sense if $R(x)$ is analytic at $x=\zeta_k$. It's possible to show that in the general situation and for $\vec{f}(x)$ from $S\begin{pmatrix}V_1,\dots,V_m\\a_1,\dots,a_m\end{pmatrix}$, the function $R(x)=\vec{P}(x)\cdot\vec{f}(x)$ is analytic at $x=a_\nu$ only in the case of $\vec{f}(x)$ uniform at $x=a_\nu$, i.e., when $V_\nu=I$.

<u>8.1.</u> In this paper we follow the classic studies of Padé approximations [20], [25] where only diagonal or subdiagonal cases were considered. In other words, we consider the cases when either $m_1=\dots=m_n=N$ or all differences $|N-m_i|$: $i=1,\dots,n$ are bounded by a constant depending only on the system of functions $\vec{f}(x)=(f_1(x),\dots,f_n(x))$.

<u>8.1.1.</u> The most important problem for applications, in particular for Number Theory, is the problem of the normality and perfectness of the Padé approximation [20] - [22]. The system of Padé approximations to $\vec{f}(x)$ corresponding to $(m_1,\dots,m_n)$ at $x=\zeta_1,\dots,\zeta_p$ is called <u>perfect</u> if for any $(m_1,\dots,m_n)$-approximation system $\vec{P}(x)=(P_1(x),\dots,P_n(x))$

to $\vec{f}(x)$ at $x=\zeta_1,\dots,\zeta_p$, we have equality in (8.0.3): $\sum_{k=1}^{p} \mathrm{ord}_{\zeta_k} R(x)=\{\sum_{i=1}^{n}(m_i+1)\}-1$. The system of Padé approximations is called "almost perfect" [20] if for an $(m_1,\dots,m_n)$-approximation $\vec{P}(x)$ to $\vec{f}(x)$ at $x=\zeta_1,\dots,\zeta_p$, we have

$$(8.1.2) \qquad |\sum_{k=1}^{p} \mathrm{ord}_{\zeta_i} R(x) - \sum_{i=1}^{h} (m_i+1) + 1| \le C(\vec{f})$$

for a constant $C(\vec{f})$ depending only on $\vec{f}(x)$ and, possibly, on $\zeta_1,\dots,\zeta_p$. There are more refined notions in the case of the one-point Padé approximation. Following [22] we call $\vec{m}=(m_1,\dots,m_n)$ a normal point of the one-point Padé approximation to $\vec{f}(x)$ at $x=\zeta_1$ if the system of $(m_1,\dots,m_n)$-Padé approximation $\vec{P}(x)=(P_1(x),\dots,P_n(x))$ to $\vec{f}(x)$ at $x=\zeta$, is unique up to a scalar factor.

There are only a few examples of normal or perfect systems of Padé approximations (see [20] - [23] for their descriptions); these well-known examples follow from the Hermite construction of the normal and perfect system of the Padé approximation to the exponential functions $(e^{\omega_1 x},\dots,e^{\omega_n x})$ at $x=0$ [23].

8.2. Results on "almost perfectness" of systems of functions satisfying linear differential equations are proved and reproved for a long time since C. Siegel (1929) work on E-functions. These results were summarized in [23] where the most general result was proved:

Proposition 8.2.1. Let $\vec{w}(x)=(w_1(x),\dots,w_n(x))$ be a solution of a system of linearly differential equations:

$$(8.2.2) \qquad \frac{d}{dx}\vec{w}(x) = \vec{w}(x)\cdot A(x)$$

for $A(x)$ from $M_n(\mathbb{C}(x))$. Let $p_1(x),\dots,p_n(x)$ be polynomials of degrees $d(p_1),\dots,d(p_n)$ and let $\vec{w}(x)$ be analytic at $x=c$ and $w_1(x),\dots,w_n(x)$ be linearly independent over $\mathbb{C}[x]$. Then for $\nu(x)=p_1(x)w_1(x)+\dots+p_n(x)w_n(x)$ we have

$$(8.2.3) \qquad \begin{aligned} \mathrm{ord}_c r(x) \le\ & n\cdot\max\{d(p_1),\dots,d(p_n)\} + \min\{\mathrm{ord}_c w_1(x),\dots,\mathrm{ord}_c w_n(x)\} + \\ & + n-1 + \frac{n(n-1)}{2} D. \end{aligned}$$

Here D is the degree of $A(x)$, i.e., for $g(x)$ being the common denominator of the elements of $A(x)$, D is the $\max\{d(g(x)), d(b_{i,j}(x))\}$ with $A(x)g(x)=(b_{i,j}(x))\varepsilon M_n(\mathbb{C}[x])$.

One of the main results of this paper is the establishment of the generalization of (8.2.3) for an arbitrary set of points in $\mathbb{CP}^1$. Also the sense of D becomes a little more clear, namely, in general, D is "the number of poles of $A(x)$ in $\mathbb{CP}^1$ counted with a multiplicity minus one" (this is a slight improvement of the definition of D in (8.2.3)). The statement 8.2.1 is very important for the applications, especially in Transcendental Number Theory. E.g., let the conditions of 8.1.1 be satisfied and $r^{(i-1)}(x)=P_{1,i}(x)\,w_1(x)+\dots+P_{n,i}(x)w_n(x)$: $i=1,\dots,n$. If $\mathrm{ord}_c r(x) \ge (n-1)\cdot\max\{d(p_1),\dots,d(p_n)\} + C'$ for C' depending on the equation (8.2.2), then linear forms $r^{(i)}(x)$ are linearly independent: $i=0,1,\dots,n-1$ and $\det(P_{i,j}(x))_{i,j=1}^{n} \ne 0$.

New developments in investigation of Padé approximations to solutions of linear differential equations belong to Ch. Osgood, W. Schmidt, who studied in detail rational approximations to algebraic functions and considered rational approximations to solutions of algebraic differential equations. Very importnat recent progress in the study of "almost perfectness" of algebraic approximants for the solutions of linear differential equations (or special nonlinear differential equations satisfied by elliptic functions) belongs to D. Brownawell and D. Masser [26]. Results of Brownawell-Masser generalized early results of Nesterenko. The estimates proved in the present paper refined considerably previous results on upper bounds of orders of rational approximations of system of functions satisfying linear differential equations.

8.3. As an example of the usefulness of our approach based on the Riemann and Fuchsian modules, we show that the system of functions $\{x^{\omega_1},\ldots,x^{\omega_n}\}$ is "perfect" with respect to any finite set in $\mathbb{C}\setminus\{0\}$. The results of [21] correspond only to the case $x=1$.

Theorem 8.3.1. Let $\omega_1,\ldots,\omega_n$ be any system of n distinct (mod $\mathbb{Z}$) complex numbers. Let $\{\zeta_1,\ldots,\zeta_p\}$ be any set of non-zero complex numbers. Then the system of function $\{x^{\omega_1},\ldots,x^{\omega_n}\}$ is "perfect" with respect to $\{\zeta_1,\ldots,\zeta_p\}$. This means that for any non-negative integers $m_1,\ldots,m_n$ and polynomials $P_1(x),\ldots, P_n(x)$ of degrees $m_1,\ldots,m_n$ we have an upper bound of the order of zero of the remainder function

(8.3.2) $$R(x) = \sum_{i=1}^{n} P_i(x)x^{\omega_i}$$

of the form

(8.3.3) $$\sum_{k=1}^{p} \mathrm{ord}_{\zeta_k}R(x) \leq \sum_{i-1}^{n} m_i + p(n-1)\ .$$

Proof. It is obvious that $R(x)$ satisfies a scalar Fuchsian linear differential equation of the order n; we denote the corresponding Fuchsian module of the solutions by $\mathfrak{A}$. Let γ be the simple closed contour in a finite part of the x-plane enclosing zero. Then continuing $R(x)$ along any closed curve homotopic in $\mathbb{CP}^1\setminus\{0,\infty\}$ to $\gamma\cdot k$, $k\in\mathbb{Z}$, we have

(8.3.4) $$R(x) \underset{\gamma\cdot k}{\longmapsto} \sum_{i=1}^{n} e^{2\pi\sqrt{-1}k\omega_i}\cdot x^{\omega_i}P_i(x)\ .$$

Let $\mu_i=e^{2\pi\sqrt{-1}\omega_i}$: $i=1,\ldots,n$ and

(8.3.5) $$R_k(x) = \sum_{i=1}^{n} \mu_i^k x^{\omega_i}P_i(x)\ :\ k\in\mathbb{Z}.$$

Then by (8.3.4) all $R_k(x)$ belong to a module $\mathfrak{A}$. Let $Q_i(x)=\sum_{k=0}^{n-1} q_{ik}x^k$ be Lagrange interpolation polynomial of the degree n-1 such that

(8.3.6) $$Q_i(\mu_j) - \delta_{i,j}\ :\ i,j-1,\ldots,n.$$

Then we have by (8.3.5) - (8.3.6)

$$P_i(x)x^{\omega_i} = \sum_{j=1}^{n} P_j(x)x^{\omega_j}Q_i(\mu_j) = \sum_{j=1}^{n}\sum_{k=0}^{n-1} q_{ik}\mu_j^k\ P_j(x)x^{\omega_j} =$$
$$=\sum_{k=0}^{n-1} q_{ik}\cdot R_k(x).$$

Thus $P_i(x)\cdot x^{\omega_i}$ belongs to a module $\mathfrak{A}$ as well. The module $\mathfrak{A}$ has real singularities at $x=0,\infty$ and a certain set Sing of apparent singularities. Since $P_i(x)x^{\omega_i}\in\mathfrak{A}$ and $d(P_i)=m_i\geq 0$, $\dim_{\mathbb{C}}\mathfrak{A}=n$.

Let $\Omega \supseteq$ Sing; $0,\infty \notin \Omega$ and $S=\Omega\cup\{0,\infty\}$; for $b\in\mathbb{C}$ we denote by $\lambda_1^{(b)},\ldots,\lambda_n^{(b)}$ the local exponents of $\mathcal{O}\!\mathcal{L}$ at $x=b$. According to (8.3.7) we have

(8.3.8) $$\lambda_i^{(0)} \geq \omega_i \;;\; \lambda_i^{(\infty)} \geq -m_i-\omega_i ,$$

for $i=1,\ldots,n$. Also for $\zeta\in\Omega$ we have by 7.5.1 the following:

(8.3.9) $$\sum_{i=1}^n \lambda_i^{(\zeta)} \geq 0+1+\ldots+(n-2) + \max_{i=1,\ldots,n} \lambda_i^{(\zeta)} \geq \frac{(n-2)(n-1)}{2} + $$

$$+\mathrm{ord}_\zeta R(x)$$

Now taking into account 7.5.2 we have

$$\sum_{\zeta\in S} \sum_{i=1}^n \lambda_i^{(\zeta)} = \frac{n(n-1)}{2} (|S| - 2).$$

$|S|-2=|\Omega|$ we get by (8.3.8) - (8.3.9) the following:

$$\frac{n(n-1)}{2} |\Omega| \geq - \sum_{i-1}^n m_i + |\Omega|\cdot \frac{(n-2)(n-1)}{2} + \sum_{\zeta\in\Omega} \mathrm{ord}_\zeta R(x).$$

Thus $\sum_{\zeta\in\Omega}\mathrm{ord}_\zeta R(x) \leq \sum_{i=1}^n m_i+|\Omega|(n-1)$ or $\sum_{\zeta\in\Omega} \{\mathrm{ord}_\zeta R(x)-n+1\} \leq \sum_{i=1}^n m_i$.

The inequality (8.3.3) can be improved by a more careful analysis. E.g. when $\mathrm{ord}_{\zeta_k}R(x) \geq n$ for all $k=1,\ldots,p$; we can change $p(n-1)$ in r.h.s. of (8.3.3) to $(n-1)$.

§9 The construction of the Padé approximation to the solutions of linear differential equations.

9.0. In this chapter we describe the explicit construction of the Padé approximation to the solutions of linear differential equations formulated in terms of monodromy data. In our presentation we follow the papers [2], [17], [18] in which the reduction of the Padé approximation problem to studies of contiguous system of functions was performed.

9.1. First of all we begin with the systems of linear differential equations in canonical regular form (1.10.1). Let's consider a basis in $\Phi(x)$ in $S\begin{pmatrix}V_1,\ldots,V_m\\a_1,\ldots,a_m\end{pmatrix}$ satisfying a canonical regular system of equations (3.1.1):

(9.1.1) $$\frac{d\Phi(x)}{dx} = \sum_{\nu=1}^m \frac{\Phi(x)U_\nu}{x-a_\nu} .$$

We assume as in 3.3.1 and 3.3.5 that for $\Phi(x)$ there exist exponential matrices W_ν at any point $x=a_\nu$: $\nu=1,\ldots,m$ including the exponent W_∞ at infinity $x=a_\infty=\infty$. The conditions for the existence of the exponential matrices are given in 3.3.5; however these can be removed because the only thing we actually need is to know the eigenvalues of W_ν defined by (3.3.6) as well as the normal form of V_ν: $\nu=1,\ldots,m,\infty$. We have by (3.3.2) - (3.3.3):

(9.1.2) $$\Phi(x) = (x-a_\nu)^{W_\nu}\cdot H_\nu(x)$$

for $H_\nu(x)$ and $H_\nu(x)^{-1}$ being analytic at $x=a_\nu$. Let $\xi_1^{(a_\nu)},\ldots,\xi_n^{(a_\nu)}$ be eigenvalues of W_ν and let $\vec{\xi}^{(a_\nu)}=(\xi_1^{(a_\nu)},\ldots,\xi_n^{(a_\nu)})$ with the order of eigenvalues corresponding to the normal form of V_ν. Then from 6.2 it follows that the matrix of the local exponents of $\Phi(x)$ has the form:

(9.1.3) $\Phi(x) \sim (\vec{\xi}^{(a_\nu)t},\dots,\vec{\xi}^{(a_\nu)t})$ at $x=a_\nu$ for $\nu=1,\dots,m,\infty$.

We consider now a fixed set S in $\mathbb{C}$ for which the Padé approximation problem will be solved. We assume that $S\subseteq\{a_1,\dots,a_m\}$ since can always consider any regular point b of (9.1.1) to be a virtual singularity of (9.1.1) by putting $U_b=0$ and $V_b=I$. In order to construct Padé approximation to $\Phi(x)$ we consider isomonodromy deformation of $\Phi(x)$ changing eigenvalues of exponential matrices W_a at $a\in S$ by rational integers.

9.2. Starting with $\Phi(x)$ and the exponential matrices W_a of $\Phi(x)$ at $x=a\in S$ we can construct different contiguous bases of $S\begin{pmatrix}V_1,\dots,V_m\\a_1,\dots,a_m\end{pmatrix}$, connected with $\Phi(x)$ by polynomial relations. For this we consider the most general case when W_a is changed to an arbitrary solution W_a' of the equation

(9.2.1) $2\pi\sqrt{-1}\cdot W_a = \mathrm{Log}\ V_a$: $a\in S$.

Then for $V_a=S_a^{-1}\cdot[J_{k_1}(\rho_1^a),\dots,J_{k_1}(\rho_1^a)]S_a$ and $k_1=k_1^a,\dots,k_l=k^a{}_{l_a}$ we obtain the following expression for W_a' from (3.4.3):

(9.2.2) $W_a' = W_a + S_a^{-1}\cdot[I_{k_1}\cdot r_1^{(a)},\dots,I_{k_l}\cdot r^{(a)}_{l_a}]\cdot S_a$

for arbitrary rational integers $r_1^{(a)},\dots,r_{l_a}^{(a)}$. We assume however

(9.2.3) $r_i^{(a)} \geq 0$: $a\in S$.

At infinity we choose an exponential matrix W_∞' with

(9.2.4) $W_\infty' = W_\infty + S_\infty^{-1}\circ[I_{k_1}\cdot m_1^{(\infty)},\dots,I_{k_{l_\infty}}\cdot m_{l_\infty}^{(\infty)}]\cdot S_\infty$.

Let all eigenvalues of W_a' be $(\xi_1^{(a)}+r_1^{(a)},\dots,\xi_n^{(a)}+r_n^{(a)})$ for $a\in S$ and let eigenvalues of W_∞' be $(\xi_1^{(\infty)}+m_1^{(\infty)},\dots,\xi_n^{(\infty)}+m_n^{(\infty)})$. If as in 6.1 $\sigma(\bar{r}^{(a)})=\sum_i r_i^{(a)}$, $\sigma(\bar{m}^{(\infty)})=\sum_i m_i^{(\infty)}$, then the formula $e^{2\pi\sqrt{-1}W_1'}\dots e^{2\pi\sqrt{-1}W_\infty'}=I$ implies:

(9.2.5) $\sum_{a\in S}\sigma(\bar{r}^{(a)}) + \sigma(\bar{m}^{(\infty)}) = 0.$

Now we choose a nonsingular matrix solution $\Psi(x)$ of the canonical regular systems having exponential matrices W_a' at $x=a\in S\cup\{\infty\}$ and W_a at finite $x=a\notin S$. This $\Psi(x)$ is a basis of $S\begin{pmatrix}V_1,\dots,V_n\\a_1,\dots,a_n\end{pmatrix}$ and its existence follows from [4] or [10]. From (1.7.2) we have

(9.2.6) $\Phi(x)\cdot P(x) = \Psi(x)$

for n x n matrix P(x) with rational function coefficients. Here we have in addition to (9.1.2):

(9.2.7) $\Psi(x) = (x-a)^{W_a'}\cdot H_a'(x)$

for $H_a'(x)$, $H_a'(x)^{-1}$ being regular at x=a and we define $W_a'=W_a$ for $a\in\mathbb{C}\setminus S$.

Theorem 9.2.8. Let us suppose that $\Phi(x)$ satisfies (9.1.1) - (9.1.3) and $\Psi(x)$ satisfies (9.2.6) - (9.2.7) with W_a' for $a\in S\cup\{\infty\}$ defined in (9.2.1) - (9.2.5). Then P(x) is a matrix with polynomial coefficients. Moreover in the notations of (9.2.4), let

(9.2.9) $N = -\min \{m_i^{(\infty)}\}$.

Then the polynomial $P(x)$ is of degree **N**.

Proof. According to (9.1.2), (9.2.6) - (9.2.7) we have $P(x)=\Phi(x)^{-1}\Psi(x)=$ $=H_a(x)^{-1}(x-a)^{-W_a}.(x-a)^{W_a'}.H_a'(x)$. Since W_a and W_a' commute (see 3.4 above) we have $P(x)=H_a(x)^{-1}.(x-a)^{W_a'-W_a}.H_a'(x)$. Because $W_a'-W_a=S_a^{-1}\cdot\mathrm{diag}(r_1^{(a)},\ldots,r_n^{(a)})\cdot S_a$ we get

(9.2.10) $P(x) = (S_a\cdot H_a(x))^{-1}\cdot\mathrm{diag}\,((x-a)^{r_1^{(a)}},\ldots,(x-a)^{r_n^{(a)}})\cdot S_a\cdot H_a'(x)$

For any finite point a the condition (9.2.3) gives $r_i^{(a)}\geq 0$ (where $r_i^{(b)}\equiv 0$ when $b\in\mathbb{C}\setminus S$) for $i=1,\ldots,n$ Thus $P(x)$ is regular in $\mathbb{C}$ and is a polynomial. On infinity we have

(9.2.11) $P(x) = (S_\infty H_\infty(x))^{-1}\cdot\mathrm{diag}\,(x^{-m_1^{(\infty)}},\ldots,x^{-m_n^{(\infty)}})\cdot S_\infty\cdot H_\infty'(x)$.

Since $H_\infty(x)^{-1}$, $H_\infty'(x)$ are holomorphic at $x=\infty$, $P(x)$ is polynomial of the degree $\max\{-m_1^{(\infty)},\ldots,-m_n^{(\infty)}\}=N$.

The theorem 9.2.8 is the basis of our construction of the Padé approximation to vector solutions of (9.1.1).

9.3. If we consider the exponential matrices W_b' for b being a regular point of (9.1.1), then $V_b=I$ and the set of all possible solutions W of $2\pi\sqrt{-1}W=\mathrm{Log}\, I$ has the form

$$W = S\cdot\mathrm{diag}\,(r_1,\ldots,r_n)\cdot S^{-1}$$

for an arbitrary matrix S and an arbitrary sequence $(r_1,\ldots,r_n)$ of rational integers.

9.4. We use Theorem 9.2.8 to construct rational approximation to a solution $\vec{y}(x)$ of (9.1.3)

(9.4.1) $\frac{d}{dx}\vec{y}(x) = \vec{y}(x)\cdot\sum_{\nu=1}^{m}\frac{U_\nu}{x-a_\nu}$.

We consider as in 9.2 Padé approximations to $\vec{y}(x)$ at a fixed set $S\subseteq\{a_1,\ldots,a_m\}$. In order to apply the definition of a Padé approximation we must assume that $\vec{y}(x)$ is analytic at any $x=a\in S$. We leave for the future the discussion of the case when the same solution $\vec{y}(x)$ has zero exponents at several real singularities $x=a_\nu$, because this implies the existence of the new relations among V_ν. Thus we assume that S contains only one, say a_1, real singularity of (9.1.1) at which monodromy is nontrivial ($V_\nu\neq I$); however S may have many regular points. We may assume, without loss of generality that $\vec{y}(x)$ corresponds to the exponent $\xi_1^{(a_1)}=0$ at $x=a_1$:

(9.4.2) $\vec{y}(x)$ is regular at $x=a_1$ and at any $x=a\in S$

According to §6 this means that 1 is an eigenvalue of V_1 and there is a simple normal factor $(\lambda-1)$ of V_1. Let S_1 be the matrix which reduces V_1 to the normal form: $V_1=$ $= S_1^{-1}[\xi_1^{(a_1)}, J_{k_2}(\rho_1^{1}),\ldots,J_{k_1}(\rho_1^{1})]\, S_1$ with a unit Jordan block for $\xi_1^{(a_1)}$. If

$S_1\Phi(x)=(\vec{y}_1^{(a_1)t},\ldots,\vec{y}_n^{(a_1)t})^t$, then we can identify $\vec{y}(x)$ with $\vec{y}_1^{(a_1)}$. We use 9.2.8 to construct a rational approximation to $\vec{y}$, which is not exactly Padé approximation since order of zero of the remainder function is less than in 8.0.1 - 8.0.3.

9.5. Let's consider the system of exponential matrices W_b' for $b\varepsilon S$ satisfying the following conditions of the type (9.2.2) - (9.2.3)

(9.5.1) $$W_b' = S_1^{-1}\cdot\text{diag }(k_b,0,0,\ldots,0)\cdot S_1 \ : \ b\varepsilon S$$

for rational integers $k_b\geq 0$: $b\varepsilon S$. Then for $x=\infty$ we define an exponential matrix W_∞' to satisfy (9.2.4), (9.2.5):

(9.5.2) $$W_\infty' = W_\infty + S_\infty^{-1}\cdot\text{diag }(-m,\ldots,-m)\ S_\infty \ \text{ and}$$

(9.5.3) $$\sum_{b\varepsilon S} k_b = nm$$

Theorem 9.5.4. Let $\Phi(x)$ have exponents W_a (for $a\varepsilon CP^1$) and let exponential matrices W_a' : $a\varepsilon S\cup\{\infty\}$ be defined as in (9.5.1) - (9.5.3) for a given sequence of integers $k_a\geq 0$ ($a\varepsilon S$). Let $\Psi(x)$ satisfy (9.2.6) - (9.2.7) for given matrices W_a' : $a\varepsilon S\cup\{\infty\}$ and $\Psi(x)=\Phi(x)\cdot P(x)$. We denote

(9.5.5) $$S_1\Phi(x) = (\vec{y}_1^{(a_1)}(x)^t,\ldots,\vec{y}_n^{(a_1)}(x)^t)^t;$$

(9.5.6) $$P(x) = (P_{ij}(x))_{i,j=1}^n,$$

(9.5.7) $$S_1\Psi(x) = (R_{i,j}(x))_{i,j=1}^n$$

Then $P_{11}(x),\ldots,P_{n1}(x)$ are polynomials of degrees at most m giving rational approximation to $\vec{y}(x)=\vec{y}_1^{(a_1)}(x)$ at $x=a$ for any $a\varepsilon S$. The function $R_{11}(x)=\sum_{i=1}^n P_{i1}(x)y_{1,i}^{(a_1)}(x)$ has order $\text{ord}_a R_{11}(x)\geq k_a$ for any $a\varepsilon S$ and (9.5.3) is satisfied.

For the proof it's enough to notice that by the choice of $\Psi(x)$ and $S_1\Psi(x)-S_1\Phi(x)P(x)$,

$$S_1\Phi(x)\cdot P(x) = \text{diag}\{x^{k_a},1,\ldots,1\}\cdot H_a'(x)$$

for $a\varepsilon S$. The bound for the degree of $P_{i1}(x)$ follows from 9.2.8.

§10. Explicit expressions and recurrent formulas for one-point Padé approximation.

10.0. We analyze below the nature of the one-point Padé approximation to solutions of linear differential equations. The explicit expressions for Padé approximants are known now only for classical orthogonal polynomials [24]. We reduce the recurrent formula for Padé approximants to Riemann's contiguous relations of the form 1.6.3 or 1.7.2. Using this reduction we propose the explicit contiguous relations for one-point Padé approximations for the solutions of matrix linear differential equations. Following Lappo-Danilevski [10] we put the point $x=\zeta$ of the Padé appoximation at infinity.

10.1. In view of §9, Padé approximations to the solutions (1.10.1) at $x=\infty$ mean that we consider contiguous systems of functions having the same exponential matrices W_ν

at the finite points $a_\nu : \nu=1,\dots,m$. The exponential matrices W_∞ is the only one that changes. We consider system (3.1.1) together with all the notations of §3, e.g. we demand the fundamental solution to be normed at x=b by $Y_b(b)=I$. Thus

$$(10.1.1) \qquad \frac{dY}{dx} = \sum_{j=1}^{m} Y\cdot\frac{U_j}{x-a_j}$$

and we assume that the normed solution $Y_b(x)$ Of (10.1.1) has exponential matrices W_ν at $x=a_\nu : \nu=1,\dots,m,\infty$. The main result in the regular case is the following.

Theorem 10.1.2. (i) Let us consider another system of the form (1.10.1):

$$(10.1.3) \qquad \frac{dY'}{dx} = \sum_{j=1}^{m} Y'\cdot\frac{U_j}{x-a_j}$$

with the solution $Y_b'(x)$ normed at x=b. The necessary and sufficient condition for $Y_b'(x)$ to have exponential matrices W_ν at $x=a_\nu$ for $\nu=1,\dots,m$ and some exponential matrix W_∞' at $x=\infty$, is the existence of the solution P(x) of

$$(10.1.4) \qquad \frac{dP}{dx} = \sum_{j=1}^{m} \frac{PU'_j-U_jP}{x-a_j}$$

such that P(x) and $P(x)^{-1}$ are polynomial matrices and P(b)=I.

Theorem 10.1.5. (i) Let us suppose that a solution P(x) of (10.1.4) which corresponds to matrices U_j': $j=1,\dots,m$ in (10.1.3) exists. Then there exists a sequence $(m_1,\dots,m_n)$ of rational integers such that

$$(10.1.6) \qquad m_1+\dots+m_n = 0,$$

and if $U_\infty=-\sum_{j=1}^{m} U_j$ has eigenvalues $(\tau_1,\dots,\tau_n)$, then the eigenvalues of $U_\infty'= =-\sum_{j=1}^{m} U_j'$ are $(\tau_1+m_1,\dots,\tau_n+m_n)$. Moreover, if V_∞ is a monodromy matrix at $x=\infty$ (common for both (10.1.1) and (10.1.3)) and S reduces V_∞ to a normal form: $V_\infty= =S\cdot[\dots]\cdot S^{-1}$, then the exponential matrix W_∞' has the form:

$$(10.1.7) \qquad W_\infty' = W_\infty + S\cdot\mathrm{diag}\,(m_1,\dots,m_n)\cdot S^{-1}.$$

(ii) The matrix polynomial P(x) has degree $\max\{-m_1,\dots,-m_n\}$ and the polynomial $P(x)^{-1}$ has degree $\max\{m_1,\dots,m_n\}$.

We can get direct proofs of Theorems 10.1.4 - 10.1.5 using Theorems 1.7.2 and 9.2.8. The matrix P(x) is defined by

$$P(x) = Y_b(x)^{-1}\cdot Y_b'(x); \quad P(x)^{-1} = Y_b'(x)^{-1}\cdot Y_b(x).$$

Then by the theorem 1.7.2, P(x) is a matrix with rational functioncoefficients, and by 9.2.8, both P(x) and $P(x)^{-1}$ are polynomial matrices. Since the monodromy matrices $V_\nu : \nu=1,\dots,m$ are the same for (10.1.1) and (10.1.3), $V_\infty=(V_1\dots V_m)^{-1}$ is the same for (10.1.1) and (10.1.3). Thus the exponential matrix W_∞' for (10.1.3) is determined by the rule (3.4.3). If S is any matrix which reduces V_∞ to the normal form:

$$(10.1.8) \qquad V_\infty = S\cdot[J_{r_1}(\omega_1),\dots,J_{r_s}(\omega_s)]\cdot S^{-1},$$

then

$$(10.1.9)\qquad W_\infty' = W_\infty + S\,[I_{r_1}\cdot n_1,\ldots,I_{r_s}\cdot n_s]\cdot S^{-1}.$$

Here in agreement with (10.1.7) we have $[I_{r_1}\cdot n_1,\ldots,I_{r_s}\cdot n_s]=\mathrm{diag}\,(m_1,\ldots,m_n)$. Applying Theorem 9.2.8 to $\Phi(x)=Y_b(x)$ we get $\deg(P(x))=\max\{-m_1,\ldots,-m_n\}$, and applying 9.2.8 to $\Phi(x)=Y_b'(x)$ we get $\deg(P(x)^{-1})=\max\{m_1,\ldots,m_n\}$.

10.2. If eigenvalues of U_j doesn't differ by rational integers, then exponents W_j exist and a system of integers $(m_1,\ldots,m_n)$ satisfying 10.1.5 exists. If $U_\infty=-\sum_{j=1}^m U_j$ also has eigenvalues distinct (mod $\mathbb{Z}$), then for a given $(m_1,\ldots,m_n)$ with (10.1.6), the solution is unique. However, as in [10] the system (10.1.3) doesn't exist necessarily for any $(m_1,\ldots,m_n)$. If U_∞ is arbitrary, then the system (10.1.3) depends not only on $(m_1,\ldots,m_n)$ but on S in (10.1.9) as well (cf. 9.3 for $V_\infty=I$).

10.3. We present now expressions for P(x) and U_j' in the case $\max(|m_1|,\ldots,|m_n|)=1$ [10]. The general case follows from this by iterations only. We assume $m_i=1$, $m_j=-1$, $m_k=0$ for $k\neq i,j$. Then $P=I+A(x-b)$, $P^{-1}=I-A(x-b)$ and $A^2=0$. Substituting expressions for $P(x)$, $P(x)^{-1}$ into (10.1.4) we get

Lemma 10.3.1. All solutions of (10.1.4) with $\deg(P(x))=\deg(P(x)^{-1})=1$ have the form

$$(10.3.2)\qquad U_j' = [I-(a_j-b)A]\cdot U_j\cdot[I+(a_j-b)A]$$

with A satisfying the system of equations

$$(10.3.3)\qquad A^2 = 0;\ A\cdot\sum_{j=1}^m U_j - \sum_{j=1}^m U_j\cdot A = A-A\cdot\sum_{j=1}^m a_jU_j\cdot A.$$

There U_j' is similar to U_j: $j=1,\ldots,m$. The matrix A has zero eigenvalues and in its normal form Jordan blocks are of sizes at most two. Let A correspond to one Jordan block $\binom{01}{00}$ at the place (i,j) and other zero eigenvalues are simple. Then (10.3.3) is equivalent to: $A^2=0$ and $[U_\infty,A]=\delta A$ with $\delta=\mathrm{tr}(SA)-1$ and $S=\sum_{j=1}^m a_jU_j$. Using the result of 4.3 we obtain:

Lemma 10.3.4. Let $U_\infty=-\sum_{\nu=1}^m U_\nu$ has a normal form $U_\infty=S_\infty^{-1}\,\mathrm{diag}(\lambda_1^{(\infty)},\ldots,\lambda_n^{(\infty)})\cdot S_\infty$ and $\sum_{\nu=1}^m a_\nu(S_\infty U_\nu S_\infty^{-1})_{ij}\neq 0$. Then the system (10.3.3) has a solution

$$(10.3.5)\qquad A(i,j) = (\lambda_i^{(\infty)}-\lambda_j^{(\infty)}+1)(\sum_{\nu=1}^m a_\nu(S_\infty\ U_\nu S_\infty^{-1})_{ij})^{-1}\cdot S_\infty^{-1}\cdot E_{ij}\cdot S_\infty$$

for $(E_{ij})_{i'j'}=\delta_{ii'}\cdot\delta_{jj'}$. This solution corresponds to a case $m_i=1$, $m_j=-1$ and U_∞' has eigenvalues $(\lambda_1^{(\infty)},\ldots,\lambda_i^{(\infty)}+1,\ldots,\lambda_j^{(\infty)}-1,\ldots,\lambda_n^{(\infty)})$.

10.4. For the most general equation (4.1.1) the formulae for one-point Padé approximations are almost the same. We must change only (10.1.4) for

$$(10.4.1)\qquad \frac{dP}{dx} = \sum_{j=1}^m\sum_{r=1}^S \frac{PU_j^{(r)'}-U_j^{(r)}P}{(x-a_j)^r}$$

and (10.3.3) for

$$A^2 = 0,\quad [A,\sum_{j=1}^m U_j^{(1)}] = A-A\cdot\sum_{j=1}^m (a_jU_j^{(1)}+U_j^{(2)})\cdot A.$$

Similarly in (10.3.5) we change $a_\nu U_\nu$ for $a_\nu U_\nu^{(1)}+U_\nu^{(2)}$.

§11. Perfectness and "almost perfectness" of the p-point Padé approximation to solutions of linear differential equations.

11.0. We study perfectness and "almost perfectness (in the sense of §8) of Padé approximations to solutions of linear differential equations with regular singularities. For this we use the results of §§5-7 on local exponents of Fuchsian modules.

11.1. The most natural system $\vec{y}(x)=(y_1(x),\dots,y_n(x))$ to be Padé approximated is the system of functions satisfying the system of linear differential equations of the first order:

$$(11.1.1) \qquad \frac{d}{dx}\vec{y}(x) = \vec{y}(x)\mathcal{A}(x)$$

for $\mathcal{A}(x)\in M_n(\mathbb{C}(x))$. Let's consider $\vec{y}(x)$ to be a row in the fundamental matrix solution $Y(x)$ of (11.1.1):

$$(11.1.2) \qquad \frac{dY(x)}{dx} = Y(x)\mathcal{A}(x) \text{ and}$$

$$(11.1.3) \qquad Y(x) = (\vec{y}_1^{\,t}(x),\dots,\vec{y}_n^{\,t}(x))^t = (\vec{f}_1^{\,t}(x),\dots,\vec{f}_n^{\,t}(x)).$$

We can consider the Padé approximation to $\vec{y}_1(x)$ (or the simultaneous system of Padé approximations to $\vec{y}_1(x),\dots,\vec{y}_n(x)$). Let $\vec{P}(x)=(P_1(x),\dots,P_n(x))$ be a system of polynomials of degrees $(m_1,\dots,m_n)$, and we introduce n remainder functions

$$(11.1.4) \qquad R_i(x) = \vec{P}(x)\cdot\vec{y}_i^{\,t}(x) = \sum_{j=1}^{n} P_i(x)y_{i,j}(x) \ : \ i=1,\dots,n.$$

11.2. The Fuchsian module associated with the system of polynomials $\vec{P}(x)=(P_1(x),\dots,P_n(x))$ and the system (11.1.1) is introduced:

$$(11.2.1) \qquad \mathcal{P} = \{\vec{P}(x)\cdot\vec{y}^{\,t}(x)\text{: for all solutions } \vec{y}(x) \text{ of } (11.1.1)\}.$$

Lemma 11.2.2. In the notations of 7.1, the $\mathbb{C}$-module $\mathcal{P}$ is the Fuchsian module $\mathcal{L}(\vec{R})$, where $\vec{R}(x)=(R_1(x),\dots,R_n(x))$.

Let Sing be the set of all singularities of (11.1.2) (i.e. of $\mathcal{A}(x)$ including ∞) and we add to Sing all the set of all apparent singularities of each of the Fuchsian modules $\mathcal{L}(\vec{f}_1),\dots,\mathcal{L}(\vec{f}_n)$ for n columns $\vec{f}_1,\dots,\vec{f}_n$ of $Y(x)$ in (11.1.3). Let Sing$=\{a_1,\dots,a_m,\infty\}$ and V_ν be a monodromy matrix of $Y(x)$ at $x=a_\nu$: $\nu=1,\dots,m,\infty$ as in (6.0.2) corresponding to a fixed basis $\gamma_1,\dots,\gamma_m,\gamma_\infty$ of $\pi^1(\mathbb{CP}^1\setminus\{a_1,\dots,a_m,\infty\})$, $\gamma_1+\dots+\gamma_m+\gamma_\infty=0$. The discussion of 6.0 - 6.1 gives us:

Lemma 11.2.3. Let the singular set Sing of (11.1.2) be $\{a_1,\dots,a_m,\infty\}$ and V_ν be the monodromy of $Y(x)$ at $x=a_\nu$: $\nu=1,\dots,m,\infty$. Then $\vec{R}(x)=(R_1(x),\dots,R_n(x))$ is an element of $S\begin{pmatrix}V_1,\dots,V_m\\ a_1,\dots,a_m\end{pmatrix}$.

Proof. By (11.1.3) - (11.1.4) we have $\vec{R}=\vec{P}\cdot Y^t$ or $\vec{R}^t=Y\cdot\vec{P}^t$. Thus by (6.0.2), after encircling a circuit γ_ν we get :

$$\vec{R}(x)^t \underset{\gamma_\nu}{\longmapsto} V_\nu\cdot Y\cdot\vec{P}^t = V_\nu\vec{R}(x)^t \ : \ \nu=1,\dots,m,\infty.$$

Since $\mathcal{P}=\mathcal{L}(\vec{R})$ is a Fuchsian module, according to 7.2.1. the set $\mathcal{P}$ is a set of all solutions of a Fuchsian linear differential equation of the order $\dim_{\mathbb{C}}\mathcal{P}\le n$. We can, following 7.4, denote by $\mathrm{Ord}_c P$ the sum of all local exponents of elements of P at $x=c$ for c in $\mathbb{CP}^1$.

11.3. Let $V_\nu = S_\nu^{-1} \cdot [J_{k_1}(\rho_i^{\nu}),\ldots,J_{k_l}(\rho_l^{\nu})] \cdot S_\nu$ as in 6.1 so that the fundamental matrix $Y(x)$ changes into $V_\nu \cdot Y(x) = Z_\nu(x)$ with $Z_\nu(x)$ having the canonical form (6.1.9) - (6.1.10). Following this structure of $Z_\nu(x)$ in 6.1 we defined the matrix of the local exponents of $Y(x)$ at $x=a_\nu$ in the form (6.1.11) (cf. (6.1.4)):

(11.3.1) $$Y(x) \sim (\vec{\lambda}_1^{(a_\nu)^t},\ldots,\vec{\lambda}_n^{(a_\nu)^t}) \text{ at } x=a_\nu;$$

(11.3.2) $$\vec{\lambda}_i^{(a_\nu)} \equiv \vec{\lambda}^{(a_\nu)} (\mathrm{mod}\ \mathbb{Z}^n) : \quad i=1,\ldots,n.$$

Lemma 11.3.3. For any finite singularity a in Sing we have

(11.3.4) $$\mathrm{Ord}_a P \geq \sum_{i=1}^n \min\{\lambda_{ji}^{(a)} : j=1,\ldots,n\},$$

where, as in (6.1.4), $\vec{\lambda}_j^{(a)} = (\lambda_{j,1}^{(a)},\ldots,\lambda_{j,n}^{(a)}) : j=1,\ldots n.$

Proof. We have $\vec{R}^t = Y \cdot \vec{P}^t$, so by (6.1.8)

(11.3.5) $$S_\nu \vec{R}^t = S_\nu \cdot Y \cdot \vec{P}^t = Z_\nu(x) \cdot \vec{P}(x)^t.$$

Since S_ν is nonsingular, $S_\nu \vec{R}^t$ is a generator of P and for $S_\nu \vec{R}^t = (r_1(x),\ldots,r_n(x))^t$,

(11.3.6) $$r_i(x) = \sum_{j=1}^n P_j(x) Z_\nu(x)_{ij} = \sum_{j=1}^n P_j(x) \cdot d_{ij}^{(a)}(x) \times (x-a)^{\lambda_{j,i}^{(a)}} \cdot \phi_{a,ij}(x)$$

with the sturcture of $d_{ij}^{(a)}(x)$ described in (6.1.10). Then from (11.3.6) it follows that $r_i(x)$ belongs to exponents $\lambda_{j,i}^{(a)} \equiv \lambda_i^{(a)} (\mathrm{mod}\ \mathbb{Z})$ and the $\mathrm{ord}_a r_i(x)$ is properly defined. Then (11.3.6) gives us for a finite a:

(11.3.7) $$\mathrm{ord}_a r_i(x) \geq \min\{\lambda_{j,i}^{(a)} : j=1,\ldots,n\}:$$

$i=1,\ldots,n$; which immediately implies (11.3.4).

Lemma 11.3.8. At infinity we have

(11.3.9) $$\mathrm{Ord}_\infty P \geq \sum_{i=1}^n \min\{\lambda_{ji}^{(\infty)} - d(P_j) : j=1,\ldots,n\}.$$

Proof. We apply (11.3.6) for $a=\infty$ and $V_\infty = \{V_1 \ldots V_m\}^{-1}$. Then expressions (6.1.10) for $d_{ij}^{(\infty)}(x)$, and the fact that for a polynomial $P(x)$, $\mathrm{ord}_\infty P = -d(P)$, give us

$$\mathrm{ord}_\infty r_i(x) \geq \min\{\lambda_{j,i}^{(\infty)} - d(P_j) : j=1,\ldots,n\}.$$

Applying Corollary 7.5.2 we obtain for any finite set Ω in $\mathbb{CP}^1$ containing $\mathrm{Sing}\cup\{\infty\}$:

$$\sum_{a\in\Omega} \mathrm{Ord}_a P \leq \frac{n(n-1)}{2} (|\Omega| - 2)$$

(since $\dim_{\mathbb{C}} P \leq n$). In particular we obtain

Corollary 11.3.10. For any set S of apparent singularities of P we have

(11.3.11) $$\sum_{b\in S} \{\mathrm{ord}_b R(x) + 1 - n\} \leq \frac{n(n-1)}{2} (|\mathrm{Sing}| - 1) - \sum_{i=1}^n \min\{\lambda_{ji}^{(\infty)} - d(P_j) : j=1,\ldots,n\} - \sum_{a\in\mathrm{Sing}} \sum_{i=1}^n \min\{\lambda_{ji}^{(a)} : j=1,\ldots n\}.$$

One should take into account that, according to 7.5.2,

$$(11.3.12)\qquad \sum_{a\in \mathrm{Sing}\cup\{\infty\}} \sum_{i=1}^{n} \lambda_{ji}^{(a)} = \frac{n(n-1)}{2}(|\mathrm{Sing}|-1)$$

for any j=1,...,n. Then choosing an appropriate Y(x) with different $\vec{\lambda}_1^{(a_\nu)},\dots,\vec{\lambda}_n^{(a_\nu)}$ we get canonical bases of contiguous functions which give normal and "perfect" Padé approximations to $\bar{y}(x)$. Results (11.3.11)-(11.3.12) are a considerable improvement over (8.2.3).

11.4. In particular we can prove normality in the case of $\{y(x),\dots,y^{(n-1)}(x)\}$ for y(x) satisfying a Fuchsian linear differential equation (5.1.1) of order n. For this we use the reduction of a scalar equation to a matrix form (5.1.4)-(5.1.8). Taking into account the computation of local exponents in 5.2, we come to the following conclusion:

Theorem 11.4.1. Let y(x) satisfy a Fuchsian linear differential equation $L\{y\} = 0$ of order n, and let $a_1,\dots,a_m$ be all finite singularities of $L\{y\} = 0$ and $P(x)=\prod_{i=1}^{m}(x-a_i)$. We consider polynomials $P_1(x),\dots,P_n(x)$ of degrees $d(P_1),\dots,d(P_n)$ and $d(P_i)=N+(n-i)(m-1)$: i=1,...,n. Then for

$$R(x)=P_1(x)y(x)+P_2(x)P(x)y'(x)+\dots+P_n(x)P(x)^{n-1}y^{(n-1)}(x)\not\equiv 0$$

we have

$$\sum_{a\in\Omega}\{\mathrm{ord}_a R+1-n\} \le \sum_{i=1}^{n} d(P_i).$$

References

1. B. Riemann, Oeuvres Mathematiques, A Blanchard, Paris, 1968.

2. G. V. Chudnovsky, Padé approximation and the Riemann problem, Lectures, Cargese summer school on mathematical physics, June 1979, Reidel Publishing Company, Dordrecht, Holland, 1979, 62 pp.

3. J. A. Lappo-Danilevski; Mat. Sbornik, 34 (1927), 113-148.

4. J. Plemelj, Problems in the sense of Riemann and Klein, John Wiley, 1964.

5. E. Hille, Ordinary differential equations in the complex domain, John Wiley, 1976.

6. A. R. Forsyth, Theory of differential equations, VI volumes, Dover, 1959.

7. I. L. Ince, Ordinary differential equations, Dover, 1959.

8. V. V. Golubev, Lectures on the analytic theory of differential equations, GTIL, Moscow, 1950.

9. L. Fuchs, Gesammelte Mathematische werke, Bd. 1-3, Berlin, 1900-1906.

10. J. A. Lappo-Danilevski, Mémoires sur la théorie des systèmes des équations différentielles linéaires, Chelsea, NY, 1953.

11. L. Markus, Group theory and differential equations, Lecture notes, University of Minnesota, 1959-1960.

12. D. V. Chudnovsky, Riemann monodromy problem, isomonodromy deformation equations and completely integrable systems, Lectures, Cargese summer school on mathematical physics, June 1979, Reidel Publishing Company, Dordrecht, Holland, 1979, 61 pp.

13. D. V. Chudnovsky, Proceedings of the International Meeting "Nonlinear evolution equations and dynamical systems," Lecce, Italy, June 1979, Lect. Not. Phys., Springer, 1979 (to appear).

14. D. V. Chudnovsky and G. V. Chudnovsky, CEN-Saclay Preprint DPh-t-79/115. Seminar on the Riemann problem.

15. G. V. Chudnovsky a) C.R. Acad. Sci. Paris, 288A (1979), pp. A-607 - A609; b) C.R. Acad. Sci., 286A (1970), pp. A-965 - A-967.

16. G. V. Chudnovsky, a) C.R. Acad. Sci. Paris, 288A (1979), pp. A-1001 - A 1004; b) J. Math. Pure Appl. v. 56 (1979), 32 pp.

17. G. V. Chudnovsky, a) Proceedings of the International Meeting "Nonlinear evolution equations and dynamical systems," Lecce, Italy, June 1979, Lect. Not. Phys., Springer, 1979 (to appear), 60 pp.; b) Lect. Not. in Math., Springer, 1979, v. 751, pp. 45 - 69.

18. G. V. Chudnovsky, C.R. Acad. Sci. 289A (1980), pp. A-15 - A-17.

19. Équations différentielles et systèmes de Pfaff dans le champ complexe, Lect. Not. Math., v. 712, Springer, 1979.

20. K. Mahler, Composito Math., 19 (1968), 95-166.

21. H. Jager, Proc. Kon. Nederl. Wet. 67 (1964), 192-249.

22. J. H. Loxton, A. J. van der Poorten, Rocky Moun. J. Math., v. 9, 3 (1979), 385-393.

23. K. Mahler, Lectures on Transcendental Numbers, Lect. Not. Math., v. 546, Springer, 1976.

24. Bateman, A. Erdelyi. Higher Transcendental Functions, v. 1-3, London, 1953.

25. D. Brownawell, D. Masser (to appear).

METHODE DE LA PHASE STATIONNAIRE ET SOMMATION DE BOREL

Bernard MALGRANGE
Laboratoire de Mathématiques Pures - Institut Fourier
dépendant de l'Université Scientifique et Médicale de
Grenoble associé au C.N.R.S.
B.P. 116, 38402 ST Martin D'Heres, France

I. - Les quelques remarques qui suivent m'ont été suggérées par le travail de Balian-Parisi-Voros sur l'oscillateur quartique [2] ; elles ont pour but d'examiner, dans un cas particulier simple, comment des corrections exponentiellement petites ("instantons"), jointes à un procédé convenable de resommation, permettent de déduire une formule exacte des développements asymptotiques obtenus par la méthode de la phase stationnaire.

Soit P un polynôme à n variables $(x_1,\dots,x_n) = x$, à coefficients réels, vérifiant les deux hypothèses suivantes :

(P 1) $P(x) \to +\infty$ si $x \to +\infty$;

(P 2) les points critiques réels $x^1,\dots,x^k$ de P sont "algébriquement isolés", c'est-à-dire que, au voisinage de x^ℓ $(\ell = 1,\dots,k)$, les équations $\frac{\partial P}{\partial x_j}(x) = 0$ n'admettent pas d'autre solution complexe que $x = x^\ell$.

Soit alors φ une fonction C^∞ à support compact sur $\mathbb{R}^n$, et considérons l'intégrale

$$I_\varphi(\tau) = \int e^{i\tau P(x)}\varphi(x)dx \;,\; (dx = dx_1\dots dx_n) \;.$$

Il est classique que, pour τ réel tendant vers $+\infty$, $I_\varphi(\tau)$ admet un développement asymptotique de la forme suivante

$$(A_\varphi) \;:\; I_\varphi(\tau) \sim \sum e^{i\tau d_\ell} f_\ell(\tau) \quad , \quad f_\ell(\tau) = \sum_{\lambda,\mu} a_{\lambda,\mu,\ell}\, \tau^\lambda (\log \tau)^\mu \;,$$

avec

i) $d_\ell = P(x^\ell)$ (les d_ℓ sont les "valeurs critiques" de P correspondant aux points critiques réels) ;

ii) les λ sont rationnels < 0 , et forment un ensemble "fini modulo $\mathbb{Z}$" (autrement dit, il existe un ensemble fini $\lambda_1,\ldots,\lambda_p$ tel que tout λ soit de la forme λ_q+r , $q = 1,\ldots,p$, r entier ≤ 0) ;

iii) $0 \leq \mu \leq n-1$.

Remarque 1. - En fait, ce résultat n'exige aucune hypothèse sur P . Il est démontré, par exemple, dans [5] dans le cas où P n'a qu'une seule valeur critique, comme conséquence d'un résultat de Jeanquartier, lui-même conséquence du théorème de désingularisation d'Hironaka. Le cas général s'en déduit facilement si l'on remarque que P n'a qu'un nombre fini de valeurs critiques (cette dernière assertion est simplement la version algébrique du théorème de Sard).

Maintenant, de l'hypothèse (P 2), on déduit facilement que les $a_{\lambda,\mu,\ell}$ pour ℓ fixé ne dépendent que des valeurs de φ et de ses dérivées au point x^ℓ (pour simplifier l'exposé, je suppose $d_1,\ldots,d_k$ distincts) ; voir par exemple l'introduction de [5] . Si donc ψ est une fonction C^∞ quelconque sur $\mathbb{R}^n$, on pourra définir "l'intégrale asymptotique" $⨍ e^{i\tau P(x)}\psi(x)dx$ comme étant, par définition, le développement asymptotique de $\int e^{i\tau P(x)}\varphi(x)dx$, φ n'importe quelle fonction C^∞ à support compact égale à ψ au voisinage des points $x^1,\ldots,x^k$.

Prenons en particulier pour ψ un polynôme Q .

LEMME 2. - <u>L'intégrale</u> $J_Q(\tau) = \int e^{i\tau P(x)}Q(x)\,dx$ <u>converge dans le demi-plan</u> $\operatorname{Im}\tau > 0$ <u>et définit une fonction holomorphe dans le demi-plan</u>.

En effet, l'hypothèse (P 1) entraine qu'il existe un $r > 0$ tel qu'on ait, pour x réel assez grand $P(x) \geq \|x\|^r$, avec $\|x\|^2 = \sum x_1^2$ (pour ce type de raisonnement, voir [4] , Appendix). Le lemme s'en déduit immédiatement.

Nous verrons un peu plus loin que la fonction I_Q se prolonge en une fonction analytique en tout point $\tau \in \mathbb{R}-\{0\}$, fonction qu'il est naturel de noter encore I_Q ; en fait, elle se prolonge même en une fonction multiforme sur

$\mathbb{C}-\{0\}$. Le problème consiste donc à étudier la relation qui existe entre $I_Q(\tau)$, $\tau \gg 0$, et le développement asymptotique (A_Q) .

Pour étudier I_Q , notons $\gamma(t)$ $(t \in \mathbb{R})$ la surface de niveau $P(x) = t$, $x \in \mathbb{R}^n$; pour $t \neq d_1,\dots,d_k$, $\gamma(t)$ est lisse ; orientons-la comme le bord de l'ensemble $P(x) < t$. L'hypothèse (P 1) implique que $\gamma(t)$ est compact pour tout $t \in \mathbb{R}$. Un calcul classique montre alors qu'on a,

$$(3) \qquad I_Q(\tau) = \int_{-\infty}^{+\infty} e^{i\tau t} dt \int_{\gamma(t)} Q \frac{dx}{dP} ,$$

l'intégrale figurant au second membre étant absolument convergente pour $\mathrm{Im}\, \tau > 0$.

II. - Pour étudier l'intégrale (3) , un point essentiel consiste à séparer les contributions des différents points critiques. On opère pour cela de la manière suivante : il existe un ensemble fini $F \subset \mathbb{C}$ tel que, en dehors de $P^{-1}(F)$, l'application $P : \mathbb{C}^n \to \mathbb{C}$ soit une fibration C^∞ localement triviale, les fibres étant non singulières (voir par exemple [3] , th. 6.13). Les points de F peuvent être de trois sortes :

i) les points $d_1,\dots,d_k$;

ii) les valeurs critiques de P , correspondant aux points critiques non réels ;

iii) les points $f \in \mathbb{C}$ tels que, au voisinage de f , P ne définisse pas à l'infini une fibration localement triviale. Un exemple de cette dernière situation est, par exemple donné par $P(x,y) = (xy-1)^2+x^2$ et $f = 0$ (je laisse le lecteur examiner les détails).

Naturellement, un point $f \in \mathbb{C}$ peut aussi appartenir à F pour plusieurs des raisons ci-dessus.

Si l'on a un chemin (C) dans $\mathbb{C}-F$, d'origine t_1 et d'extrémité t_2, la fibration se trivialise au-dessus de ce chemin ; on en déduit, en particulier, un isomorphisme $H_{n-1}(P^{-1}(t_1),\mathbb{C}) \simeq H_{n-1}(P^{-1}(t_2),\mathbb{C})$, dont on montre qu'il ne dépend pas de la trivialisation choisie, mais qui peut dépendre de (C) . Si donc on a une classe d'homologie $\delta(t_1) \in H_{n-1}(P^{-1}(t_1),\mathbb{C})$, on en déduit une classe $\delta(t_2) \in H_{n-1}(P^{-1}(t_2),G)$, dite "déduite de $\delta(t_1)$ par continuité le long de (C) ".

Cela étant, supposons qu'on ait $d_1 < \ldots < d_k$, et examinons de ce point de vue le cycle $\gamma(t)$ défini plus haut avec $t \in \mathbb{R}-F$. Pour $t < d_1$, on a évidemment $\gamma(t) = 0$. Lorsque t varie entre $d_{\ell-1}$ et d_ℓ , il est facile de voir que la classe d'homologie de $\gamma(t)$ (que nous noterons encore $\gamma(t)$), varie par continuité au sens précédent, tant qu'on ne passe pas un point $f \in F \cap]d_{\ell-1}, d_\ell[$; au passage d'un tel point f , $\gamma(t)$ varie continuement à condition de "contourner" f , par exemple dans le demi-plan supérieur (cf. figure 1)

fig.1

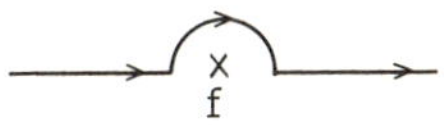

En effet, soit D un voisinage suffisamment petit de f : quoique l'application P ne se trivialise pas sur $P^{-1}(D)$, elle se trivialise au voisinage de $\gamma(f)$ d'une manière compatible avec la conjugaison complexe et cela suffira pour obtenir le résultat cherché.

Reste à examiner ce qui se passe lorsque t passe la valeur d_ℓ . Soient t' , $t'' \in \mathbb{R}$, assez voisins de d_ℓ , avec $t' < d_\ell$, $t'' > d_\ell$, et soit $\bar{\gamma}(t'') \in H_{n-1}(P^{-1}(t''), \mathbb{C})$ la classe obtenue par continuité à partir de $\gamma(t')$ en contournant d_ℓ du côté $\operatorname{Im} t > 0$, c'est-à-dire le long d'un chemin du type indiqué à la figure 2 . Posons $\gamma_\ell(t'') = \gamma(t'') - \bar{\gamma}(t'')$.

fig.2

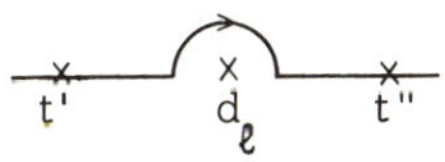

LEMME 4. - γ_ℓ <u>est un cycle évanescent au point</u> x^ℓ (autrement dit, on peut trouver un représentant de $\gamma_\ell(t'')$ de support arbitrairement voisin de x^ℓ , pourvu que t'' soit assez voisin de d_ℓ) .

Nous ne donnerons pas la démonstration détaillée de ce lemme, qui est essentiellement "bien connu". Il est conséquence facile, des résultats de Milnov sur les singularités isolées d'hypersurfaces [6] ; en gros, soit B une boule ouverte de rayon assez petit centrée en x^ℓ ; il faut montrer que, pour t'' assez voisin de d_ℓ , l'image de $\gamma_\ell(t'')$ dans $H_{n-1}(P^{-1}(t'') \bmod P^{-1}(t'') \cap B, \mathbb{C})$ est nulle, ce qui se fait en trivialisant convenablement la restriction de P à un voisinage de $\gamma(d_\ell)-B$ dans $\mathbb{C}^n-B$.

Ceci étant, soit (C_ℓ) un chemin infini obtenu à partir de $]d_\ell, +\infty[$ en contournant dans le demi-plan $\operatorname{Im} t > 0$ les points de $F \cap]d_\ell, +\infty[$; soit $\gamma_\ell(t)$ le cycle obtenu par continuité à partir de γ_ℓ le long de ce chemin ; l'intégrale $\int_{(C_\ell)} e^{i\tau t} dt \int_{\gamma_\ell(t)} Q \frac{dx}{dP}$ est absolument convergente pour $\operatorname{Im} \tau > 0$: au point $t = d_\ell$, cela résulte de la "positivité" des exposants caractéristiques (voir [5]) ; à l'infini, cela résulte des propriétés de croissance des intégrales $\int_{\gamma_\ell(t)} \frac{dx}{dP}$ ("théorème de régularité") que nous rappellerons dans un instant.

Alors, la formule (3) et le lemme 4 nous donnent la formule suivante

$$I_Q(t) = \sum_\ell \int_{(C_\ell)} e^{i\tau t} dt \int_{\gamma_\ell(t)} Q \frac{dx}{dP} . \tag{5}$$

III. - Soit $\mathbb{P}_1(\mathbb{C})$ l'espace projectif complexe à une dimension ("sphère de Riemann") ; le "théorème de régularité" de Nilsson-Griffiths-Katz-Deligne (cf. [3]) nous affirme en particulier ceci : sur $\mathbb{P}_1(\mathbb{C}) - F \cup \{\infty\}$, les $\int_{\gamma_\ell(t)} Q \frac{dx}{dP}$ se prolongent en des fonctions multiformes de détermination finie (i.e. les différentes déterminations au voisinage d'un point forment un espace vectoriel de dimension finie), les différentes déterminations étant à croissance modérée dans tout secteur de sommet un point singulier, i.e. un point de $F \cup \{\infty\}$. Une autre manière de dire la même chose est la suivante : ces fonctions multiformes satisfont, sur $\mathbb{P}_1(\mathbb{C})$, à une équation différentielle dont tous les points singuliers sont "singuliers réguliers", ou "du type de Fuchs".

Ceci nous permet d'abord d'effectuer le prolongement analytique de $I_Q(t)$ pour τ réel > 0 ; soit en effet (C'_ℓ) une demi-droite d'origine d_ℓ, située dans le demi-plan $\operatorname{Im} t > 0$, et faisant avec $[d_\ell, +\infty[$ l'angle α ; si α est suffisamment petit on aura

$$\int_{(C'_\ell)} e^{i\tau t} dt \int_{\gamma_\ell(t)} Q \frac{dx}{dP} = \int_{(C_\ell)} e^{i\tau t} dt \int_{\gamma_\ell(t)} Q \frac{dx}{dP}$$

pour $\operatorname{Im} \tau > 0$ et $\operatorname{Im} (\tau e^{i\alpha}) > 0$, et la première intégrale converge pour $\operatorname{Im}(\tau e^{i\alpha}) > 0$; d'où le prolongement cherché.

Cette formule n'est plus exacte si α est tel que le secteur compris entre (C_ℓ) et (C'_ℓ) contient des points de F (on suppose que (C'_ℓ) n'en contient pas d'autre que d_ℓ) ; il est nécessaire de la corriger en ajoutant

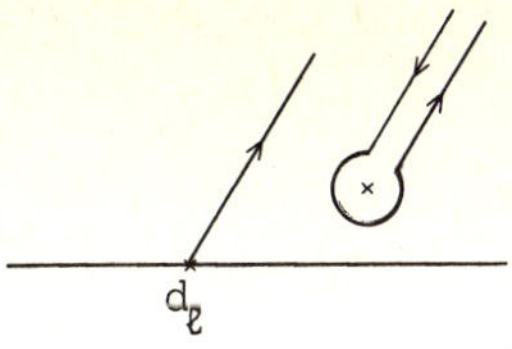

fig. 3

des termes correspondant à des <u>chemins complémentaires</u> entourant les points en question (cf. figure 3). Par ce procédé, en augmentant d indéfiniment, on prolonge analytiquement $I_Q(\tau)$ en une fonction multiforme sur $\mathbb{C}-\{0\}$.

PROPOSITION 5. - *Pour τ réel, tendant vers $+\infty$, $I_Q(\tau)$ admet pour développement asymptotique l'expression $⨍ e^{i\tau P(x)} Q(x) dx$ définie au* § I .

Donnons une esquisse de la démonstration. Il suffit d'établir le résultat suivant : soit (C'_ℓ) un chemin comme ci-dessus, avec $0 < \alpha < \pi$, et soit d'autre part φ une fonction C^∞ à support compact contenu dans un voisinage assez petit de x^ℓ , égale à 1 au voisinage de x^ℓ ; alors la différence

$$\int_{(C'_\ell)} e^{i\tau t} dt \int_{\gamma_\ell(t)} Q \frac{dx}{dP} - \int_{\mathbb{R}^n} e^{i\tau P(x)} Q(x) \varphi(x) dx$$

est à décroissance rapide pour $\tau \to +\infty$. Ceci se voit en reprenant les calculs de [5] , § 6 et 7 et en montrant que le cycle γ_ℓ coïncide avec celui qui est introduit dans cet article, § 7 sous le nom de γ^+ .

IV. - Le but de cet exposé est de montrer que les développements asymptotiques précédents, convenablement resommés, permettent d'obtenir la valeur exacte de $I_Q(\tau)$ pour $\tau \gg 0$, à condition d'ajouter éventuellement des corrections exponentiellement petites correspondant aux "chemins complémentaires" considérés fig. 3. Ces corrections sont elle-mêmes susceptibles du même traitement que nous allons donner maintenant. Tout ceci n'est en fait qu'un cas particulier de la théorie de la resommation des solutions des équations différentielles au voisinage d'un point singulier irrégulier ; voir à ce propos la conférence de J.-P. Ramis.

Rappelons le résultat suivant (Nörlund [7] ; nous nous contenterons ici d'un énoncé relativement grossier ; des estimations plus précises se trouvent chez divers auteurs ; cf. Ramis).

THEOREME 6. - <u>Soit</u> $A > 0$, <u>et soit</u> $D_A \subset \mathbb{C}$ <u>la réunion du disque</u> $|t| < A$ <u>et de la demi-bande</u> : $\mathrm{Re}\, t > 0$, $|\mathrm{Im}\, t| < A$. <u>Soit</u> f <u>holomorphe dans</u> D_A , <u>et telle que</u>, <u>pour un</u> $k > 0$ <u>convenable</u>, $f(t)e^{-kt}$ <u>soit bornée dans</u> $D_A \cap \{\mathrm{Re}\, t \geq 1\}$. <u>Alors</u>, <u>pour</u> $\omega \geq \frac{\pi}{2A}$, <u>la transformée de Laplace</u> $F(\sigma) = \int_0^{+\infty} e^{-\sigma t} f(t)dt$ <u>est développable en série de factorielles</u> $a_o(\omega) + \sum \frac{a_n(\omega)n!}{\frac{\sigma}{\omega}(\frac{\sigma}{\omega}+1)\ldots(\frac{\sigma}{\omega}+n)}$ <u>uniformément convergente dans un demi-plan</u> $\mathrm{Re}\, \sigma \gg 0$.

<u>Remarque 7</u>. - La connaissance de développement asymptotique de F pour $\sigma \to +\infty$, ou plus généralement pour σ tendant vers l'infini sur une demi-droite $\mathrm{Im}\, \sigma = \lambda\, \mathrm{Re}\, \sigma$, $\mathrm{Re}\, \sigma > 0$, permet de calculer de proche en proche les $a_n(\omega)$; par conséquent, ce résultat peut s'interpréter comme un procédé de re-sommation de séries asymptotiques, d'ailleurs cas particulier du procédé de sommation de Borel ; si f est holomorphe dans un ouvert plus grand, d'autres variantes de la méthode de Borel peuvent être aussi envisagées.

Pour pouvoir être appliqué ici, ce théorème doit être un peu généralisé : soit f une fonction multiforme sur $D_A - \{0\}$, de détermination finie ; on fait les hypothèses suivantes :

i) il existe $k > 0$ tel que toutes les déterminations de $e^{-kt}f$ restent bornées dans $D_A \cap \{\mathrm{Re}\, t \geq 1\}$;

ii) dans tout secteur de sommet 0 , toute détermination de f est à croissance modérée au voisinage de 0 ,

iii) n'importe quelle détermination de f sur $]0,+\infty[$ est intégrable au voisinage de 0 .

Soit alors f_o une telle détermination, et posons $F(\sigma) = \int_0^{+\infty} f_o(t)e^{-\sigma t}dt$; on peut établir que F s'écrit comme une somme finie $\sum \sigma^{\alpha}(\log \sigma)^k F_{d,k}(\sigma)$ (avec $\alpha < 0$, k entier ≥ 0) , les $F_{d,k}$ étant justiciables du théorème 6 . Bien entendu, les d et les k sont liés de manière simple au comportement asymptotique de f_o quand $t \to 0$; ce point est laissé au lecteur. Quant à la remarque 7, elle s'appliquera encore ici sans modification.

Si C'_ℓ ne contient pas d'autre point de F que d_ℓ , les résultats précédents s'appliquent au calcul des intégrales $\int_{(C'_\ell)} e^{i\tau t} dt \int_{\gamma_\ell(t)} Q \frac{dx}{dP}$, moyennant le changement de coordonnées $\tau = i\sigma e^{-i\alpha}$ (α étant l'angle de (C'_ℓ) avec $[d_\ell, +\infty[$) ; la connaissance supposée du développement asymptotique de cette intégrale permet donc d'en obtenir la valeur exacte pour les grandes valeurs de τ.

Si l'on prend α assez petit, ceci donnera donc la valeur de $I_Q(\tau)$, pour τ réel $\gg 0$ (je dois cette remarque à L. Boutet de Monvel). Cependant, en pratique, il peut être plus intéressant de prendre α plus grand, par exemple voisin de $\pi/2$. Il intervient alors des intégrales sur des "chemins complémentaires"; théoriquement, elles sont justiciables des mêmes méthodes ; cependant, leur calcul <u>effectif</u> présente au moins une difficulté supplémentaire : même dans le cas où tous les points de F proviennent de points critiques complexes de nature assez simple (par exemple de points de Morse), la connaissance du cycle sur lequel on doit intégrer pose ici un problème de <u>monodromie globale</u>, du système local $t \mapsto H_{n-1}(P^{-1}(t), \mathbb{C})$ associé à la fibration $P : \mathbb{C}^n - P^{-1}(F) \to \mathbb{C} - F$.

A l'heure actuelle, il faut bien dire que l'on sait fort peu de choses sur ce sujet ; cette remarque, un peu désabusée, me servira de conclusion.

Je remercie J.-P. Ramis de m'avoir signalé que l'utilisation de la sommabilité de Borel et les séries de factorielles permettaient dans l'exemple considéré, d'arriver à des formules exactes ; dans une première version, je m'étais contenté, comme [2], d'examiner ce que Poincaré appelle la "sommation au sens des astronomes", c'est-à-dire, le procédé qui consiste à tronquer une série asymptotique au plus petit terme.

BIBLIOGRAPHIE

[1] V.I. ARNOL'D - Remarques sur la méthode de la phase stationnaire (en russe), Uspekhi Mat. Nauk 28-5 (1973), pp.17-45.

[2] R. BALIAN, G. PARISI, A. VOROS - Quartic oscillator, Math. Problems in Feynman path integrals, Springer Lect. Notes in Math (1979).

[3] P. DELIGNE - Equations différentielles à points singuliers réguliers, Springer Lect. Notes in Math n°163 (1970).

[4] L. HÖRMANDER - Linear partial differential operators, Springer (1963).

[5] B. MALGRANGE - Intégrales asymptotiques et monodromie, Ann. E.N.S. 4-7 (1974) pp.405-430.

[6] J. MILNOR - Singular points of complex hypersurfaces, Ann. Math. Studies n°61, Princeton (1968).

[7] N.E. NÖRLUND - Leçons sur les séries d'interpolation, Collection Borel, Gauthiers Villars (1926).

LES SERIES k - SOMMABLES ET LEURS APPLICATIONS

par J. P. RAMIS
INSTITUT DE RECHERCHE MATHEMATIQUE AVANCEE
Laboratoire Associé au C.N.R.S.
Université Louis Pasteur
7, rue René Descartes
67084 STRASBOURG CEDEX

Faire du neuf avec du vieux.
(G. COLOMB)

INTRODUCTION

On trouvera ci-dessous la rédaction de l'exposé fait au Colloque des Houches et quelques compléments. Nous avons essayé de dégager en quelques pages les points essentiels de la théorie qui sera exposée dans une série d'articles en préparation (RAMIS [34], [35], [36], [37], [39] ; RAMIS-THOMANN [40]); on trouvera un résumé plus détaillé dans [33] et dans [38] .

Cette rédaction se divise en deux parties (I à V) et (VI à VIII). La première traite des fondements de la théorie. La seconde de son application aux équations différentielles. (Il y a d'autres applications, en particulier aux équations linéaires aux différences, que nous n'aborderons pas ici.)

La théorie des développements asymptotiques Gevrey se rattache étroitement à un certain nombre de travaux anciens (EULER, STIRLING, LAGUERRE, STIELTJES, POINCARE, BOREL, LEROY, MAILLET, WATSON, F. NEVANLINNA, CARLEMAN, NÖRLUND, PERRON, VALIRON,...), que nous avons essayé de renouveler en utilisant, entre autres, l'approche cohomologique récemment introduite dans la théorie classique par MALGRANGE et SIBUYA.

I. SERIES FORMELLES AVEC CONDITIONS DE GEVREY.

J'ai réintroduit cette notion dans [33], dans une perspective d'interpolation. Elle est due à BOREL, LEROY, PERRON [5], [18], [27], [28].

Soit $\hat{f}(x) = \sum_{n \geq 0} a_n x^n \in \mathbb{C}[[x]]$. Si $\hat{f}$ est le jet à l'origine d'une fonction entière f , les formules :

$$1/\rho = \operatorname*{Liminf}_{n\geq o}(-\operatorname{Log}|a_n|/n \operatorname{Log} n) \quad \text{et}$$

$$(\tau e \rho)^{1/\rho} = \operatorname*{Lim\,sup}_{n\geq o} n^{1/\rho}|a_n|^{1/n} ,$$

définissent l'ordre ρ de f, et, ρ étant fixé, le type τ de f.
En utilisant ces formules on étend la définition de l'ordre ρ et du type τ à une série formelle $\hat{f} \in \mathbb{C}[[x]]$ quelconque ; ρ décrit $]0^+,+\infty]\cup[-\infty,0^-[/\{\pm\infty\}$ (muni de la relation d'ordre évidente) et τ décrit $[-\infty,+\infty]$.

On a le

<u>LEMME</u> 1.1.

Soient $\rho \in]-\infty,0^-[$ et $\tau \in [-\infty, 0[$. On pose $k = -\rho > 0$ et $\tau A''^k = -1$, $A''^k = ekA'^k = kA^k (A,A',A''>0)$. Pour $\hat{f} \in \mathbb{C}[[x]]$, les conditions suivantes sont équivalentes $(\hat{f}(x) = \sum_{n\geq o} a_n x^n)$:

(i) $\hat{f}$ est d'ordre $\leq \rho$ et de type $\leq \tau$.

(ii) Pour tout $\varepsilon>0$, il existe $C_\varepsilon>0$, tel que

$$|a_n| < C_\varepsilon (n!)^{1/k}(A+\varepsilon)^n .$$

(iii) Pour tout $\varepsilon>0$, il existe $C'_\varepsilon>0$, tel que

$$|a_n| < C'_\varepsilon\ n^{n/k}(A'+\varepsilon)^n .$$

(iv) Pour tout $\varepsilon>0$, il existe $C''_\varepsilon>0$, tel que

$$|a_n| < C''_\varepsilon \Gamma(1+{}^n/k)(A''+\varepsilon)^n .$$

DEFINITION 1.2.

Si $\hat{f} \in \mathbb{C}[[x]]$ vérifie les conditions du lemme 1.1., on dira que $\hat{f}$ est Gevrey d'ordre s et d'ordre précisé $(s,A+)$; $s = 1+1/k$. On notera $\hat{f} \in \mathbb{C}[[x]]_{(s,A+)}$.

Si on a $|a_n| < C(n!)^{1/k} A^n$, on dira que $\hat{f}$ est d'ordre précisé (s,A). On notera $\hat{f} \in \mathbb{C}[[x]]_{s,A}$.

On pose $\mathbb{C}[[x]]_s = \bigcup_{A>0} \mathbb{C}[[x]]_{s,A}$ et $\mathbb{C}[[x]]_{(s)} = \bigcap_{A>0} \mathbb{C}[[x]]_{s,A}$.

(Cf. RAMIS [33].)

On vérifie que $\mathbb{C}[[x]]_{s,A}, \mathbb{C}[[x]]_{(s,A+)}, \mathbb{C}[[x]]_s$ et $\mathbb{C}[[x]]_{(s)}$ sont des sous $\mathbb{C}$-algèbres de $\mathbb{C}[[x]]$ contenant $\mathbb{C}\{x\}$; $\mathbb{C}[[x]]_{(s,A+)}, \mathbb{C}[[x]]_s$ et $\mathbb{C}[[x]]_{(s)}$ sont stables par dérivation ; $\mathbb{C}[[x]]_s$ et $\mathbb{C}[[x]]_{(s)}$ sont stables par composition ; $\mathbb{C}[[x]]_{s,A}$ est stable par composition par une fonction entière.

II. DEVELOPPEMENTS ASYMPTOTIQUES GEVREY.

On va considérer des "secteurs" du revêtement universel $\mathbb{R} \times \mathbb{R}_*^+$ de $\mathbb{C}^*$. On note $\underline{x} = (\theta, r)$ le point courant de $\mathbb{R} \times \mathbb{R}_*^+$ et $p(\underline{x}) = x = re^{i\theta}$ sa projection sur $\mathbb{C}^*$. On munit $\mathbb{R} \times \mathbb{R}^+$ de sa structure naturelle de variété analytique réelle à bord, et on prolonge p en $\mathbb{R} \times \mathbb{R}^+ \xrightarrow{p} \mathbb{C}$, que l'on factorise en $\mathbb{R} \times \mathbb{R}^+ \to \mathbb{T} \times \mathbb{R}^+ \xrightarrow{\pi} \mathbb{C}$ ($\mathbb{T} = \mathbb{R}/2\pi\mathbb{Z}$) ; on note $\pi^{-1}(0) = \mathbb{T} \times \{0\} = S$. On munit enfin $\mathbb{R} \times \mathbb{R}_*^+ \cup \{0\}$ de la topologie quotient de celle de $\mathbb{R} \times \mathbb{R}^+$.

Soient $\alpha \in \mathbb{R}$, $\beta \in \mathbb{R}_*^+$, $R \in \mathbb{R}_*^+ \cup \{+\infty\}$. On pose

$$\underline{V}_{\alpha,\beta,R} = \{\bar{x} = (\theta, r)/\alpha - \beta/2 < \theta < \alpha + \beta/2 \, ; \, 0 < r < R\} \, ;$$

$\overline{\underline{V}}_{\alpha,\beta,R}$ est l'adhérence de $\underline{V}_{\alpha,\beta,R}$ dans $\mathbb{R} \times \mathbb{R}_*^+ \cup \{0\}$.

Si $\overline{\underline{W}} - \{0\} \subset \underline{V}$, on note $\underline{W} < \underline{V}$.

Si $\overline{\underline{V}}$ est un secteur fermé, on désigne par $\mathcal{C}(\underline{V})$ l'espace de Banach des fonctions continues sur $\overline{\underline{V}}$ (à valeurs complexes), par $\mathcal{O}(\underline{V})$ l'espace de Fréchet des fonctions holomorphes sur $\underline{V}$ et par $B(\underline{V})$ l'espace de Banach $B(\underline{V}) = \mathcal{C}(\overline{\underline{V}}) \cap \mathcal{O}(\underline{V})$.

Si $f \in \mathcal{O}(\underline{V})$ admet un développement asymptotique sur $\underline{V}$, uniforme sur tout $\underline{W} < \underline{V}$, on note $f \in \mathcal{A}(\underline{V})$.

On rappelle la définition (RAMIS [33]) :

DEFINITION 2.1.

Soit $s > 1$. Soit $f \in \mathcal{O}(\underline{V})$; on dira que f est Gevrey-Roumien d'ordre s si, pour tout $\underline{W} < \underline{V}$, on a

$$\sup_{\underline{x}\in\underline{W}} |f^{(p)}(\underline{x})| < C_{\underline{W}}(p!)^s A_{\underline{W}}^p \quad (C_{\underline{W}}>0\,,\, A_{\underline{W}}>0 \text{ convenables}).$$

On note alors $f \in \mathcal{G}_s(\underline{V}) \subset \mathcal{G}(\underline{V})$. On désigne par $J(f) = \hat{f}$ le jet de f à l'origine, par $\mathcal{G}_{o,s}(\underline{V})$ le noyau de l'application $\mathcal{G}_s(\underline{V}) \xrightarrow{J} \mathbb{C}[[x]]_s$.

La théorie des développements asymptotiques Gevrey est, pour l'essentiel, due à WATSON [49], [50], [51]. Elle a été améliorée par NEVANLINNA [24] (cf. aussi CARLEMAN [7]). Elle présente des avantages considérables par rapport à la théorie classique des développements asymptotiques (POINCARE [31]), grâce en particulier aux deux propriétés suivantes :

a) La propriété de "sommation au plus petit terme" fournit des résultats d'approximation numérique très précis ;

b) La théorie est "non quasi-analytique" pour les "petits ouverts" (ouverture $< \frac{\pi}{k}$) et "quasi-analytique" pour les "gros ouverts" (ouverture $> \frac{\pi}{k}$).

Ainsi la théorie Gevrey est formellement voisine de la théorie classique pour les "petits ouverts" (dans la théorie classique les ouverts sont tous "petits"!) tandis qu'on a un phénomène complètement nouveau de "resommation" pour les "gros ouverts". Plus précisément on a les deux résultats suivants :

THEOREME 2.2. (BOREL-RITT avec majorations GEVREY).

Soit $s>1$. Soit $\underline{V}$ un secteur ouvert d'ouverture $\leq \pi/k$.

On a une suite exacte : $\mathcal{G}_s(\underline{V}) \xrightarrow{J} \mathbb{C}[[x]]_s \to 0$.

C'est la situation "non quasi-analytique".

THEOREME 2.3. Soit $s>1$. Soit $\overline{V}$ un secteur fermé d'ouverture $\geq \pi/k$.

On a une suite exacte : $0 \to \mathcal{G}_s(\underline{V}) \xrightarrow{J} \mathbb{C}[[x]]_s$.

C'est la situation "quasi-analytique".

Le théorème 2.3. est dû à NEVANLINNA [24], et, sous une forme moins précise à WATSON [48].

Les fonctions de $\mathcal{G}_{o,s}$ se caractérisent par une propriété de décroissance exponentielle d'ordre k (RAMIS [33], [34], [38]).

Soit $\varphi(t) = \sum_{n \geq o} b_n t^n \in \mathbb{C}\{t\}$ de rayon de convergence $R > 0$. On définit "sa" transformée de Leroy incomplète d'ordre k par :

$$\ell_{k,r}(\varphi)(\underline{x}) = \frac{k}{x^k} \int_o^r \varphi(t)\, e^{-t^k/x^k}\, t^{k-1}\, dt$$

$(0 < r < R)$. (Si on change r, on modifie $\ell_{k,r}(\varphi)$ par une fonction de $\mathcal{G}_{o,s}$.)
On établit le Théorème 2.2. en utilisant $\ell_{k,r}(\varphi)$.

On a les caractérisations suivantes des fonctions de $\mathcal{G}_s(\underline{V})$:

THEOREME 2.4. Soit $k > 0$ $(s = 1+1/k)$. Pour $f \in \mathcal{G}(\underline{V})$, les conditions suivantes sont équivalentes $(J(f)(x) = \hat{f}(x) = \sum_{n \geq o} a_n x^n)$:

(i) Pour tout $\underline{W} < \underline{V}$, il existe $C_{\underline{W}} > 0$ et $A_{\underline{W}} > 0$:

$$\sup_{\underline{x} \in \underline{W}} |f^{(n)}(\underline{x})| \leq C_{\underline{W}} (n!)^s A_{\underline{W}}^n ;$$

(ii) Pour tout $\underline{W} < \underline{V}$, il existe $K_{\underline{W}} > 0$ et $B_{\underline{W}} > 0$:

$$\sup_{\underline{x} \in \underline{W}} |x|^{-n} |f(\underline{x}) - \sum_{p = o,\ldots,n-1} a_p x^p| \leq K_{\underline{W}} (n!)^{1/k} B_{\underline{W}}^n ;$$

Si l'ouverture de $\underline{V}$ est $< \frac{\pi}{k}$ et sa bissectrice le demi-axe réel positif, ces conditions sont de plus équivalentes aux suivantes :

(iii) $\varphi(t) = \sum_{n \geq o} \frac{a_n}{\Gamma(1+n/k)} t^n \in \mathbb{C}\{t\}$, de rayon de convergence $R > 0$ et pour un $0 < r < R$, on a :

$$f(\underline{x}) - \ell_{k,r}(\varphi)(\underline{x}) \in \mathcal{G}_{o,s}(\underline{V}) ;$$

(iv) φ est convergente et il existe $L_{\underline{W}} > 0$, $K'_{\underline{W}} > 0$, $C'_{\underline{W}} > 0$ $(\underline{W} < \underline{V})$ tels que :

$$|f(\underline{x}) - \sum_{p < L_{\underline{W}} |x|^{-k}} a_p x^p| \leq C'_{\underline{W}}\, e^{-K_{\underline{W}} |x|^{-k}},$$

pour $\underline{x} \in \underline{W}$.

Les conditions (iii) et (iv) et le principe de leur équivalence avec (i) et (ii) m'ont été signalées par MALGRANGE (juin 1979). La condition (iv) est du type "sommation au plus petit terme" (cf. POINCARE [32]).

III. SERIES k - SOMMABLES.

Nous introduisons d'abord la notion de série k - sommable sous forme "abstraite". Nous indiquerons ensuite brièvement quelques méthodes "concrètes" de resommation.

DEFINITION 3.1. Soient $k>0$ et $\hat{f} \in \mathbb{C}[[x]]$.

(i) Soit $\alpha \in S$. On dira que $\hat{f}$ est k - sommable (resp. fortement k - sommable) dans la direction α, s'il existe un secteur(*) $V_{\underline{\alpha},\beta,R}$, avec $\beta = \pi/k$ (resp. $\beta > \pi/k$) et une fonction $f \in \mathcal{G}_s(\overline{V}_{\underline{\alpha},\beta,R})$ telle que $J(f) = \hat{f}$. On dira alors que f est la somme de $\hat{f}$ dans la direction α(**). (On notera éventuellement $f = f_\alpha$.)

(ii) On dira que $\hat{f}$ est k - sommable si elle est k - sommable dans toutes les directions, sauf éventuellement, un nombre fini, formant le "support singulier" $\Sigma(\hat{f})$ de $\hat{f}$. Dans le cas (ii), on note $\hat{f} \in \mathbb{C}\{x\}_s$ $(s = 1+1/k)$. On a les inclusions strictes de $\mathbb{C}$ - algèbres stables par dérivation :

$$\mathbb{C}\{x\} \subset \mathbb{C}\{x\}_s \subset \mathbb{C}[[x]]_s .$$

Si $\hat{f} \in \mathbb{C}\{x\}_s$ et si $\Sigma(\hat{f}) = \emptyset$, $\hat{f} \in \mathbb{C}\{x\}$.

Si $\hat{f}$ est k - sommable dans toute direction $\alpha \in]\alpha_1,\alpha_2[$, elle est fortement k - sommable dans toute direction $\alpha \in]\alpha_1,\alpha_2[$; de plus les différentes sommes $f_\alpha (\alpha \in]\alpha_1,\alpha_2[)$ se recollent. (Ces résultats s'établissent en utilisant la transformation de Leroy.)

En utilisant l'interprétation par formules intégrales de l'isomorphisme de MALGRANGE (cf. infra), on établit la

PROPOSITION 3.2. Toute série k - sommable $\hat{f}$ de support singulier $\Sigma(\hat{f}) = \{\alpha_1,\ldots,\alpha_\ell\}$ s'écrit de manière unique (modulo $\mathbb{C}\{x\}$) :

(*) On "identifie" $\underline{\alpha}$ et α.

(**) f est déterminé de façon unique (Théorème 2.3.).

$$\hat{f} = \hat{f}_o + \hat{f}_1 + \dots + \hat{f}_\ell \text{ , avec } \hat{f}_o \in \mathbb{C}\{x\} \text{ , } \hat{f}_i \in \mathbb{C}\{x\}_s$$

$(i = 1,\dots,\ell)$ et $\Sigma(\hat{f}_i) = \{\alpha_i\}$ $(i = 1,\dots,\ell)$.

Si $\hat{f}$ est k - sommable dans la direction du demi-axe réel positif (on se ramène éventuellement à ce cas par rotation), la somme $f = f_o$ de $\hat{f}$ se "calcule" en utilisant la transformation de Leroy :

$f(x) = \frac{k}{x^k} \int_o^{+\infty} \varphi(t) e^{-t^k/x^k} dt$ (cf. RAMIS [34], [38]) .

Si k est entier et si $\hat{f}$ est k - sommable dans la direction α et les directions qui s'en déduisent par multiplication par les racines k - ièmes de l'unité, la somme $f = f_\alpha$ de $\hat{f}$ se calcule en utilisant des "séries de facultés généralisées" convergentes :

$$\sum_{p=o,\dots,k-1} \frac{1}{z^p} \sum_{n \geq o} \frac{b_{p,n+1} \omega^n e^{ik\alpha(n+1)}}{z^k (z^k + \omega e^{ik\alpha}) \dots (z^k + n\omega e^{ik\alpha})}$$

$(\omega > \omega_o$ convenable) ; cf. RAMIS [34], [38].

On étent cette dernière méthode au cas au cas où $k \in \mathbb{Q}^+$.

On peut inversement caractériser les séries "sommables Leroy" ou "sommables en séries de facultés généralisées" par des propriétés voisines de la k - sommabilité (WATSON [51], NEVANLINNA [24], RAMIS [34], [38]).

La méthode utilisant les séries de facultés peut être rendue raisonnablement "effective" ; son aspect numérique est discuté dans RAMIS-THOMANN [40].

IV. LA TRANSFORMATION DE CAUCHY-HEINE ET L'ISOMORPHISME DE MALGRANGE.

La transformation de Cauchy-Heine est une variante de la transformation de Cauchy : $T \mapsto f(x) = \frac{1}{2i\pi} < T_\zeta , \frac{1}{\zeta - x} >$. (Nous l'utilisons dans un esprit voisin de celui dans lequel STIELTJES utilisait la transformation de Cauchy dans [44].) Cette transformation a été introduite par HEINE, puis PINCHERLE [29], [30] . Nous en utiliserons en fait une variante "locale" (au sens "homologie locale").

Soient V un secteur ouvert de sommet 0 , γ un chemin différentiable par morceaux issu de 0 (non tangent aux côtés de V à l'origine).

DEFINITION 4.1. Si φ est continue le long de γ , sa transformée de Cauchy-Heine est :

$$H_\gamma(\varphi)(x) = \frac{1}{2i\pi}\int \frac{\varphi(\zeta)d\zeta}{\zeta - x} = \frac{1}{2i\pi x} * \mu_\gamma(\varphi) .$$

La fonction $H_\gamma(\varphi)$ est holomorphe pour $x \in \mathbb{C} - \mathrm{Im}\gamma$.

Posons $a_n = \frac{1}{2i\pi}\int_\gamma \frac{\varphi(\zeta)d\zeta}{\zeta^{n+1}}$ et $\hat{H}_\gamma(\varphi)(x) = \Sigma\, a_n x^n \in \mathbb{C}[[x]]$;

$$R_n(x) = \frac{1}{2i\pi}\int_\gamma \frac{\varphi(\zeta)d\zeta}{\zeta^n(\zeta - x)} .$$

PROPOSITION 4.2. Si $\mathrm{Sup}|\zeta^{-n}\varphi(\zeta)| < M_n$ $(n \in \mathbb{N})$, pour $x \in V_1 (\mathrm{Im}\gamma \cap V_1 = \emptyset)^{(*)}$

(i) $|a_n| < C\, M_{nt1}$;

(ii) $|x^{-n}||R_n(x)| < C' M_{n+1}$; où $C, C' > 0$ sont indépendants de n et φ .

Ainsi $H_\gamma(\varphi) \in \mathcal{G}(V_1)$ et $J(H_\gamma(\varphi)) = \hat{H}_\gamma(\varphi)$.

La situation ci-dessus est vérifiée pour $\varphi \in \mathcal{G}_o(V)$ et, en particulier pour $\varphi \in \mathcal{G}_{o,s}(V)$. (Des renseignements "asymptotiques" sur φ fournissent de plus des renseignements "asymptotiques" sur sa transformée de Cauchy-Heine formelle $\hat{H}_\gamma(\psi)$) .

Dans [21] MALGRANGE construit une application naturelle $\mathbb{C}[[x]]/\mathbb{C}\{x\} \xrightarrow{\mu} H^1(S;\mathcal{G}_o)$. Il montre que c'est un isomorphisme. Par restriction à $\mathbb{C}[[x]]_s/\mathbb{C}\{x\}$, on obtient une injection μ_s à valeurs dans $H^1(S;\mathcal{G}_{o,s})$ (cf. RAMIS [33]). En utilisant la transformation de Cauchy-Heine on redémontre très simplement la surjectivité de μ ; on démontre également celle de μ_s . (L'idée de cette méthode a été obtenue indépendamment par MALGRANGE (non publié) et SIBUYA (cf. [43]).)

On peut caractériser de façon simple l'image de $\mathbb{C}\{x\}_s/\mathbb{C}\{x\}$ dans $H^1(S;\mathcal{G}_{o,s})$ par l'application μ_s (RAMIS [34], [38]). (Les considérations précédentes ont des applications intéressantes à l'étude des équations linéaires aux différences

---

(*) γ non tangent aux côtés de V_1 à l'origine.

à coefficients rationnels.) On en déduit (RAMIS [34], [38]) :

PROPOSITION 4.3. Soit $D \in \mathbb{C}[x][\frac{d}{dn}]$. Il existe une base de $\mathrm{Ker}(\mathbb{C}[[x]]/_{\mathbb{C}\{x\}} \xrightarrow{D} \mathbb{C}[[x]]/_{\mathbb{C}\{x\}})$ formée de séries formelles $\hat{f}(x) = \Sigma\, a_n x^n$; a_n ayant un "développement asymptotique" en $\frac{1}{n^{1/q}}$:

$$a_n = C(n!)^{1/k} e^{\Lambda_o n + \Lambda_1 n^{q-1/q} + \dots + \Lambda_{q-1} n^{1/q} + \Lambda_q \log n} \left(1 + \frac{b_1}{n^{1/q}} + \dots +\right),$$

où $k \in \mathbb{Q}^+$; $\Lambda_o, \dots, \Lambda_q \in \mathbb{C}$; $q \in \mathbb{N}$; $b_1, \dots \in \mathbb{C}$.

Cette Proposition précise des résultats de PERRON [27], VALIRON [47], RAMIS [33]. On utilise dans sa démonstration la "formule de Stirling" pour la fonction "Γ généralisée" :

$$G_P(u) = \int_o^{+\infty} t^{u-1}\, e^{-P(t)}\, dt \quad (P \in \mathbb{C}[t]) \; ;$$

cf. DE BRUIJN [9], A. DUVAL [10].

V. L'ISOMORPHISME DE MALGRANGE-SIBUYA : CAS NON ABELIEN.

Soit m un entier fixé. On désigne par $GL(m;\mathbb{C}[[x]]) = \hat{G}$ l'ensemble des matrices carrées d'ordre m , à coefficients dans $\mathbb{C}[[x]]$, inversibles. De la même façon on considère $G = GL(m;\mathbb{C}\{x\})$, $\hat{G}_s = GL(m;\mathbb{C}[[x]]_s)$, $G_s = GL(m;\mathbb{C}\{x\}_s)$. On a, par ailleurs (cf. MALGRANGE [21]) des faisceaux sur S : $\Lambda = GL(m;\mathcal{Q})$, $\Lambda_s = GL(m;\mathcal{Q}_s)$, $\Lambda_{s,A+} = GL(m;\mathcal{Q}_{s,A+})$. On note Λ_o le sous-faisceau de Λ formé des matrices asymptotiques à l'identité ; $\Lambda_{o,s} = \Lambda_o \cap \Lambda_s$. MALGRANGE définit dans [21] une application

$$\hat{G}/G \xrightarrow{\nu} H^1(S;\Lambda_o)$$

(analogue à l'application μ de IV).

La restriction de ν à $\hat{G}_s/G$ fournit une application ν_s à valeurs dans $H^1(S;\Lambda_{o,s})$; la restriction à $\hat{G}_{s,A+}/G$ fournit une application $\nu_{s,A+}$ à valeurs dans $H^1(S;\Lambda_{o,s,A+})$.

On a le résultat fondamental :

THEOREME 5.1.

(i) L'application ν est bijective.

(ii) L'application ν_s est bijective.

(iii) L'application $\nu_{s,A+}$ est bijective.

De plus on peut donner une caractérisation simple de l'image de G_s/G dans $H^1(S;\Lambda'_{o,s})$.

L'assertion (i) est due à MALGRANGE (non publié ; cf. [21]) et SIBUYA [42]. Les démonstrations de MALGRANGE et SIBUYA sont assez différentes : l'une utilise une "Dolbeaultisation", l'autre une variante du "Lemme de Cartan" [8]. On peut étendre (avec des modifications convenables) chacune de ces démonstrations(*) au cas Gevrey, ce qui établit (ii) et (iii) (la variante k - sommable se déduit de la démonstration "à la SIBUYA").

Le Théorème 5.1. joint au "Théorème fondamental des développements asymptotiques Gevrey" (version non linéaire ; cf. infra) permet de donner une variante Gevrey des problèmes de modules traités par MALGRANGE dans [21] (cf. aussi BALSER-JURKAT-LUTZ [2], [16]). Plus précisément on obtient une "stratification" de l'espace des modules de MALGRANGE (liée à l'"iterierte Blockstruktur" de BALSER-JURKAT-LUTZ [1], [2], [16]).

VI. LE THEOREME FONDAMENTAL DES DEVELOPPEMENTS ASYMPTOTIQUES GEVREY.

Nous considérerons des opérateurs différentiels $D \in \mathbb{C}\{x\}[\frac{d}{dx}]$ (ou plus généralement $D \in \mathbb{C}\{x\}[x^{-1}][\frac{d}{dx}]$) et des "systèmes d'ordre un", de la forme $\Delta = x^{k_o+1}\frac{d}{dx} - A(x)$ (où $A(x)$ est une matrice (m,m) à coefficients dans $\mathbb{C}\{x\}$; $k_o \in \mathbb{N}$). Le "dictionnaire" de passage de D à Δ (et inversement) est bien connu. Plus généralement nous aurons à considérer $D \in \mathbb{C}[[x]][x^{-1}][\frac{d}{dx}]$ et Δ avec $A \in \mathrm{End}(m;\mathbb{C}[[x]])$. A un tel D (resp. Δ), on associe son polynôme de Newton $N(D)$ (resp. $N(\Delta)$; cf. RAMIS [33]. Si le polynôme de Newton $N(D)$ a un unique côté (issu de 0), nous dirons que D est "de première espèce". On lui associe un "polynôme déterminant" (MALGRANGE [21], page 9),

(*) Je remercie SIBUYA pour l'aide qu'il m'a apportée dans l'extension de mon résultat.

qui n'a pas de racine nulle.

Si $\Delta = x^{k_o+1} \frac{d}{dx} - A(x)$, et si $A(0)$ n'a pas de valeur propre nulle, nous dirons que Δ est "de première espèce". On lui associe un "polynôme caractéristique" qui est le polynôme caractéristique de Δ . Le polynôme de Newton $N(\Delta)$ a alors un unique côté (issu de 0) , de pente k_o , et k_o est l'invariant de Katz de Δ .

Nous noterons $k(D)$ (resp. $k(\Delta)$) l'invariant de Katz de D (resp. Δ) ; i.e. la pente de côté de plus grande pente de $N(D)$ (resp. $N(\Delta)$) .

A $\Delta = x^{k_o+1} \frac{d}{dx} - A(x)$ nous associerons le système

$\mathrm{End}\,\Delta = x^{k_o+1} \frac{d}{dx} + [.,A]$, d'ordre m^2 .

Plus généralement, pour D (resp. Δ) à invariant de Katz entier $k_1 = k(D)$ (resp. $k_1 = k(\Delta)$) , on peut considérer le "polynôme déterminant" de D (resp. le "polynôme caractéristique" de Δ) relatif au côté de pente k_1 de $N(D)$ (resp. $N(\Delta)$) . Les racines de ce polynôme sont non nulles et notées $\Lambda_1,\ldots,\Lambda_r$ $(r \leq m)$. On pose $\Lambda = \inf_{i=1,\ldots,r} |\Lambda_i|$ et $B(D) = \Lambda^{-1/k}$ (resp. $B(\Delta) = \Lambda^{-1/k}$) .

Les nombres complexes x de module un vérifiant l'une des conditions $\Lambda_i/x^k \in \mathbb{R}^-$ donnent des points de S formant, par définition, le spectre singulier $\Sigma(D)$ (resp. $\Sigma(\Delta)$) de D (resp. Δ). Ce spectre singulier est invariant par $\mathbb{Z}/k\mathbb{Z}$.

Dans le cas général on se ramène au cas où $k(D)$ est entier par un revêtement $x = t^q$ $(q \in \mathbb{N})$. On étend alors les notions précédentes.

On a la version linéaire du "Théorème fondamental des développements asymptotiques Gevrey" :

THEOREME 6.1. Soit $k > 0$. Soit D (resp. Δ) d'invariant de Katz k_1 . Soit $\underline{V}$ un secteur ouvert, d'ouverture $\leq \pi/k_1$. On a la suite exacte :

$$\mathcal{A}_{o,s}(\underline{V}) \xrightarrow{D} \mathcal{A}_{o,s}(\underline{V}) \to 0$$

(resp. la suite exacte :

$$(\mathcal{A}_{o,s}(\underline{V}))^m \xrightarrow{\Delta} (\mathcal{A}_{o,s}(\underline{V}))^m \to 0) .$$

La démonstration de ce résultat est élémentaire mais assez longue et délicate ; cf. RAMIS [35], [36].

THEOREME 6.2. Soit D (resp. Δ) de première espèce et d'invariant de Katz k_o. Alors l'application $(s_o = 1+1/k_o)$:

$$\mathbb{C}[[x]]_{s_o} \xrightarrow{D} \mathbb{C}[[x]]_{s_o} \quad (\text{resp. } (\mathbb{C}[[x]]_{s_o})^m \xrightarrow{\Delta} \mathbb{C}[[x]]^m_{s_o})$$

est un isomorphisme.

Plus généralement, si $B > B(D)$ (resp. $B(\Delta)$), l'application

$$\mathbb{C}[[x]]_{s_o,B} \xrightarrow{D} \mathbb{C}[[x]]_{s_o,B} \; (\text{resp. } (\mathbb{C}[[x]]_{s_o,B})^m \xrightarrow{\Delta} (\mathbb{C}[[x]]_{s_o,B})^m)$$

est un isomorphisme.

L'injectivité est facile. La surjectivité se déduit des résultats de RAMIS [33], [39].

LEMME 6.3. Soit D (resp. Δ) d'invariant de Katz k_1. Soit $\alpha \in S$, $\alpha \notin \Sigma(D)$ (resp. $\Sigma(\Delta)$). Soit $\underline{V}$ un secteur ouvert de bissectrice α et d'ouverture $\geq \pi/k_1$ (*). Alors on a une suite exacte :

$$0 \to \mathcal{A}_{o,s_1}(\underline{V}) \xrightarrow{D} \mathcal{A}_{o,s_1}(\underline{V})$$

(resp. $0 \to (\mathcal{A}_{o,s_1}(\underline{V}))^m \xrightarrow{\Delta} (\mathcal{A}_{o,s_1}(\underline{V}))^m$). Si $\text{ouv}(\underline{V}) = \pi/k_1$ (**), on a un isomorphisme

$$\mathcal{A}_{o,s_1}(\underline{V}) \xrightarrow{D} \mathcal{A}_{o,s_1}(\underline{V})$$

$$\text{resp. } (\mathcal{A}_{o,s_1}(\underline{V}))^m \xrightarrow{\Delta} (\mathcal{A}_{o,s_1}(\underline{V}))^m) .$$

(*) A α fixé ceci reste vrai pour $\text{ouv}(\underline{V}) \geq \pi/k_1 - \varepsilon_\alpha$, avec ε_α assez petit.

(**) Ou plus généralement $\pi/k_1 - \varepsilon_\alpha \leq \text{ouv}(\underline{V}) \leq \pi/k_1$; ε_α assez petit.

PROPOSITION 6.4. Soit D (resp. Δ) de première espèce et d'invariant de Katz k_o.

(i) Soit $\alpha \notin \Sigma(D)$ (resp. $\Sigma(\Delta)$). Alors D réalise un automorphisme de l'espace des fonctions fortement k_o-sommables dans la direction α.

(ii) L'opérateur D (resp. Δ) fournit un isomorphisme

$$\mathbb{C}\{x\}_{s_o} \xrightarrow{D} \mathbb{C}\{x\}_{s_o} \quad (\text{resp. } (\mathbb{C}\{x\}_{s_o})^m \xrightarrow{\Delta} (\mathbb{C}\{x\}_{s_o})^m).$$

Plus précisément si $D\hat{f} = \hat{g}$, avec $\hat{g} \in \mathbb{C}\{x\}_{s_o}$, on a $\hat{f} \in \mathbb{C}\{x\}_{s_o}$ et $\Sigma(\hat{f}) \subset \Sigma(\hat{g}) \cup \Sigma(D)$ (resp. $\Sigma(\hat{g}) \cup \Sigma(\Delta)$). En particulier si $\hat{g} \in \mathbb{C}\{x\}$, $\Sigma(\hat{f}) \subset \Sigma(D)$.

PROPOSITION 6.5. Soit D (resp. Δ) d'invariant de Katz k_1.

(i) Soit $\alpha \notin \Sigma(D)$ (resp. $\Sigma(\Delta)$). Soient $\hat{g} \in \mathbb{C}[[x]]_{s_1}$, fortement k_1-sommable dans la direction α et $\hat{f} \in \mathbb{C}[[x]]_{s_1}$, avec $D\hat{f} = \hat{g}$ (resp. $\hat{g} \in (\mathbb{C}[[x]]_{s_1})^m, \dots, \hat{f} \in (\mathbb{C}[[x]]_{s_1})^m$, avec $\Delta\hat{f} = \hat{g}$). Alors $\hat{f}$ est fortement k_1-sommable dans la direction α et $Df_\alpha = g_\alpha$ (resp. $\Delta f_\alpha = g_\alpha$).

(ii) Soient $\hat{g} \in \mathbb{C}\{x\}_{s_1}$ et $\hat{f} \in \mathbb{C}[[x]]_{s_1}$, avec $D\hat{f} = \hat{g}$ (resp. $\hat{g} \in (\mathbb{C}\{x\}_{s_1})^m$ et $\hat{f} \in ([[x]]_{s_1})^m$, avec $\Delta\hat{f} = \hat{g}$). Alors $\hat{f} \in \mathbb{C}\{x\}_{s_1}$ (resp. $(\mathbb{C}\{x\}_{s_1})^m$).

On a des versions non linéaires, à coefficients Gevrey ou k-sommables, des résultats précédents.

On va considérer des équations non linéaires de la forme suivante :

$$(E) \qquad x^{k_o+1} \frac{dY}{dx} = \Phi(x,Y) = \Phi_o(x) + A(x)Y + \Psi(x,Y)(Y,Y) \; ;$$

où $k_o \in \mathbb{N}$, $A(x)Y$ est linéaire en Y, $\Psi(x,Y)(Y,Y)$ est une forme quadratique en Y (à coefficients dépendant de x et Y).

A une telle équation on associera $\Delta = x^{k_o+1} \frac{d}{dx} - A(x)$. On transpose à (E) les notions introduites pour Δ (invariant de Katz, polygone de Newton, première espèce,...).

On a d'abord le Théorème fondamental des développements asymptotiques :

THEOREME 6.6. Soit (E) $x^{k_o+1}\frac{dY}{dx} = \Phi(x,Y)$, d'invariant de Katz k_1 . Soit $k>0$. Soit $\underline{V}$ un secteur ouvert, avec $\mathrm{ouv}(\underline{V}) \leq \pi/k_1$ et π/k . On suppose que $\Phi(x,Y)$ est holomorphe en (x,Y) (au voisinage de $(0,Y_o)$) , ou que $\Phi(x,Y)$ est polynomiale en Y , à coefficients dans $\mathcal{A}_s(\underline{V})$. Soit $\hat{Y} \in (\mathbb{C}[[x]]_s)^m$ une solution formelle de (E) (avec $\hat{Y}(0) = Y_o$) . Alors il existe, quitte à diminuer le rayon de $\underline{V}$, $Y \in (\mathcal{A}_s(\underline{V}))^m$, solution de (E) , avec $J(Y) = \hat{Y}$.

Ce résultat se démontre suivant un argument classique en se ramenant (en utilisant le Théorème 2.2.) au :

THEOREME 6.7. Soit (E) $x^{k_o+1}\frac{dY}{dx} = \Phi_o(x)+A(x)Y+\Psi(x,Y)(Y,Y)$, d'invariant de Katz k_1 . Soit $k>0$. Soit $\underline{V}$ un secteur ouvert, avec $\mathrm{ouv}(\underline{V}) \leq \pi/k_1$ et π/k. On suppose que $\Phi_o \in (\mathcal{A}_{o,s}(\underline{V}))^m$, $A \in \mathrm{End}(m;\mathcal{A}_s(\underline{V}))$, et que $\Psi(x,Y)$ est analytique en (x,Y) (au voisinage de $(0,0)$) , ou polynomiale en Y , à coefficients dans $\mathcal{O}(V)$, bornés sur tout $\underline{W} < \underline{V}$(*) . Alors il existe $Y \in (\mathcal{A}_{o,s}(\underline{V}))^m$ solution de (E) (quitte à diminuer le rayon de $\underline{V}$) .

On a la version non linéaire du Lemme 6.3. :

PROPOSITION 6.8. Soit (E) $x^{k_o+1}\frac{dY}{dx} = A(x)Y + \Psi(x,Y)(Y,Y)$, d'invariant de Katz k_1 . Soit $\alpha \subset S$, $\alpha \notin \Sigma(E)$. Soit $\underline{V}$ un secteur ouvert de bissectrice α et d'ouverture $\geq \pi/k_1 - \varepsilon_\alpha$, ($\varepsilon_\alpha$ assez petit). Alors l'unique solution de (E) dans $(\mathcal{A}_{o,s_1}(\underline{V}))^m$ est la solution banale $Y = 0$.

THEOREME 6.9. Soit (E) $x^{k_o+1}\frac{dY}{dx} = \Phi_o(x) + A(x)Y + \Psi(x,Y)(Y,Y)$, de première espèce (d'invariant de Katz k_o) . On suppose que $\Psi(0,Y) = 0$. On suppose également que $\left| \begin{array}{l}\Phi_o(x) \in \mathbb{C}[[x]]_{s_o} \\ A \in \mathrm{End}(m; \mathbb{C}[[x]]_{s_o})\end{array}\right.$ et que $\Psi(x,Y)$ est polynomial en Y , à coefficients dans $\mathbb{C}[[x]]_{s_o}$. Alors (E) admet une solution formelle unique $\hat{Y} \in (\mathbb{C}[[x]])^m$ et $\hat{Y} \in (\mathbb{C}[[x]]_{s_o})^m$. Si $B > B(E)$,

---

(*) On peut prendre ces coefficients dans $\mathcal{A}(\underline{V})$.

si $\left| \begin{array}{l} \Phi_o \in \mathbb{C}[[x]]_{s_o,B} \\ A \in \mathrm{End}(m, \mathbb{C}[[x]]_{s_o,B}) \end{array} \right.$ et si Ψ est à coefficients dans $\mathbb{C}[[x]]_{s_o,B}$, on a

$\hat{Y} \in (\mathbb{C}[[x]]_{s_o,B})^m$.

THEOREME 6.10. Soit (E) $x^{k_o+1} \frac{dY}{dx} = \Phi_o(x) + A(x)Y + \Psi(x,Y)(Y,Y)$, de première espèce (d'invariant de Katz k_o) . On suppose que $\Psi(0,Y) = 0$.

(i) Soit $\alpha \notin \Sigma(E)$. On suppose que Φ_o est k_o - sommable dans la direction α , que A est à coefficients k_o - sommables dans la direction α ainsi que Ψ supposée polynomiale en Y . Alors (E) admet une solution formelle unique $\hat{Y}$; $\hat{Y}$ est k - sommable dans la direction α et sa somme Y_α est solution de (E) .

(ii) On suppose que $\Phi_o \in (\mathbb{C}\{x\}_{s_o})^m$, que $A \in \mathrm{End}(m; \mathbb{C}\{x\}_{s_o})$ et que Ψ est polynomiale en Y , à coefficients dans $\mathbb{C}\{x\}_{s_o}$. Alors (E) admet une solution formelle unique $\hat{Y}$; $\hat{Y} \in (\mathbb{C}\{x\}_{s_o})^m$ et plus précisément $\Sigma(\hat{Y})$ est contenu dans la réunion des spectres singuliers de Φ_o et des coefficients de A , Ψ , et de $\Sigma(E)$.

L'assertion (i) généralise un résultat de HARRIS [13].

THEOREME 6.11. Soit (E) $x^{k_o+1} \frac{dY}{dx} = \Phi_o(x) + A(x)Y + \Psi(x,Y)(Y,Y)$, d'invariant de Katz k_1 .

(i) Soit $\alpha \notin \Sigma(E)$. On suppose que Φ_o et les coefficiants de A , Ψ (supposée polynomiale en Y) sont k_1 - sommables dans la direction α . Si $\hat{Y} \in (\mathbb{C}[[x]]_{s_1})^m$ est solution formelle de (E) , $\hat{Y}$ est k_1 - sommable dans la direction α ; $\hat{Y}_\alpha$ est solution de (E) .

(ii) On suppose que $\Phi_o \in (\mathbb{C}\{x\}_{s_1})^m$, $A \in \mathrm{End}(m; ((\mathbb{C}\{x\})_{s_1})^m$ et que Ψ est polynomiale en Y à coefficients dans $\mathbb{C}\{x\}_{s_1}$. Si $\hat{Y} \in (\mathbb{C}[[x]]_{s_1})^m$ est solution formelle de E , $\hat{Y} \in (\mathbb{C}\{x\}_{s_1})^m$.

En reprenant un argument de MALGRANGE [21] (Théorème 10.1), on déduit du Théorème 6.6. le :

THEOREME 6.12. Soit $\underline{V}_o$ un secteur ouvert. Soit $k > 0$. Soit $\Phi(x,Y,Z)$ polynôme en $Y,Z \in \mathbb{C}^n$ à coefficients dans $\mathcal{G}_s(\underline{V}_o)$, ou analytique en (x,Y,Z) au voisinage de $(0,Y_o,Z_o) \in \mathbb{C}^{2m+1}$. Supposons qu'il existe $\hat{Y} \in (\mathbb{C}[[x]]_s)^m$ vérifiant $\Phi(x,\hat{Y}, \frac{d\hat{Y}}{dx}) = 0$ $(\hat{Y}(0) = Y_o , \frac{d\hat{Y}}{dx}(0) = Z_o)$. Supposons que

$$\frac{\partial\hat{\Phi}}{\partial Z}(x,\hat{Y}, \frac{d\hat{Y}}{dx}) \in GL(m;\mathbb{C}[[x]]_s[x^{-1}]) .$$

Alors il existe $k_o > 0$, tel que, pour tout $\underline{V} \subset \underline{V}_o$, avec ouv $\underline{V} \leq \pi/k$ et π/k_o, il existe $Y \in (G_s(\underline{V}))^m$, solution de $\Phi(x,Y, \frac{dY}{dx}) = 0$, avec $J(Y) = \hat{Y}$.

COROLLAIRE 6.13. Soit $G(x,y_o,\dots,y_m) \in \mathbb{C}\{x\}[y_o,\dots,y_m]$. Soit $\hat{f} \in \mathbb{C}[[x]]$ une solution formelle de $G(x,f,f',\dots,f^{(m)}) = 0$. Alors il existe $k_o, k > 0$, tels que, pour tout $\underline{V}$ d'ouverture $< \pi/k_o$ et π/k, et de rayon assez petit, il existe $f \in G_s(\underline{V})$, solution de $G(x,f,f',\dots,f^{(m)})$, avec $J(f) = \hat{f}$.

Ainsi, pour une solution formelle d'équation non linéaire analytique la "sommation au plus petit terme" (au sens donné plus haut) est valable. Le corollaire 6.13. se déduit du Théorème 6.12. et d'un résultat de MAILLET [53], MAHLER [54].

(En général on a $k_o \geq k$ et l'on dispose d'un calcul précis de k_o tandis que l'on n'a que des estimations inférieures assez grossières de k ; on est donc loin de la resommation.)

VII. THEOREMES D'ALGEBRISATION DES SYSTEMES GEVREY OU k-SOMMABLES.

Considérons le système $\Delta Y = x^{k_o+1}\frac{dY}{dx} - A(x)Y$.

La transformation $Y = PZ$ (P inversible) le transforme en le système $\Delta^P Y = x^{k_o+1}\frac{dZ}{dx} - B(x)Z$, avec $A^P = B = -x^{k_o+1}P^{-1}\frac{dP}{dx} + P^{-1}AP$. Cette dernière relation s'écrit $x^{k_o+1}\frac{dP}{dx} = AP - PB$. Si $A = B$, on retrouve End Δ. Si P a des pôles, on remplace éventuellement k_o par k_1. On a les généralisations suivantes du Théorème "d'algébrisation" de BIRKHOFF [4] :

THEOREME 7.1. Soit $k > 0$. Soit $\Delta = x^{k_o+1}\frac{d}{dx} - A$, avec $A \in End(m;\mathbb{C}[[x]]_s)$. Il existe $P \in GL(m;\mathbb{C}[[x]]_s[x^{-1}])$ transformant Δ en un système algébrique

$$\Delta^P = x^{k_1+1}\frac{d}{dx} - B , \text{ avec } B \in End(m;\mathbb{C}[x]) .$$

On établit ce résultat (suivant une méthode qui m'a été indiquée par MALGRANGE (juin 79)) en utilisant le Théorème 2.2. (BOREL-RITT avec majoration Gevrey), le Théorème 6.6. (Théorème fondamental des développements asymptotiques) et l'isomorphisme du Théorème 5.1. (ii). On obtient alors une transformation $Q \in GL(m;\mathbb{C}[[x]]_s)$ transformant Δ en un système analytique : $A^Q \in GL(m;\mathbb{C}\{x\})$. On utilise ensuite le résultat de BIRKHOFF [4] pour algébriser.

Pour $k \in \mathbb{Q}^+$ (qui est le cas intéressant en pratique) MALGRANGE donne dans [23] une démonstration différente du Théorème 7.1., en utilisant les opérateurs pseudodifférentiels.

THEOREME 7.2. Soit $k > 0$. Soit $\Delta = x^{k_o+1}\frac{d}{dx} - A$, avec $A \in \mathrm{End}(m;\mathbb{C}\{x\}_s)$. Il existe $P \in GL(m;\mathbb{C}\{x\}_s[x^{-1}])$ transformant Δ en un système algébrique $\Delta^P = x^{k_1+1}\frac{d}{dx} - B$, avec $B \in \mathrm{End}(m;\mathbb{C}[x])$.

VIII. RESOMMATION DE LA SOLUTION FONDAMENTALE FORMELLE D'UN SYSTEME ANALYTIQUE.

Nous esquissons ci-dessous la solution complète (dans le cas le plus général) du problème de "resommation" des solutions formelles des équations (ou des systèmes) à coefficients analytiques. Nous décrivons le résultat pour un système. Pour une équation on obtient une "resommation" de la décomposition en "branches formelles" donnée par BALDASSARI [52], MALGRANGE [22], ROBRA [41] (i.e. une "Théorie de HENSEL resommable") ; cf. RAMIS [37], [38].

Le premier exemple de resommation a été donné par EULER [11]. Le problème de resommation est posé par BOREL dans [5], [6]. La conviction qu'il est résoluble "par séries de facultés" apparaît chez NORLUND. Un pas décisif est fait par HORN [24], [25], qui le résoud dans certains cas (de première espèce, "générique", invariant de KATZ égal à un). Des progrès sont faits ensuite par TURRITIN [46], TRJITZINSKY [45] ; TRJITZINSKY remarque que les séries de facultés ordinaires $(k = 1)$ ne suffisent pas. EVANS attaque le problème dans [12] (mais sa démonstration est fausse). Plus récemment, le problème est repris par KOHNO-OHTOMO [17] et LUTZ [19](*). (Cf. aussi WASOW [49] chapitre XI).

Dans le cas général la nécessité de "mélanger" divers processus de k - sommation (avec des k différents) est a priori évidente pour des raisons algébriques élémentaires (et compte tenu du fait qu'une série k_1 - sommable n'est pas en général k_2 - sommable pour $k_1 \neq k_2$) . Il est curieux que ce point n'ait pas été dégagé par mes prédecesseurs dans l'étude de ce problème...

PROPOSITION 8.1. Soit $\Delta = x^{k_o+1} \frac{d}{dx} - A$, avec $A = \begin{pmatrix} A_{11} & A_{12} \\ A_{21} & A_{22} \end{pmatrix}$, $A_{12}(0) = 0$, $A_{22}(0) = 0$; $A_{11}(0)$ et $A_{22}(0)$ n'ayant par ailleurs aucune valeur propre commune. Alors :

(i) Si $A \in End(m; \mathbb{C}[[x]]_{s_o})$, il existe $P \in GL(m; \mathbb{C}[[x]])$ unique de la forme $P = I + \begin{pmatrix} 0 & P_{12} \\ P_{21} & 0 \end{pmatrix}$ telle que $A^P = B = \begin{pmatrix} B_{11} & 0 \\ 0 & B_{22} \end{pmatrix}$. De plus $P \in GL(m; \mathbb{C}[[x]]_{s_o})$.

(ii) Si $A \in End(m; \mathbb{C}\{x\}_{s_o})$ (en particulier si $A \in End(m; \mathbb{C}\{x\})$, il existe P unique comme ci-dessus. De plus $P \in GL(m; \mathbb{C}\{x\}_{s_o})$.

On détermine P de façon classique(**) en déterminant P_{12} (resp. P_{21}) comme solution d'une équation non linéaire. Cette équation est de première espèce et d'invariant de KATZ k_o . On obtient alors (i) et (ii) en appliquant respectivement les Théorèmes 6.9. et 6.10. (ii).

(La Proposition 8.1. a été obtenue indépendamment par MALGRANGE [23].)

En utilisant le Théorème 7.1. (resp. 7.2) d'algébrisation(***), on peut ensuite trouver dans le cas (i) (resp. (ii)) des matrices

---

(*) LUTZ traite le cas "générique" dans un esprit voisin du nôtre. Il nous semble toutefois que sa démonstration dans [19] présente une lacune. Par ailleurs, le fait d'utiliser les résultats de NEVANLINNA [24] (tenant compte de l'apport de NÖRLUND [25], [26]) au lieu de ceux de WATSON [51] simplifie et améliore l'approche de LUTZ [19].

(**) Pour la récurrence qui suit, nous nous limiterons à une "analytisation" qui permet d'éviter l'introduction de pôles.

(***) Méthode due à SIBUYA (Cf. WASOW [48]).

$$Q_1 \in GL(m_1; \mathbb{C}[[x]]_{s_o}[x^{-1}]) \quad \text{et} \quad Q_2 \in GL(m_2; \mathbb{C}[[x]]_{s_o}[x^{-1}])$$

(resp. $Q_1 \in GL(m_1; \mathbb{C}\{x\}_{s_o}[x^{-1}])$ et $Q_2 \in GL(m_2; \mathbb{C}\{x\}_{s_o}[x^{-1}])$) , telles que $B_{11}^{Q_1}$ et $B_{22}^{Q_1}$ soient à coefficients algébriques. Remarquons que si dans (i) on part de $A \in End(m; \mathbb{C}\{x\}_{s_o})$, la transformation faisant passer de A à $\begin{pmatrix} B_{11}^{Q_1} & 0^{Q_2} \\ 0 & B_{22} \end{pmatrix}$ qui est a priori à coefficients dans $\mathbb{C}[[x]]_{s_o}$ est en fait à coefficients dans $\mathbb{C}\{x\}_{s_o}$ (Théorème 6.11.) ; ce qui permet d'éviter l'emploi du Théorème 7.2. si on le désire.

On peut alors "casser" par double récurrence (sur l'ordre et l'invariant de Katz) le système Δ en systèmes "non couplés" suivant la méthode classique (WASOW [48], MALGRANGE [20]). Il est d'ailleurs plus agréable de passer avant à un revêtement $x = t^q$ convenable et de faire une transformation rationnelle approchant convenablement (au sens m-adique) en transformation de réduction formelle ; l'invariant de Katz décroît alors à chaque cassure et il suffit de faire une récurrence sur l'ordre.

On obtient alors le Théorème de resommation :

Soit $\Delta = x^{k_o+1}\frac{d}{dx} - A$, avec $A(x) \in End(m; \mathbb{C}\{x\})$.

Elle admet une solution fondamentale formelle de la forme $\hat{H}(t)t^{\Lambda}e^{Q(\frac{1}{t})}$; où $x = t^q$ ($q \in \mathbb{N}$ convenable ; $q \le$ p.p.c.m.$(1,2,\dots,m) \le m!$) , Λ est une matrice constante, $Q(\frac{1}{t})$ est une matrice diagonale dont les coefficients sont des polynômes en t (notés $q_1(t),\dots,q_\ell(t)$, et, s'il y a lieu $q_{\ell+1}(t) = 0$) . (On note $k_1,\dots,k_r$ $(k_1 < \dots < k_r)$ les degrés des $q_{j_1} - q_{j_2}$ non nuls ($k_i \in \mathbb{N}$; $i = 1,\dots,r$) ; k_r est l'invariant de KATZ de Δ(*)). Enfin $\hat{H}(t) \in GL(m; \mathbb{C}[[t]])$.

THEOREME 8.2. Dans les conditions ci-dessus, il existe une décomposition :

$$\hat{H}(t) = \hat{H}_o(t)\hat{H}_{k_r}(t)\cdots\hat{H}_{k_1}(t) \text{ , avec } \hat{H}_o(t) \in GL(m; \mathbb{C}[t][t^{-1}]) \text{ ,}$$

$\hat{H}_{k_i}(t) \in GL(m; \mathbb{C}\{t\}_{s_i})$ ($i = 1,\dots,r$; $s_i = 1+1/k_i$) . Le support singulier de

---

(*) Les k_i/q sont les pentes des côtés de $N(End\,\Delta)$.

$\hat{H}_{k_i}$ étant contenu dans un ensemble fini $\Sigma_i(\Delta)$ (qui ne dépend que de $Q(\frac{1}{t})^{(*)}$).
De plus, pour $\alpha \in S$, $\alpha \notin \bigcup_{i=1,\dots,r} \Sigma_i(\Delta)$, si on note $H_\alpha = H_o H_{k_r\alpha} \cdots H_{k_1\alpha}$ ($H_{k_i\alpha}$ étant la somme de $\hat{H}_{k_i}$ dans la direction α), $H_\alpha t^\Lambda e^{Q(\frac{1}{t})}$ est une solution fondamentale de Δ sur un "oeil ouvert" $L_\alpha (0 \in \partial L_\alpha)$. On peut interpréter les coefficients de H_α comme combinaisons linéaires de produits de sommes de "k-séries de facultés".

Le Théorème 8.2. est une version "concrète" du résultat "abstrait" de BALSER-JURKAT-LUTZ [1], [2], [19] qui fournit une solution "canonique" dans toute direction "générique" α par un procédé inspiré de BIRKHOFF. (Dans le cas "générique" H_α est bien canonique ; dans le cas général il y a priori un nombre fini de décompositions distinctes $\hat{H} = \hat{H}_o \hat{H}_{k_r} \dots \hat{H}_{k_1}$, liées à des choix d'ordres ; je ne pense pas que les H_α correspondants coïncident.)

On peut préciser le résultat du Théorème 8.2. en donnant pour chaque $\hat{H}_{k_i}$ une décomposition en blocs liée à la "structure par blocs itérée" de BALSER-JURKAT-LUTZ [1], [2], [19]. Ceci est relié à la "stratification" de l'espace des modules de MALGRANGE [21] (cf. aussi BERTRAND [3]).

Pour terminer je voudrais signaler l'aide considérable que m'a apportée B. MALGRANGE dans la préparation de ce travail et dont les références précises que j'ai citées ne donnent qu'une faible idée.

BIBLIOGRAPHIE

[1] BALSER W. - Einige Beiträge zur Invariantentheorie meromorpher Differentialgleichungen. (Habilitationsschrift 1978).

[2] BALSER W., JURKAT W., LUTZ P.A. - A general theory of invariants for meromorphic differential equations ; Part II, Proper Invariants (à paraître)

[3] BERTRAND D. - Travaux récents sur les points singuliers des équations différentielles linéaires. Séminaire Bourbaki 1978-79, n°538 (01-15).

[4] BIRKHOFF G. - The generalized Riemann problem for linear differential equations. Proc. An. A. Sc. 49 (1913), 521-568.

[5] BOREL E. - Mémoire sur les Séries Divergentes. Annales Sc. de l'ENS, 3ème série, T. 16, 1899 (9-136).

[6] BOREL E. - Leçons sur les Séries Divergentes - Deuxième édition, 1928. (Gauthier-Villars, Paris)

[7] CARLEMAN T. - Les Fonctions Quasi-Analytiques. (Gauthier-Villars, Paris 1926).

[8] CARTAN H. - Sur les matrices holomorphes de n variables complexes, J. Math. pures et appl., 9ème série, t.19, 1940, p.1-26.

(*) Plus précisément $\Sigma_i(\Delta)$ ne dépend que des termes de plus haut degré k_i des q_{j1} correspondants.

[9] DE BRUIJN N.G. - Asymptotic Methods in Analysis. (North-Holland Publ. Co., Amsterdam 1961).
[10] DUVAL A. - Etude asymptotique d'une intégrale analogue à la fonction "Γ modifiée". Preprint I.R.M.A. Strasbourg (1980).
[11] EULER L. - De seriebus divergentibus, commentationes analyticae, ad Theoriam Serierum Infinatorum Pertinentes, volumen primum (C. Brehm et G. Faber ed. 1925 ; p.601-602).
[12] EVANS R.L. - Asymptotic and convergent factorial series in the solution of linear ordinary differential equations. Proceedings A.M.S., 5, 1954.
[13] HARRIS W.A. - Laplace integrals and factorial series on singular differential equations. Notices A.M.S. 1979, 192, vol.16.1.
[14] HORN J. - Fakultätenreihen in der Theorie der linearen Differentialgleichungen. Math. Ann. 71 (1912), p.510-532.
[15] HORN J. - Integration linearer differentialgleichungen durch Laplacesche Integrale un Fakultätenreihen. Jahresb. deutsch. Math. - Ver. 24 (1915), p. 309-329.
[16] JURKAT W.B. - Meromorphe Differentialgleichungen. (Lecture Notes 637, Springer-Verlag 1978).
[17] KOHNO M., OHTOMO M. - A note on generalized factorial series expansions. Hiroshima Math. J. 7 (1977), 605-628.
[18] LEROY E. - Sur les séries divergentes et les fonctions définies par un développement de Taylor. Ann. Fac. Sc. Université de Toulouse 1900, p.317-430.
[19] LUTZ P.A. - On the summation of formal solutions at an irregular singular point by factorial series. International conference on Differential Equations. Acad. Press Inc., 1975, p.489, 509.
[20] MALGRANGE B. - Sur les points singuliers des équations différentielles linéaires. Enseign. Math., 20 (1970), 147-176.
[21] MALGRANGE B. - Remarques sur les équations différentielles à points singuliers irréguliers. Equations différentielles et systèmes de Pfaff dans le champ complexe. R. Gérard et J.P. Ramis éd. (Lecture Notes 712, Springer Verlag 1979).
[22] MALGRANGE B. - Sur la réduction formelle des équations différentielles à singularités irrégulières. Preprint, Grenoble 1979.
[23] MALGRANGE B. - Cours de 3ème cycle à Grenoble (1979-80).
[24] NEVANLINNA F. - Zur Theorie der Asymptotischen Potenzreihen. Annales Academiae Scientiarum Fennicae, ser. A, Fom.XII. Helsinski 1919.
[25] NÖRLUND N.E. - Sur les séries de facultés. Acta Math. 37 (1914), p.327-387.
[26] NÖRLUND N.E. - Differenzenrechnung. (Die Grundlehren ..., Bd XIII, Springer Verlag 1924).
[27] PERRON O. - Über lineare Differenzengleichungen. Acta Math. 34 (1910),p.109-137.
[28] PERRON O. - Über lineare Differentialgleichungen mit rationalen Koeffizienten, Acta Math. 34 (1910), p.139-163.
[29] PINCHERLE S. - Sopra una transformazione della equazioni differentiali lineari in equazioni alle differenze e viceversa - Rend. ist. Lomb. (2) 19 (1886) p;559-562.
[30] PINCHERLE D. - Sur la génération des systèmes récurrents au moyen d'une équation différentielle. Acta Math. 16 (1892), p.341-363.
[31] POINCARE H. - Sur les intégrales irrégulières. Acta Math. 8 (1886), p.295-344.
[32] POINCARE H. - Les méthodes nouvelles de la mécanique Céleste, T.II.
[33] RAMIS J.P. - Dévissage Gevrey. Astérisque S.M.F. 59-60 (1978),p.173-204.
[34] RAMIS J.P. - Développements asymptotiques Gevrey et séries k-sommables. En préparation.
[35] RAMIS J.P. - Le théorème fondamental des développements asymptotiques Gevrey I. (L'équation d'Euler). En préparation.
[36] RAMIS J.P. - Le théorème fondamental des développements asymptotiques Gevrey II. En préparation.
[37] RAMIS J.P. - La théorie de Hensel des opérateurs différentiels et le problème de resommation. En préparation.
[38] RAMIS J.P. - Développements asymptotiques Gevrey, séries k-sommables et applications aux équations différentielles et aux différences. Preprint I.R.M.A. Strasbourg (mars 1980).

[39] RAMIS J.P. - Théorème d'indice pour les équations différentielles et aux différences linéaires. En préparation.

[40] RAMIS J.P., THOMANN - Remarques sur l'utilisation numérique des séries de factorielles. Preprint I.R.M.A. Strasbourg (Mars 1980).

[41] ROBBA P. - Lemmes de Hensel pour les opérateurs différentiels. Application à la réduction formelle des équations différentielles. (Preprint, Paris 1979).

[42] SIBUYA Y. Linear differential equations in the complex domain, problems of analytic continuation. Kinokuniya, Tokyo (1976) ; en japonais.

[43] SIBUYA Y. - A theorem concerning uniform simplification at a transition point ; conférence à Leiden (29/2/1980).

[44] STIELTJES - Recherches sur les fractions continues Ann. Fac. Sc. Université de Toulouse, T. VIII, IX, 1894, 1895.

[45] TRJITZINSKY W.J. - Laplace intégrales and factorial seriesin the theory of linear differential and difference equations. Trans. Am. Math. Soc.37, 80-146 (1935).

[46] TURRITIN H.L. - Convergent solutions of ordinary linear homogeneous differential equations in the neighborhood of an irregular singular point. Acta Math. 93, 27-66 (1955).

[47] VALIRON G. - Lectures on the general theory of integral functions. Chelrea Publ. Co (1949).

[48] WASOW W. - Asymptotic expansions for ordinary differential equations (Interscience publishers 1965).

[49] WATSON G.N. - A theory of asymptotic series. Philosophical transactions of the royal society of London (Ser.A), vol. CCXI (1911), p.279-313.

[50] WATSON G.N. - The characteristics of asymptotic series. Quarterly journal of Mathematics, vol. XLIII (1911-1912), p.65-77.

[51] WATSON G.N. - The transformation of an asymptotic series into a convergent series of inverse factorials. Cir. Mat. Palermo, Rend. 34 (1912),p.41-88.

[52] BALDASSARI F. - Differential modules and singular points of p-adic differential equations. (Preprint).

[53] MAILLET E. - Sur les séries divergentes et les équations différentielles. Annales E.N.S., Paris, 1903, p.487-518.

[54] MAHLER K. - On formal power series as integrals of algebraic differential equations. Lincei, 36, vol.I, 1971, p.76-89.

La rédaction de cet article était terminée lorsque Y. SIBUYA nous a signalé les travaux suivants (Cf. [13]) :

[55] BRAAKSMA B.L.J. - Laplace integrals, factorial series and singular differential equations. Proc. Bicentennial congress of the Wiskundig Genootschap, Amsterdam 1978.

[56] BRAAKSMA B.L.J. - Laplace integrals in singular differential and difference equations. (Preprint).

[57] BRAAKSMA B.L.J., HARRIS Jr., W.A. - Laplace integrals and factorial series in singular functional differential systems. Applicable analysis, 1978, Vol.8, 21-45. (Gordon and Breach, 1978).

On comparera ces résultats à notre théorème 6.10 (i). Le cas traité par BRAAKSMA et HARRIS correspond à notre cas 1 - sommable (Gevrey 2 sur des secteurs fermés d'ouverture $\geq \pi$). Les méthodes sont différentes : BRAAKSMA et HARRIS se ramènent à des équations intégrales alors que nous utilisons une méthode de "globalisation" (sans référence à la transformation de Laplace).

Reflection of Analytic Singularities

by

Johannes Sjöstrand

Let M be an n-dimensional real analytic manifold with boundary, P a second order differential operator on M with analytic coefficients and with real principal symbol p. Assume that P is of principal type and that near any point of the boundary ∂M, there are local coordinates $(x_1,\dots,x_n) = (x',x_n)$ such that (up to a non-vanishing factor)

$$p = \xi_n^2 + r(x,\xi') , \qquad M: x_n \geq 0 ,$$

where $d_{x',\xi'}r(x',0,\xi')$, $\sum_1^{n-1} \xi_j dx_j$ are linearly independent. This condition can be formulated invariantly ([5]), and is satisfied by the wave operator in a cylinder, $\mathbb{R}_t \times \Omega_x$, $\Omega \subset \mathbb{R}^{n-1}$.

We decompose

$$T^*\partial M\backslash 0 = \mathcal{E} \cup \mathcal{G} \cup \mathcal{H}$$

where $\mathcal{E}$, $\mathcal{G}$, $\mathcal{H}$ are given by $r_o > 0$, $r_o = 0$, $r_o < 0$ respectively, and $r_o(x',\xi') = r(x',0,\xi')$. At a point of $\mathcal{G}$, the complement $\complement M$ is strictly convex along the bicharacteristics iff $\partial r/\partial x_n < 0$ and we let this inequality define a subset $\mathcal{G}_+ \subset \mathcal{G}$. Let $\Sigma_b = \mathcal{H} \cup \mathcal{G} \cup \left(p^{-1}(0)|_{\overset{\circ}{M}}\right)$ with the natural topology.

<u>Definition</u>. An analytic ray is a continuous curve $\gamma : I \to \Sigma_b$, where $I \subset \mathbb{R}$ is an interval, such that for every $t_o \in I$:

1° If $\gamma(t_o) \in p^{-1}(0)|_{\overset{\circ}{M}}$, then γ is differentiable at t_o and

$\dot{\gamma}(t_o) = H_p(\gamma(t_o))$, where H_p is the Hamilton field of

$$p = \xi_n^2 + r(x,\xi').$$

2° If $\gamma(t_o) \in \mathcal{H}$ then $\gamma(t) \in p^{-1}(0)|_{\overset{\circ}{M}}$ for $|t-t_o| > 0$ small enough.

3° If $\gamma(t_o) \in G$ and we write $\gamma(t) = (x(t),\xi(t))$, then $(x(t),\xi'(t))$ is differentiable at t_o and $\dot{x}_n(t_o) = 0$, $(\dot{x}'(t_o),\dot{\xi}'(t_o)) = H_{r_o}$.

Recall from Melrose-Sjöstrand [6], that the corresponding definition for C^∞-rays is obtained by replacing $\mathcal{H}$ by $\mathcal{H} \cup G_+$ in 2°. Thus contrary to a C^∞-ray, an analytic ray may glide along a strictly convex part of an obstacle (thinking of the wave operator in a cylinder as our main example).

We write

$u \in \mathcal{D}'(M)$, if u is an extendable distribution on $\overset{\circ}{M}$,

$u \in a(M)$, if u is real-analytic near M,

$v \in a(\partial M)$, if v is real-analytic on ∂M.

If $u \in \mathcal{D}'(M)$, $Pu \in a(M)$ we define

$$WF_{ba}(u) = WF_a\left(u\big|_{\overset{\circ}{M}}\right) \cup \left(WF_a\left(u\big|_{\partial M}\right) \cup WF_a\left(\partial_{x_n} u\big|_{\partial M}\right)\right)$$

$$\subset (T^*\overset{\circ}{M}\setminus 0) \cup (T^*\partial M\setminus 0) \quad ,$$

where WF_a denotes the analytic wavefront set.

If $u \in \mathcal{D}'(M)$ and

$$\text{(1)} \qquad Pu \in a(M) \; , \qquad u\big|_{\partial M} \in a(\partial M)$$

it follows from results of Schapira [10] that $WF_{ba}(u)$ is a closed conic subset of Σ_b.

Our main result is,

<u>Theorem 1. ([11].)</u> If $u \in \mathcal{D}'(M)$ satisfies (1) and $\rho \in WF_{ba}(u)$, then there exists a maximally extended analytic ray passing through ρ and contained in $WF_{ba}(u)$.

In $T^*\overset{\circ}{M}\setminus 0$ this is due to Sato-Kawai-Kashiwara [9], Hörmander [3] and near $\mathcal{H}$ the result is due to Schapira [10].

This result is quite satisfactory away from G_+ , since analytic rays cannot split there. Near G_+ it is natural to ask what unions of analytic rays can be carriers of the analytic singularities of a solution to (1). The following result eliminates the case of a single gliding ray in G_+ :

Theorem 2. ([12].) Let $u \in \mathcal{D}'(M)$ satisfy (1), let $\rho \in G_+$, and for $\delta > 0$ small enough let $\gamma:[-\delta,\delta] \to \Sigma_b$ be the unique C^∞ ray with $\gamma(0) = \rho$, $\gamma(t) \in p^{-1}(0)|_{\overset{\circ}{M}}$, $t \neq 0$. If $\gamma(\delta), \gamma(-\delta) \notin WF_a\left(u|_{\overset{\circ}{M}}\right)$, then $\rho \notin WF_{ba}(u)$.

After this conference we received a preprint of Kiyômi Kataoka, proving essentially also Theorem 2.

In the case of

$$P = \sum_2^n \frac{\partial^2}{\partial x_j^2} - (1 + x_n) \frac{\partial^2}{\partial x_1^2} .$$

Friedlander-Melrose [2] showed for a particular solution that propagation along non-C^∞ rays takes place. A similar result was obtained by J. Rauch [7] in the following situation:

Let $\Omega \subset \mathbb{R}^n$ be open with analytic boundary and assume for simplicity that $\complement\Omega$ is compact. Let $x_o \in \Omega$ and let $u \in \mathcal{D}'(\mathbb{R}_t \times \bar{\Omega})$ be the solution to the mixed problem:

$$(2) \qquad \Box u = 0 , \quad u|_{\mathbb{R} \times \partial\Omega} = 0$$

$$u(0,x) = 0 , \quad \partial_t u(0,x) = \delta(x-x_o) .$$

Using a simple and elegant argument based on Holmgren's uniqueness theorem and the property of finite speed of propagation, Rauch proved that if $x_1 \in \bar{\Omega}$, $x_1 \neq x_o$ then $(\pm\ell,x_1)$ belong to the analytic singular support of u , if ℓ is the distance <u>in $\bar{\Omega}$</u> from x_o to x_1. In particular there are analytic singularities in the C^∞-shadow. One might expect from this that all analytic rays, passing over $(0,x_o)$ are contained in $WF_{ba}(u)$, but there is a difficulty in carrying out Rauch's argument for analytic rays whose x-space projection is not length minimizing. However, using Theorem 1 and the techniques of

Rauch [7] we were able to show,

<u>Theorem 3. (Rauch-Sjöstrand [8]).</u> Let $\Omega \subset \mathbb{R}^2$ have analytic boundary and assume that $\complement\Omega$ is convex and bounded. If u is the solution of (2), then $WF_{ba}(u)$ is the union of all analytic rays passing over $(0,x_o)$.

The idea of the proof is to introduce the universal covering space $\pi\colon \hat{\Omega} \to \Omega$. Then over $\mathbb{R}\times\hat{\Omega}$, all rays which hit G_+ have length-minimizing projections in $\hat{\Omega}$. Let $\hat{u}$ solve the "lifted" problem

$$(\hat{2}) \qquad \hat{\Box}\hat{u} = 0 \ , \quad \hat{u}\big|_{\mathbb{R}\times\partial\hat{\Omega}} = 0 \ , \quad \hat{u}(0,\hat{x}) = 0 \ , \ \partial_t\hat{u}(0,\hat{x}) = \delta(\hat{x}-\hat{x}_o) \ ,$$

where $\pi(\hat{x}_o) = x_o$. Then using Theorem 1, the techniques of Rauch [7] and a refined version of Holmgren's uniqueness theorem due to Kashiwara [9], we can first determine $WF_{ba}(\hat{u})$ completely. Then we recover $u(t,x)$ by the formula

$$u(t,x) = \sum_{\pi(\hat{x})=x} \hat{u}(t,\hat{x}) \ ,$$

and using the convexity assumption and Theorem 1 we can verify that no cancellation of analytic singularities takes place.

The proofs of Theorems 1, 2 combine the ideas of energy estimates used in [5], [6] and by Andersson [1], with the techniques of complex phase functions, resolutions of the identity, etc. In a way, the proof of Theorem 1 involves a complex canonical transformation, which transforms (1) into an elliptic boundary value problem. More recent work based on this point of view seems to give considerable generalizations of Theorem 1, to higher order equations and more general boundary conditions. In particular in the interior we would recover a result of Kashiwara-Kawai [4] on the propagation of analytic singularities for microhyperbolic operators.

REFERENCES

1. K.G. Andersson, Régularité jusqu'au bord pour des problèmes mixtes. C.R.A.S. 282 (2 février 1976), 275-277.

2. F.G. Friedlander, R.B. Melrose, The wavefront set of the solution of a simple initial-boundary problem with glancing rays, II. Math. Proc. Camb. Phil. Soc. (1977), 81, 97-120.

3. L. Hörmander, Uniqueness theorems and wavefront sets for solutions of linear differential equations with analytic coefficients. C.P.A.M., 23 (1970), 329-358.

4. M. Kashiwara, T. Kawai, Microhyperbolic pseudo-differential operators I. Journal of Math. Soc. Japan, 27 (1975), 359-404.

5. R.B. Melrose, J. Sjöstrand, Singularities of boundary value problems I. C.P.A.M., 31 (1978), 593-617.

6. R.B. Melrose, J. Sjöstrand, Singularities of boundary value problems II. In preparation, but see Sém. Goulaouic-Schwartz, 1977-78, no. 15 for the essential results.

7. J. Rauch, The leading wavefront for hyperbolic mixed problems. Bull. Soc. R. Sci. de Liège, 46, no. 5-8 (1977), 156-161.

8. J. Rauch, J. Sjöstrand, Propagation of analytic singularities along diffracted rays. To appear.

9. M. Sato, T. Kawai, M. Kashiwara, Microfunctions and pseudo-differential equations. Springer Lecture Notes in Mathematics, no. 287.

10. P. Schapira, Propagation at the boundary and reflection of analytic singularities of solutions of linear partial differential equations, I. Publ. RIMS, Kyoto Univ., 12 Suppl. (1977), 441-453.

11. J. Sjöstrand, Propagation of analytic singularities for second order Dirichlet problems, I. Comm. P.D.E., to appear.

12. J. Sjöstrand, Propagation of analytic singularities for second order Dirichlet problems, II. Comm. P.D.E., to appear.

LES OPERATEURS METADIFFERENTIELS

A. Unterberger
Département de Mathématiques
Université de Reims
B.P. 347, 51062-Reims-Cedex

A la suite de Maslov, le professeur Leray a insisté, dans ses travaux sur l'Analyse lagrangienne [8], pour que les outils définis en vue d'applications à la mécanique quantique soient invariants par les changements de repères symplectiques de l'espace de phase $R^{2\nu}$. Nous adoptons ici le point de vue qu'un repère $\omega = (Y,\theta)$ est constitué par la donnée d'un point $Y = (y,\eta)$ de l'espace de phase et d'une norme symplectique $\| \ \|_\theta$ sur celui-ci, entendant par là toute norme euclidienne se déduisant de la norme canonique de $R^{2\nu}$ par une transformation symplectique ; il revient au même de se donner la "transformation de Fourier" $\mathfrak{F}_\omega$ définie ($Op_{\frac{1}{2}}$ étant la quantification de Weyl-Wigner) par

$$\mathfrak{F}_\omega = Op_{\frac{1}{2}}(2^{\nu/2}\, e^{\frac{i\pi\nu}{4}} \exp - 2i\pi \|X-Y\|_\theta^2) ,$$

ou encore l'oscillateur harmonique $\Lambda_\omega = Op_{\frac{1}{2}}(\pi \|X-Y\|_\theta^2) - \frac{\nu}{2}$, ou encore l'espace propre de niveau d'énergie zéro de celui-ci, ou enfin le point Y et la structure complexe sur l'espace de phase pour laquelle la forme R-bilinéaire B définie par

$$B(X_1,X_2) = (X_1,X_2)_\theta + i[X_1,X_2]$$

($[\ ,\ ]$ étant la forme symplectique) est sesquilinéaire. La représentation de Siegel permet d'expliciter un état fondamental φ_ω de Λ_ω, les autres états propres φ_ω^α ($\alpha \in N^\nu$) s'obtiennent à partir de φ_ω en faisant agir des opérateurs de création: ce procédé est bien connu des physiciens [5].

Etant donné un ensemble Ω de repères muni d'une mesure $d\omega$, une condition appropriée permettra d'écrire les éléments de $L^2(R^\nu)$ comme sommes convenables de fonctions φ_ω^α ($\omega \in \Omega$), avec $|\alpha| \leq n$ ne dépendant que de Ω : on parvient ainsi à la notion d'espaces de repères, le procédé que nous venons de décrire pouvant être rattaché à la représentation de Fock. Signalons qu'une décomposition de ce genre, mais ne faisant intervenir que des gaussiennes et par conséquent entachée d'un terme d'erreur, a été décrite par Cordoba et Fefferman [6] dans une situation adaptée aux opérateurs pseudo-différentiels classiques. Les opérateurs métadifférentiels associés à un couple (Ω,Ω') d'espaces de repères sont enfin définis comme ceux susceptibles de s'écrire comme sommes de "projecteurs-obliques" $u \longmapsto (u,\varphi)\psi$, où φ et ψ sont des gaussiennes, le noyau de cette décomposition étant concentré de telle sorte que le couple (φ,ψ) reste en moyenne proche de couples du genre $(\varphi_{\omega'},\varphi_\omega)$ ($\omega' \in \Omega'$, $\omega \in \Omega$). Des opérateurs de ce type sont éventuellement susceptibles d'intervenir dans la discussion de l'équation de Schrödinger ; des versions particulières comprennent des classes très générales d'opérateurs pseudo-différentiels, ou des opérateurs proches d'opérateurs intégraux de Fourier généralisés.

Pour faciliter l'utilisation de ces opérateurs, il convient d'exhiber des conditions sur un symbole permettant de fournir une représentation métadifférentielle de l'opérateur associé : ceci nécessite de décrire complétement les fonctions de Wigner $H(\varphi_{\omega}^{\alpha}, \varphi_{\omega'}^{\beta})$ d'états propres de deux oscillateurs harmoniques distincts : c'est la partie la plus technique de ce travail.

Bien entendu, il y a une quantité de travaux qui utilisent l'oscillateur harmonique : citons par exemple le traitement par Melin [11] de l'inégalité de Gårding ; signalons également que Ralston [12] a récemment utilisé des gaussiennes dans des problèmes de propagation des singularités. Le fait que les normes symplectiques jouent un rôle dans la description des opérateurs pseudo-différentiels (les symboles de ceux-ci étant définis comme ayant des dérivées normalisées bornées, la normalisation dépendant du point et d'une norme attachée à ce point) a été mis en évidence par Beals [2], Hörmander [7] ; dans [16], nous avons abandonné la technique traditionnelle d'étude des opérateurs pseudo-différentiels (l'intégration par parties) pour la remplacer par celle, plus efficace, qui consiste à décomposer l'opérateur comme somme de ramollisseurs harmoniques ; la décomposition comme somme de projecteurs-obliques a été introduite dans [17] ; ces deux dernières méthodes sont toutes deux employées dans le présent travail.

Plan de l'article :

1. Généralités sur le calcul de Weyl-Wigner
2. Extensions du lemme de Cotlar
3. Sur l'équation de Schrödinger
4. Un oscillateur harmonique
5. Deux oscillateurs harmoniques
6. Ramollisseurs harmoniques
7. Normes de créabilité-annihilabilité relatives à un oscillateur ou une paire d'oscillateurs
8. Espaces de repères
9. Opérateurs métadifférentiels.

1. Généralités sur le calcul de Weyl-Wigner

On considère l'espace R^ν, l'espace de phase $R^{2\nu}$ muni de la forme symplectique définie par

(1.1) $$[(x,\xi),(y,\eta)] = -< x,\eta > + < y,\xi >,$$

et la quantification de Weyl $Op_{\frac{1}{2}}$ définie, pour $a \in \mathcal{S}(R^{2\nu})$ et $u \in \mathcal{S}(R^\nu)$, par

(1.2) $$Op_{\frac{1}{2}}(a)\, u(x) = \iint a(\tfrac{x+y}{2},\xi)\, e^{2i\pi< x-y,\xi>} u(y)\, dy\, d\xi .$$

On sait que, pour tout $a \in \mathcal{S}(R^{2\nu})$, $Op_{\frac{1}{2}}(a)$ opère de $\mathcal{S}'(R^\nu)$ dans $\mathcal{S}(R^\nu)$, et, pour tout $a \in \mathcal{S}'(R^{2\nu})$, $Op_{\frac{1}{2}}(a)$ opère de $\mathcal{S}(R^\nu)$ dans $\mathcal{S}'(R^\nu)$; a est le symbole de $Op_{\frac{1}{2}}(a)$.

On a aussi, pour $u, v \in \mathcal{S}(R^\nu)$:

(1.3) $$(Op_{\frac{1}{2}}(a)u,v) = \iint a(x,\xi)\, H(u,v,x,\xi)\, dx\, d\xi,$$

où $H(u,v)$ est la fonction de Wigner définie par

(1.4) $$H(u,v,x,\xi) = 2^\nu \int u(x+z)\, \bar{v}(x-z)\, e^{-4i\pi< z,\xi >} dz.$$

La fonction $H(u,v)$ appartient à $\mathcal{S}(R^{2\nu})$ (resp. $\mathcal{S}'(R^{2\nu})$) si u, v appartiennent toutes deux à $\mathcal{S}(R^\nu)$ (resp. $\mathcal{S}'(R^\nu)$). Il est classique, et fondamental, que la fonction de Wigner $H(\varphi,\psi)$ peut aussi s'interpréter comme le symbole du projecteur oblique $\mathcal{H}(\varphi,\psi)$ défini par

(1.5) $$\mathcal{H}(\varphi,\psi)u = (u,\psi)\varphi .$$

On conservera par la suite cette notation :

(1.6) $$\mathcal{H}(\varphi,\psi) = Op_{\frac{1}{2}}(H(\varphi,\psi)).$$

Les grandes lettres X, Y, Z,... désigneront toujours des points de l'espace de phase $R^{2\nu}$. On convient dans cet article, dans le cas de symboles dépendant de paramètres appartenant à $R^{2\nu}$, d'attribuer toujours à X le rôle de variable opératoire, c'est-à-dire que, par convention,

(1.7) $$Op_{\frac{1}{2}}(a(X,Y,Z,...)) = Op_{\frac{1}{2}}(X \longmapsto a(X,Y,Z,...)).$$

Rappelons la formule de composition des symboles : si a et b appartiennent à $\mathcal{S}(R^{2\nu})$, on a

(1.8) $$Op_{\frac{1}{2}}(a)\, Op_{\frac{1}{2}}(b) = Op_{\frac{1}{2}}(a \# b),$$

avec

(1.9) $$(a \# b)(X) = 2^{2\nu} \iint a(Y)\, b(Z)\, e^{-4i\pi[Y-X,Z-X]}\, dY\, dZ.$$

Nous aurons fréquemment à considérer des opérateurs agissant sur les fonctions définies sur l'espace de phase. Le symbole d'un tel opérateur est donc une fonction définie sur l'ensemble des points (X,Ξ) de $R^{2\nu} \times (R^{2\nu})^*$. On évitera d'identifier $(R^{2\nu})^*$ à $R^{2\nu}$, mais on utilisera souvent, en revanche, l'isomorphisme σ de $(R^{2\nu})^*$ sur $R^{2\nu}$ défini par l'identité

$$(1.10) \qquad \langle X, \Xi \rangle = -[X, \sigma\Xi].$$

Si a, b, f $\in \mathcal{S}(R^{2\nu})$, on a, pour tout $X \in R^{2\nu}$:

$$\begin{aligned}(a \# f \# b)(X) &= 2^{4\nu}\iiiint a(Z_1)f(Y)b(T_2)e^{-4i\pi[Z_1-T_1,Y-T_1]}e^{-4i\pi[T_1-X,T_2-X]}\, dT_1\, dT_2\, dY\, dZ \\ &= 2^{2\nu}\iint a(Z_1)f(Y)b(X+Y-Z_1)e^{-4i\pi[Z_1,Y]}e^{4i\pi[X,Y-Z_1]}dZ_1\, dY \\ &= \iint a(\tfrac{X+Y+Z}{2})f(Y)b(\tfrac{X+Y-Z}{2})\, e^{-2i\pi[X-Y,Z]}\, dY\, dZ,\end{aligned}$$

d'où il résulte que le symbole de l'opérateur $f \longmapsto a \# f \# b$ est la fonction

$$(1.11) \qquad (X,\Xi) \longmapsto a(X + \tfrac{\sigma\Xi}{2})\, b(X - \tfrac{\sigma\Xi}{2}).$$

Rappelons qu'il existe un homomorphisme unique $M \longmapsto \tilde{M}$, de noyau réduit à deux éléments, du groupe métaplectique $Mp(\nu)$ sur le groupe symplectique $Sp(\nu)$ tel que, pour tout $a \in \mathcal{S}'(R^{2\nu})$, on ait

$$(1.12) \qquad M\, Op_{\frac{1}{2}}(a)\, M^{-1} = Op_{\frac{1}{2}}(a \circ \tilde{M}^{-1}).$$

Si B est une matrice réelle symétrique de format $\nu \times \nu$, si A est réelle, inversible de format $\nu \times \nu$, les transformations M et N définies ci-dessous sont métaplectiques :

$$(1.13) \qquad \begin{aligned}(Mu)(x) &= e^{i\pi(Bx,x)}u(x). \\ (Nu)(x) &= |\det A|^{-\frac{1}{2}}u(A^{-1}x).\end{aligned}$$

Leurs images dans $Sp(\nu)$ sont données, en matrices-blocs, par

$$(1.14) \qquad \tilde{M} = \begin{pmatrix} 1 & 0 \\ B & 1 \end{pmatrix}, \quad \tilde{N} = \begin{pmatrix} A & 0 \\ 0 & A'^{-1} \end{pmatrix}.$$

Pour permettre au groupe symplectique affine d'opérer sur les symboles, introduisons également les <u>translations de phase</u> τ_Z définies, pour $Z = (z,\zeta) \in R^{2\nu}$, par

$$(1.15) \qquad (\tau_Z u)(x) = u(x-z)\, e^{2i\pi\langle x - \frac{z}{2}, \zeta \rangle}.$$

Ces transformations vérifient les identités

$$(1.16) \qquad (\tau_Z)^* = \tau_{-Z} = \tau_Z^{-1},$$

$$(1.17) \qquad \tau_{Z_1}\tau_{Z_2} = e^{i\pi[Z_1,Z_2]}\,\tau_{Z_1+Z_2},$$

et

$$(1.18)\qquad H(\tau_Z u,\tau_Z v,X) = H(u,v,X-Z).$$

Cette dernière identité est équivalente à

$$(1.19)\qquad \tau_Z \mathrm{Op}_{\frac{1}{2}}(a)\,\tau_{-Z} = \mathrm{Op}_{\frac{1}{2}}(a(X-Z))$$

(cf. (1.7)).

On se servira également de la formule, dont la vérification est immédiate :

$$(1.20)\qquad H(\tau_{-Z}u,\tau_Z v,X) = e^{4i\pi[X,Z]}\,H(u,v,X).$$

En combinant (1.18) et (1.20), on obtient

$$(1.21)\qquad H(\tau_Y u,\tau_{Y'} v,X) = e^{-i\pi[Y,Y']}\,e^{2i\pi[X,Y'-Y]}\,H(u,v,X-\frac{Y+Y'}{2}),$$

autrement dit, en introduisant des translations de phase sur l'espace de phase :

$$(1.22)\qquad H(\tau_Y u,\ \tau_{Y'} v) = \tau_{\frac{Y+Y'}{2},\sigma^{-1}(Y-Y')}\,H(u,v).$$

2. Extensions du lemme de Cotlar

Ces deux lemmes sont des formes assez précises du lemme de Cotlar-Stein sur les "sommes de presque-projecteurs-orthogonaux" dont l'emploi dans la discussion de la continuité des opérateurs pseudo-différentiels est décisif depuis le travail de Calderon-Vaillancourt [4] ; il sont un peu plus généraux que les deux lemmes que nous avions prouvés dans [17].

Il faut auparavant préciser la définition suivante relative aux noyaux : soient Ω et Ω' deux espaces munis de mesures positives σ-finies $d\omega$ et $d\omega'$, et soit $k(\omega,\omega')$ une fonction mesurable sur $\Omega\times\Omega'$; nous dirons que k est le noyau d'un opérateur borné de $L^2(\Omega')$ dans $L^2(\Omega)$ si les conditions suivantes sont remplies :

(i) pour toute $f\in L^2(\Omega')$, on a $\int |k(\omega,\omega')|\ |f(\omega')|\ d\omega' < \infty$ pour presque tout ω.

(ii) la fonction $(Kf)(\omega) = \int k(\omega,\omega)\ f(\omega')\ d\omega'$ ainsi définie appartient à $L^2(\Omega)$.

(iii) on a $\|Kf\| \le C\ \|f\|$ avec une constante $C>0$ indépendante de f.

Naturellement, K s'appelle l'opérateur de noyau k. La composition des noyaux marche bien dans le cas où les valeurs absolues des noyaux fournissent également des opérateurs bornés : elle marche aussi dans le cas suivant (non comparable au précédent) dont nous aurons davantage l'usage :

<u>Lemme 2.1</u> : Soient Ω, Ω', Ω'' trois espaces munis de mesures positives σ-finies $d\omega$, $d\omega'$ et $d\omega''$; soient $k(\omega,\omega')$ et $\ell(\omega'',\omega')$ deux noyaux définissant des opérateurs bornés (entre les espaces L^2 correspondants) K et L ; on suppose de plus que pour presque tout ω on a $\int |k(\omega,\omega')|^2\ d\omega' < \infty$ et que pour presque tout ω'' on a $\int|\ell(\omega'',\omega')|^2 d\omega'<\infty$.

Si l'on pose (pour presque tout (ω,ω'')) :

$$m(\omega,\omega'') = \int k(\omega,\omega')\, \overline{\ell}(\omega'',\omega')\, d\omega',$$

la fonction m vérifie la condition $\int |m(\omega,\omega'')|^2\, d\omega'' < \infty$ pour presque tout ω, et elle constitue le noyau de l'opérateur borné $KL^* : L^2(\Omega'') \to L^2(\Omega)$.

Preuve : posons, pour presque tout ω, $k_\omega(\omega') = k(\omega,\omega')$; la fonction k_ω appartient à $L^2(\Omega')$ pour presque tout ω et l'on a, pour presque tout ω'' :

$$(L\,\overline{k}_\omega)(\omega'') = \int \ell(\omega'',\omega')\, \overline{k}(\omega,\omega')\, d\omega'$$

d'où $m(\omega,\omega'') = \overline{L\,\overline{k}_\omega(\omega'')}$, ce qui prouve le premier point. On a par ailleurs, pour toute fonction $g \in L^2(\Omega'')$, et presque tout ω :

$$\begin{aligned}(KL^*g)(\omega) &= \int k(\omega,\omega')(L^*g)(\omega')\, d\omega' \\ &= \int g(\omega'')\, \overline{L\,\overline{k}_\omega(\omega'')}\, d\omega'' = \int m(\omega,\omega'')\, g(\omega'')\, d\omega''.\end{aligned}$$

Lemme 2.2 : Soient Ω et Ω' deux espaces munis de mesures positives σ-finies $d\omega$ et $d\omega'$, et soient $L_{\omega'}$, M_ω et $A_{\omega,\omega'}$ des opérateurs linéaires bornés sur $L^2(\mathbb{R}^\nu)$, dépendant mesurablement des variables spécifiées. On pose

$$k(\omega,\omega') = \|A_{\omega,\omega'}\| \;,\; k_1(\omega,\omega_1) = \|M_{\omega_1} M^*_\omega\| \text{ et } k_2(\omega',\omega'_1) = \|L_{\omega'_1} L^*_{\omega'}\| \;,$$

toutes normes représentant la norme dans l'espace des endomorphismes bornés de $L^2(\mathbb{R}^\nu)$. On suppose que k, k_1 et k_2 sont les noyaux d'opérateurs bornés de $L^2(\Omega')$ dans $L^2(\Omega)$, de $L^2(\Omega)$ dans $L^2(\Omega)$ et de $L^2(\Omega')$ dans $L^2(\Omega')$ respectivement, de normes respectives C, C_1 et C_2. Alors l'opérateur A sur $L^2(\mathbb{R}^\nu)$ défini par

$$A = \iint_{\Omega\times\Omega'} M^*_\omega\, A_{\omega,\omega'}\, L_{\omega'}\, d\omega\, d\omega'$$

et borné sur $L^2(\mathbb{R}^\nu)$, de norme moindre que $C(C_1C_2)^{\frac{1}{2}}$.

Preuve : d'après [17], on a, pour tout $u \in L^2(\mathbb{R}^\nu)$, les inégalités

$$\int \|M_\omega u\|^2\, d\omega \le C_1\, \|u\|^2$$

et

$$\int \|L_{\omega'} u\|^2\, d\omega' \le C_2\, \|u\|^2.$$

Alors

$$\begin{aligned}|(Au,v)| &= \left|\iint (A_{\omega,\omega'} L_{\omega'} u, M_\omega v)\, d\omega\, d\omega'\right| \\ &\le \iint k(\omega,\omega')\, \|L_{\omega'} u\|\, \|M_\omega v\|\, d\omega\, d\omega' \\ &\le C\left(\int \|L_{\omega'} u\|^2\, d\omega'\right)^{\frac{1}{2}} \left(\int \|M_\omega v\|^2\, d\omega\right)^{\frac{1}{2}}\end{aligned}$$

$$\leq C(C_1C_2)^{\frac{1}{2}} \|u\| \|v\| .$$

Lemme 2.3 : Soient Ω et Ω' deux espaces munis de mesures positives σ-finies $d\omega$ et $d\omega'$. Soit $k(\omega,\omega')$ un noyau mesurable sur $\Omega \times \Omega'$, et $\omega \longmapsto \psi_\omega$ (resp. $\omega' \longmapsto \varphi_{\omega'}$) une fonction mesurable sur Ω (resp. Ω') à valeurs dans $L^2(\mathbb{R}^\nu)$. On pose $k_1(\omega,\omega_1) = (\psi_{\omega_1},\psi_\omega)$ et $k_2(\omega',\omega'_1) = (\varphi_{\omega'_1},\varphi_{\omega'})$. Soient K, K_1 et K_2 les opérateurs de noyaux respectifs k, k_1 et k_2. Alors si ces opérateurs sont bornés entre les espaces L^2 correspondants, K_1 et K_2 sont autoadjoints positifs, et l'opérateur A sur $L^2(\mathbb{R}^\nu)$ défini faiblement par

$$(Au,v) = \iint_{\Omega\times\Omega'} k(\omega,\omega')(u,\varphi_{\omega'})(\psi_\omega,v)\, d\omega\, d\omega'$$

est borné, et vérifie l'égalité $\|A\| = \|K_1^{\frac{1}{2}} K K_2^{\frac{1}{2}}\|$.

Preuve : Soit (E_n) une suite croissante de parties de mesure finie de Ω, telle que $\Omega = \cup E_n$, et $\|\psi_\omega\| \leq n$ pour $\omega \in E_n$.
Si l'on définit $M_n : L^2(\mathbb{R}^\nu) \to L^2(E_n)$ par $(M_n u)(\omega) = (u,\psi_\omega)$, on a

$(M_n^* f)(x) = \int_{E_n} f(\omega)\, \psi_\omega(x)\, d\omega$ pour toute $f \in L^2(E_n)$, et

$$(M_n M_n^* f)(\omega) = \int_{E_n} k_1(\omega,\omega_1)\, f(\omega_1)\, d\omega_1 ,$$

d'où $\|M_n M_n^*\| \leq C$ avec C indépendante de n puisque K_1 est borné. Il en résulte que $\|M_n u\| \leq C\|u\|$, et par suite l'opérateur $M : L^2(\mathbb{R}^\nu) \to L^2(\Omega)$ défini par $(Mu)(\omega) = (u,\psi_\omega)$ est borné.
Quelles que soient f et g appartenant à $L^2(\Omega)$, on peut alors écrire

$$(MM^* f,g) = \int (M^* f,\psi_\omega)\, g(\omega) d\omega = \int (f,M\psi_\omega)\, \overline{g}(\omega)\, d\omega$$
$$= \int \overline{g}(\omega) d\omega \int f(\omega_1)(\psi_{\omega_1},\psi_\omega)\, d\omega,$$

d'où $MM^* = K_1$.

On définit de même $L : L^2(\mathbb{R}^\nu) \to L^2(\Omega')$ par $(Lu)(\omega') = (u,\varphi_{\omega'})$, d'où $LL^* = K_2$.
On a enfin, quelles que soient u et $v \in L^2(\mathbb{R}^\nu)$, l'identité

$$(M^* KLu,v) = (KLu,Mv) = \int (\psi_\omega,v)\, d\omega \int k(\omega,\omega')(u,\varphi_{\omega'})\, d\omega' .$$

D'où $A = M^* K L$, et (puisque $\|M^* f\|^2 = (K_1 f,f)$ pour toute $f \in L^2(\Omega)$) :

$$\|A\| = \|M^* KL\| = \|K_1^{\frac{1}{2}} KL\| = \|L^* K^* K_1^{\frac{1}{2}}\| = \|K_2^{\frac{1}{2}} K^* K_1^{\frac{1}{2}}\| = \|K_1^{\frac{1}{2}} K K_2^{\frac{1}{2}}\| .$$

On écrira aussi

$$Au = \iint k(\omega,\omega')(u,\varphi_{\omega'})\, \psi_\omega\, d\omega\, d\omega'$$

ou même (cf.(1.5)) :

$$A = \iint k(\omega,\omega')\, \mathcal{H}(\psi_\omega,\varphi_{\omega'})\, d\omega\, d\omega' .$$

3. Sur l'équation de Schrödinger

Soit $A = Op_{\frac{1}{2}}(a)$ un opérateur linéaire symétrique sur $L^2(\mathbb{R}^\nu)$, de domaine $\mathcal{S}(\mathbb{R}^\nu)$: a est donc réel, et l'on supposera que A opère de $\mathcal{S}(\mathbb{R}^\nu)$ dans $\mathcal{S}(\mathbb{R}^\nu)$. Un problème fondamental est la recherche de conditions pour qu'il existe un groupe fortement continu

$$(3.1) \qquad R_t = e^{2i\pi tA}$$

d'opérateurs unitaires de $L^2(\mathbb{R}^\nu)$ dont le générateur infinitésimal soit une extension de $2i\pi A$; le cas échéant, il faut voir également si ce groupe est unique.

On sait, en vertu d'un théorème de Stone, que le premier problème équivaut à la recherche d'extensions autoadjointes de A ; la seconde question posée est de savoir si A, de domaine $\mathcal{S}(\mathbb{R}^\nu)$, est essentiellement autoadjoint : on y répond par l'affirmative si l'on sait construire un groupe R_t qui vérifie la propriété supplémentaire que, pour tout t, R_t opère de $\mathcal{S}(\mathbb{R}^\nu)$ dans lui-même. On peut, en pensant aux représentations de Schrödinger et d'Heisenberg de la mécanique ondulatoire, étudier la construction d'un groupe R_t en deux temps. Le premier (sur lequel le présent papier ne contient pas d'information) va consister en la construction du groupe $\widetilde{R}_t$ défini par

$$(3.2) \qquad Op_{\frac{1}{2}}(\widetilde{R}_t b) = R_t \, Op_{\frac{1}{2}}(b) \, R_t^{-1}$$

pour tout symbole $b \in L^2(\mathbb{R}^{2\nu})$; compte-tenu du fait que les éléments de $L^2(\mathbb{R}^{2\nu})$ sont exactement les symboles des opérateurs de Hilbert-Schmidt, on voit aisément que (3.2) définit effectivement un groupe unitaire fortement continu si R_t a déjà été construit (mais c'est le passage dans l'autre sens qui nous intéresse) ; le générateur infinitésimal de ce groupe doit être une extension de l'opérateur

$$(3.3) \qquad 2i\pi \, \widetilde{A} = \mathrm{ad}\; 2i\pi \, a$$

défini pour $b \in \mathcal{S}(\mathbb{R}^{2\nu})$ par

$$(3.4) \qquad (\mathrm{ad}\; 2i\pi \, a)b = 2i\pi(a \# b - b \# a) \ :$$

cette formule résulte naturellement de (3.2)et de la formule

$$(3.5) \qquad \frac{d}{dt}(R_t \, u)\Big|_{t=o} = 2i\pi \, Au$$

valable si $u \in \mathcal{S}(\mathbb{R}^\nu)$.

D'après (1.11), on peut donc écrire

$$(3.6) \qquad \widetilde{A} = Op_{\frac{1}{2}}(\widetilde{a})$$

avec

$$(3.7) \qquad \widetilde{a}(X,\Xi) = a(X + \frac{\sigma\Xi}{2}) - a(X - \frac{\sigma\Xi}{2}).$$

Remarquons d'ailleurs que $\tilde{A}$ opère toujours de $\mathcal{S}(\mathbb{R}^{2\nu})$ dans $\mathcal{S}(\mathbb{R}^{2\nu})$ puisque, ainsi qu'il a été remarqué dans [15] (preuve du théorème 4.2), chacun des deux termes de sa décomposition est le conjugué, par une certaine transformation métaplectique, de l'opérateur $A \otimes I$ opérant sur les fonctions définies sur $\mathbb{R}^{2\nu} \times \mathbb{R}^{2\nu}$; ce résultat est également annoncé dans [1]. A titre de curiosité, ces symboles $\tilde{a}$ vérifient la propriété que l'opérateur $Op_{\frac{1}{2}}(\tilde{a})$ coïncide avec l'opérateur $Op(\tilde{a})$ obtenu en faisant agir les convolutions avant les multiplications : plus généralement, $\tilde{a}$ est invariant par le groupe J^t défini dans l'exercice 1.3 de [14], dont la valeur en $t = \frac{1}{2}$ connecte les deux calculs symboliques considérés.

Le problème étudié ici consiste à remonter du groupe $\tilde{R}_t = e^{2i\pi t\tilde{A}}$, supposé construit, à un groupe R_t. Naturellement, si l'on est optimiste, on peut penser, après développement de Taylor au premier ordre dans (3.7), que l'opérateur $b \longmapsto b \circ \Phi_t$, où Φ_t est le flot hamiltonien de a, peut dans certains cas être le premier terme d'une approximation convenable de $\tilde{R}_t$; si l'on est pessimiste, on retiendra quand même que l'on a $\tilde{R}_t = \Phi_t$ exactement si a est un polynôme d'ordre 2, le théorème qui suit fournissant alors une nouvelle construction des opérateurs du groupe métaplectique, où l'on n'éprouve pas les difficultés usuelles relatives aux singularités de Maslov des phases (deux gaussiennes n'étant jamais orthogonales) : le but essentiel de ce théorème est pour nous de montrer l'intérêt qui peut s'attacher d'une part aux opérateurs définis comme sommes de projecteurs-obliques, d'autre part aux décompositions de l'identité d'un genre qui sera étudié plus loin.

<u>Théorème 3.1</u> : soit a un symbole réel tel que $Op_{\frac{1}{2}}(a)$ définisse un opérateur A de $\mathcal{S}(\mathbb{R}^{\nu})$ dans $\mathcal{S}(\mathbb{R}^{\nu})$; soit $\tilde{A}$ l'opérateur de $\mathcal{S}(\mathbb{R}^{2\nu})$ dans $\mathcal{S}(\mathbb{R}^{2\nu})$ défini par (3.6) et (3.7). Soient D un sous-ensemble de $\mathcal{S}(\mathbb{R}^{2\nu})$ et $t_0 > 0$. On suppose que pour tout $b \in D$ il existe une unique fonction de classe C^1 : $t \longmapsto b^t$ de $]-t_0, t_0[$ dans $L^2(\mathbb{R}^{2\nu})$, prenant en fait ses valeurs dans $\mathcal{S}(\mathbb{R}^{2\nu})$, qui soit solution du problème de Cauchy

$$\begin{cases} \frac{d}{dt} b^t = 2i\pi\, \tilde{A}\, b^t & (|t| < t_0) \\ b^0 = b. \end{cases}$$

Soient par ailleurs Ω un ensemble muni d'une mesure positive σ-finie $d\omega$, $\omega \longmapsto \psi_\omega$ une application mesurable de Ω dans la sphère unité de $L^2(\mathbb{R}^{\nu})$, et k une fonction à valeurs complexes sur $\Omega \times \Omega$, vérifiant les propriétés suivantes :

(i) pour tout $\omega \in \Omega$, $\Psi_\omega = H(\psi_\omega, \psi_\omega) \in D$

(ii) l'opérateur de noyau $(\psi_\omega, \psi_{\omega'})$ sur $L^2(\Omega)$ et ceux de noyaux $k(\omega,\omega')$ et $\bar{k}(\omega',\omega)$ sont bornés et vérifient l'hypothèse supplémentaire du lemme 2.1 qui permet la composition des noyaux.

(iii) l'opérateur $\iint_{\Omega\times\Omega} k(\omega,\omega')\,\mathcal{H}(\psi_\omega, \psi_{\omega'})\, d\omega\, d\omega'$ (borné sur $L^2(\mathbb{R}^{\nu})$ d'après le lemme 2.3) coïncide avec l'identité.

On suppose enfin la condition suivante vérifiée :

(iv) pour tout $\omega \in \Omega$, et tout $t \in]-t_o, t_o[$, on a $(\Psi_\omega^t, \Psi_\omega) \neq 0$.

Posons, pour $\omega \in \Omega$ et $|t| < t_o$:

$$\mu_\omega(t) = \int_o^t \frac{(\Psi_\omega^s \# H(2i\pi A\, \psi_\omega, \psi_\omega), \Psi_\omega)}{(\Psi_\omega^s, \Psi_\omega)}\, ds ,$$

et

$$\psi_\omega^t = e^{\mu_\omega(t)}\, Op_{\frac{1}{2}}(\Psi_\omega^t)\, \psi_\omega .$$

La famille (R_t) $(|t| < t_o)$ d'opérateurs définis par

$$R_t = \iint k(\omega,\omega')\, \mathcal{H}(\psi_\omega^t, \psi_{\omega'})\, d\omega\, d\omega'$$

s'étend alors en un groupe fortement continu d'opérateurs unitaires sur $L^2(R^\nu)$; le domaine du générateur infinitésimal de ce groupe contient l'espace vectoriel engendré par les fonctions ψ_ω $(\omega \in \Omega)$, et sur ce dernier espace le générateur infinitésimal coïncide avec $2i\pi A$. Enfin, si l'on a $R_t(\mathcal{S}(R^\nu)) \subset \mathcal{S}(R^\nu)$ pour $|t| < t_o$, l'opérateur A est essentiellement autoadjoint sur $\mathcal{S}(R^\nu)$.

<u>Preuve</u> : elle sera décomposée en lemmes pour plus de clarté ; signalons avant tout que, pour tout ω, $\psi_\omega \in \mathcal{S}(R^\nu)$ puisque $\Psi_\omega \in \mathcal{S}(R^{2\nu})$.

<u>Lemme 3.1</u> : pour $|t| < t_o$, et tout $\omega \in \Omega$, Ψ_ω^t est le symbole d'un projecteur orthogonal de rang 1.

<u>Preuve</u> : posant $c(X,\Xi) = 2i\pi\, \tilde{a}(X,\Xi)$, on a $\bar{c}(X,\Xi) = c(X,-\Xi)$: cette propriété, comme le montre un calcul élémentaire, signifie que l'opérateur $2i\pi\, \tilde{A}$ commute avec la conjugaison complexe ; l'équation différentielle

(3.8) $$\frac{d}{dt} \Psi_\omega^t = 2i\pi\, \tilde{A}\, \Psi_\omega^t$$

montre alors que Ψ_ω^t est réel pour tout t.
Cette équation fournit également

(3.9) $$\|\Psi_\omega^t\| = 1 \quad \text{pour tout } t.$$

Par ailleurs,

$$\frac{d}{dt}(\Psi_\omega^t \# \Psi_\omega^t) = 2i\pi[(\tilde{A}\, \Psi_\omega^t) \# \Psi_\omega^t + \Psi_\omega^t \# \tilde{A}\, \Psi_\omega^t]$$

$$= 2i\pi[(a \# \Psi_\omega^t - \Psi_\omega^t \# a) \# \Psi_\omega^t + \Psi_\omega^t \# (a \# \Psi_\omega^t - \Psi_\omega^t \# a)]$$

$$= 2i\pi\, \tilde{A}(\Psi_\omega^t \# \Psi_\omega^t).$$

Jointe à la condition initiale $\Psi_\omega^o \# \Psi_\omega^o = \Psi_\omega^o$, cette équation différentielle entraîne $\Psi_\omega^t \# \Psi_\omega^t = \Psi_\omega^t$. Nous avons donc prouvé que Ψ_ω^t est le symbole d'un projecteur orthogonal de norme de Hilbert-Schmidt égale à 1, ce qui établit le lemme.

Lemme 3.2 : ψ_ω^t vérifie le problème de Cauchy

$$\begin{cases} \frac{d}{dt} \psi_\omega^t = 2i\pi\, A\, \psi_\omega^t & (|t| < t_o) \\ \psi_\omega^o = \psi_\omega \end{cases}$$

Preuve : le second point est évident.
Utilisant le lemme 3.1, posons provisoirement

(3.10) $\quad \Psi_\omega^t = H(\varphi_\omega^t, \varphi_\omega^t).$

On a alors, pour $|t| < t_o$:

(3.11) $\quad (\psi_\omega, \varphi_\omega^t) \neq 0,$

(3.12) $\quad \mu_\omega'(t) = \dfrac{(2i\pi\, A\psi_\omega, \varphi_\omega^t)}{(\psi_\omega, \varphi_\omega^t)}$

et

(3.13) $\quad \psi_\omega^t = e^{\mu_\omega(t)}\, (\psi_\omega, \varphi_\omega^t)\, \varphi_\omega^t .$

Partant de la définition originale de ψ_ω^t , on obtient

$$\begin{aligned} \frac{d}{dt} \psi_\omega^t &= \frac{(2i\pi\, A\psi_\omega, \varphi_\omega^t)}{(\psi_\omega, \varphi_\omega^t)} \psi_\omega^t + e^{\mu_\omega(t)} \frac{d}{dt}(Op_{\frac{1}{2}}(\Psi_\omega^t)\, \psi_\omega) \\ &= \frac{(2i\pi\, A\psi_\omega, \varphi_\omega^t)}{(\psi_\omega, \varphi_\omega^t)} \psi_\omega^t + 2i\pi\, e^{\mu_\omega(t)} (A\, Op_{\frac{1}{2}}(\Psi_\omega^t) - Op_{\frac{1}{2}}(\Psi_\omega^t)A)\, \psi_\omega \\ &= 2i\pi\, A\psi_\omega^t + \frac{(2i\pi\, A\psi_\omega, \varphi_\omega^t)}{(\psi_\omega, \varphi_\omega^t)} \psi_\omega^t - 2i\pi\, e^{\mu_\omega(t)} (A\psi_\omega, \varphi_\omega^t)\, \varphi_\omega^t , \end{aligned}$$

ce qui fournit l'équation différentielle en se servant à nouveau de (3.13).

Lemme 3.3 : $\Psi_\omega^t = H(\psi_\omega^t, \psi_\omega^t)$.

Preuve : c'est vrai lorsque $t = 0$; par ailleurs, d'après le lemme 10.2, on a

$$\begin{aligned} \frac{d}{dt} H(\psi_\omega^t, \psi_\omega^t) &= H(2i\pi\, A\, \psi_\omega^t, \psi_\omega^t) + H(\psi_\omega^t, 2i\pi\, A\, \psi_\omega^t) \\ &= 2i\pi(a \# H(\psi_\omega^t, \psi_\omega^t) - H(\psi_\omega^t, \psi_\omega^t) \# a) \\ &= 2i\pi\, \tilde{A}\, H(\psi_\omega^t, \psi_\omega^t). \end{aligned}$$

Lemme 3.4 : les opérateurs R_t $(|t| < t_o)$ sont unitaires sur $L^2(\mathbb{R}^\nu)$.

Preuve : en utilisant le lemme 3.2, on voit que $(\psi_\omega^t, \psi_{\omega'}^t) = (\psi_\omega, \psi_{\omega'})$, et le lemme 2.3 montre alors que les opérateurs R_t sont bornés.

On peut en outre écrire, compte tenu du lemme 2.1 :

$$R_t^* R_t = \int \bar{k}(\omega_1,\omega_1')k(\omega,\omega')(\psi_\omega^t,\psi_{\omega_1}^t)\mathcal{H}(\psi_{\omega'_1},\psi_{\omega'})\, d\omega\, d\omega'\, d\omega_1\, d\omega_1'$$

et

$$I \circ I = \int \bar{k}(\omega_1,\omega_1')\, k(\omega,\omega')(\psi_\omega,\psi_{\omega_1})\mathcal{H}(\psi_{\omega_1'},\psi_{\omega'})\, d\omega\, d\omega'\, d\omega_1\, d\omega_1' ,$$

d'où $\quad R_t^* R_t = I.$

Lemme 3.5 : Les conditions $|s| < t_o$, $|t| < t_o$ et $|s+t| < t_o$ entraînent $R_t R_s = R_{s+t}$.

Preuve : à l'aide du lemme 3.2, on voit que, quels que soient ω et ω_1 , on a

$$\frac{d}{dt}(\psi_{\omega_1}^{t+s},\ \psi_\omega^t) = 0 \quad , \text{ d'où } \quad (\psi_{\omega_1}^{t+s},\psi_\omega^t) = (\psi_{\omega_1}^s,\psi_\omega)$$

et

$$R_t^*\ \psi_{\omega_1}^{t+s} = \iint \bar{k}(\omega,\omega')(\psi_{\omega_1}^{t+s},\psi_\omega^t)\ \psi_{\omega'}\ d\omega\, d\omega'$$

$$= \iint \bar{k}(\omega,\omega')(\psi_{\omega_1}^s,\psi_\omega)\ \psi_{\omega'}\ d\omega\, d\omega' = \psi_{\omega_1}^s .$$

Par suite $\quad R_t\ \psi_{\omega_1}^s = \psi_{\omega_1}^{t+s}$; le lemme en résulte immédiatement.

Lemme 3.6 : si R_t conserve $\mathcal{S}(R^\nu)$ pour $|t| < t_o$, on a $AR_t = R_t A$.

Preuve : quels que soient ω et ω_1, on a

(3.14) $\qquad (A\ \psi_{\omega_1}^t,\psi_\omega^t) = (A\ \psi_{\omega_1},\psi_\omega)$:

c'est une conséquence du lemme 3.2.

On peut alors écrire

$$R_t^*(A\ \psi_{\omega_1}^t) = \iint \bar{k}(\omega,\omega')(A\ \psi_{\omega_1}^t,\psi_\omega^t)\ \psi_{\omega'}\ d\omega\, d\omega'$$

$$= \iint \bar{k}(\omega,\omega')(A\ \psi_{\omega_1},\psi_\omega)\ \psi_{\omega'}\ d\omega\, d\omega'$$

$$= A\ \psi_{\omega_1}.$$

Il en résulte que

(3.15) $\qquad A\ \psi_{\omega_1}^t = R_t(A\ \psi_{\omega_1}) .$

Remarquons maintenant que l'hypothèse (iii) implique que les fonctions ψ_ω $(\omega \in \Omega)$ forment un système total dans $L^2(R^\nu)$ (en effet, si u et $v \in L^2(R^\nu)$, la condition $v \perp \psi_\omega$ pour tout ω entraîne $(u,v) = 0$) ; également, $R_t^*\ \psi_{\omega_1} = \psi_{\omega_1}^{-t} \in \mathcal{S}(R^\nu)$.

Pour prouver le lemme, il suffit donc de montrer que les conditions $u \in \mathcal{S}(R^\nu)$, $v \in \mathcal{S}(R^\nu)$ et $R_t^*\ v \in \mathcal{S}(R^\nu)$ entraînent $(R_t u, Av) = (R_t A\ u,\ v)$.

Or on a

$$
\begin{aligned}
(R_t u,Av) &= \iint k(\omega,\omega')(u,\psi_{\omega'})(\psi_\omega^t,Av)\,d\omega\,d\omega' \\
&= \iint k(\omega,\omega')(u,\psi_{\omega'})(A\,\psi_\omega^t,v)\,d\omega\,d\omega' \\
&= \iint k(\omega,\omega')(u,\psi_{\omega'})(R_t A\psi_\omega,v)\,d\omega\,d\omega' \\
&= \iint k(\omega,\omega')(u,\psi_{\omega'})(\psi_\omega,AR_t^* v)\,d\omega\,d\omega' \\
&= (u,AR_t^* v) = (Au,R_t^* v) = (R_t Au,v).
\end{aligned}
$$

<u>Preuve du théorème 3.1</u> :

Pour tout $\omega \in \Omega$, et tout $t \in]0,t_o[$, on a d'après le lemme 3.2 et la preuve du lemme 3.5 :

$$
\frac{R_t\,\psi_\omega-\psi_\omega}{t} = \frac{\psi_\omega^t-\psi_\omega}{t} = \int_o^1 2i\pi\,A\,\psi_\omega^{\lambda t}\,d\lambda = \int_o^1 2i\pi\,R_{\lambda t}(A\,\psi_\omega)\,d\lambda, \tag{3.16}
$$

d'où $\quad \| R_t\,\psi_\omega-\psi_\omega\| \leq t\,\|2i\pi\,A\,\psi_\omega\|.$

Il en résulte que le groupe (R_t) est fortement continu (puisque les ψ_ω forment un système total) ; par suite $\|R_t A\,\psi_\omega - A\,\psi_\omega\|$ tend vers 0 lorsque $t \to 0$ pour tout ω, et le générateur infinitésimal du groupe (R_t) coïncide bien avec $2i\pi$ A sur les ψ_ω. Enfin, pour prouver la dernière assertion, il reste à montrer (cf. [13], th. VIII.10) que moyennant l'hypothèse supplémentaire que $R_t(\mathcal{S}(R^\nu)) \subset \mathcal{S}(R^\nu)$, on a

$$
\frac{R_t u - u}{t} \longrightarrow 2i\pi\,Au \qquad \text{pour tout } u \in \mathcal{S}(R^\nu) \text{ lorsque } t \to 0.
$$

Or, pour tout $u \in \mathcal{S}(R^\nu)$ et tout $v \in \mathcal{S}(R^\nu)$ tel que $R_t^* v \in \mathcal{S}(R^\nu)$ et $\|v\| \leq 1$, on peut écrire

$$
\begin{aligned}
\left(\frac{R_t u-u}{t},v\right) &= \iint k(\omega,\omega')(u,\psi_{\omega'})\left(\frac{\psi_\omega^t-\psi_\omega}{t},v\right)d\omega\,d\omega' \\
&= \int_o^1 d\lambda \iint k(\omega,\omega')(u,\psi_{\omega'})(2i\pi\,R_{\lambda t}A\psi_\omega,v)\,d\omega\,d\omega' \\
&= \int_o^1 d\lambda \iint k(\omega,\omega')(u,\psi_{\omega'})(2i\pi\,\psi_\omega^{\lambda t},Av)\,d\omega\,d\omega' \\
&= 2i\pi \int_o^1 (R_{\lambda t}u,Av)\,d\lambda \\
&= 2i\pi \int_o^1 (R_{\lambda t}Au,v)\,d\lambda.
\end{aligned}
$$

Par suite

$$
\begin{aligned}
&\left|\left(\frac{R_t u-u}{t} - 2i\pi\,Au,v\right)\right| \\
&= \left|\int_o^1 ((R_{\lambda t}-I)\,2i\pi\,Au,v)\,d\lambda\right| \\
&\leq \int_o^1 \|(R_{\lambda t}-I)\,2i\pi\,Au\|\,d\lambda
\end{aligned}
$$

tend vers 0 quand $t \to 0$, ce qui termine la preuve du théorème 3.1.

4. Un oscillateur harmonique

Soit B une forme bilinéaire réelle symétrique non dégénérée sur l'espace de phase $R^{2\nu}$, et soit par ailleurs une structure complexe sur $R^{2\nu}$ définie par un endomorphisme J de $R^{2\nu}$ de carré -1 ; nous dirons que la forme et la structure complexe sont compatibles (relativement à la structure symplectique de $R^{2\nu}$) si la forme complexe $\tilde{B}$ définie par

$$(4.1) \qquad \tilde{B}(X_1,X_2) = B(X_1,X_2) + i[X_1,X_2]$$

est sesquilinéaire.

Soit B_o la matrice symétrique telle que $B(X_1,X_2) = \langle X_1,B_oX_2 \rangle$. La condition de compatibilité est l'identité $B(X_1,X_2) = [JX_1,X_2]$, et s'exprime donc par l'égalité

$$(4.2) \qquad J = \sigma B_o ,$$

qui montre que la donnée de J ou celle de B sont équivalentes.

Définition 4.1 : soit $X \longmapsto \|X\|$ une norme euclidienne sur l'espace de phase $R^{2\nu}$; nous dirons que cette norme est symplectique si la forme quadratique B associée par polarisation est compatible avec la structure complexe définie par un endomorphisme J de $R^{2\nu}$ symplectique de carré -1.

Posant $\|X\|^2 = \langle X,B_oX \rangle$ avec B_o symétrique, on demande que $J = \sigma B_o$ soit symplectique. Comme, dans la base canonique de $R^{2\nu}$ et la base duale de $(R^{2\nu})^*$, σ est représentée par la matrice $\begin{pmatrix} 0 & -1 \\ 1 & 0 \end{pmatrix}$, il revient au même d'exiger que la matrice θ représentant B_o^{-1} dans ces bases soit symplectique. On peut alors choisir V symplectique avec $VV' = \theta$, d'où $\|X\|^2 = |V^{-1}X|^2$, où $X \longmapsto X$ est la norme canonique de $R^{2\nu}$. Nous reviendrons incessament à cette décomposition, basée sur la représentation de Siegel. Ce qui précède montre qu'une norme euclidienne est symplectique si et seulement si elle admet une base orthonormale dont la matrice de passage, relativement à la base canonique de $R^{2\nu}$, est symplectique. Autrement dit, l'ensemble des normes symplectiques s'identifie à l'espace homogène des classes à gauche $Sp(\nu)/Sp(\nu)\cap O(2\nu)$. Nous l'identifierons de préférence à l'espace P_ν des matrices θ symplectiques symétriques positives sur $R^{2\nu}$, par la correspondance définie par $\|X\|_\theta^2 = (\theta^{-1}X,X)$, où (,) désigne le produit scalaire canonique sur $R^{2\nu}$. Dans ce qui suit, tout ce qui concerne l'espace P_ν et la représentation de Siegel est emprunté au livre de Maaβ [10]. Il convient d'observer, cependant, que la représentation de l'espace des normes symplectiques par P_ν, puis celle de Siegel, dépendent du choix d'un repère privilégié de $R^{2\nu}$, commode pour effectuer les calculs ; mais les résultats finaux de cet article, comme on pourra s'en convaincre, ne dépendent que de la donnée de l'espace de phase sous la forme du produit d'un espace vectoriel réel de dimension ν par son dual.

Définition 4.2 : soit $\omega = (Y,\theta) \in R^{2\nu} \times P_\nu$.
Nous appellerons oscillateur harmonique associé l'opérateur $L_\omega = Op_{\frac{1}{2}}(\pi\|X-Y\|^2_\theta)$. On posera également $\Lambda_\omega = L_\omega - \frac{\nu}{2}$.
Avec $\theta = VV'$, $V \in Sp(\nu)$, on a donc

$$(4.3) \qquad L_\omega = Op_{\frac{1}{2}}(\pi|V^{-1}(X-Y)|^2),$$

et les formules (1.12) et (1.19) montrent que L_ω est unitairement équivalent à l'oscillateur harmonique canonique $L_o = Op_{\frac{1}{2}}(\pi|X|^2)$ dont la résolution spectrale est donnée par $1 = \sum_{j\geq 0} P_j$, $L_o = \sum_{j\geq 0} (\frac{\nu}{2} + j)P_j$, où l'image du projecteur P_o est engendrée par $\varphi(x) = 2^{\nu/4}e^{-\pi|x|^2}$, celle de P_j par les fonctions d'Hermite $\varphi^\alpha(|\alpha| = j)$; la description de celles-ci à l'aide des opérateurs de création est bien connue (cf. par exemple [5]), mais nous allons la rappeler pour l'oscillateur général L_ω. Soient $Y = (y,\eta)$, et $\theta = \begin{pmatrix} A & B \\ B' & C \end{pmatrix}$; le fait que θ appartient à P_ν s'exprime (cf. [10]) par les conditions $B'C = CB$, $AB' = BA$, $AC - B^2 = I$, et l'on a $\theta = VV'$, avec

$$(4.4) \qquad V = \begin{pmatrix} 1 & 0 \\ A^{-1}B & 1 \end{pmatrix} \begin{pmatrix} A^{\frac{1}{2}} & 0 \\ 0 & A^{-\frac{1}{2}} \end{pmatrix} :$$

c'est une formule de Siegel.
Un état fondamental de l'oscillateur $Op_{\frac{1}{2}}(\pi\|X\|^2_\theta)$ est alors, d'après (1.12), (1.13), et (1.14), donné par

$$(4.5) \qquad \varphi_\theta(x) = (\det A)^{-\frac{1}{4}}\, 2^{\nu/4}\, e^{-\pi(A^{-1}x,x)}\, e^{i\pi(A^{-1}Bx,x)}.$$

Enfin, d'après (1.19), un état fondamental de l'oscillateur L_ω est donné par

$$(4.6) \qquad \varphi_\omega = \tau_Y\, \varphi_\theta .$$

Cette formule a déjà été donnée donnée dans [16], ainsi que la définition suivante :

Définition 4.3 : On appellera μ l'application de P_ν dans l'espace $\mathfrak{S}_\nu$ des matrices symétriques complexes de format $\nu \times \nu$ dont la partie réelle est définie positive, donnée par

$$(4.7) \qquad \mu\begin{pmatrix} A & B \\ B' & C \end{pmatrix} = A^{-1} - i\, A^{-1}B.$$

La formule (4.5) exprime donc que, à normalisation près, un état fondamental de L_θ est donné par $\exp -\pi(\mu(0)x,x)$. Pour construire les autres états propres, on a besoin des opérateurs de création et d'annihilation :

Définition 4.4 : lorsque f parcourt l'espace des formes linéaires complexes sur $R^{2\nu}$, antiholomorphes (i.e. antilinéaires) pour la structure complexe sur $R^{2\nu}$ associée à θ, l'opérateur $Op_{\frac{1}{2}}(f(X-Y))$ décrit l'espace des opérateurs de création associés à $\omega = (Y,\theta)$, et l'opérateur $Op_{\frac{1}{2}}(\overline{f}(X-Y))$ décrit l'espace des opérateurs d'annihilation associés à ω.

Plus spécifiquement, toujours avec $\theta = VV'$, où V est définie par (4.4), voici des bases d'opérateurs de création et d'annihilation : lorsque $\omega = (0,I)$, on définit $(1 \leq k \leq \nu)$

$$a_o^{(k)}(x,\xi) = \pi^{\frac{1}{2}}(x_k - i\,\xi_k). \tag{4.8}$$

Pour $\omega = (Y,\theta)$, on pose

$$a_\omega^{(k)}(X) = a_o^{(k)}(V^{-1}(X-Y)), \tag{4.9}$$

et

$$A_\omega^{(k)} = Op_{\frac{1}{2}}(a_\omega^{(k)}),\ (A_\omega^{(k)})^* = Op_{\frac{1}{2}}(\bar{a}_\omega^{(k)}) . \tag{4.10}$$

Les $A_\omega^{(k)}$ (resp. $(A_{(\omega)}^{(k)})^*$) constituent donc une base de l'espace des opérateurs de création (resp. annihilation) associés à ω. Soit $M \in Mp(\nu)$ telle que $\tilde{M} = V$ (cf. 1.12)). En posant $N = \tau_Y M$, on a donc $A_\omega^{(k)} = N\, Op_{\frac{1}{2}}(a_o^{(k)})\, N^{-1}$, ce qui permet de réduire au cas de l'oscillateur canonique L_o la vérification, alors immédiate, des résultats qui suivent :

<u>Proposition 4.1</u> : On a les relations $[A_\omega^{(j)}, A_\omega^{(k)}] = 0$ (crochet de deux opérateurs et non forme symplectique !) ; $[(A_\omega^{(j)})^*, (A_\omega^{(k)})^*] = 0$; $[A_\omega^{(j)}, (A_\omega^{(k)})^*] = -\delta_{jk}$;

$\Lambda_\omega = \Sigma A_\omega^{(k)}(A_\omega^{(k)})^*$; $\Lambda_\omega + \nu = L_\omega + \frac{\nu}{2} = \Sigma(A_\omega^{(k)})^* A_\omega^{(k)}$; $L_\omega A_\omega^{(k)} = A_\omega^{(k)} L_\omega + A_\omega^{(k)}$.

Cette proposition montre d'une part que, pour tout multi-indice $\alpha = (\alpha_1,\alpha_2,\ldots,\alpha_\nu)$, la fonction $A_\omega^\alpha \varphi_\omega = (A_\omega^{(1)})^{\alpha_1} \ldots (A_\omega^{(\nu)})^{\alpha_\nu} \varphi_\omega$ est un état propre de niveau d'énergie $|\alpha|$ (i.e. correspondant à la valeur propre $\frac{\nu}{2} + |\alpha|$) de l'oscillateur L_ω , d'autre part que, dans l'espace engendré par les opérateurs de création et d'annihilation, les opérateurs d'annihilation peuvent être caractérisés comme ceux qui annihilent l'état fondamental φ_ω . Si $\nu = 1$, on a $(A\varphi, A\varphi) = (A^*A\varphi,\varphi) = ((\Lambda+1)\varphi,\varphi)$ pour tout φ, ce qui prouve que $\|A^j\varphi\|^2 = j!$. Dans le cas général, on normalisera donc les fonctions $A_\omega^\alpha\ \varphi_\omega$ en posant

$$\varphi_\omega^\alpha = (\alpha!)^{-\frac{1}{2}} A_\omega^\alpha\, \varphi_\omega . \tag{3.11}$$

En désignant par 0 le point privilégié $(0,I)$ de $R^{2\nu} \times P_\nu$, on a alors, avec N défini plus haut, les relations $\varphi_\omega^\alpha = N\, \varphi_o^\alpha$, à condition que M ait été défini, conformément au procédé qui a conduit à (4.5), par la formule

$$(Mu)(x) = (\det A)^{-\frac{1}{4}}\, e^{i\pi(A^{-1}Bx,x)} u(A^{-\frac{1}{2}}x). \tag{4.12}$$

Pour ω fixé, les diverses fonctions φ_ω^α sont orthogonales entre elles puisque, dans le cas où $\omega = (0,I)$, on se ramène au cas $\nu = 1$, où ces diverses fonctions correspondent alors à des niveaux d'énergie différents.

Il est utile de transporter sur l'espace des opérateurs d'annihilation la structure hermitienne héritée, par passage aux symboles, de celle de dual complexe de l'espace de phase : se ramenant au cas où $\omega = (0,I)$ on vérifie que l'on a, pour deux tels opérateurs F et G,

(4.13) $$(F,G) = \pi(G^* \varphi_\omega \,,\, F^* \varphi_\omega).$$

5. Deux oscillateurs harmoniques

Définition 5.1 : Soient $\omega = (Y,\theta)$ et $\omega' = (Y',\theta')$ appartenant à $R^{2\nu} \times P_\nu$. On définit $\omega \circ \omega'$ appartenant à $R^{2\nu} \times (R^{2\nu})^* \times P_{2\nu}$ par $\omega \circ \omega' = (\frac{Y+Y'}{2}, \sigma^{-1}(Y-Y'),\Theta)$,

avec
$$(\Theta^{-1}(X,\Xi),(X,\Xi)) = \|X + \frac{\sigma\Xi}{2}\|^2_\theta + \|X - \frac{\sigma\Xi}{2}\|^2_{\theta'} .$$

On définit ainsi une composition "extérieure", mais il convient de vérifier que la norme définie par Θ est symplectique : c'est bien le cas lorsque $\theta = \theta' = I$ (on espère que, θ étant par définition symétrique, le lecteur tolérera la notation θ' pour une autre matrice symétrique indépendante de θ) ; posant $\|X\|^2_\theta = |V^{-1}X|^2$ et $\|X\|^2_{\theta'} = |W^{-1}X|^2$, et résolvant les équations

$$Y + \frac{\sigma H}{2} = V^{-1}(X + \frac{\sigma\Xi}{2})$$
$$Y - \frac{\sigma H}{2} = W^{-1}(X - \frac{\sigma\Xi}{2}),$$

on voit qu'il faut vérifier le caractère symplectique de la matrice

$$\begin{pmatrix} \frac{V^{-1}+W^{-1}}{2} & \frac{V^{-1}-W^{-1}}{4}\sigma \\ \sigma^{-1}(V^{-1}-W^{-1}) & \sigma^{-1}\frac{V^{-1}+W^{-1}}{2}\sigma \end{pmatrix} = \begin{pmatrix} A & B \\ C & D \end{pmatrix},$$

autrement dit les équations

$A'C = C'A$, $B'D = D'B$, $A'D - C'B = I$.

Tenant compte des équations $V'^{-1} = \sigma^{-1}V\sigma$ et $W'^{-1} = \sigma^{-1}W\sigma$ qui expriment que V et W sont symplectiques, cette vérification résulte immédiatement des formules

$$A' = \frac{1}{2}\sigma^{-1}(V+W)\sigma \,,\; B' = -\frac{1}{4}(V-W)\sigma,\; C' = -\sigma^{-1}(V-W),\; D' = \frac{1}{2}(V+W).$$

Proposition 5.1 : avec les notations de la définition 5.1, appelons, pour tout $X \in R^{2\nu}$, $f(X)$ l'unique Y tel que $\|X-Y\|^2_\theta + \|Y\|^2_{\theta'}$ soit minimum, et posons

$$\|X\|^2_{\theta,\theta'} = 2(\|X-f(X)\|^2_\theta + \|f(X)\|^2_{\theta'}),$$

et $$Q_{\theta,\theta'}(X) = \lceil X,f(X)\rceil.$$

Au sens de la définition 4.3, $\mu(\Theta)$ est alors donné par

$$(\mu(\Theta)X,X) = 2\|X\|^2_{\theta,\theta'} - 4i\, Q_{\theta,\theta'}(X).$$

Preuve : signalons d'abord que la norme $\| \ \|_{\theta,\theta'}$ diffère par le facteur 2 de la définition que nous en avions donnée dans [16] et [17] (ceci pour assurer $\| \ \|_{\theta,\theta} = \| \ \|_{\theta}$).

En utilisant, comme dans ([16], lemme A 2), une base symplectique dans laquelle on ait

$$\|x,\xi\|_{\theta}^2 = \Sigma\, x_j^2 + \xi_j^2\ ,$$
$$\|x,\xi\|_{\theta'}^2 = \Sigma\, \lambda_j\, x_j^2 + \lambda_j^{-1}\, \xi_j^2\ ,$$

on a alors

$$\|x,\xi\|_{\theta,\theta'}^2 = 2\ \Sigma\ \frac{\lambda_j}{1+\lambda_j}\, x_j^2 + \frac{1}{1+\lambda_j}\, \xi_j^2$$

et

$$Q_{\theta,\theta'}(x,\xi) = \Sigma\ \frac{1-\lambda_j}{1+\lambda_j}\, x_j\, \xi_j\ .$$

Avec $X = (x,\xi)$, et $\sigma\Xi = Z = (z,\zeta)$, on a, en négligeant l'indice j :

$$(\Theta^{-1}(X,\Xi),(X,\Xi)) = \|X + \frac{Z}{2}\|_{\theta}^2 + \|X - \frac{Z}{2}\|_{\theta'}^2$$

$$= (x+\frac{z}{2})^2 + \lambda(x-\frac{z}{2})^2 + (\xi+\frac{\zeta}{2})^2 + \lambda^{-1}(\xi-\frac{\zeta}{2})^2,$$

d'où

$$\begin{pmatrix} 1 & 0 \\ 0 & \sigma \end{pmatrix} \Theta^{-1} \begin{pmatrix} 1 & 0 \\ 0 & \sigma^{-1} \end{pmatrix} = \begin{pmatrix} 1+\lambda & 0 & \frac{1-\lambda}{2} & 0 \\ 0 & \frac{\lambda+1}{\lambda} & 0 & \frac{\lambda-1}{2\lambda} \\ \frac{1-\lambda}{2} & 0 & \frac{1+\lambda}{4} & 0 \\ 0 & \frac{\lambda-1}{2\lambda} & 0 & \frac{1+\lambda}{4\lambda} \end{pmatrix},$$

et par suite $\Theta = \begin{pmatrix} A & B \\ B' & C \end{pmatrix}$ avec

$$A = \begin{pmatrix} \frac{1+\lambda}{4\lambda} & 0 \\ 0 & \frac{1+\lambda}{4} \end{pmatrix} \quad \text{et} \quad B = \begin{pmatrix} 0 & \frac{1-\lambda}{2\lambda} \\ \frac{1-\lambda}{2} & 0 \end{pmatrix}.$$

Il en résulte que

$$\mu(\Theta)) = A^{-1} - i\, A^{-1}B = \begin{pmatrix} \frac{4\lambda}{1+\lambda} & 0 \\ 0 & \frac{4}{1+\lambda} \end{pmatrix} - i \begin{pmatrix} 0 & 2\,\frac{1-\lambda}{1+\lambda} \\ 2\,\frac{1-\lambda}{1+\lambda} & 0 \end{pmatrix},$$

ce qui termine la preuve de la proposition 5.1.

Proposition 5.2 : avec les notations de la définition 5.1 , la fonction $\Phi_{\omega,\omega'} = H(\varphi_{\omega},\varphi_{\omega'})$ est un état fondamental normalisé de l'oscillateur $L_{\omega,\omega'}$.

Preuve : l'égalité $\|\Phi_{\omega,\omega'}\| = 1$ résulte de (1.6) et de l'égalité entre la norme L^2 d'un symbole et la norme de Hilbert-Schmidt de l'opérateur associé. Le cas où $\omega = (0,\theta)$ (on écrira $\omega = \theta$ par abus d'écriture commode) résulte de la proposition 5.1 et du lemme A2 de [16]. Dans le cas général, grâce à (1.22), on a

$$H(\varphi_\omega,\varphi_{\omega'}) = H(\tau_Y\,\varphi_\theta,\tau_{Y'}\varphi_{\theta'}) = \tau_{\frac{Y+Y'}{2},\ \sigma^{-1}(Y-Y')}\,\Phi_{\theta,\theta'}\ ,$$

ce qui termine la preuve de la proposition.

On passe maintenant à la description des états propres non fondamentaux de l'oscillateur $L_{\omega\sigma\omega'}$. Pour tout $\omega = (Y,\theta) \in R^{2\nu} \times P_\nu$, introduisons les coordonnées complexes (holomorphes pour la structure complexe définie par θ) $X_\omega^{(k)}$ définies par

$$(5.1)\qquad \pi^{\frac{1}{2}}\, X_\omega^{(k)} = \bar{a}_\omega^{(k)}(X),$$

où $a_\omega^{(k)}$ est défini par (4.9).

Ceci permet de définir également les opérateurs $\dfrac{\partial}{\partial X_\omega^{(k)}}$ et $\dfrac{\partial}{\partial \bar{X}_\omega^{(k)}}$.

<u>Définition 5.2</u> : Pour tout $\omega \in R^{2\nu} \times P_\nu$, on définit les 4ν opérateurs différentiels suivants opérant sur les fonctions définies sur l'espace de phase :

$$\mathcal{A}_\omega^{(k)} = \pi^{\frac{1}{2}}\, \bar{X}_\omega^{(k)} - \frac{1}{2\pi^{\frac{1}{2}}}\frac{\partial}{\partial X_\omega^{(k)}}$$

$$\bar{\mathcal{A}}_\omega^{(k)} = \pi^{\frac{1}{2}}\, X_\omega^{(k)} - \frac{1}{2\pi^{\frac{1}{2}}}\frac{\partial}{\partial \bar{X}_\omega^{(k)}}$$

$$(\mathcal{A}_\omega^{(k)})^* = \pi^{\frac{1}{2}} X_\omega^{(k)} + \frac{1}{2\pi^{\frac{1}{2}}}\frac{\partial}{\partial \bar{X}_\omega^{(k)}}$$

$$(\bar{\mathcal{A}}_\omega^{(k)})^* = \pi^{\frac{1}{2}}\, \bar{X}_\omega^{(k)} + \frac{1}{2\pi^{\frac{1}{2}}}\frac{\partial}{\partial X_\omega^{(k)}}\ .$$

Rappelons que $X_\omega^{(k)} = \bar{a}_o^{(k)}(V^{-1}(X-Y))$ (cf. 4.9)), avec V défini par (4.4).

Le symbole de l'opérateur $\dfrac{1}{2i\pi}\dfrac{\partial}{\partial X_\omega^{(k)}}$ est la fonction

$$\frac{1}{2i\pi}\frac{\partial}{\partial X_\omega^{(k)}}\left(e^{2i\pi< X,\Xi >}\right)(X=0) = \frac{\partial}{\partial X_\omega^{(k)}} < X,\Xi >$$

$$= \frac{\partial}{\partial X_o^{(k)}} < VX,\Xi > = -\frac{\partial}{\partial X_o^{(k)}}\,[X,V^{-1}\sigma\,\Xi].$$

Avec $V^{-1}\sigma\Xi = (z,\zeta)$, cela s'écrit $\frac{1}{2}(\zeta_k + i\, z_k) = \frac{i}{2}(z_k - i\zeta_k)$

$$-\frac{i}{2\pi^{\frac{1}{2}}}\, a_o^{(k)}(V^{-1}\sigma\,\Xi).$$

Il en résulte que le symbole de $\mathcal{A}_\omega^{(k)}$ est la fonction

$$a_o^{(k)}(V^{-1}(X-Y)) + \frac{1}{2}\, a_o^{(k)}(V^{-1}\,\sigma\,\Xi) = a_\omega^{(k)}(X + \frac{\sigma\Xi}{2}).$$

Avec trois autres calculs analogues, on obtient :

<u>Proposition 5.3</u> : avec les notations de la définition 5.2, on a les identités

$$\mathcal{A}_\omega^{(k)} = \mathrm{Op}_{\frac{1}{2}}(a_\omega^{(k)}(X + \frac{\sigma\Xi}{2}))$$
$$\bar{\mathcal{A}}_\omega^{(k)} = \mathrm{Op}_{\frac{1}{2}}(\bar{a}_\omega^{(k)}(X - \frac{\sigma\Xi}{2}))$$
$$(\mathcal{A}_\omega^{(k)})^* = \mathrm{Op}_{\frac{1}{2}}(\bar{a}_\omega^{(k)}(X + \frac{\sigma\Xi}{2}))$$
$$(\bar{\mathcal{A}}_\omega^{(k)})^* = \mathrm{Op}_{\frac{1}{2}}(a_\omega^{(k)}(X - \frac{\sigma\Xi}{2})).$$

<u>Proposition 5.4</u> : quelles que soient φ et $\psi \in \mathcal{S}(R^\nu)$, on a les identités (où $A_\omega^{(k)}$ a été défini en 4. 10) :

$$H(A_\omega^{(k)}\varphi,\psi) = \mathcal{A}_\omega^{(k)} H(\varphi,\psi)$$
$$H(\varphi,A_\omega^{(k)}\psi) = \bar{\mathcal{A}}_\omega^{(k)} H(\varphi,\psi)$$
$$H(A_\omega^{(k)*}\varphi,\psi) = (\mathcal{A}_\omega^{(k)})^* H(\varphi,\psi)$$
$$H(\varphi,(A_\omega^{(k)})^*\psi) = (\bar{\mathcal{A}}_\omega^{(k)})^* H(\varphi,\psi).$$

<u>Preuve</u> : démontrons par exemple la seconde identité :
d'après (1.6), $H(\varphi,A_\omega^{(k)}\psi)$ est le symbole de l'opérateur

$$u \longmapsto (u,A_\omega^{(k)}\psi)\,\varphi = ((A_\omega^{(k)})^* u,\psi)\,\varphi\ ,$$

c'est-à-dire que l'on a

$$\mathcal{H}(\varphi,A_\omega^{(k)}\psi) = \mathcal{H}(\varphi,\psi) \circ (A_\omega^{(k)})^*.$$

Comme $(A_\omega^{(k)})^* = \mathrm{Op}_{\frac{1}{2}}(\bar{a}_\omega^{(k)})$,

il résulte alors de (1.11) que l'on a

$$H(\varphi,A_\omega^{(k)}\psi) = \mathrm{Op}_{\frac{1}{2}}(\bar{a}_\omega^{(k)}(X - \frac{\sigma\Xi}{2}))\, H(\varphi,\psi),$$

ce qui prouve la proposition.

<u>Définition 5.3</u> : avec les notations des définitions 5.1 et 5.2, on pose

$$\Phi_{\omega,\omega'}^{\alpha,\beta} = H(\varphi_\omega^\alpha, \varphi_{\omega'}^\beta) = (\alpha!)^{-\frac{1}{2}}(\beta!)^{-\frac{1}{2}} H(A_\omega^\alpha \varphi_\omega\ , A_{\omega'}^\beta \varphi_{\omega'})$$
$$= (\alpha!)^{-\frac{1}{2}}(\beta!)^{-\frac{1}{2}} \mathcal{A}_\omega^\alpha\, \bar{\mathcal{A}}_{\omega'}^\beta\, H(\varphi_\omega\ , \varphi_{\omega'}).$$

<u>Proposition 5.5</u> : on a l'identité

$$\Lambda_{\omega\circ\omega'} = L_{\omega\circ\omega'} - \nu = \Sigma\, \mathcal{A}_\omega^{(k)}(\mathcal{A}_\omega^{(k)})^* + \bar{\mathcal{A}}_{\omega'}^{(k)}(\bar{\mathcal{A}}_{\omega'}^{(k)})^*.$$

Preuve : quelles que soient φ et $\psi \in \mathcal{S}(R^\nu)$, on a

$$(5.2)\qquad \Sigma \mathcal{A}_\omega^{(k)}(\mathcal{A}_\omega^{(k)})^* H(\varphi,\psi) = \Sigma H(A_\omega^{(k)}(A_\omega^{(k)})^* \varphi,\psi) = H((L_\omega - \tfrac{\nu}{2})\varphi,\psi),$$

où l'on a utilisé la proposition 4.1.

D'après (1.11), le symbole de l'opérateur $\Sigma \mathcal{A}_\omega^{(k)}(\mathcal{A}_\omega^{(k)})^*$ est donc $\pi\|X + \frac{\sigma\Xi}{2} - Y\|_\theta^2 - \frac{\nu}{2}$; on vérifie de même que le symbole de $\Sigma\bar{\mathcal{A}}_{\omega'}^{(k)}(\bar{\mathcal{A}}_\omega^{(k)})^*$ est $\pi\|X - \frac{\sigma\Xi}{2} - Y'\|_{\theta'}^2 - \frac{\nu}{2}$.

Enfin, d'après la proposition 5.1, le symbole de l'opérateur $L_{\omega\circ\omega'}$ est

$$(5.3)\qquad \pi\|X - \frac{Y+Y'}{2} + \frac{1}{2}\sigma(\Xi - \sigma^{-1}(Y-Y'))\|_\theta^2 + \pi\|X - \frac{Y+Y'}{2} - \frac{1}{2}\sigma(\Xi - \sigma^{-1}(Y-Y'))\|_{\theta'}^2$$

$$= \pi\|X + \frac{\sigma\Xi}{2} - Y\|_\theta^2 + \pi\|X - \frac{\sigma\Xi}{2} - Y'\|_{\theta'}^2 .$$

Proposition 5.6 : les fonctions $\Phi_{\omega,\omega'}^{\alpha,\beta}$ constituent une base orthonormale d'états propres de l'oscillateur $L_{\omega\circ\omega'}$, la fonction $\Phi_{\omega,\omega'}^{\alpha,\beta}$ ayant le niveau d'énergie $|\alpha| + |\beta|$.

Preuve : d'après (1.3) et (1.6), on a

$$(H(\varphi_\omega^\alpha,\varphi_{\omega'}^\beta),\ H(\varphi_\omega^\lambda,\varphi_{\omega'}^\mu)) = \int H(\varphi_\omega^\alpha,\varphi_{\omega'}^\beta,X)\, \overline{H(\varphi_{\omega'}^\mu,\varphi_\omega^\lambda,X)}\, dX$$

$$= (\varphi_{\omega'}^\mu,\varphi_{\omega'}^\beta)(\varphi_\omega^\alpha,\varphi_\omega^\lambda) = \delta_{\beta,\mu}\ \delta_{\alpha,\lambda} .$$

Par ailleurs, compte tenu de la proposition 4.1, on a (pour φ, $\psi \in \mathcal{S}(R^\nu)$) les identités

$$\Lambda_{\omega\circ\omega'}\mathcal{A}_\omega^{(k)}H(\varphi,\psi) = \Lambda_{\omega\circ\omega'}\, H(A_\omega^{(k)}\varphi,\psi) = H(\Lambda_\omega A_\omega^{(k)}\varphi,\psi) + H(A_\omega^{(k)}\varphi,\Lambda_{\omega'}\psi)$$

$$= H((A_\omega^{(k)}\Lambda_\omega + A_\omega^{(k)})\varphi,\psi) + H(A_\omega^{(k)}\varphi,\Lambda_{\omega'}\psi) - \mathcal{A}_\omega^{(k)}\Lambda_{\omega\circ\omega'}H(\varphi,\psi) + \mathcal{A}_\omega^{(k)}H(\varphi,\psi),$$

est l'identité analogue relative à $\bar{\mathcal{A}}_\omega^{(k)}$.

Ces identités prouvent que $\Phi_{\omega,\omega'}^{\alpha,\beta}$ est une fonction propre de $\Lambda_{\omega\circ\omega'}$ de niveau d'énergie $|\alpha| + |\beta|$.

Compte tenu des propositions 5.4 et 5.6, il est clair que les 2ν opérateurs $\mathcal{A}_\omega^{(k)}$ et $\bar{\mathcal{A}}_{\omega'}^{(k)}$ (resp. $(\mathcal{A}_\omega^{(k)})^*$ et $(\bar{\mathcal{A}}_{\omega'}^{(k)})^*$) constituent une base de l'espace des opérateurs de création (resp. annihilation) relatifs à l'oscillateur $L_{\omega\circ\omega'}$. Rappelons que sur l'espace des opérateurs d'annihilation existe une norme euclidienne canonique : la formule (4.13) fournit

$$(5.4)\qquad ((\mathcal{A}_\omega^{(j)})^*,(\mathcal{A}_\omega^{(k)})^*) = \pi(\mathcal{A}_\omega^{(k)}\Phi_{\omega,\omega'},\mathcal{A}_\omega^{(j)}\Phi_{\omega,\omega'}) = \pi(\Phi_{\omega,\omega'}^{(k),0},\Phi_{\omega,\omega'}^{(j),0}) = \delta_{jk} ,$$

et deux égalités analogues, qui montrent que la base constituée des opérateurs $(\mathcal{A}_\omega^{(1)})^*,\dots,(\mathcal{A}_\omega^{(\nu)})^*$, $(\bar{\mathcal{A}}_{\omega'}^{(1)})^*,\dots,(\bar{\mathcal{A}}_{\omega'}^{(\nu)})^*$ est orthonormale.

6. Ramollisseurs harmoniques

Définition 6.1 : nous appellerons ramollisseur harmonique attaché à $\omega=(Y,\theta)\in R^{2\nu}\times P_\nu$ le semi-groupe $e^{-t\Lambda_\omega}=e^{-t(L_\omega-\frac{\nu}{2})}$ $(t\in R^+)$. On pose également

$$e^{-t\Lambda_\omega}=Op_{\frac{1}{2}}(\lambda_\omega^t)\ .$$

D'après [16], on a

$$(6.1)\qquad \lambda_\omega^t(X)=2^\nu(1+e^{-t})^{-\nu}\exp-2\pi(\text{th}\,\tfrac{t}{2})\,\|X-Y\|_\theta^2$$

et en particulier

$$(6.2)\qquad \lambda_\omega^\infty(X)=2^\nu\exp-2\pi\,\|X-Y\|_\theta^2\ .$$

Proposition 6.1 : avec les notations de la définition 5.1, on a, pour tout $a\in\mathcal{S}(R^{2\nu})$, et tout $t\geq 0$,

$$(6.3)\qquad e^{-t\Lambda_{\omega\circ\omega'}}a=\lambda_\omega^t\#a\#\lambda_{\omega'}^t\ .$$

Preuve : d'après (1.11), le symbole de l'opération $a\longmapsto\lambda_\omega^t\#a\#\lambda_{\omega'}^t$ est

$$(6.4)\qquad \lambda_\omega^t(X+\tfrac{\sigma\Xi}{2})\,\lambda_{\omega'}^t(X-\tfrac{\sigma\Xi}{2})=$$

$$2^{2\nu}(1+e^{-t})^{-2\nu}\exp-2\pi(\text{th}\,\tfrac{t}{2})\{\|X+\tfrac{\sigma\Xi}{2}-Y\|_\theta^2+\|X-\tfrac{\sigma\Xi}{2}-Y'\|_{\theta'}^2\},$$

ce qui prouve la proposition à l'aide de (5.3) et (6.1).

Proposition 6.2 : avec les notations de la définition 5.1, soit $\{\lambda_1,\ldots,\lambda_\nu,\lambda_1^{-1},\ldots,\lambda_\nu^{-1}\}$ l'ensemble des valeurs propres de la forme quadratique $\|X\|_\theta^2$ par rapport à la forme quadratique $\|X\|_{\theta'}^2$; posons $\mu_j=\lambda_j^{\frac{1}{2}}+\lambda_j^{-\frac{1}{2}}$ $(1\leq 1\leq\nu)$.

La norme sur $L^2(R^\nu)$ de l'opérateur $e^{-t\Lambda_\omega}e^{-t\Lambda_{\omega'}}$ est alors donnée par

$$(6.5)\qquad \|e^{-t\Lambda_\omega}e^{-t\Lambda_{\omega'}}\|=2^\nu(1+e^{-2t})^{-\nu}\,\Pi(\mu_j\,\text{th}\,t+(\mu_j^2\,\text{th}^2t+\text{ch}^{-4}t)^{\frac{1}{2}})^{-\frac{1}{2}}$$

$$\exp-\tfrac{\Pi}{2}(\text{th}\,t)\,\|Y'-Y\|_{\theta,\theta'}^2\ .$$

Preuve : d'après des calculs assez laborieux effectués dans [17], on a, si $\varepsilon=2\delta(1+\delta^2)^{-1}$ et

$$B_\omega^\varepsilon=Op_{\frac{1}{2}}(2^{\nu/2}(1+\delta^2)^{\nu/2}\,e^{-2\pi\delta\,\|X-Y\|_\theta^2}),$$

l'égalité

$$(6.6)\qquad \|B_\omega^\varepsilon B_{\omega'}^\varepsilon\|=2^\nu\,\Pi(\mu_j\varepsilon+(\mu_j^2\,\varepsilon^2+(\varepsilon^2-1)^2)^{\frac{1}{2}}\exp-\tfrac{\pi\varepsilon}{2}\,\|Y'-Y\|_{\theta,\theta'}^2\ .$$

Il suffit alors de poser $\delta=\text{th}\,\frac{t}{2}$, $\varepsilon=\text{th}\,t$, d'où $e^{-t\Lambda_\omega}=(1+e^{-2t})^{-\nu/2}B_\omega^\varepsilon$.

On se contentera dans les applications de l'inégalité (qui résulte de (6.5) et de l'inégalité $\mu_j\geq 2$) :

(6.7) $\left\| e^{-t\Lambda_\omega} e^{-t\Lambda_{\omega'}} \right\| \le 2^\nu \prod (1+\mu_j \text{th } t)^{-\frac{1}{2}} \exp - \frac{\pi}{2}(\text{th } t) \| Y'-Y\|^2_{\theta,\theta'}$.

7. Normes de créabilité-annihilabilité relatives à un oscillateur ou une paire d'oscillateurs.

Dans ce paragraphe, on étudie la possibilité de représenter une fonction u comme somme de translatées de phase de l'état fondamental φ_ω d'un oscillateur harmonique ; remontant sur l'espace de phase, on étudiera ensuite la représentation d'un symbole comme somme de translatées de phase de l'état $\Phi_{\omega,\omega'}$, relatif à une paire d'oscillateurs.

On posera $\omega = (Y,\theta)$ et $\theta = VV'$ avec V défini par (4.4) dans la première partie ; dans la seconde, on posera $\omega' = (Y',\theta')$ et $(\omega \circ \omega') = (\frac{Y+Y'}{2}, \sigma^{-1}(Y-Y'), \Theta)$.

Lemme 7.1 : pour tout $u \in L^2(\mathbb{R}^\nu)$, on a

$$\|u\|^2 = \int |(u,\tau_Z \varphi_\omega)|^2 \, dZ .$$

Preuve : $\int |(u,\tau_Z \varphi_\omega)|^2 \, dZ = (Au,u)$,

avec

(7.1) $A = \int (\tau_Z\varphi_\omega , \tau_Z \varphi_\omega) \, dZ = Op_{\frac{1}{2}}(a)$,

où

(7.2) $a(X) = \int H(\tau_Z \varphi_\omega, \tau_Z \varphi_\omega, X) \, dZ = \int \Phi_{\omega,\omega}(X-Z) dZ = \mathrm{Tr}\, \mathcal{H}(\varphi_\omega,\varphi_\omega) = 1.$

Lemme 7.2 : si $u = \int f(Z) \, \tau_Z \varphi_\omega \, dZ$, avec $f \in L^2(\mathbb{R}^{2\nu})$, on a

$$\|u\|^2 \le \int |f(Z)|^2 \, dZ$$

Preuve : $\|u\|^2 = \int f(Z)(\tau_Z \varphi_\omega, u) \, dZ$

$$\le (\int |f(Z)|^2 dZ)^{\frac{1}{2}} (\int |(\tau_Z \varphi_\omega, u)|^2 \, dZ)^{\frac{1}{2}} = (\int |f(Z)|^2 \, dZ)^{\frac{1}{2}} \|u\| .$$

Lemme 7.3 : on a les identités

$$(A_\omega^{(k)})^* \, \tau_Z \varphi_\omega = \pi^{\frac{1}{2}} \, z_\theta^{(k)} \, \tau_Z \varphi_\omega ,$$

où la coordonnée complexe $z_0^{(k)}$ est définie par (5.1).

Preuve : avec $\bar{a}_o^{(k)}(x,\xi) = \pi^{\frac{1}{2}} (x_k + i\xi_k)$, on a

(7.3) $(A_\omega^{(k)})^* = Op_{\frac{1}{2}} (\bar{a}_o^{(k)}(V^{-1}(X-Y)))$,

d'où, d'après (1.19)

(7.4) $\tau_{-Z}(A_\omega^{(k)})^* \, \tau_Z = Op_{\frac{1}{2}}(\bar{a}_o^{(k)}(V^{-1}(X+Z-Y)))$, et

$$(7.5)\qquad (A_\omega^{(k)})^* \tau_Z \varphi_\omega = \tau_Z((A_\omega^{(k)})^* + \bar{a}_o^{(k)}(V^{-1}Z))\varphi_\omega$$

$$= \bar{a}_o^{(k)}(V^{-1}Z)\ \tau_Z \varphi_\omega = \bar{a}_\theta^{(k)}(Z)\ \tau_Z \varphi_\omega = \pi^{\frac{1}{2}} Z_\theta^{(k)}\ \tau_Z \varphi_\omega .$$

Définition 7.1 : Soit $k = 0,1,2,\ldots$, et soit $u \in \mathcal{S}'(R^\nu)$. On pose

$$\|u\|_{\omega,\mathcal{H}^k} = (\int (1+\|Z\|_\theta^2)^k\ |(u,\tau_Z \varphi_\omega)|^2\ dZ)^{\frac{1}{2}}.$$

On appelle $\mathcal{H}^k$ l'espace vectoriel (indépendant de ω ainsi qu'il résulte des lemmes qui suivent) des $u \in \mathcal{S}'(R^\nu)$ pour lesquels cette norme est finie.

Lemme 7.4 : il existe une constante $C > 0$, ne dépendant que de k et de ν, telle que l'on ait pour tout $u \in \mathcal{H}^k$:

$$C^{-1}\|u\|^2_{\omega,\mathcal{H}^k} \leq \sum_{|\alpha|\leq k} \|A_\omega^\alpha u\|^2 \leq C\ \|u\|^2_{\omega,\mathcal{H}^k} .$$

Preuve : d'après le lemme 7.1,

$$\sum_{|\alpha|\leq k} \|A_\omega^\alpha u\|^2 = \sum_{|\alpha|\leq k} \int |(A_\omega^\alpha u,\tau_Z \varphi_\omega)|^2\ dZ = \sum_{|\alpha|\leq k} \int |(u,(A_\omega^\alpha)^*\ \tau_Z \varphi_\omega)|^2\ dZ$$

$$= \sum_{|\alpha|\leq k} \pi^{|\alpha|} \int |Z_\theta^\alpha|^2\ |(u,\tau_Z \varphi_\omega)|^2\ dZ \sim \|u\|^2_{\omega,\mathcal{H}^k} .$$

Lemme 7.5 : il existe une constante C, ne dépendant que de k et de ν, telle que l'on ait, pour tout $u \in \mathcal{H}^k$:

$$\sum_{|\alpha|\leq k} \frac{1}{\alpha!}\|(A_\omega^\alpha)^* u\|^2 \leq \sum_{|\alpha|\leq k} \frac{1}{\alpha!}\|A_\omega^\alpha u\|^2 \leq C \sum_{|\alpha|\leq k} \frac{1}{\alpha!}\|(A_\omega^\alpha)^* u\|^2 .$$

Preuve : commençons par remarquer que l'on a

$$k! \sum_{|\alpha|=k} \frac{1}{\alpha!} A_\omega^\alpha (A_\omega^\alpha)^* = \Lambda_\omega(\Lambda_\omega-1)\ldots(\Lambda_\omega-k+1).$$

En effet, pour $k=1$, ce fait a été remarqué dans la proposition 4.1 ; on procède ensuite par récurrence, en utilisant la relation $[(A_\omega^{(j)})^*, A_\omega^{(k)}] = \delta_{jk}$, et en remarquant que la relation à prouver peut s'écrire aussi

$$\Lambda_\omega(\Lambda_\omega-1)\ldots(\Lambda_\omega-k+1) = \sum A_\omega^{(j_o)}\ldots A_\omega^{(j_{k-1})}(A_\omega^{j_o})^*\ldots(A_\omega^{j_{k-1}})^*,$$

la somme étant étendue à toutes les suites $(j_o,\ldots,j_{k-1})$ de k entiers ≥ 0.

On montre de même que

$$k! \sum_{|\alpha|=k} \frac{1}{\alpha!} (A_\omega^\alpha)^* A_\omega^\alpha = (\Lambda_\omega + \nu)(\Lambda_\omega + \nu + 1)\ldots(\Lambda_\omega + \nu + k-1).$$

Il suffit alors de remarquer qu'il existe $(C_o,\ldots,C_k)$ tels que l'on ait l'identité entre polynômes $\in R[\Lambda]$:

(7.6) $$(\Lambda+\nu)(\Lambda+\nu+1)\ldots(\Lambda+\nu+k-1) = C_o + \sum_{j=1}^{k} C_j\, \Lambda(\Lambda-1)\ldots(\Lambda-j+1).$$

Les lemmes qui précèdent montrent que $u \in \mathcal{H}^k$ si et seulement si l'on a $x^\alpha D^\beta u \in L^2$ pour tout couple (α,β) tel que $|\alpha| + |\beta| \leq k$.

Proposition 7.1 : la norme $\|u\|_{\omega,\mathcal{H}^k}$ reste dans un rapport borné (par une constante ne dépendant que de ν et de k) avec la borne inférieure des expresssions

(7.7) $$\left(\int (1+\|Z\|_\theta^2)^k\, |f(Z)|^2 dZ\right)^{\frac{1}{2}}$$

étendue aux fonctions $f \in L^2(R^{2\nu})$ vérifiant

(7.8) $$u = \int f(Z)\, \tau_Z\, \varphi_\omega\, dZ.$$

Preuve : Si $\|u\|_{\omega,\mathcal{H}^k} \leq 1$, on peut écrire,d'après (7.2) :

(7.9) $$u = \int (u, \tau_Z\, \varphi_\omega)\, \tau_Z\, \varphi_\omega\, dZ,$$

et l'on a alors

$$\int (1+\|Z\|_\theta^2)^k\, |(u,\tau_Z\, \varphi_\omega)|^2\, dZ \leq 1.$$

Réciproquement, si l'on a (7.7) avec

(7.10) $$\int (1+\|Z\|_\theta^2)^k |f(Z)|^2 dZ \leq 1,$$

on peut alors écrire, pour $|\alpha| \leq k$:

(7.11) $$(A_\omega^\alpha)^* u = \int f(Z)(A_\omega^\alpha)^*\, \tau_Z\, \varphi_\omega\, dZ = \pi^{|\alpha|/2} \int Z_\theta^\alpha\, f(Z)\, \tau_Z\, \varphi_\omega\, dZ\ ,$$

d'où

$$\|(A_\omega^\alpha)^* u\|^2 \leq \pi^{|\alpha|} \int |Z_\theta^\alpha|^2\, |f(Z)|^2\, dZ$$

d'après le lemme 7.2.

Les lemmes 7.4 et 7.5 permettent alors de conclure.

On passe maintenant sur l'espace de phase, en appliquant les résultats qui précèdent au cas de l'oscillateur associé à $\omega \circ \omega'$.

Proposition 7.2 : pour $a \in \mathcal{H}_{\omega\circ\omega',k}$, la norme $\|a\|_{\omega\circ\omega',\mathcal{H}^k}$ reste dans un rapport borné avec la borne inférieure des expressions

(7.12) $$\left(\iint (1+\|Z\|_\theta^2 + \|Z'\|_{\theta'}^2)^k\, |K(Z,Z')|^2\, dZ\, dZ'\right)^{\frac{1}{2}}$$

étendue aux fonctions $K \in L^2(R^{4\nu})$ vérifiant l'identité

(7.13) $$a(X) = \iint K(Z,Z')\, H(\tau_Z\varphi_\omega,\ \tau_{Z'}\varphi_{\omega'}, X)\, dZ\, dZ'\ .$$

Preuve : d'après (1.22), on a

$$H(\tau_Z \varphi_\omega , \tau_{Z'} \varphi_{\omega'}) = \tau_{\frac{Z+Z'}{2}, \sigma^{-1}(Y-Y')} H(\varphi_\omega, \varphi_{\omega'}),$$

ce qui permet d'écrire (7.13) sous la forme

$$(7.14) \qquad a(X) = \iint K(S + \tfrac{\sigma\Sigma}{2}, S - \tfrac{\sigma\Sigma}{2})(\tau_{(S,\Sigma)} \Phi_{\omega o \omega'})(X)\, dS\, d\Sigma .$$

D'après la proposition 7.1, $\|a\|_{\omega o \omega', \mathcal{H}^k}$ reste donc dans un rapport borné avec la borne inférieure, pour les K vérifiant (7.13), de l'expression

$$(7.15) \qquad (\iint (1+ \|(S,\Sigma)\|^2_\Theta)^k \, |K(S + \tfrac{\sigma\Sigma}{2}, S - \tfrac{\sigma\Sigma}{2})|^2 \, dS\, d\Sigma)^{\frac{1}{2}}$$

$$= (\iint (1 + \|S + \tfrac{\sigma\Sigma}{2}\|^2_\theta + \|S - \tfrac{\sigma\Sigma}{2}\|^2_{\theta'})^k \, |K(S + \tfrac{\sigma\Sigma}{2}, S - \tfrac{\sigma\Sigma}{2})|^2 \, dS\, d\Sigma)^{\frac{1}{2}}$$

$$= (\iint (1 + \|Z\|^2_\theta + \|Z'\|^2_{\theta'})^k \, |K(Z,Z')|^2 \, dZ\, dZ')^{\frac{1}{2}}.$$

8. Espaces de repères

Dans cette section nous allons considérer un ensemble Ω et une application $\rho : \Omega \to R^{2\nu} \times P_\nu$. Pour tout $\omega \in \Omega$ on posera toujours, tacitement, $\rho(\omega) = (Y,\theta)$ (et de même $\rho(\omega')=(Y',\theta')$); pour simplifier, on désignera par L_ω , Λ_ω , $\omega \circ \omega'$, etc ..., ce qui, conformément aux notations employées jusqu'à maintenant, devrait être noté $L_{\rho(\omega)}$, $\Lambda_{\rho(\omega)}$, $\rho(\omega) \circ \rho(\omega')$, etc ..., et ceci bien que l'application ρ ne soit pas nécessairement injective.

Quels que soient θ et $\theta' \in P_\nu$, on désigne par $\det_{\theta,\theta'}$ le déterminant de la forme quadratique $\|X\|^2_{\theta,\theta'}$ par rapport à une base symplectique de $R^{2\nu}$; les nombres μ_j définis dans la proposition 6.2 permettent d'écrire

$$(8.1) \qquad \det_{\theta,\theta'} = 2^{2\nu}(\Pi\, \mu_j)^{-2} .$$

Définition 8.1 : On appellera espace de repères la donnée d'un espace Ω muni d'une mesure positive σ-finie $d\omega$, et d'une application mesurable $\rho : \Omega \to R^{2\nu} \times P_\nu$, vérifiant les propriétés suivantes :

(i) il existe $N > 0$ tel que la fonction

$$(\omega,\omega') \longmapsto \det^{\frac{1}{4}}_{\theta,\theta'} (1 + \|Y-Y'\|^2_{\theta,\theta'})^{-N}$$

soit le noyau d'un opérateur borné sur $L^2(\Omega, d\omega)$.

(ii) il existe une application mesurable $\omega \longmapsto a_\omega$ de Ω dans $\mathcal{S}(R^{2\nu})$ telle que d'une part la norme $\|a_\omega\|_{\omega o \omega, \mathcal{H}^k}$ (cf. section 7) soit bornée pour tout k, et que d'autre part on ait l'identité $1 = \int a_\omega(X)\, d\omega$ au sens faible suivant : quels que soient u et $v \in L^2(R^\nu)$, on a

$$(u,v) = \int d\omega \, (\int a_\omega(X)\, H(u,v,X)\, dX).$$

Remarque : il convient, pour justifier cette définition, de prouver la convergence de l'intégrale

$$\int d\omega \left| \int a_\omega(X)\, H(u,v,X)\, dX \right| = \int |(Op_{\frac{1}{2}}(a_\omega)\, u,v)|\, d\omega.$$

D'après la proposition 7.2, pour tout $k = 0, 1, 2, \ldots$, il existe $C > 0$ et, pour tout $\omega \in \Omega$, une décomposition de a_ω de la forme

$$(8.2) \qquad a_\omega(X) = \iint K(\omega,Z,Z')\, H(\tau_Z\, \varphi_\omega\,,\ \tau_{Z'}\, \varphi_\omega,\ X)\, dZ\, dZ'$$

avec

$$(8.3) \qquad \left(\iint |K(\omega,Z,Z')|^2 (1 + \|Z\|_\theta^2 + \|Z'\|_\theta^2)^k\, dZ\, dZ'\right)^{\frac{1}{2}} \le C.$$

D'après (8.2) et (8.3), on a

$$(8.4) \qquad |(Op_{\frac{1}{2}}(a_\omega)u,v)| \le C\left(\iint (1+\|Z\|_\theta^2 + \|Z'\|_\theta^2)^{-k}\, |(u,\tau_{Z'}\varphi_\omega)|^2\, |(v,\tau_Z\varphi_\omega)|^2\, dZ\, dZ'\right)^{\frac{1}{2}},$$

et l'on est ramené à prouver la convergence, pour $u \in L^2(\mathbb{R}^\nu)$ et k assez grand, de l'intégrale

$$\iint (1+\|Z\|_\theta^2)^{-k/2}\, |(u,\tau_Z\varphi_\omega)|^2\, dZ\, d\omega\,.$$

Ce résultat est conséquence de la proposition plus précise suivante :

Proposition 8.1 : Soit $(\Omega, d\omega, \rho)$ vérifiant la condition (i) de la définition 8.1. Soit N_o la borne inférieure des $N \ge \frac{\nu}{2}$ tels que la fonction

$$\lambda_N(\omega,\omega') = (1 + \|Y-Y'\|_{\theta,\theta'}^2)^{-N+\frac{\nu}{2}} \prod_{j=1}^{\nu} (\mu_j + \|Y-Y'\|_{\theta,\theta'}^2)^{-\frac{1}{2}}$$

(où les $\mu_j = \mu_j(\theta,\theta')$ ont été définis dans la proposition 6.2) soit le noyau d'un opérateur borné sur $L^2(\Omega)$.

Alors, quel que soit $N' > N_o$, il existe $C > 0$ tel que, quel que soit $u \in L^2(\mathbb{R}^\nu)$, on ait

$$\iint (1 + \|Z\|_\theta^2)^{-N'}\, |(u,\tau_Z\, \varphi_\omega)|^2\, dZ\, d\omega \le C\, \|u\|^2\,.$$

Remarque : d'après(8.1), on a

$$\lambda_N(\omega,\omega') \le 2^{-\nu/2}\, \det{}^{\frac{1}{4}}_{\theta,\theta'}\, (1+ \|Y-Y'\|_{\theta,\theta'}^2)^{-N+\frac{\nu}{2}},$$

ce qui justifie la définition de N_o.

Preuve de la proposition 8.1 :

C'est essentiellement la preuve du théorème 3.1 de [17], dans un cadre différent : on choisit $N \in]N_o, N'[$ et l'on pose

$$(8.5) \qquad \|A_1 u\|^2 = \iint (1+ \|Z\|_\theta^2)^{-N'}\, |(u,\tau_Z\varphi_\omega)|^2\, dZ\, d\omega$$
$$\le C \int_0^1 \varepsilon^{N'-1}\, (K_\varepsilon u,u)\, d\varepsilon,$$

avec

(8.6) $\qquad (K_\varepsilon u,u) = \iint e^{-2\pi\varepsilon \|Z\|^2_\theta} |(u,\tau_Z\,\varphi_\omega)|^2\, dZ\, d\omega$.

L'opérateur K_ε a pour symbole

(8.7) $\qquad 2^\nu \iint e^{-2\pi\varepsilon\|Z\|^2_\theta}\, e^{-2\pi\|X-Y-Z\|^2_\theta}\, dZ\, d\omega = (1+\varepsilon)^{-\nu} \int \exp - \frac{2\pi\varepsilon}{1+\varepsilon} \|X-Y\|^2_\theta\, d\omega$

et il s'agit de prouver que la norme de $\varepsilon^N K_\varepsilon$ est bornée quand $\varepsilon \to 0$.

Compte-tenu de (6.1), on peut, dans ce problème, remplacer K_ε par

(8.8) $\qquad K'_\varepsilon = \int e^{-2\varepsilon\Lambda_\omega}\, d\omega$.

D'après le lemme 2.1 de [17] (déjà utilisé dans la preuve du lemme 2.2 du présent papier), il suffit de prouver que la fonction

(8.9) $\qquad k_\varepsilon(\omega,\omega') = \varepsilon^N \| e^{-\varepsilon\Lambda_\omega}\, e^{-\varepsilon\Lambda_{\omega'}} \|$

est, pour $0 < \varepsilon \leq 1$, le noyau d'un opérateur borné en norme sur $L^2(\Omega)$ par une constante indépendante de ε.

D'après (6.7), on a

(8.10)
$$\begin{aligned} k_\varepsilon(\omega,\omega') &\leq C\, \varepsilon^N \Pi(1+\mu_j \text{ th } \varepsilon)^{-\frac12} \exp - \frac{\pi}{2}(\text{th } \varepsilon) \|Y'-Y\|^2_{\theta,\theta'} \\ &\leq C\, \varepsilon^N \Pi(1+\mu_j \text{ th } \varepsilon)^{-\frac12} (1+(\text{th } \varepsilon)\|Y'-Y\|^2_{\theta,\theta'})^{-N} \\ &\leq C\, \varepsilon^N \{\Pi(1+\mu_j \text{ th } \varepsilon)(1+(\text{th } \varepsilon)\|Y'-Y\|^2_{\theta,\theta'})\}^{-\frac12} (1+(\text{th } \varepsilon)\|Y'-Y\|^2_{\theta,\theta'})^{-N+\frac{\nu}{2}} \\ &\leq C\, (\mu_j + \|Y'-Y\|^2_{\theta,\theta'})^{-\frac12} \quad (1+\|Y'-Y\|^2_{\theta,\theta'})^{-N+\frac{\nu}{2}} . \end{aligned}$$

Ceci termine la preuve de la proposition 8.1.

<u>Théorème 8.1</u> : Soit $(\Omega, d\omega, \rho)$ un espace de repères. Pour tout $n = 0, 1, 2, \ldots$, il existe $C > 0$ telle que, quel que soit $u \in L^2(R^\nu)$, on ait

$$\sum_{|\alpha|\leq n} \int |(u,\varphi^\alpha_\omega)|^2\, d\omega \leq C\, \|u\|^2 .$$

De plus, pour n assez grand, il existe $C > 0$ telle que l'on ait

$$\sum_{|\alpha|\leq n} \int |(u,\varphi^\alpha_\omega)|^2\, d\omega \geq C^{-1} \|u\|^2 .$$

<u>Preuve</u> : Quel que soit $\varepsilon > 0$ et quel que soit $u \in L^2(R^\nu)$, on a

(8.11)
$$\begin{aligned} \int \sum_{j\geq 0} e^{-2j\varepsilon} \sum_{|\alpha|=j} |(u,\varphi^\alpha_\omega)|^2\, d\omega &= \int (e^{-2\varepsilon\Lambda_\omega} u,u)\, d\omega \\ &= (K'_\varepsilon u,u) \leq C\, \varepsilon^{-N} \|u\|^2 , \end{aligned}$$

où l'on a utilisé les notations employées dans la preuve de la proposition 8.1 : la première partie du théorème 8.1 en résulte.

Par ailleurs, d'aprés (8.4), on a

(8.12) $\|u\|^2 \leq C \|A_1 u\|^2$,

d'où, d'aprés (8.5) et la preuve de la proposition 8.1 :

(8.13) $\|u\|^2 \leq C \int_\delta^1 \varepsilon^{N'-1}(K_\varepsilon u,u)\, d\varepsilon$

pour $\delta > 0$ assez petit indépendant de u.

Comme $(K_\varepsilon u,u)$ est une fonction décroissante de ε, on a aussi, pour δ assez petit :

(8.14) $\|u\|^2 \leq C \int (Op_{\frac{1}{2}}(2^\nu e^{-2\pi\delta\|X-Y\|_\theta^2})u,u)\, d\omega$.

La deuxième partie du théorème 8.1 se démontre alors exactement comme le théorème 4.1 de [16].

Théorème 8.2 : Soit $(\Omega, d\omega, \rho)$ un espace de repères. Il existe un entier $n \geq 0$ et des fonctions $k_{\alpha,\beta}$ ($|\alpha| \leq n$, $|\beta| \leq n$) définies sur $\Omega \times \Omega$, constituant les noyaux d'opérateurs $K_{\alpha\beta}$ bornés sur $L^2(\Omega)$, telles que l'opérateur identité de $L^2(\mathbb{R}^\nu)$ puisse s'écrire

$$I = \sum_{\substack{|\alpha| \leq n \\ |\beta| \leq n}} \iint k_{\alpha,\beta}(\omega,\omega')\mathcal{H}(\varphi_\omega^\beta, \varphi_{\omega'}^\alpha)\, d\omega\, d\omega'.$$

Preuve : remarquer d'abord que le membre de droite est bien défini en vertu du lemme 2.3.

Posons $\Omega_n = \sum_{|\alpha| \leq n} \int \mathcal{H}(\varphi_\omega^\alpha, \varphi_\omega^\alpha)\, d\omega$:

c'est, d'après le théorème 8.1, un opérateur borné et inversible sur $L^2(\mathbb{R}^{2\nu})$ pour n assez grand. On écrit alors

$$I = \sum_\alpha \int \mathcal{H}(\Omega_n^{-1} \varphi_{\omega'}^\alpha, \varphi_{\omega'}^\alpha)\, d\omega' = \sum_\beta \int \mathcal{H}(\varphi_\omega^\beta, \Omega_n^{-1} \varphi_\omega^\beta)\, d\omega$$

$$= \sum_{\alpha,\beta} \iint k_{\alpha,\beta}(\omega,\omega')\mathcal{H}(\varphi_\omega^\beta, \varphi_{\omega'}^\alpha)\, d\omega\, d\omega'$$

avec

$$k_{\alpha,\beta}(\omega,\omega') = (\Omega_n^{-1} \varphi_{\omega'}^\alpha, \Omega_n^{-1} \varphi_\omega^\beta).$$

Remarquons également que d'une part (cf. lemme 2.3) on peut écrire l'opérateur associé à $k_{\alpha,\beta}$ sous la forme ML^* avec $(Mu)(\omega) = (u, \Omega_n^{-1} \varphi_\omega^\beta)$ et $(Lu)(\omega') = (u, \Omega_n^{-1}\varphi_{\omega'}^\alpha)$, et que d'autre part $k_{\alpha,\beta}$ vérifie l'hypothèse du lemme 2.1 relatif à la composition des noyaux.

F. Bruyant [3] a donné des conditions pour que Ω_n^{-1} soit un opérateur pseudo-différentiel ; des applications sont données dans ce même travail, ainsi que dans celui de N. Lerner [9].

Bien qu'il soit quelquefois techniquement nécessaire de sortir de ce cadre, l'exemple de loin le plus important d'espaces de repères est constitué par un triple $(R^{2\nu}, dX, \rho)$, où dX est la mesure de Lebesgue sur l'espace de phase $R^{2\nu}$, et $\rho : R^{2\nu} \to R^{2\nu} \times P_\nu$ a pour première coordonnée l'identité ; autrement dit, un tel espace est défini par un champ $Y \mapsto \| \ \|_Y$ de normes symplectiques sur $R^{2\nu}$; l'hypothèse (ii) de la définition 8.1 est vérifiée lorsqu'il existe $C > 0$ telle que, quels que soient X,Y et $Z \in R^{2\nu}$, l'inégalité $\|Y-X\|_X \leq C^{-1}$ entraîne $\|Z\|_Y \leq C\|Z\|_X$, comme on le voit en utilisant la partition de l'unité définie par (3.2) dans [16]. Dans le cas où il existe une transformation canonique κ telle qu'on ait l'identité $\|X\|_Y = |\kappa'(Y)X|$, la structure quasi-complexe définie par (4.2) est intégrable, et l'on obtient une structure kählérienne sur $R^{2\nu}$: cette structure très riche sera étudiée ailleurs. Signalons enfin que si l'on pose (avec $Y = (y,\eta)$, $0 \leq \delta < 1$)

$$(8.15) \qquad \|Z\|_Y^2 = (1+|\eta|)^{2\delta} |z|^2 + (1+|\eta|)^{-2\delta} |\zeta|^2 ,$$

ce qu'il est convenable de faire si l'on s'intéresse aux classes $S_{\delta,\delta}$ de Hörmander, la condition (i) de la définition 8.1 est vérifiée dès que $N > \nu$.

<u>Proposition 8.2</u> : Soit $(R^{2\nu}, dY, \rho)$ l'espace de repères défini par la donnée d'un champ $\rho : Y \mapsto \| \ \|_Y$ de normes symplectiques sur l'espace de phase. Pour tout $M \in R$, soit E_M l'espace des fonctions a mesurables sur $R^{2\nu}$ telles que $\int (1+|Y|)^{2M} |a(Y)|^2 < \infty$. On suppose la condition suivante vérifiée :

(iii) pour tout M_1, il existe k et M_2 tels que l'opérateur de noyau $K(Y,Y') = (1+\|Y-Y'\|_Y)^{-k}$ opère de E_{M_2} dans E_{M_1}.

Alors, quel que soit $M_1 \in R$, il existe $k > 0$ tel que l'intégrale

$$\iint (1+\|Z\|_Y)^{-k} (1+|Y|)^{M_1} |(u, \tau_Z\, \varphi_Y)|^2 \, dY\, dZ$$

reste bornée lorsque u parcourt une partie bornée de $\mathcal{S}(R^\nu)$.

<u>Preuve</u> : cette intégrale s'écrit (Au,u), où le symbole a de A est donnée par

$$a(X) = 2^\nu \int (1+\|Z\|_Y)^{-k} (1+|Y|)^{M_1} e^{-2\pi \|X-Y-Z\|_Y^2} \, dY\, dZ$$

$$\leq C \int (1+|Y|)^{M_1} (1+\|X-Y\|_Y)^{-k} \, dY.$$

L'hypothèse (iii), traduite en termes relatifs au transposé de l'opérateur de noyau K, montre que cette intégrale est majorée par $C(1+|X|)^{M_2+2\nu+1}$, ce qui prouve que a est tempérée et termine la preuve de la proposition.

On peut voir facilement que (iii) est vérifiée si d'une part la plus grande des valeurs propres de la forme $\| \ \|_Y^2$ est une fonction à croissance lente de Y, et si d'autre part il existe $M > 0$ et $C > 0$ tels que, quels que soient X et Y $\in R^{2\nu}$, on ait l'inégalité

$$1 + \| X-Y\|_X \leq C(1+\|X-Y\|_Y)^M.$$

<u>Proposition 8.3</u> : Soit ($R^{2\nu}$,dY,ρ) l'espace de repères défini par la donnée d'un champ $\rho : Y \longmapsto \| \ \|_Y$ de normes symplectiques sur l'espace de phase. Supposons que la plus grande des valeurs propres de la forme $\| \ \|_Y^2$ soit une fonction à croissance lente de Y. Soit $R > 0$.

Si $u \in \mathcal{S}'(R^\nu)$ est telle que, quel que soit $M_1 > 0$, on ait

$$\iint e^{-2\pi R\|Z\|_Y^2} (1+|Y|)^{M_1} |(u,\tau_Z\, \varphi_Y)|^2 \, dY\, dZ < \infty,$$

alors $u \in \mathcal{S}(R^\nu)$; si $E \subset \mathcal{S}(R^\nu)$ est telle que, pour tout $M_1 > 0$, cette intégrale reste bornée lorsque u parcourt E, alors E est une partie bornée de $\mathcal{S}(R^\nu)$.

<u>Preuve</u> : il suffit de prouver le second point. D'après le théorème 8.1, on a, pour n assez grand :

$$\|u\|_0^2 \sim \sum_{|\alpha|\leq n} \int |(u,\varphi_Y^\alpha)|^2 \, dY.$$

D'après l'hypothèse faite, les opérateurs de création et d'annihilation canoniques sont combinaisons linéaires, à coefficients moindres que $C(1+|Y|)^M$ pour un certain $M \geq 1$, de l'identité et des opérateurs de création et d'annihilation relatifs à l'oscillateur associé à $(Y,\| \ \|_Y)$. Il en résulte que pour tout entier $N \geq 0$, on a

$$\|u\|_{\mathcal{H}^N}^2 \leq C \int (1+|Y|)^{2MN} \sum_{|\alpha|\leq n+N} |(u,\varphi_Y^\alpha)|^2 \, dY.$$

Pour tout $r < 1$, on a donc

$$\|u\|_{\mathcal{H}^N}^2 \leq C \int (1+|Y|)^{2MN} (Op_{\frac{1}{2}}(e^{-2\pi r\|X-Y\|_Y^2})u,u) \, dY.$$

Enfin, d'après la formule (8.7), cette expression s'écrit aussi

$$\|u\|_{\mathcal{H}^N}^2 \leq C \iint e^{-\frac{2\pi r}{1-r}\|Z\|_Y^2} (1+|Y|)^{2MN} |(u,\tau_Z\, \varphi_Y)|^2 \, dY\, dZ.$$

<u>Proposition 8.4</u> : supposons les hypothèses de la proposition 8.3 satisfaites, et soit F une partie bornée de $\mathcal{S}'(R^\nu)$. Il existe $M_1 \in R$ et $k > 0$ tels que l'intégrale

$$\iint (1+\|Z\|_Y)^{-k}(1+|Y|)^{-M_1} |(v,\tau_Z\, \varphi_Y)|^2 \, dY\, dZ$$

est bornée lorsque v parcourt F.

<u>Preuve</u> : soient $C > 0$ et N un entier ≥ 0 tels que, pour tout $v \in F$ et toute $u \in \mathcal{S}(R^\nu)$, on ait $|(v,u)| \leq C\|u\|_{\mathcal{H}^N}$, norme de création-annihilation canonique.

L'argument utilisé dans la preuve de la proposition 8.3 permet d'écrire

$$\|\tau_Z \varphi_Y\|_{\mathcal{H}^N} \le C(1+|Z|+|Y|^M)^N \le C(1+\|Z\|_Y)^N(1+|Y|)^{M'},$$

d'où
$$\iint (1+\|Z\|_Y)^{-k}(1+|Y|)^{-M_1}|(v,\tau_Z\varphi_Y)|^2\, dY\, dZ \le C \iint (1+\|Z\|_Y)^{-k+2N}(1+|Y|)^{-M_1+2M'} dY\, dZ$$

9. Opérateurs métadifférentiels

Soient $(\Omega, d\omega, \rho)$ et $(\Omega', d\omega', \rho')$ deux espaces de repères ; soit N_o le nombre défini dans la proposition 8.1, et soit N'_o le nombre défini de la façon analogue relativement au deuxième espace de repères.

Définition 9.1 : soit $k > 2(N_o+N'_o)$. On appellera opérateur métadifférentiel d'ordre 0 et de classe $\mathcal{H}^k$ (relatif à (Ω,Ω')) tout opérateur A pouvant s'écrire

$$A = \iint Op_{\frac{1}{2}}(a_{\omega,\omega'})\, d\omega\, d\omega',$$

où $(\omega,\omega') \longmapsto a_{\omega,\omega'}$ est une application mesurable de $\Omega \times \Omega'$ à valeurs dans $\mathcal{H}^k(R^{2\nu})$, telle que $\|a_{\omega,\omega'}\|_{\omega\circ\omega',\mathcal{H}^k}$ soit le noyau d'un opérateur borné de $L^2(\Omega')$ dans $L^2(\Omega)$.

Il convient de préciser le sens de l'opérateur A, ce qui est l'objet du théorème suivant : rappelons que les normes $\|a\|_{\omega\circ\omega',\mathcal{H}^k}$ ont été définies dans la section 7 ; on pose $k = 2(N_1+N'_1)$, avec $N_1 > N_o$ et $N'_1 > N'_o$.

Théorème 9.1 : supposons les hypothèses de la définition 9.1 vérifiées. Il existe alors $C > 0$ telle que, quelles que soient u et $v \in L^2(R^\nu)$, l'intégrale

$$(Au,v) = \iint (Op_{\frac{1}{2}}(a_{\omega,\omega'})u,v)\, d\omega\, d\omega'$$

soit absolument convergente, et de plus vérifie

$$|(Au,v)| \le C\, \|u\|\, \|v\| .$$

Preuve : d'après la proposition 7.2, on peut pour tout $(\omega,\omega') \in \Omega \times \Omega'$, trouver une décomposition

$$a_{\omega,\omega'}(X) = \iint K_{\omega,\omega'}(Z,Z')\, H(\tau_Z\varphi_\omega, \tau_{Z'}\varphi_{\omega'}, X)\, dZ\, dZ'$$

avec

$$\left(\iint (1+\|Z\|^2_\theta + \|Z'\|^2_{\theta'})^k |K_{\omega,\omega'}(Z,Z')|^2\, dZ\, dZ'\right)^{\frac{1}{2}} \le C\|a_{\omega,\omega'}\|_{\omega\circ\omega',\mathcal{H}^k}.$$

on a alors

$$|(Au,v)| \le \iiiint |K_{\omega,\omega'}(Z,Z')|\ |(v,\tau_Z\varphi_\omega)|\ |(u,\tau_{Z'}\varphi_{\omega'})|\, dZ\, dZ'\, d\omega\, d\omega'$$
$$\le \iint \|a_{\omega,\omega'}\|_{\omega\circ\omega',\mathcal{H}^k}\, d\omega\, d\omega'$$
$$\left(\int (1+\|Z\|^2_\theta)^{-N_1} |(v,\tau_Z\varphi_\omega)|^2 dZ\right)^{\frac{1}{2}} \left(\int (1+\|Z'\|^2_{\theta'})^{-N'_1} |(u,\tau_{Z'}\varphi_{\omega'})|^2\, dZ'\right)^{\frac{1}{2}}.$$

Il suffit alors d'utiliser la proposition 8.1.

Remarque 9.1 : considérons le cas où $(\Omega, d\omega) = (\mathbb{R}^{2\nu}, dY)$, et où le champ de normes symplectiques $Y \mapsto \| \ \|_Y$ définissant l'espace vérifie la propriété, déjà mentionnée, que, pour une certaine constante $C > 0$, l'inégalité $\|Y-X\|_X \le C^{-1}$ entraîne $\|Z\|_Y \le C\|Z\|_X$; alors les opérateurs pseudo-différentiels d'ordre 0 définis dans [16] sont effectivement métadifférentiels au sens de la définition 9.1 : en effet, une partition de l'unité banale (celle définie par (3.2) dans [16]) permet d'écrire l'opérateur A sous la forme $\int Op_{\frac{1}{2}}(a_\omega)\, d\omega$, avec $\|a_\omega\|_{\omega\infty\omega, \mathcal{H}^k} < C$; pour obtenir une fonction $a_{\omega,\omega'}$ plutôt qu'une mesure concentrée sur la diagonale, il suffit de remplacer $a_Y(X)$ par $\int a_Y(X) f(\|Y-Y'\|_Y^2) dY'$, où $f \in \mathcal{D}(\mathbb{R})$ est à support dans $]-\infty, C^{-1}[$ et vérifie $\int f(|Z|^2)\, dZ = 1$.

Remarque 9.2 : si $(\Omega, d\omega) = (\mathbb{R}^{2\nu}, dY)$ et $(\Omega', d\omega') = (\mathbb{R}^{2\nu}, dY')$, et si de plus $a_{\omega,\omega'} = a_{Y,Y'}$ est concentré, en un certain sens, près du graphe d'une certaine transformation canonique, alors l'opérateur métadifférentiel A est proche d'une version généralisée (il n'y a pas ici d'hypothèse d'homogénéité) d'un opérateur intégral de Fourier.

Lemme 9.1 : quels que soient $N > \nu$, et $N' < N$, il existe $C > 0$ telle que, quels que soient (Y,θ) et $(Y',\theta') \in \mathbb{R}^{2\nu} \times P_\nu$, on ait

$$\iint (1+\|Z\|_\theta^2)^{-N} \exp -\pi \|Y+Z-Y'-Z'\|_{\theta,\theta'}^2 (1+\|Z'\|_{\theta'}^2)^{-N} dZ\, dZ'$$

$$\leqslant C(1+\|Y-Y'\|_{\theta,\theta'}^2)^{-N'+\nu}.$$

Preuve : soit E l'ensemble des couples (Z,Z') tels que

$$\|Z-Z'\|_{\theta,\theta'} \ge \frac{1}{2}\|Y-Y'\|_{\theta,\theta'} .$$

L'intégrale sur le complémentaire de E ne pose pas de difficulté ; avec $\mu = \nu + N - N' > \nu$, l'intégrale sur E est majorée par

$$\iint_E (1+\|Z\|_\theta^2)^{-\mu}(1+\|Z'\|_\theta^2)^{-\mu}(1+\|Z\|_\theta^2 + \|Z'\|_{\theta'}^2)^{-N'+\nu} dZ\, dZ'$$

$$\le C(1+ \inf_E(\|Z\|_\theta^2 + \|Z'\|_{\theta'}^2))^{-N'+\nu} \le C(1 + \frac{1}{8}\|Y-Y'\|_{\theta,\theta'}^2)^{-N'+\nu}$$

(cf. prop. 5.1).

Théorème 9.2 : Soient $(\Omega, d\omega, \rho)$, $(\Omega', d\omega', \rho')$ et $(\Omega'', d\omega'', \rho'')$ trois espaces de repères; soit N' tel que la fonction

$$\lambda(\omega_1', \omega') = \det{}^{\frac{1}{4}}_{\theta_1',\theta'} (1+\|Y'_1-Y'\|_{\theta_1',\theta'}^2)^{\frac{1}{2}(-N'+\nu)}$$

soit le noyau d'un opérateur borné sur $L^2(\Omega')$(cf. définition 8.1). Soit A un opérateur métadifférentiel d'ordre 0 et de classe $\mathcal{H}^{k_1}$ relativement à (Ω,Ω') et soit B un opérateur métadifférentiel d'ordre 0 et de classe $\mathcal{H}^{k_1}$ relativement à (Ω',Ω'') ;

alors, si $k_1 - k = N > \max(N', \nu)$ et si k vérifie l'hypothèse de la définition 9.1 relativement à (Ω, Ω''), AB est un opérateur métadifférentiel d'ordre 0 et de classe $\mathcal{H}^k$ relativement à (Ω, Ω'').

Preuve : à l'aide de la proposition 7.2, écrivons

$$A = \int K_{1,\omega,\omega'}(Z,Z')\mathcal{H}(\tau_Z \varphi_\omega, \tau_{Z'} \varphi_{\omega'})\, dZ\, dZ'\, d\omega\, d\omega'$$

et

$$B = \int K_{2,\omega'_1,\omega''}(Z'_1,Z'')\mathcal{H}(\tau_{Z'_1} \varphi_{\omega'_1}, \tau_{Z''} \varphi_{\omega''})\, dZ'_1\, dZ''\, d\omega'_1\, d\omega''$$

d'où

$$AB = \int K_{\omega,\omega''}(Z,Z'')\mathcal{H}(\tau_Z \varphi_\omega, \tau_{Z''} \varphi_{\omega''})\, dZ\, dZ''\, d\omega\, d\omega'',$$

avec

$$K_{\omega,\omega''}(Z,Z'') = \int K_{1,\omega,\omega'}(Z,Z')K_{2,\omega'_1,\omega''}(Z'_1,Z'')(\tau_{Z'_1}\varphi_{\omega'_1}, \tau_{Z'}\varphi_{\omega'})dZ'dZ'_1\, d\omega'\, d\omega_1. \tag{9.1}$$

Posons

$$M_{\omega,\omega''}(Z,Z'') = (1+\|Z\|^2_\theta + \|Z''\|^2_{\theta''})^{k/2} K_{\omega,\omega''}(Z,Z''),$$

$$M_{1,\omega,\omega'}(Z,Z') = (1+\|Z\|^2_\theta + \|Z'\|^2_{\theta'})^{\frac{k_1}{2}} K_{1,\omega,\omega'}(Z,Z'),$$

$$M_{2,\omega'_1,\omega''}(Z'_1,Z'') = (1+\|Z'_1\|^2_{\theta'_1} + \|Z''\|^2_{\theta''})^{\frac{k_1}{2}} K_{2,\omega'_1,\omega''}(Z'_1,Z''),$$

ainsi que

$$K(\omega,\omega'') = \|M_{\omega,\omega''}\|_{L^2(dZ\, dZ'')},$$

$$K_1(\omega,\omega') = \|M_{1,\omega,\omega'}\|_{L^2(dZ\, dZ')}$$

et

$$K_2(\omega'_1,\omega'') = \|M_{2,\omega'_1,\omega''}\|_{L^2(dZ'_1\, dZ'')}$$

On peut supposer que K_1 est le noyau d'un opérateur borné de $L^2(\Omega')$ dans $L^2(\Omega)$ et que K_2 est le noyau d'un opérateur borné de $L^2(\Omega'')$ dans $L^2(\Omega')$, et il s'agit de montrer que K est le noyau d'un opérateur borné de $L^2(\Omega'')$ dans $L^2(\Omega)$.

Pour estimer le produit scalaire intervenant dans (9.1), on utilise la proposition 6.2 avec $t = \infty$, ce qui donne

$$|(\tau_{Z'_1} \varphi_{\omega'_1}, \tau_{Z'} \varphi_{\omega'})| = \det{}^{\frac{1}{4}}_{\theta'_1,\theta'} \exp - \frac{\pi}{2}\|Y'_1 + Z'_1 - Y' - Z'\|^2_{\theta'_1,\theta'} .$$

Partant de (9.1), on obtient

$$|M_{\omega,\omega''}(Z,Z'')| \leq \int (1+\|Z\|^2_\theta + \|Z''\|^2_{\theta''})^{k/2} (1+\|Z\|^2_\theta + \|Z'\|^2_{\theta'})^{-k_1/2} (1+\|Z_1'\|^2_{\theta_1'} + \|Z''\|^2_{\theta''})^{-k_1/2}$$

$$\det{}^{\frac{1}{4}}_{\theta_1',\theta'} \exp - \frac{\pi}{2} \| Y_1' + Z_1' - Y' - Z \|^2_{\theta_1',\theta'}$$

$$|M_{1,\omega,\omega'}(Z,Z')| \, |M_{2,\omega_1',\omega''}(Z_1',Z'')| \, dZ' dZ_1' \, d\omega' \, d\omega_1'$$

$$\leq \int (1+\|Z'\|^2_{\theta'})^{-N/2} (1+\|Z'_1\|^2_{\theta_1'})^{-N/2}$$

$$\det{}^{\frac{1}{4}}_{\theta_1',\theta'} \exp - \frac{\pi}{2} \| Y_1' + Z_1' - Y' - Z' \|^2_{\theta_1',\theta'}$$

$$|M_{1,\omega,\omega'}(Z,Z')| \, |M_{2,\omega_1',\omega''}(Z_1',Z'')| \, dZ' dZ_1' \, d\omega' \, d\omega_1' .$$

A l'aide de l'inégalité de Schwarz (seulement par rapport à $dZ' \, dZ_1'$) et du lemme 9.1, on obtient

$$|M_{\omega,\omega''}(Z,Z'')| \leq C \int d\omega' \, d\omega_1' \; \det{}^{\frac{1}{4}}_{\theta_1',\theta'} (1+\|Y_1'-Y'\|^2_{\theta_1',\theta'})^{\frac{1}{2}(-N'+\nu)}$$

$$\left(\int |M_{1,\omega,\omega'}(Z,Z')|^2 \, |M_{2,\omega_1',\omega''}(Z_1',Z'')|^2 \, dZ' \, dZ_1'\right)^{\frac{1}{2}}$$

$$= C \int \lambda(\omega_1',\omega') \, d\omega' \, d\omega_1'$$

$$\left(\int |M_{1,\omega,\omega'}(Z,Z')|^2 \, |M_{2,\omega_1',\omega''}(Z_1',Z'')|^2 \, dZ' \, dZ_1'\right)^{\frac{1}{2}}.$$

A l'aide de l'inégalité de Minkowski intégrale sur $(L^2(dZ \, dZ''))$, on obtient

$$K(\omega,\omega'') \leq C \int \lambda(\omega_1',\omega') \, d\omega' \, d\omega_1' \; K_1(\omega,\omega') \, K_2(\omega_1',\omega''),$$

soit enfin $\quad K \leq C \, K_1 \circ \lambda \circ K_2$

au sens de la composition des noyaux, ce qui termine la preuve du théorème 9.2.

<u>Théorème 9.3</u> : si A est un opérateur métadifférentiel d'ordre 0 et de classe $\mathcal{H}^k$ relativement à (Ω,Ω'), A^* est un opérateur métadifférentiel d'ordre 0 et de classe $\mathcal{H}^k$ relativement à (Ω',Ω).

<u>Preuve</u> : c'est évident.

Nous allons étudier pour terminer l'action sur $\mathcal{S}(\mathbb{R}^\nu)$ d'opérateurs métadifférentiels d'ordre éventuellement non nul, en nous limitant toutefois à des espaces de repères définis par des champs de normes symplectiques sur l'espace de phase.

<u>Proposition 9.1</u> : soient $(\mathbb{R}^{2\nu}, dY, \rho)$ et $(\mathbb{R}^{2\nu}, dY', \rho')$ les espaces de repères définis par la donnée de deux champs de normes symplectiques sur l'espace de phase ; on suppose que le premier vérifie l'hypothèse de la proposition 8.3, et que le second vérifie

la propriété (iii) de la proposition 8.2.

Soit $(Y,Y')\longmapsto a_{Y,Y'}$ une application mesurable de $\mathbb{R}^{2\nu}\times\mathbb{R}^{2\nu}$ dans $\mathcal{S}(\mathbb{R}^{2\nu})=\bigcap_k \mathcal{H}^k(\mathbb{R}^{2\nu})$.

On suppose les deux conditions suivantes remplies :

a) il existe $C>0$ et $M>0$ tels que $a_{Y,Y'}$ ne puisse être non nul que si l'on a $1+|Y|\leq C(1+|Y'|)^M$.

b) il existe $M_2>0$ tel que la fonction

$$(Y,Y')\longmapsto(1+|Y'|)^{-M_2}\|a_{Y,Y'}\|_{Y\circ Y',\mathcal{H}^k}$$

soit, pour tout k, le noyau d'un opérateur borné sur $L^2(\mathbb{R}^{2\nu})$.

Alors l'opérateur $A=\iint Op_{\frac{1}{2}}(a_{Y,Y'})\,dY\,dY'$ opère de $\mathcal{S}(\mathbb{R}^\nu)$ dans $\mathcal{S}(\mathbb{R}^\nu)$.

Preuve : remarquons d'abord que les symboles pseudo-différentiels a d'ordre tempéré quelconque peuvent effectivement s'écrire sous la forme $a=\iint a_{Y,Y'}\,dY\,dY'$ de façon que les conditions a) et b) soient remplies ; pour les opérateurs métadifférentiels, la condition a) est une exigence raisonnable, mais non vide.

Il s'agit, pour $u\in\mathcal{S}(\mathbb{R}^\nu)$ et $v\in\mathcal{S}'(\mathbb{R}^\nu)$, de donner une définition sesquilinéaire de (Au,v) telle que (Au,v) reste borné quand u parcourt une partie bornée E de $\mathcal{S}(\mathbb{R}^\nu)$ et v parcourt une partie bornée F de $\mathcal{S}'(\mathbb{R}^\nu)$. Posons

$$a_{Y,Y'}=\iint K_{Y,Y'}(Z,Z')\,H(\tau_Z\,\varphi_Y,\tau_{Z'}\,\varphi_{Y'})\,dZ\,dZ'$$

conformément à la proposition 7.2. On peut alors supposer que

$$K(Y,Y')=(1+|Y'|)^{-M_2}\left(\iint|K_{Y,Y'}(Z,Z')|^2(1+\|Z\|_Y^2+\|Z'\|_{Y'}^2)^k\,dZ\,dZ'\right)^{\frac{1}{2}}$$

est le noyau d'un opérateur borné sur $L^2(\mathbb{R}^\nu)$(dépendant de k, qui peut être choisi arbitrairement grand). L'inégalité de Schwarz fournit

$$|(Au,v)|\leq\iint(1+|Y'|)^{M_2}K(Y,Y')\,dY\,dY'$$

$$\left(\int(1+\|Z\|_Y^2)^{-k/2}\,|(v,\tau_Z\,\varphi_Y)|^2\,dZ\right)^{\frac{1}{2}}$$

$$\left(\int(1+\|Z'\|_{Y'}^2)^{-k/2}\,|(u,\tau_{Z'}\,\varphi_{Y'})|^2\,dZ'\right)^{\frac{1}{2}}.$$

Choisissons M_1 et k (on pourra augmenter ce dernier plus loin) conformément à la proposition 8.4. Il suffit alors, pour que (Au,v) reste borné, qu'il en soit ainsi de l'intégrale

$$\iint(1+|Y'|)^{2M_2+MM_1}(1+\|Z'\|_{Y'})^{-k}\,|(u,\tau_{Z'}\,\varphi_{Y'})|^2\,dY'\,dZ',$$

ce qui (M,M_1 et M_2 étant connus) est assuré pour k assez grand par la proposition 8.2.

REFERENCES

(1) M.A. ANTONETS - The classical limit for Weyl quantization, Letters in Math. Physics 2 (1978), 241-245.

(2) R. BEALS - A general calculus of pseudo-differential operators, Duke Math. J. 42 (1975), 1-42.

(3) F. BRUYANT - Estimations pour la composition d'un grand nombre d'opérateurs pseudo-différentiels et applications, Thèse de 3ème cycle, Univ. de Reims (1979).

(4) A.P. CALDERON & R. VAILLANCOURT - A class of bounded pseudo-differential operators, Proc. Nat. Acad. Sci. USA 69 (1972), 1185-1187.

(5) P. CARTIER - Quantum mechanical commutation relations and theta functions, Proc. Symp. Pure Math. IX, A.M.S. (1966).

(6) A. CORDOBA & C. FEFFERMAN - Wave packets and Fourier integral operators, Comm. Part. Diff. Eq.3, 11 (1978), 979-1006.

(7) L. HORMANDER - The Weyl calculus of pseudo-differential operators, CPAM 23,3 (1979), 359-443.

(8) J. LERAY - Analyse lagrangienne et mécanique, Séminaire au Collège de France (1976-77).

(9) N. LERNER - Sur les espaces de Sobolev généraux associés aux classes récentes d'opérateurs pseudo-différentiels, Thèse de 3ème cycle, Univ. de Reims (1979).

(10) H. MAAS - Siegel's modular forms and Dirichlet Series, Lecture Notes in Math. 216 (1971), Springer-Verlag.

(11) A. MELIN - Lower bounds for pseudo-differential operators, Arkiv för Mathematik 9, 1 (1971), 117-140.

(12) J. RALSTON - Gaussian beams and the propagation of singularities (preprint).

(13) M. REED & B. SIMON - Fourier Analysis, Self-Adjointness, Ac. Press (1975).

(14) A. UNTERBERGER - Pseudo-differential operators and applications : an introduction, Lecture Notes, Aarhus (1976).

(15) A. UNTERBERGER - Encore des classes de symboles, Séminaire Goulaouic-Schwartz (1977-78), Ecole Polytechnique, Paris.

(16) A. UNTERBERGER - Oscillateur harmonique et opérateurs pseudo-différentiels, Annales de l'Institut Fourier 29, 3 (1979), 201-221.

(17) A. UNTERBERGER - Sur la continuité sur L^2 des opérateurs pseudo-différentiels, Séminaire Kree sur les équations aux dérivées partielles en dimension infinie, 4ème année, Institut Henri Poincaré (1979).

PART III

GENERAL STRUCTURE OF GREEN FUNCTIONS AND COLLISION AMPLITUDES

ASYMPTOTIC BEHAVIOUR OF FEYNMAN INTEGRALS

M.C. BERGERE

DPh-T, CEN Saclay, BP n°2, 91190 Gif-sur-Yvette, France

I - INTRODUCTION

The description of physical observables in various kinematic domains is one of the ultimate purpose of quantum field theory. Among these various domains, the asymptotic domains where some kinematic variables become large represent one of the best testing ground between theory and experiment (large energy-momentum behaviour of Green functions; scattering of two particles at large energy, large transfert and fixed angle; Regge phenomenology; deep inelastic scattering; etc...).

The predictions of quantum field theory on this subject were for a long time rather small, although several attempts and various methods were constructed to achieve this goal ; let us for instance quote the original work of Gell-Mann and Low[1] on the renormalization group; let us also remind the enormous amount of work on the Bethe-Salpeter[2] kernel and its application to Regge theory. It is only in the 70's that a large step forward has been performed by the introduction of modern renormalization group techniques[3], differential equations[4], short distance and light-cone expansions[5].

However, these techniques do not directly describe the asymptotic behaviour of Green functions and somewhere an approximation, according to what asymptotic behaviour we are looking for, has to be performed; such an approximation is obtained from a power counting theorem[6] which is proved to be valid in perturbation, on Feynman amplitudes, and the hypothesis is made that what is leading in perturbation remains leading on Green functions.

In these lecture notes, we wish to describe how to obtain the asymptotic behaviour of Feynman amplitudes; this technique has been already applied in several cases[7], but the general solution for any kind of asymptotic behaviour has not yet been found. From the mathematical point of view, the problem to solve is close to the following problem : find the asymptotic expansion at large λ of the integral

$$\int \dots \int \{dx\}\, e^{-\lambda P\{x\}} \qquad (I.1)$$

where $P\{x\}$ is a polynomial of several variables. Theoretically this problem is solved[8] but first, Feynman amplitudes are not as simple as the above integral, and second, we wish to compute the coefficients of the expansion so that we may sum these coefficients over all Feynman amplitudes which describe a physical process. The first part of these notes (section II) introduces the Feynman amplitudes and its Mellin transform in regards to some asymptotic variable ; in section III and IV we prove the following theorem[9]:

Given a Feynman amplitude $I_G\{q\}$ attached to a Feynman graph G with external momenta q_i satisfying $\sum_{i=1}^{N} q_i = 0$; we denote by $\{q\}$ the set of all Lorentz invariants $(q_{i_1}+q_{i_2}+\ldots+q_{i_p})^2$ with $p < N$. Let us now scale by a parameter λ the invariants which belong to a given subset $\{q'\}$; then $I_G[\{\lambda q'\},\{q''\}]$ has the asymptotic behaviour

$$I_G[\{\lambda q'\},\{q''\}] = \sum_{p=\Omega}^{-\infty} \lambda^p \sum_{r=o}^{r_{max}(p)} (\mathrm{Log}\lambda)^r \; g_{pr}\{q\} \quad . \tag{I.2}$$

The above theorem is proved in Euclidean space for any asymptotic behaviour and it remains valid in Minkowski space for a subclass of asymptotic behaviours.

Finally in section V we show how to compute the coefficients $g_{pr}\{q\}$ in the simpler case.

II - FEYNMAN AMPLITUDE AND MELLIN TRANSFORM

For sake of simplicity, we shall only introduce here Euclidean scalar Feynman amplitudes, ultraviolet and infrared convergent. What follows may be extended to other kind of Feynman amplitudes encountered in quantum field theory.

Given a Feynman graph G with ℓ internal lines, we define a Feynman amplitude in dimension D as

$$I(\{q\}) = \int_o^\infty \prod_{a=1}^{\ell} d\alpha_a \; \frac{e^{-\sum_{a=1}^{\ell} \alpha_a m_a^2} \; e^{-\frac{V(\alpha,\{q\})}{P(\alpha)}}}{P(\alpha)^{D/2}} \tag{II.1}$$

where one variable α is attached to each internal line and where the polynomials $P(\alpha)$ and $V(\alpha,\{q\})$ are defined as follows :

Definition : - A one-tree T_1 is a connected tree graph (no loop) which contains all the vertices of the graph G.

- A two-tree T_2 is a tree graph with two connected components, and which contains all the vertices of the graph G.

A two-tree is obtained from a one-tree by omiting one line. A two-tree splits the set of external momentum $\{q_1,\ldots q_r\}$ into two sets $\{q_{i_1},\ldots q_{i_s}\}$ and $\{q_{i_{s+1}}\ldots q_{i_r}\}$

such that

$$\left(\sum_{j=1}^{s} q_{i_j}\right)^2 = \left(\sum_{j=s+1}^{r} q_{i_j}\right)^2 \tag{II.2}$$

by energy-momentum conservation.

Then,

$$P_G(\alpha) = \sum_{T_1} \prod_{a\notin T_1} \alpha_a \quad , \tag{II.3a}$$

$$V(\alpha,\{q\}) = \sum_{T_2} \left\{ \prod_{a\notin T_2} \alpha_a (\Sigma q)^2_{T_2} \right\} \quad . \tag{II.3b}$$

The notation $\{q\}$ stands for the set of all possible squares $(\Sigma q)^2$. Several two-trees T_2 define the same square $(\Sigma q)^2$ and partial summation of the two-trees for a given square can be performed. We define a cut c as a set of lines such that if we cut these lines the graph G is split into two connected parts R and L and such that no subset of these lines have the same property. A cut defines a square $(\Sigma q)^2_c$. Then $V(\alpha,\{q\})$ may also be expressed as

$$V(\alpha,\{q\}) = \sum_{c} \left[\prod_{a\in c} \alpha_a \, P_R(\alpha) P_L(\alpha) (\Sigma q)^2_c \right] \tag{II.4}$$

The polynomials $P(\alpha)$ and $V(\alpha,\{q\})$ are positive when all α'_s are positive and vanishes when some α'_s vanish (see later). The ratio $V(\alpha,\{q\})/P$ is regular for $\alpha'_s \geq 0$.

Suppose now that some (not necessary all) invariants are scaled to infinity ($\{q\} \to \lambda\{q\}$), then V becomes

$$V \to V' + \lambda W \tag{II.5}$$

and we study the large λ behaviour of $I(\lambda)$. We introduce the Mellin transform

$$M(x) = \int_0^\infty d\lambda \, \lambda^{-x-1} I(\lambda)$$

which is found to be

$$M(x) = \Gamma(-x) \int_0^\infty \prod_{a=1}^{\ell} d\alpha_a \, e^{-\sum_{a=1}^{\ell} \alpha_a m_a^2} \left(\frac{W}{P}\right)^x \frac{e^{-\frac{V'}{P}}}{P^{D/2}} \tag{II.6}$$

for $\Omega < \mathrm{Re}x < 0$ where Ω is the largest singularity in x of the integrals α. We now study these integrals.

Definition of Hepp's sectors

We define a sector S as an ordering of the α'_s :

$$0 \leq \alpha_{a_1} \leq \alpha_{a_2} \leq \ldots \leq \alpha_{a_\ell} \tag{II.7}$$

There are $\ell!$ sectors and the union over all sectors is the integration domain in α'_s.
A sector defines a set of nested subgraphs $R_i = \{a_1, a_2, \ldots a_i\}$, $R_i \supset R_{i-1}$, $R_\ell = G$.

In each sector, we perform the change of variables

$$\alpha_{a_i} = \prod_{j=i+1}^{\ell} \beta_j \; \beta_i \qquad \begin{cases} 0 \leq \beta_\ell \leq \infty \\ 0 \leq \beta_{i<\ell} \leq 1 \end{cases} \tag{II.8}$$

so that,

$$\text{the Jacobian} \quad \prod_{a=1}^{\ell} d\alpha_a \to \prod_{i=1}^{\ell} \left[\beta_i^{\ell(R_i)-1} \; d\beta_i \right] \quad \text{with} \quad (\ell(R_i) = i) \tag{II.9}$$

$$P_G(\alpha) \to \prod_{i=1}^{\ell} \beta_i^{L(R_i)} \; [1 + Q(\beta)] \tag{II.10}$$

where $L(R_i)$ are the number of independent loops of the subgraphs R_i, and $Q(\beta)$ is a non-negative polynomial.

Definition : A function $f(\alpha_a)$ is said to be FINE if in each sector

$$f(\alpha_a) \to \prod_{i=1}^{\ell} \beta_i^{\nu_i} \; g(\beta_i) \tag{II.11}$$

where $g(\beta_i)$ has a simultaneous Taylor expansion in all β'_s around the point $\beta_1 = \beta_2 = \ldots = \beta_\ell = 0$.

In general $W(\alpha, \{q\})$ is NON-FINE and transforms into a sum of FINE functions. Nevertheless, many asymptotic behaviours have the property to possess a FINE $W(\alpha, \{q\})$ function and we shall first study this simpler case.

III - THE CASE $W(\alpha, \{q\})$ FINE

The function $W(\alpha, \{q\})$ satisfies the property that under the change of variables (II.8), we get :

$$W(\alpha, \{q\}) \to \prod_{i=1}^{\ell} \beta_i^{\nu_i} \; W'(\beta_i, \{q\}) \quad , \quad \begin{cases} \nu_\ell = L(G)+1 \quad , \\ \nu_i \geq L(R_i) \end{cases} \tag{III.1}$$

where W' has a simultaneous Taylor expansion in all β'_s around the point $\beta_1 = \beta_2 = \ldots = \beta_\ell = 0$, and where ν_i are non-negative integers. If we decompose $M(x)$ over the sectors, we obtain after integration over β_ℓ ,

$$M_S(x) = \Gamma(-x)\Gamma\left(x - \frac{\omega(G)}{2}\right) \int_0^1 \prod_{i=1}^{\ell-1} \left[d\beta_i \; \beta_i^{-\frac{\omega(R_i)}{2} + [\nu_i - L(R_i)]x - 1} \right] h(\beta_i, x, \{q\}, m_a^2) \tag{III.2}$$

where, from the property of ultraviolet convergence,

$$\omega(R_i) = L(R_i)D - 2\ell(R_i) < 0 \quad . \tag{III.3}$$

($\omega(R_i)$ is called the superficial degree of convergence of R_i), and where $h(\beta_i,x)$ is analytic in x and has a simultaneous Taylor expansion in all β'_s around $\beta_1 = \beta_2 = \ldots = \beta_{\ell-1} = 0$.

The structure of the β-integrals and the Euler functions Γ show that $M_S(x)$ is a meromorphic function. The poles of $M_S(x)$ are generated from the subgraphs R_i such that $\nu_i > L(R_i)$. Such subgraphs are called essential subgraphs. The essential subgraphs are those which determine the asymptotic behaviour of $I(\lambda)$. The poles of $M(x)$, apart from those at $x = 0,1,2,\ldots$ which comes from $\Gamma(-x)$, are found at negative rational values of x (if the space-time dimension D is rational)

$$x_k^S = \frac{\omega(S) - 2k}{2[\nu_S - L(S)]} \qquad \text{for} \qquad \begin{cases} k = 0,1,2,\ldots \\ S \text{ essential} \end{cases} \tag{III.4}$$

where ν_S is given by

$$W(\rho\alpha_{a\in S},\alpha_{a\notin S},\{q\}) \underset{\rho\to o}{\sim} \rho^{\nu_S} \; [W^o(\alpha_a,\{q\}) + \mathcal{O}(\rho)] \tag{III.5}$$

The leading pole is found to be

$$\Omega = \sup_{\substack{S \\ \text{essential}}} \left\{ \frac{\omega(S)}{2[\nu_S - L(S)]} \right\} \tag{III.6}$$

and we call leading subgraphs, those essential subgraphs such that $x_o^S = \Omega$.

The function $\frac{M(x)}{\Gamma(-x)}$ can be written as an infinite sum of multiple poles at $x = x_k^S$ and from the inverse Mellin transform

$$I(\lambda) = \frac{1}{2i\pi} \int_{\sigma-i\infty}^{\sigma+i\infty} d(\mathrm{Im}x)\, \lambda^x \, M(x) \qquad \Omega < \sigma < 0 \tag{III.7}$$

we may push the contour σ towards the left (the convergence at $|\mathrm{Im}x| \to \infty$ is ensured by the function $\Gamma(-x)$) and we obtain, by Cauchy's theorem, the asymptotic behaviour

$$I_G(\lambda) = \sum_{p=\Omega}^{-\infty} \lambda^p \sum_{r=o}^{r_{max}(p)} (\mathrm{Log}\lambda)^r \, g_{pr}(\{q\}) \tag{III.8}$$

where p runs over the rational values x_k^S and where r, for a given p, runs over a finite number of non-negative integers.

IV - THE CASE $W(\alpha,\{q\})$ NON FINE

The form obtained for the asymptotic behaviour and given at the end of section III remains valid when $W(\alpha,\{q\})$ is NON FINE. We now explain how to obtain this result.

The function $W(\alpha,\{q\})$ can be decomposed in many ways as

$$W(\alpha,\{q\}) = \sum_{j=1}^{N} W_j(\alpha,\{q\}) \tag{IV.1}$$

where each function $W_j(\alpha,\{q\})$ is FINE. Then we may transform a sum into a product by using a representation of the type

$$\Gamma(-x)(A+B)^x = \frac{1}{2i\pi} \int_{\sigma-i\infty}^{\sigma+i\infty} d\mathrm{Im}z\ \Gamma(-z)\Gamma(z-x)A^z B^{x-z} \tag{IV.2}$$

with $\mathrm{Re}x < \sigma < 0$, $\mathrm{Re}A > 0$, $\mathrm{Re}B > 0$. This leads directly to the introduction of a multiple Mellin representation. From

$$e^{-\frac{W_j}{P}} = \frac{1}{2i\pi} \int_{\sigma_j-i\infty}^{\sigma_j+i\infty} d\mathrm{Im}x_j\ \Gamma(-x_j) \left(\frac{W_j}{P}\right)^{x_j} , \quad \sigma_j < 0 \tag{IV.3}$$

we obtain the multiple Mellin representation as

$$M(x_j) = \prod_{j=1}^{N} \Gamma(-x_j) \int_0^\infty \prod_{a=1}^{\ell} d\alpha_a\ e^{-\sum_{a=1}^{\ell} \alpha_a m_a^2} \prod_{j=1}^{N} \left(\frac{W_j}{P}\right)^{x_j} \frac{e^{-\frac{V'}{P}}}{P^{D/2}} . \tag{IV.4}$$

The above α-integrand is FINE and the technique of section III may be applied. For each sector, we find that $M_S(x_j)$ is analytic provided $\sigma_j = \mathrm{Re}x_j$ satisfies the inequalities

$$\left\{ \begin{array}{ll} \sigma_j < 0 & \text{for all } j \ , \\ \sum_{j=1}^{N} n_{ij}\sigma_j - \frac{1}{2}\,\omega(R_i) > 0 & \text{for all } i \ . \end{array} \right. \tag{IV.5}$$

where $n_{ij} = \nu_{ij} - L(R_i) \geq 0$ and

$$W_j(\rho\alpha_{a\in R_i}, \alpha_{a\notin R_i}, \{q\}) \underset{\rho\to o}{\sim} \rho^{\nu_{ij}} [W_j^o(\alpha_a,\{q\}) + \mathcal{O}(\rho)] \tag{IV.6}$$

The above conditions over σ_j define a closed convex polyhedron $\mathcal{P}_S$ for each sector. This polyhedron is not empty because the amplitude is ultraviolet convergent. The inverse multiple Mellin transform defines

$$I_S(\lambda) = \int_{\sigma_j-i\infty}^{\sigma_j+i\infty} \frac{d\mathrm{Im}x_j}{2i\pi}\ \lambda^{\sum_{j=1}^{N} x_j}\ M_S(x_j) \tag{IV.7}$$

with $\sigma_j \in P_S$. Because $\bigcap_S P_S \neq \phi$, it is also possible to define an inverse multiple Mellin transform for $I(\lambda)$.

The behaviour of $I_S(\lambda)$ at large λ is given by the point (or points) such that $\left(\sum_{j=1}^{N} x_j\right)$ is minimum in P_S ;

$$I_S(\lambda) \underset{\lambda\to\infty}{\sim} \lambda^{\Omega_S} \quad \text{up to powers of Log}\,\lambda \quad , \tag{IV.8}$$

where

$$\Omega_S = \operatorname*{Inf}_{P_S} \left(\sum_j x_j\right) \tag{IV.9}$$

If $\Omega = \sup_S \Omega_S$, $I(\lambda) \underset{\lambda\to\infty}{\sim} \lambda^{\Omega}$ up to powers of Log λ.
(it must be noted that $\Omega \leq \operatorname*{Inf}_{\bigcap_S P_S} \left(\sum_j x_j\right)$).

The singularities of $M_S(x_j)$ are on the varieties

$$\left\{ \begin{array}{l} \sigma_j = k_j \\ \sum_{j=1}^{N} n_{ij}\sigma_j - \frac{1}{2}\,\omega(R_i) = -k_i \end{array} \right. \tag{IV.10}$$

where k_i or k_j are non-negative integers. The functions $M_S(x_j)$ can be written (up to $\Gamma(-x_j)$ functions) as an infinite sum of terms of the type

$$\frac{R(\{q\})}{\prod_{i=1}^{\ell} \left[\sum_{j=1}^{N} n_{ij}x_j - \frac{\omega(R_i)}{2} + k_i \right]} \quad . \tag{IV.11}$$

We may now use the inverse multiple Mellin transform, and push the contour along a path where $\left(\sum_{j=1}^{N} \sigma_j\right)$ is monotonically decreasing ; if we use Cauchy's theorem as many times as needed we obtain for $I_S(\lambda)$ and consequently for $I(\lambda)$ an asymptotic expansion of the type described before in (III.8).

Although, there are many ways of splitting $W(\alpha,\{q\})$ into FINE functions $W_j(\alpha,\{q\})$, two ways seem to be topologically interesting :

a) split according to the cuts c_j which define the large invariants ; then

$$W_j(\alpha,\{q\}) = \prod_{a\in c_j} \alpha_a \; P_R(\alpha)\; P_L(\alpha)\; (\Sigma q)^2_{c_j} \quad . \tag{IV.12}$$

In this case, the coefficients n_{ij} corresponding to a graph R_i and a cut c_j are equal to the difference between the number of connected parts when R_i is cut by c_j and the original number of connected parts of R_i .

b) split according to all two-trees ; we decompose the function V according to (II.3b),

$$V_j(\alpha,\{q\}) = \prod_{a \notin T_2^j} \alpha_a \, (\Sigma q)^2_{T_2^j} \tag{IV.13}$$

In this case we may write

$$[P(\alpha)]^{-\sum_{j=1}^{N} x_j - D/2} = \frac{1}{\Gamma\left[-\sum_{j=1}^{N} x_j - D/2\right]} \int_{\sigma_k - i\infty}^{\sigma_k + i\infty} \frac{d\mathrm{Im}y_k}{2i\pi} \Gamma(-y_k) \, [P_k(\alpha)]^{y_k} \, \delta\left[\sum_{j=1}^{N} x_j + \sum_{k=1}^{M} y_k + \frac{D}{2}\right] \tag{IV.14}$$

where we split $P(\alpha)$ over its one-tree components $P_k(\alpha) = \prod_{a \notin T_1^k} \alpha_a$. It is then possible to integrate over all α'_s and we obtain the complete Mellin representation of Feynman amplitudes as obtained by C.de Calan and A.P.C.Malbouisson[10]

$$I(\{q\}) = \int_{\sigma_j - i\infty}^{\sigma_j + i\infty} \frac{d\mathrm{Im}x_j}{2i\pi} \Gamma(-x_j) \int_{\sigma_k - i\infty}^{\sigma_k + i\infty} \frac{d\mathrm{Im}y_j}{2i\pi} \Gamma(-y_j) \, \frac{\delta\left[\sum_{j=1}^{N} x_j + \sum_{k=1}^{M} y_k + D/2\right]}{\Gamma\left(\sum_{k=1}^{M} y_k\right)} \prod_j \left[(\Sigma q)^2_{T_2^j}\right]^{x_j}$$

$$\times \prod_{a=1}^{\ell} (m_a^2)^{-\phi_a} \, \Gamma(\phi_a) \tag{IV.15}$$

where the forms ϕ_a are linear

$$\phi_a = \sum_{j=1}^{N} n_{aj} x_j + \sum_{k=1}^{M} m_{ak} y_k + 1 \tag{IV.16}$$

and the coefficients n_{aj}, m_{ak} are 0 or 1 depending whether the line a belongs to the two-tree T_2^j or one-tree T_1^k , or not. The point $\{\sigma_j,\sigma_k\}$ must be chosen on the intersection of the variety M

$\left[\sum_{j=1}^{n} x_j + \sum_{k=1}^{M} y_k + \frac{D}{2}\right]$ with the closed convex polyhedron P

$$\begin{cases} \sigma_j < 0 \quad , \quad \sigma_k < 0 \\ \phi_a > 0 \end{cases} \tag{IV.17}$$

and any asymptotic behaviour in $\{q\}$ may be obtained by minimizing in $M \cap P$ a given linear form of the x'_js.

Let us end this section by mentioning that the asymptotic behaviour obtained (in III.8) for $I(\lambda)$ is valid for amplitudes with spinor and coupling derivatives, for renormalized amplitudes and also for infrared (zero mass) asymptotic behaviour. If we try to extend the theorem to Minkowski space, two cases appear : either the Wick rotation transform the functions $g_{pr}\{q\}$ into distributions (as it does for the Feynman amplitude) and the asymptotic behaviour remains valid, in this sense, or the functions g_{pr} become infinite because the Landau singularities perturb deeply the asymptotic behaviour.

V - THE COEFFICIENTS $g_{pr}\{q\}$ IN THE CASE $W(\alpha,\{q\})$ FINE

A way to obtain the coefficients $g_{pr}\{q\}$ is to compute the residues of the poles in each Hepp's sector. The problem in doing so is, first, that it is very tedious to calculate over $\ell!$ sectors, second, that it completely destroys the topological structure of the subgraphs and reduced graphs, third, that it does not lead to an analytic form of $g_{pr}\{q\}$ which may be easily summed over all graphs contributing to the Green function.

We present here an alternative method which can be applied, step by step, on the α-integrals in its general form and which avoids the three above disadvantages. We first give a one-dimensional example.

Definition : We introduce the Taylor operators acting over a function f(x) infinitely differentiable at x = 0, as

$$\begin{cases} T^n_x f(x) = \sum\limits_{p=o}^{n} \dfrac{x^p}{p!} f^{(p)}(0) , & \text{for } n \geq 0 \\ T^n_x f(x) = 0 & \text{for } n < 0 \end{cases} \tag{V.1}$$

We now extend this definition to functions f(x) such that $x^{-\nu} f(x)$ (ν complex) is infinitely differentiable at x = 0 ; then, we define the operator τ^n_x as

$$\tau^n_x f(x) = x^\nu T^{n-E'(\nu)} \{x^{-\nu} f(x)\} \tag{V.2}$$

where $E'(\nu)$ is the smallest integer $\geq \mathrm{Re}\nu$. One property of this operator is that

$$(1-\tau^n_x) f(x) \underset{x\to o}{\sim} x^{n+\varepsilon} , \quad \varepsilon > 0 \tag{V.3}$$

and consequently, if $\int\limits_o dx\, f(x)$ is not integrable, $\int\limits_o dx\, (1-\tau^{-1}_x) f(x)$ is integrable $\forall\nu$.

It is important to note that the number of subtractions is dependent with ν and change by one whenever ν increases and becomes integer. As an application we may obtain analytic continuation of functions defined from an integral.

Example :

$$\Gamma(x) = \int_o^\infty d\lambda\, (1-\tau^{-1}_\lambda) \{\lambda^{x-1} e^{-\lambda}\} \tag{V.4}$$

is defined for all $x \neq -k$, k = 0,1,2,... .

We now generalize to several variables.

Given a function $f(\alpha)$ which is FINE, and given a subgraph S, we define

$$\tau_S^n f(\alpha) = \left[\tau_\rho^n f(\alpha) \Big|_{\alpha_a \to \rho\alpha_a \forall a \in S} \right]_{\rho=1} \tag{V.5}$$

Then, if

$$\begin{aligned} (1-\tau^n)\, f(\alpha) &= g(\alpha) \quad , \\ g[\rho\alpha_{a\in S}, \alpha_{a\notin S}] &\underset{\rho\to o}{\sim} \rho^{n+\varepsilon} \quad , \qquad \varepsilon > 0 \quad . \end{aligned} \tag{V.6}$$

We define the operator

$$R = \prod_{\underline{S\subset G}} (1-\tau_S^{-\ell(S)}) \tag{V.7}$$

where $\ell(S)$ is the number of lines of the subgraph S.

Although the operators τ_S do not commute when relative to overlapping subgraphs, in $Rf(\alpha)$, the product which defines R may be taken in any order if $f(\alpha)$ is FINE.

<u>Theorem 1</u> [11] :

Given the meromorphic function

$$M(x) = \Gamma(-x) \int_o^\infty \prod_{a=1}^{\ell} d\alpha_a \, e^{-\sum_{a=1}^{\ell} \alpha_a m_a^2} \left(\frac{W}{P}\right)^x \frac{e^{-\frac{V'}{P}}}{P^{D/2}} \tag{V.8}$$

where $W(\alpha)$ is FINE , the above integral is convergent for Rex > Ω and M(x) has multiple poles at $x = x_S^k$ for $k = 0,1,2...$ and S essential. The α-integrals can be analytically continued in x for Rex < Ω and away from the poles. We obtain the absolutely convergent integral representation

$$M(x) = \Gamma(-x) \int_o^\infty \prod_{a=1}^{\ell} d\alpha_a \, R \left\{ e^{-\sum_{a=1}^{\ell} \alpha_a m_a^2} \left(\frac{W}{P}\right)^x \frac{e^{-\frac{V'}{P}}}{P^{D/2}} \right\} \tag{V.9}$$

Again, the number of subtractions performed by R depends of the band (between two consecutive poles) where stands the point x in the complex x plane.

<u>Theorem 2</u> [12] : In the function

$$\bar{M}(x) = \Gamma(-x) \int_o^\infty \prod_{a=1}^{\ell} d\alpha_a \, e^{-\sum_{a=1}^{\ell} \alpha_a m_a^2} R \left\{ \left(\frac{W}{P}\right)^x \frac{e^{-\frac{V'}{P}}}{P^{D/2}} \right\} \tag{V.10}$$

the α-integrals are absolutely convergent $\forall$x.

The function $\bar{M}(x)$ is not continuous on the right of the lines $x_S^k + iy$ (y from $-\infty$ to $+\infty$).

To compute the coefficients g_{pr} we use these theorems in the following manners : we are looking for a multiple pole of $M(x)$ at $x = p$. On each side of the line $\mathrm{Re}\,x = p$, we define the bands $\mathcal{B}_-$ and $\mathcal{B}_+$ which separate the pole $x = p$ from its neighbours. The coefficient g_{pr} may be obtained by taking a Cauchy contour around $\mathrm{Re}\,x = p$ taking into account the definition of $M(x)$ in $\mathcal{B}_-$ and in $\mathcal{B}_+$ as given in V.9. Another way of calculating g_{pr} is to define the function $M^{\mathcal{B}_-}(x)$ which is equal to $\bar{M}(x)$ in the band $\mathcal{B}_-$ and which is analytic for all x away from the poles which are at the left of $\mathcal{B}_-$. Then, the difference $[M(x)-M^{\mathcal{B}_-}(x)]$ contains all the meromorphic structure at $\mathrm{Re}\,x = p$. In both cases, we are led to compute the action of τ_S operators over the α-integrand. In doing so, factorization properties occur between a function attached to the subgraph S and a function attached to the reduced graph $[G/S]$; the coefficients g_{pr} are found to be a sum of expressions, each being attached to a forest F (set of non-overlapping subgraphs) of essential subgraphs. For each forest, we obtain a factorization of coefficients (defined by their integral representation) attached to reduced graphs $[S]_F$ where $S \in F$ and where $[S]_F$ means the graph obtained from S when all subgraphs inside S and which belong to F are shrinked into points.

It is this quasi-geometric structure of the coefficients g_{pr} which explains why, in some cases, the asymptotic behaviour of $I_G(\lambda)$ for a given graph G may be summed over all graphs of a given Green function.

Let us mention that we have applied efficiently this technique to determine the scaling[7] and the deep-inelastic behaviour of scalar field theories where the α-integrand of the Mellin transform is FINE (and where of course, we use a formalism adapted to renormalized Feynman amplitude). We also determined the Regge pole behaviour (planar graphs) of scalar field theories[7] where the α-integrand is not FINE but where nevertheless, the operator R has been shown to determine the leading coefficients $g_{\Omega r}$ $\forall r$.

On the other hand, the construction of a generalized R operator which could determine the coefficients g_{pr} in the case NON FINE is definitely needed, for instance, if we wish to handle the infrared behaviour of quantum electro-(chromo) dynamics.

REFERENCES

[1] A.Peterman, E.C.G.Stückelberg, Helv.Phys.Acta, 26, 499 (1953).
M.Gell-Mann, F.E.Low, Phys.Rev. 95, 1300 (1954).

[2] N.Nakanishi, Suppl.Prog.Theor.Phys. 43, 1 (1969).

[3] K.G.Wilson, Phys.Rev. B.4, 3174 (1971) ; 3184 (1971).
K.G.Wilson, J.B.Kogut, Phys.Reports 12C, 75 (1974).

[4] C.G.Callan, Phys.Rev. D.2, 1541 (1970).
K.Symanzik, Comm.Math.Phys. 18, 227 (1970) ; 23, 49 (1971).

[5] K.G.Wilson, W.Zimmermann, Comm.Math.Phys. 24, 87 (1972).
W.Zimmermann, Ann.Physics 77 (1973) 570.
S.A.Anikin and O.I.Zavialov, Ann.Physics 116, 135 (1978).

[6] S.Weinberg, Phys.Rev. 118, 838 (1960).
P.Cvitanovic, T.Kinoshita, Phys.Rev. D.10, 3978, 3991 (1974).

[7] M.C.Bergère, Y.M.P.Lam, Comm.Math.Phys. 39, 1 (1974) ; "Asymptotic expansion of Feynman amplitudes, part.II. The divergent case". Freie Universität Berlin HEP (May 74/9).
M.C.Bergère, C.Gilain, J.Math.Phys. 19, 1495 (1978).
M.C.Bergère, C.de Calan, "Regge pole behaviour from perturbative scalar field theories" DPh-T preprint n°79/7, to be published in Phys.Rev.

[8] A.N.Vartchenko, Functional analysis and its applications 10, 13 (1976).
A.G.Kouchnirenko, Inventiones Mathematicae 32, 1 (1976).

[9] M.C.Bergère, C.de Calan, A.P.C.Malbouisson, Comm.Math.Phys. 62, 137 (1978).

[10] A.P.C.Malbouisson, "Comportement Asymptotique des Amplitudes de Feynman", Doctoral Thesis, Paris VI (1978).
C.de Calan, A.P.C.Malbouisson, "Complete Mellin representation and asymptotic behaviours of Feynman amplitudes", Ecole Polytechnique preprint A.318 (Sept. 1979).

[11] M.C.Bergère, F.David, J.Math.Phys. 20, 1244 (1979).

[12] M.C.Bergère, Y.M.P.Lam, J.Math.Phys. 17, 1546 (1976).

INTEGRAL RELATIONS IN COMPLEX SPACE AND THE GLOBAL ANALYTIC AND MONODROMIC STRUCTURE OF GREEN'S FUNCTIONS IN QUANTUM FIELD THEORY : SOME GENERAL IDEAS AND RECENT RESULTS

J. BROS

DPh-T, CEN Saclay, BP 2, 91190 Gif-sur-Yvette, France

In this lecture, we wish to present some of the ideas of a global consistent approach to the analytic and monodromic structure of Green's functions and scattering amplitudes of elementary particles on the basis of general quantum field theory. This approach, which originates in Symanzik's pioneer program [1] called "Many-Particle Structure Analysis" (in brief M.P.S.A.) has already been presented by us in its complex momentum space formulation at several places [2][3].

It integrates in an organized scheme concepts which pertain to three kinds of standard investigations concerning analyticity in particle physics namely :

(i) Analyticity properties of Green's functions $H^{(n)}$ and scattering amplitudes* on the basis of the axioms of local fields [4][5].

(ii) Analytic structure of scattering amplitudes in axiomatic S-matrix theory [6]

(iii) Analytic structure of the Feynman amplitudes in the perturbative treatment of Quantum Field Theory [7][8][9][10].

The foundations of the M.P.S.A. approach are the axioms used in (i), supplemented by an extra-axiom called "asymptotic completeness" (A.C): the latter is actually equivalent to the property of "unitarity of the S-matrix" which is one of the basic ingredients of (ii). The method which we use in M.P.S.A. is based on the study of certain integral relations in complex momentum space which have similarities and closed links with the Feynman integrals ; the technique which we apply to studying the generation of complex singularities through the integration process and the results which we obtain are comparable to those of a standard approach of (iii) [8].

From the framework of (i), the M.P.S.A. approach keeps the benefit of well-defined global analyticity domains for the Green functions $H^{(n)}(k_1,\ldots,k_n)$ in $\mathbb{C}^{4(n-1)} = \{k=(k_1,\ldots k_n)\ ;\ k_1+\ldots+k_n = 0\}$; these domains - called "primitive analyticity domains" and denoted by D_n - can be enlarged by holo-convex completion** (i.e. by computing their holomorphy envelopes or parts of the latter) , but they

---

* Let us recall that the scattering amplitudes for the $m\to n-m$ particle processes are appropriate real boundary values of the restriction of Green's function $H^{(n)}$ to a certain complex manifold $M_n^{(c)}$ called the "complex mass shell".

** For example, for n=4, this yields the domains of "dispersion relations" for the two-particle scattering amplitudes of certain reactions.

obey strong limitations : they are "schlicht" somains in $\mathbb{C}^{4(n-1)}$ which are bounded by certain "cuts" accross which it would be impossible to expect analytic continuation for the $H^{(n)}$, were the A.C. axiom not postulated.

Exploitation of the A.C. axiom allows one to investigate the monodromic structure of Green's functions around well defined singular sets, called "Landau varieties". Investigations of this type are in some sense analogous to those which can be carried out in the S-matrix framework (ii) by exploiting the unitarity integral relations. In (ii) however, no primitive global analytic structure based on physical principles has been proposed up to now (except the one which was based "a priori" on requirements of "maximal analyticity"[11]) so that the monodromic structure of the scattering amplitudes can only be studied there in the framework of microlocal analysis, the latter having being justified on physical grounds by the macrocausality assumptions and through the method of local Fourier transformation[6].

The integral relations of M.P.S.A. through which the A.C. axiom can be efficiently exploited are exact counterparts of certain identities which relate appropriate formal series of Feynman amplitudes in the framework of perturbation theory. The various terms of these integral relations can be represented by graphs which obey similar integration prescriptions to those of Feynman amplitudes, but whose vertices are associated with "general n-point functions" instead of constants or monomials.

The "general n-point functions" which occur in these relations are not only Green's functions $H^{(n)}$ themselves, but other quantities which are called "ℓ-particle irreducible n-point functions" ($\ell=1,2,\ldots,\infty$) and which have an intuitive interpretation in perturbation theory[1]. These functions, whose definition involve a recursive algorithm starting from the lowest values of the integer ℓ, enjoy the following important property : for increasing values of ℓ they are analytic in increasingly larger domains of $\mathbb{C}^{4(n-1)}$ which all contain the primitive domain of $H^{(n)}$; in the axiomatic M.P.S.A. approach which we consider, these analyticity properties of the ℓ-particle irreducible functions are expected to be obtained as consequences of the A.C. axiom which supplement the primitive analyticity properties entailed by (i)*. At the present stage of our program these properties have only been established for $\ell=1,2,3$ but some steps have also been taken in the treatment of the general case.

From considering the integral relations of M.P.S.A., it emerges that the global analytic and monodromic structure of Green's functions should be completely governed by the two following operations :

----

*Here, it is interesting to notice that in some tractable models of Quantum Fields which have been studied in the constructivist approach ($P(\varphi)_2$ at small coupling), the required analyticity properties of ℓ-particle irreducible functions can be checked directly [12][13].

(I) Integration of analytic functions over prescribed classes of cycles in definite complex domains.

(II) Fredholm inversion of analytic kernels depending analytically on parameters over prescribed classes of cycles in definite complex domains.

In the recursive process of M.P.S.A., one can indeed expect that to each Green function $H^{(n)}$ and integer ℓ, there corresponds a definite domain $D_n^{(\ell)}$ in complex four-momentum space $\mathbb{C}^{4(n-1)}$ in which the above operations (I) and (II) should determine exhaustively the Landau singularities and the associated monodromic structure of $H^{(n)}$ implied by the completeness of the $1,2,\ldots,\ell-1$, ℓ-particle asymptotic states of the theory (in brief : the "$1,2,\ldots,\ell-1$, ℓ-particle structure"). The domains $D_n^{(\ell)}$ should be determined by the primitive analyticity domains of the corresponding ℓ-particle irreducible n-point functions, ; their size, governed by the production threshold of $(\ell+1)$-particle states should increase with ℓ, and to the extent that in the limit $\ell \to \infty$, $D_n^{(\ell)}$ fills up the whole complex four-momentum space $\mathbb{C}^{4(n-1)}$, the M.P.S.A. program may have - in principle - the ambition of characterizing at a global level in $\mathbb{C}^{4(n-1)}$ the singularities and the organization of the complete monodromic structure of Green's functions.

At the present stage, the obtained results (for $\ell=1,2,3$) are indications in favour of this wishful description.

Before coming to a brief account of some results which have been recently obtained for the two and three-particle structure of the six-point Green function $H^{(6)}$ and of the three-particle scattering amplitude, we shall make a few comments on the way in which the mathematical operations (I) and (II) are acting in the M.P.S.A. approach.

(I) Integration in complex space : $t=(t_1,\ldots,t_q) \in \mathbb{C}^q$, $z=(z_1,\ldots,z_n) \in \mathbb{C}^n$.

$$I(t) = \int_{\Gamma_t} f(t,z)\, dz_1 \wedge \ldots \wedge dz_n \quad , \tag{1}$$

f is analytic in a domain $D = \bigcup_{t\in\Delta} (t,D_t)$, and $\Gamma_t \subset D_t$ is an n-dimensional cycle which has to "vary continuously" with t(see [8]).

A typical example which illustrates the relevance of this operation to M.P.S.A. is the "triangle term" which contributes to the r.h.s. of the structural equation (9) (given below) for $H^{(6)}$, namely :

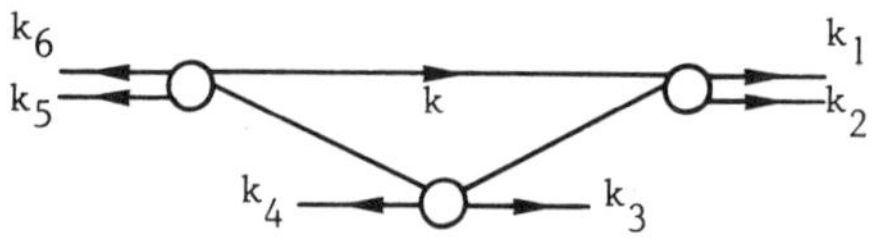

in the latter the bubble vertices represent the Green function* $H^{(4)}$, and the associated contribution to the six-point function is the integral :

---

* More precisely, a suitably modified form of $H^{(4)}(k_1k_2k_3k_4)$ which is regularized at infinity and "amputated" from its two initial poles $k_1^2 = m^2$, $k_2^2 = m^2$ (see [2],[3]).

$$H^{(6)}$$ 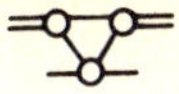 $$(k_1,\ldots,k_6)=\int_{\Gamma_{(k_1\ldots k_6)}} H^{(4)}(k_5,k_6,k,-k-k_5-k_6)H^{(4)}(k_1,k_2,-k,k-k_1-k_2)\times \ldots$$

$$\ldots \times H^{(4)}(k_3,k_4,k_5+k_6+k,k_1+k_2-k)\ d_4k \qquad (2)$$

$\Gamma_{(k_1\ldots k_6)}$ being a well-defined class of four-dimensional cycles.

At the preliminary step of M.P.S.A. - at which $H^{(4)}$ is only known to be analytic in its primitive domain D_4 - the integral (2) can be shown to define a function which is analytic in the primitive domain D_6 : this is in fact the content of the "G -convolution theorem" proved in [14], in the case when G is the triangle graph, here considered.

At the step $\ell=2$ (namely, the exploitation of the two-particle structure) $H^{(4)}$ can be shown to be analytic in a larger domain which is a two-sheeted Riemann surface with branching singularity $s=(k_1+k_2)^2=(2m)^2$ (m being a fundamental mass parameter in the theory). Correspondingly, the integral (2) can be shown to have an analytic continuation on a certain Riemann surface which contains the so-called "three-particle elastic scattering region on the mass shell" (defined by $k_i^2=m^2$, $i=1,\ldots,6$; $(3m)^2 \leq (k_1+k_2+k_3)^2 < 4m^2$). This Riemann surface has common features with that of the Feynman amplitude corresponding to the Feynman graph  , but it is confined over a certain definite domain : this restriction expresses the fact that *only* the two-particle structure of the theory has been taken into account at this stage. As in the perturbative case, the Landau singularities of $H^{(6)}$  are produced by the pinches of the integration cycle $\Gamma_{(k_1\ldots k_6)}$ by certain "vanishing cells"[8][15][17] of the singular set S of the integrand; here S is the union of the three polar manifolds $k^2=m^2$, $(k_1+k_2-k)^2=m^2$, $(k_5+k_6+k)^2=m^2$, and of the branching manifold $(k_1+k_2+k_3-k)^2=(2m)^2$ (notice that the latter is absent for the corresponding Feynman integrand).

(II) <u>Fredholm inversion in complex space or "Bethe-Salpeter inversion"</u> :

One is interested in the analyticity properties (in(t,z,z',λ)) of the Fredholm resolvent F of a given analytic kernel G, satisfying the equation :

$$\forall\lambda\in\mathbb{C},\ F(t,z,z',\lambda) = G(t,z,z')+\lambda\int_{\Gamma_t} F(t,z,z'',\lambda)G(t,z'',z')\ dz''_1\wedge\ldots\wedge dz''_n \quad ; \qquad (3)$$

G is analytic in a domain $D=\bigcup_{t\in\Delta}(t,D_t\times D'_t)$ and $\Gamma_t\subset D_t\cap D'_t$ is an n-dimensional cycle which has to vary continuously with t (see [16] for a systematic study of the analytic and monodromic structure of F, under suitable simple assumptions).

A typical example which illustrates the relevance of this operation to M.P.S.A. is the Bethe-Salpeter equation which expresses $H^{(4)}$ as the Fredholm resolvent of the two-particle irreducible four-point function $L^{(4)}$:

$$H^{(4)}(K,z,z') = L^{(4)}(K,z,z')+\frac{1}{2}\int_{\Gamma_K} H^{(4)}(K,z,z'')\ L^{(4)}(K,z'',z')dz'' \qquad (4)$$

or graphically :

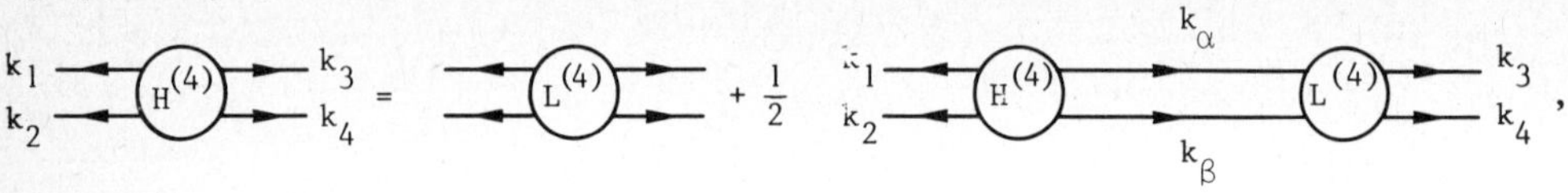

where $K=k_3+k_4 = -(k_1+k_2)$, $z = \frac{k_2-k_1}{2}$, $z' = \frac{k_3-k_4}{2}$, $z'' = \frac{k_\alpha - k_\beta}{2}$, and Γ_K is defined as the "euclidean subspace"* $(i\,\mathbb{R})\times \mathbb{R}^3$.

Equation (4) is actually the starting point of the step $\ell=2$** of M.P.S.A[2],since it accounts for the generation of the two-sheeted Riemann surface of $H^{(4)}$ (quoted in the previous paragraph), once the primitive "schlicht" analyticity domain of $L^{(4)}$ is given. The latter includes the two-particle elastic scattering region on the mass shell ($k_i^2 = m^2$, $i = 1,\dots,4$, $(2m)^2 \leq (k_1+k_2)^2 < (4m)^2$), and the branching singularity $s=K^2=(k_1+k_2)^2 = (2m)^2$ of $H^{(4)}$ is produced by the pinch of the integration cycle Γ_K by the vanishing cell of the polar set S of $L^{(4)}$, namely :

$$S = \{(K,z'') \; ; \; (\tfrac{K}{2}+z'')^2 = m^2\} \cup \{(K,z'');(\tfrac{K}{2}-z'')^2 = m^2\} \quad .$$

Let us point out that if the primitive ("schlicht") domain of $H^{(4)}$ is distinguished (as usual) as the "physical sheet" of $H^{(4)}$, the second sheet or "unphysical sheet" is a domain of *meromorphy* for $H^{(4)}$, because the solution of (4) can be written in the whole domain of $H^{(4)}$:

$$H^{(4)}(K,z,z') \quad = \quad \frac{N^{(4)}(K,z,z')}{D^{(2)}(K)} \quad ,$$

where $N^{(4)}$ is a general four-point function and where $D^{(2)}$ is a general two-point function [2] which may have zeros in the considered second-sheet domain : such poles of $H^{(4)}$ have the physical interpretation of new particles which are either bound states or unstable particles according to whether their position lies on the real axis of the s-plane or not.

At a further step,Bethe-Salpeter equations of the form

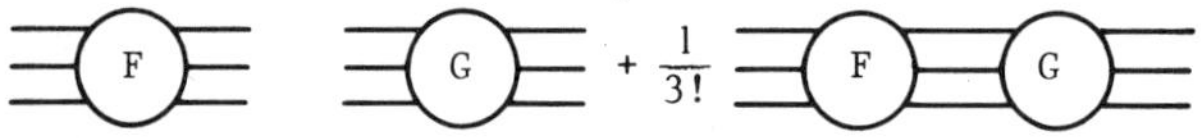

(5)

have to be considered, and for such equations of type (3) new difficulties arise as far as the analysis of the Landau singularities of F are concerned. These difficulties are linked with the occurrence of singular sets S for G involving "non-separable components" with respect to z and z' - variables (for instance : poles of the form $1/(z_i-z'_j)^2-m^2)$. The latter generate infinite sequences of Landau varieties whose elements are singularities for the successive terms of the Neumann series of equation (3).

----

* Here, a Lorentz frame has to be chosen : $K = (K^{(o)},\vec{K})$, and the Minkowski metric $K^2=K^{(o)2}-\vec{K}^2$ defines a euclidean metric in $(i\,\mathbb{R})\times\mathbb{R}^3$.

** Equation (4), which we quote here, actually corresponds to the case of even quantum field theories : in the latter the step $\ell=1$ (... ℓ odd) does not exist for $H^{(4)}$.

In the physical case of equations of the type (5), such singularities accumulate to the threshold singularity $(k_1+k_2+k_3)^2 = (3m)^2$, and a careful analysis of the various sheets of the resolvent has to be done. The results for $H^{(6)}$ which we shall now state are valid in a domain which is bounded by the cut : $s=(k_1+k_2+k_3)^2=(3m)^2+\rho$ $(\rho \geq 0)$; in other words the monodromic structure around the three-particle threshold has not yet been explored there. In order to describe properly these results, we first need to specify some notations.

We consider the case of even Quantum Field Theories with a single fundamental mass m.

We put $H^{(6)}(k_1,\ldots,k_6) = H^{(6)}(K,z,z')$, where $K=k_1+k_2+k_3=-(k_4+k_5+k_6)$, $z=(\underline{k}_4,\underline{k}_5,\underline{k}_6)$ with : $\underline{k}_i=k_i+\frac{K}{3}$, i=4,5,6 ; $z'=(\underline{k}_1,\underline{k}_2,\underline{k}_3)$ with $\underline{k}_i=k_i-\frac{K}{3}$, i=1,2,3. K is kept in the manifold $K=(K^{(o)},\vec{0})$, and $K^{(o)}$ varies in a domain $\Delta_M^{cut}=\Delta_M-\{K^o\geq 3m\}$,

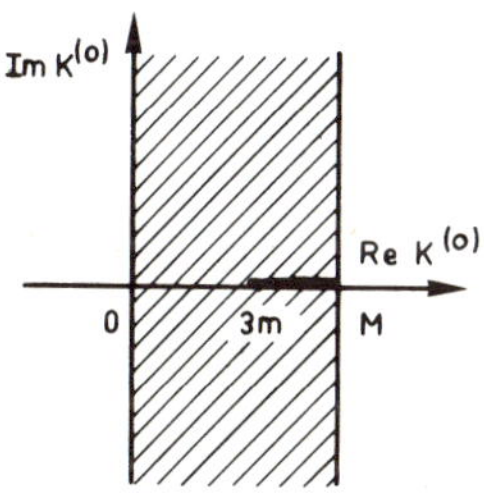

Fig.1 : The domain Δ_M^{cut}

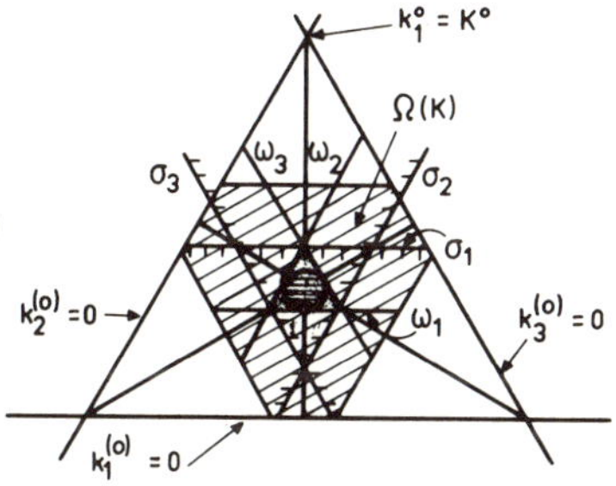

Fig.2a. $K^{(o)} > 3m$

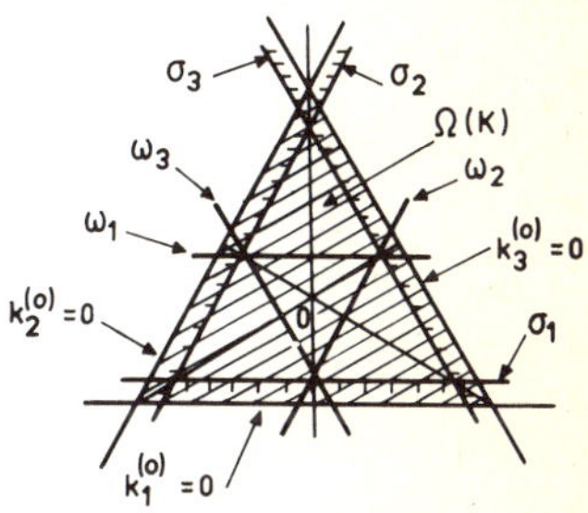

Fig.2b. $K^{(o)} < 3m$

where $\Delta_M = \{K^{(o)};0 < \text{Re}\,K^o < M$, with $3m < M \leq 5m\}$ (Fig.1). Fig.2 represents the situation in Re $z'^{(o)}$- plane ; ω_h is the polar manifold $k_h^2 = m^2$, σ_h is the branch manifold $s_{ij}=(k_i+k_j)^2 = 4m^2$ $(\{i,j,h\}=\{1,2,3\})$; the dark disk at the center of fig.2a) represents in projection the physical region of the reaction in which the incoming (resp. outgoing) particles are labelled by the set {4,5,6} (resp.{1,2,3}).

For every $K^{(o)}$ in Δ_M, one defines in z-space a domain D_K of the following form :

$$D_K = \{z=(z^{(o)},\vec{z})\ ;\ \text{Im}\,z^{(o)}\in\mathbb{R}^2,\ \text{Re}\,z^{(o)}\in\Omega(K),\ \text{Re}\,\vec{z}\in\mathbb{R}^6; |\text{Im}\vec{z}| < \varepsilon\ (\text{Re}\,z^{(o)})\}$$

where the domain $\Omega(K)$ is represented in light hatchings on Fig.2.

The following domains $\mathcal{D}_M, \mathcal{D}_M^{cut}$ are used in the results below :

$$\mathcal{D}_M(\text{resp.}\,\mathcal{D}_M^{cut}) = \{(K^{(o)},z,z');K^{(o)}\in\Delta_M(\text{resp.}\ \Delta_M^{cut})\ ;\ z\in D_K;\ z'\in D_K\}\ .$$

One-and-two particle dressing of $H^{(6)}$:

The aim of this step is to express* $H^{(6)}$ in terms of a suitable irreducible part $G^{(6)}$ and of auxiliary kernels which only involve $H^{(4)}$ and $L^{(4)}$. We first define the

----

*Actually, $H^{(6)}$, $H^{(4)}$, $H^{(2)}$ are replaced here by the corresponding regularized and amputated forms (see one of our previous footnotes).

latter, namely :

T_{in} is the tree-graph product $\sum_{h,n}$ 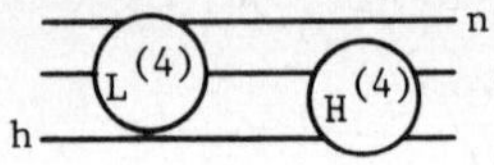 and similarly

$T_{out} = \sum_{hn}$ 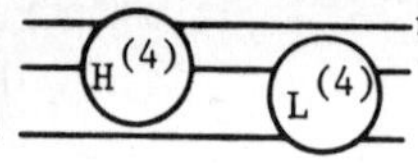 ; they admit B.S. inverses (in the sense of Eq.(5)) which we call respectively U_{in} and U_{out}. We also introduce the operators

$\Lambda = \mathbb{1} - \frac{1}{6} \sum_{hn}$ 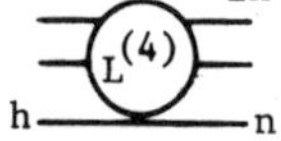 and $V = \mathbb{1} + \frac{1}{6} \sum_{hn}$ 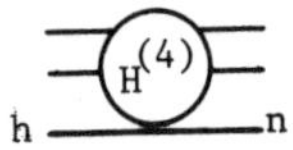 which satisfy :

$$\Lambda\, V = \mathbb{1} - \frac{1}{6} T_{in},\quad V\,\Lambda = \mathbb{1} - \frac{1}{6} T_{out}.$$

The B.S. inversion of T_{in}, T_{out} allows us to define the inverse of Λ as :

$$\Lambda^{-1} = V.(1 + \frac{1}{6} U_{in}) = (1 + \frac{1}{6} U_{out}).V.$$

We now introduce the six-point function $G^{(6)}$ through the equation :

$$G^{(6)} = \Lambda H_1^{(6)} \Lambda - \Lambda T_{out} \tag{6}$$

(note that $\Lambda T_{out} = T_{in} \Lambda$) ,where

$$H_1^{(6)} = H^{(6)} - \quad \text{[diagram]} \quad . \tag{7}$$

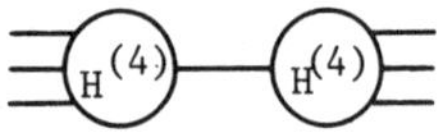

Suitable one and two-particle irreducibility properties can be proved for $G^{(6)}$, which imply in particular the following analyticity property of $\hat{G}^{(6)} = \prod_{i=1}^{6} (k_i^2 - m^2) G^{(6)}$.

<u>*Lemma*</u> : $\hat{G}^{(6)}$ *is analytic in a domain* $\mathcal{D}_M^{cut}$, *such that* $M = 11m/3$.

Now the equation (6) which defines $G^{(6)}$ can be inverted and yields the following "two-particle dressing equation"

$$H_1^{(6)} = (1 + \frac{1}{6} U_{out})\; V\, G^{(6)}\; V(1 + \frac{1}{6} U_{in}) + U_{out}\; V \quad . \tag{8}$$

By using the analyticity properties of $H^{(4)}$ and $G^{(6)}$, and by exploiting in complex space some standard combinatorics of Fredholm theory, one can prove

<u>*Theorem*</u> = *In* $\mathcal{D}_M^{cut}$ *(with* $M = 11m/3$*) the following decomposition of* $\hat{H}^{(6)}$ *holds* :

$$\hat{H}_1^{(6)} = \text{[diagram]} + \sum_{h,n} \text{[diagram]} + \sum_{h,n} \text{[diagram]} + \ldots \tag{9}$$

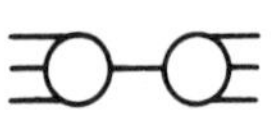 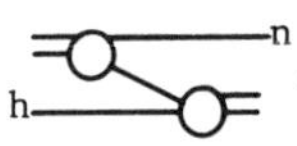 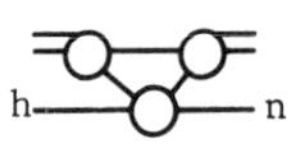

$$\ldots + \sum_{h,n} \text{[diagram]} + \sum_{h,n} \psi_{hn}$$

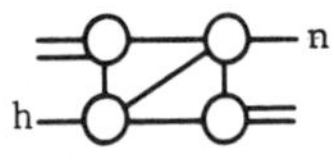

in the latter, all the "bubbles" of the G-convolution products represent $H^{(4)}$ *and each analytic function* ψ_{hn} *of the residual sum* ($h \in \{4,5,6\} n \in \{1,2,3\}$ *has the following properties* :

i) ψ_{hn} *is analytic in a "first-sheet domain" which is* $\mathcal{D}_M^{(cut)} - (\bar{\sigma}_h \cup \bar{\sigma}_n)$; *here the sets* $\bar{\sigma}_h$, $\bar{\sigma}_n$ *denote the following "cuts"* :

$\bar{\sigma}_h = \{(K,z,z'); s_{ij}(K,z) = 4m^2+\rho; \rho \geq 0\}$ *and* $\bar{\sigma}_n = \{(K,z,z'); s_{\ell m}(K,z') = 4m^2+\rho; \rho \geq 0\}$.

ii) ψ_{hn} *admits a* <u>*local*</u> *analytic continuation across the sets* $\overline{\sigma}_h$ *and* $\overline{\sigma}_n$ *on the Riemann surface associated with the Feynmann graph* 

Besides each G-convolution term of this decomposition is analytic in $\mathcal{D}_M^{cut}$ *minus the Landau singular set associated with the corresponding graph G ; it moreover admits a local analytic continuation on the Riemann surface associated with G.*

<u>*Remark*</u> : The only singularities of $\hat{H}^{(6)} = \prod_{i=1}^{6} (k_i^2 - m^2)\ H^{(6)}$ which are produced in $\mathcal{D}_M^{cut} - \bigcup_{h,n} (\overline{\sigma}_h \cup \overline{\sigma}_n)$, considered as a first sheet defined by the extension of the primitive analyticity domain, are the Landau singularities of the 0-loop, 1-loop and two-loop "truss-bridge graphs" ; the ℓ-loop graphs ($\ell > 2$)  are potentially present in the Neumann expansions of U_{in}, U_{out} and illustrate the two-particle dressing of $H_1^{(6)}$ from both sides of $G^{(6)}$ in formula (8) ; however, it is a consequence of the previous theorem that their Landau singularities are *effective only in other sheets.*

By taking the restriction of $\hat{H}^{(6)}$ to the complex mass-shell $M_6^{(c)}$, one obtains the following corollary of the previous theorem :

<u>*Corollary*</u> : *In* $\mathcal{D}_M^{cut} \cap M_6^{(c)}$, *the* $3 \to 3$ *particle scattering amplitude admits the following decomposition as a sum of analytic functions :*

$S_{IJ}^{(6)} = \sum_{h,n}$ 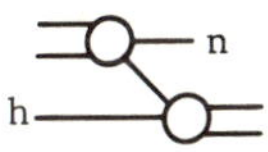 $+ \sum_{h,n}$ 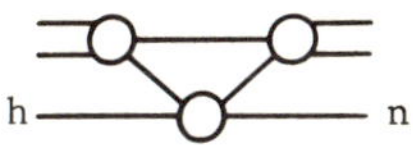 $+ \sum_{h,n} \hat{\psi}_{hn}$,

where each function $\hat{\psi}_{hn}$ *is analytic in* $\mathcal{D}_M^{cut} \cap M_6^{(c)} - (\overline{\sigma}_h \cup \overline{\sigma}_n)$; *across each cut* $\overline{\sigma}_h, \sigma_n$ $\hat{\psi}_{hn}$ *admits a local analytic continuation which spreads in a two-sheeted Riemann surface around each threshold* σ_h, σ_n.

As far as the *three-particle structure of* $H^{(6)}$ is concerned, we can only say here that :

(i) a three-particle irreducible kernel $L^{(6)}$ has been introduced on the basis of three-particle asymptotic completeness (see [3b],[17]).

(ii) By using a Bethe-Salpeter equation which relates $G^{(6)}$ to $L^{(6)}$, it is then possible to show that $G^{(6)}$, and $H^{(6)}$ can be analytically continued at least locally across the cut $s=(3m)^2+\rho$ ($\rho > 0$). The global study of this monodromic structure remains to be done.

<u>*Remark*</u> : This result contains as a subproduct the local analyticity properties which are sufficient to ensure macrocausality conditions* for the $3 \to 3$ scattering process in the corresponding low-energy region.

---

*in the sense of Stapp-Iagolnitzer's S.matrix theory [6].

REFERENCES

[1] K. SYMANZIK, J. Math. Phys. 1, 249 (1960) and in "Symposium on Theoretical Physics", 3, 121, New York, Plenum Press (1967)

[2] J. BROS in "Analytic Methods in Math. Physics", p.85, Gordon and Breach, New York (1970)
J. BROS, M. LASSALLE, Comm. Math. Phys. 54, 33 (1977)

[3]a) J. BROS, M. LASSALLE, in "Structural Analysis of Collision Amplitudes" p.97 North Holland, Amsterdam (1976)

b) J. BROS, in A.M.P. Conference, Lausanne, August 1979

[4] R. OMNES in "Relations de Dispersion et Particules Elémentaires", Hermann, Paris (1960), and references quoted therein

[5] H. EPSTEIN in "Axiomatic Field Theory", vol.1, p3-133, Gordon and Breach, New York (1965) and references quoted therein

[6] D. IAGOLNITZER, The S-Matrix, North Holland, Amsterdam (1978) and references quoted therein ; and D. Iagolnitzer in this volume.

[7] L.D. LANDAU, Nuclear Physics, 13, 181 (1959) ;
R.J. EDEN, P.V. LANDSHOFF, D.I. OLIVE, J.C. POLKINGHORNE, "The analytic S-matrix", Cambridge Univ. Press (1966) and references quoted therein

[8] D. FOTIADI, M. FROISSART, J. LASCOUX, F. PHAM, Topology (Gr. Brit) 4, 159 (1965); F. PHAM : "Introduction à l'étude topologique des singularités de Landau", Gauthier-Villars, Mémorial des Sc. Math. 164 (1967)

[9] G. PONZANO, T. REGGE, E.R. SPEER, M.J. WESTWATER, Comm. Math. Phys. 15, 83 (1969)

[10] M. KASHIWARA, T. KAWAI, T. OSHIMA, Comm. Math. Phys. 60, 97 (1978) and references quoted therein.
A. VAN DEN ESSEN, in this volume

[11] G.F. CHEW, The Analytic S-matrix, W.A. Benjamin, New York (1966) and references quoted therein

[12] T. SPENCER, F. ZIRILLI, Comm. Math. Phys. 49, 1 (1976)

[13] M. COMBESCURE, F. DUNLOP, "n-particle irreducible functions in euclidean Q.F.T." Preprint IHES, Bures-sur-Yvette (1978)

H. KOCH, Thesis, Genève Univ. (1979)

[14] M. LASSALLE, Comm. Math. Phys. 34, 185 (1974)

[15] J. LERAY, Bull. Soc. Math. de France 87, 81 (1959)

[16] J. BROS, D. PESENTI, "Fredholm theory in complex manifolds ..." to be published in Journal de Math. Pures et Appliquées (Preprint Orsay University, 1979)

[17] J. BROS, "Cours de 3ème cycle", Lausanne University, May 1979

MICROCAUSALITY, MACROCAUSALITY AND THE PHYSICAL REGION (MICRO)ANALYTIC S-MATRIX

D. IAGOLNITZER

DPh-T, CEN Saclay, BP 2, 91190 Gif-sur-Yvette, France

ABSTRACT

Recent works on the physical region analytic structure of multiparticle collision amplitudes in relativistic quantum theory are presented. First, the structure that can be expected and which is the expression, in terms of general essential support or microanalyticity properties, of macrocausality and macrocausal factorization, is described. It is shown that, taken together, these properties are equivalent to decompositions of the S-matrix, in bounded parts of the physical region, in terms of generalized Feynman integrals. Derivations of this structure obtained recently for $3 \to 3$ processes below the four-particle threshold both in S-matrix theory (without recourse to the crucial ad hoc assumption of "separation of singularities" in unitarity equations used previously) and in axiomatic field theory are then reviewed. It is finally explained how this structure applies in two-dimensional space-time and yields factorization of the multiparticle S-matrix itself for a class of models.

CONTENTS

1. INTRODUCTION

The purpose of these lectures is to present recent results on the derivation, in relativistic quantum theory, of the physical-region analytic structure of multiparticle collision amplitudes between sets of massive particles with short-range interactions. The unitary collision operator S, usually called the S matrix, and collision amplitudes are introduced in Sect.2 (see Ch.I of [1] for details).

Analyticity was recognized long ago as an important notion in particle physics[2], particularly in the study of strong (= short-range) interactions, and important progress was made in the sixties in the understanding and derivation of general momentum space analyticity properties, on the complex mass-shell, for two-body processes (two initial and two final particles). In the multiparticle case, however, progress has been made mainly in the study of analyticity properties in the neighborhood of the real *physical regions*. This study has appeared to have its own physical and mathematical interest and to be possibly a starting point for further developments.

A satisfactory understanding of the structure that can be expected in the physical region, and which is in fact equivalent to physical properties of *macrocausality* and *macrocausal factorization* (= asymptotic properties of transition amplitudes between appropriate sequences of initial and final wave functions, corresponding respectively to *non causal* and *causal* configurations of displaced particles or subgroups of particles), was first achieved at the end of the sixties[3,4] and in a more precise and complete way in the seventies (see Ch.II of [1]). The above properties are stated mathematically in the form of *essential support* properties which, by the general results of essential support theory (see App.1), are *equivalent*, for each given physical process, to physical-region analyticity, or more generally microanalyticity, properties of the scattering function f of the process considered (= its momentum-space connected collision amplitude, after factorization of its energy-momentum conservation δ^4-function : see Sect.1). Away from certain $+\alpha$-Landau points, macrocausality is in particular equivalent to decompositions of f, in bounded parts of the physical region, as a sum of contributions f_β associated with the various $+\alpha$-Landau surfaces $L^+(G_\beta)$ of connected graphs G_β encountered, each f_β being analytic outside $L^+(G_\beta)$ and being along $L^+(G_\beta)$ the boundary value of an analytic function from the "plus $i\varepsilon$" directions dual, at each point P of $L^+(G_\beta)$, to the causal directions $\hat{u}_+(P;G_\beta)$. Macrocausality by itself yields no further information on the structure of each f_β . Macrocausal factorization then yields (under some conditions), and is in fact equivalent in the regions considered to similar decompositions in which each f_β is a "generalized Feynman integral" associated with G_β . Equivalently, in usual cases (and for "simple" graphs : see Sect.3), the discontinuity of f_β around $L^+(G_\beta)$ is the integral, over internal on-mass-shell four-momenta, of the product of connected S-matrix kernels associated with each vertex of G_β .

These results are described in Sect.3, with emphasis on aspects relevant later or not treated earlier, and illustrated in Sect.4 in the case, of interest below, of the three-body (3→3) S-matrix below the 4-particle threshold. A brief introduction to the notion of essential support, microanalyticity and related results, which play a crucial role throughout these lectures, is given in App.1.

Many efforts have been devoted on the other hand to the derivation of the above physical-region structure from basic principles. However, as recalled below, the results obtained up to a few years ago either made use in S-matrix theory of a crucial, ad hoc assumption, or were in axiomatic field theory, very far from the desired ones. Progress has then been achieved recently in both theories for the three-body S matrix below the four-particle threshold. It is presented in Sect.5 and 6 respectively.

The various approaches proposed in S-matrix theory are based on the study of the possible singularities of the "bubble diagram functions" that occur in equations derived from unitarity and the decomposition of S into connected components, and of the way these singularities have to cancel among themselves. A first approach, proposed in the sixties[5], starts from the principle of "maximal analyticity" which, applied to the physical region, is interpreted as providing analyticity of individual scattering functions outside the singularities of phase-space integrals (= Landau surfaces, or in a more useful way modified Landau surfaces[6]), and the plus $i\varepsilon$ rule at $+\alpha$-Landau points. The aim was then to establish analyticity outside the $+\alpha$-parts of the Landau surfaces and discontinuity formulae around the $+\alpha$-Landau singularities. The "proofs" of [5] rely crucially,however, on some assertions that were unproved and , as recognized recently, cannot be expected to be correct in any reasonable framework (see Remark 4 in Sect.5).Results were obtained on the other hand at the end of the sixties[7] and in a more complete and satisfactory way in the seventies (see [8,9] and Ch.III of [1]) in the related approach that starts, besides unitarity, from macrocausality. In view of the results described in Sect.3, macrocausality can be considered as a precise and strengthtened version of the analyticity properties associated previously in the physical region with "maximal analyticity" . As recalled above, the results that one aims to derive are then equivalent to macrocausal factorization. However, while this latter property has a satisfactory physical interpretation, it already contains detailed information, and does not seem, in the quantum case, to be independent from macrocausality and unitarity. It is therefore desirable, to derive it, if possible, from these more general and basic physical principles.

Besides macrocausality and unitarity, the above results are based, however, in a crucial way (see Remark 4 in Sect.5) on an ad hoc assumption of "separation of singularities" in unitarity equations according to their underlying "topological structure" (see App.1). "Separation of singularities" is believed to hold, but it has no à priori basis and should thus itself be derived. Proofs that do not use this assumption have then been achieved partly in [10], and in a complete way in [11], for the three-body S-matrix below the four-particle threshold (and away from M_0 points such that some initial, or some final, four-momenta are colinear). Macrocausality provides (see above) decompositions of the three-body scattering function f, in the region considered, as a sum of "plus iε" boundary values of analytic functions associated with the surfaces $L^+(G_\beta)$ encountered (= here surfaces of graphs G_β with one internal line and triangle graphs). Possible pathologies of these boundary values are first eliminated by a "no sprout assumption", which slightly completes the analyticity properties derived for f from macrocausality(*). The use of unitarity equations then allows one, as explained in Sect.5, to show that they necessarily coincide with the generalized Feynman integrals introduced in Sect.3 ."Separation of singularities" is established when needed in the course of the derivation, and can then be derived in other cases (in the region considered) from the results obtained, following [9] and more generally App.B of Ch.III of [1].

In the course of the derivation a refined version of macrocausality (see the end of Sect.3.1), which provides the regularity property R (see mathematical lecture) for the connected S matrix is also used to solve[12] the crucial "u=0" problem (see Sect.5) encountered for some of the bubble diagram functions. For the alternative approach to the u=0 problem developed in the framework of the theory of holonomic microfunctions, see the lectures by Kashiwara-Kawaï.

The approach of axiomatic field theory presented here (Sect.6) is that of [13]. The alternative approach developed by J. Bros[14] and coworkers , with related results, is described in his lectures. As first recalled in Sect.6, *microcausality* and the *spectral condition* (on the spectrum of the energy-momentum operator) yield general information on the essential support of collision amplitudes in their physical region[15]. This information is, however, much weaker than that associated with macrocausality. We emphasize that the properties of microcausality, which applies to the *fields* on the microscopic space-time level, and of macrocausality, which is expressed in terms of *particles* at the macroscopic level, are of a different nature and that microcausality therefore does not imply by itself macrocausality. We shall then explain, following [13], how the essential support properties associated with macrocausality and macrocausal factorization do follow for the three-body S matrix below the four-particle threshold (and up to some limitations) from a further exploitation of the "off-shell" structure of the theory, together with the axiom of *asymptotic completeness* and regularity assumptions needed to avoid à priori possible à la Martin pathologies.

(*) It is a weak aspect of the idea of "maximal analyticity", or of holonomy, and can possibly be derived from the refined macrocausality condition mentioned below.

We conclude these lectures by outlining in Sect.7 how the general structure described in Sect.2 applies in *two-dimensional* space-time. Whereas macrocausality and macrocausal factorization are, as mentioned above, asymptotic properties of transition amplitudes between initial and final wave functions, for non causal or causal configurations of displaced particles, we shall see how they yield in this case[16,17] an important factorization property of the momentum-space multiparticle S matrix itself for a class of models.

Sect.2 to 6 are presented for simplicity for a theory with only one type of particle, a "boson" of mass $\mu > 0$, and in four-dimensional space-time. The statements and results extend, however, to more general theories, including different types of particles with strictly positive masses, and to other space-time dimensions, with only minor changes. The results of Sect.5 apply, however, only for space-time dimensions strictly larger than two. Needed modifications to Sect.2 and 3 in two-dimensional space-time, and for theories with several equal-mass particles, are outlined in Sect.7.

2. THE S-MATRIX

As announced in Sect.1 we consider below, for simplicity, a theory with only one type of particle, a "boson" of spin zero and mass $\mu > 0$. The Hilbert space of free-particle states is then defined as :

$$\mathcal{H} = \bigoplus_{m \geq 1} \mathcal{H}_m$$

where $\mathcal{H}_1 = L^2(M)$ is the space of "wave" functions φ of an energy-momentum variable $p = (p_o, \vec{p})$ (where p_o is the energy and $\vec{p} = \{p_\mu\}, \mu = 1,2,3$, is the momentum), which are defined on the positive-energy, real mass-shell manifold M $(p^2 = p_o^2 - \vec{p}^2 = \mu^2,\ p_o > 0)$ and are square integrable with respect to the invariant measure $d\mu(p) = \delta(p^2-\mu^2)\theta(p_o)\ d^4p \equiv \frac{d^3\vec{p}}{2\omega(\vec{p})}$, $\omega(\vec{p}) = (\vec{p}^2+\mu^2)^{1/2}$. $\mathcal{H}_m$ is the symmetrized tensorial product of m spaces $\mathcal{H}_1$.

Each physical system, which is considered here with all its evolution, is assumed to be characterized before and after interactions by well defined rays in $\mathcal{H}$ (= non zero vectors up to multiplication by a complex $\lambda \neq 0$). General principles (= conservation of probabilities) then ensure the existence of a unitary *collision operator* S from $\mathcal{H}$ to $\mathcal{H}$, usually called the S matrix, which, to each vector $|\varphi\rangle$ representing the state of a system before interactions associates the vector $S|\varphi\rangle$, which represents its state after interactions. For unit norm $|\varphi\rangle$ and $|\psi\rangle$, $|\langle\psi|S|\varphi\rangle|^2$ is the transition probability from $|\varphi\rangle$ to $|\psi\rangle$, i.e. the probability of detecting the final state of the system in the state $|\psi\rangle$, if its initial state is $|\varphi\rangle$. $\langle\psi|S|\varphi\rangle$ is called the transition amplitude from $|\varphi\rangle$ to $|\psi\rangle$. If $|\varphi\rangle \in \mathcal{H}_m$, $|\psi\rangle \in \mathcal{H}_n$, and if $|\varphi\rangle$ and $|\psi\rangle$ are both symmetrized tensorial products (over the values k=1 , ...,m and $k = m+1,\dots,m+n$) of wave functions φ_k associated with each initial and final particle, the transition amplitude $S(\{\varphi_k\}) = \langle\psi|S|\varphi\rangle$ can be written formally as :

$$S(\{\varphi_k\}) = \int S_{m,n}(p_1,\dots,p_m\,;\,p_{m+1},\dots,p_{m+n}) \times \prod_{i=1}^{m} \varphi_i(p_i) \prod_{j=m+1}^{m+n} \overline{\varphi}_j(p_j) \times \prod_{k=1}^{m+n} d\mu(p_k)\ , \qquad (1)$$

where $\overline{\varphi}$ is the complex conjugate of φ and $d\mu(p_k) = \delta(p_k^2-\mu^2)\theta((p_k)_o)\, d^4p_k$.

The kernel $S_{m,n}(p_1,\dots,p_{m+n})$ (which is symmetric with respect to the initial, and the final, variables) is called the momentum-space transition, or collision amplitude of the process $m \to n$. Since S is unitary, it is the kernel of a bounded operator $S_{m,n}$ from $\mathcal{H}_m$ to $\mathcal{H}_n$, and is thus in particular (by Schwartz's nuclear theorem) a well defined tempered distribution on the space $M_{m,n}$ of all real points $p=(p_1,\dots,p_{m+n})$ such that $p_k^2 = \mu^2$, $(p_k)_o > 0$, $\forall k$.

From general principles, the operator S is known to satisfy *energy-momentum conservation*. Correspondingly, one has (away from points p such that all p_k are colinear) :

$$S_{m,n}(p_1,\dots,p_{m+n}) = s_{m,n} \times \delta^4\left(\sum_1^m p_i - \sum_{m+1}^{m+n} p_j\right) \qquad (2)$$

where $s_{m,n}$ is a distribution defined on the *physical-region* $\mathcal{M}_{m,n}$ of the process $m\to n$ $(=\{p=(p_1\dots p_{m+n}),\ p_k^2=\mu^2, (p_k)_o > 0,\ \forall k,\ \Sigma p_i = \Sigma p_j\})$.

The *connected* parts of the S matrix, which will correspond physically to "interactions involving all particles together", are defined by induction through the formulae :

$$S_{m,n} = S^c_{m,n} + \sum_{K,N(K)>1} \left(\bigotimes_{K\in K} S^c_{m_K,n_K} \right)_K \qquad (3)$$

where the sum Σ runs over all partitions K of the sets of initial indices $1,\dots m$ and final indices $m+1,\dots,m+n$ into subsets (I_K,J_K), $K=1,\dots,N(K)$, m_K, n_K are the numbers of indices is I_K and J_K , and $(\bigotimes S^c_{m_K,n_K})_K$ is a tensorial product of operators $S^c_{m_K,n_K}$ associated with the partition K in the appropriate way. We recall that $S_{1,1} = S^c_{1,1} = 1_{1,1}$, $S_{1,n} \equiv 0$ if $n \neq 1$, $S_{m,1} \equiv 0$ if $m \neq 1$. In view of (3), $S^c_{m,n}$ is, like $S_{m,n}$ a bounded operator which satisfies also energy-momentum conservation and whose kernel can thus be written in the same form as (2) (away from points p such that all p_k are colinear) :

$$S^c_{m,n}(p_1,\dots,p_{m+n}) = f_{m,n} \times \delta^4(\Sigma p_i - \Sigma p_j) \ . \qquad (4)$$

The distribution $f_{m,n}$ is again defined on $\mathcal{M}_{m,n}$ and is usually called the *scattering function* of the process $m \to n$.

General arguments allow one to conclude that the connected S-matrix kernels, in contrast to the non connected ones, include no contribution containing δ-functions of energy-momentum conservation between subgroups of initial and final particles. This is, in particular, a consequence of the macrocausality condition introduced in Sect.3 (in the form of the factorization property for non causal configurations of subgroups of displaced particles), which will give, moreover, further information on the analytic structure of $f_{m,n}$.

3. THE MACROCAUSAL S-MATRIX

We first mention some geometrical facts that will be useful below. The variables dual to the energy-momentum variables are identified physically with space-time variables. Being given a point $p=(p_1,\ldots,p_{m+n})$ in $R^{4(m+n)}$ and a point $u=(u_1,\ldots,u_{m+n})$ in the dual space, where $u_k = ((u_k)_o, \vec{u}_k)$ is a point with time component $(u_k)_o$ and space component $\vec{u}_k$, the scalar product will be defined as

$$\langle u,p\rangle = \sum_k \varepsilon_k p_k \cdot u_k ,$$

where $\varepsilon_k = +1$ if $k=1,\ldots,m$, $\varepsilon_k=-1$ if $k=m+1,\ldots,m+n$, $p_k.u_k = (p_k)_o(u_k)_o-\vec{p}_k.\vec{u}_k$.

A point in the cotangent space $T_P^* M_{m,n}$, resp. $T_P^* \mathcal{M}_{m,n}$, at a point $P=\{P_k\}$ of $M_{m,n}$ is thus a point $u=(u_1,\ldots,u_{m+n})$ defined modulo addition for each k of $\lambda_k P_k$, where λ_k is any real scalar, resp. is a point u defined modulo addition for each k of $\lambda_k P_k + a$, where a is an arbitrary space-time vector independent of k. It can be characterized by the configuration of trajectories (= straight lines) (P_k,u_k) in space-time which, for each k, are parallel to P_k and pass through u_k, resp. by the relative configuration, defined modulo global space-time translations, of these trajectories.

3.1. Macrocausality[3,1]

It will be convenient, in this section and in Sect.3.2, to consider sequences of initial and final wave functions of the form :

$$\varphi_{k,\tau}(p_k) = \chi_k(p_k)\, e^{-\gamma\tau(\vec{p}_k-\vec{P}_k)^2}\, e^{-i\, p_k.\tau u_k} \tag{5}$$

where $\vec{P}_k$ is a given momentum, χ_k is analytic in the neighborhood of $P_k=((P_k)_o,\vec{P}_k)$, $(P_k)_o = (\vec{P}_k+\mu^2)^{1/2}$, u_k is a space-time vector, $\gamma > 0$ and τ is a positive scalar that will tend to infinity. The factor $e^{ip_k.\tau u_k}$ corresponds to space-time translation of particle k by τu_k . These wave functions correspond to particles that are asymptotically localized, in the $\tau\to\infty$ limit and in a space-time coordinate system scaled to τ (i.e. each point τx in space-time is replaced by x), along the classical trajectories (P_k,u_k), modulo well specified exponential fall-off properties.

Apart from exceptional cases (not encountered later) in which further limiting procedures[6,12] might have to be considered (see App.2) , macrocausality states the exponential fall off (in a well specified sense) in the $\tau\to\infty$ limit of transition probabilities between initial and final wave functions of the form (5), for *non causal* configurations of the initial and final trajectories, i.e. if there exists no classical multiple scattering diagram $\mathcal{D}_+(P,u)$ (with possibly[6,12] vertices at infinity(*)) whose external trajectories are those determined by (P_k,u_k) for each k. This statement corresponds physically to the above mentioned localization properties of the initial and final particles and to the short-range character of the interactions, which is interpreted as inducing exponential fall-off with distance (i.e. with τ in actual space-time) of all transfers of energy-momentum that cannot be attributed to actual stable particles. The absence of pathologies, such as the existence of sequences of unstable particles with arbitrarily small rates of exponential decay, by which energy-momentum might be transferred, is also assumed.

The above property yields trivially a similar exponential fall-off property of the transition amplitude $S(\{\varphi_{k,\tau}\})$, which can be stated, in terms of essential supports as :

$$u \notin ES_P(S) \text{ if there exists no } \mathcal{D}_+(P,u) \quad , \tag{6}$$

where u is defined here modulo addition of $\lambda_k P_k$ for each k. An analogous (and equivalent : see App.1) statement has been proposed independently as a "microanalyticity postulate" by Pham and Sato[18] in terms of the notion of singular spectrum.

A slightly stronger form of macrocausality, which will be that *needed* in Sect.5 (see Remark 3 in that section) says this : if the only possible diagrams $\mathcal{D}_+(P,u)$ are composed of several disconnected parts connecting together given subgroups K of initial and final particles, then the overall transition probability is equal, modulo a remainder to which the previous exponential fall-off properties apply, to the product of the partial transition probabilities associated with each subgroup. It can be shown[4] that the transition amplitude $S(\{\varphi_{k,\tau}\})$ then satisfies a similar factorization property, which can be stated in terms of essential supports as :

$$u \notin ES_P\,(S - \bigotimes_K S_K) \tag{7}$$

where $\bigotimes_K S_K$ is the (tensorial) product of S-matrix kernels associated with each subgroup of initial and final variables, and that (7) is itself equivalent, apart from exceptional situations(**), if they exist, not encountered later, to :

$$u \notin ES_P\,(S^c) \text{ if there exists no } \textit{connected}\ \mathcal{D}_+(P,u)\ . \tag{7'}$$

(*) Vertices "at infinity" in some direction are allowed when all incoming and outgoing trajectories at that vertex are parallel. These trajectories then need not coincide, but are required, following [6], to satisfy "angular-momentum conservation".

(**) Even if there is no connected $\mathcal{D}_+(P,u)$ (7') does not necessarily follow from (7) if there exist different disconnected diagrams $\mathcal{D}_+(P,u)$, $\mathcal{D}'_+(P,u)$,..., composed of parts that relate together different subgroups of initial and final particles : see [4].

In (7) and (7'), u is defined again modulo $\{\lambda_k P_k\}$. By the general results of essential support theory (or hyperfunction theory), (7') is in turn equivalent to :

$$E\, S_P(f) \subset C_+(P) \quad , \tag{8}$$

where $C_+(P)$ is the subset of $T^*_P M$ corresponding to all *relative* configurations of external trajectories of connected diagrams $\mathcal{D}_+(P)$ with external four-momenta P_k.

The set $C_+(P)$ is the union of the sets $C_+(P;G)$ corresponding to connected diagrams $\mathcal{D}_+(P;G)$ whose topological structure is a given multiple scattering connected graph G. The $+\alpha$-*Landau surface* $L^+(G)$ is the subset of the physical region M composed of points P such that $C_+(P;G)$ is non empty. It is known[19] that these surfaces have at most codimension one, and are not dense, in M. Away from M_o *points* (= points P such that some initial, or some final, P_k are colinear) and from points P at which several surfaces are *related*(*), (which belong in both cases to subsets of codimension > 1) the surfaces $L^+(G)$ of connected graphs are (if non empty) real analytic codimension one submanifolds, and $C_+(P;G)$ is correspondingly composed, at each P of $L^+(G)$, of a unique direction $\hat{u}_+(P;G)$. The set associated with a class of related surfaces at a non M_o point P is still a closed convex salient cone. (See example at the end of Sect.3.2).

If we leave aside the M_o points and the points P that lie on several non related $+\alpha$ - Landau surfaces, the set $C_+(P)$ is thus known to be either empty (at non $+\alpha$-Landau points), or to be a closed convex salient cone, which reduces in general to a unique direction $\hat{u}_+(P;G)$. The points that have been left out lie in subsets of M of codimension strictly larger than one. In view of the general results of essential support theory, (8) therefore ensures that f is analytic outside the $+\alpha$-Landau surfaces of connected graphs and is at almost all $+\alpha$-Landau points the boundary value of a well defined function $\underline{f}$, analytic in a domain of the complexified manifold $\underline{M}$ of M, from the *plus* $i\varepsilon$ directions dual at each point P of a surface $L^+(G)$ to $\hat{u}_+(P;G)$, or to the cone $C_+(P)$ if P lies on several related surfaces.

At non M_o points P that lie on several *non related* surfaces, $C_+(P)$ is no longer contained in general in a closed convex salient cone : see Sect.4 ; f can then be decomposed locally, as a *sum* of plus $i\varepsilon$ boundary values of analytic functions $\underline{f}_\beta$, each of which being associated with one of the surfaces, or classes of related surfaces involved at P. If the cones associated at P with each class of related surfaces

----

(*) Surfaces $L^+(G')$, $L^+(G'')$, ... are related at P if P lies on these surfaces or their closures, and also on the surface $L^+(G)$, or its closure, of a graph G such that G',G'',... are contractions of G. A contraction of G is either G or any graph obtained by removing certain internal lines and identifying the vertices where they were incoming and outgoing.

Note that graphs that differ only by trivial changes which do not change the surface (such as the inclusion of vertices on certain lines or sets of lines) are identified.

are *disjoint* (apart from the origin), each f_β is well defined locally modulo an analytic function. There are à priori ambiguities otherwise (see Examples 2 and 3 in Sect.4). If P is on the other hand an M_o point, $C_+(P)$ is not contained in a closed convex salient cone. There is no natural decomposition of $C_+(P)$ into a union of closed convex salient cones, and hence of f into a sum of boundary values of analytic functions.

Finally, the general decomposition theorems of essential support theory, or hyperfunction theory, also yield various decompositions of f, in bounded parts of the physical region, as a sum of contributions associated with the various $+\alpha$-Landau surfaces encountered. There are à priori a number of ambiguities that we shall not discuss here in general. See the example of Sect.4, where they are eliminated, in the region considered, by the "no sprout" assumption.

Recent studies about macrocausality have been made in connection with the problems encountered in Sect.5. On the one hand, the situation at M_o points has been analyzed[6,12] : see App.3. On the other hand, a refined version of macrocausality has been proposed[12]. It gives information on the way rates of exponential fall-off tend to zero, in certain situations, when causal directions (= directions in the essential support) are approached. It follows, in the cases that have been considered, essentially from the same physical arguments as the previous version of macrocausality, and provides the regularity property R (see mathematical lecture) for the connected S matrix. For conciseness, a further discussion is omitted here, the reader being referred to [12].

3.2. Macrocausal factorization

We now consider again wave functions of the form (5), but points (P,u) such that there may exist diagrams $\mathcal{D}_+(P,u)$. For simplicity, we shall restrict our attention to cases when these diagrams include not set of two or more internal trajectories outgoing and incoming at common vertices, and we shall consider here cases (which are the usual ones in a theory with only one type of particle - see Sect.7 for an extension) in which there is only one $\mathcal{D}_+(P,u)$, or such that all possible $\mathcal{D}'_+(P,u)$ can be obtained from a unique $\mathcal{D}_+(P,u)$ by replacing certain interactions at certain vertices of $\mathcal{D}_+$ by no interaction, or by several subinteractions : see example in Fig.1. In this example, $\mathcal{D}_+$ is such that (i) the external trajectories 9, 10 coincide with the internal trajectories incoming at c and (ii) $P_4+P_5=P_7+P_8$. The trajectories 9,10 in $\mathcal{D}'_+$ are issued respectively from b and a : although they meet at the space-time position of c, there is no vertex there. On the other hand, 4,5,7,8 have in $\mathcal{D}'_+$ a separate interaction at a vertex b' which is different from b, although b and b' have the same space-time position.

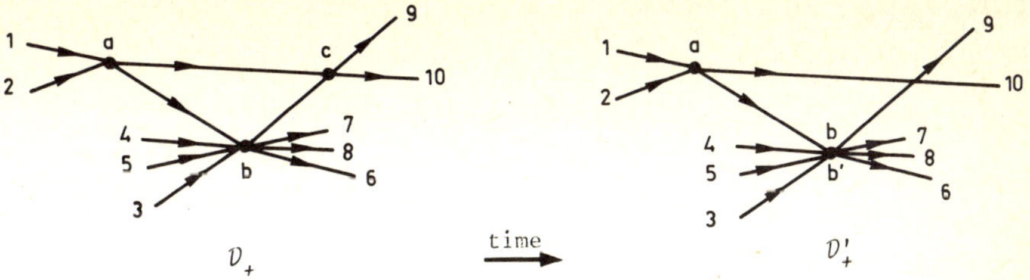

Fig. 1

Apart from exceptional cases (see App.2), *macrocausal factorization* is then the assertion that the transition amplitude $S(\{\varphi_{k,\tau}\})$ is asymptotically equal, modulo a remainder that falls off exponentially in the $\tau \to \infty$ limit (in a sense similar to that of Sect.3.1), to the term $D(\{\varphi_{k,\tau}\})$, where D is the integral, over all possible internal *on-mass-shell* four-momenta associated with the internal lines, of the product of S-matrix kernels associated with each vertex of the graph G that represents the topological structure of $\mathcal{D}_+(P,u)$. For instance, if G = 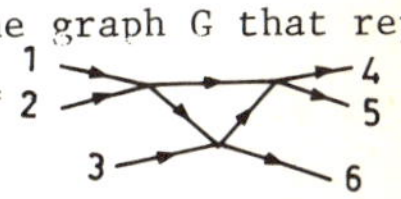
then

$$D(p_1,p_2,p_3;p_4,p_5,p_6) = \int S_{2,2}(p_1,p_2,k_1,k_2)\; S_{2,2}(k_2,p_3;k_3,p_6)\; S_{2,2}(k_1,k_3,p_4,p_5)$$

$$\prod_{\ell=1,2,3} \delta(k_\ell^2-\mu^2)\; \theta((k_\ell)_o)\; d^4k_\ell \tag{9}$$

or in a diagrammatical notation :

D = 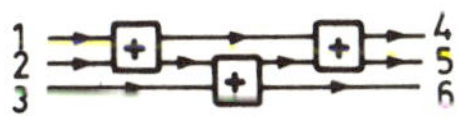 (9')

$D(\{\varphi_{k,\tau}\})$ is defined in the same way as $S(\{\varphi_{k,\tau}\})$ in Eq.(1). The terms D are always well defined as kernels of bounded operators (see[12]), hence as tempered distributions, and contain an overall energy-momentum conservation δ-function.

In terms of essential supports, macrocausal factorization is stated as :

$$u \notin ES_P\,(S - D) \quad . \tag{10}$$

where u is, as in (6) (7), defined modulo $\{\lambda_k P_k\}$.

In view of (7),(10) can be restricted to the case of *connected* diagrams $\mathcal{D}_+(P,u)$. It can then be checked in usual (not always trivial : see example in Ch.III of [1]) cases, such as all those encountered in Sect.4, that (10) then entails also a corresponding property for the *connected* S matrix S^c :

$$u \notin ES_P \left(S^c - \sum_\beta D^c_\beta\right) \tag{11}$$

where there is now one term D^c_β for each possible *connected* diagram $\mathcal{D}_+(P,u)$ or $\mathcal{D}'_+(P,u)$, and where D^c_β is defined in the same way as D, except that *connected* S-matrix kernels are now associated with each vertex. For instance, if we consider the simplified version of Fig.1, in which 4,5,7,8 are left out and 9,10 are relabelled 4,5, and if there is no other connected $\mathcal{D}'_+(P,u)$:

$u \notin ES_P \Big(S^c_{3,3}$ − 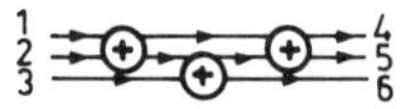 − 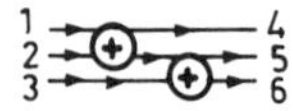 $\Big)$ (12)

where  stands for connected S-matrices $S^c_{2,2}$.

Each term D^c_β can be written, as D, (away from points p such that all p_k are colinear) in the form :

$$D^c_\beta(p_1,\ldots,p_m;p_{m+1},\ldots,p_{m+n}) = d^c_\beta \times \delta^4(\Sigma p_i - \Sigma p_j) \tag{13}$$

where d^c_β is defined on $M_{m,n}$; (11) is then equivalent to :

$$u \notin ES_P \left(f - \sum_\beta d^c_\beta\right) \tag{14}$$

where u is defined modulo $\lambda_k P_k + a$ for each k.

In order to study the implications of macrocausal factorization, we first introduce *generalized Feynman integrals* associated with graphs G_β .

Let us consider a codimension one surface $L^+(G_\beta)$ of a "simple" graph G_β(=graph with no set of more than one line between two vertices), which, in a given domain R_β of the physical region, is related to no other +α-surface. In usual cases, such as all those encountered in Sect.4, the structure theorem on bubble diagram functions (see App.1) allows one to show that the essential support of d^c_β in R_β is (contained in) the union of the closures of the sets $C_+(G_\beta)$ and $\widehat{C}(G_\beta)$ which correspond respectively to relative configurations of external trajectories of diagrams $\mathcal{D}_+(G_\beta)$ and of other diagrams $\mathcal{D}$, and which are disjoint (apart from the origin) at any P in R_β . The decomposition theorems of essential support theory (or hyperfunction theory) then provide a decomposition of d^c_β in R_β of the form :

$$d^c_\beta = (d^c_\beta)_+ + \widehat{d^c_\beta} \quad , \tag{15}$$

where the only possible directions in the essential supports of $(d^c_\beta)_+$ and $\widehat{d^c_\beta}$ are those of $C_+(G_\beta)$ and $\hat{C}(G_\beta)$ respectively, and where $(d^c_\beta)_+$ and $\widehat{d^c_\beta}$ are well defined modulo analytic backgrounds in R_β . The essential support of $(d^c_\beta)_+$ is empty (apart from the origin) outside $L^+(G_\beta)$ and has the unique direction $\hat{u}_+(P;G_\beta)$ at any point P of $L^+(G_\beta)$ in R_β ; $(d^c_\beta)_+$ is correspondingly analytic outside $L^+(G_\beta)$ and is along $L^+(G_\beta)$ the boundary value of an analytic function from the plus $i\varepsilon$ directions dual at each P to $\hat{u}_+(P;G_\beta)$.

The distribution $(d^c_\beta)_+$ thus defined can be considered as a physical-region generalization, modulo an analytic background, of the Feynman integral associated with G_β in R_β : see discussion in [11] and references therein. If $L^+(G_\beta)$ is not related, outside R_β , to other surfaces $L^+(G_{\beta'})$ encountered in R_β , then $(d^c_\beta)_+$ is, roughly speaking, the analogue of the Feynman integral associated with G_β , the constants at each vertex being replaced by off-shell analytic continuations of the scattering functions associated with these vertices. (Appropriate cut-off factors have also to be introduced in the integrand). This will be the case for instance in Sect.4. Then $(D^c_\beta)_+ = (d^c_\beta)_+ \times \delta^4(\Sigma p_i - \Sigma p_j)$ will be written diagrammatically in the same way as D^c_β , but with + signs above each internal line (see example in Sect.4). If $L^+(G_\beta)$ is related to some other surface $L^+(G_{\beta'})$ outside R_β , then $(D^c_\beta)_+$ as defined above corresponds to the replacement of the constants at some vertices by off-shell continuation of the scattering functions, from which relevant singularities have first been removed.(Otherwise $(D^c_\beta)_+$ would be singular also along $L^+(G_{\beta'})$ in R_β . One generalized Feynman integral can alternatively be used for a given class of related surfaces. In this connection, see the example given at the end of this section).

Remark : in usual cases, again such as those of Sect.4, the structure theorem (App.1) shows that $E\,S_P\,(d^c_\beta)$, at any point P belonging to $L^+(G_\beta)$ in R_β , is composed of $\hat{u}_+(P,G_\beta)$ and of the opposite direction $\hat{u}_-(P;G_\beta)$ (i.e. $\hat{C}$ reduces, if $P \in L^+(G_\beta)$, to $\hat{u}_-(P;G_\beta)$. It is then easily checked that d^c_β is in the neighborhood of P the discontinuity of $(d^c_\beta)_+$ around $L^+(G_\beta)$, i.e. the difference of $(d^c_\beta)_+$ and of the "minus $i\varepsilon$" boundary value d^c_β of its minus $i\varepsilon$ analytic continuation around $L^+(G_\beta)$ from the "non physical" side of $L^+(G_\beta)$ in R_β (where $d^c_\beta \equiv 0$) : see details e.g. in Ch.III of [1]. In the above mentioned case, $(d^c_\beta)_+$ can as a matter of fact be defined equivalently modulo an analytic background in R_β as the distribution which is analytic outside $L^+(G_\beta)$ and whose discontinuity around $L^+(G_\beta)$ is d^c_β -

In the neighborhood of a point P of $L^+(G_\beta)$ that lies on no other $+\alpha$-Landau surface or its closure, macrocausal factorization entails that :

$$f \simeq (d^c_\beta)_+ \quad , \tag{16}$$

where $\simeq$ means equality modulo an analytic background. In bounded domains R of the physical region in which all surfaces $L^+(G_\beta)$ encountered have the properties listed above (15), (14) entails more generally the following decomposition (see example in Sect.4) in R to which it is in fact equivalent :

$$f \simeq \sum_\beta (d^c_\beta)_+ \tag{17}$$

Proof : We prove below that (14) entails (17).(The converse is proved similarly). It is sufficient to prove that $f - \sum_\beta (d^c_\beta)_+$ is analytic at any point P in R , i.e. equivalently that $ES_P(f-\sum_\beta (d^c_\beta)_+)$ is empty (apart from the origin). Being given P, let us divide the set of surfaces $L^+(G_\beta)$ that contain P (if there are) into subsets for which the directions $\hat{u}_+(P;G_\beta)$ coincide. Let $\hat{u}^{(1)}_+$, $\hat{u}^{(2)}_+$,... be the directions thus obtained. If $\hat{u} \neq \hat{u}^{(1)}_+$, $\hat{u}^{(2)}_+$,... then $\hat{u} \notin ES_P(f)$ by macrocausality and $\hat{u} \notin ES_P((d^c_\beta)_+)$ (see above). Hence $\hat{u} \notin ES_P\ (f - \sum_\beta (d^c_\beta)_+)$. If $\hat{u}$ is for instance $\hat{u}^{(1)}_+$, then one may write :

$$f - \sum_\beta (d^c{}_\beta)_+ = (f - \sum_{(1)} d^c_\beta) + \sum_{(1)} \widehat{d^c_\beta} - \sum_{(2)} (d^c_\beta)_+ - \ldots$$

where the sums $\sum_{(1)}$, $\sum_{(2)}$, ... refer to each subset and where (15) has been used.

In view of (14), $\hat{u}^{(1)}_+ \notin ES_P\ (f - \sum_{(1)} d^c_\beta)$. On the other hand, in view of the essential support properties previously described of $\widehat{d^c_\beta}$ and $(d^c_\beta)_+$, $\hat{u}^{(1)}_+ \notin ES_P(\sum_{(1)} \widehat{d^c_\beta})$, $\hat{u}^{(1)}_+ \notin ES_P(\sum_{(2)} (d^c_\beta)_+)$, ... Hence $\hat{u}^{(1)}_+ \notin ES_P(f - \sum_\beta (d^c_\beta)_+)$ Q.E.D.

The extension of the definitions and results to regions that contain points P at which several surfaces are related on the one hand, and in which surfaces of graphs with sets of several lines between some vertices are encountered on the other hand, is possible in principle, although there are difficulties. There is , roughly speaking, one generalized Feynman integral, in a given domain R, for each class of related surfaces. We only give here a simple example.

Example : Let P be a point of the surfaces $L^+(G)$, $L^+(G_1)$, $L^+(G_2)$, where :

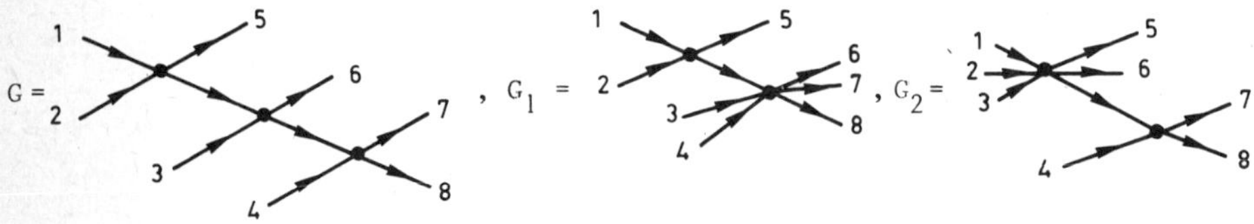

The surfaces $L^+(G_1)$, $L^+(G_2)$ are codimension one surfaces in $M_{4,4}$ defined by the equations $k_1^2 = \mu^2$, $(k_1)_o > 0$ and $k_2^2 = \mu^2$, $(k_2)_o > 0$ respectively, where $k_1 = p_1 + p_2 - p_5$, $k_2 = p_7 + p_8 - p_4$. The surface $L^+(G)$ is here the intersection of $L^+(G_1)$, $L^+(G_2)$. Being given a point $P \in L^+(G)$ that lies for simplicity on no surface (or its closure) other than $L^+(G_1)$, $L^+(G_2)$, the vertices a,b,c of any diagram $\mathcal{D}_+(P;G)$ satisfy $b-a = \lambda_1 K_1$, $c-b = \lambda_2 K_2, \lambda_1 > 0, \lambda_2 > 0$ and there is one direction $\hat{u}_+(P;G)$ for each ratio of λ_2/λ_1 . In the limits when λ_2/λ_1 tends to zero and to infinity, the directions $\hat{u}_+(P;G_1)$ and $\hat{u}_+(P;G_2)$ are obtained. $C_+(P)$ is the closed convex salient cone composed of all directions $\hat{u}_+(P;G)$ and of $\hat{u}_+(P;G_1)$, $\hat{u}_+(P;G_2)$.

Macrocausal factorization ensures that all directions $\hat{u}_+(P;G)$ are outside $ES_P(f-d^c)$ and that $\hat{u}_+(P;G_i) \notin ES_P\,(f-d_i^c)$, i=1,2... On the other hand $\hat{u} \notin ES_P(f)$ if $\hat{u} \notin C_+(P)$. These essential support properties determine f in a unique way in the neighborhood of P, modulo an analytic background, as the following generalized Feynman integral :

$$f \simeq \frac{a(p_1,\ldots,p_8)}{(k_1^2-\mu^2+i\varepsilon)(k_2^2-\mu^2+i\varepsilon)} \tag{18}$$

where a is any locally analytic function such that

$$a\Big|_{k_1^2=k_2^2=\mu^2} = f_{2,2}(p_1,p_2;p_5,k_1)f_{2,2}(k_1,p_3;k_2,p_6)f_{2,2}(k_2,p_4;p_7,p_8)\ .$$

(See e.g. App.C in Ch.III of [4]). In a diagrammatical notation :

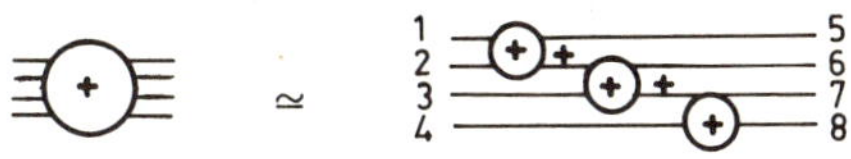

(18')

4. THE THREE-BODY S-MATRIX BELOW THE FOUR-PARTICLE THRESHOLD

In this section we consider a 3 → 3 process in the region

$$\mathcal{R} = \{p;p \in M_{3,3}\ ,\ (3\mu)^2 < s < (4\mu)^2\ ,\ \ p \notin M_o\} \tag{19}$$

where $s = (\sum_1^3 p_i)^2 = (\sum_4^6 p_j)^2$.

In $\mathcal{R}$, one encounters 18 surfaces $L^+(G_\beta)$ of connected graphs G_β, including 9 surfaces of graphs with one internal line, such as G_1 = 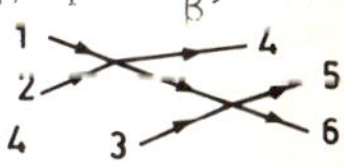

and 9 surfaces of triangle graphs, such as G_2 = 

. These surfaces are codimension one real analytic submanifolds of $\mathcal{R}$ and are not related at any point P in $\mathcal{R}$ or outside $\mathcal{R}$.

Macrocausality yields decompositions in $\mathcal{R}$ of the form :

$$f = \sum_\beta f_\beta \tag{20}$$

where each f_β is analytic outside $L^+(G_\beta)$ and is along $L^+(G_\beta)$ the boundary value of an analytic function $\underline{f}_\beta$ from the plus iε directions dual, at each point P of $L^+(G_\beta)$, to $\hat{u}_+(P;G_\beta)$.

We give below some examples of situations described in Sect.3.1

Example 1 : A simple example (see Ch.II of [1]) of points P where f is *not* locally the boundary value of a unique analytic function are all points P such that $P_1=P_6$, $P_2=P_4$, $P_3=P_5$. These points lie on six surfaces $L^+(G_\beta)$ of graphs G_β with one internal line, which can be grouped in pairs in a way such that the directions $\hat{u}_+(P;G_\beta)$ are *opposite* in each pair. This is the case for instance for the graphs 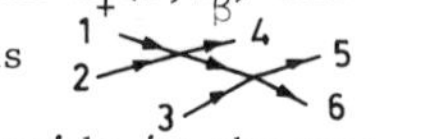

and 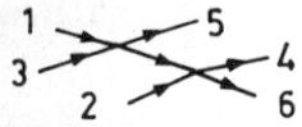

(The trajectories 1,6, as also 2,4 and 3,5 coincide in the two diagrams $\mathcal{D}_+$, the internal trajectory coinciding with 1,6).

Example 2 : An example of points P such that the directions $\hat{u}_+(P;G_\beta)$ now *coincide*, and where the contributions f_β to f in (20) are not à priori well defined, locally, modulo analytic backgrounds, is obtained[12] by considering the two graphs G_1,G_2 mentioned at the beginning of this section, and any point P of the (non empty) subset ω of $L^+(G_2)$ defined as follows. Being given any $P'=(P_1',\ldots,P_6')$ of $L^+(G_2)$ in $\mathcal{R}$, the internal on-mass-shell four-momenta of $\mathcal{D}_+(P';G_2)$ are well determined. Then all points P obtained from P' by replacing P_4', P_5' by the two internal four-momenta of the lines incoming at the vertex where 4,5 are outgoing belong to ω . The diagrams $\mathcal{D}_+(P;G_1)$ and $\mathcal{D}_+(P;G_2)$ are similar to the diagrams $\mathcal{D}_+'$ and $\mathcal{D}_+$ of Fig.1 in Sect.3.3, except that the external trajectories 4,5,7,8 of Fig.1 are left out, and one has clearly $\hat{u}_+(P;G_1) = \hat{u}_+(P;G_2)$. (These points play an important role in Sect.5 in the course of the derivation of the contributions associated with triangle graphs since they always occur in integration domains : see [11]).

Example 3 : Examples of points P in which both phenomena occur are all points P such that $P_1=P_6, P_2=P_4, P_3=P_5$ and such that P_1,P_2,P_3 lie moreover in a common plane ($\Sigma\,\lambda_i P_i=0$, $\lambda_1,\lambda_2,\lambda_3 \neq 0$). These points lie on eight surfaces $L^+(G_\beta)$ tangent at P, four of which give rise to the same direction $\hat{u}_+(P;G_\beta)$, the four others giving rise to the opposite direction. Each group contains three surfaces of graphs with one internal line and one surface of a triangle graph. The diagrams $\mathcal{D}_+(P;G_\beta)$ corresponding in each group to the triangle graph are shown in Fig.2. Those corresponding to the graphs with one internal line are obtained by removing one of the vertices.

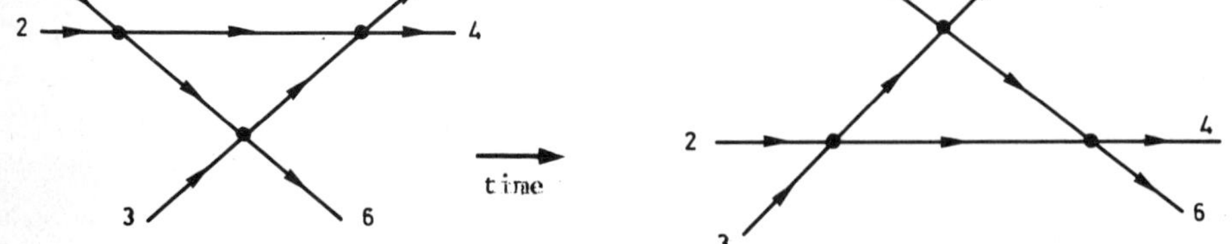

Fig. 2

Macrocausal factorization is equivalent to the decomposition (17) in R in terms of generalized Feynman integrals associated with the 18 surfaces $L^+(G_\beta)$ encountered, which can be written in a diagrammatical form (see Sect.3.2) as :

 (21)

i=1,2,3 j=4,5,6 — i=1,2,3 j=4,5,6

where, by definition, each term involved in the right-hand side is equal to one of the terms $(D^c_\beta)_+ = (d^c_\beta)_+ \times \delta^4 (p_1+p_2+p_3-p_4-p_5-p_6)$ and where $\simeq$ means as previously equality modulo an analytic background in R (after factorization of the δ^4-functions).

5. S-MATRIX THEORY

Theorem[11]

"Refined macrocausality, unitarity and the no sprout assumption (see below) imply the decomposition (21) of the three-body S-matrix in R in terms of generalized Feynman integrals".

We first give some explanations on the assumptions. The method is based on the exploitation of the following equations, which directly follow from $SS^{-1} = \mathbb{1}$, or $S^{-1}S = \mathbb{1}$, and from the definition (3) of the connected parts of the S-matrix :

 (22)

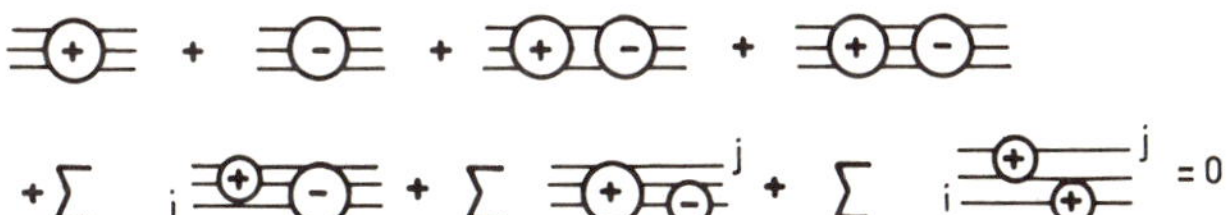 (23)

in the respective regions $s < (3\mu)^2$, and $s < (4\mu)^2$. The bubbles  , and  stand for connected operators $S^c_{m,n}$, and $(S^{-1})^c_{m,n}$ respectively, or their kernels. The "bubble diagram functions" F_B associated with each bubble diagram B are defined, in a way similar to the terms D^c in Sect.3.3, as integrals, over all possible on-mass-shell four-momenta associated with the internal lines, of the product of kernels associated with each bubble. They are always well defined as kernels of bounded operators (e.g. 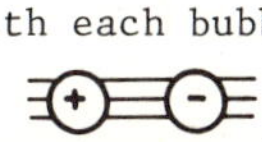 is the kernel of the product of $S^c_{3,3}$ and $(S^{-1})^c_{3,3}$), hence as tempered distributions, and contain, as in Sect.3.3. an overall δ^4-function (if all p_k are not colinear) :

$$F_B = f_B \times \delta^4 (\Sigma p_i - \Sigma p_j) \quad . \tag{24}$$

Refined macrocausality (see the end of Sect.3.1) is the essential support (7), used here for $S^c_{2,2}$ in the region $s < (3\mu)^2$ and for $S^c_{2,3}$, $S^c_{3,2}$, $S^c_{3,3}$ in the region $(3\mu)^2 < s < (4\mu)^2$, plus the regularity property R which will be applied to $S^c_{3,3}$ at $\mathcal{M}_o$ points in the region $(3\mu)^2 < s < (4\mu)^2$ (see below).

Unitarity ($S^{-1} = S^{\dagger}$) is used here to ensure that the essential support of the minus bubbles is opposite to that of the plus bubbles at any physical-region point (and to ensure the regularity property R for $(S^{-1})^c_{3,3}$).

The essential support properties (7) and the corresponding essential support properties of the minus bubbles, together with the regularity property R, will yield information on the essential support of the terms F_B (see App. 1). Although we are interested only in the region $\mathcal{R}$, the essential support properties (7) of $S^c_{3,3}$ are needed in the full region $(3\mu)^2 < s < (4\mu)^2$ since $\mathcal{M}_o$ points are always encountered in integration domains for the individual bubbles, for instance for the term 
The regularity property R is needed because all points $P=(p_1,\ldots,p_6)$ are u=0 points for this term ; hence, the standard results of essential support, or hyperfunction, theory, which apply only at u≠0 points, give no information on its essential support, or singular spectrum. This absence of information would completely disrupt the proofs. The u=0 problem arises, in the region $(3\mu)^2 < s < (4\mu)^2$, from the occurrence, just mentioned, of $\mathcal{M}_o$ points in integration domains.

<u>No sprout assumption</u>

Let us consider the decompositions (20) of f provided in $\mathcal{R}$ by macrocausality. Being given a system of real analytic local coordinates $q_1,\ldots,q_{14}$ of $\mathcal{M}$ at a point P of $L^+(G_\beta)$ in $\mathcal{R}$, chosen such that $L^+(G_\beta)$ is represented locally by $q_1=0$ and $\hat{u}_+(P;G_\beta)$ is the direction of the positive u_1-axis, and being given any open cone Γ with apex at the origin whose closure is contained (apart from the origin) in the half-space $q_1 > 0$, the analyticity properties of f_β stated below (20) ensure the existence of a complex neighborhood $\underline{\omega}$ of P such that $\underline{f}_\beta$ is analytic in $\underline{\omega} \cap \{|\mathrm{Im}\underline{q}_1 \in \Gamma\}$, where $\underline{q}$ denote the complexified variables of q. By definition, f_β satisfies the *no sprout property* at P if there exists $\underline{\omega}$ such that $\underline{f}_\beta$ is analytic in $\underline{\omega} \cap \{\mathrm{Im}\underline{q}_1 > 0\}$.

A slight extension of the decomposition (20) in the region $(3\mu)^2 < s < (4\mu)^2$ will be useful. Namely, macrocausality ensures decompositions in that region of the form :

$$f = \sum_\beta f_\beta + a \tag{25}$$

where a is analytic in $\mathcal{R}$ and where each f_β has the same properties as those stated below (20) in a region $\mathcal{R}_\beta$ slightly larger than $\mathcal{R}$. For the graphs G_1, G_2 of Fig.1, $\mathcal{R}_\beta$ is obtained by removing only from the region $(3\mu)^2 < s < (4\mu)^2$ the points p of $L^+(G_1)$, resp. $L^+(G_2)$, such that $p_1=p_2$ or $p_5=p_6$, resp. $p_1=p_2$ or $p_4=p_5$.

No sprout assumption : "There exist decompositions of the form (25) such that each f_β satisfies the no sprout property at each point P of $L^+(G_\beta)$ in R_β ".

Lemma (see [11] : "the terms f_β in the decompositions provided by the no sprout assumption are uniquely defined in R_β modulo addition of analytic functions".

This lemma is an easy consequence of Bremerman's continuity theorem.

Proof of the theorem

We now give some hints on the proof of the theorem. In view of the lemma just quoted, one has to show that $f_\beta \simeq (d^c_\beta)_+$ in R . We outline the proof for the term f_1 associated with the graph G_1 of Fig.1, following [11].

Eq.(23) can be written in the form :

$[F_1$ +  + 3 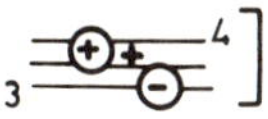 $]$ + H_1 = 0 (26)

where $F_1 = f_1 \times \delta^4(\Sigma p_i - \Sigma p_j)$,  is defined as previously, as an integral over internal on-mass-shell four-momenta(*), 3 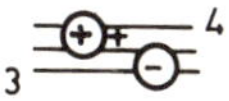 is defined from 3 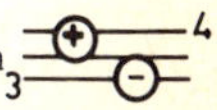 in the same way as $(D^c)_+$ is defined from D^c in Sect.3.3, and H_1 is the sum of all terms in the l.h.s. of (23), except that 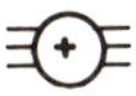 ,  and 3 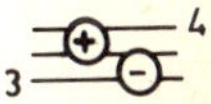 are replaced by $\sum_{\beta\neq 1} F_\beta + A$, $\sum_{\beta\neq 1}$  +  and 3  = 3 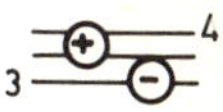 - 3 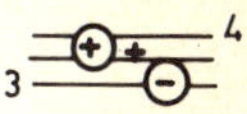

($A = a \times \delta^4(\Sigma p_i - \Sigma p_j)$ where a is introduced in Eq.(25)).

The two terms singled out with F_1 in the bracket of (26) are those which (after factorization of their e.m.c. δ^4-function) have at most, like f_1, $\hat{u}_+(P;G_1)$ in their essential support along $L^+(G_1)$ and are analytic in R_1 outside $L^+(G_1)$. Moreover, their singularities are those that correspond to diagrams $\mathcal{D}_B$ with a related topological structure (see App.1). According to the general idea of "separation of singularities in unitarity equations" (see App .1), their singularities are expected to cancel among themselves. We shall *prove* below that this is the case, i.e. more precisely that h_1 is analytic in R_1, where $H_1 = h_1 \times \delta^4(\Sigma p_i - \Sigma p_j)$.

Once this is achieved, the desired result ($f_1 \simeq (d^c_1)_+$) is obtained by replacing each term  in (26) by  (again in the sense of integrals over internal on-mass-shell four-momenta), where  is as in Sect.3 the non connected S matrix $S_{2,2}$. In view of (22), the sum of the first two terms in (26) is

----

(*) F_1 is no longer known to be the kernel of a bounded operator. However, this term, as also similar ones encountered below, are well defined by the standard u≠0 results on products and integrals of distributions.

transformed into F_1 and the third term is transformed (see [11] into $-$ 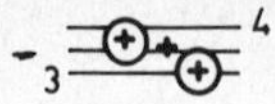
Hence :

$$F_1 \;=\;$$ 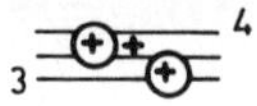 $-$  (27)

The analyticity of h_1 in R_1 yields analyticity of the last term (after factorization of its e.m.c. δ^4-function), and the result is therefore proved.

Finally, the analyticity of h_1 in R_1 is proved as follows :

(i) A detailed analysis [10,11] of each term contained in h_1 allows once to show that $\hat{u}_+(P;G_1) \notin ES_P(h_1)$ if $P \in L^+(G_1)$ lies outside the union $\Omega_+ \cup N$ of an open subset Ω_+ of $L^+(G_1)$ and of other lower-dimensional submanifolds .

The equality (26) and the essential support properties of the three terms in the first bracket of (23) (see above) then imply that $ES_P(h_1)$ is empty, i.e. that h_1 is analytic along $L^+(G_1)$ outside $\Omega_+ \cup N$.

(ii) It can be checked that the three terms in the bracket of (26) (after factorization of their e.m.c. δ^4-functions), and hence h_1 by (26), satisfy, like f_1, the no sprout property at any point P of $L^+(G_1)$ in R_1. Hence by an application of Bremerman's continuity theorem (see [10, 11]) h_1, being analytic at some points of $L^+(G_1)$ is necessarily analytic all along $L^+(G_1)$ in R_1 . Q.E.D.

Remarks : 1) the elimination in step (ii) of the possible $\hat{u}_+$-singularities remaining in step (i) is crucial. Otherwise these $\hat{u}_+$-singularities would propagate all along $L^+(G_1)$ for the term  and hence no result would be obtained anywhere. We emphasize also that step (ii) applies to the *sum* H_1, but *not* to the individual terms in H_1 : some of them, such as  , 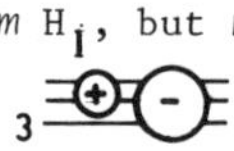 , 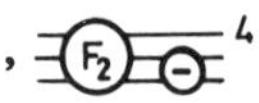 , are certainly expected to have $\hat{u}_+(P;G_1)$ in their essential support along Ω_+ and other submanifolds (after factorization of their e.m.c. δ^4-functions).

2) $(d_1^c)_+$ and hence $f_1 - (d_1^c)_+$, satisfies the no sprout property, as easily checked, at all points P of $L^+(G_1)$ in R_1. However, unless "separation of singularities in unitarity equations" is *assumed*, this fact cannot be used to prove *directly* from Eq.(27) that $f_1-(d_1^c)_+$ is analytic along $L^+(G_1)$: the argument that would be similar to step (ii) above is not possible because several terms in 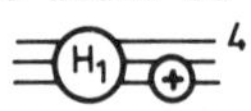 such as 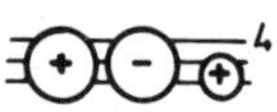 or 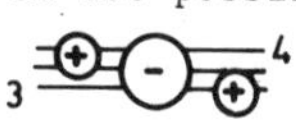 have $\hat{u}_+(P;G_1)$ in their essential support at *all* points of $L^+(G_1)$, (after factorization of their e.m.c. δ^4-functions).

3) the proof would be completely disrupted if macrocausality was used in its form (6) rather than (7) : for instance it would not be possible to prove that

$\hat{u}_+(P;G_1)$ is absent from the essential support of  (after factorization of its e.m.c.δ^4-function) at any point P of $L^+(G_1)$.

4) Previous proofs of the above result ($f_1 \simeq (d_1^c)_+$) all start essentially modulo unessential variations, from the following equations, which replace (26)(27):

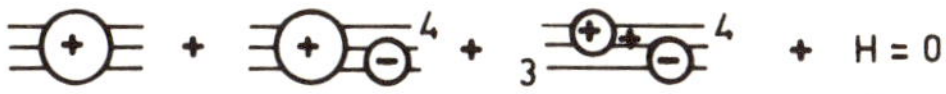 (26')

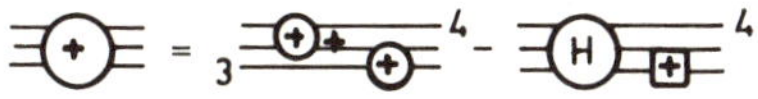 (27')

where H is here simply the sum of all terms in the l.h.s. of (23) except that  and  have been removed and 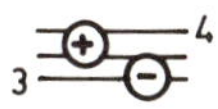 is replaced by 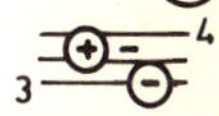

(i) The "proof" of [5] relies on the assertion that all terms in H are analytic, or like , *minus* iε boundary values, all along $L^+(G_1)$, after factorization of their e.m.c. δ^4-functions. This is however not correct (see Remark 1). The proof of [5] is then completely ruined.

(ii) The proofs using the assumption of separation of singularities start from Eq.(27'). The only terms whose singularities correspond to diagrams with a structure related to $\mathcal{D}_+(P;G_1)$ = 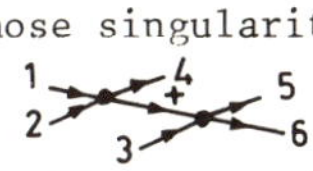 are F_1 and . Hence the result is automatic if this assumption is used. Otherwise no result is obtained for the same reason as in Remark 2) above.

(iii) The proof of [10] is close to that presented above, but there are some complications and a supplementary step is needed, in particular because h, in contrast to h_1, cannot be expected to be analytic along Ω_+. ($H-H_1$ contains the term  which has $\hat{u}_+(P,G_1)$ in its ess.support along Ω_+) -

Once the above result on f_1 ($f_1 \simeq (d_1^c)_+$ in R_1) is obtained, a similar analysis allows one to determine the contributions f_β associated with triangle graphs, such as f_2. One starts again from (23) and one now singles out the contribution F_2 to 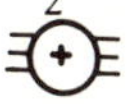, together with the terms  and  whose singularities along $L^+(G_2)$ correspond to diagrams with a related topological structure. The arguments are similar to above, although somewhat more subtle : see details in [11].

6. AXIOMATIC FIELD THEORY

The basic quantities of axiomatic field theory for the purposes of this section are the connected "chronological functions" τ_c which are, for each n, well defined distributions in the space R^{4n} of n space-time vector variables $x_1,\dots,x_n$ and are in fact (possibly regularized) connected, "amputated" vacuum expectation values of the chronological product $T(x_1,\dots,x_n)$ of n field operators $A(x_1),\dots,A(x_n)$. Their Fourier transforms $\tilde{\tau}_c$ are correspondingly defined in the space R^{4n} of energy-momentum variables $p_1,\dots,p_n$. In view of the translation invariance of τ_c under translation of all x_k by a common space-time vector a, $\tilde{\tau}_c$ contains an energy-momentum conservation δ^4-function $\delta^4(\sum_{k=1}^{n} p_k)$. It can be shown (see below), as first established in [20], that, being given any process $m \to m'$, with $m+m' = n$, the distributions $\tilde{\tau}_c$ can be restricted (as distributions) to the mass-shell $M_{m,m'} = \{p=(p_1,\dots,p_n)\ ;\ p_k^2 = \mu^2,\ \forall k,\ (p_k)_o < 0,\ k=1,\dots,m\ ,\ (p_k)_o > 0,\ k=m+1,\dots,n\}$, and the following relation holds :

$$S^{c}_{m,m'}(p_1,\dots,p_m;p_{m+1},\dots,p_n) = \tilde{\tau}_c(-p_1,\dots,-p_m;p_{m+1},\dots,p_n)\big|_{M_{m,m'}} \qquad (28)$$

We shall also consider functions $(\tau_c)_I$, where I is a subset of $(1,\dots,n)$, which are defined in a way similar to τ_c, except that $T(x_1,\dots,x_n)$ is replaced by the product of $T(x(J))$ and $T(x(I))$, where J is the complement of I in $(1,\dots,n)$ and $x(I)$, resp. $x(J)$, denote the sets of points x_k , $k \in I$, resp. $k \in J$.

Microcausality

It asserts that the commutator $[A(x),A(y)]$ of two field operators vanishes if $(x-y)^2 < 0$, where $x^2 = x_o^2 - \vec{x}^2$, i.e. if x-y is space-like, and entails (see [15]) that :

$$\tau_c(x_1,\dots,x_n) = (\tau_c)_I\,(x_1,\dots,x_n) \qquad \text{if } (x_1,\dots,x_n) \notin \Sigma_I \qquad (29)$$

where $\Sigma_I = \{(x_1,\dots,x_n),\ x_j - x_i \gtrsim 0$ for any pair of points x_i in $x(I)$ and x_j in $x(J)\}$. The condition $x_j - x_i \gtrsim 0$ means either that $x_j - x_i$ is space-like ($(x_j-x_i)^2 < 0$), or belongs to the cone $\overline{V}^+$ (= set of points x in space-time such that $x^2 \geq 0$, $x_o \geq 0$) .

The *spectral condition* is the assertion that the spectrum of the energy-momentum operator of the theory is contained (for a theory with only one type of particle of mass $\mu > 0$) in the union of the origin (corresponding to the "vacuum"), of the positive-energy mass-shell hyperboloïd $V^+(\mu)$ ($p^2 = \mu^2, p_o > 0$), corresponding to one-particle states, and of the continuum $\overline{V}^+(2\mu)$ ($p^2 \geq (2\mu)^2$, $p_o > 0$). It entails[15] that $(\tilde{\tau}_c)_I$ has its support in the region

$$\begin{aligned} \sum_{j\in J} p_j \ (= -\sum_{i\in I} p_i) &\in \overline{V}^+(\mu) \cup \overline{V}^+(2\mu), \ \text{if } 1 < |J| < n-1 \\ &\in \overline{V}^+(2\mu) \qquad , \ \text{if } |J| = 1 \text{ or } n-1 . \end{aligned} \qquad (30)$$

Being given a point P in $M_{m,m'}$, let S_P be the set of all I such that $(\tilde{\tau}_c)_I$ vanishes in the neighborhood of P. It is then easily seen that[15] :

$$ES_P\,(\tilde{\tau}_c) \subset \bigcap_{I \in S_P} \Sigma_I \quad . \tag{31}$$

Proof : By (29), $\tau_c-(\tau_c)_I = 0$ if $(x_1,\dots,x_n) \notin \Sigma_I$. Hence, for any P, $ES_P(\tilde{\tau}_c-(\tilde{\tau}_c)_I) \subset \Sigma_I$ (see lemma below). If $I \in S_P$, $(\tilde{\tau}_c)_I = 0$ locally. Hence $ES_P((\tilde{\tau}_c)_I)$ is empty (apart from the origin) and $ES_P(\tilde{\tau}_c) \subset \Sigma_I$, since $\tilde{\tau}_c=(\tilde{\tau}_c-(\tilde{\tau}_c)_I) + (\tilde{\tau}_c)_I$.

We have used the following elementary lemma (see e.g.[21]) : if the Fourier transform $\tilde{f}$ (here $\tau -(\tau_c)_I$) of a distribution f (here $\tilde{\tau}_c-(\tilde{\tau}_c)_I$) vanishes outside a cone C with apex at the origin, then the generalized Fourier transform of f at any point P falls off exponentially in the appropriate sense outside C, i.e. $ES_P(f)\subset C$. (Note that the analogues of the variables x of the mathematical lecture are here the energy-momentum variables $p_1,\dots,p_n$, and not the space-time variables).

By standard results of essential support theory, (31) entails that the restriction of $\tilde{\tau}_c$ to $M_{m,m'}$ does exist (away from M_o points), and it gives information on the essential supports of $S^c_{m,m'}$, hence of the scattering function $f_{m,m'}$ at any point P of $M_{m,m'}$: they are contained respectively in the sets associated with all possible configurations of trajectories (P_k,x_k) corresponding to all points $(x_1,\dots,x_n)$ in $ES_P(\tilde{\tau}_c)$, resp. relative configurations of such trajectories.

Four-point function (two-body processes)

For a two-body process (m=m'=2,n=4), the result (31) reduces, at any physical region point P, to :

$$ES_P(\tilde{\tau}_c) \subset \{x=(x_1,\dots,x_4),\ x_1=x_2,\ x_3=x_4,\ x_3-x_1 \in \overline{V}^+\} \tag{32}$$

where $\overline{V}^+$ is the set of points x in space-time such that $x^2 \geq 0, x_o \geq 0$. By Lorentz invariance, this result can moreover be improved, namely the condition $x_3-x_1 \in \overline{V}^+$ can be replaced by $x_3-x_1=\lambda(P_1+P_2)$, $\lambda \geq 0$. The possible points in $ES_P(\tilde{\tau}_c)$ and $ES_P(S^c_{2,2})$ are represented in Fig.3

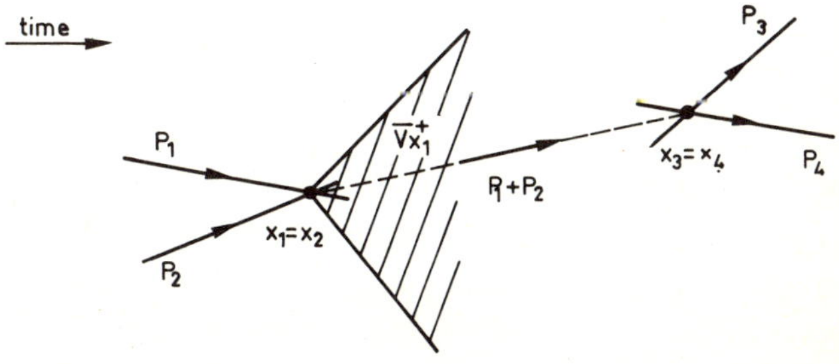

Fig. 3

Let us consider a point P in the region $(2\mu)^2 < s < (3\mu)^2$. There, a better result is expected according to the macrocausality ideas, namely $x_1=x_2=x_3=x_4$, since energy-momentum cannot be transferred from $x_1=x_2$ to $x_3=x_4$ by stable particles if $x_3 \neq x_1$. This result can be obtained (see [13] and references) by a regularity assumption which eliminates à priori possible à la Martin pathologies, corresponding physically to the possibility of having sequences of unstable particles, with arbitrarily small rates of exponential decay, by which energy-momentum might be transferred. (The exclusion of such pathologies is also part of the physical arguments upon which macrocausality is based : see Sect.3.1).

We admit below correspondingly that :

$$ES_P(\tilde{\tau}_c) \subset \{(x_1,x_2,x_3,x_4),\ x_1=x_2=x_3=x_4\},\ \forall\, P \text{ s.t.} (2\mu)^2 < s < (3\mu)^2 \quad . \tag{33}$$

Equivalently, $\tilde{\tau}_c$, after factorization of its e.m.c. δ^4-function, and $f_{2,2}$ are analytic at P.

Five-point function (2→3 or 3→2 processes)

We next consider a 2→3 process in the region $s<(4\mu)^2$ $(s=(p_1+p_2)^2=(p_3+p_4+p_5)^2)$. In this case, (31) entails that :

$$ES_P(\tilde{\tau}_c) \subset C_3 \cup C_4 \cup C_5 \quad , \tag{34}$$

where 3,4,5 denote as previously the final particles and

$$C_k = \{x=(x_1,\dots,x_5); x_1=x_2, x_i=x_j,\ x_i-x_k \in \overline{V}^+,\ x_k-x_1 \in \overline{V}_+\} \tag{35}$$

k=3,4,5, (i,j,k) = (3,4,5). The cone C_3 is represented in Fig.4a)

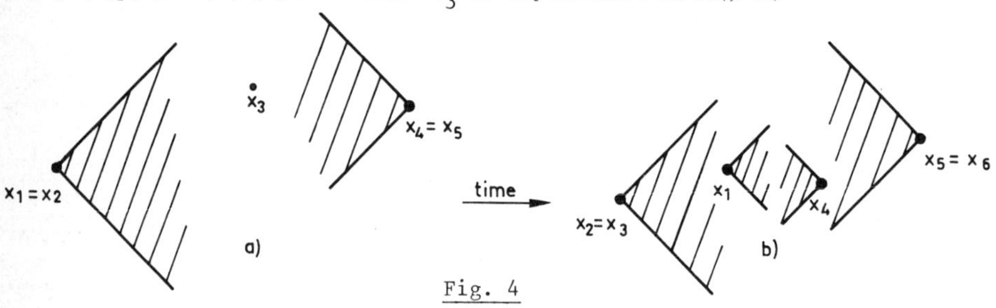

Fig. 4

This result, which cannot be improved in a simple way by Lorentz invariance, is very far from the essential support expected from the macrocausality ideas, which corresponds in the region $s < 4\mu^2$, and away from M_o points, to $x_1=x_2=x_3=x_4=x_5$. We explain below how the result (34) can be improved by using the further axiom of *asymptotic completeness*, together with the previous result (33) on the four-point function.

Let $(\tilde{\tau}_c)^-_{i,j} = \tilde{\tau}_c - (\tilde{\tau}_c)_I$, where I = (1,2,k), (i,j,k)=(3,4,5) and let 

and 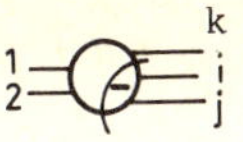 represent diagrammatically $\tilde{\tau}_c$ and $(\tilde{\tau}_c)^{-}_{i,j}$ respectively. Microcausality and the spectral condition yield the following information, which is the analogue of (34) :

$$ES_P \left(\text{[diagram]} \right) \subset C_i \cup C_j \cup C_k^- \qquad (36)$$

where $C_k^- = \{x=(x_1,\ldots,x_5),\ x_1=x_2,\ x_i=x_j,\ x_k-x_1 \in \overline{V}^+,\ x_k-x_i \in \overline{V}^+\}$.

Asymptotic completeness yields on the other hand the following relations in the region $s < (4\mu)^2$, obtained by introducing in the appropriate place a complete set of intermediate states and by standard analysis of axiomatic field theory (see e.g. [22]) :

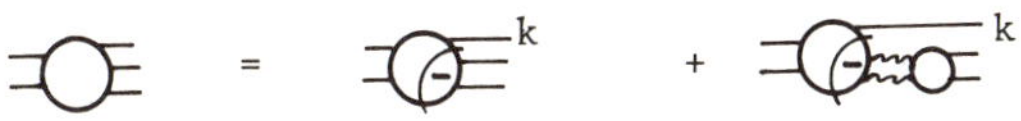

(37)

where the last term in the right-hand side is defined in the same way as bubble diagram functions in Sect.5, as a convolution integral over internal *on-mass-shell* four-momenta (with the measure $\delta(k_\ell^2-\mu^2)\theta(k_\ell)_o)d^4k_\ell$ for each internal line). The only difference is that the external four-momenta are not restricted to the mass-shell. In view of (33) (36), the standard $u\neq0$ results of essential support theory (or hyperfunction theory) on products and integrals (see App.1) show that this integral is well defined and give information on its essential support. Moreover, a systematic use of the relations (37) for the various values of k=3,4,5, yields :

<u>*Theorem*</u> [13] "ES_P () $\subset C = \{x; x_1=x_2, x_3=x_4=x_5, x_3-x_1 \in \overline{V}^+\}$ (38)

at any point P of $M_{2,3}$ s.t. $s < (4\mu)^2$, $(P_i+P_j-P_k)^2 \neq \mu^2$ where i,j,k is any permutation of (3,4,5)".

The exclusion of the codimension one submanifolds $(P_i+P_j-P_k)^2=\mu^2$ (which contain the M_o points) in the results obtained so far in [13], is due to reasons that we shall not discuss here in detail. Lorentz invariance implies now, as for the four-point function, an improvement of (38) in which the condition $x_3-x_1 \in \overline{V}^+$ is replaced by $x_3-x_1=\lambda(P_1+P_2)$, $\lambda \geq 0$. The situation is then the same as that shown in Fig.3 for the 2→2 processes, except that $x_3=x_4$ is replaced by $x_3=x_4=x_5$. To reobtain macrocausality (see above), a regularity assumption for 2→3 processes would again be needed to avoid the à priori possible à la Martin pathologies.

<u>Six-point function</u> (3→3 processes)

We shall now consider, as in Sect.5, a 3→3 process in the region $s<(4\mu)^2$.

Generalized Feynman integrals, such as 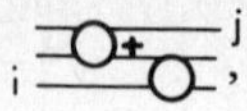, 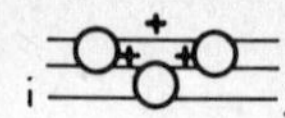 are defined in a natural way in axiomatic field theory[14,13] where the bubbles  are not defined à priori on-mass-shell, by associating Feynman-type propagators to each internal line. It can be checked that they coincide, after restriction to the mass-shell, and modulo analytic backgrounds, with the $(D^c_\beta)_+$ introduced previously.

The analogue of (34), which follows again from microcausality and the spectral condition, is here :

$ES_P \Big($  $- \sum_{\substack{i=1,2,3\\ j=4,5,6}}$ 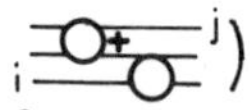 $\Big) \subset \bigcup_{\substack{s=1,2,3\\ t=4,5,6}} C^t_s$ (39)

where e.g. $C^4_1 = \{x, x_2=x_3, x_5=x_6, x_1-x_2 \in \overline{V}^+, x_5-x_4 \in \overline{V}^+, x_4-x_1 \in \overline{V}^+\}$ (Fig.4b)).

The asymptotic completeness equations are the three equations (37) (with three lines instead of two on the left of the 3→3 bubbles) and three analogous equations involving terms 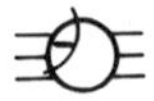 (i,j,k) = (1,2,3), k=1,2,3. As previously, microcausality and the spectral condition yield information also on 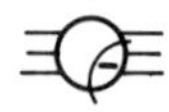, or 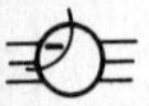 The full use of all previous informations, together with the standard u≠0 results, then yields :

<u>*Theorem*</u> [13] " $ES_P \Big($  $- \sum$ 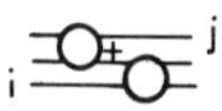 $- \sum$ 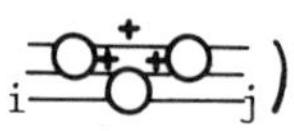 $\Big)$

$$\subset \{x;\ x_1=x_2=x_3,\ x_4=x_5=x_6,\ x_4-x_1 \in \overline{V}^+\} \qquad (40)$$

at any point P of $M_{3,3}$ s.t. $s < (4\mu)^2$, $(P_i+P_j-P_k)^2 \neq \mu^2$, where (i,j,k) is any permutation of (1,2,3) or of (4,5,6)".

The condition $x_4-x_1 \in \overline{V}^+$ can, as previously be replaced by $x_4-x_1=\lambda(P_1+P_2+P_3)$, $\lambda \geq 0$ by Lorentz invariance. A supplementary regularity assumption (for 3→3 processes) would be needed to avoid à la Martin pathologies, i.e. to replace this latter condition by $x_1= \dots = x_6$, and thus to reobtain the decomposition (21), equivalent to macrocausality and macrocausal factorization, on the mass-shell. As in the case of the five-point function, the submanifolds $(P_i+P_j-P_k)^2=\mu^2$ are excluded so far in the results of [13].

<u>*Remark*</u> : Although this has not yet been established so far, it is believed that the terms , or 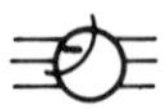, after restriction to the mass-shell and factorization of their e.m.c. δ^4-function, can be obtained from the scattering

function $f_{3,3}$ by analytic continuation, on the complex mass-shell, around two-particle thresholds.

In S-matrix theory, these analytic continuations, and the mass-shell versions of Eq.(37) or of its analogues for 3→3 processes are not known at the outset, although one would like to reobtain them at a later stage. Hence only physical on-mass-shell unitarity has been used in Sect.5.

7. TWO-DIMENSIONAL SPACE-TIME[16,17]

We first consider, as in [16], a theory with only one type of particle and a 3→3 process. The physical-region can be divided into sectors, each of which corresponding to a given relative position of the three initial two-momenta and of the three final two-momenta (in two-dimensional energy-momentum space). M_o points lie on the boundary of these sectors. In each sector, one encounters eight +α-Landau surfaces of connected graphs associated with graphs with one internal line and triangle graphs, which all coincide with the real analytic codimension one surface Σ of the physical region defined by requiring that the sets of initial and of final two-momenta should coincide, e.g. $p_1=p_6$, $p_2=p_4$, $p_3=p_5$ in one of these sectors.(These conditions reduce to only one condition in $M_{3,3}$ in view of the mass-shell constraints and of energy-momentum conservation : $p_1+p_2+p_3 = p_4+p_5+p_6$). The situation is then the same as that described in Example 3 of Sect.4, except that these +α-Landau points P belong now to a codimension one surface and not to a low-dimensional surface of the physical region as in four-dimensional space-time. The directions $\hat{u}_+(P;G_\beta)$ and hence the plus iε directions associated with two different subsets of these surfaces are opposite all along Σ and the scattering function $f_{3,3}$ is thus not along Σ the boundary value of an analytic function.

In contrast to the four-dimensional case, the structure of the connected S matric is *not* simpler in two-dimensional space-time than that of the non connected S matrix. This is related to the fact that(non colinear)space-time trajectories necessarily meet in this latter case . In the neighborhood of a point P of a given sector that may lie on Σ but lies on no other +α-Landau surface, the structure of $S_{3,3}$ implied by macrocausality and macrocausal factorization, can be determined as follows. The terms D_1, D_2 associated with the two diagrams of Fig.2 of Sect.4 coincide (and vanish outside Σ) : $D_1=D_2=D$ with

$$D = \mathbb{1}_{3,3}(p_1,\ldots,p_6) \times \prod_{\substack{i<j \\ i,j=1,2,3}} s_{2,2}(p_i,p_j;p_{i+3},p_{j+3}) \tag{41}$$

where two-body kinematics in two-dimensional space-time ($\delta^2(p_1+p_2-p_3-p_4)$ reduces to the kernel $\mathbb{1}_{2,2}$ $(p_1,\ldots,p_4)$ of the identity) has been used. Macrocausality (Eq.(6) for 2→2 processes) ensures that each $s_{2,2}$ is locally analytic.

Macrocausality (Eq.(6)) and macrocausal factorization (Eq.(10)) then imply in a straightforward way[16] that $ES_P(s_{3,3}-d)$ is empty (apart from the origin), i.e. equivalently :

$$S_{3,3}(p_1,\ldots,p_6) = \mathbb{I}_{3,3}(p_1,\ldots,p_6) \times \prod_{\substack{i<j \\ i,j=1,2,3}} s_{2,2}(p_1,p_j,p_{i+3},p_{j+3}) + R(p_1,\ldots,p_6) \tag{42}$$

where $R = r \times \delta^2(p_1+p_2+p_3-p_4-p_5-p_6)$ and r is locally *analytic* in the neighborhood of P : i.e. $S_{3,3}$ "factorizes into a product of two-body S matrices" modulo an analytic background.

Let us now consider the two following properties that are established, or believed to hold, in a class of models : (i) absence of particle production (i.e. $S_{m,n} = 0$ if $m \neq n$) and (ii) coincidence of the sets of initial and of final two-momenta (i.e. for the three-body case $S_{3,3} = 0$ outside Σ). These properties cannot be expected in four-dimensional space-time, but can be expected in two-dimensional space-time. They are also related to each other by (heuristic) arguments. Property (i) entails that the only relevant $+\alpha$-Landau surfaces are those mentioned above in each sector. Property (ii) then implies that r is zero outside Σ and is thus identically zero since it is analytic in the neighborhood of P. Hence one obtains exact factorization.

This result can first be extended by induction to the multiparticle S matrix $S_{m,m}$, $m > 3$, in models in which properties (i) and (ii) are satisfied : see [16] . It can on the other hand be extended to models with several types of equal-mass particles as we now explain for the three-body S matrix. Being given the external particles, there can exist several diagrams $\mathcal{D}_+$ for each one of the diagrams of Fig.2, depending on the types of particles associated with the internal lines. The term D_1, D_2 involved in each case in the statement of macrocausal factorization are then sums of the corresponding contributions and are no longer à priori equal. However, it is easily seen that macrocausality and macrocausal factorization, together with properties (i) and (ii), imply that $D_1 = D_2$ and $S_{3,3} = D_1 = D_2$ in each sector : see [17] . The equality $D_1 = D_2$ entails *consistency relations*, also called factorization equations, between sums of products of (non connected) scattering functions, such as those used in the lectures by Berg and references therein.

APPENDIX 1 - ESSENTIAL SUPPORT, STRUCTURE THEOREM AND RELATED TOPICS

The essential support[21] of a distribution f defined in $\mathbb{R}^n$ (resp. in a real analytic manifold M) is, at each point X, a cone with apex at the origin in the dual space (resp. in $T^*_X M$) composed of "singular directions" along which a generalized Fourier transform $F_X(f)$ of f at X does not decrease exponentially in a well specified sense ; f is analytic at X, resp. is at X the boundary value of an analytic function from the directions of an open cone Γ (in the space of imaginary parts of the variables), if and only if $ES_X(f)$ is empty (apart from the origin), resp. is contained in the closed convex salient dual cone C of Γ . The notion of essential support characterizes more generally possible (in general non unique) decompositions of f, in the neighborhood of a point, or over real regions, into sums of boundary values of analytic functions from specified directions (which depend on the real point considered) : see [21] and references therein.

This characterization coincides with that obtained independently in hyperfunction theory[23] in terms of the notion of singular spectrum, except that boundary values occurring in general in decompositions of f may be hyperfunctions in [23] even when f itself is a distribution. It is, however, proved in [24] that the two notions do coincide. Following the language of [23], we say by definition that f is *microanalytic* at the point X and in a given direction $\hat{u}$ (of $T^*_X M$) if $\hat{u} \notin ES_X(f) \equiv SS_X(f)$.

It follows from the definition of the essential support that $ES_X(f_1+f_2) \subset ES_X(f_1) + ES_X(f_2)$, and that $ES_X(f) \subset ES_X(f_1) \cap ES_X(f_2)$ if $f=f_1=f_2$ in the neighborhood of X. This latter result is a useful generalization of the edge-of-the-wedge theorem. These two results are used extensively in Sect.3 to 7.

<u>Structure theorem on bubble diagram functions</u> (S-matrix theory).[8,12]

Being given bubble diagram function F_B such as those encountered in Sect.3 to 5 (= integrals, over internal on-mass-shell four-momenta, of products of distributions  , or  , or $F_1, F_2, \ldots$ associated with each bubble and defined on-the-mass-shell), the general $u \neq 0$ results of essential support (or hyperfunction) theory on products and integrals of distributions entail that their possible singularities are associated with space-time diagrams, as explained below. Being given the bubble diagram B, let us first define diagrams ε_B : they are collections of subdiagrams ε_b that are associated with each bubble b of B and "fit together". More precisely, each ε_b is a configuration of incoming and outgoing trajectories associated with the incoming and outgoing lines of b and representing (see the beginning of Sect.3) a possible point in the essential support of F_b. The various ε_b fit together if, being given any internal line ℓ of B, which is outgoing from a bubble b_1 and incoming at a bubble b_2, the two trajectories associated with ℓ in ε_{b_1} and ε_{b_2} respectively coincide. They are then identified as an internal line of ε_B .

The various ε_b encountered in applications are, by macrocausality, configurations of external trajectories of diagrams $\mathcal{D}_+$ if b is a bubble 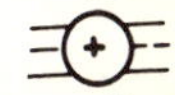 , or a bubble $F_1, F_2, \ldots$, or (by unitarity : $S^{-1} = S^{\dagger}$) of opposite diagrams $\mathcal{D}_-$ if b is a minus bubble. The configurations of external trajectories of diagrams ε_B will thus be configurations of external trajectories of diagrams $\mathcal{D}_B$ whose internal lines are either the original internal lines of ε_B or are internal lines of subdiagrams $\mathcal{D}_b$.

A point $P = \{P_k\}$ (which is a set of external four-momenta) is a u=0 point of B if there exist "u=0" diagrams ε_B , or $\mathcal{D}_B$, whose all external trajectories pass through a common point, e.g. the origin, (and have the respective four-momenta P_k), whereas some internal trajectories of ε_B do not pass through this point. Away from u=0 points, the essential support of F_B is (contained in) the set determined by all possible configurations *of external trajectories* of diagrams ε_B, or $\mathcal{D}_B$. At u=0 points, all directions may à priori be singular and F_B is not necessarily à priori well defined. On the other hand F_B is well defined (see [12]) if the bubbles are kernels of bounded operators, i.e. for instance if B =  . If the regularity property R holds for each F_b, then the essential support is contained in the set determined by considering all possible limit configurations of external trajectories of sequences of modified diagrams in which each internal trajectory is replaced by a pair of trajectories (one for each bubble) which must tend to each other in the limit. (The internal trajectories that are considered here are à priori those of ε_B . They may also be the internal trajectories of subdiagrams $\mathcal{D}_b$ if the approach of [12] to macrocausality is adopted, but their doubling does not seem to be required apart from exceptional cases : see App.2).

Configurations of external trajectories of different diagrams $\mathcal{D}_B$ may coincide, for a common or for different bubble diagrams B. At u=0 points of a given B, all directions should on the other hand be associated with the above "u=0" diagrams $\mathcal{D}_B$, if u=0 results are not known. "*Separation of singularities*" in unitarity equations is the assertion in all cases that singular contributions to the various bubble diagram functions that correspond to diagrams $\mathcal{D}_B$ with a related structure (see details in Ch.III of [1]) should cancel among themselves, independently of the other possible singularities corresponding to diagrams with a non related structure.

Structure theorem (field theory)

The bubble diagram functions encountered in field theory are again integrals over internal *on-mass-shell* four-momenta (see Sect.6), but the external four-momenta are not à priori restricted to the mass-shell.

A diagram ε_B is then a collection of subdiagrams ε_b, each of which is now a configuration (= set) of *points* in space-time associated with the incoming and outgoing lines of b, this configuration representing a possible point in the essential support of F_b for given values of the external four-momenta and given values of the internal(on-mass-shell) four-momenta K_ℓ in the integration domain. The ε_b must moreover fit together in the sense that, being given any internal line ℓ of B, outgoing from b_1 and incoming at b_2, the points $(x_\ell)_1$ and $(x_\ell)_2$ that represent it in ε_{b_1} and ε_{b_2} must satisfy :

$$(x_\ell)_2 - (x_\ell)_1 = \lambda_\ell K_\ell \tag{43}$$

where λ_ℓ is an arbitrary real scalar.

A u=0 point $P = \{P_k\}$ of B is now a point such that there exists a diagram ε_B whose all "external" points have a common space-time position (e.g. the origin), whereas some "internal" points are not there.

Away from u=0 points, F_B is a well defined distribution and its essential support is (contained in) the set corresponding to all possible configurations of external points of diagrams ε_B .

APPENDIX 2 - COMPLEMENTS ON MACROCAUSALITY AND MACROCAUSAL FACTORIZATION

Macrocausality

Exponential decay cannot be expected if there exists a diagram $\mathcal{D}_+(P,u)$, *or* if the trajectories (P_k,u_k) can be obtained as limits of the external trajectories of a sequence of modified diagrams, in which some of the constraints of the diagrams $\mathcal{D}_+$ are relaxed, if the violations tend to zero in the limit. These considerations are at the origin of the introduction of vertices at infinity, satisfying angular-momentum conservation, in [6] , where certain violations to ordinary diagrams $\mathcal{D}_+$ (with no vertex at infinity) are considered by analogy with results obtained there on phase-space integrals (= violations of the loop equations and of the mass-shell conditions). The consideration of limiting procedures is then no longer necessary (each sequence of modified diagrams giving rise to a limit diagram $\mathcal{D}_+(P,u)$), away from the special class of M_o points P, if they exist, for which there exist "u=0" diagrams $\mathcal{D}_+(P)$ whose all external trajectories pass through the origin (and have the four-momenta P_k), but such that the non parallel vertices do not all lie at the origin. (A non parallel vertex is a vertex at which the incoming and outgoing trajectories are not all parallel).

A further physical discussion of macrocausality, with similar conclusions, is given in [12], where the possibility of vertices at infinity is accepted from the outset, and in which, in addition to violations of strict locality at each vertex, a doubling of the internal trajectories is considered. This doubling accounts physically for the fact that, in the quantum case, internal particles cannot be strictly localized along classical trajectories : the consideration of all possible classical trajectories (K_ℓ,W_ℓ) for each internal line ℓ should thus be replaced by the consideration of all possible wave functions of the form (5) corresponding to all possible trajectories (K_ℓ,W_ℓ), in agreement with the fact that these wave functions, with $\chi_\ell = 1$ and appropriate normalization factors $N(\gamma,\tau)$, form a complete set of intermediate states :

$$\int d\vec{K}_\ell \, d\vec{W}_\ell \, \left|\varphi_{\ell,\tau;K_\ell,W_\ell}\right\rangle \left\langle\varphi_{\ell,\tau;K_\ell,W_\ell}\right| = \mathbb{1} \tag{44}$$

where $W_\ell=(0,\vec{W}_\ell)$. $(K_\ell,W_\ell$ are for each $\varphi_{\ell,\tau}$ the analogues of P_k , u_k in (5)).

Macrocausal factorization

We consider, as in Sect.3.2, cases in which there exists only one $\mathcal{D}_+(P,u)$ and we shall denote here by (K_ℓ,W_ℓ) the well determined internal trajectories of this diagram. The variables K_ℓ,W_ℓ in (44) will thus be replaced by K'_ℓ,W'_ℓ. We denote below, on the other hand, by $\hat{\mathcal{D}}(P,u)$ (non causal) diagrams $\mathcal{D}$ (if they exist) with the same external trajectories (P_k,u_k), obtained in the same way as the diagrams $\mathcal{D}_B$ in App.1 by replacing here each vertex v of $\mathcal{D}_+(P,u)$ by a subdiagram $\mathcal{D}_+$ (not necessarily reduced to a single vertex). The well determined internal trajectories of $\hat{\mathcal{D}}(P,u)$ that replace the internal trajectories of $\mathcal{D}_+(P,u)$ will be denoted $(\hat{K}_\ell,\hat{W}_\ell)$. If there is no set of more than one internal trajectory between two common vertices of $\mathcal{D}_+(P,u)$, there is in general no $\hat{\mathcal{D}}(P,u)$ whose trajectories $(\hat{K}_\ell,\hat{W}_\ell)$ lie, for each ℓ, in a given neighborhood of (K_ℓ,W_ℓ) (*).

Macrocausal factorization is then the physical requirement that $S(\{\varphi_{k,\tau}\})$ is asymptotically equal, modulo exponential fall-off in the $\tau \to \infty$ limit, to the integral, over sufficiently small neighborhoods N_ℓ of (K_ℓ,W_ℓ) in (K'_ℓ,W'_ℓ)-space, for each ℓ, of the product of transition amplitudes $S(\{\varphi_{k,\tau};\varphi_{\ell,\tau},K'_\ell,W'_\ell\}|_v)$ associated with each vertex v of $\mathcal{D}_+(P,u)$.

In usual cases, such as all those encountered in Sect.4, there exists no $\hat{\mathcal{D}}(P,u)$ at all. The methods of [12] then allow one to show that the integral over the complement of the product of the neighborhoods N_ℓ falls off exponentially with τ in the appropriate sense. The statement of macrocausal factorization given in Sect.3.2 follows, in view of (44).

REFERENCES

[1] - D. IAGOLNITZER, *The S Matrix*, North-Holland, Amsterdam (1978) (or for Chs.I and II, the earlier version of this book : *Introduction to S Matrix Theory*, ADT, Paris (1973)).

[2] - G.F. CHEW, *The Analytic S Matrix*, W.A. Benjamin, New-York (1966)

[3] - D. IAGOLNITZER, H.P. STAPP, Comm. Math. Phys. 14, 15 (1969) This work is a development of several earlier works, including in particular : C. CHANDLER, H.P. STAPP, J. Math. Phys. 90, 826 (1969).

[4] - D. IAGOLNITZER, in *Lectures in Theoretical Physics*, Vol.XI D, ed. by K.T. Mahanthappa and W.E. Brittin, Gordon and Breach, New-York (1969), p.229

(*) This is no longer true if there are sets of $\alpha > 1$ lines between some vertices. Then, one is led to associate with these vertices modified S-matrices, corresponding to remove the interactions between the particles of these sets, and to add a new vertex on each set, to which an actual S-matrix is associated. In usual cases, this procedure leads alternatively to the statement (10) with D being defined as in Sect.3.2 with actual S-matrices associated to each vertex, except that a new vertex is introduced as above on each set, to which the inverse S_α^{-1} of the restriction of S to $\mathcal{H} = \bigoplus_{m \geq \alpha} \mathcal{H}_m$ is associated.

[5] - R.J. EDEN, P.V. LANDSHOFF, D.I. OLIVE, J.C. POLKINGHORNE, *The Analytic S Matrix* Cambridge Univ. Press. (1966), ch.IV.
M. BLOXHAM, D.I.OLIVE, J.C. POLKINGHORNE, J. Math. Phys. 10, 494, 545, 553 (1969)
D.I. OLIVE, in *Hyperfunctions and Theoretical Physics*, Lecture Notes in Mathematics, Springer-Verlag, Heidelberg (1975).

[6] - M. KASHIWARA, T. KAWAI, H.P. STAPP, Comm. Math. Phys. 66, 95 (1979).

[7] - J. COSTER, H.P. STAPP, J. Math. Phys. 10, 371 (1969), 11, 1441 (1970), 11, 2753 (1970).

[8] - D. IAGOLNITZER, Comm. Math. Phys. 41, 39 (1975).

[9] - H.P. STAPP, in *Structural Analysis of Collision Amplitudes*, ed. by R. Balian and D. Iagolnitzer, North-Holland, Amsterdam (1976), p.275.

[10] - D. IAGOLNITZER, H.P. STAPP, Comm. Math. Phys. 57, 1 (1977).

[11] - D. IAGOLNITZER, Macrocausality, Unitarity and Physical-Region Structure of the Multiparticle S Matrix, to be published in Comm. Math. Phys. (1980).

[12] - D. IAGOLNITZER, Comm. Math. Phys. 63, 49 (1978).

[13] - H. EPSTEIN, V. GLASER, D. IAGOLNITZER, in preparation.

[14] - J. BROS, Cours de Lausanne (May 1979), and in preparation.

[15] - J. BROS, H. EPSTEIN, V. GLASER, Helv. Phys. Acta 45, 149 (1972). The direct proof given here of (31) is that of D. IAGOLNITZER, App.C in Part IV of Ref.1.

[16] - D. IAGOLNITZER, Phys. Rev. D, 18, 1275 (1978).

[17] - D. IAGOLNITZER, Phys. Lett. 76B, 207 (1978).

[18] - F. PHAM, in *Hyperfunctions and Theoretical Physics*, op. cit. in Ref.5 and M. SATO.

[19] - C. CHANDLER, H.P. STAPP, Ref.3.

[20] - K. HEPP, in *Lectures in Particle Symmetries and Axiomatic Field Theory*, vol.1, ed. by M. Chretien and S. Deser, Gordon and Breach, New-York (1966).

[21] - D. IAGOLNITZER, in *Structural Analysis of Collision Amplitudes*, op. cit. in Ref.9, p.295.

[22] - J. BROS, M. LASSALLE, Comm. Math. Phys. 43, 279 (1975) and references therein.

[23] - M. SATO, T. KAWAI, M. KASHIWARA, in *Microfunctions and Pseudo-differential Equations*, Lecture Notes in Mathematics, Springer-Verlag, Heidelberg (1973), p.265.

[24] - J.M. BONY, in Goulaouic - Schwartz Séminaire, Paris (1976).

ANALYTIC 2-PARTICLE STRUCTURE
AND CROSSING CONSTRAINTS

GUSTAV SOMMER
Fakultät für Physik, Universität Bielefeld, Germany

ABSTRACT : Some aspects of 2-particle structure and crossing symmetry in general local quantum field theory are reviewed.

1. TWO-PARTICLE-STRUCTURE IN GENERAL QUANTUM FIELD THEORY AND CROSSING SYMMETRY

The n-particle structure of momentum space Green's functions in general quantum field theory has been studied extensively by Bros[1)] and by Bros and Lassalle[2)]. The starting point was to define for euclidean momenta a Feynman convolution corresponding to a definite diagram with vertex functions having the analytic properties of general quantum field theory. This allows for an extension to the primitive holomorphy domain and for the definition of p-particle-irreducible Green's functions through certain structure equations. For details see an article in these Proceedings[3)].

Here we review what is known in the simplest case of many-particle structure, i.e. for the 2-particle structure (n=2) of 4-point functions[4)]. The aim then is to study the Green's functions far away from the euclidean region and from the primitive domain, and in fact near the intersection of two physical energy regions where extra constraints result from crossing symmetry and asymptotic completeness.

The frame for the discussion of 2-particle structure given below is axiomatic Local Quantum Field Theory[5)] (LQFT) together with Asymptotic Completeness (AC). By LQFT we understand here that there are given

a) a unitary representation of the (proper) Poincaré group on a Hilbert space with an invariant vector Ω and such that the translation part of the representation has the support of its measure in $(\{(p^o)^2-\vec{p}^2= o\}\cup\{(p^o)^2-\vec{p}^2=m^2>o\}\cup\{(p^o)^2-\vec{p}^2= =4m^2+\mathbb{R}_+\})\cap\{p^o\geq o\}$ (spectral condition), and

b) a Poincaré-covariant operator-valued tempered distribution (the "field") $A(x)$, $(x^o,\vec{x})\in\mathbb{R}^4$, acting cyclically on Ω and such that two fields $A(x)$, $A(y)$ commute for space-like separations $(x-y)^2< o$ (microcausality condition).

By AC we denote the assumption that the underlying Hilbert space is equal to the Hilbert spaces spanned by the incoming resp. outgoing states, these two spaces in fact being the same.

The assumptions of LQFT guarantee that functions $H_{(\sigma)}(k)$ defined in terms of the Fourier transforms of 4-point generalized retarded functions of the type

$$r_{(\sigma)}(x_1x_2x_3x_4):=(\Omega|\sum_{\pi=\pi(1,2,3,4)}\prod_{j=1}^{4}\sigma_j^{\pi}\,\theta(\sigma_j^{\pi}(x^o_{\pi(j-1)}-x^o_{\pi j}))A(x_{\pi 1})A(x_{\pi 2})A(x_{\pi 3})A(x_{\pi 4})|\Omega)$$

$$H_{(\sigma)}(k_1k_2k_3k_4)\delta(\sum_{j=1}^{4}k_j):=\prod_{j=1}^{4}(k_j^2-m^2)\cdot(\mathcal{F}r_{(\sigma)})(k_1k_2k_3k_4)$$

are holomorphic in certain tubes. In these definitions $(\sigma)=\{(\sigma_1^{\pi},..\sigma_4^{\pi})\}$ are arbitrary sign factors $\sigma_j^{\pi}=\pm 1$ fulfilling $\sigma_j^{\pi_k\pi}=\sigma_j^{\pi}\ \forall\ j\neq k$ if π_k is a transposition between positions (k-1) and (k) and π an arbitrary permutation of (1,2,3,4). Taking into account the spectral condition all such functions $H_{(\sigma)}(k)$ can be continued through parts of the real faces of the tubes ("coincidence regions") and are in fact one single analytic function $H(k)$ holomorphic in the "primitive domain" which is the union of

$$\mathcal{T}:=\{(k_1,k_2,k_3,k_4)|\sum_j k_j=o,k_j=p_j+iq_j,p_j\in\mathbb{R}^4,(q_{\pi_1},q_{\pi_2},q_{\pi_3})\in V_+^{\otimes 3}\cup V_-^{\otimes 3}$$
$$\text{or }(q_{\pi_1}+q_{\pi_2}+q_{\pi_3},q_{\pi_2}+q_{\pi_4},q_{\pi_3}+q_{\pi_4})\in V_+^{\otimes 3}\cup V_-^{\otimes 3}\}$$

with the coincidence regions in the boundaries of the tubes. From Poincaré of the fields the function H on the manifold $\Gamma=\{(k);\sum_j k_j=o\}$ may as well be considered as a function

$H(s,t|\underset{\sim}{\zeta},\underset{\sim}{\zeta}')=H(k_1,k_2,-k_1',-k_2')$ of the Lorentz invariants

$$\zeta_i=k_i^2,\ \zeta_i'=k_i'^2\ (i=1,2)\ ,\ \underset{\sim}{\zeta}=\binom{\zeta_1}{\zeta_2},\underset{\sim}{\zeta}'=\binom{\zeta_1'}{\zeta_2'},\ s=(k_1-k_1')^2,t=(k_1+k_2)^2$$

and is an analytic function also in these variables in the envelope of holomorphy $\mathcal{E}(\mathcal{D})$ of a domain $\mathcal{D}$ corresponding to $\mathcal{T}$. From AC there exists a unitary S-operator whose matrix elements between two-particle states can be expressed in terms of the function H. Also from AC, the discontinuities A across certain cut-singularities of H can be represented by symmetric Green's functions of the form $(\Omega,R^+R\Omega)$, partly reexpressable by H itself, and have positivity relations of the type[6)]

$$(1)\qquad \int_{\omega_r\times\omega_r} A(K+l,K-l,-K+l',-K-l')\,\bar{\psi}(l)\psi(l')\,d^4l\,d^4l'\geq o$$

for all test functions $\psi\in\mathcal{D}(\omega_r)$, $\omega_r\subset\mathbb{R}^4$ and K a fixed 4-vector. Using these properties, AC can help in various ways to elucidate the analytic structure of the Green's function H: (i) enlargement of the holomorphy domain, (ii) monodromy structure, (iii) crossing constraint equations on boundary values and the existence and uniqueness of solutions. Being interested primarily in (iii), further results of the types (i) and (ii) are required for this purpose.

Starting point for the considerations is the fact, proven by Bros[1)2)], that AC allows one to define by the Bethe-Salpeter type of equation

$$\text{(2)}\quad H(k_1,k_2,-k_1',-k_2') = I_t(k_1,k_2,-k_1',-k_2') + \frac{1}{2}\int I_t(k_1,k_2,-k_1'',-k_2'')\prod_{i=1}^{2}(k_i''^2-m^2)^{-1} H(k_1'',k_2'',-k_1',-k_2')\, d^4k_1'' d^4k_2''$$

or, graphically:

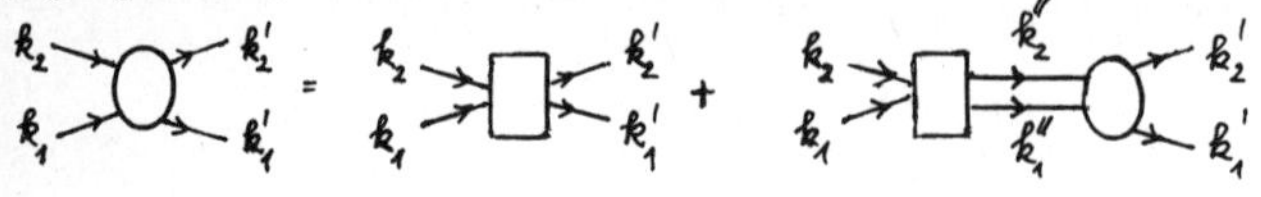

$$H \;=\; I_t \;+\; I_t \odot_t H$$

the so-called t-channel 2-particle irreducible Green's function I_t which with H has the primitive analytic structure connected to the domain $\mathcal{S}$ and moreover is irreducible with respect to 2-particle intermediate states in the sense that it has vanishing discontinuity across $C_t^{el} := \{(k)\in\Gamma \,|\, 4m^2 \leq (k_1+k_2)^2 \equiv t < 16m^2\}$ which is a cut-singularity of H . Inside the primitive analyticity domain $\mathcal{S}$ resp. $\mathcal{D}$, equation (2) may as well be formulated in terms of the invariant variables:

$$\text{(2')}\quad H(s,t|\underset{\sim}{\zeta},\underset{\sim}{\zeta}') = I_t(s,t|\underset{\sim}{\zeta},\underset{\sim}{\zeta}') + \int ds'ds''d^2\underset{\sim}{\zeta}'' K(s,s',s'',t\ \underset{\sim}{\zeta},\underset{\sim}{\zeta}'',\underset{\sim}{\zeta}') I_t(s',t|\underset{\sim}{\zeta},\underset{\sim}{\zeta}'')\cdot \cdot H(s'',t|\underset{\sim}{\zeta}'',\underset{\sim}{\zeta}')$$

where K is a well-defined kernel function. Apart from the fact that equs.(2) or (2') define I_t , in either case H can be represented as a Fredholm series of iterated "Feynman convolutions" $I_t \odot_t I_t \odot_t \cdots \odot_t I_t$. On the other hand, from the existence of fields commuting at space-like separations and from the resulting algebraic structure of retarded functions, H has a property known as crossing symmetry

$$\text{(3)}\quad H\left(s,t\left|\begin{smallmatrix}\zeta_1\zeta_1'\\ \zeta_2\zeta_2'\end{smallmatrix}\right.\right) = H\left(t,s\left|\begin{smallmatrix}\zeta_1\zeta_2\\ \zeta_1'\zeta_2'\end{smallmatrix}\right.\right) = H\left(u,t\left|\begin{smallmatrix}\zeta_1\zeta_2'\\ \zeta_2\zeta_1'\end{smallmatrix}\right.\right), \quad u = \sum_j(\zeta_j+\zeta_j') - s - t,$$

and expressing the fact that the boundary values of H in the physical regions corresponding to the 2-particle processes

$$(1)+(1') \rightarrow (2)+(2')\ ,\ (1)+(\bar{2}) \rightarrow (\bar{1}')+(2'), (1)+(\bar{2}') \rightarrow (2)+(\bar{1}')$$

have to be equal (in the sense of distributions). Now this symmetry shows that H can, e.g., also be represented by a Fredholm series of iterations $I_s \odot_s I_s \odot_s \cdots \odot_s I_s$ of s-channel Feynman convolutions $\odot_s$ of s-channel 2-particle irreducible Green's functions I_s. This fact strongly constraints the boundary values of H , and, in particular, on those parts of the boundary where two physical energy regions intersect. To exploit such a constraint effectively, the simultaneous validity of the two Fredholm series not only in the primitive domain is needed, but rather near the interesting boundary point. While any term in the series is holomorphic (or at least meromorphic) in $\mathcal{E}(\mathcal{D})$, it is not at all trivial that any such term can be represented by an iterated convolution (with integrations in the analyticity domain) when analytically continued to points which can be reached only by a complex completion of the primitive domain or by positivity arguments. Of particular interest in this connection is the discontinuity equation which results if the discontinuity (Δ_s) across $C_s := \{ (k)\in\Gamma \,|\, 4m^2 \leq (k_1-k_1')^2 \equiv s < \infty \}$ of equation (2) or (2') is taken. Up to some symmetrization (indicated by $\overset{*}{=}$) it has the form

(4) $\Delta_s H \overset{*}{=} \Delta_s I_t + \Delta_s I_t *_t \Delta_s H$ or graphically

with $*_t$ a convolution with compact domain of integration, so that the iteration solution of this equation has only a finite number of terms at any value of s. In the s-channel elastic region $C_s^{el} := \{ (k)\in\Gamma \,|\, 4m^2 \leq (k_1-k_1')^2 \equiv s < 16m^2 \}$, the discontinuity $\Delta_s I_t$ of the 2-particle irreducible kernel I_t of the t-channel completeness equation (2), i.e. kernel function of equation (4), can be expressed by H via the on-the-mass-shell $\{ k_i^2 \equiv \zeta_i = m^2 \}$ s-channel completeness relation which is just the elastic unitarity equation (with ^ indicating on-shell integration)

(5) $\Delta_s H = \Delta_s I_t = H \,\hat{\odot}_s\, H \qquad s \in C_s^{el}$

or graphically

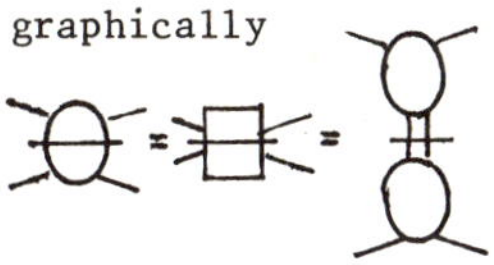

Whenever for the values of $\underset{\sim}{\zeta}, \underset{\sim}{\zeta}'$ considered and for $s \in C_s^{el}$ the function H has cut-plane analyticity in t, then the $\hat{\odot}_s$-convolution in (5) can even be reexpressed by $\Delta_t H$ with another type of composition $\hat{\otimes}_s$ with

(6) $H \,\hat{\odot}_s\, H \equiv \Delta_t H \,\hat{\otimes}_s\, \Delta_t H \qquad s \in C_s^{el}$

so that $\Delta_s H$ and the crossed channel discontinuity $\Delta_t H$ are connected via the constraint

(7) $\Delta_s H \overset{*}{=} \Delta_t H \,\hat{\otimes}_s\, \Delta_t H \qquad s \in C_s^{el}$.

In this case the function H can also be continued towards its double cut singularity at $s \in C_s$, $t \in C_t^{el}$ leading in fact to two different constraints for the double discontinuity $\Delta_t(\Delta_s H)$ (with $\hat{*}_t$ the restriction of $*_t$ to the mass-shell)

(8a) $\Delta_t(\Delta_s H) \overset{*}{=} \Delta_s I_t \,\hat{*}_t\, \Delta_s H + \Delta_s I_t *_t \Delta_t (\Delta_s H)$

(8b) $\Delta_t(\Delta_s H) \overset{*}{=} \Delta_s H \,\hat{*}_t\, \Delta_s H + \Delta_t(\Delta_s H) \,\hat{*}_t\, \Delta_t(\Delta_s H)$

They correspond to two different ways of continuing to the double cut, in case of eq.(8a) along

$$\{ s\notin C_s,\ t\notin C_t\} \to \{s\in C_s,\ t\notin C_t\} \to \{ s\in C_s,\ t\in C_t^{el}\}$$

using the completeness equation (2), and in case of eq.(8b) along

$$\{ s\notin C_s,\ t\notin C_t\} \to \{ s\notin C_s,\ t\in C_t^{el}\} \to \{ s\in C_s,\ t\in C_t^{el}\}$$

using crossed channel elastic unitarity (mass shell completeness relation).

Before explaining in Sections 3 and 4 to what extent the continuation to the different boundaries and the derivation of the different constraints which combine completeness and crossing symmetry has been achieved, we shall discuss briefly in Section 2 possible applications in connection with Banach space fixed point methods.

2. CROSSING SYMMETRIC BETHE-SALPETER-EQUATIONS AND BANACH SPACE METHODS

In this Section we assume that the set

$$(9)\quad V := \left\{ (s,t,\underset{\sim}{\zeta},\underset{\sim}{\zeta}')\in \mathbb{C}^6 \;\middle|\; \begin{array}{l} t\in \mathcal{N}([o,16m^2])\cap(\mathbb{C}\setminus\{4m^2+\mathbb{R}_+\}) \\ (\underset{\sim}{\zeta},\underset{\sim}{\zeta}')\in\mathcal{N}(\tilde{\Pi}^t\times\tilde{\Pi}^t) \\ s\in\mathbb{C}\setminus(\{4m^2+\mathbb{R}_+\}\cup\{\sum\limits_{j=1}(\zeta_j+\zeta_j')-t-4m^2-\mathbb{R}_+\}) \end{array} \right\}$$

with $\mathcal{N}$ denoting complex neighbourhoods and with

$$(10)\quad \tilde{\Pi}^t := \{ \underset{\sim}{\zeta}\in\mathbb{C}^2 \mid \underset{\sim}{\zeta} = \tfrac{1}{2}\binom{\zeta_+ + \zeta_-}{\zeta_+ - \zeta_-},\ \zeta_+\in\mathbb{R},\ \zeta_-\in i\,\mathbb{R},\ o\leq -\zeta_-^2 \leq t(t-2\zeta_+)\}$$

belongs to the holomorphy domain of H and I_t, leaving open for a while the question whether analyticity in this domain follows from LQFT and AC (what is expected, however). Then all the constraint relations of Section 1 can be shown to hold[7]. From this, e.g., a system of nonlinear integral equations can be derived which in an abbreviated form ($\overset{*}{=}$) with $s\in[4m^2,\infty)$, $t\in[o,16m^2)$, $(\underset{\sim}{\zeta},\underset{\sim}{\zeta}')\in\tilde{\Pi}^t\times\tilde{\Pi}^t$ reads:

$$\text{(EU)}\quad \Delta_s H \overset{*}{=} \Delta_t H \,\hat{\otimes}_s\, \Delta_t H \qquad\qquad \text{on } \{ s\in[4,16m^2)\}$$

$$\text{(BS)}\quad \left[\begin{array}{l} \Delta_s H \overset{*}{=} \Delta_s I_t + \Delta_s I_t * \Delta_s H - \Delta_s I_t \,\hat{*}\, \Delta_t(\Delta_s H) \\ \Delta_t(\Delta_s H) \overset{*}{=} \Delta_s I_t \,\hat{*}\, \Delta_s H + \Delta_s I_t * \Delta_t(\Delta_s H) \end{array}\right.$$

$$\text{(DD)}\quad \left[\begin{array}{l} \Delta_s I_t \overset{*}{=} \Delta_s H - \Delta_s I_t * \Delta_s H + \Delta_s I_t \,\hat{*}\, \Delta_s H \,\hat{*}\, \Delta_s H + \Delta_s I_t \,\hat{*}\, \Delta_t(\Delta_s H) \,\hat{*}\, \Delta_t(\Delta_s H) \\ \Delta_s I_t \overset{*}{=} \Delta_s H - \Delta_s I_t * \Delta_s H + \Delta_s I_t \,\hat{*}\, \Delta_t(\Delta_s H) + \Delta_t(\Delta_s H) - \Delta_s H \,\hat{*}\, \Delta_s H - \Delta_t(\Delta_s H) \,\hat{*}\, \Delta_t(\Delta_s H) \end{array}\right.$$

(IR) $\Delta_s H \overset{*}{=} \Delta_s I_t + \Delta_s I_t *_t \Delta_s H - \Delta_s I_t \hat{*}_t \Delta_t (\Delta_s H) + Z_s$

with

$$Z_s \overset{*}{=} \Delta_s I_{stu} + \Delta_s I_t *_t \Delta_s H - \Delta_s I_t \hat{*}_t \Delta_t (\Delta_s H) - \Delta_s I_t$$

Here I_{stu} is a linear combination of Green's functions which are 2-particle irreducible in 2 channels:

(11) $I_{stu} := I_{st} - I_{su} + I_{ut}$

(12) $I_{st} := I_s + I_t - H$

In all equations symmetrization with respect to the components of $\underset{\sim}{\zeta}$ and of $\underset{\sim}{\zeta}'$ is understood. The different equations represent constraints from elastic unitarity (EU), from the general structure equation of the Bethe-Salpeter type (BS), from the double discontinuity (DD) and from the inelastic energy region (IR). In the last equation other expressions for Z_s could have been used.

The set of equations (EU), (BS), (DD), (IR) can be made a well-defined map in a suitable Banach space $\mathcal{B}$ of functions $F\ (s,t,\underset{\sim}{\zeta},\underset{\sim}{\zeta}')$, Hölder-continuous in s,t, continuous or analytic in $\underset{\sim}{\zeta},\underset{\sim}{\zeta}'$, and with growth properties such that e.g.

$$| s^{\nu}(\log s)^{1+\omega}(\zeta\zeta')^{1-\rho}\ F(\cdot)| < \infty \quad , \quad o < \nu < 1 \ , \quad o < \omega \ , \ \rho < 1$$

The application of the Banach contraction theorem and the Schauder fixed point theorem thus enables us to study the solubility of the system of integral equations when the 2-channel 2-particle irreducible Green's function or rather $\Delta_s I_{stu}$ is considered as given. Then it results:

Theorem: If $\Delta_s I_{stu}$ is given with $a' \leq \|\Delta_s I_{stu}\| \leq a$, where $\|\cdot\|$ is the norm in , then one unique solution for $\Delta_s H$ and $\Delta_s I_t$ exists with $b' \leq \|\Delta_s H\| \leq b$ (b,b' suitable) and no solution with $o \leq \|\Delta_s H\| \leq b'$, and different $\Delta_s I_{stu}$ yield different $\Delta_s H$ and $\Delta_s I_t$.

There exist also other results for other choices of Z_s or of the assumed growth properties.

We also remark that though invariant variables $s,t,\underset{\sim}{\zeta},\underset{\sim}{\zeta}'$ have been used there are restrictions to the dimension d of the underlying space-time. For Banach spaces of the type described necessarily $d \geq 4$ must hold. That this is so follows from the use of functions with continuity properties and the fact that in $d < 4$ dimensions the constraints effectively restrict some of the integrations in the integral equations to measure 0 sets.

3. TWO-PARTICLE STRUCTURE AND STABLE DOMAINS

As pointed out before, each term in the iterative solution of eq.(2) with respect to H or I_t is analytic at least in the envelope of holomorphy of the primitive domain of H. It is, however, not clear from the beginning that outside the primitive domain the analytic continuation of each term $H \odot_t H \odot_t \cdots \odot_t H$ is again of the same type, i.e. can again be represented as in iterated convolution. On the other hand, such a property is necessary for applications as indicated in Section 2, where all variables $(s,t,\zeta_1,\zeta_2,\zeta_1',\zeta_2')$ are allowed to stay simultaneously at or near real positive values. Therefore the question arises to what extent at such points H and I_t or their discontinuities can still be determined from each other without continuing back to the primitive domain. In fact, in order to determine either the numerical values or analytic structure of the convolution $(F_1 *_t F)(s,t,\underset{\sim}{\zeta},\underset{\sim}{\zeta}')$ at the point $(s,t,\underset{\sim}{\zeta},\underset{\sim}{\zeta}')$ one has to know the numerical values or analytic structure, respectively, of $F_1(s_1',t,\underset{\sim}{\zeta},\underset{\sim}{\zeta}'')$ and $F_2(s_2',t,\underset{\sim}{\zeta}'',\underset{\sim}{\zeta}')$ at all points of the integration regions $\{(s_i',\underset{\sim}{\zeta}'') \mid \sqrt{s_i'} \in [2m,\sqrt{s}-2m],\ \underset{\sim}{\zeta}'' \in \tilde{I}^t_{s,\underset{\sim}{\zeta},\underset{\sim}{\zeta}'}\}$ with $\tilde{I}^t_{s,\underset{\sim}{\zeta},\underset{\sim}{\zeta}'}$ well-defined compact, but parameter-dependent domains in $\mathbb{C}^2$. Therefore, to determine $\{\Delta_s H\}$ and $\{\Delta_s I_t\}$ for some $t \in \mathbb{R}_+$ from each other via the Bethe-Salpeter type of equation (4) it is necessary (and sufficient) that a domain $\tilde{I}^t \subset \mathbb{C}^2$ exists which is stable under the map

$$*_t : (F_1,F_2) \mapsto F_1 *_t F_2$$

in the following sense: We call $*_t$-stable (stable under the convolution $*_t$) a domain $\tilde{I}^t$ such that

$$(\underset{\sim}{\zeta},\underset{\sim}{\zeta}') \in \tilde{I}^t \times \tilde{I}^t \Rightarrow \bigcup_{s\in[4m^2,\infty)} \tilde{I}^t_{s,\underset{\sim}{\zeta},\underset{\sim}{\zeta}'} \subset \tilde{I}^t$$

To find such a stable domain is a purely geometrical problem. Now for the purpose of applications as in Section 2 it turns out that it is sufficient to find and to consider such stable domains only on the "diagonal" manifold

$$(13)\quad \tilde{Z}_D := \{ \underset{\sim}{\zeta} \in \mathbb{C}^2 \mid \underset{\sim}{\zeta} = \tfrac{1}{2}\binom{\zeta_+ + \zeta_-}{\zeta_+ - \zeta_-},\ \zeta_+ \in \mathbb{R},\ \zeta_- \in \sqrt{-1}\,\mathbb{R} \},$$

i.e. to consider only $Z_D = \tilde{Z}_D \times \tilde{Z}_D$ in the space $Z = \mathbb{C}^4$ of mass variables $(\zeta_1, \zeta_2,\zeta_1',\zeta_2')$. In fact, for $s \in \{4m^2 + \mathbb{R}_+\}$, $t \in [4m^2,16m^2]$ it holds that $(\underset{\sim}{\zeta},\underset{\sim}{\zeta}') \in \tilde{Z}_D \times \tilde{Z}_D$ implies $\tilde{I}^t_{s,\underset{\sim}{\zeta},\underset{\sim}{\zeta}'} \subset \tilde{Z}_D$. To find then a stable domain $\tilde{I}^t \restriction \tilde{Z}_D \equiv \tilde{\Pi}^t$ on the diagonal manifold still involves 10 real variables. Nevertheless it is possible to prove the following.

Lemma: For any $t \in (0,16m^2)$ the smallest $*_t$-stable domain $\tilde{\Pi}^t$ on $\tilde{Z}_D$, different fom the mass-shell $\zeta_1{}^2 = \zeta_2{}^2 = m^2$ and containing at least one real point $\zeta_1{}^2 = \zeta_2{}^2 \leq m^2$, is given by $\tilde{\Pi}^t := \{ (\zeta_1,\zeta_2) \mid \zeta_1 = \zeta_2{}^*,\ \zeta_1 \in \Pi^t\}$ with

(14) $\quad \Pi^{\tau} := \{ \zeta \in \mathbb{C} \mid o \leq (\frac{1}{\sqrt{\tau}} \operatorname{Im}\zeta)^2 \leq \frac{\tau}{4} - \operatorname{Re} \zeta \}$

The stable domain $\tilde{\Pi}^t$ agrees with eq.(10). The knowledge of these domains, invariant under the 2-particle structure map $*_t$, enables us to determine the analyticity properties of $\Delta_s I_t$ if those of $\Delta_s H$ are given in $\tilde{\Pi}^t \times \tilde{\Pi}^t$, and vice versa. Because $*_t$ involves only compact integration regions this immediately yields:

Theorem: (a) If $\Delta_s H(s,t,\underset{\sim}{\zeta},\underset{\sim}{\zeta}')$ for all $s \in [4m^2,\infty)$ is holomorphic in all points of

(15) $$V^{\tau}_{\varepsilon} := \{ (t,\underset{\sim}{\zeta},\underset{\sim}{\zeta}') \in \mathbb{C}^5 \;\Big|\; \begin{array}{l} t \in \{ |\frac{t}{m^2} - \tau + \varepsilon| < \varepsilon \} \cap (\mathbb{C} \setminus \{ 4m^2 + \mathbb{R}_+ \}) \\ \underset{\sim}{\zeta} \in \mathcal{N}(\tilde{\Pi}^{\tau}) \;,\; \underset{\sim}{\zeta}' \in \mathcal{N}(\tilde{\Pi}^{\tau}) \end{array} \}$$

where $o < \tau < 16$, then $\Delta_s I_t(s,t,\underset{\sim}{\zeta},\underset{\sim}{\zeta}')$ is also holomorphic in $V^{\tau}_{\varepsilon'} \cup \{ t,\underset{\sim}{\zeta},\underset{\sim}{\zeta}' \mid t \in \{ 4m^2, \tau m^2 \} \;,\; (\underset{\sim}{\zeta},\underset{\sim}{\zeta}') \in \tilde{\Pi}^t \times \tilde{\Pi}^t \}$ for some $\varepsilon' > o$.

(b) If $\Delta_s I_t$ for all $s \in [4m^2,\infty)$ is holomorphic in V^{τ}_{ε}, then $\Delta_s H$ is holomorphic in $V^{\tau}_{\varepsilon'}$ for some ε'.

This theorem supplies the information desired in applications of the type discussed in Section 2 provided the hypothesis of the theorem can be proved. The present state is discussed in the following Section.

4. ANALYTIC PROPERTIES OF TWO-PARTICLE IRREDUCIBLE GREEN'S FUNCTIONS

In this Section we briefly review the present situation concerning the holomorphy properties of the 2-particle irreducible Green's function I_t as they are needed or desired in applications of the type described in Section 2.

First we quote for the full Green's function the following (partial) result:

Theorem 1: $H(s,t,\underset{\sim}{\zeta},\underset{\sim}{\zeta}')$ is holomorphic in a quasi-topological product of the cut s-plane and a neighbourhood of

(16) $$\Delta^4 := \{ t \in [0,4m^2) \;,\; (\underset{\sim}{\zeta},\underset{\sim}{\zeta}') \in \tilde{\Pi}^4 \times \tilde{\Pi}^4 \}$$

The proof uses

(i) old results leading to the well-known on-the-mass-shell analyticity properties,

(ii) the fact that ellipse type domains of holomorphy in t of $\Delta_s H$ for $(\underset{\sim}{\zeta},\underset{\sim}{\zeta}') \in \tilde{\Pi}^4 \times \tilde{\Pi}^4$ (corresponding to completions of the primitive domain along $\operatorname{Im}(k_1-k_1')=o, \operatorname{Im} k_1=o$) are such that their real sections for

$$\zeta = \begin{pmatrix} \zeta_+ + \zeta_- \\ \zeta_+ - \zeta_- \end{pmatrix} \;,\; \zeta' = \begin{pmatrix} \zeta'_+ + \zeta'_- \\ \zeta'_+ - \zeta'_- \end{pmatrix} \;,\quad \zeta_+,\zeta'_+ \in \mathbb{R} \;,\; \zeta_-,\zeta'_- \in i\,\mathbb{R}$$

contain the corresponding real sections for $\zeta_- = \zeta'_- = o$,

(iii) an analytic completion along $\mathrm{Im}(k_1+k_2)=o$ and $\mathrm{Im}\,k_1=o$ or $\mathrm{Im}\,k_1'=o$ for $|s|\leq\varepsilon$, $t\in[o,4m^2)$ yielding holomorphy of H in $\tilde{\Pi}^4\times\tilde{\Pi}^4$, and

(iv) the fact that positivity of $\Delta_s H$ for $(\underset{\sim}{\zeta},\underset{\sim}{\zeta}')\in\tilde{\Pi}^4\times\tilde{\Pi}^4$ is implied by positivity of $\Delta_s H$ for $\zeta_-=\zeta_-'=o$.

Property (iv) is an application of the Glaser-Positivity-Theorem[6]:

If (a) $F(k,l)$ is analytic in $\Omega\times\Omega^*\subset\mathbb{C}^N\times\mathbb{C}^N$, and

(b) an open $\omega_r\subset\mathbb{R}^N\cap\Omega$ exists with

$$\int_{\omega_r\times\omega_r}F(k,l)\,\overline{\psi(k)}\,\psi(l)dkdl\geq o\qquad\forall\,\psi\in\mathcal{D}(\omega_r)$$

Then: (A) for all open $\omega\subset\Omega$ with $\overline{\omega}\subset\Omega$

$$\int_{\omega\times\omega}F(k,l)\,\overline{\varphi(k)}\,\varphi(l)\,\frac{d\bar{k}\wedge dk}{(2i)^N}\,\frac{d\bar{l}\wedge dl}{(2i)^N}\geq o\qquad\forall\,\varphi\in L^2(\omega),\text{ and}\tag{17}$$

(B) a sequence of functions $f_\nu(k)$ analytic in Ω exists with

$$F(k,l)=\sum_{\nu=1}^{\infty}f_\nu(k)\,\overline{f_\nu}(l)\quad\text{uniformly in }\Omega\times\Omega\ ,$$

and (A) and (B) are equivalent statements.

Because of the Theorem of Section 3 the results of Theorem 1 are sufficient to prove some of the desired properties of I_t. In fact it holds:

Theorem 2 : For all $s\in[4m^2,\infty)$ the discontinuity $\Delta_s I_t$ of the t-channel 2-particle irreducible Green's function I_t is holomorphic in a neighbourhood of

$$\Delta:=\{\,t\in[0,4m^2),(\underset{\sim}{\zeta},\underset{\sim}{\zeta}')\in\tilde{\Pi}^t\times\tilde{\Pi}^t\}\tag{18a}$$

and therefore the dispersion relation for I_t continues to Δ.

Knowing that the primitive domain for I_t is bigger than that of H by a completion along $\mathrm{Im}\,k_1-k_2')=o$ Theorem 2 can be supplemented by

Theorem 3: For all $s\in[4m^2,\infty)$ the 2-particle irreducible function I_t is holomorphic at all points of

$$\Delta':=\{\,t\in\mathbb{R}_+,\ \sqrt{t}<\min\,[4m,\sqrt{s-36m^2}-\sqrt{s}]\ ,\ (\underset{\sim}{\zeta},\underset{\sim}{\zeta}')\in\tilde{\Pi}^t\times\tilde{\Pi}^t\}\tag{18b}$$

and $\Delta_s H$ is holomorphic in $\Delta'\cap(\mathbb{C}^5\setminus\{\,t=4m^2+\mathbb{R}_+\})$.

Again the Theorem of Section 3 has been used, here, however, in order to improve the information on $\Delta_s H$ from the knowledge of the bigger domain of $\Delta_s I_t$, while before it was applied to improve the information on $\Delta_s I_t$ from the positivity properties of $\Delta_s H$ which possibly do not hold for $\Delta_s I_t$.

One has good reasons to believe that also holomorphy in $\{\,t\in[4m^2,16m^2),$

$(\underset{\sim}{\zeta},\underset{\sim}{\zeta}') \in \tilde{\Pi}^4 \times \tilde{\Pi}^4 \} \setminus \Delta'$ can be shown using positivity from completeness of the sub-Hilbert space of (n > 2)-particle states. In this case the structure considerations as indicated in Section 2 could be put on a completely firm basis. In particular, if Glaser positivity, eq.(17), could be established for $\Delta_s I_t$, then the desired properties would result rather easily. This is because the $*_t$-convolutiom has the following Glaser-positivity preserving property:

Theorem 4: If F_1 and F_2 for $s \in [4m^2,\infty)$, $(t,\zeta,\zeta') \in \Delta$, eq.(18a), have Glaser positivity, eq.(17), then also $F_1 *_t F_2$ has Glaser positivity.

REFERENCES

1) J. Bros, in: Analytic Methods in Mathematical Physics, New York 1970

2) M. Lassalle, Commun. math. Phys. 36, 184 (1974)
J. Bros, M. Lassalle, Commun. math. Phys. 43, 279 (1975), 54, 33 (1977)
J. Bros, M. Manolessou-Grammaticou, Saclay preprint DPh-T/78-109

3) J. Bros, Proceedings Les Houches Summer Institute 1979

4) G. Sommer, in preparation

5) R.F. Streater, A.S. Wightman, PCT, Spin & Statistics, and All That, NY. 1964
Res Jost, The General Theory of Quantized Fields, Providence 1965
H. Epstein, in: Axiomatic Field Theory, ed. M. Chretien and S. Deser, NY. 1966

6) V. Glaser, R.C.P. 25, Vol. 16, Strasbourg 1973

7) G. Sommer, in: Proceed. Steklov Institute of Math. 1 (1978), 139.

ON THE SCATTERING MATRIX WITH INDEFINITE METRIC

Xia Daoxing

Research Institute of Mathematics, Fudan University,
Shanghai, China

1. Quantum theory with indefinite metric.

Some physicists [1] developed a theory by using negative metric to get over the difficulty of divergence in the theory of relativistic quantum fields. This scattering problem concerning indefinite metric can be formulated as follows.

Let K be a linear space with indefinite inner product $[x,y]$ for $x,y \in K$. We suppose that there are two linear subspaces K_+ and K_- in K, such that $K = K_+ + K_-$, the subspace K_+ and K_- are mutually perpendicular with respect to this indefinite inner product and become Hilbert spaces, if they are endowed with inner products $[x_+,y_+]$, $x_+,y_+ \in K$ and $-[x_-,y_-]$, $x_-,y_- \in K_-$ respectively.

The subspace K_+ is the space of all states of free particles with positive metric and the subspace K_- is the space of states in which the fictitious particles appear. According to this decomposition of K, there is a natural inner product defined in K,

$$\langle x_+,x_- | y_+,y_- \rangle = [x_+,y_+] - [x_-,y_-]$$

for all $x_+,y_+ \in K_+$ and $x_-,y_- \in K_-$. If the projection from K onto $K_\pm$ is denoted by $P_\pm$ and $J = P_+ - P_-$, then the indefinite inner product $[x,y]$ is $\langle x|J|y\rangle$.

The Hamiltonian of a system is an operator H self-adjoint with respect to the indefinite metric. Let H^* be the operator adjoint of H with respect to the inner product. Then $H^* = JHJ$. If we write the vector $x_- \oplus x_+$, $x_\pm \in K_\pm$ in a column as

$$\begin{pmatrix} x_- \\ x_+ \end{pmatrix}$$

then the operator H can be written in the matrix form

$$H = \begin{pmatrix} H_- & \Gamma \\ -\Gamma^* & H_+ \end{pmatrix},$$

where $H_\pm$ are the selfadjoint operators in $K_\pm$ respectively, and Γ is an operator from K_+ to K_-.

Now we shall consider the space of all physical states. A vector $x \in K$ is called with real local spectrum, if there is a vector-valued analytic function $x(.)$ defined on the whole complex plane except the real axis, such that $(H-\lambda I)\, x(\lambda) = x$ and for any $a \in K$, there is a function $f_\pm(\omega,a) \in L^1$ such that

$$\langle a|x(\lambda)\rangle = \frac{1}{2i\pi}\int \frac{f_\pm(\omega,a)\, d\omega}{\omega - \lambda}, \quad \forall\lambda \gtrless 0.$$

Let K' be the space of vectors in K with real local spectrum. In the following, we only consider the case when K' is a closed subspace of K. And we shall adopt K' as the space of physical states. We denote the physical Hamiltonian $H|_{K'}$ by H'.

2. Formulation of scattering problem.

The free Hamiltonian is a selfadjoint operator H_o in K. The wave operator $W_{\pm} = W_{\pm}(H',H_o)$, if it exists, is an operator from K_+ to K, defined by

$$\lim_{t\to\pm\infty} \| e^{-itH'} W_{\pm}(H',H_o)x - e^{-itH_o} x\| = 0, \quad \text{for } x \in K_+$$

and the scattering operator is $\hat{S} = W_-(H',H_o)^{-1} W(H',H_o)$. Since the operators H_+ and H_o are selfadjoint in Hilbert space K_+, the existence, unitarity and the expression of the wave operators $W_{\pm}(H_+,H_o)$ do not relate to negative metric and are considered in the ordinary quantum mechanics. On the otherhand,

$$W_{\pm}(H',H_o) = W_{\pm}(H',H_+)\, W_{\pm}(H_+,H_o).$$

Thus $\hat{S} = W_-(H,H_o)^{-1}\, S\, W_+(H_+,H_o)$, where

$$S = W_-(H',H_+)^{-1}\, W_+(H',H_+)$$

Thus we may suppose that $H_+ = H_o$ and we shall only consider the existence, unitarity and expression of this scattering operator S.

The main tool in our consideration is the following characteristic function. Let σ be the spectrum of H_+. For $\lambda \bar{\in} \sigma$, we construct the operator

$$h(\lambda) = \lambda I - H_- + \Gamma(\lambda I\ H)^{-1}\, \Gamma^*$$

in K_-. This operator-valued analytic function h(.) is called characteristic function and it depends upon the Hamiltonian H and the decomposition $K = K_+ + K_-$.

3. Some assumptions.

Suppose that K_+ is the space of strongly measurable and quadratic integrable vector-valued functions with values in a separable Hilbert space $\mathcal{E}$ and for $\varphi(.) \in K_+$,

$$\|\varphi\|^2 = \int_\sigma \|\varphi(\omega)\|^2 d\omega, \qquad (H_+\varphi)(\omega) = \omega\varphi(\omega) ,$$

where the measure $d\omega$ is the Lebesgue measure. Suppose further that there is an operator-valued function $\zeta(\omega)$ such that $\zeta(\omega)$ is the operator from $\mathcal{E}$ to K_-, for $\omega \in \sigma$ and

$$\Gamma\varphi = \int_\sigma \zeta(\omega)\ \varphi(\omega)\ d\omega$$

the operator H_- is a bounded selfadjoint operator.

4. The main theorem.

Let $h(\omega \pm io) = \underset{\varepsilon\to o+}{\text{s-lim}}\ h(\omega \pm io)$ be the boundary value in σ of the analytic function $h(\lambda)$.

Theorem. If $h(\omega \pm io)$ and $h(\omega \pm io)^{-1}$ exist for almost all real ω and

$$\int \| h(\omega \pm io)^{-1}\ \zeta(\omega) \|^2\ d\omega < \infty .$$

Then the $[x,x] > 0$ for all $x \in K'$ and $x \neq 0$. The space K' endowed with $[x,y]$ becomes Hilbert space. The wave operators $W_\pm = W_\pm(H', H_+)$ exist, are unitary from K_+ onto K' and

$$W_\pm \varphi = \begin{pmatrix} \int_\sigma h(\omega \pm io)^{-1}\ \zeta(\omega)\ \varphi(\omega)\ d\omega \\ \varphi(.) + i\zeta(.)^* \int \dfrac{h(t \pm io)^{-1}\ \zeta(t)\ \varphi(t)\ dt}{t - (. \pm io)} \end{pmatrix}, \quad \text{for } \varphi \in K_+ .$$

The scattering $S = W_-^{-1}\ W_+$ is a unitary operator in K_+ and

$$(S\varphi)(\omega) = \left(I - 2\pi i \zeta(\omega)^*\ h(\omega + io)^{-1}\ \zeta(\omega)\right) \varphi(\omega) \quad , \qquad \varphi \in K_+ .$$

The method in the proof of this theorem is the complex analysis of operator-valued analytic function defined in the upper or lower half-plane.

5. The complex Lee and Wick's model.

As an example, we consider Lee and Wick's model. The Hamiltonian of the system is

$$H = m_V^o\ V^*V + \int \omega(\vec{k})\ a^*(\vec{k})\ a(\vec{k})\ d^3\vec{k} + \int \varphi_o(\vec{k})(V^*Na(\vec{k}) - VN^*a^*(\vec{k}))\ d^3\vec{k}$$

where $\varphi_o(\vec{k})$ is a real function, m_V^o is the rest mass of V-particle and $\omega(k) = \sqrt{\mu^2 + \vec{k}^2}$, $(\mu > 0)$. The operators V, N, a, V^*, N^* and a^* are the annihilation and production operators of V-particle, N-particle and θ-particle respectively and they satisfy the commutation relation

$$[a(\vec{k}), a^*(\vec{k}')] = \delta(\vec{k} - \vec{k}')$$

and anti-commutation relation

$$\{N, N^*\} = 1, \quad \{V, V^*\} = -1$$

respectively. Let the vacuum be denoted by Φ_o, $[\Phi_o, \Phi_o]$ $\langle \Phi_o | \Phi_o \rangle = 1$. Then the state in which the V-particle occurs is of negative metric, and $[V^*\Phi_o, V^*\Phi_o] = -1$. The space $K^{(n)}$ of all states $[V + (n-1)\theta] \oplus [N + n\theta]$ is a reduced subspace of the Hamiltonian and $K^{(n)} = K_+^{(n)} \oplus K_-^{(n)}$ where $K_+^{(n)}$ is the space of all states

$$\Phi_\varphi = N^* \int \varphi(\vec{k}_1, \ldots, \vec{k}_n)\ a^*(\vec{k}_1) \ldots\ a^*(\vec{k}_n) d^3\vec{k}_1 \ldots d^3\vec{k}_n \Phi_o$$

with $\int|\varphi|^2\, d\vec{k}_1\ldots d\vec{k}_n<\infty$ and $K_-^{(n)}$ is the space of all states

$$\Psi_\psi = V^*\int\psi(\vec{k}_1,\ldots,\vec{k}_{n-1})\; a^*(\vec{k}_1)\ldots a\;(\vec{k}_{n-1})d^3\vec{k}_1\ldots d^3\vec{k}_{n-1}\Phi_o$$

with $\int|\psi|^2\, d\vec{k}_1\ldots d\vec{k}_n<\infty$. We identify the states Φ_φ and Ψ_ψ with φ and ψ respectively. The space $K_+^{(n)}$ is the $N\,\overbrace{\theta\ldots\theta}^{n}$ sector. The indefinite metric is positive in $K_+^{(n)}$ and negative in $K_-^{(n)}$. In this case,

$$H_+\varphi = \sum_1^n \omega(\vec{k}_\nu)\;\varphi(\vec{k}_1,\ldots,\vec{k}_n)$$

$$H_-\psi = (m + \sum_1^{n-1}\omega(\vec{k}_\nu))\;\psi(\vec{k}_1,\ldots,\vec{k}_n)$$

$$\Gamma\varphi = n\int\varphi(\vec{k}_1,\ldots,\vec{k}_{n-1},\vec{k}_n)\;\varphi_o(\vec{k}_n)\;d^3\vec{k}_n$$

6. The structure of the S-matrix in Nθθθ-sector.

Lee and Wick have already obtained the explicit form of elements of scattering matrix in $K^{(1)}$ (the case n = 1, Nθ sector) by solving a function equation and in $K^{(2)}$ (the case n = 2, Nθθ sector) by solving a singular integral equation with one variable. But for $n\geq 3$, the problem comes to solving singular integral equation with several variables, and certainly we'll meet with some essentially new difficulty. The method used in [2] may be extended to the case n = 4 and will be omitted here owing toits complexity. We'll only consider the case n = 3 and find the concrete form of elements of scattering matrix in Nθθθ sector. In this case we solve singular integral equation of two variables by the method of complex analysis to obtain the concrete form of the operator $h(\omega+io)^{-1}$. This singular integral equation is

$$\beta(\omega_1+\omega_2)y(\omega_1,\omega_2) - \int_\mu^\omega \frac{y(\omega_1,\omega')\alpha(\omega_2)\alpha(\omega') + y(\omega',\omega_2)\alpha(\omega')\alpha(\omega_1)}{\omega'+\omega_1+\omega_2-\omega-io}\,d\omega'$$

$$= \delta(\omega_1-a_1)\;\delta(\omega_2-a_2),$$

where α and β are given functions determined by m_v^o, μ and $\varphi_o(.)$ and $y(\omega_1,\omega_2)$ is unknown function of two variables.

We denote the vector $\vec{k}$ in R^3 by $\vec{k}=\omega\vec{\tau}$, where $\omega=|\vec{k}|$, τ is a unit vector. Then the S-matrix element is

$$S(\vec{k}_1,\vec{k}_2,\vec{k}_3;\vec{k}_1',\vec{k}_2',\vec{k}_3') = S(\omega_1,\omega_2,\omega_3;\omega_1',\omega_2',\omega_3')\;\delta(\vec{\tau}_1,\vec{\tau}_2,\vec{\tau}_3,\vec{\tau}_1',\vec{\tau}_2',\vec{\tau}_3')$$

and

$$S(\omega_1,\omega_2,\omega_3;\omega_1',\omega_2',\omega_3') = e^{-i\sum_1^3(\eta(\omega_j)+\eta(\omega_j'))}\Big\{\delta(\omega_1,\omega_2,\omega_3;\omega_1',\omega_2',\omega_3') +$$

$$\delta(\omega_1+\omega_2+\omega_3-\omega_1'-\omega_2'-\omega_3')\prod_j\nu(\omega_j)\nu(\omega_j')\sum_{\ell,j=1}^3\frac{1}{L(\omega_j')}\left(\frac{\delta(\omega_j-\omega_j')}{f(\omega_j')}+M(\omega_j,\omega_j')\right)\Big\}$$

where η is real and η, ν, L, f, M are functions which can be expressed by α and β.

References

1. Lee, T.D. & Wick, G.D. : Nucl. Phys. B., 9, 2 (1969), 209-243 ; 10, 1 (1969), 1-10 ; Phys. Rev. D., 6(1970), 1033-1048.

2. Xia, D.X., Scienta Sinica, 18 (1975), 165-183.

3. Xia, D.X., Acta Math. Sinica, 19 (1976), 39-51.

ON THE REPRESENTATION OF THE LOCAL CURRENT ALGEBRA

Xia Daoxing

Research Institute of Mathematics, Fudan University
Shanghai, China.

In the course of the past few years, several physicists have become attracted to the possibility that one can write down a complete physical theory in which the fundamental dynamical variables are local observables, such as current densities ([1], [2], [3]) instead of the canonical fields. For simplicity, at present time, we only consider the non-relativistic field. In mathematics, it reduces to the representation of the group of diffeomorphisms.

Let $Diff(R^n)$ be the group of all C^∞-diffeomorphic bijuctions φ in R^n, which is identical mapping beyond a compact set K_φ, and with composition

$$o : (\varphi,\psi) \mapsto \varphi(\psi(.))$$

as group operation. The group $Diff(R^n)$ is naturally equipped with Schwartz's topology to be a topological group. Let $D_o(R^n)$ be the connected component containing identity of $Diff(R^n)$. It is an infinite-dimensional Lie group. Let K_n^n be the Schwartz's space of all C^∞-mappings $g : R^n \to R^m$ with compact supports. Then K_n^n becomes infinite-dimensional algebra with the following non-associative multiplication :

$$[g,h](x) = \Sigma_j \left(h^j(x) \frac{\partial g(x)}{\partial x^j} - g^j(x) \frac{\partial h(x)}{\partial x^j} \right) .$$

For every $g \in K_n^n$, the solution φ_t^g of the differential equation

$$\frac{d}{dt} \varphi_t^{g\,'} = g o \varphi_t^g \quad , \qquad \varphi_o^g = I$$

is denoted by $\exp(tg)$. Then

$$\exp : g \mapsto \exp(tg)$$

is the exponential mapping from the Lie algebra K_n^n to the Lie group.

Let $\rho(x)$ and $J^j(x)$, $j = 1,2,\ldots n$ be the particle density and flux density. These are operator-valued distributions. We take $f \in K_n^I$ and $g \in K_n^n$ and construct the smeared currents

$$\rho(f) = \int \rho(x) f(x) dx \quad , \quad J(g) = \sum_{j=1}^{n} \int J^j(x) g_j(x) dx.$$

These are self-adjoint operators in the Hilbert space $\mathcal{H}$ of all physical states.

The commutational relation of the current algebra $\{J(x),\rho(x)\}$ can be written as

$$[\rho(f_1),\rho(f_2)] = 0,$$
$$[\rho(f),J(g)] = i\rho\left(\sum_{j=1}^{n} g_j \frac{\partial f}{\partial x^j}\right),$$
$$[J(g_1)J(g_2)]\ iJ([g_1,g_2]).$$

Since $\rho(f)$ and $J(g)$ are unbounded operators, we shall use the unitary operators $e^{it\rho(f)}$ and $e^{itJ(g)}$ instead of them. If we denote

$$U(tf) = e^{it\rho(f)} \quad , \quad V(\exp(tg)) = e^{itJ(g)} \ ,$$

Then the commutational relation of the local current algebra is equivalent to the following

$$U(f_1)U(f_2) = U(f_1 + f_2) \ , \tag{1}$$
$$V(\varphi_1)V(\varphi_2) = V(\varphi_1 \circ \varphi_2) \ , \tag{2}$$
$$V(\varphi)U(f) = U(f \circ \varphi)V(\varphi) \ , \tag{3}$$

where f_1, f_2, $f \in K_n^1$ and φ_1, φ_2, $\varphi \in D_o(R^n)$. If we consider K_n^1 as an additive group, then (1) means that

$$U : f \longmapsto U(f)$$

is a representation of K_n^1 in $\mathcal{H}$. Also, (2) means that $V : \varphi \longmapsto V(\varphi)$ is a representation of group $D_o(R^n)$ in $\mathcal{H}$.

The problem of finding the representations of the local current algebra is equivalent to determine those pairs of representations U of group K_n^1 and V of group $D_o(R^n)$, which satisfy the commutational relation

$$V(\varphi)U(f) = U(f \circ \varphi)V(\varphi). \tag{3}$$

In the following, we shall consider a kind of such representations which are closely related to the quasi-invariant measure and is seemly important in the quantum physics.

Let K_n^* be the dual of K_n^1 (i.e. Schwartz's space of distributions), $\mathcal{B}$ be the Borel field of sets in K_n^*, generated by the class of weakly closed sets in K_n^*. For any $\varphi \in \mathrm{Diff}(R^n)$, the distribution $F(\varphi^{-1}(x))D(x,\varphi^{-1})$ is denoted by $\varphi^* F$, where $D(x,\varphi)$ is the Jacobian of φ at x. The mapping $\varphi \to \varphi^*$ is a representation of $\mathrm{Diff}(R^n)$. Let μ be a measure on the space $(K_n^*, \mathcal{B})$ for every $\varphi \in D_o(R^n)$, we consider the measure μ_φ on $(K_n^*, \mathcal{B})$:

$$\mu_\varphi(E) = \mu(\varphi^* E),$$

where $\varphi^{*}E = \{\varphi^{*}F \mid F \in E\}$ for $E \in \mathcal{B}$. If $\mu_{\varphi} \sim \mu$ for all $\varphi \in D_o(R^n)$, then we say that μ is a quasi-invariant measure with respect to the group $D_o(R^n)$.

We consider a measure μ on $(K_n^{*}, \mathcal{B})$ quasi-invariant with respect to $D_o(R^n)$, and the Hilbert space $\mathcal{H} = L^2(K_n^{*}, \mathcal{B}, \mu)$. We define

$$(U(f)\Psi)(F) = e^{i(F,f)}\Psi(F) \quad , \quad F \in K_n^{*}, \Psi(.) \in \mathcal{H}$$

(in this case $\rho(x) = F(x)$)

$$(V(\varphi)\Psi(F) = \Psi(\varphi^{*}F)\left(\frac{d\mu(\varphi^{*}F)}{d\mu(F)}\right)^{1/2} \quad , \quad F \in K_n^{*}, \Psi(.) \in \mathcal{H} \quad ,$$

for $f \in K_n^1$ and $\varphi \in D_o(R^n)$, where $\frac{d\mu(\varphi^{*}F)}{d\mu(F)}$ is the Radon-Nikodym derivative of $\mu(\varphi^{*}(.))$ with respect to $\mu(.)$. Evidently, $U : f \longmapsto U(f)$ and $V : \longmapsto V(f)$ are group representations of K_n^1 and $D_o(R_n)$ in $\mathcal{H}$ respectively, and satisfy the commutational relation (3).

The main problem is to determine the measure μ (which is related the vacuum state) such that in this representation we have the prescribed well-defined Hamiltonian.

Following [1-3], we considered the case, when

$$H_K = \frac{1}{8}\Sigma_j \int dx\, K^j(x) \frac{1}{\rho(x)} K^j(x) \ ,$$

where $K^j(x) = \frac{\partial}{\partial x^j}\rho(x) + 2i\, J^j(x)$,

$$H_p = V_o + \sum_{m=1}^{\infty} \int V_m(x_1,\dots,x_m)\rho(x_1)\dots\rho(x_m)dx_1\dots dx_m$$

and $H = K_k + H_p$.

Firstly, we consider the general form of Radon-Nikodym derivative. Under certain conditions, we called that $S(g,.) = d\mu(\varphi^{*}(.))/d\mu(.)$ is an analytic functional[5].

Theorem[5] if $S(g,.)$ is an analytic functional, then

$$\frac{d\mu(\varphi^{*}F)}{d\mu(F)} = \exp\left\{C(\int Fdx)(F, \ln D(x,\varphi)) + P(\varphi^{*}F) - P(F)\right\} . \tag{4}$$

where $C(.)$ is an analytic function, P

$$P(F) = \sum_{m=1}^{\infty} \int P(x_1,\dots,x_m)F(x_1)\dots F(x_m)dx_1\dots dx_m .$$

Example. We consider the Poisson measure[6]. Let $f(x)$ be a fixed smooth function on R^n, $0 < f(x) < \infty$ and m be the measure on R^n, : $m(U) = \int_U f(x)dx$. If $\{x_n\}$ is a sequence of points in R^n without limiting point and $x_m \neq x_{m'}$ for $m \neq m'$, then we construct the corresponding distribution $\sum_{m=1}^{\infty} \delta(x-x_m)$. Let $\mathcal{M}$ be the set of all such distributions. There is a unique probability measure μ_o on $(K_n^*, \mathcal{B})$ which concentrates on $\mathcal{M}$ such that

$$\mu_o\left\{\Sigma_m \delta(x-x_m) \in \mathcal{M} \mid \text{there are just n points of } \{x_m\} \text{ in } U\right\}$$
$$= e^{-\lambda m(U)} \frac{(\lambda m(U))^n}{n!},$$

for any open set U in R^n, where λ is a fixed positive parameter.

This measure μ_o is called Poisson measure. In this case,

$$\frac{d\mu_o(\varphi^* F)}{d\mu_o(F)} = \exp\left(F, \ln(D(x,\varphi) \frac{f(\varphi(x))}{f(x)}\right) .$$

If we take a measurable functional P(F) such that $0 < \exp(P(F)-(F,\ln f) < \infty$, for almost all $F \in K_n^*$ and define a measure μ as

$$\mu(E) = \int_E \exp(P(F)-(F,\ln f) d\mu_o(F),$$

Then

$$\frac{d\mu(\varphi^* F)}{d\mu(F)} = \exp\{(F, \ln D(x,\varphi)) + P(F^*)-P(F)\}$$

and formally

$$S(g,F) = \sum_{j=1}^{n}\left[\left(\frac{\partial}{\partial x^j} F, g^j\right) + \left(F \frac{\partial}{\partial x^j} \frac{\delta P(F)}{\delta F(x)}, g(x)\right)\right].$$

In [1], [2], [3] Menikoff and Sharp investigated the problem to find representation space of a current algebra. Now we have to find out quasi-invariant measures μ such that we can give a definite interpretation of the Hamiltonian H such that H becomes a non-negative self-adjoint operator and $H\Omega = 0$, where Ω is the vacuum state 1 in $L^2(K_n^*, \mathcal{B}, \mu)$.

Suppose that μ satisfies the conditions in our Theorem. Let e_j be the unit vector whose j-th component is 1 and the other components are zero. Let

$$R_j(x,F) = S(e_j\delta(.-x),F) + \frac{\partial}{\partial x^j} F(x).$$

Then the problem of determining μ is reduced to solving the equation

$$\int e^{i(F,f)}\left(\int \left[\frac{1}{8} \frac{\sum_1^n R_j(x,F)^2}{F(x)} + \frac{i}{4} R_j(x,F)\partial_j f\right] dx + H_p(F)\right) d\mu(F) = 0, \quad f \in K_n^1 \tag{5}$$

which is simply $H\Omega = 0$. Now we consider the case that $C(.) \equiv 1$ in (4). Hence

$$R_j(x,F) = F(x) \frac{\partial}{\partial x^j} \frac{\delta P(F)}{\delta F(x)}$$

We can prove that in this case (5) becomes

$$H_p(F) = \int \left[\frac{1}{8} F(x) \sum_{j=1}^{n} \left(\frac{\partial}{\partial x^j} \frac{\delta P(F)}{\delta F(x)}\right)^2 + \frac{1}{2} \frac{\delta P(F)}{\delta F(x)} \sum_{j=1}^{n} \frac{\partial}{\partial x^j} F\right] dx$$

References

[1] Goldin, G. Grodnik, J., Power, R. and Sharp, D.H., J. Math. Phys. 15 (1974), 88.

[2] Menikoff, R., J. Math. Phys. 15 (1974), 1138, 1394 ; 16 (1976) 234, 2353.

[3] Menikoff, R., Sharp, D.H., J. Math. Phys. 19 (1978), 135.

[4] Xia Dao-xing, Measure of Integration Theory on Infinite-Dimensional spaces (translated by E.J. Brody), Acad. Press. N.Y. § London (1972).

[5] Xia Dao-xing, On the Representation of Local Current Algebra and the Group of Diffeomorphism (I) to appear in Scientia Sinica (1979).

[6] Viechik, A.M., Gelford, I.M., Graev, I.M, (1975), 6, 1.

PART IV

TWO-DIMENSIONAL MODELS AND RELATED DEVELOPMENTS

S MATRIX THEORY OF THE MASSIVE THIRRING MODEL

B. BERG

II. Institut für Theoretische Physik, Universität Hamburg.

CONTENTS

Abstract:

The S matrix theory of the massive Thirring model, describing the exact quantum scattering of solitons and their boundstates, is reviewed. Treated are: Factorization equations and their solution, boundstates, generalized Jost functions and Levinson's theorem, scattering of boundstates, "virtual" and anomalous thresholds.

I. Introduction

For a variety of two dimensional relativistic field theories physicists know now the exact quantum S matrix. I like to call these models "quantum field theories with soliton behaviour". Examples are the Sine-Gordon theory (alias massive Thirring model)/1,2/, the O(N) non-linear σ-model /3/, the Gross Neveu model /4/, supersymmetric extensions of the last two models /5/ and the chiral invariant SU(N) Thirring model /6,7/.

I will up to some side remarks concentrate on the massive Thirring model (MTM), because it is the prototype of a field theory with soliton behaviour and it's S matrix theory is with respect to a number of properties best understood. Someone who understands this model should have no difficulties with the original literature concerning the other models. I also have tried to be complementary to the review of Zamolodchikov and Zamolodchikov /8/, which relies on the results of ref./1-4/ and previous literature.

Following the original development, we use the method of factorization equations to derive the S matrix and the spectrum of the MTM. This is "old fashioned", because it is now possible to get the same results in the more fundamental Hamiltonian approach by means of the Bethe ansatz. This is treated in the lectures of Honerkamp. So far one has, however, not succeeded to solve the Bethe ansatz for all models where the exact S matrix is known.

Our lecture is concerned with details of the MTM S matrix. The S matrix is calculable, because in the MTM infinitely many higher conserved classical currents survive quantization /9/ and imply /1o/ factorization of the S matrix (section II). Although there is only elastic scattering, nevertheless the S matrix exhibits a suprisingly rich structure. After solving the factorization equations for soliton-antisoliton scattering (section III) one discovers the boundstate(breather) spectrum, which was previously known from the quasiclassical treatment/11/, to be described through simple poles in the physical sheet of the forward scattering amplitude. In the nonrelativistic limit this is consistent with the two dimensional version of Levinson's theorem. More generally it can be shown /12/, that for the relativistic S matrix Jost functions exist in close analogy to potential scattering (section IV).The distinction between boundstate and CDD poles is also discussed in section IV.

The complete S matrix of the MTM /13/ includes the amplitudes for soliton-breather and breather-breather scattering (section V). In these channels of the complete S matrix one finds double poles, which populate the physical sheets of the forward scattering amplitudes. They have a natural interpretation as anomalous thresholds /14/ (section VI). Another interesting feature arises from the fact, that breather-breather boundstates have to leave the spectrum of physical particles at a value of the coupling, where their mass is two times the soliton mass and still less than the sum of the masses of their breather constituents. This gives rise to a "virtual threshold" (section VI).

In summary the S matrix theory of the MTM is well understood and provides a lot of material, which could be used for illustration in an introductory course about S matrix theory. An outlook is given in the final section VII.

II. The factorization equations

The MTM is described by the Lagrangian

$$\mathcal{L} = \bar{\Psi} i \gamma_\mu \partial^\mu \Psi - m \bar{\Psi}\Psi - \tfrac{1}{2} g\, j_\mu j^\mu \quad \text{with} \quad j_\mu = \bar{\Psi}\gamma_\mu \Psi \qquad (1.a)$$

of a selfinteracting fermi field in 1 + 1 dimensions. By Coleman's /15/ equivalence the MTM is known to be equivalent to the Sine Gordon (SG) theory described by the Langrangian

$$\mathcal{L} = \partial_\mu \varphi \partial^\mu \varphi + \frac{\alpha}{\beta^2}\left(\cos(\beta\varphi) - 1\right) \qquad (1.b)$$

of aselfinteracting bose field. The correspondence is such that the elementary fermions $f, \bar{f}$ of the MTM correspond to the solitons $s, \bar{s}$ of the SG theory. The elementary SG boson corresponds in the MTM to the lowest $f\bar{f}$ boson boundstate b_1.

The coupling constants have to be defined in terms of physical quantities. Using the parametrization of /15/ the coupling constants are related by

$$\gamma = \pi \left(\frac{4\pi}{\beta^2} - 1 \right) \tag{2}$$

α has dimension $(\text{mass})^2$ and is related to the MTM fermion mass.

The SG theory is classically completely integrable by means of the inverse scattering method /16/. This implies the existence of an infinite set of higher conservation laws /16/ which govern the dynamics of the model. Their physical meaning is conservation of higher powers of in- and outgoing momenta, i.e.

$$(p_1^{in})^m + (p_2^{in})^m + \ldots + (p_n^{in})^m = (p_1^{out})^m + (p_2^{out})^m + \ldots + (p_{n'}^{out})^m \quad (m=1,3,5..)$$

for $n \rightarrow n'$ particle scattering, similar higher conserved currents exist in the classical (Grassmann) MTM /17/. In the quantized theory these conservation laws survive to all orders of perturbation theory /9/ and have the following severe consequences for particle scattering:

a) There is no production or annihilation.

b) The sets of incoming and outgoing momenta are identical $\{p_i^{in}, \ldots, p_n^{in}\} = \{p_1^{out}, \ldots, p_n^{out}\}$.

Relying on axiomatic S matrix theory it has been proven rigorously by Iagolnitzer /10/ that properties a) and b) imply factorization of the S matrix in terms of two particle scattering amplitudes. Roughly:

$$S_n\,(p_1 \ldots p_n) = \prod_{i<j} S_2\,(p_i, p_j) \tag{3}$$

This factorization has first been recognized for the SG boson in SG perturbation theory ($1/\beta^2$ small) /18/ and then also been verified for the MTM fermion in MTM perturbation theory (g small) /19/.

As it stands equation (3) is not well defined. Let us consider in detail elastic fermion-fermion and fermion-antifermion 2- particle scattering. Energy momentum conservation $p_1+p_2 = p_1' + p_2'$ and the mass shell condition $p_i^2 = p_i'^2 = m^2$ $(i = 1,2)$ imply that only forward ($p_1=p_1'$, $p_2 = p_2'$) and (for distinguishable particles) backward ($p_1 = p_2'$, $p_2=p_1'$) scattering is possible. The S matrix acts on 2-particle states by

$$\left|p_1,p_2\right\rangle^{in} = S\left|p_1,p_2\right\rangle^{out} = u(\hat{s})\left|p_1,p_2\right\rangle^{out} \tag{4.a}$$

and

$$\left|p_1,\bar{p}_2\right\rangle^{in} = S\left|p_1,\bar{p}_2\right\rangle^{out} = t(\hat{s})\left|p_1,\bar{p}_2\right\rangle^{out} + r(\hat{s})\left|\bar{p}_1,p_2\right\rangle^{out} \tag{4.b}$$

Here $\hat{s} = (p_1 + p_2)^2$; particles with momentum p_i are denoted p_i and anti-particles with momentum p_i are denoted $\bar{p}_i$.

The transmission amplitudes $u = u(s)$, $t = t(s)$ and the reflection amplitude $r = r(s)$ are supposed to be real meromorphic functions $\bar{f}(s)=f(\bar{s})$ ($f = u,t,s$) of the complex Mandelstam variables, which are defined on an (possibly) infinitely many sheeted Riemann surface with branching points at o and $4m^2$. By standard convention the cuts are choosen from $-\infty$ to o and from $4m^2$ to ∞. Because of the conservation laws no other singularities are required. The assumption is that only the minimal singularities are realized. By putting the considerations about Levinson's theorem (section IV) and anomalous thresholds (section VI) at the beginning one would have good arguments for this. The physical values $\hat{s}$ of s are real and above the elastic threshold: $\hat{s}= (p_1+p_2)^2 \geq 4m^2$. Following the standard convention, the physical values of the amplitudes are obtained as limit

$$\lim_{\varepsilon \to 0^+} f(\hat{s} + i\varepsilon) \qquad (f=u,t,s)$$

on a certain Riemannian sheet, which is called the physical sheet.

The further discussion is much simplified by introducing the complex rapidity variable θ defined by

$$\operatorname{ch} \theta = \frac{s-2m^2}{2m^2} \tag{5}$$

The variable θ plays the role of a global uniformizing variable of the amplitudes. The conformal mapping (5) is one-to-one between the infinitely many sheeted complex s-plane and the complex θ-plane. The physical sheet is required to be mapped into the strip $0 < \operatorname{Im}\theta < \pi$ of the θ-plane. All other sheets are mapped into similar strips above or below the physical strip (cf.Fig.1.a,b)). The amplitudes are defined as functions of θ by $f(\theta) = f(s(\theta))$. The Hermitian analyticity $\bar{f}(s)=f(\bar{s})$ carries over into $\bar{f}(\theta)=f(-\bar{\theta})$.

In the s-plane the crossing and unitarity relations of the amplitudes are

$$\lim_{\varepsilon\to 0^+} u(s+i\varepsilon) = \lim_{\varepsilon\to 0^+} t(s-4m^2-i\varepsilon)$$

$$\lim_{\varepsilon\to 0^+} r(s+i\varepsilon) = \lim_{\varepsilon\to 0^+} r(s-4m^2-i\varepsilon)$$

and

$$\lim_{\varepsilon\to 0^+} u(s+i\varepsilon)\cdot u(s-i\varepsilon) = 1$$

$$\lim_{\varepsilon\to 0^+}\left[t(s+i\varepsilon)\cdot t(s-i\varepsilon) + r(s+i\varepsilon)\cdot r(s-i\varepsilon)\right] = 1$$

$$\lim_{\varepsilon\to 0^+}\left[t(s+i\varepsilon)\cdot r(s-i\varepsilon) + r(s+i\varepsilon)\cdot t(s-i\varepsilon)\right] = 0$$

Using the Schwarz reflection principle we obtain from these relations, that the following equations hold for all complex θ.

Crossing:
$$u(\theta) = t(i\pi-\theta) \tag{6.a}$$
$$r(\theta) = r(i\pi-\theta) \tag{6.b}$$

Unitarity:
$$u(\theta)\cdot u(-\theta) = 1 \tag{7.a}$$
$$t(\theta)\cdot t(-\theta) + r(\theta)\cdot r(-\theta) = 1 \tag{7.b}$$
$$t(\theta)\cdot r(-\theta) + r(\theta)\cdot t(-\theta) = 0 \tag{7.c}$$

We are now prepared for stating precisely the factorization equations. The 2-particle S matrix is defined to act on a multiparticle state by means of

$$S_2(p_i,p_j)\,|p_1\ldots p_i\ldots p_j\ldots p_n\rangle^{out} = u(\hat{s}_{ij})\,|p_1\ldots p_i\ldots p_j\ldots p_n\rangle^{out}$$

and

$$S_2(p_i,p_j)\,|p_1\ldots p_i\ldots \bar{p}_j\ldots p_n\rangle^{out} = t(\hat{s}_{ij})\,|p_1\ldots p_i\ldots \bar{p}_j\ldots p_n\rangle^{out}$$
$$+\, r(\hat{s}_{ij})\,|p_1\ldots \bar{p}_i\ldots p_j\ldots p_n\rangle^{out}$$

where $\hat{s}_{ij} = (p_i-p_j)^2$.

In lowest order MTM perturbation theory (cf.e.g./19/) one may check easily, that the reflection amplitude does not vanish and the 2-particle S matrix is really a matrix. The 2-particle S matrices of different arguments do not commute and therefore the factorization equation (3) makes only sense if we specify the order of the factors on the r.h.s. For this task it is useful to consider elastic scattering of well localized wave packets such that all the $\frac{1}{2}n(n-1)$ 2-particle interaction points are far apart from another. Let us assume $p_1^1 > p_2^1 > \ldots > p_n^1$, then an order which corresponds to a physically possible scattering process is

$$S_n(p_1 \ldots p_n) = \prod_{i=1}^{n-1} \left(\prod_{j=i+1}^{n} S_2(p_i, p_j) \right) \tag{8}$$

The product factors have to be written down from left to right. For 3-particle scattering (cf.fig.2a) this reads

$$S_3(p_1,p_2,p_3) = S_2(p_1,p_2)\, S_2(p_1,p_3)\, S_2(p_2,p_3) \tag{9}$$

By parallel shifting of particle lines ordering procedures equivalent to (8) can be obtained, which have to give the same result for the n-particle S matrix. Already the restriction obtained from 3-particle scattering enables factorization of n-particle scattering. By shifting one line the equivalent equation to (9)(cf.fig.2b)) implies

$$S_2(p_1,p_2)\, S_2(p_1,p_3)\, S_2(p_2,p_3) = S_2(p_2,p_3)\, S_2(p_1,p_3)\, S_2(p_1,p_2) \tag{1o}$$

Equations of this type are known for a long time /2o/ as consistency equations for factorization of multiparticle amplitudes in non-relativistic δ-potential scattering. It has, however, only been recognized much later by Karowski et al /2/, that one may go the other way round and use these equations as input for the calculation of exact (even relativistic) S matrices.

Applying (1o) to $|p_1,p_2,\bar{p}_3\rangle^{in}$ and looking for the final state $|p_1,\bar{p}_2,p_3\rangle^{out}$ the following severe restriction on the scattering amplitudes is obtained:

$$h(\theta_{31}+\theta_{12}) = h(i\pi+\theta_{31})\cdot h(\theta_{12}) + h(\theta_{31})\cdot h(i\pi-\theta_{12})$$

where $h(\theta)=\frac{t(\theta)}{r(\theta)}$ and $\theta_{31} = -\theta_{13}$. For $i<j$ the rapidity variables θ_{ij} are defined by $ch\,\theta_{ij}=\frac{s_{ij}-4m^2}{2m^2}$ (cf.(5)).

III. The exact S matrix

We are now able to derive the main result. Let us first note, that equation (11) has the unique solution

$$h(\theta) = \frac{\mathrm{sh}\lambda\theta}{\mathrm{sh}\, i\pi\lambda} \tag{12}$$

here λ is an arbitrary real parameter, which will become related to the coupling constant. The proof of (12) is similar to that of the addition theorem for the exponential function although slightly more complicated. λ is real because $\bar{h}(\theta) = h(-\bar{\theta})$.

Let us assume the crossing and unitarity relations (6),(7) and equation (12). Under the following additional two assumptions the scattering amplitudes become uniquely determined:

(a) The phase $\delta(\theta)$ defined by $u(\theta) = |u(\theta)|e^{i2\delta(\theta)}$ is bounded in the physical strip.

(b) There exists a repulsive region in the parameter λ, where $t(\theta)$ has no poles and zeros in the physical strip and for arbitrary λ $t(\theta;\lambda)$ can be obtained by analytic continuation from the repulsive region.

The explicit results for the amplitudes are:

$$t(\theta) = \frac{F(\theta)}{F(2\pi i-\theta)} \tag{13.a}$$

where

$$F(\theta) = \prod_{k=1}^{\infty}\prod_{l=0}^{\infty} \frac{\left(2l+\frac{k}{\lambda}+\frac{\theta}{i\pi}\right)\left(2l+\frac{k-1}{\lambda}+\frac{\theta}{i\pi}\right)}{\left(2l-1+\frac{k}{\lambda}+\frac{\theta}{i\pi}\right)\left(2l+1+\frac{k-1}{\lambda}+\frac{\theta}{i\pi}\right)} .$$

One could write $F(\theta)$ as infinite product of Γ-functions, but this is not really a simplification. The other amplitudes are

$$r(\theta) = \frac{t(\theta)}{h(\theta)} = \frac{\sin\pi\lambda}{\sin\pi\lambda\frac{\theta}{i\pi}}\, t(\theta) \tag{13.b}$$

and

$$u(\theta) = t(i\pi-\theta) = \frac{\sin\pi\lambda\left(1-\frac{\theta}{i\pi}\right)}{\sin\pi\lambda\frac{\theta}{i\pi}}\, t(\theta) \tag{13.c}$$

Remark:

For the MTM the region with $g < o$ is known to be repulsive (cf.e.g./15/). The assumptions of a bounded phase and the absence of poles and zeros of $t(\theta)$ in the physical stripe for the repulsive region are required by Levinson's theorem (cf.next section).

Proof of (13.a):

Unitarity and crossing relations yield

$$\frac{h(\theta)}{h(i\pi-\theta)} = t(\theta)\, t(i\pi+\theta) \qquad (14)$$

Let C denote a contour, which is identical with the boundary of the physical strip , except that possible poles or zeros of $t(\theta)$ on the boundary of the physical stripe are circumvented by small half-circles (lying inside the physical strip) and let θ be an element of the physical strip . Under the done assumptions we have in the repulsive region the integral representation

$$\ln t(\theta) = \frac{1}{2\pi i}\int_C \frac{dz}{\mathrm{sh}(z-\theta)} \ln t(z) = \frac{1}{2\pi i}\int_{-\infty}^{+\infty} \frac{dx}{\mathrm{sh}(x-\theta)} \ln\left(\frac{\mathrm{sh}\lambda x}{\mathrm{sh}\lambda(i\pi-x)}\right)$$

The last equality is obtained by using (12), (14).
After some calculation one obtains

$$t(\theta) = \exp i\int_0^\infty \frac{dx}{x}\,\frac{\mathrm{sh}\frac{x}{2}(1-\lambda)}{\mathrm{sh}\frac{x}{2}\,\mathrm{ch}\frac{x}{2}\lambda}\,\sin\frac{x\lambda}{\pi}(i\pi-\theta)$$

Using Malmstén's formula /21/, we finally arrive at (13.a).

Discussion of the result:

From (13a) we can read off the poles and zeros of $t(\theta)$ by looking for the zeros of the factors. One immediately recognizes, that all poles and zeros are lying on the imaginary θ-axis. Simple linear relations between $1/\lambda$ and Re φ with $\varphi = \frac{\theta}{i\pi}$ are obtained. They are drawn in fig.3. In the physical strip there are only simple poles given by

$$\theta_k = i\pi\left(1-\frac{k}{\lambda}\right) \qquad (k=1,2,\dots,[\lambda]) \qquad (15.a)$$

$[\lambda]$ is the largest integer smaller then λ . The poles describe $f\bar{f}$ boundstates (bosons). By translating the θ-dependence back in the s-plane we obtain a spectrum of quasiclassical type /11/

$$m_k = 2m\sin\left(\pi\frac{k}{2\lambda}\right) \qquad (k=1,2,\dots,[\lambda]) \qquad (15.b)$$

With the relationship

$$\lambda = 1+\frac{2g}{\pi} = \frac{8\pi}{\beta^2}-1 \qquad (16)$$

between λ and the MTM coupling g (SG coupling β, cf.(2)) the quasiclassical spectrum becomes exact as has been conjectured in ref./11/. Before the exact S matrix was known, the spectrum (15) has also been

derived by lattice methods /22/ and by assuming factorisation for the SG boson S matrix /23/.
In contrary to the spectrum the quasiclassical S matrix /24/ is only exact for integer λ. These are precisely the points, where the reflection amplitude (13.b) vanishes and a new particle (cf.(15.a)) enters the physical spectrum. Zamolodchikov /1/ first obtained the exact MTM S matrix by looking for the simplest analytic interpolation between these points, which has a nonvanishing reflection amplitude. It has been checked /25/ that $t(\theta)$ agrees up to third order in g with MTM perturbation theory.

IV Boundstates and Levinsons's theorem

In the previous section we have associated simple poles (we consider always poles and zeros in the physical strip now) of the amplitude $t(\theta)$ with $f\bar{f}$ boundstates. It is, however, known in potential scattering, that only those simple poles of the S matrix describe boundstates, which correspond to zeros in a related Jost function. In the nonrelativistic limit of the MTM $f\bar{f}$ scattering is known to be described by a smooth potential /26/.One may therefore conjecture, that results similar to potential scattering hold for the relativistic MTM S matrix. This question has been answered affirmative by explicit inspection of the S matrix /12/.

Let us first review potential scattering in one space dimension (e.g. /27/) for a well behaved symmetric potential V(x):

$$\int_{-\infty}^{+\infty} dx\ (1+|x|)\ |V(x)| < \infty \qquad \text{and} \qquad V(x) = V(-x).$$

The Jost solutions f(x,k) of the Schrödinger equation

$$-\frac{1}{2m}\frac{d^2}{dx^2}\psi(x) + V(x)\psi(x) = E\,\psi(x), \qquad E=\frac{k^2}{2m}$$

are defined by the boundary condition

$$f(x,k) \approx e^{ikx} \qquad \text{as} \qquad x \to \infty$$

f(x,k) is analytic in the open and continuous on the closed upper half of the complex k-plane. The scattering solution of the Schrödinger equation corresponding to an incoming plane wave e^{-ikx} from the right is given by

$$\Psi_r(x,k) = f(x,-k) + r(k)\cdot f(x,k) = t(k)\cdot f(x,-k)$$

where t and r are the transmission and reflection coefficients. The parity even (+) and odd (-) channels of the S matrix are

$$s_{\pm}(k) = t(k) \pm r(k) .$$

They obey the unitarity condition $|s_{\pm}(k)| = 1$ for k real. The Jost functions of the parity $\pm$ channels are defined to be ($f_x = \frac{df}{dx}$)

$$f_+(k) = \frac{f_x(0,k)}{ik} \qquad \text{and} \qquad f_-(k) = f(0,k) .$$

It can be proved (e.g. /12,27/), that the Jost functions are related to the amplitudes by

$$s_{\pm}(k) = \frac{f_{\pm}(-k)}{f_{\pm}(k)} \tag{17}$$

and

$$t(k) = \frac{1}{f_+(k)\cdot f_-(k)} \tag{18}$$

Furthermore it can be proved, that the Jost functions $f_{\pm}(k)$ have in the upper half complex k-plane no poles and only a finite number of simple zeros. The zeros are located on the imaginary k-axis. They are different for f_+ and f_- and correspond to boundstates with even or odd wave functions respectively. If a zero is located at $k=i\varkappa$, $\varkappa > 0$ the energy of the corresponding boundstate is $E = -\varkappa^2$. We recognize that all poles in t(k) correspond to physical boundstates, whereas in the $s_{\pm}$ channels redundant poles are possible.

We now state Levinson's theorem. For real k the phase shifts defined modulo π by

$$s_{\pm}(k) = e^{2i\delta_{\pm}(k)}$$

are real functions. For the Jost functions $\bar{f}_{\pm}(k) = f_{\pm}(-\bar{k})$ holds and the phase shifts are determined modulo 2π (cf. (17)) from

$$f_{\pm}(k) = |f_{\pm}(k)|\; e^{-i\delta_{\pm}(k)} .$$

The functions $f_{\pm}(k)$ are analytic in the upper half plane and we can apply the argument principle to get the numbers of parity $\pm$ boundstates

$$n_{\pm} = \frac{1}{2\pi i} \int_C dk \frac{f'_{\pm}(k)}{f_{\pm}(k)} = \frac{1}{\pi} (\delta_{\pm}(o) - \delta_{\pm}(\infty)) - \frac{1}{2}\alpha_{\pm} \tag{19.a}$$

Here the contour C encloses the upper half plane, avoiding k=o on a small half circle inside the upper half plane, where the behaviour $f(k) \approx k^{\alpha_\pm}$ for $k \to o$ is assumed. The integers $\alpha_{\pm}$ can be obtained (e.g./12/) from the threshold behaviour of $s_{\pm}$ via

$$\alpha_{\pm} = \pm \frac{1}{2} (s_{\pm}(o) - 1) \tag{19.b}$$

Equations (19) are Levinson's theorem for one dimensional potential scattering.

In the nonrelativistic limit of the MTM the physical strip of the complex θ-plane becomes the upper half complex k-plane. Let us now turn to the relativistic case.

Because of CPT and T invariance parity $\pm$ eigenstates in the MTM are identical with C=$\pm$ 1 eigenstates

$$|\pm\rangle = \frac{1}{2} (|p_1,\bar{p}_2\rangle \pm |\bar{p}_1,p_2\rangle) \tag{21}$$

On these states the two-particle S matrix becomes diagonal

$$S_2 = \begin{pmatrix} s_+ & \\ & s_- \end{pmatrix} \qquad \text{with} \qquad s_{\pm} = t \pm r \; .$$

In the $s_{\pm}$ channels the unitarity relations (7.b,c) read

$$s_+(\theta)\cdot s_+(-\theta) = s_-(\theta)\cdot s_-(-\theta) = 1 \; .$$

Explicit formulas for $s_{\pm}$ are

$$s_+(\theta) = - \frac{\mathrm{sh}\frac{\lambda}{2} (\theta+i\pi)}{\mathrm{sh}\frac{\lambda}{2} (\theta-i\pi)} t(i\pi -\theta) \tag{22.a}$$

and

$$s_-(\theta) = - \frac{\mathrm{ch}\frac{\lambda}{2} (\theta+i\pi)}{\mathrm{ch}\frac{\lambda}{2} (\theta-i\pi)} t(i\pi -\theta) \tag{22.b}$$

We note from these equations that the $s_{\pm}$ amplitudes have in the physical strip only poles for certain imaginary θ and no zeros. Let us exclude

λ =integer for the moment, then the poles are given graphically as function of λ in fig.4a,b). The boson boundstates (cf.(15)) energe as simple poles in s_- for k=odd and in s_+ for k=even. Therefore the boundstates have parity $P=(-1)^k$. This is consistent with the opinion that the lowest (=lightest) boundstate b_1 is a pseudoscalar, which corresponds to the elementary SG field.

It is not difficult to compute the residua of the poles with the result

$$(+i)\cdot \operatorname*{Res}_{\theta=\theta_k} s_+ = R_k > o \qquad \text{for} \quad P=+1 \qquad (\text{i.e. k even}) \tag{23.a}$$

$$(+i)\ \operatorname*{Res}_{\theta=\theta_k} s_- = R_k < o \qquad \text{for} \quad P=-1 \qquad (\text{i.e. k odd}) \tag{23.b}$$

Here $\operatorname*{Res}_{\theta=\theta_k} f(\theta) = +\lim_{\theta\to\theta_k} (\theta-\theta_k)\cdot f(\theta)$. This is in agreement with the result$^{+)}$:

a) Symmetric wave functions correspond to negative and antisymmetric wave functions correspond to positive residua. This is required by general S matrix theory (e.g./28/); poles with the wrong residuum give rise to ghosts with negative norm.
b) Fermion-antifermion boundstates of parity P=-1 have symmetric wave functions and of parity P=+1 have antisymmetric wave functions (e.g./29/). Putting a) and b) together we precisely arrive at (23).

Besides the physical boundstate poles we notice in $s_\pm$(fig.4) further poles which are redundant. Following the suggestions of potential scattering (17),(18) one can introduce "generalized Jost functions" by the ansatz

$$s_\pm(\theta) = \frac{g_\pm(-\theta)}{g_\pm(\theta)} \tag{24}$$

and

$$t(\theta) = \frac{1}{g_+(\theta)\cdot g_-(\theta)} \tag{25}$$

$^{+)}$Formulated in the s-plane. Residua in s- and θ-plane are related by

$(i\alpha)\ \operatorname*{Res}_{\theta_k} = \operatorname*{Res}_{s_k}$ with $\alpha > o$.

and finds the solution

$$g_+(\theta) = \prod_{l=o}^{\infty}\prod_{k=1}^{\infty} \frac{(2l-1+\frac{2k}{\lambda}+\frac{\theta}{i\pi})\cdot(2l+1+\frac{2k-1}{\lambda}+\frac{\theta}{i\pi})}{(2l+\frac{2k}{\lambda}+\frac{\theta}{i\pi})\cdot(2l+\frac{2k-1}{\lambda}+\frac{\theta}{i\pi})}$$

$$\times \frac{(2l+2+\frac{2k+1}{\lambda}-\frac{\theta}{i\pi})\cdot(2l+2+\frac{2k-2}{\lambda}-\frac{\theta}{i\pi})}{(2l+1+\frac{2k+1}{\lambda}-\frac{\theta}{i\pi})\cdot(2l+3+\frac{2k-2}{\lambda}-\frac{\theta}{i\pi})} \tag{26}$$

and $g_-(\theta)=g_+(\theta-\frac{i\pi}{\lambda})$. The generalised Jost functions have no poles and for imaginary θ a finite number of simple zeros in the physical strip . Parity $\pm$ boundstates correspond presisely to these zeros of $g_\pm$.

The functions $g_\pm(\theta)$ can be used to obtain the generalized version of Levinson's theorem. The numbers of parity $\pm$ boundstates are

$$n_\pm = \frac{1}{2\pi i}\int_C d\theta \frac{g'_\pm(\theta)}{g_\pm(\theta)} =$$

$$\frac{1}{\pi}(\delta_\pm(o)-\delta_\pm(\infty)) + \frac{1}{\pi}(\delta(o)-\delta(\infty)) - \frac{1}{2}\alpha_\pm \tag{27.a}$$

Here the contour C is identical with the boundary of the physical strip , except that $\theta=o$ is avoided on a small half circle inside the physical strip , where we have the behaviour $g(\theta)\approx\theta^{\alpha_\pm}$ for $\theta\to o$. The integers $\alpha_\pm$ can be obtained from the threshold behaviour of $s_\pm$ via

$$\alpha_\pm = \mp\frac{1}{2}(s_\pm(o)-1) \tag{27.b}$$

The sign difference compared with (19.b) relies on the different threshold behaviour of bosons (potential scattering) and fermions (present case), cf./12/. Equations (27) constitute the generalized Levinson theorem. By explicit calculation of the phase shifts we may obtain the already known results

$$n_+ = [\lambda/2]\ ,\qquad n_- = [(\lambda+1)/2]\ ,\qquad n_+ + n_- = [\lambda]$$

The term $\frac{1}{\pi}(\delta(o)-\delta(\infty))$, which reflects crossing, is defined by

$u(\theta) = |u(\theta)| \cdot e^{2i\,\delta(\theta)}$ and gives the contribution $\frac{1}{4} \cdot (\lambda - 1)$.

In the semiclassical limit all results agree with those computed previously /3o/. This shows again, that the semiclassical method is reliable in the SG theory.

Let us close this section with some remarks on the so far excluded case λ =integer. For this case the reflection amplitude vanishes and we have

$$s_+(\theta) = s_-(\theta) = t(\theta)$$

Therefore poles with the wrong residuum (ghosts) appear in the $s_\pm$ channels. Following Karowski /31/ a consistent particle interpretation is guaranteed if there are no transitions between genuine particles and ghosts. Let us denote the ghost with mass m_n by b_n'. Anticipating method and notation of the next section the f $b_n \rightarrow$ f b_n' transition amplitude is

$$^{out}\langle b_n' f | b_n f \rangle^{in} = u_{23} \cdot t_{13} - t_{23} \cdot u_{13}$$

This vanishes for $u = \pm t$, as is indeed the case for λ integer (13.c). Similarly all other transitions between genuine particles and ghosts vanish.

For S matrices with higher symmetries (e.g. the chiral SU(n) Thirring model /6/) analog considerations are quite restrictive and imply, that only a finite number of boundstate poles can be introduced consistently. Further there are poles in the physical strip possible, which are not interpreted as boundstates or ghosts. They are called CDD /32/ poles. If such poles are allowed, i.e.assumption b) of section III is given up, the solution of the factorization equations for the transmission amplitude becomes

$$T(\theta) = \prod_{i=1}^{L} \frac{\mathrm{sh}\frac{1}{2}(\theta+\theta_i) \cdot \mathrm{ch}\frac{1}{2}(\theta-\theta_i)}{\mathrm{sh}\frac{1}{2}(\theta-\theta_i) \cdot \mathrm{ch}\frac{1}{2}(\theta+\theta_i)} \cdot t(\theta)$$

where $t(\theta)$ is the previous (minimal) solution and the θ_i are imaginary. If the relativistic model (like the SG theory) is described by a well behaved potential in the nonrelativistic limit, CDD poles are excluded /33/, because they would be in contradiction to the existence of Jost functions.

V The complete S matrix

The complete S matrix includes scattering of boundstates. Relying on ref./13/, this section is devoted to the derivation of the boundstate scattering amplitudes.

The $f\bar{f}$ boson boundstate of mass m_k as given by (15.b) is denoted b_k. There is the possibility of boson-boson scattering $b_k b_l$ and of boson-fermion scattering $b_k f$. For the description of scattering of particles with different masses m_k, m_l we use the rapidity variable Θ_{kl}, defined analog to (5) by

$$s_{kl} = m_k^2 + m_l^2 + 2m_k m_l \operatorname{ch}(\Theta_{kl}). \tag{28}$$

s_{kl} is the complex Mandelstam variable, which for physical values reduces to $\hat{s}_{kl}=(P_k+P_l)^2$. The l.h. cut in the s_{kl}-plane starts at $s_{kl}=(m_k-m_l)^2$ and the r.h. cut at $s_{kl}=(m_k+m_1)^2$. We further use the variable

$$\varphi_{kl} = \frac{\Theta_{kl}}{i\pi}. \tag{29}$$

If no misunderstanding is possible the indices k,l are dropped.

The fermion-fermion amplitude $u(\Theta)$ (13.c) is continuous at the boundstate poles Θ_k as may be seen from fig.3. Therefore and from the discussion of the $s_{\pm}$ amplitudes (cf. (22) ff.) we conclude, that in the fermion sectors the residua of the S matrix at $s_{12}= (P_1+P_2)^2=m_k^2$ (P_1 and P_2 are at unphysical values) act as projection operators. More presisely, the operators

$$P_{12}^k \overset{\text{def}}{=} \frac{1}{R_k} \operatorname*{Res}_{s_{12}=m_k^2} S(p_1,p_2) \tag{30}$$

project for $\lambda \neq$ integer onto chargeless states

$$\frac{1}{\sqrt{2}}\left(|p_1,\bar{p}_2\rangle + (-1)^k |\bar{p}_1,p_2\rangle \right)$$

with momentum $p=p_1+p_2$, mass m_k and parity $(-1)^k$. They are identified with the boundstates b_k.

Let us first consider the S matrix for boundstate-fermion scattering. It is defined by

$$S(p_1+p_2,p_3)\Bigg|_{\substack{(p_1+p_2)^2=m_k^2 \\ p_3^2 = m^2}} = \frac{1}{R_k} \operatorname*{Res}_{s_{12}=m_k^2} S_{12}S_{13}S_{23} \tag{31.a}$$

Using the factorization equation (1o) and the definition of the projector we find

$$\frac{1}{R_k} \operatorname*{Res}_{s_{12}=m_k^2} S_{12}S_{13}S_{23} = P^k_{12}S_{13}S_{23} = S_{23}S_{13}P^k_{12} = P^k_{12}S_{23}S_{13}P^k_{12} \tag{31.b}$$

Formulas (31) have a graphical interpretation as given in fig.5a).One can associate with b_k a fermion and antifermion with "parallel" momenta p_1 and p_2. Since for real and parallel momenta the mass shell condition $p_i^2=m^2$ implies $(p_1+p_2)^2=4m^2$, the momenta have to be complex and only the real parts are parallel. $p=p_1+p_2$ is real, because it is the physical momentum of b_k. A convenient choice is $p_1=(p^o,\ \mathrm{Re}p^1+i\,\mathrm{Im}p^1)$ and $p_2=(p^o,\ \mathrm{Re}p^1-i\mathrm{Im}p^1)$ with p^oreal. $(p_1+p_2)^2$ can take all values between o and $4m^2$ for convenient choices of $\mathrm{Im}p^1$.

We now calculate the amplitudes explicitly. There is no reflection, since the projector P^k_{12}appears on both sides of equation (31b). It is convenient to use amplitudes defined as functions of φ (cf.(29)) by $f(\varphi)\equiv f\ (\varphi(\theta))$. The boson-fermion scattering amplitude $t_{b_k f}(\varphi)$ becomes now

$$t_{b_k f}(\varphi) = {}^{out}\langle b_n f | S_{23}S_{13} | b_n f\rangle^{out}$$

$$= \frac{1}{2}\ (t_{23}u_{13} + (-1)^n r_{23}r_{13} + u_{23}t_{13})$$

where $\quad t_{23} = t(\varphi_{23}) \quad$ etc.

and $\quad \varphi_{13} = \varphi + \frac{1}{2}(1-\frac{n}{\lambda}), \qquad \varphi_{23} = \varphi - \frac{1}{2}(1-\frac{n}{\lambda}).$

Using equations (13) for t,r and u one finds after some calculation

$$t_{b_k f}(\varphi) = (-1)^k \frac{\sin\frac{\pi}{2}(\varphi + \frac{k}{2\lambda} + \frac{1}{2}) \ \sin\frac{\pi}{2}(\varphi - \frac{k}{2\lambda} + \frac{1}{2})}{\sin\frac{\pi}{2}(\varphi - \frac{k}{2\lambda} - \frac{1}{2}) \ \sin\frac{\pi}{2}(\varphi + \frac{k}{2\lambda} - \frac{1}{2})}$$

$$\times \left(\prod_{j=1}^{k-1} \frac{\sin\frac{\pi}{2}(\varphi + \frac{k-2j}{2\lambda} + \frac{1}{2})}{\sin\frac{\pi}{2}(\varphi - \frac{k-2j}{2\lambda} - \frac{1}{2})} \right)^2 \qquad (32)$$

Similarly boson-boson scattering can be treated. The S matrix for scattering $b_n b_m$ is defined by

$$S(p_1+p_2, p_3+p_4) \Bigg|_{\substack{(p_1+p_2)^2=m_k^2 \\ (p_3+p_4)^2=m_l^2}} =$$

$$\frac{1}{R_k} \operatorname*{Res}_{s_{12}=m_k^2} \frac{1}{R_l} \operatorname*{Res}_{s_{34}=m_l^2} S_{12}S_{13}S_{14}S_{23}S_{24}S_{34}$$

cf.fig.6b).Again there is only forward scattering. After some calculation one finds for $k \leq l$:

$$t_{b_k b_l}(\varphi) = \frac{\operatorname{tg}\frac{\pi}{2}(\varphi + \frac{k+l}{2\lambda}) \ \operatorname{tg}\frac{\pi}{2}(\varphi + \frac{l-k}{2\lambda})}{\operatorname{tg}\frac{\pi}{2}(\varphi - \frac{k+l}{2\lambda}) \ \operatorname{tg}\frac{\pi}{2}(\varphi - \frac{l-k}{2\lambda})}$$

$$\times \left(\prod_{j=1}^{k-1} \frac{\operatorname{tg}\frac{\pi}{2}(\varphi + \frac{k+l-2j}{2\lambda})}{\operatorname{tg}\frac{\pi}{2}(\varphi - \frac{k+l-2j}{2\lambda})} \right)^2 \qquad (33)$$

A consistency check is to build up the $t_{b_k b_l}$ amplitude from $b_k f\bar{f}$ scattering. The result is the same. This and analog checks were carried out by Karowski and Thun /13/. All amplitudes are in agreement with previous calculations /24/ in the semiclassical limit.

Particularly simple is the scattering amplitude for two elementary SG bosons b_1:

$$t_{b_1b_1}(\varphi) = \frac{\sin\pi\varphi + \sin\frac{\pi}{\lambda}}{\sin\pi\varphi - \sin\frac{\pi}{\lambda}}$$

At $\lambda=3$ or equivalently $\beta^2=2\pi$, where $m_1=m$ and $m_2=m\sqrt{3}$, one recognizes /34/

$$-u(\Theta) = -t(\Theta) = t_{b_1f}(\Theta) = t_{b_1\bar{f}}(\Theta) = t_{b_1b_1}(\Theta)$$

and

$$t_{b_2f}(\Theta) = t_{b_2\bar{f}}(\Theta) = t_{b_1b_2}(\Theta) \ .$$

This can be interpreted as a hidden SU(2) symmetry. The particles $f,b_1,\bar{f}$ are in an isotriplet and b_2 is in an isosinglet. Using different methods this symmetry was first observed by Coleman /35/ in his investigation of the massive Schwinger model.

Finally let us consider the poles (there are no zeros) of t_{b_kf} and $t_{b_kb_l}$ in the physical strip . In the boson fermion amplitude t_{b_kf} there are two simple poles at

$$\varphi = \frac{\pm k}{2\lambda} + \frac{1}{2} \qquad (34.a)$$

and double poles at

$$\varphi = \frac{\pm j}{2\lambda} + \frac{1}{2} \qquad (34.b)$$

with j=o,2,.. ,k-2 if k is even and j=1,3.., k-2 if k is odd.

Let us choose $k<l$ for the boson-boson amplitude $t_{b_kb_l}$.
Then we find simple poles at

$$\varphi = \frac{k+l}{2\lambda}, \qquad \varphi = \frac{l-k}{2\lambda} \qquad (35.a)$$

and double poles at

$$\varphi = \frac{k+l-2j}{2\lambda} \qquad (j=1,\ldots,k-1) \qquad (35.b)$$

Further we encounter the poles required by crossing symmetrie at $1-\varphi$.

For the special cases $t_{b_4 f}$ and $t_{b_2 b_4}$ the poles are drawn in fig. 6a) and b). At the first look it is surprising, that besides simple poles double poles are found. This and related questions are considered in the next section.

The simple pole at $\varphi = (k+1)/2\lambda$ in the boson-boson amplitude $t_{b_k b_1}$ has the interpretation as boson boundstate b_{k+1}. This is seen by calculating the associated mass and checking the residuum.

Also the simple pole in the boson-fermion amplitude $t_{b_k f}$ has a natural interpretation. It's associated mass is precisely m. The meaning is, that the fermion f can be interpreted as $b_k f$ boundstate. This is consistent with the picture /36/ of the SG soliton being a coherent boundstate of SG bosons. The soliton may "eat" a boson and remains a soliton. Again the check of the residuum is consistent.
The sign of the residuum R_b of a boundstate b is given by the compact formula (b=boundstate)

$$R_b \eta_b / (\eta_\alpha \eta_\beta) < 0 \tag{36}$$

where $\eta_{b_k} = (-1)^k$, $\eta_f = \eta_{\bar{f}} = i$ and α, β are the boundstate constituents.

More recently /37/ the interpretation of the fermion as boson-fermion boundstate has been used to reconstruct the MTM fermion S matrix from the boundstate S matrices. This method can be used /38/, to obtain the kink S matrix of the Gross Neveu model, where the boundstate S matrix is known /4/.

VI Anomalous and virtual thresholds

The double poles (34.b), (35.b) of the previous section turn out /14/ to be anomalous threshold. We follow closely ref./14/. Let us first recall what an anomalous threshold is.

For physical external momenta normal Landau singularities of the S matrix occur for those values of the momenta, for which one can draw a space-time graph of a process involving classical particles, which are all on mass shell, all moving forward in time, and interacting only through energy-momentum conserving interactions localized at a space-time point. The discontinuity of the singularity is obtained by the Cutkosky rules /39/. One evaluates the space-time graph as if it were a Feynman graph, with two exceptions:

a) The point interactions are replaced by actual S matrix elements.
b) Feynman propagators are replaced by mass-shell δ-functions, $\theta(p^{o})\,\delta\,(p^2-m^2)$.

Singularities below threshold are called anomalous singularities. In this case one may choose the time components of the external momenta to be real and the space components to be imaginary. Such a two-vector is called pseudomomentum. The rules for finding an anomalous singularity are /4o/:

(1) The graph must be a geometrical figure in Euclidean space.
(2) With every internal line there must be associated a pseudomomentum of lenght m, the mass of the particle associated with this line.
(3) The pseudomomentum must be parallel to the associated line.
(4) All pseudomomenta at a given vertex must sum to zero.

Thus, locating an anomalous singularity becomes a problem in Euclidean plane geometry. As an example let us consider the graph of fig.7).

The Cutkosky rules show that in four dimensions the singularity is a branch point. We have an eight-dimensional integral over six δ-functions; thus we obtain a finite discontinuity over the cut.
In two dimensions we have only a four-dimensional integral. Thus we

have two δ-functions left after integration. The associated singularity is a double pole. This is best shown by considering the singularity of the Feynman diagram associated with fig.7. By momentum conservation, the two virtual particles carry equal momenta, which we denote p_c. Thus the Feynman integral contains two factors $1/(p_c^2-m_c^2)$ and can be written as derivative with respect to m_c of an integral with only one such factor. The discontinuity of the new integral is $(p_c^2-m_c^2)$. Thus the singularity of the new integral is proportional to $1/(p_c^2-m_c^2)$ and the singularity of its derivative is a double pole.

It remains to show, that for each double pole (34.b), (35.b) there exists an associated diagram and vice versa. For the decay of a particle A the angle $\varphi(a,c;A)$ defined in fig.8 is related to the rapidity of a,c:

$$m_A^2 = m_a^2 + m_c^2 + 2m_am_c\cos\{\pi\varphi(a,c;A)\} .$$

For fig.9 to be a possible graph, the lines a and b must intersect, thus

$$\varphi(c,a;A) + \varphi(b,c;B) < 1 \tag{37}$$

The angle φ of fig.9 is the location of the singularity in the t_{AB} amplitude. By elementary geometry it is determined to be

$$\varphi = 2 - \varphi(A,c;a) - \varphi(B,c;b) \tag{38}$$

These equations are all we need to locate the double poles for any two dimensional forward scattering process corresponding to fig.7. All we have to do is to plug in the particle masses.

For the SG spectrum (15.b) the relevant angles of the three particle vertices are readily computed to be

$$\varphi(f,\bar{f};b_k) = 1 - \frac{k}{\lambda} \quad (39.a) \qquad \varphi(b_k,f;\bar{f}) = \frac{1}{2}\left(1+\frac{k}{\lambda}\right) \quad (39.b)$$

$$\varphi(b_k,b_1;b_{k+1}) = \frac{k+1}{2\lambda} \quad (39.c) \qquad \varphi(b_1,b_{k+1};b_k) = 1-\frac{k}{2\lambda} \quad (39.d)$$

Using these rules all double poles can be explained. For example fermion-boson scattering: Let A be f and B be b . While the initial fermion can only decay into a fermion and a boson, the initial boson may decay into either a fermion-antifermion pair or two bosons. We will for illustration only consider the case, where the initial boson decays into a fermion-antifermion pair. Therefore $a=b_1$, $b=\bar{f}$ and $c=f$ in fig.7. By equation (37) and (38) we obtain $2k-1>\lambda$. And by equation (38) the locations of the double poles are given by

$$\varphi = \frac{1}{2} + \frac{j}{2\lambda}$$

with $j=2l-k$, $l=1,\dots$ $[2k-\lambda]$ this is seen to describe a part of the range given in (34.b). The other part of the range is obtained by considering the decay of the initial boson into bosons.

The discussion can be completed /14/ and explains all double poles in $t_{b_k f}$ and $t_{b_k b_l}$. Especially there are no anomalous thresholds in the ff and $f\bar{f}$ amplitudes, because (37) cannot be fulfilled for this case. The simple pole at $\varphi=\frac{1-k}{2\lambda}$ (35.a) is explained by the tree graph of fig.9.

Let us finally consider fig.6. The pole given by $\varphi=\frac{6}{2\lambda}$ corresponds for $\lambda>6$ to the mass m_6 of the b_2b_4 boundstate b_6. At $\lambda=6$ the boundstate b_6 ceases to be present in the MTM spectrum of $f\bar{f}$ boundstates (15). Precisely at this point the pole at $\varphi=6/2\lambda$ and the crossing symmetric pole give a double pole. For $4<\lambda<6$ we have again two simple poles in the physical strip . The poles have, however, changed the sign of their residuum, such that an interpretation as boundstate is no longer natural. In a theory with production, there would be a branch point corresponding to the anomalous threshold $b_2b_4\rightarrow f\bar{f}$, the boundstate pole would move into the second sheet and become a resonance. This is impossible here, because of the conservation laws. The remnant is a virtual threshold in form of a double pole. Analog considerations apply to the b_{l+k} boundstate in the $t_{b_k b_l}$ amplitude.

VII Summary and outlook

The S matrix theory of the MTM is a well understood and fairly closed topic. A challenging open problem is the complete solution of the MTM and other models with soliton behaviour. Complete solution means construction of the Green's functions. Different approaches to this problem are treated at this conference in the lectures of a) Honerkamp, b) Jimbo, Miwa, Sato and c) Karowski. The bootstrap program outlined in the lecture of Karowski tries to construct Green's functions from known exact S matrices and is therefore closely related to the present lecture. So far the bootstrap program has lead to a simple construction /41/ of the Green's functions of the Ising model in the scaling limit.

Acknowledgement

It is a pleasure to acknowledge the kind hospitality of Prof.Iagolnitzer during this conference. I also would like to thank M.Karowski, H.J.Thun and P.Weisz for useful discussions.

References:

1.) A.B.Zamolodchikov, Commun.math. Phys.55 (1977) 183

2.) M.Karowski, H.J.Thun, T.T.Truong and P.Weisz, Phys.Lett. 67B (1977) 321

3.) A.B.Zamolodchikov and Aj.B.Zamolodchikov, Nucl.Phys. B133 (1978) 525

4.) A.B.Zamolodchikov and Aj.B.Zamolodchikov, Phys.Lett. 72B (1978) 481

5.) R.Shankar and E.Witten, Phys.Rev. D17 (1978) 2194

6.) B.Berg and P.Weisz, Nucl.Phys. B146 (1978) 2o5; E.Abdalla,B.Berg and P.Weisz, Nucl.Phys. B157 (1979) 387

7.) V.Kurak and J.A.Swieca, Phys.Lett. 82B (1979) 289

8.) A.B.Zamolodchikov and Aj.B.Zamolodchikov, Ann.Phys. (N.Y.) 12o (1979) 253

9.) M.Lüscher, Nucl.Phys. B117 (1976) 475;
P.P.Kulish and E.R.Nissimov, JETP Lett. 24 (1976) 22o;
B.Berg, M.Karowski and H.J.Thun, Nuovo Cim. 38A (1977) 11;
R.Flume and S.Meyer, Nuovo Cim. Lett. 18 (1977) 238;
J.Lowenstein and E.Speer, Commun.math. Phys. 63 (1978) 97

1o.) D.Iagolnizer, Phys.Rev.D18 (1978) 1275; Phys.Lett. 76B (1978) 2o7

11.) R.J.Dashen, B.Hasslacher and A.Neveu, Phys.Rev. D11 (1975) 3424;
V.E.Korepin and L.D.Faddeev, Theor.Math.Phys. 25 (1975) 1o39

12.) B.Berg, M.Karowski, W.R.Theis and H.J.Thun, Phys.Rev. D17 (1978) 1172

13.) M.Karowski and H.J.Thun, Nucl.Phys. B13o (1977) 295;
A.B.Zamolodchikov, Preprint ITEP-12 (1977) unpublished

14.) S.Coleman and H.J.Thun, Commun.math.Phys. 61 (1978) 31

15.) S.Coleman, Phys.Rev. D11 (1975) 2o88
For a review see: J.A.Swieca, Fortsch.Phys. 25 (1977) 3o3

16.) M.J.Ablowitz, D.J.Kaup, A.C.Newell and H.Segur, Phys.Rev.Lett.31 (1973) 152;
L.A.Takhtadzhyan and L.D.Faddeev, Theor.Math.Phys. 21 (1975) 1o46.
Infinitely many higher conservation laws were first obtained by M.D.Kruskal and D.Wiley, American Mathematical Society, Summer seminar on nonlinear wave motion, ed. A.C.Newell, Potsdam, N.Y., July 1972. For a review see A.C.Scott, F.Y.F. Chu and D.W. McLaughlin, Proc.IEEE 61 (1973) 1443

17.) P.P. Kulish and E.R. Nissimov /9/ and references given therein.

18.) I. Ya. Araf'eva and V. E. Korepin, JETP Lett. 2o (1974) 312

19.) B. Berg, Nuovo Cimento 41A (1977) 58 and references given therein.

2o.) J.B. McGuire, J. Math. Phys. 5 (1964) 622;
E. Brezin and J. Zinn-Justin, Compt. Rend. Acad. Sci. Paris B263 (1966) 67o;
C.N. Yang, Phys. Rev. Lett. 19 (1967) 1312 and Phys. Rev 168 (1968) 192o

21.) I.S. Gradshteyn and I.M. Ryzhik, "Table of Integrals Series and Products", Academic Press (N.Y.1965), 8.341(7), p.939

22.) A. Luther, Phys. Rev. B14 (1977) 2153

23.) S. Nussinov, Phys. Rev. D14 (1976) 647;
B. Schroer, T.T. Truong and P.H. Weisz, Phys. Lett. 63B (1976) 422;
cf. also M. Karowski /37/.

24.) V. E. Korepin and L.D. Faddeev /11/

25.) P.H. Weisz, Nucl. Phys. B122 (1977) 1

26.) V.E. Korepin, JETP Lett. 23 (1976) 224; for a review see /8/.

27.) L.D. Faddeev, Am. Math. Soc. Trans. 65 (1967) 139

28.) W. Brenig and R. Haag, Fortschr. Phys. 7 (1959) 183

29.) J.D. Bjorken and S.D. Drell, "Relativistische Quantenfeldtheorie", B.I. Hochschultaschenbücher 1o1/1o1a^{+}, S.124

3o.) R. Jackiw and G. Woo, Phys. Rev. D12 (1975) 1649;
S. Coleman, Phys. Rev. D12 (1975) 165o

31.) M. Karowski, Nucl. Phys. B153 (1979) 244

32.) L. Castillejo, R.H. Dalitz and F.J. Dyson, Phys. Rev. 1o1 (1956) 453

33.) R. Haag, Nuovo Cimento 5 (1957) 2o3

34.) S. Meyer and P.H. Weisz, Phys. Lett. 68B (1977) 471

35.) S. Coleman, Ann. Phys. (N.Y.) 1o1 (1976) 239

36.) R. Jackiw, "Collective Phenomena in Quantum Field Theory", Cracow lectures 1975, Acta Physica Polonica B6 (1975) 919

37.) M. Karowski, Kaiserslautern lectures 1979, to be published by Plenum Press (London)

38.) M. Karowski and H.J. Thun, to be published

39.) R.J. Eden, P.V. Landshoff, P.I. Olive and J.C. Polkinghorne, "The analytic S matrix", Cambridge University Press (Cambridge 1966), p.87 ff. and p.11o ff.

4o.) J.D. Bjorken and S.D. Drell /29/ S.232 ff.

41.) B. Berg M. Karowski and P.H. Weisz, Phys. Rev. D19 (1979) 2477

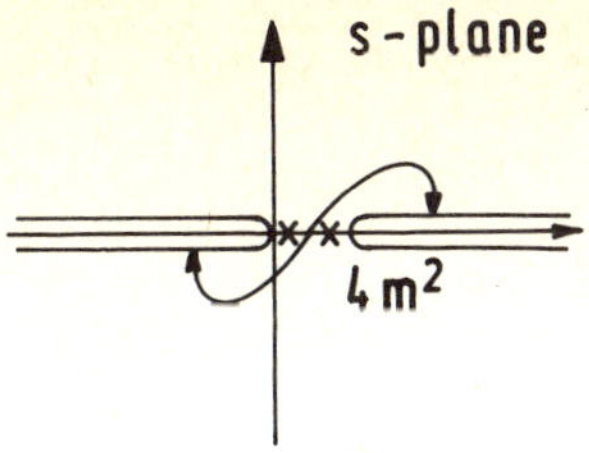

s-plane

Fig. 1a

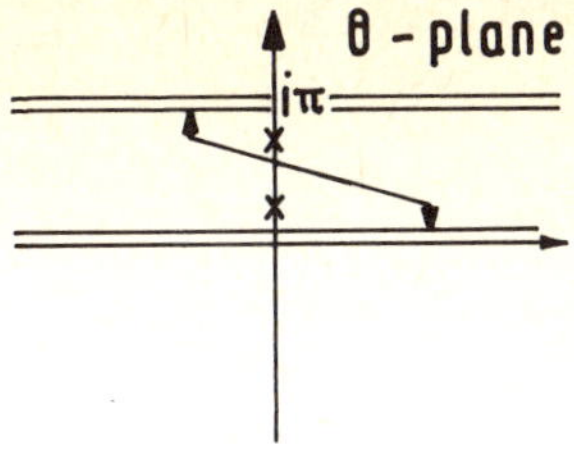

θ-plane

Fig. 1b

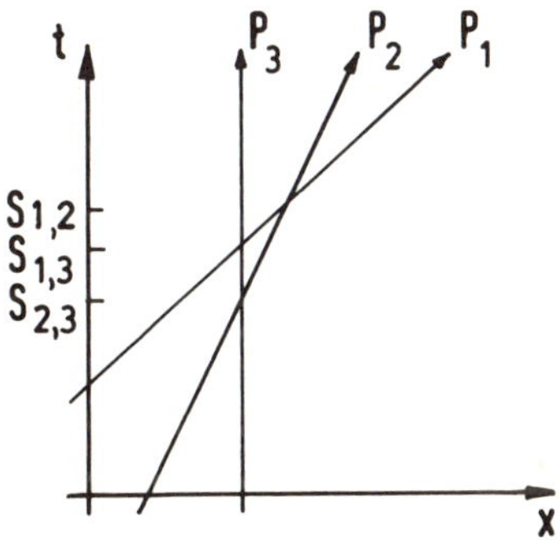

Fig. 2a

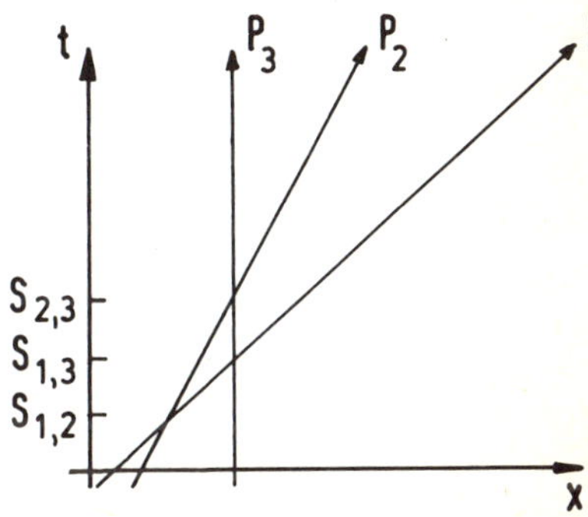

Fig. 2b

Equivalent possibilities for factorization of 3-particle scattering.

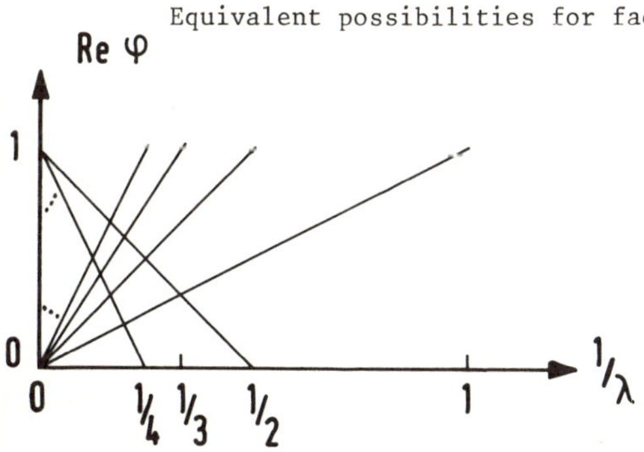

Fig. 4a

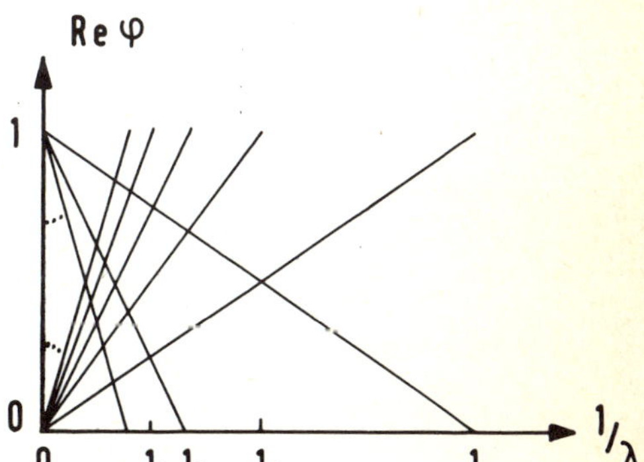

Fig. 4b

a) Poles and zeros of the amplitude $s_+ = t+r$.

b) Poles and zeros of the amplitude $s_- = t-r$.

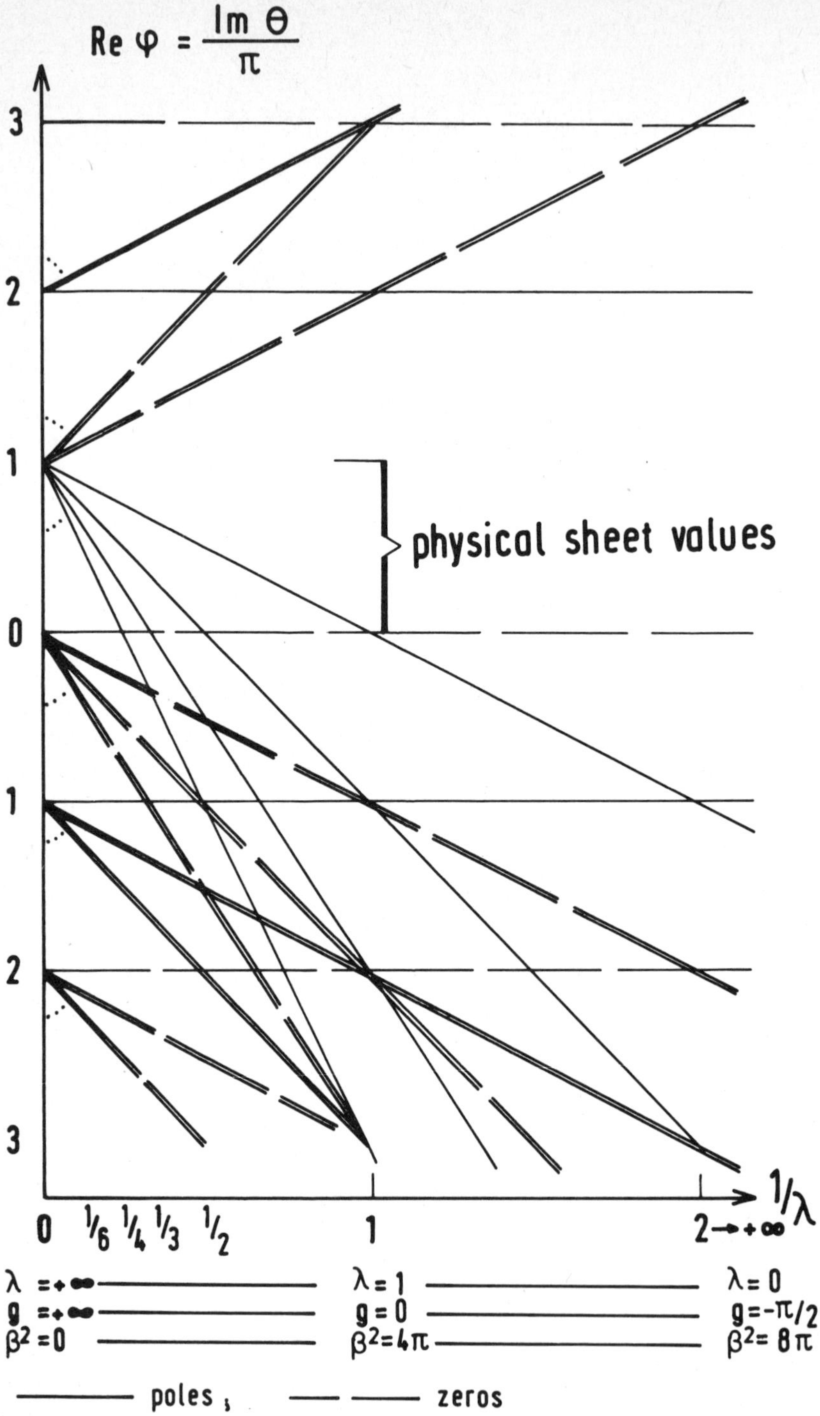

Fig. 3

Poles (———) and zeros (-----) of $t(\theta)$ for arbitrary coupling.

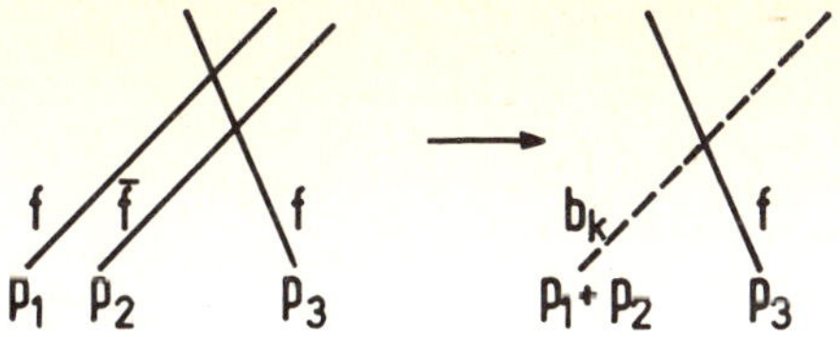

Fig.5a - Composition of a fermion-antifermion pair to a boson b_k.

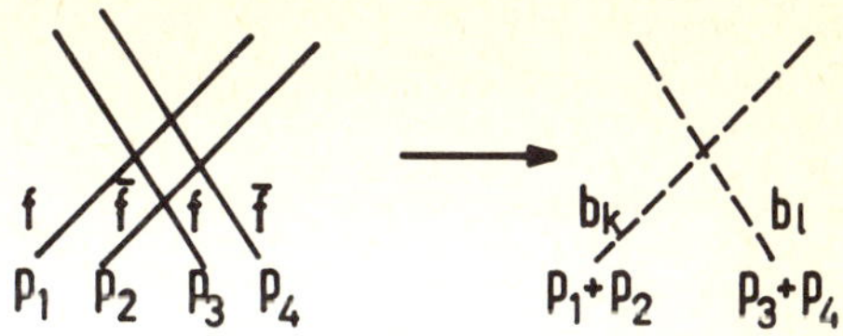

Fig.5b - Composition of two $f\bar{f}$ pairs to bosons b_k, b_l.

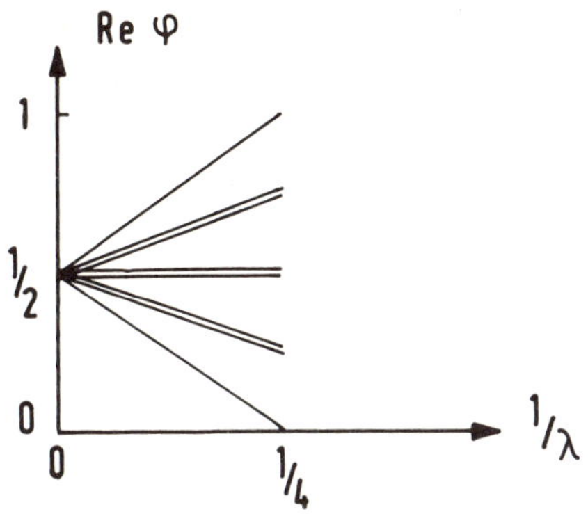

Fig.6a - Poles (——) and double poles (═══) of $t_{b_4 f}$.

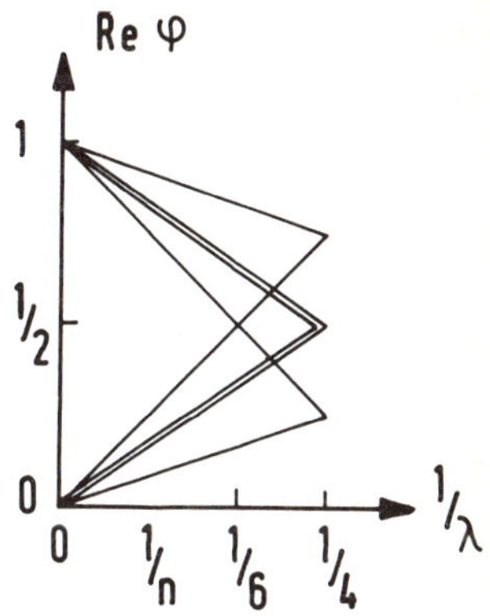

Fig.6b - Poles and double poles of $t_{b_4 b_2}$.

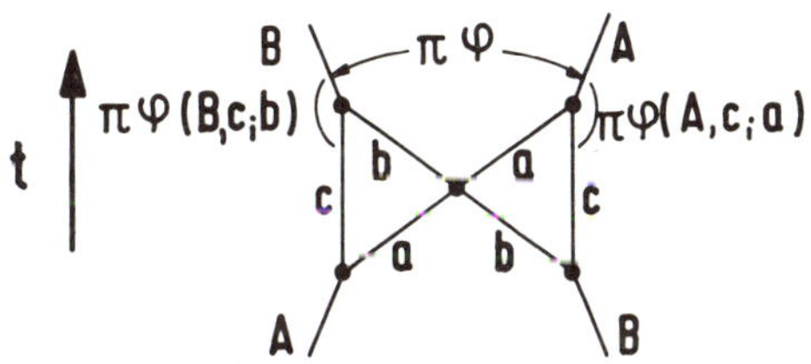

Fig.7 - The graph that produces double poles as anomalous thresholds in two dimensions.

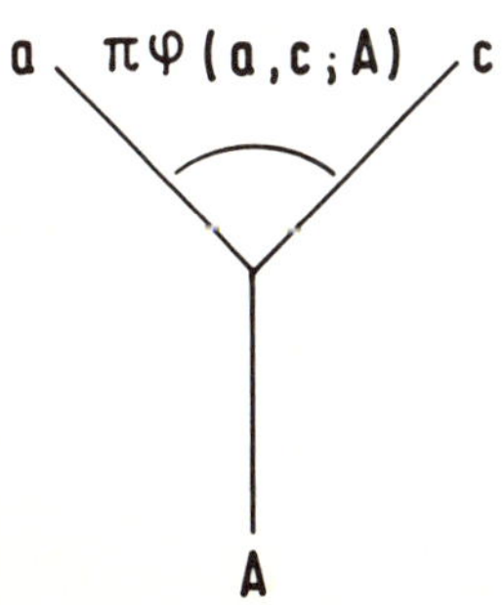

Fig.8 - Graphical representation of the angle $\varphi(a,c;A)$.

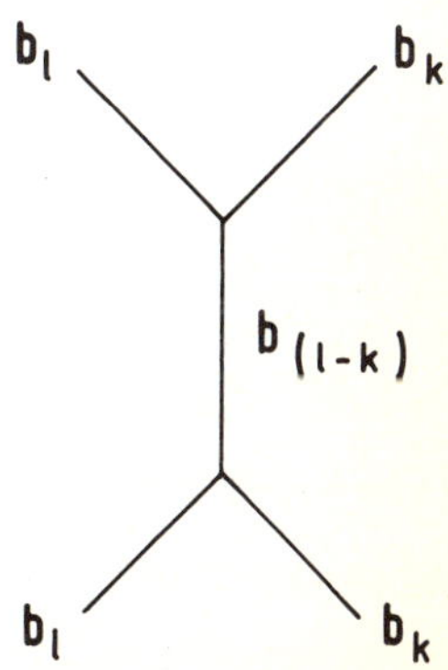

Fig.9 - Graph explaining a simple pole.

FIELD THEORIES IN 1+1-DIMENSIONS WITH SOLITON BEHAVIOUR :

FORM FACTORS AND GREEN'S FUNCTIONS+)

M. KAROWSKI

Institut für Theoretische Physik, Freie Universität Berlin

D-1000 Berlin 33, Arnimallee 3

Abstract : For a number of field theoretic models with soliton behaviour the on-shell solutions have been obtained in the last few years. In this paper it is shown how to calculate offshell quantities like form factors and Green's Functions by means of the known S-matrices, general properties of quantum field theory and minimality assumptions.

1. Introduction

Various 1+1 dimensional field theoretic models with soliton behaviour have been investigated recently, e.g. the sine-Gordon alias the massive Thirring model [1], the O(N) nonlinear σ-model and the Gross-Neveu model [2] , the chiral SU(N) model [3] and a Z(N)-Ising model in the scaling limit [4]. The later will be used in this paper in order to demonstrate the bootstrap program for the construction of offshell quantities like form factors and Green's functions from the known exact S-matrix, general properties of quantum field theory and some minimality assumptions. A Z(N)-Ising model in the scaling limit is defined by taking the correlation functions of a Z(N)-Ising model on a 2-dimensional lattice in the limit

$$\lim_{T\to T_c} \left|1-\frac{T}{T_c}\right|^{-n\beta} \langle\sigma_{R_1}\cdots\sigma_{R_n}\rangle = S(r_1\ldots r_n)$$

with $R_i \to \infty$ and $r_i = R_i|1-T/T_c|$ fixed. By analytic continuation one obtains from the euclidian Schwinger functions the Green's functions in Minkowski space. For $N > 2$ there exist several Z(N)invariant nearest neighbour interactions. Let us make the assumption that one of these interactions enforces soliton behaviour for the model in scaling limit.

For field theories with soliton behaviour the dynamics is governed by an infinite number of higher conservation laws. In particular for scattering processes they imply the absence of particle production and factorization of the n-particle S-matrix [5]

$$S^{(n)}(p_1\ldots p_n) = \prod_{i=1}^{n-1}\prod_{j=i+1}^{n} S^{(2)}(p_i,p_j) \tag{1}$$

where the two-particle S-matrix $S^{(2)}(p_i p_j)$ applied to an n-particle state acts on the particles with momenta p_i and p_j

$$S^{(2)}(p_i,p_j)\,|\ldots\alpha_i(p_i)\ldots\alpha_j(p_j)\ldots\rangle^{in} = |\ldots\alpha'_i(p_i)\ldots\alpha'_j(p_j)\ldots\rangle^{in}\, S_{\alpha'_i\alpha'_j\,\alpha_i\alpha_j}(p_i,p_j) \tag{2}$$

The α's label the kinds of the particles. Together with unitarity, crossing and minimality assumptions this property allows the exact calculation of the S-matrix

[1,2,3,4] . Let us repeat briefly the argument for the Z(N)-Ising model with soliton behaviour [4] . Assume the existence of an elementary boson b_1 corresponding to the order variable $\sigma(x)$ and the existence of a two-particle bound state $b_2 = (b_1 b_1)$. Then [6] we get the series of bound states b_k with masses

$$m_k = m_1 \frac{\sin ak}{\sin a} \tag{3}$$

Since the order variable σ assumes the values - the N roots of one - we have

$$\sigma^+ = \sigma^{N-1} \tag{4}$$

which suggests the assumption that the antiparticle $\bar{b}_1$ is identical to the bound state b_{N-1}. Then from $m_1 = m_{N-1}$ we get for the parameter a the value

$$a = \frac{\pi}{N} \tag{5}$$

Minimality means that the transmission amplitude for the scattering of two particles b_1 has only the pole corresponding to the bound state b_2. This implies uniquely [4]

$$S_{11}(\theta) = \frac{\operatorname{sh} \frac{1}{2} (\theta + \frac{2\pi i}{N})}{\operatorname{sh} \frac{1}{2} (\theta - \frac{2\pi i}{N})} \tag{6}$$

where θ is the rapidity difference defined by

$$p_1 p_2 = m^2 \operatorname{ch} \theta . \tag{7}$$

From eq.(6) we obtain for the scattering of the boundstates b_j and b_k

$$S_{jk}(\theta) = \exp 4 \int_0^\infty \frac{dx}{x} \frac{\operatorname{ch} x (1 - \frac{|j-k|}{N}) - \operatorname{ch} x (1 - \frac{j+k}{N})}{\operatorname{sh} x \operatorname{th} \frac{x}{N}} \operatorname{ch} x \frac{\theta}{i\pi} . \tag{8}$$

This formula is consistent with crossing , the above assumption $\bar{b}_1 = b_{N-1}$ and more general even with $\bar{b}_k = b_{N-k}$ since

$$S_{j\bar{k}}(\theta) = S_{jk}(i\pi - \theta) = S_{j\, N-k}(\theta) \tag{9}$$

This model is simple because of absence of backward scattering which is a consequence of $\bar{b}_1 = b_{N-1}$ [4] .

For N=2 the model is much more simple since there is only one kind of particles $b = \bar{b}$ and the two-particle S-matrix is

$$S^{(2)} = -1 \tag{10}$$

because of this simplification the bootstrap program has been carried through for this model [8] .

2. Watson's Theorem [7]

For form factors, i.e. matrix elements of local operators, we derive a set of equations which follow from general principles of quantum field theory, maximal analyticity, and the S-matrix factorization. For simplicity we first consider the case where we have only one kind of bosons in the model and a hermitean operator $O(x)$. If we define the function $F(\theta)$ (for θ c.f. eq.(7)) by $\langle o|O(o)|p_1 p_2\rangle^{in} = F(\theta)$ it follows from CPT-invariance $\langle o|O(o)|p_1 p_2\rangle^{out} = F(-\theta)$, unitarity and factorization $F(\theta) = \langle o|O(o)|p_1 p_2\rangle^{out}\, S(\theta)$, and crossing $\langle p_1|O(o)|\hat{p}_2\rangle = F(i\pi-\theta)$.
Hence we have Watson equations:

$$\begin{aligned} F(\theta) &= F(-\theta)S(\theta) \\ F(i\pi-\theta) &= F(i\pi+\theta) \end{aligned} \qquad (11)$$

The corresponding equations for the general case where the α's denote the different kinds of particles read

$${}^{out}\langle \alpha_1(p_1)\ldots\alpha_m(p_m)\,|O(o)|\ldots\alpha_n(p_n)\rangle^{in} = F_{\alpha_1\ldots\alpha_n}(\theta_{ij}, i\pi-\theta_{rs}, \theta_{k\ell})$$

$$= S^{(m)}_{\alpha_1\ldots\alpha_m\;\alpha'_1\ldots\alpha'_m}(\theta_{ij})\; F_{\alpha'_1\ldots\alpha'_n}(-\theta_{ij}, i\pi+\theta_{rs}, -\theta_{k\ell})\; S^{(n-m)}_{\alpha'_{m+1}\ldots\alpha'_n\;\alpha_{m+1}\ldots\alpha_n}(\theta_{k\ell})$$

where $1 \leq i < j \leq m$, $1 \leq r \leq m < s \leq n$, $m < k < \ell \leq n$ and $o \leq m \leq n$.

Solutions of these equations are only known for simple cases.

3. Construction of Form Factors

The construction of form factors has been performed for models with an arbitrary number of kinds of particles for the case n=2. Hence one particle expectation values of local operators, the usual form factors are known exactly for various models [7]. Absence of backward scattering enables us to make proposals for more general form factors with n>2, e.g. for the sine-Gordon [7] and the Z(N)-Ising model. Furthermore arbitrary n-particle matrix elements are known for the Z(2)-Ising model in the scaling limit [8].

There are two distinct problems in this procedure.

a) the "matrix problem": The set of matrix equations (12) has not been solved for the general case. For n=2 we diagonalize the two-particle S-matrix and get for each eigenvalue the simple equations (11).

Theorem [7]

An analytic function F(θ) fulfilling eqs.(11) is (up to a normalization) uniquely determined by the positions of the poles (and zeros) at $\theta=i\alpha_k$ in the physical strip $0< \mathrm{Im}\,\theta < \pi$:

$$F(\theta) = K(\theta)\, F^{min}(\theta) \qquad (13)$$

where $K(\theta) = \text{const} \left[\prod_k \mathrm{sh}\,\frac{1}{2}(\theta-i\alpha_k)\ \mathrm{sh}\,\frac{1}{2}(\theta-i\alpha_k)\right]^{-1} \qquad (13a)$

is a solution of eqs.(11) with S=1 and

$$F^{min}(\theta) = \exp \frac{1}{4\pi i} \int_{-\infty}^{\infty} \frac{dz}{\mathrm{sh}\,\frac{1}{2}(z-\theta)} \frac{\mathrm{ch}\,\frac{1}{2}\theta}{\mathrm{ch}\,\frac{1}{2}z}\ \ell n\, S(z) \qquad (13b)$$

For vanishing reflèction $S^{(n)}$ is diagonal,and a solution of the general equations (12) is

$$F_{\alpha_1..\alpha_n}(\theta_{12}...) = K_{\alpha_1..\alpha_n}(\theta_{12}...) \prod_{i<j} F^{min}_{\alpha_i\alpha_j}(\theta_{ij}) \qquad (14)$$

where the function K is a solution of eqs.(12) with S=1, this means it has no branch cuts in terms of the variables p_ip_j .

b) The "pole problem": The poles in the physical region, of the function

$$F_{\alpha_1..\alpha_n}(\theta_{12}...) = \langle o|O(o)|\alpha_1(p_1)...\alpha_n(p_n)\rangle^{in}$$

which are contained in the function K, are determined by one-particle states in all subchannels $\alpha_{i_1}...\alpha_{i_j}$. If we make the minimality assumption that there are no further so-called redundant poles and no zeros in the physical region, we can make proposals for exact form factors, which can be checked by perturbation theory.

4. Examples

a) The Z(N)-Ising model in the scaling limit with soliton behaviour.
The minimal form factor function corresponding to the external legs b_j and b_k for the Z(N)-Ising model follows from eqs.(8,13b):

$$F^{min}_{jk}(i\pi-\theta)=\exp \int_0^{\infty} \frac{dx}{x} \frac{\mathrm{ch}\, x(1-\frac{|j-k|}{N}) - \mathrm{ch}\, x(1-\frac{j+k}{N})}{\mathrm{sh}^2 x\ \ \mathrm{th}\,\frac{x}{N}} \left(1-\mathrm{ch}\, x\,\frac{\theta}{i\pi}\right) \qquad (15)$$

Since there is no backward scattering in this model the general form factor is given by eq.(14). The determination of K as a rational function of $p_i p_j$ is in general a complicated problem. Let us consider in more detail the case N=3. Then we have two particles $b_1, b_2=\bar{b}_1$ with mass m and the two-particle S-matrix

$$S_{11}(\theta)=S_{12}(i\pi-\theta) = \frac{\mathrm{sh}\,\frac{1}{2}(\theta+\frac{2\pi i}{3})}{\mathrm{sh}\,\frac{1}{2}(\theta-\frac{2\pi i}{3})} \tag{16}$$

The minimal form factor functions are

$$F_{11}^{min}(i\pi-\theta) = \exp\int_0^\infty \frac{dx}{x}\,\frac{\mathrm{sh}\,\frac{x}{3}}{\mathrm{sh}^2 x}\,(1-\mathrm{ch}\,x\frac{\theta}{i\pi})$$
$$F_{12}^{min}(i\pi-\theta) = \exp\int_0^\infty \frac{dx}{x}\,\frac{\mathrm{sh}\,\frac{2x}{3}}{\mathrm{sh}^2 x}\,(1-\mathrm{ch}\,x\frac{\theta}{i\pi}) \tag{17}$$

The matrix elements of the order variable $\sigma(x)$ must have poles corresponding to the bound state $b_2=\bar{b}_1$ of two particles b_1 . Hence we make the proposals:

$$\langle o|\sigma(o)|b_2(p_1)b_2(p_2)\rangle^{in} = \mathrm{const}\,\frac{1}{(p_1+p_2)^2-m^2}\,F_{11}^{min}(\theta_{12}) \tag{18a}$$

$$\langle o|\sigma(o)|b_1(p_1)b_1(p_2)b_2(p_3)\rangle^{in} = \mathrm{const}\,\frac{(p_1+p_2)^2}{(p_1+p_2)^2-m^2}\,\frac{1}{(p_1+p_2+p_3)^2-m^2}\prod_{i<j}F_{\alpha_i\alpha_j}^{min}(\theta_{ij}) \tag{18b}$$

$$\langle o|\alpha(o)|b_1(p_1)b_1(p_2)b_1(p_3)b_1(p_4)\rangle^{in} = \mathrm{const}\left[(p_1+p_2+p_3+p_4)^2-m^2\right]\prod_{i<j}\frac{F_{\alpha_i\alpha_j}^{min}(\theta_{ij})}{(p_i+p_j)^2-m^2} \tag{18c}$$

The numerators in eqs.(18b,c) prevent reflection in two-particle scattering and particle production 2→3, respectively. The proposals (18) fulfil the consistency conditions

$$\mathop{\mathrm{Res}}_{\theta_{12}=\frac{2\pi i}{3}}\langle o|\sigma(o)|b_1(p_1)b_1(p_2)b_2(p_3)\rangle^{in} = \mathrm{const}\,\langle o|\sigma(o)|b_2(p_1+p_2)b_2(p_2)\rangle^{in} \tag{19a}$$

$$\mathop{\mathrm{Res}}_{\theta_{12}=\frac{2\pi i}{3}} \langle o|\sigma(o)|b_1(p_1)b_1(p_2)b_1(p_3)b_1(p_4)\rangle^{in} = \text{const} \langle o|\sigma(o)|b_2(p_1+p_2)b_1(p_3)b_1(p_4)\rangle^{in} \tag{19b}$$

Furthermore from the four point function (18b) one gets back the two-particle S-matrix eq.(16):

$$S_{11}(\theta_{12}) = 1 + \frac{1}{4m^2 i\,sh\theta_{12}} \left[\left((p_1+p_2+p_3)^2-m^2\right)\langle b_1(-p_2)|\sigma(o)|b_1(p_1)b_1(p_2)\rangle^{in}\right]_{(p_1+p_2+p_3)^2=m^2} \tag{20}$$

We have up to now no other possibilities to check these results, because we do not know, as mentioned in the introduction, which interaction is that one which enforces the soliton behaviour for the Z(3)-Ising model in the scaling limit.

b) The Z(2)-Ising model.

For the Z(2)-Ising model in the scaling limit the problem of calculating arbitrary matrix elements of the order variable σ

$$\langle o|\sigma(o)|p_1 \ldots p_n\rangle^{in} = F^{(n)}(\theta_{12})\ldots) \tag{21}$$

has been solved completely [8]. The minimal solution of Watson's equations (11) with $S^{(2)}=-1$ is obtained from eq.(15)

$$F^{min}(\theta) = i\,sh\,\frac{\theta}{2} \tag{22}$$

Since there is no reflection we can use formula (14). The function $K^{(n)}$ contains the poles from one-particle states in all subchannels. Because there is only one kind of bosons b with mass m, we obtain for n=3

$$K^{(3)} = \frac{\text{const}}{(p_1+p_2+p_3)^2-m^2} = \frac{Z^{(3)}}{ch\,\frac{\theta_{12}}{2}\;ch\,\frac{\theta_{13}}{2}\,ch\,\frac{\theta_{23}}{2}}\;. \tag{23}$$

This proposal fulfills the consistency condition that $F^{(3)}$ with the leg corresponding to $\sigma(x)$ on-shell, as in eq.(20), has to reproduce the correct S-matrix $S^{(2)}$. From this condition we also obtain the constant to be $Z^{(3)}=2$.

For n≥5 the arguments are just slightly more involved. The function $K^{(n)}$ must have poles at $(p_i+p_j+p_k)^2=m^2$ where any set of particles i,j,k is combined to give a one - particle state. However, the residues of "higher poles" where more than three particles build up a one-particle state must vanish, since the presence of such poles would be in contradiction to the absence of particle production. Hence the minimality hypothesis implies that $K^{(n)}$ has the form

$$K^{(n)} = \frac{\text{const}\left[\prod_{i<j}(p_i+p_j)^2\right]^{\frac{n-3}{2}}}{\prod_{i<j<k}\left[(p_i+p_j+p_k)^2-m^2\right]} = \frac{Z^{(n)}}{\prod_{i<j}\text{ch}\,\frac{\theta_{ij}}{2}} \tag{24}$$

The numerator is necessary in order to modify the severe singularity of the denominator at $p_i+p_j=o$ into simple poles. The constant $Z^{(n)}$ can again be determined to be $Z^{(n)}=2^{\frac{n-1}{2}}$ by crossing and taking the off-shell leg $\sigma(x)$ on-shell. Finally we obtain

$$\langle o|\sigma(o)|p_1\ldots p_n\rangle^{in} = (2i)^{\frac{n-1}{2}} \prod_{i<j} \text{th}\,\frac{\theta_{ij}}{2} \tag{25}$$

for odd n. The matrix-elements vanish for even n because of Z(2)-invariance.

5. Green's Functions

Till now the problem of constructing Green's functions has been solved only for the simplest field theory with soliton behaviour, the Z(2)-Ising model in the scaling limit $T \to T_c$ [8]. Having deduced all matrix elements of the field $\sigma(x)$, we can immediately write down expressions for the Green's functions. For example the two-point function is

$$\tau(p) = \int d^2x\; e^{ipx} \langle T\sigma(x)\,\sigma(o)\rangle = i\int d\kappa^2\, \frac{\rho(\kappa^2)}{p^2-\kappa^2+i\varepsilon} \tag{26}$$

with the spectral function ρ given by

$$\rho(p^2) = \sum_{n\ \text{odd}} \frac{1}{n!} \int \frac{d\theta_1}{4\pi} \ldots \int \frac{d\theta_n}{4\pi}\; 2\pi\,\delta^{(2)}(p-\Sigma p_i) \prod_{1\leq i<j\leq n} \text{th}^2\,\frac{\theta_{ij}}{2} \tag{27}$$

There are similar formulae for the n-point Green's functions which are all in agreement with those derived in Ref. [9] by different methods. This agreement supports the validity of the minimality hypothesis and gives some confidence in the feasibility of the bootstrap program. The determination of the correlation functions for one of the more involved soliton field theories remains, however, a challenging open problem.

Acknowledgement:

This talk is based on collaborations at the Institut für Theoretische Physik FU Berlin, with B.Berg, V.Kurak, B.Schroer, R.Seiler, H.J.Thun, T.T.Truong and P.Weisz. I would like to thank them for many discussions.

References

[1] A.B.Zamolodchikov, Comm.Math.Phys.55 (1977)183;
M.Karowski, H.J.Thun, Nucl.Phys.B 130(1977)295;
M.Karowski, H.J.Thun, T.T.Truong, P.Weisz, Phys.Lett.67B(1977)321.

[2] A.B.and Al.B.Zamolodchikov, Nucl.Phys.B133(1978)525; Phys.Lett.72B(1978)72B;
B.Berg, M.Karowski, V.Kurak, P.Weisz, Phys.Lett.76B(1978)502.

[3] B.Berg, M.Karowski, V.Kurak, P.Weisz, Nucl.Phys.B134(1978)125;
B.Berg, P.Weisz, Nucl.Phys.B 146(1978)205;
V.Kurak, J.A.Swieca, Phys.Lett.

[4] R.Köberle, J.A.Swieca, Factorizable Z(N)Models, S.Carlos preprint (1979).

[5] P.P.Kulish, Theor.Math.Phys.26(1976)198;
M.Karowski, H.J.Thun, ref. 1 ;
D.Jagolnitzer, Phys.Rev.D18(1978)1257;
S.Parke, Harvard preprint HUTP-79/A037.

[6] B.Schroer, T.Truong, P.Weisz, Phys.Lett.63B(1976)422;
M.Karowski, FU-Berlin preprint HEP 7/79

[7] M.Karowski, P.Weisz, Nucl.Phys.B139(1978)455.

[8] B.Berg, M.Karowski, P.Weisz, Phys.Rev.D19 (1979)2477.

[9] R.Z.Bariev, Phys.Lett.55A(1976)456;
B.McCoy , C.A.Tracy, T.T.Wu, Phys.Rev.Lett.38(1977)783;
M.Sato, T.Miwa, M.Jimbo, Proc.Japan Acad.53A(1977)6.

+) Supported by Deutsche Forschungsgemeinschaft, contract Schr 160/5 and Schr 415.

ONE AND MULTIDIMENSIONAL COMPLETELY INTEGRABLE SYSTEMS ARISING FROM THE ISOSPECTRAL DEFORMATION

D.V. Chudnovsky
CEN-Saclay
Boite Postale No. 2
91190 Gif-sur-Yvette, France

Introduction. In mathematical physics there is one subject that now attracts the attention of mathematicians working in different fields. This is the study of different "solvable" or "completely integrable" models, some of which are very important field theory models. Perhaps for the first time since Jacobi and Weierstrass, the whole flow of integrable systems of mechanical and physical importance appeared in direct connection with the problems of functional analysis and algebraic and differential geometry. Different mathematical methods such as the inverse scattering method [59], the monodromy theory [18], the Riemann boundary value problem [43], the Riemann-Roch theorem [46], classification of vector bundles, etc. are the natural tools for the modern study of completely integrable systems. Recently new developments have appeared, such as: explicit examples of complete integrability in 3 and 4 dimensions (cf. [5], [17], [19]); the quantum inverse scattering method [39], [40], [56]; the inverse spectral problem for operators of an arbitrary order; and relation with the monodromy properties for irregular singularities (Stokes phenomena). The latter direction is definitely very promising; we devote a separate paper to the discussion of this problem. In the present paper you can find all the information necessary to begin the corresponding discussion for the matrix linear differential operators of an arbitrary order. The present paper is devoted only to the linear differential operators of orders one and two. We cannot cover the subject of completely integrable systems even in a big treatise, though there is none. For this reason we have decided to restrict this paper to the consistent representation of a very general class of completely integrable systems (containing almost all the practically interesting ones) together with the exposition of an entirely new description of their Hamiltonian structure. We consider here classical completely integrable systems and simultaneously, their operator generalizations [13], where the solutions are operators on a Hilbert space H depending on time-space variables x,t,y. The most interesting among these systems are operator generalizations of the sin-Gordon equation (§§2,5), containing all the SO(N) nonlinear σ-models and principal chiral fields (§§1-2). This operator formalism allows us to establish our main result. We show how two-dimensional completely integrable systems arising from the isospectral deformation for linear differential operators of order one and two can be decomposed into the common action of two simple commuting one dimensional Hamiltonians. In general, these Hamiltonians arise as natural generalizations of a very special and important completely integrable Hamiltonian system

$$f_{i,xx} = \lambda_i f_i + f_i \sum_{j=1}^{n} f_j g_j;\ g_{i,xx} = \lambda_i g_i + g_i \sum_{j=1}^{n} f_j g_j:$$

$i = 1,\dots,n$. This system, studied for a century in [35], [41], [51], [23-25] is examined in §6, where following [35], we call it the Russian Chain [14], [16], [17]. The infinite-component and operator generalizations of the Russian Chain play the crucial role in the decomposition of two and three dimensional completely integrable systems into commuting one dimensional Hamiltonians. The grounds for this decomposition are connected with a very peculiar phenomenon: the coefficients of a linear differential operator $L = \sum_{j=1}^{n} u_j \frac{d^j}{dx_j}$ can be expressed in terms of products of eigenfunctions of the direct $L\psi = k^n\psi$ and conjugate $L^*\phi = k^n\phi$ eigenvalue problem for L. For $n = 2$ and the Schrödinger operator $-y'' + u(x)y = -\lambda^2 y$ this was observed for the first time in [15], [23-26]; or in [42], [52], [30]. In §11 we establish the representation $U = -2\int_\Omega \omega\check{\phi}_\omega d\Sigma_\omega \hat{\phi}_\omega$ for an operator spectral measure $d\Sigma_\omega$ associated with the eigenvalue problem $\check{\phi}_{\omega,xx} = U\check{\phi}_\omega + \omega^2\check{\phi}_\omega$; $\hat{\phi}_{\omega,xx} = \hat{\phi}_\omega U + \omega^2\hat{\phi}_\omega$ in the case of an operator U. We develop the general form of such a representation in §§9-10 using the resolvent expansion [32], [37] and the matrix moment problem. The moment problem associated with the linear operator (cf. the Green formula) imposes interesting constraints on eigenfunctions $\check{\phi}_\omega$, $\hat{\phi}_\omega$ in the form of algebraic relations such as $\int_\Omega \check{\phi}_\omega d\Sigma_\omega \hat{\phi}_\omega = 1$. These constraints (§§§7, 9, 10) lead to the conclusion that two and three dimensional completely integrable systems which arise from the isospectral deformation can be viewed as a free motion constraint to an infinite dimensional symplectic manifold, explicitly defined by the intersections of quartics (cf. the discussion in §§7, 10). From the three dimensional systems we have chosen only the equations of the Kadomtsev-Petviashvili type [59], §§1, 12: $3u_{yy} = (4u_t + u_{xxx} - 12uu_x)_x$. In this case, according to §12, $u = u(x,t,y)$ can be represented as $u = -\int_\Omega \mathrm{tr}[\psi_\omega^+ d\Sigma_\omega \psi_\omega]$ for $\psi_\omega = \psi_\omega(x,t,y)$ satisfying an operator Russian chain in x and two commuting systems in y and t.

In this presentation we restrict ourselves only to isospectral deformation equations. The more general approach is that of the isomonodromy deformation treated in [18], [19]. It should be noted that the Russian chain equation is a very particular case of the general Schlesinger isomonodromy deformation equations [16], [35]. We also do not consider here explicit solutions of completely integrable systems in terms of θ-functions [4], [33], [49]. Nevertheless, such θ-function objects as Baker functions on Riemann surfaces [4], [10], [18], [46] immediately lead to the Russian chain and its generalizations.

This paper is organized as follows. In §1 we define the most general class of isospectral deformation equations in dimensions from one to four. §2 is devoted to sin-Gordon and SO(N) and SU(N) σ-model equations. In §3 and §4 we explicitly define the spectral transformation associated with the operator Schrödinger equation. §5 contains the general form of isospectral deformation equations together with numerous examples of operator two-dimensional equations. We would like to attract attention to the most promising example of the operator sin-Gordon (5.11) or (5.14), containing all SO(n) σ-models. In §6 the Russian chain and its history is described, and a detailed study of its relationship with the resolvent expansion and the higher KdV equations is presented. In §7 we study the constraints associated with the Russian chain and their appearance in the SU(N) σ-model [26]. §8 deals with the formal quantization of the Russian chain. In §§9-10 we describe isospectral deformation equations generated by linear differential equations of the first order. We decompose this system into a common action of commuting one dimensional Hamiltonians using the operator analog of the Russian chain (systems (10.7), (10.8)). In §11, on the basis of the investigation of the operator Russian chain [14], we study the decomposition of operator equations of the KdV type. Three dimensional operator systems of the Kadomtsev-Petviashvili type [17] are introduced in §12, where they are decomposed into a common action of three commuting one dimensional operator Hamiltonians. This paper is concluded by a short discussion of the operators of order three.

The author is deeply thankful to Professor D. Iagolnitzer and to CEN-Saclay for the kind invitation to present this paper. I thank the members of the Theoretical Division of CEN-Saclay for stimulating discussion and support. The partial support from ONR and NSF is gratefully acknowledged.

§1. Isospectral deformation equation in two, three and four dimensions. Chiral field theories and different σ-models.

1.0. We present here the description of two-dimensional systems of nonlinear equations associated with commuting linear operators (isospectral deformation equations). Special attention is devoted to chiral and gauge two-dimensional field theories.

1.1. The most general approach to the completely integrable systems is based on the representation of the system of nonlinear equations as the condition of commutatively of linear differential operators in partial derivatives. The first such representation in one-dimensional case belongs to Burchuall and Chaundy [10] (1922). The corresponding idea is presented here along the lines of the Zaharov-Shabat approach [59], [27]. For the (x,t)-space-time dimensional case, $x_1 = x$, $x_2 = t$, we use light-cone coordinates $2\xi = t - x$, $2\eta = t + x$.

The most general compatibility condition in (x,t)-dimensions has the following form

(1.1) $$i\psi_\xi = U\psi; \quad i\psi_\eta = V\psi$$

for (complex)operators U, V and eigenfunction ψ depending on ξ, η. Then (1.1) is equivalent to

(1.2) $$U_\eta - V_\xi + i[U,V] = 0.$$

The system (1.2) itself is trivially solvable if there are no restrictions placed on U, V. The general solution has the form

$$U = i\psi_\xi\psi^{-1}, \quad V = i\psi_\eta\psi^{-1}.$$

The system (1.2) becomes, however, nontrivial if the spectral parameter λ is introduced, in which case U and V must be rational (or in general meromorphic) functions in λ.

Now assume that according to the Zakharov-Mikhailov-Shabat scheme [59], [27], U and V, as functions of λ, have poles of fixed orders $m_1,\ldots,m_n$ at fixed points $\overset{o}{\lambda}_1,\ldots,\overset{o}{\lambda}_n$ of a λ-plane. Then the system (1.2) becomes a "completely" integrable system of nonlinear equations on the coefficients of the U and V: i.e., on the residues of U and V at $\lambda = \overset{o}{\lambda}_j$.

The Lax representation [47]

(1.3) $$\frac{dL}{dt} = [A,L]$$

for two linear differential operators A and L in $\frac{d}{dx}$ with coefficients depending on x and t, is imbedded into the scheme. To get the Lax representation (1.3) from (1.2) we demand U and V to be matrices having poles only at one fixed λ_o (say, $\lambda_o = \infty$) with orders of poles at $\lambda = \lambda_o$ for U and V equal to the orders of operators L and A.

The most interesting for the gauge theory is the case of U and V being rational functions with only one pole for U and one pole for V.

In the simplest case one can set:

$$U = \frac{A}{\lambda + 1}, \quad V = -\frac{B}{\lambda - 1}$$

for A, B independent of λ. Then from (1.2) the main system of equations is obtained [58], [59], [27], [20]:

$$(1.4) \qquad A_\eta = \frac{i}{2}[A,B], \quad B_\xi = -\frac{i}{2}[A,B].$$

This system of equations (1.4) is equivalent to the condition of consistency for the system

$$(1.5) \qquad i\psi_\xi = \frac{A}{\lambda + 1}, \quad i\psi_\eta = -\frac{B}{\lambda - 1}\psi.$$

1.2. The system (1.4) leads to field theories, e.g., to principle chiral fields and σ-models, connected with Lie groups.

Let us suppose that we have at any point (ξ,η) some element $g(\xi,\eta)$ of Lie group G. We consider the following equations of motion

$$(1.6) \qquad g_{\xi\eta} = \frac{1}{2}(g_\xi g^{-1} g_\eta + g_\eta g^{-1} g_\xi)$$

and the corresponding action

$$S = \int d\xi d\eta \, \frac{1}{2} \, \mathrm{Tr}\left(\frac{\partial}{\partial\xi} g \, \frac{\partial}{\partial\eta} g^{-1}\right).$$

The equation (1.6) can be reduced to the form (1.4). We define:

$$(1.7) \qquad A = ig_\xi g^{-1}, \quad B = ig_\eta g^{-1}.$$

Then in the notations (1.7), the system (1.6) is equivalent to the system (1.4):

$$(1.4') \qquad A_\eta - B_\xi - i[A,B] = 0, \quad A_\eta + B_\xi = 0.$$

The system (1.6) is called principle chiral field theory associated with G.

The most interesting from physical and mathematical points of view are two cases $G = SO(N)$ and $G = SU(N)$.

1.3. Moreover, it is often necessary to consider also some reduction of the principle field g in (1.6). For example, the most important systems, σ-models, correspond to restrictions:

$$g^2 = 1. \tag{1.8}$$

Under the constraints (1.8) the system (1.6) is reduced to the system:

$$g_{\xi\eta} = \frac{1}{2}(g_\xi g g_\eta + g_\eta g g_\xi). \tag{1.9}$$

The restriction (1.8) is consistent with the equation (1.9).

Let us see to what systems of equations the systems (1.8) - (1.9) can be reduced for $G = SO(N)$, $SU(N)$. We present g in the form

$$g = 1 - 2P \tag{1.10}$$

where $P^2 = P$, i.e., P is an operator of projecting, we can put

$$g = 1 - 2P_k, \tag{1.10'}$$

where k is the dimension of the image of P, the number k is unchanged in the evolution. In other words, the system (1.9) with the constraints (1.8) is indeed a system of equations:

$$[(P_k)_{\xi\eta}, P_k] = 0\colon\ k = 1,\dots,[\tfrac{N}{2}] \tag{1.11}$$

with the action [58]:

$$S_k = \frac{1}{2}\int d\xi d\eta\ \mathrm{Tr}(P_{k\xi}P_{k\eta}).$$

In other words one can consider each of the systems (1.11) as the field defined on the space of projectors of a given dimension k. This space in the real case is the Grassman real variety $\Gamma^{\mathbb{R}}_{N,k}$ and in the complex case is the complex Grassman variety $\Gamma^{\mathbb{C}}_{N,k}$:

$$\Gamma^{\mathbb{R}}_{N,k} = \frac{SO(N)}{O(k)SO(N-k)}, \quad \Gamma^{\mathbb{C}}_{N,k} = \frac{SU(N)}{U(k)SU(N-k)}.$$

The most important case is $k = 1$, when we have $\mathbb{R}P^{N-1}$ or $\mathbb{C}P^{N-1}$. In this case the projector P_1 can be written in the form

(1.12) $$(P_1)_{ij} = n_i\bar{n}_j \quad \text{and} \quad \sum_{i=1}^{N} n_i\bar{n}_i = 1.$$

Thus we have two nonlinear σ-models [54], [58], [20], [27].

A) <u>Real $\mathbb{R}P^{N-1}$ or SO(N) σ-model</u>

(1.13) $$\vec{n}_{\xi\eta} + (\vec{n}_\xi,\vec{n}_\eta)\vec{n} = 0; \quad (\vec{n},\vec{n}) = 1;$$

B) <u>Complex $\mathbb{C}P^{N-1}$ or SU(N) σ-model</u>

(1.14) $$\vec{n}_{\xi\eta} + \frac{1}{2}((\vec{n}_\xi,\vec{n}_\eta) + (\vec{n}_\eta,\vec{n}_\xi) + (\vec{n}_{\xi\eta},\vec{n}) - (\vec{n},\vec{n}_{\xi,\eta}))\vec{n}$$

$$- (\vec{n},\vec{n}_\xi)\vec{n}_\eta - (\vec{n},\vec{n}_\eta)\vec{n}_\xi + 2(\vec{n},\vec{n}_\eta)(\vec{n},\vec{n}_\xi)\vec{n} = 0$$

and

$$(\vec{n},\vec{n}) = 1.$$

Here we put $\vec{n} = (n_1,\ldots,n_N)$ and $(\vec{a},\vec{b}) = \sum_{i=1}^{N} a_i\bar{b}_i$ in the complex case.

It is clear from (1.13), (1.14) that both SO(N) and SU(N) nonlinear σ-models can be generalized to infinite-dimensional case, when $\vec{n}$ belongs to an infinite-dimensional Hilbert space H. This generalization is given below in §3.

In §2 we give a different form of equations of SO(N) nonlinear σ-model with examples for the case N = 3,4. The study of nonlinear σ-models is continued later in §§2, 5, 7.

Other two-dimensional examples of systems (1.2), (1.3) will be presented in §5, where we associate with the linear problem (1.1) the spectral and inverse scattering problems. Equations (1.2), (1.3) are usually called isospectral deformation equations (in (x,t)-dimensions).

<u>1.4</u>. General commutativity conditions can be used also to write three- and four-dimensional equations suspicious for being "completely integrable". These three-dimensional (x,y,t)-systems are of Zakharov-Shabat [59] type. They can be written as:

(1.15) $$\frac{\partial}{\partial y} L_n - \frac{\partial}{\partial t} L_m = [L_m,L_n]$$

for matrix linear differential operators L_n, L_m of orders n and m in $\frac{d}{dx}$ with coefficients being functions of x, y, t. Such a system is the condition of compatibility for

$$L_m\psi = \frac{\partial\psi}{\partial y}, \quad L_n\psi = \frac{\partial\psi}{\partial t}.$$

In the scalar case ($n = 2$, $L_2 = -d^2/dx^2 + u$, L_m is of third order) we get the so-called Kadomtsev-Petviashvili equation that will be treated later:

$$3u_{yy} = \frac{\partial}{\partial x}(4u_t + u_{xxx} - 12uu_x).$$

Four-dimensional systems that can be considered along the lines of compatibility conditions are those closely related to four-dimensional self-dual Yang-Mills (Belavin-Zakharov [5] representations) and have the form

$$\nabla_1\psi = -\lambda\nabla_2\psi; \quad \nabla_1^*\psi = \lambda^{-1}\nabla_2^*\psi$$

or $$[\nabla_1 + \lambda\nabla_2, \nabla_1^* - \lambda^{-1}\nabla_2^*]$$

for $$\nabla_1 = \frac{\partial}{\partial z_1} + B_1,\ \nabla_2 = \frac{\partial}{\partial z_2} + B_2,\ \nabla_1^* = -\frac{\partial}{\partial \bar{z}_1} + B_1^*,\ \nabla_2^* = -\frac{\partial}{\partial \bar{z}_2} + B_2^*.$$

§2. Examples of principle chiral fields, SO(N) nonlinear σ-models and sin-Gordon equations.

2.0. We study here in detail the equations of SU(N), SO(N) principle chiral fields and SO(N) nonlinear σ-models. All these equations are considered as generalizations of sin-Gordon equation $\alpha_{\xi\eta} = \sin\alpha$ [1], [59]. We present different forms of the equations defining these chiral (or gauge) theories and the Lax representation of an arbitrary SO(N) σ-model. Later, in §3 and §5 we define the most general operator sin-Gordon equation. The SO(N), SU(N) principle chiral fields and SO(N) nonlinear σ-models are particular cases of the operator sin-Gordon equation of §5.

2.1. Sin-Gordon equation [1], [59]:

$$\alpha_{\xi\eta} = \sin\alpha \quad \text{or}$$
(2.1)
$$\alpha_{xx} - \alpha_{tt} = \sin\alpha.$$

This is SO(2) principle chiral field. At the same time this is the SO(3) nonlinear σ-model; or the SU(2)(CP^1) σ-model.

2.2. The following system is known under different names:

Complex sin-Gordon equation, or
Pohlmeyer-Lund-Regge [54], [58] system, or
Getmanov's Lorentz invariant system [58].

This is a system of equations for two scalar fields α and β:

$$\beta_{\xi\eta} + \frac{1}{\sin\alpha}(\alpha_\eta\beta_\xi + \alpha_\xi\beta_\eta) = 0;$$

(2.2)

$$\alpha_{\xi\eta} + \sin\alpha - \frac{\sin(\alpha/2)}{2\cos^3(\alpha/2)}\beta_\xi\beta_\eta = 0.$$

For $\beta = 0$ one gets sin-Gordon. The system (2.2) is simultaneously SU(2) principle chiral field and at the same time SO(4) nonlinear σ-model.

2.3. Now it is possible to write for any $n \geq 3$ the equation of SO(n) nonlinear σ-model as the system of relativistic SO(n-2) covariant differential equations involving n-2 scalar fields. We follow here K. Pohlmeyer and K.H. Rehren [54].

Pohlmeyer [54] makes the following transformation. For SO(n) σ-model (1.13):

$$\bar{n}_{\xi\eta} + (\bar{n}_\xi, \bar{n}_\eta)\bar{n} = 0 : \|\bar{n}\| = 1,$$

$\|\bar{n}_\xi\| = \|\bar{n}_\eta\| = 1$ we consider an orthonormal basis

$$b_{-1} = \bar{n},\ b_o = \bar{n}_\xi,\ b_1 = \frac{\bar{n}_\eta - \cos\alpha\,\bar{n}_\xi}{\sin\alpha},\ b_k : k = 1,\dots,n-2$$

in $\mathbb{R}^n$ if $\alpha = \arccos(\bar{n}_\xi, \bar{n}_\eta)$. Let $S^+_{jk} = (\bar{b}_{j\xi}, \bar{b}_k)$; $S^-_{jk} = (\bar{b}_{j\eta}, \bar{b}_k)$. Now $S^+_{ij} = (\bar{f}_{i\xi}, \bar{f}_j)$; $S^-_{ij} = (\bar{f}_{i\eta}, \bar{f}_j)$ for $f_1,\dots,f_{n-2}$ forming an orthonormal basis in $\mathbb{R}^{n-2}$.

Now defining $\bar{\phi}$ by

$$\sin\alpha \cdot \bar{f} = \bar{\phi}$$

we reduce the equation of SO(n) σ-model to (n-2)-component equations without constraints. This equation has the form

$$\bar{\phi}_{\xi\eta} + \frac{(\bar{\phi}, \bar{\phi}_\eta)\bar{\phi}_\xi}{1 - \|\bar{\phi}\|^2} + \sqrt{1 - \|\bar{\phi}\|^2}\,\bar{\phi} = 0. \tag{2.3}$$

Of course there exists a symmetric equation corresponding to another choice of the basis b_{-1}, b_0, b_1:

$$\bar{\chi}_{\xi\eta} + \frac{(\bar{\chi}, \bar{\chi}_\xi)\bar{\chi}_\eta}{1 - \|\bar{\chi}\|^2} + \sqrt{1 - \|\bar{\chi}\|^2}\,\bar{\chi} = 0.$$

2.4. There exists another form representing SO(n) σ-models [54]. In the SO(6) σ-model we have the following 4-component field theory in $\mathbb{R}^4$. Let $[\cdot,\cdot,\cdot]$ be the vector product in $\mathbb{R}^4$:

$$[A,B,C]_i = \varepsilon_{ijk\ell}A_j B_k C_\ell.$$

Then the equation takes the form:

$$(2.4)\quad \bar{\psi}_{\xi\eta} + \frac{(\bar{\psi},\bar{\psi}_\xi)\bar{\psi}_\eta + (\bar{\psi},\bar{\psi}_\eta)\bar{\psi}_\xi - (\bar{\psi}_\xi,\bar{\psi}_\eta)\bar{\psi}}{1 - \|\bar{\psi}\|^2} - [\bar{\psi},\bar{\psi}_\xi,\bar{\psi}_\eta]/1 - \|\bar{\psi}\|^2 + (1-\|\bar{\psi}\|^2)\bar{\psi} = 0,$$

which is already symmetric in ξ and η. This is a 4-component equation of the SO(6) nonlinear σ-model.

2.5. We can call equations (2.2) - (2.4) to be generalizations of sin-Gordon equation (2.1). These generalizations can be written down for arbitrary compact Lie group G and are naturally connected with the principle chiral fields. For G = SU(N), G = SO(N) these generalizations belong to Budagov and Tahtadzan [8] (or [58]).

Let A,B,R,W be diagonal matrices; A,B,R be hermitian and W anti-hermitian. Then the "sin-Gordon equation for the group SU(N)" takes the form:

$$(2.5)\qquad (\mathfrak{A}^{-1}\mathfrak{A}_\xi)_\eta = [\mathfrak{A}^{-1}B\mathfrak{A},A] + i[\mathfrak{A}^{-1}\mathfrak{A}_\xi,R_\eta] + iR_{\xi\eta}.$$

Another equivalent representation for the "sin-Gordon equation for the group SU(N)" is:

$$(2.6)\qquad (\mathfrak{A}^{-1}\mathfrak{A}_\xi)_\eta - (\mathfrak{A}^{-1}W_\xi\mathfrak{A})_\eta = [\mathfrak{A}^{-1}B\mathfrak{A},A].$$

In the case of SO(N) we have W = 0 and for SO(N) matrices there exist the similar equation:

$$(2.7)\qquad (\mathfrak{A}^T\mathfrak{A}_t)_t - (\mathfrak{A}^T\mathfrak{A}_x)_x = [\mathfrak{A}^T\mathfrak{A}_t,\mathfrak{A}^T\mathfrak{A}_x] + 4[A,\mathfrak{A}^TB\mathfrak{A}] = 0.$$

All these equations (2.5) - (2.7) are very special examples of operator (matrix) equations of the order two (§3). They are particular cases of the "Operator sin-Gordon equation" proposed below in §5.

We note that equations (2.5) - (2.7) are equivalent to SU(N) and SO(N) principle chiral field equations (see §1). More precisely, Budagov [8] and Zakharov-Mikhailov [58] have proved the equivalence (guage equivalence) of the Lax representations for "sin-Gordon for the group G" with the Lax representation for the principle chiral field equation on G, both for G = SU(N) (equations (2.5), (2.6)) and G = SO(N): (2.7) cases.

2.6. There is one way to write down the Lax representation and isospectral datae corresponding to SO(N) nonlinear σ-models. This form is preferable to that of §1 because this was one obtains a very simple spectral problem. The difference between two spectral problems (1.4) and (2.8) is the same as between $u_{\xi\eta} = \sin u$ and $u_{xx} - u_{tt} = \sin u$ equations (look at [1] and [59]).

We consider SO(n) nonlinear σ-model as (n-2)-component system with components $\phi_1,\dots,\phi_{n-2}$ in the form (2.3) (without constraints):

$$\bar{\phi}_{\xi\eta} + \bar{\phi}_\xi(\bar{\phi}_\eta,\bar{\phi})/(1-\|\bar{\phi}\|^2) + \bar{\phi}\sqrt{1 - \|\bar{\phi}\|^2} = 0.$$

The essential point is Pohlmeyer's method for the introduction of matrix Riccati equations [54]. Let Γ^i: i = 1,...,n-2 form the basis elements of the

Clifford algebra C_{n-2}:

$$\{\Gamma^i,\Gamma^j\} = 2\delta^{ij}\mathbf{1}.$$

The lowest-dimensional representation of C_{n-2} are in $2^{[(n-2)/2]} \times 2^{[(n-2)/2]}$ matrices. Then the Lax pair of SO(n) σ-model equation has the following form [54]:

$$(2.8)\quad \begin{cases} \begin{pmatrix} \mathbf{I} & 0 \\ 0 & -\mathbf{I} \end{pmatrix} Y_\xi - \frac{1}{2}\begin{pmatrix} 0 & \Phi_\xi\Big/\sqrt{1-\|\bar{\Phi}\|^2} \\ \Phi_\xi\Big/\sqrt{1-\|\bar{\Phi}\|^2} & 0 \end{pmatrix} Y = \lambda Y; \\[2ex] Y_\eta = (4\lambda)^{-1}\begin{pmatrix} -\sqrt{1-\|\phi\|^2} & \Phi \\ \Phi & \sqrt{1-\|\bar{\Phi}\|^2} \end{pmatrix} Y. \end{cases}$$

Here

$$(2.9)\qquad \Phi = \sum_{j=1}^{n-2} \phi_j \Gamma^j .$$

The conditions of consistency of two equations (2.8) is exactly the SO(n) system of equations (2.3) for $\{\phi_1,\ldots,\phi_{n-2}\}$.

Equivalent representation of commuting linear differential operators arises from Clifford algebra C_{n-1}: $\tilde{\Gamma}^j = \begin{pmatrix} 0 & \Gamma^j \\ \Gamma^j & 0 \end{pmatrix}$: $j = 1,\ldots,n-2$; $\tilde{\Gamma}^{n-1} = \begin{pmatrix} \mathbf{I} & 0 \\ 0 & -\mathbf{I} \end{pmatrix}$.

Then two linear systems have the form:

$$\frac{\partial}{\partial\xi} Y = UY; \qquad \frac{\partial}{\partial\eta} Y = VY$$

and

$$U = \lambda\tilde{\Gamma}^{n-1} + \frac{1}{2}\sum_{j=1}^{n-2} \tilde{\Gamma}^{n-1}\tilde{\Gamma}^j \phi_{j\xi}\Big/\sqrt{1-\|\bar{\Phi}\|^2} ;$$

$$V = (4\lambda)^{-1}\left(-\sqrt{1-\|\bar{\Phi}\|^2}\,\tilde{\Gamma}^{n-1} + \sum_{j=1}^{n-2} \tilde{\Gamma}^j \phi_j\right)$$

§3. Operator generalizations of classical two-dimensional equations.

3.0. One of the main purposes of this paper is to consider operator generalizations of two and three dimensional completely integrable systems and to solve them in terms of one dimensional operator Hamiltonian systems. In Chapters 3 - 5 we describe operator generalizations of two dimensional equarions and the most general form of operator nonlinear evolutionary equations, associated with isospectral deformation for linear differential equations of order two with operator function coefficients.

3.1. We can put in correspondence with any equation of Zakharov-Mikhailov-Shabat type an operator equation [3]. The most general such equation is the condition of compatibility of two linear problems [59], [27].

$$\frac{\partial\psi}{\partial\xi} = U(\lambda)\psi, \quad \frac{\partial\psi}{\partial\eta} = V(\lambda)\psi,$$

where $U(\lambda)$, $V(\lambda)$ are rational functions in the λ-plane. In the classical cases $U(\lambda)$, $V(\lambda)$ are $N \times N$ matrices and all such systems are characterized by: i) the positions λ_j^0 of poles of $U(\lambda)$, $V(\lambda)$ that are fixed, and ii) orders of poles of $U(\lambda)$, $V(\lambda)$ at $\lambda = \lambda_j^0$ that are constant.

In order to replace classical systems by an operator one changes $U(\lambda)$, $V(\lambda)$ to be operators keeping the same poles λ_j^0 and order of poles fixed [13]. In general in the quantum case there appears also the necessity to rearrange the order of terms in $U(\lambda)$, $V(\lambda)$ and the equations themself. Hence, in each concrete case this must be done separately.

One should not think that the tranformation from usual equations to operator ones is immediate and that this operation merely involves the replacement in the equation of c-functions by operators. First, for certain equations it is simply impossible, e.g., in the sin-Gordon $\psi_{xt} = \sin\psi$ the function ψ cannot be substituted by an operator, if complete integrability is to be maintained; instead the form of the equation must be changed (see §§2,5). Second, there appears the problem of ordering of operators. The only way to get the true operator generalization is to return back to isospectral equations and then generalize them to the operator case.

3.2. As an example of operator generalization we can consider the equation (1.4) or principle chiral field equation (1.6) in the case of infinite-dimensional group G.

For this we consider the most general σ-model for an operator-function $g = g(\xi,\eta)$ being an operator on $L^2(\Omega,d\sigma)$ and satisfying

$$g_{\xi\eta} = \frac{1}{2}(g_\xi g^{-1} g_\eta + g_\eta g^{-1} g_\xi)$$

together with the condition

$$g^2 = 1.$$

We consider only the case when $g = 1 - 2P_1$, where $P_1^2 = P_1$ is one-dimensional projector corresponding to two vectors $\bar{n}$ and $\bar{m}$ from $L^2(\Omega,d\sigma)$:

$$(P_1\phi)(\omega) = \bar{n}(\omega)\int_\Omega \bar{m}(\omega_1)\phi(\omega_1)d\sigma_1.$$

Then equations for σ-model are equivalent to the equations

$$\begin{aligned}(3.1)\quad &\bar{n}_{\xi\eta} + \{(\bar{n}_\xi,\bar{m}_\eta) + (\bar{n}_\eta,\bar{m}_\xi) + (\bar{n},\bar{m}_{\xi\eta})\}\bar{n} + (\bar{n},\bar{m}_\xi)\bar{n}_\eta + (\bar{n},\bar{m}_\eta)\bar{n}_\xi + 2(\bar{n},\bar{m}_\eta)(\bar{n},\bar{m}_\xi)\bar{n} = 0;\\ &\bar{m}_{\xi\eta} + \{(\bar{n}_\xi,\bar{m}_\eta) + (\bar{n}_\eta,\bar{m}_\xi) + (\bar{n},\bar{m}_{\xi\eta})\}\bar{m} - (\bar{n},\bar{m}_\xi)\bar{m}_\eta - (\bar{n},\bar{m}_\eta)\bar{m}_\xi + 2(\bar{n},\bar{m}_\eta)(\bar{n},\bar{m}_\xi)\bar{m} = 0\end{aligned}$$

together with constraints:

$$(\bar{n},\bar{m}) \equiv 1.$$

In the case $\dim L^2(\Omega, d\sigma) = n$ the condition $\bar{n} = \bar{m}$ leads us to $SO(n)$ σ-model ($\mathbb{RP}^{n-1}$ case) and the condition $\bar{m} = \bar{n}^*$ is equivalent to $SU(n)$ σ-model ($\mathbb{CP}^{n-1}$ σ-model). In both cases we take $\bar{n}$, $\bar{m}$ from ℓ^2_n with the scalar product in the sense of ℓ^2_n.

3.3. For these lectures we decided to restrict ourselves only to those operator nonlinear evolutionary equations that are associated with linear differential operators of order at most two with operator function coefficients. In principle linear differential operators of any order can be treated along the same lines. This is shown briefly later for linear differential operators of order three.

We call [13], [14] an isospectral deformation equation of order two (or at most two) if this equation admits the Lax representation (1.3):

$$\frac{dL}{dt} = [A, L]$$

where A is a differential (or integro-differential) operator in $\frac{d}{dx}$ and L is a differential operator of order at most two in $\frac{d}{dx}$. In other words, either

(i) $L = A\frac{d}{dx} + V$, where A is a diagonal operator, $V = [A, U]$;

or

(ii) $L = \frac{d^2}{dx^2} + U$.

Both cases are considered here. The case (i) is treated later and chapters 3-5 are devoted to (ii).

Of course, one can reformulate the definition of our class of equation in terms of the Zakharov-Shabat scheme. If we consider two dimensional nonlinear operator equations arising from the scheme of §1:

$$\psi_{x_1}(x_1, x_2, \lambda) = U(\lambda)\psi;$$

$$\psi_{x_2}(x_1, x_2, \lambda) - V(\lambda)\psi$$

then we call the corresponding nonlinear equation

$$U_{x_2} - V_{x_1} + [U, V] = 0$$

to be of order two if each of the operators $U(\lambda)$, $V(\lambda)$ have only one pole in λ-plane, but orders of these poles can be arbitrary.

In this class of equations one includes the majority of known two-dimensional completely integrable systems: Kd V, m Kd V, nonlinear Schrödinger, sin-Gordon, chiral fields, σ-models, Gross-Neveu model, Thirring model, etc.

One of the interesting features of equations of order two is the existence of a large class of solutions (so called finite band potentials) expressed in terms of θ-functions of hyperelliptic curves. The presence of hyperelliptic curves is another common feature for the class of equations, called "of order two". For a degenerate hyperelliptic curve one gets just soliton solutions.

The theory of matrix Lax systems of order two was developed by Gelfand and Dikij [36], who introduced Hamiltonian two dimensional structure on these systems and gave recurrent formulae for conservation laws. However, we need some other simple way to present equations of order two and to connect them directly with the evolution of scattering data. This will be done in §5 and for this we use in §5 certain single integro-differential operator L associated with potential U(x). Such integro-differential operator L had been introduced by Ablowitz, Kaup, Newell, Segur [1] and then by Calogero and Degasperis [11], [31] in the matrix case.

§4. Scattering data for operator Schrödinger equation.

4.0. In this chapter we describe briefly the inverse scattering data for the Schrödinger equation with operator-function coefficients. In the appendix we illustrate how isospectral deformation equations (of order two) are linearized via inverse scattering method. The corresponding scheme is due to Kruscal-Miura-Gardner-Green and Zakharov-Shabat.

One can propose a scheme of inverse scattering for the operator Schrödinger equation on a Hilbert space H. Not too much is different from an ordinary scalar case. Classical operator approach in the arbitrary Banach space belongs to Krein [45], Berezansky [6], [7], Nizhnik [53], Levitan [48] and in the matrix case from the point of view of inverse scattering to Calogero, Degasperis [11], [31], Wadati, Kamijo.

4.1. We assume that U(x) is an operator on H with the norm $\|U(x)\|$ vanishing asymptotically exponentially or faster (as $|x| \to \infty$).

We consider Jost functions being operator solutions of (two) Schrödinger equations with the potential U(x), corresponding to the continuous spectrum:

$$\psi_{xx}(x,k) = U(x)\psi(x,k) - k^2\psi(x,k) \tag{4.1}$$

and

$$\tilde{\psi}_{xx}(x,k) = \tilde{\psi}(x,k)U(x) - k^2\tilde{\psi}(x,k): \quad k \geq 0 \tag{4.2}$$

together with the boundary conditions:

$$\psi(x,k) \to T(k)\exp(-ikx): \quad x \to -\infty \tag{4.3}$$

$$\psi(x,k) \to \exp(-ikx) + R(k)\exp(ikx): \quad x \to +\infty \tag{4.4}$$

and

$$\begin{aligned} &\tilde{\psi}(x,k) \to \tilde{T}(k)\exp(-ikx): \quad x \to -\infty; \\ &\tilde{\psi}(x,k) \to \exp(-ikx) + \tilde{R}(k)\exp(ikx) \text{ if now } x \to +\infty. \end{aligned} \tag{4.5}$$

Here R(k) is the reflection operator and T(k) is the transmission operator. Of course we have

$$\tilde{R}(k) = R(k). \tag{4.6}$$

The most important property of scattering data is the analytic continuation of the $R(k)$, $T(k)$ into k-plane. Under the conditions of fast asymptotic vanishing of $U(x)$, both $R(k)$, $T(k)$ are meromorphic in the whole k-plane.

In this case, $R(k)$ may have N simple poles at the values χ_j: $j = 1,\dots,N$. Then χ_j^2, $j = 1,\dots,N$ are exactly discrete eigenvalues of the problem (4.1). Of course, χ_j^2: $j = 1,\dots,N$ are also eigenvalues of (4.2). There are the corresponding eigenfunctions

$$\psi_{xx}^{(j)} = U(x)\psi^{(j)} - \chi_j^2\psi^{(j)}$$

$$\bar{\psi}_{xx}^{(j)} = \bar{\psi}^{(j)}U(x) - \chi_j^2\bar{\psi}^{(j)}: \; j = 1,\dots,N \tag{4.7}$$

Scattering data corresponding to the eigenvalue χ_j^2 have the following form:

$$\lim_{k\to\chi_j} \{[k - \chi_j]R(k)\} = P_j: \quad j = 1,\dots,N .$$

The system

$$S = \{R(k);\chi_j,P_j: \quad j = 1,\dots,N\}$$

is called the system of scattering data associated to the potential $U(x)$.

In the case of nondegenerate eigenvalues all P_j are projection operators of rank one.

4.2. Now one can generalize in a normal way inverse scattering method consisting of reconstruction of the potential $U(x)$ from the scattering data S.

The best tool for this is Gelfand-Levitan equation. First of all we construct initial data associated with S

$$F(y) = \frac{1}{2\pi}\int_{-\infty}^{\infty} R(k)\exp(iky)\,dk + \sum_{i=1}^{N} P_i\exp(-\chi_i y).$$

Then we write Gelfand-Levitan equation as the Fredholm operator integral equation

$$K(x,x_1) + F(x+x_1) + \int_x^{+\infty} K(x,z)F(z+x_1)\,dz = 0: \; x_1 \le x.$$

Now the potential $U(x)$ is reconstructed from $K(x,x_1)$ in a very simple way

$$U(x) = -2\frac{d}{dx}K(x,x).$$

In the case of $U(x)^+ = U(x)$ the situation is simplified considerably.

We have in the case $U(x)^+ = U(x)$ only finitely many eigenvalues that are discrete. Also

$$R^+(k) = R(-k^*)$$

and

$$\tilde{\psi}(x,k) = [\psi(x,-k^*)]^+.$$

Independent of the condition $U^+ = U$ we have the classical formulae for scattering data

$$\tilde{T}(-k)T(k) + \tilde{R}(-k)R(k) = 1.$$

To establish this identity we look at the Wronskian (which is constant):

$$W(x,k) = \tilde{\psi}(x,-k)\psi_x(x,k) - \tilde{\psi}_x(x,-k)\psi(x,k).$$

In the case of hermitian U ($U^+ = U$) and real k we have the well-known unitarity equation:

$$T^+(k)T(k) + R^+(k)R(k) = 1.$$

Appendix to §4. Inverse scattering method as a "nonlinear Fourier transform".

The essence of the inverse scattering method is very simple: one reduces a complicated nonlinear evolution equation on $U(x,t)$ to a linear equation of the evolution of scattering data for $U(x,t)$. Of course, these equations (isospectral deformation equations) are rather special.

The process of solving Cauchy problem can be explained as follows:

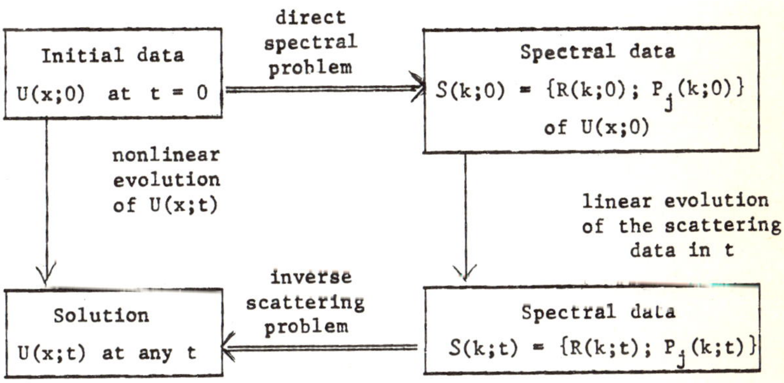

Such a scheme has obvious anisotropy in x and t.

Remark. This simplicity is the consequence of the restrictive condition $U(x) \to 0$ as $|x| \to \infty$.

§5. Two-dimensional operator systems of "order two".

5.0. We describe here the general form of two-dimensional operator evolutionary equations associated with the inverse scattering method for Schrödinger operator. These equations (of order two in the terminology of §3) are described in Theorem 5.1. Most of this chapter is devoted to examples of equations of order two [13], [14]: operator Korteweg-de Vries, operator mKdV, operator nonlinear Schrödinger, and operator sin-Gordon equations. We consider, as a particular case, a multicomponent nonlinear Schrödinger equation [15] and its stationary version - a Russian chain

[16], [25], [35]. This Russian chain and its operator version become the subject of our studies in §§6 - 9. The most interesting in this chapter is, perhaps, the introduction of the operator sin-Gordon equation.

5.1. We consider now the most general two-dimensional operator evolutionary equations for U(x,t) associated with the operator Schrödinger equation:

$$\frac{d^2\psi}{dx^2} = U\psi - k^2\psi; \qquad \psi = \psi(x,t,k)$$

(5.1)

$$\frac{d^2\tilde{\psi}}{dx^2} = \tilde{\psi}U - k^2\tilde{\psi}; \qquad \tilde{\psi} = \tilde{\psi}(x,t,k) \ .$$

We consider equations of order two (in the sense of §3) for which the evolution in t of the scattering coefficient R(k,t) of (5.1) is linear. For this we consider the integro-differential operator L (see [1], [11], [31]):

$$4LF(x) = F_{xx}(x) - 2\{U(x,t),F(x)\} + G\int_x^{+\infty} dx'F(x') \ ,$$

(5.2)

$$GF(x) = \{U_x(x,t),F(x)\} + [U(x,t), \int_x^{+\infty} dx'[U(x',t),F(x')]].$$

Now the class of the equations under consideration can be described by iterative applications of the operator L. We remind that $[A,B] = AB - BA$, $\{A,B\} = AB + BA$.

Theorem 5.1 (cf. [11], [31], [14] and [1], [52]). For fixed entire functions $\alpha(z)$, $\beta(z)$ and fixed constant operators M, N the following nonlinear operator equation for U(x,t):

(5.3) $$U_t(x,t) = \alpha(L)[N,U(x,t)] + \beta(L)\cdot G\cdot M$$

is equivalent to the linear differential eequation for the scattering coefficient R(k,t) of (5.1):

$$R_t(k,t) = \alpha(-k^2)[N,R(k,t)] + 2ik\beta(-k^2)\{M,R(k,t)\}.$$

Of course, this is not the complete picture of evolution for scattering data, since we need to know the evolution for a discrete part of the spectrum. The evolution of a discrete part of the spectrum can be obtained from the evolution of the continuous spectrum, and we have

$$x_{j,t} = 0: \ j = 1,\dots,N$$

$$P_{j,t} = \alpha(-x_j^2)[N,P_j] + 2ix_j\beta(-x_j^2)\{M,P_j\}.$$

The importance of operator L for our purposes is very simple. The operator L acts in the following way on the product $\psi\tilde{\psi}(x,k)$ of the left and right eigenfunctions of the operator Schrödinger equation:

(5.4) $$L\psi\tilde{\psi} = -k^2\psi\tilde{\psi}.$$

Remark 5.2. The action of L on the momentae

$$\mathcal{O}\!\ell_n = \int k^n\psi(x,k)\tilde{\psi}(x,k)\,dk$$

is very simple:

$$L\cdot\mathcal{O}\!\ell_n = -\mathcal{O}\!\ell_{n+2}.$$

Let us now present examples of operator equations of order two, following the statement of Theorem 5.1.

5.2. Examples.

1) The operator Kd V equation [13]:

(5.4) $$\phi_t = \phi_{xxx} + 3(\phi_x\phi + \phi\phi_x).$$

Here $M = I$, $\beta(L) = L$; $\alpha(L) = 0$.

2) The operator modified Kd V equation [13]:

(5.5) $$\psi_t = \psi_{xxx} - 3(\psi_x\psi^2 + \psi^2\psi_x).$$

Here

$$U = \begin{pmatrix} \psi^2 & \psi_x \\ \psi_x & \psi^2 \end{pmatrix}, \quad M = \begin{pmatrix} I & 0 \\ 0 & I \end{pmatrix}, \quad \beta(L) = L;\ \alpha(L) = 0.$$

3) The operator coupled nonlinear Schrödinger equation [13], [14]:

(5.6) $$i\psi_t = \psi_{xx} + \psi\phi\psi; \quad -i\phi_t = \phi_{xx} + \phi\psi\phi$$

and nonlinear Schrödinger equation

(5.7) $$i\psi_t = \psi_{xx} + \psi\psi^+\psi$$ (suggested by A. Neveu [60]).

Here

$$U = \begin{pmatrix} \psi\phi & \psi \\ \phi & \phi\psi \end{pmatrix}, \quad N = \begin{pmatrix} I & 0 \\ 0 & -I \end{pmatrix}, \quad \alpha(L) = L,\ \beta(L) = 0.$$

Operator equations 2) and 3) (but not 1)) give us multi (infinite-) component equations if one considers operators ψ of rank one [13], [14]. We consider H to be $L^2(\Omega, d\mu)$ with the scalar product: $(a,b) = \int_\Omega a(\omega)b(\omega)^* d\mu$. Then, as a particular case of the operator nonlinear Schrödinger equation we get a multicomponent nonlinear Schrödinger equation

(5.8) $$i\bar{\phi}_t = \bar{\phi}_{xx} + \bar{\phi}\cdot(\bar{\phi},\bar{\phi}).$$

and a multicomponent modified Kd V equation:

$$\bar{\phi}_t = \bar{\phi}_{xxx} - 3\,\bar{\phi}_x\cdot(\bar{\phi},\bar{\phi}) - 3\bar{\phi}\cdot(\bar{\phi}_x,\bar{\phi}) \tag{5.9}$$

for $\bar{\phi} = \bar{\phi}(x,t)$ from $H = L^2(\Omega,d\mu)$.

As we see in §12, the combination of these two equations gives us Kadomtsev-Petviashvili equations in (x,y,t)-dimensions.

In the case of finite-dimensional H, dim(H) = n, we obtain the n-component coupled nonlinear Schrödinger equation:

$$\begin{cases} i\phi_{\ell,t} = \sum_{j=1}^{n} \phi_j\psi_j\cdot\phi_\ell - \phi_{\ell,xx}, \\ -i\psi_{\ell,t} = \psi_\ell\cdot\sum_{j=1}^{n}\phi_j\psi_j - \psi_{\ell,xx}:\ \ell = 1,\dots,n. \end{cases}$$

The most interesting case is that of the stationary equation, when

$$\phi_\ell = \phi_\ell(x)e^{i\lambda_\ell t},\ \psi_\ell = \psi_\ell(x)e^{-i\lambda_\ell t}.$$

We obtain in this case the following completely integrable Hamiltonian system called Russian chain [14], [15], [16], [18], [19](first appearing in [35], [22]-[26]):

$$\begin{cases} \phi_{\ell,xx} = \sum_{j=1}^{n} \phi_j\psi_j\cdot\phi_\ell + \lambda_\ell\phi_\ell, \\ \psi_{\ell,xx} = \sum_{j=1}^{n} \phi_j\psi_j\cdot\psi_\ell + \lambda_\ell\psi_\ell:\ \ \ell = 1,\dots,n \end{cases} \tag{5.10}$$

with arbitrary $\lambda_1,\dots,\lambda_n$. For the detailed description of the Russian chain see §6.

<u>5.3.</u> The most interesting example of equation of order two is the operator sin-Gordon (or nonlinear σ-models).

4) <u>Sin-Gordon:</u> $\phi_{xt} = \sin\phi,\ \ \phi(x,t) \to 0(\text{mod } 2\pi):\ |x| \to \infty$.

Here

$$H = \mathbb{R}^2,\ U = \begin{pmatrix} -\phi_x^2/4 & -\frac{1}{2}\phi_{xx} \\ \frac{1}{2}\phi_{xx}, & \phi_x^2/4 \end{pmatrix},\ \alpha(L) = 0,\ \beta(L) = L^{-1},\ M = \mathbb{I}.$$

The operator, even matrix generalization of sin-Gordon equation is of rather complicated form. The natural operator analog of sin-Gordon equation corresponds to the space $H \times H$ and potential of the form

$$U = \begin{pmatrix} \frac{1}{4}P^2 & \frac{1}{2}P_x \\ \frac{1}{2}P_x & \frac{1}{4}P^2 \end{pmatrix}$$

The corresponding operator equation of the order two arises in the situation $\alpha(L) = 0$, $\beta(L) = L^{-1}$, $M = \begin{pmatrix} I & 0 \\ 0 & I \end{pmatrix}$. Then "The Operator Sin-Gordon" equation takes the form:

(5.11) $$P_{xt} = -\frac{1}{4}\{P, \int_{-\infty}^{x} (P^2)_t dx'\} + P.$$

In the scalar case for $P = \phi_x$ we get the usual sin-Gordon equation $\phi_{xt} = \sin\phi$. Here is another form for the "operator sin-Gordon equation". We take, as in the previous case,

(5.12) $$U = \frac{1}{2}\begin{pmatrix} P^2/2 & P_x \\ P_x & P^2/2 \end{pmatrix} = V^2 + V_x$$

for $V = \frac{1}{2}\begin{pmatrix} 0 & P \\ P & 0 \end{pmatrix}$. Here the potential U corresponds to the Schrödinger equation

$$(-\frac{d^2}{dx^2} + U)\phi = E\phi$$

and the potential V corresponds to a Dirac equation

$$\sigma_3 \frac{d}{dx}\psi = V\psi + \lambda\psi$$

for $\sigma_3 = \begin{pmatrix} I & 0 \\ 0 & -I \end{pmatrix}$. Then the simplest equation in (x,t)-dimensions corresponding to the choice $\alpha(z) = z^{-1}$, $\beta(z) = 0$ in Theorem 5.1 is presented as the following (x,t) equations on the operator P:

(5.13) $$\begin{cases} 2P_{xt} = \{P, T\}; \\ 2T_x = -(P^2)_t \, . \end{cases}$$

This system is equivalent to the "operator sin-Gordon" proposed above in (5.11):

(5.14) $$P_{xt} = -\frac{1}{4}\{P, \int_{\gamma}^{x} (P^2)_t dx_1\}.$$

This operator sin-Gordon equation (5.11), (5.14) contains "sin-Gordon equations for SU(N), SO(N)" from §2, (2.5)-(2.7).

Let's show how SO(n) nonlinear σ-model is obtained as a trivial reduction of this general sin-Gordon equation. For this, one uses Clifford algebra approach (of §2, 2.5) and imbeds SO(n) nonlinear σ-model in the form (2.3) into systems (5.12), (5.13).

Let Γ^j: $j = 1,\dots,n-2$ be generators of the Clifford algebra C_{n-2}. We assume now that

(5.15) P belongs to a vector space, generated by matrices Γ^j: $j = 1,\dots,n-2$.

If

(5.16) $$P = \sum_{j=1}^{n-2} r_j \Gamma^j;$$

then by the equations (5.13):

$$2T_x = -\sum_{j=1}^{n-2} 2r_j r_{jt}\mathbf{I}$$

or we can put

$$T = \int_{\gamma}^{x} \sum_{j=1}^{n-2} r_j r_{jt} dx_1 \mathbf{I} .$$

Other equations are

$\frac{r_{jxt}}{r_j} = T$: $j = 1,\ldots,n-2$ and then putting $r_j = \phi_{jx}\Big/\sqrt{1 - \|\overline{\phi}\|^2}$, we get $T = \sqrt{1 - \|\overline{\phi}\|^2}$, which is equivalent to the system of equations (2.3). This system of equations (2.3) is equivalent to an SO(n) σ-model by §2, 2.3.

Of course, isospectral problem (5.1) for this particular choice of U in (5.12) is equivalent to a Lax representation of SO(n) nonlinear σ-model in §2, 2.5.

§6. The Russian chain in the scalar case.

6.0. In this chapter we discuss the Russian chain (5.10) (the stationary multicomponent nonlinear Schrodinger). We decided to deduce this system from the isomonodromy deformation for the Fuchsian linear differential equations [18]. Indeed, as it was shown in [16], [18], [19] the isomonodromy deformation equations are closely related to classical completely integrable systems. In the one-dimensional case isospectral deformation equations are equivalent to simplified Schlesinger equations (see (A) below). In multidimensional cases we later see how isospectral deformation equations in two- or three-dimensions can be reduced to two or three commuting Hamiltonians of the Russian chain type.

6.1. We can start to consider the Riemann monodromy problem [18], [19], [28] immediately from Fuchsian linear differential equations. For this we consider the simplest linear differential equations:

$$\frac{d\bar{y}}{dx} = \sum_{i=1}^{n+2} \bar{y}\,\frac{A_i}{x-t_i} \tag{6.1}$$

for $\bar{y} = (y_1,\ldots,y_m)$ and A_i being $m \times m$ constant matrices: $i = 1,\ldots,m$ and n+2 singularities $t_1,\ldots,t_{n+2}$ (where, e.g., $t_{n+1} = 0$, $t_{n+2} = 1$).

One of the solutions of the Riemann monodromy problem is based on isomonodromy deformation equations, an example of which is given by Painlevé transcendent VI [18] (the case $m = 2$, $n = 1$). The general isomonodromy equations were written by Schlesinger [55] who used Riemann theorem [28] that the system of functions satisfying Fuchsian linear differential equations with the same monodromy group are linearly dependent over $\mathbb{C}[z]$. In other words, Schlesinger used the "consistency condition" as the form of representation of a system of nonlinear "completely integrable" equations.

Schlesinger's Theorem 6.1. ([55], [18], [16]) Let $Y = Y(x_o;x)$ be the fundamental matrix solution of (6.1); $Y(x_o;x_o) = I$. The necessary and sufficient condition for A_i: $i = 1,\dots,n+2$ as the function of the parameters $t_1,\dots,t_{n+2},x_o$ to have the fixed monodromy group for $Y(x)$, is the following completely integrable system of total differential equations

$$dA_j = -\sum_{i\neq j}[A_j,A_i]d\log\frac{t_j-t_i}{x_o-t_i}: \quad j = 1,\dots,n+2.$$

This Schlesinger system can be written in a classical form [55] (1908-1912):

$$\frac{\partial A_j}{\partial t_i} = [A_j,A_i]\left(\frac{1}{t_j-t_i} - \frac{1}{x_o-t_i}\right), \quad j \neq i$$

$$\frac{\partial A_i}{\partial t_i} = -\sum_{j\neq i}[A_i,A_j]\frac{1}{t_i-t_j}, \tag{6.2}$$

$$\frac{\partial A_i}{\partial x_o} = \sum_{j\neq i}[A_i,A_j]\frac{1}{x_o-t_j}: \quad i = 1,\dots,n.$$

This system of equations is the source of "classical" completely integrable systems like stationary KdV, e.g.

There is, of course, a natural temptation to identify the deformation equations of the type (6.2) with the Lax type of isospectral deformation. This is not the case at all! The deformation equations (6.2) are reduced in the most interesting cases to the equations, differential with respect to a spectral parameter [18], while Lax's equations represent equations rational with respect to a spectral parameter.

In the Schlesinger system the following transformation due to Painlevé and Garnier [35] is performed

$$t_i \to a_i + \varepsilon t_i$$

$$A^i \to \varepsilon^{-1}A^i \quad : i = 1,\dots,n+2$$

for constants a_i, where, in particular, $a_{n+1} = 0$, $a_{n+2} = 1$. If we put $\varepsilon \to 0$ then the system (6.2) takes the following form [16], [18], [19], [35]:

$$\frac{\partial A^j}{\partial t_i} = \frac{[A^i,A^j]}{a_j-a_i} : \quad j \neq i \tag{A}$$

$$\sum_{j=1}^{n+2}\frac{\partial A^j}{\partial t_i} = 0 \quad : i = 1,\dots,n.$$

The system (A) is indeed completely integrable and, moreover, can be solved in Abelian integrals (of the first and second kind) corresponding to a certain algebraic curve [16], [18], [35].

Moreover, any completely integrable one-dimensional system connected with the algebra of commuting differential operators can be represented in a form (A) [16], [18], [19]. We call the system (A) a "simplified Schlesinger equation" [16].

<u>6.2</u>. Let us consider the simplest example of simplified Schlesinger system (A) which has a rather long history.

The case n = 0 is trivial (no deformation), but the case $m \geq 2$; $\underline{n = 1}$ is already extremely important. In this case we obtain the following system of non-linear differential equations equivalent to (A) for $n = 1$, $m \geq 2$:

(R) or (6.3)
$$\begin{cases} \ddot{f}_i = \lambda_i f_i - 2f_i \sum_{j=2}^{m} f_j g_j; \\ \ddot{g}_i = \lambda_i g_i - 2g_i \sum_{j=2}^{m} f_j g_j : i = 2,\ldots,m. \end{cases}$$

for m-2 arbitrary constants $\lambda_2,\ldots,\lambda_m$.

This system was derived from Schlesinger's equations by R. Garnier [35].

The system with different constraints was studied beginning in 1830, especially for m = 3 (cf. Jacobi [41]); C. Neumann [51] (m = 3,4).

R. Garnier [25] gave the solution of (R) in terms of hyperelliptic integrals, and he emphasized the achievements of the Russian school.

Since S. Kovalevski [44] (1880) the system (R) was studied by Russians in relation the motion of a heavy rigid body. Some of the people who studied (R) in the XIX century are: S. Tchapligin, G. Kolosoff, D. Gorjachev, V. Steklov, G. Appelrot, N. Delone, P. Nekrasov [3].

Because of their work we decided to call (R) the "Russian chain".

R. Garnier was the first who remarked (in 1919) that (R) was possibly connected with the system of equations written by J. Drach [32] (1919). As we now know, J. Drach indeed discovered a stationary KdV equation [15].

Without knowing about Garnier's remark, D.V and G.V Chudnovsky walked into (R) in 1976-1977 and indeed proved the equivalence of (R) to the stationary KdV equations in 1977 in a series of papers [22]-[25]. Our attention to this problem was attracted by Professors V. Glaser and A. Martin to whom D.V. and G.V. Chudnovsky are deeply grateful.

<u>Remark</u>. Of course, the system (R) or (6.3) arises not only from isomonodromy deformation equations (simplified Schlesinger system (A)) but also from isospectral deformation equations (of §5). Indeed the system (R) is identical to (5.10): to the stationary multicomponent nonlinear Schrödinger equation (5.8).

Now we know that this system and its infinite-component and operator generalizations contain all two-dimensional nonlinear systems associated with inverse scattering method. In other words, these two-dimensional completely integrable systems (such as KdV, nonlinear Schrödinger, sin-Gordon, σ-models) can be decomposed into common action of two commuting Hamiltonians on an infinite-dimensional symplectic manifold; one of these Hamiltonians is just an operator Russian chain and another is one of its first integrals [14], [19].

6.3. Infinite-component Russian chain.

In order to prepare for operator generalizations of the Russian chain (6.3) we consider a Russian chain on the Hilbert space $L^2(\Omega,d\mu_k)$ corresponding to a measurable function $\lambda:\Omega \to \mathbb{C}$. Let us define the following Hamiltonian system for $f(k) = f(x,k)$; $g(k) = g(x,k)$:

$$(6.4)\qquad \begin{cases} f(k)_{xx} = -2\int_\Omega f(\ell)g(\ell)d\mu_\ell \cdot f(k) - \lambda(k)f(k); \\ g(k)_{xx} = -2\int_\Omega f(\ell)g(\ell)d\mu_\ell \cdot g(k) - \lambda(k)g(k). \end{cases}$$

This system has the Hamiltonian

$$(6.5)\qquad H_I = \int_\Omega f_x(k)g_x(k)d\mu_k + \int_\Omega \lambda(k)f(k)g(k)d\mu_k + (\int_\Omega f(k)g(k)d\mu_k)^2,$$

where $g_x(k)$ is a variable conjugate to $f(k)$ and $f_x(k)$ is a variable conjugate to $g(k)$. The system (6.4) possesses a lot of internal symmetries. Their existence can be explained because (6.4) (or (6.5)) admits a large group of symmetry transformations of Darboux-Bäcklund type. These transformations will be described elsewhere, but they are simple consequences of the Darboux-Bäcklund transformation for the second order linear differential operator [29], [62]. Here (6.4) *shows* that $f(k)$, $g(k)$ are eigenfunctions of the Schrödinger operator with potential $V = -2\int_\Omega f(\ell)g(\ell)d\mu_\ell$, corresponding to the eigenvalue $\lambda(k)$.

There are different ways to represent internal symmetries of (6.4) (cf. below); the simplest is to write them as quartics in canonical variables.

Theorem 6.1. We have the following first integrals of the system (6.4)

$$(6.6)\qquad \begin{aligned} K[k] &= f_x(k)g(k) - f(k)g_x(k); \\ C[k] &= f_x(k)g_x(k) + \lambda(k)f(k)g(k) + f(k)g(k)\int_\Omega f\cdot g d\mu_\lambda - \fint \frac{d\mu_\ell}{\lambda(\ell)-\lambda(k)} \times \\ &\quad \times (f_x(\ell)g(k) - f(\ell)g_x(k)) \times (g_x(\ell)f(k) - g(\ell)f_x(k)). \end{aligned}$$

The first integrals $K[k]$; $C[k]$ are in involution, and independent. There are additional series of first integrals:

$$\begin{aligned} C_2[k] &= f_x(k)^2 + \lambda(k)f(k)^2 + f(k)^2\int_\Omega fgd\mu_\lambda - \\ &\quad - \fint \frac{d\mu_\ell}{\lambda(\ell)-\lambda(k)} (f_x(\ell)f(k) - f(\ell)f_x(k))(g_x(\ell)f(k) - g(\ell)f_x(k)); \\ C_3[k] &= g_x(k)^2 + \lambda(k)g(k)^2 + g(k^2)\int_\Omega fgd\mu_\lambda - \\ &\quad - \fint \frac{d\mu_\ell}{\lambda(\ell)-\lambda(k)} (f_x(\ell)g(k) - f(\ell)g_x(k))(g_x(\ell)g(k) - g(\ell)g_x(k)). \end{aligned}$$

There are certain relations between $K[k]$, $C[k]$, $C_i[k]$, e.g., we have $\{K[k'], C_j[k]\} = 0$ if $k' \neq k$; $\{K[k'], K[k]\} = 0$; $\{K[k], C_j[k]\} = 2C_j[k]$: $j = 2,3$.

The proof of Theorem 6.1 follows by direct computation; however, time can be saved considerably if you notice that expressions like $(f_x(\ell)g(k) - f(\ell)g_x(k))$ are Wronskians of two solutions of linear differential equations of the second order.

New Hamiltonians, commuting with H_I can be represented in the following form:

$$H_{h(k)} = \int h(k)C[k]d\mu_k .$$

In particular, $H_I = \int C[k]d\mu_k$. Momentae Hamiltonians $H_{\lambda(k)^n}$ are the most important and are connected with resolvent expansion for the Sturm-Liouville problem.

6.4. Resolvent expansion and the Russian chain.

The Russian chain (6.4) represents equations for eigenfunctions f(k), g(k) of the Schrodinger operator

$$(6.7)\qquad -y'' + [u(x) + \zeta]y = 0$$

with the potential

$$(6.8)\qquad u(x) = -2\int_{\Omega} f(k)g(k)d\mu_k$$

and the eigenvalue $\zeta = -\lambda(k)$. The resolvent expansion of (6.7) gives the interpretation to momentae $\int_{\Omega} \lambda(k)^n f(k)g(k)d\mu_k$ in terms of differential polynomials in u(x), u'(x),... . We briefly state basic facts on the resolvent expansion of (6.7) following Drach [32] and [37], [15] and explain its relation with higher Korteweg-deVries equations [37], [38], [15], [24].

The equation for the resolvent $\overline{R}(x,z) \in C[[z^{-1}]]:z^2 = \zeta$ of the Sturm-Liouville problem (6.7) has the form [32], [37], [38]:

$$(6.9)\qquad -\overline{R}''' + 4\,(u+\zeta)\overline{R}' + 2u'\overline{R} = 0;$$

$$\overline{R} = \sum_{k=o}^{\infty} \overline{R}_k \zeta^{-k-1/2},$$

where $\overline{R}_k = \overline{R}_x[u,u',u'',...]$ are differential polynomials in u(x) (i.e., polynomials in $u, u_x, u_{xx}, ...$).

After the integration of the equation (6.9) J. Drach obtains the equation of the second order

$$(6.10)\qquad -2\,\overline{R}\,\overline{R}'' + (\overline{R}')^2 + 4(u+\zeta)\overline{R}^2 = c(\zeta)$$

for $c(\zeta) \in C[[\zeta^{-1}]]$ with constant coefficients. For $c(\zeta) \equiv 1$ in (6.10) we obtain

$$(6.11)\qquad R(x,z) = \sum_{k=o}^{\infty} R_k[u]\zeta^{-k-1/2}$$

Here

$$R_o = 1/2,\ R_1 = -\ 1/4u,\ R_2 = 1/16(3u^2-u''),\dots .$$

As a corollary of (6.9) we have one of the most famous identities ("Drach-Burchnall-Chaundy-Lenard"):

(6.12) $$-R_k''' + 4uR_k' + 4R_{k+1}' + 2u'R_k = 0.$$

These quantities $R_k[u]$ are the grounds for the proof of the complete integrability of the Korteweg-deVries equation [37], [47]

$$u_t = 6uu_x - u_{xxx}.$$

According to [47] the KdV equation is equivalent to the following identity

$$\frac{dL_2}{dt} = [L_3, L_2]$$

for $L_2 = -d^2/dx^2 + u(x,t)$, $L_3 = -4d^3/dx^3 + 3(u\frac{d}{dx} + \frac{d}{dx}u)$. In general the functional $I_n = \int R_n[u,u',\ldots]dx$ defines a nonlinear (x,t) Hamiltonian system

(6.13) $$u_t = -4\frac{\partial}{\partial x}R_n[u]$$

commuting with the KdV equation. This equation (6.13) is called (n-1)th KdV equation (cf. $n = 1,2$) and is equivalent to the Lax representation

$$\frac{dL_2}{dt} = [L_{2n+1}, L_2]$$

for an operator L_{2n+1} of order $2n+1$ in $\frac{d}{dx}$.

<u>6.5</u>. We may now ask, when does the potential $u(x)$ admit the representation (6.8):

$$u(x) = -2\int_\Omega f(k,x)g(k,x)d\mu_k$$

for eigenfunctions $f(k,x)$, $g(k,x)$ of (6.7): $-y'' + uy = k^2y$. We consider this representation in §7. E.g., for the potential rapidly decreasing on infinity $u(x)$ the measure $d\mu_k$ is a combination of a Lebesque measure on $[0,\infty)$ with a finite number of δ-functions corresponding to discrete eigenvalues $-x_i^2$ [15]. For periodic C^∞-potentials $u(x)$, $d\mu_k$ is singular with support at points of periodic or antiperiodic spectrum so-called ends of lacunae.

The representation (6.8) or even

(6.14) $$u = -2\int_\Omega f^2(k,x)d\mu_k$$

for

(6.15) $$f(k)_{xx} = uf(k) - k^2f(k)$$

is not unique (cf. §7 for periodic potentials $u(x)$).

However, the measure $d\mu_k$ (corresponding to the choice $\lambda(k) = k^2$) can be fixed if one demands certain norming conditions for representation of the coefficients $R_n[u]$ of the resolvent expansion in terms of momentae $\int k^{2n}f(x,k)g(x,k)d\mu_k$ (cf. [26], [14]).

We change slightly the norming of eigenfunctions, from $f(x,k)\cdot g(x,k)$ to $\lambda(k)\phi(x,k)\cdot\psi(x,k)$. In this case by the classical moment problem [2] the following representation can be proved: (for the general statement see §9).

There exists a measure $d\mu_k$ (corresponding to $\lambda(k)$) such that

$$(6.16) \qquad R_n[u] = 1/2\int_\Omega \lambda(k)^n \phi(x,k)\psi(x,k)\,d\mu_k$$

for all $n = 0,1,2,\ldots$ and $\phi(x,k)$, $\psi(x,k)$ are the eigenfunctions of the Schrödinger operator corresponding to the eigenvalue $\lambda(k)$ and the potentials:

$$(6.17) \qquad u(x) = -2\int_\Omega \lambda(k)\phi(x,k)\cdot\psi(x,k)\,d\mu_k.$$

Moreover, for real $u(x)$ and self-adjoint $-\frac{d}{dx^2} + u(x)$ we can choose $f \equiv g$, so (6.14)-(6.15) are satisfied.

<u>Main Statement</u> ([15], [26]). For self-adjoint L and the spectral measure $d\mu_k$ corresponding to $L\phi = -k^2\phi$ there exists a system $\phi(x,k)$ of eigenfunctions of

$$(6.18) \qquad (-\frac{d^2}{dx^2} + u(x))\phi = -k^2\phi$$

such that

$$(6.19) \qquad R_n[u] = 1/2\int k^{2n}\phi^2(x,k)\,d\mu_k$$

for all $n = 0, 1, 2, \ldots$. In particular,

$$(6.20) \qquad 1 = \int\phi^2(x,k)\,d\mu_k;$$

$$(6.21) \qquad -\frac{u}{2} = \int k^2\phi^2(x,k)\,d\mu_k, \ldots$$

Now the measure $d\mu_k$ and the system of eigenfunctions $\phi(x,k)$ is fixed by the conditions (6.19).

<u>Remark</u>: If we demand only

$$R_n[u] = \int P_n(k^2)\phi^2(x,k)\,d\mu_k' + C_n$$

for polynomials $P_n(k^2)$ of degree n, then the measure $d\mu_k'$ is not unique at all (we know only that supp $d\mu_k' \subseteq$ supp $d\mu_k$).

<u>6.6</u>. Now let us explain the important relation between higher kdV equations and the Hamiltonian flows $H_{h(k)}$ commuting with the Hamiltonian (6.5) H_I of the Russian chain.

We consider potential $u(x)$ in the Sturm-Liouville problem $-y_k'' + uy_k = k^2 y_k$ and the spectral measure $d\mu_k$ corresponding to $\lambda(k) = k^2$. Now by section 6.5:

$$R_n[u] = 1/2\int_\Omega k^{2n}\phi(x,k)\psi(x,k)\,d\mu_k \quad : n = 0, 1, 2, \ldots,$$

for $\phi(x,k)_{xx} = u(x)\phi(x,k) - k^2\phi(x,k)$; $\psi(x,k)_{xx} = u(x)\psi(x,k) - k^2\psi(x,k)$. If now we change $k\phi(x,k)$ to $f(x,k)$, and $k\psi(x,k)$ to $g(x,k)$ we come to a Russian chain (6.4) with $\lambda(k) = k^2$ having constraints

$$R_n[u] = 1/2\int k^{2n-2} f(x,k)g(x,k)d\mu_k: \quad n = 0, 1, 2, \ldots, \tag{6.22}$$

where

$$u(x) = -2\int_\Omega f(x,k)g(x,k)d\mu_k. \tag{6.23}$$

Let us now consider the Hamiltonian $H_{h(k)} = \int_\Omega h(k^2)C[k]d\mu_k$. Let $t = t_h$ be the time corresponding to the evolution according to $H_{h(k)}$. It is easy to compute the evolution of u(x) in (6.23) when (f_k, g_k) evolves according to $H_{h(k)}$:

$$u_{t_h} = -2\left(\int_\Omega h(k^2)f(x,k)g(x,k)d\mu_k\right)_x.$$

In particular, for $h(k^2) = k^{2n}$, $H_n = \int k^{2n}C[k]d\mu_k$ we get the evolution according to H_n: $n \geq 0$ in the following form, taking into account (6.22):

$$u_{t_n} = -4\frac{\partial}{\partial x}R_{n-1}[u]: \quad n \geq 1. \tag{6.24}$$

This is equivalent to an n-th KdV equation (6.13). In other words, for u(x,t) in the form (6.23), the evolution of u according to an n-th KdV equation is equivalent to the evolution of (f_k, g_k) according to H_I ($= H_o$) in the x-direction and according to H_n in the t-direction.

7. The Russian chain with constraints.

7.1. While the Russian chain itself is a completely integrable system, the most interesting problem is the existence of invariant constraints naturally arising from resolvent relations with momentae (see section 6.4). We change (as in section 6.5) our notations, multiplying $f(x;k)\cdot g(x;k)$ by $\lambda(k)$. In this case we obtain the following Canonical Russian Chain with constraints on $f_\alpha = f(x;\alpha)$ and $g_\alpha = g(x;\alpha)$ corresponding to $\lambda(\alpha)$ and Hilbert space $L^2(\Omega;d\mu_\alpha)$:

$$\begin{aligned} f_{\alpha,xx} &= -2f_\alpha\int_\Omega \lambda(\beta)f_\beta g_\beta d\mu_\beta + \lambda(\alpha)f_\alpha; \\ g_{\alpha,xx} &= -2g_\alpha\int_\Omega \lambda(\beta)f_\beta g_\beta d\mu_\beta + \lambda(\alpha)g_\alpha. \end{aligned} \tag{7.1}$$

The simplest constraint is the equation $(\vec{f},\vec{g}) = 1$ for $\vec{f} = (f_\alpha:\alpha \in \Omega)$, $\vec{g} = (g_\alpha:\alpha \in \Omega)$:

$$\int_\Omega f_\alpha g_\alpha d\mu_\alpha \equiv 1. \tag{7.2}$$

If we consider the potential u(x) defined by (6.17):

$$u(x) = -2\int_\Omega \lambda(\alpha)f_\alpha g_\alpha d\mu_\alpha, \tag{7.3}$$

then the set of constraints on (f_α, g_α) is given by [15], [26]:

$$R_n[u] = 1/2\int\lambda(\alpha)^n f_\alpha g_\alpha d\mu_\alpha: \tag{7.4}$$

$n = 0, 1, 2, \ldots$. The first two constraints ($n = 0, 1$) are (7.2), (7.3). The complete set of constraints can be deduced from (7.4).

Then the constraints for the system (7.1) come from the equation for resolvent (6.10). These constraints are exactly the following:

$$\sum_{\ell=0}^{\infty} \zeta^{-\ell} \int \lambda(\alpha)^{\ell} f_\alpha g_\alpha d\mu_\alpha + \tag{7.5}$$

$$+ \frac{1}{4} \zeta^{-1} \left\{ \sum_{\ell=0}^{\infty} \zeta^{-\ell} \int \lambda(\alpha)^{\ell} (f_\alpha g_\alpha)' d\mu_\alpha \right\}^2 -$$

$$- \zeta^{-1} \cdot \sum_{\ell=0}^{\infty} \zeta^{-\ell} \int \lambda(\alpha)^{\ell} f_\alpha g_\alpha d\mu_\alpha \cdot \sum_{\ell=0}^{\infty} \zeta^{-\ell} \times$$

$$\times \int \lambda(\alpha)^{\ell} f'_\alpha g'_\alpha d\mu_\alpha \equiv 1.$$

Here are the first constraints:

$$\int f_\alpha g_\alpha d\mu_\alpha \equiv 1; \quad \int (f_{\alpha x} g_\alpha + f_\alpha g_{\alpha x}) d\mu_\alpha \equiv 0; \tag{7.6}$$

$$\int \lambda(\alpha) f_\alpha g_\alpha d\mu_\alpha \equiv \int f_{\alpha x} g_{\alpha x} d\mu_\alpha; \tag{7.7}$$

$$\int \lambda(\alpha)^2 f_\alpha g_\alpha d\mu_\alpha \equiv \int \lambda(\alpha) f_{\alpha x} g_{\alpha x} d\mu_\alpha + \left(\int \lambda(\alpha)\, f_\alpha g_\alpha d\mu_\alpha\right)^2. \tag{7.8}$$

It is possible to verify directly that, in the most interesting case,

$$f_\alpha \equiv g_\alpha; \ \alpha \in \Omega$$

our system of constraints imply that all the first integrals $C[\alpha]$ of the Russian chain are zeros! In particular, the Hamiltonian is zero.

So far, one can call this system "a harmonic oscillator" with constraints; though constraint to the "zero ground state".

In section 7.3 we examine in detail the system $f_{k,xx} = u(x) f_k - \lambda(k) f_k$ together with the restriction

$$\int_\Omega f_g^2 d\mu_k \equiv 1 \tag{7.9}$$

(corresponding to (7.2) for $f_k \equiv g_k$). We found that then the corresponding equations for f_k are, in fact, equivalent to canonical Russian chain (7.1) with constraints (7.2), (7.3). In other words, systems of Jacobi-Neumann type with constraints (7.9) are indeed imbedded into Russian chain (7.1) with constraints (7.6)-(7.7).

7.2. Let us show how the Russian chain arises from stationary solutions of σ-models (§§1-3).

While σ-models themselves are connected with the inverse scattering method for

Laplace operator or a general $(\frac{\partial}{\partial x},\frac{\partial}{\partial t})$ linear partial differential operator of the second order; the stationary solutions of σ-models are connected with the "Russian chain" or inverse scattering method for Sturm-Liouville problem.

Let us present the description of all stationary solutions of different σ-models following the paper [26]. We consider the most general operator nonlinear σ-model equation (3.1) for $\bar{n},\bar{m}$ in $L^2(\Omega,d\sigma_\alpha)$. We can write the stationary solution of equations of σ-model in the form

(7.10)
$$(\bar{n})_\alpha = f_\alpha(\xi+\eta)e^{\rho(\alpha)(\xi-\eta)};$$
$$(\bar{m}) = g_\alpha(\xi+\eta)e^{-\rho(\alpha)(\xi-\eta)};\quad \alpha\in\Omega,$$

where

$$(\bar{f},\bar{g}) \equiv 1.$$

Substituting (7.10) in (3.1) we obtain a system of equations on f_α, g_α. Then our system of equations for $\bar{f} = (f_\alpha:\alpha\in\Omega)$, $\bar{g} = (g_\alpha:\alpha\in\Omega)$ is equivalent to the following

(7.11)
$$\begin{cases} f_{\alpha,\xi\xi} = \rho(\alpha)^2 f_\alpha + Vf_\alpha; \\ g_{\alpha,\xi\xi} = \rho(\alpha)^2 g_\alpha + Vg_\alpha \end{cases}$$

for

$$V = -\int_\Omega f_{\alpha,\xi}g_{\alpha,\xi}d\sigma_\alpha - \int_\Omega \rho(\alpha)^2 f_\alpha g_\alpha d\sigma_\alpha$$

together with constraints

(7.12)
$$\int_\Omega f_\alpha g_\alpha d\sigma_\alpha \equiv 1;\quad \int_\Omega \rho(\alpha) f_\alpha g_\alpha d\sigma_\alpha \equiv 0.$$

This is one of the versions of the "Russian chain" (7.1) We can turn it into a better known form by taking into account that

(7.13)
$$V = -2\int_\Omega \rho(\alpha)^2 f_\alpha g_\alpha d\sigma_\alpha + C.$$

If now $\mu(\alpha)^2 = \rho(\alpha)^2 + C$, then we have the following system (which is the Russian chain (7.1)):

(7.14)
$$\begin{cases} f_{\alpha,\xi\xi} = \mu(\alpha)^2 f_\alpha - 2\int_\Omega \mu(\beta)^2 f_\beta g_\beta d\sigma_\beta\cdot f_\alpha; \\ g_{\alpha,\xi\xi} = \mu(\alpha)^2 g_\alpha - 2\int_\Omega \mu(\beta)^2 f_\beta g_\beta d\sigma_\beta\cdot g_\alpha, \end{cases}$$

togehher with constraints (7.12) and (7.7) and

$$\int_\Omega f_{\alpha\xi}g_{\alpha\xi}d\sigma_\alpha \equiv \int_\Omega \rho(\alpha)^2 f_\alpha g_\alpha d\sigma_\alpha .$$

We note that a change of $\rho(\alpha)^2$ to $\mu(\alpha)^2$ corresponds to a gauge transformation (cf. [20]) of $\vec{n}$ and $\vec{m}$ in (7.10).

The same reduction of the stationary solutions of σ-models to a Russian chain is true if we take stationary solutions in the form

$$(\vec{n})_\alpha = f_\alpha(\xi+\gamma\eta)e^{\rho(\alpha)(\xi-\gamma\eta)};$$

$$(\vec{m})_\alpha = g_\alpha(\xi+\gamma\eta)e^{-\rho(\alpha)(\xi-\gamma\eta)}$$

for $\gamma \neq 0$. For $\gamma = 0$ we get simply instanton solutions [20], [26].

7.3. Here we investigate the relation between the canonical Russian chain (7.1) with constraints (7.6)-(7.7) and the Jacobi-Neumann systems of harmonic oscillator constraint to a sphere. We show that these systems are indeed imbedded into the Russian chain with constraints and correspond to the zero values of first integrals (e.g., of Hamiltonian) of (7.1). That is why we call the systems (7.1) or similar to (7.1) "free motion" with constraints rather than "harmonic oscillator". We consider the Sturm-Liouville equations

$$f(\omega;x)_{xx} = u(x)f(\omega;x) + \lambda(\omega)f(\omega;x) \tag{7.15}$$

restricted to a sphere of $L^2(\Omega,d\sigma_\omega)$:

$$\int_\Omega f(\omega;x)^2 d\sigma \equiv 1. \tag{7.16}$$

Of course, the condition (7.16) itself does not define an invariant manifold. In order to write down equations for an invariant manifold which are the consequences of (7.15)-(7.16), it is necessary to differentiate (7.15) once to obtain

$$\int_\Omega f(\omega;x)f(\omega;x)_x d\sigma = 0. \tag{7.17}$$

Differentiating (7.17) once more and using (7.15), we come to

$$\int_\Omega f(\omega;x)_x^2 d\sigma + u\int_\Omega f(\omega;x)^2 d\sigma + \int_\Omega \lambda(\omega)f(\omega;x)^2 d\sigma = 0.$$

In other words, first of all we get

$$u = -\int_\Omega f(\omega;x)_x^2 d\sigma - \int_\Omega \lambda(\omega)f(\omega;x)^2 d\sigma \tag{7.18}$$

and the condition (7.16) is equivalent to the system

$$\int_\Omega f(\omega;x)^2 d\sigma = 1, \quad \int_\Omega f(\omega;x)f(\omega;x)_x d\sigma = 0. \tag{7.19}$$

Now the system (7.15), (7.18) takes the Jacobi-Neumann form

$$f(\omega;x)_{xx} = \left\{-\int_\Omega f(\omega_1;x)_x^2 d\sigma_1 - \int_\Omega \lambda(\omega_1)f(\omega_1;x)^2 d\sigma\right\} \times f(\omega;x) + \lambda(\omega)f(\omega;x). \tag{7.20$_\lambda$}$$

where the invariant restrictions (7.19) should be imposed.

It is clear that if one denotes invariant submanifold (7.19) by $\mathfrak{M}_1$, then the system (7.20), restricted to $\mathfrak{M}_1$, is Hamiltonian with the following Hamiltonian

(7.21) $$H_1 = \frac{1}{2}\int_\Omega p(\omega)^2 d\sigma - \frac{1}{2}\int_\Omega \lambda(\omega)q(\omega)^2 d\sigma + \\ + \frac{1}{2}\left(\int_\Omega q(\omega)^2 d\sigma - 1\right)\left\{\int_\Omega \lambda(\omega)q(\omega)^2 d\sigma + \int_\Omega p(\omega)^2 d\sigma\right\}$$

Here it is assumed, of course, that $q(\omega)$ stands for $f(\omega;x)$ and $p(\omega)$ stands for $f(\omega;x)_x$.

Equations of evolution for the Hamiltonian H_1 take the form:

$$q(\omega)_x = p(\omega) \pmod{\mathfrak{M}_1};$$

$$p(\omega)_x = \lambda(\omega)q(\omega) - q(\omega)\left\{\int_{\Omega_1} \lambda(\omega_1)q(\omega_1)^2 d\sigma_1 + \int_{\Omega_1} p(\omega_1)^2 d\sigma_1\right\} \pmod{\mathfrak{M}_1},$$

which on $\mathfrak{M}_1$ is equivalent to the equations $(7.20)_\lambda$. Now the equations $(7.20)_\lambda$ can be naturally imbedded into the Russian chain (7.1).

Lemma 7.1. The system of equations $(7.20)_\lambda$ on $\mathfrak{M}_1$ with given λ and given value of energy H_1 is equivalent to the Russian chain (7.1) with $f \equiv g$:

$(7.23)_{\lambda'}$ $$f(\omega;x)_{xx} = -2\int f(\omega_1;x)^2 d\sigma_{\omega_1} \cdot f(\omega;x) + \lambda'(\omega)f(\omega;x)$$

for the function $\lambda'(\omega) = \lambda(\omega) + 2H_1$ on the manifold $\mathfrak{M}_1$.

Proof. Taking into account the form of Hamiltonian H_1 we conclude that on $\mathfrak{M}_1$:

$$H_1 = \frac{1}{2}\left\{\int_\Omega p(\omega)^2 d\sigma - \int_\Omega \lambda(\omega)q(\omega)^2 d\sigma\right\} \pmod{\mathfrak{M}_1} \text{ - is a constant.}$$

In other words,

(7.24) $$\int_\Omega p(\omega)^2 d\sigma - \int_\Omega \lambda(\omega)q(\omega)^2 d\sigma - 2H_1\int_\Omega q(\omega)^2 d\sigma = 0 \pmod{\mathfrak{M}_1};$$

$$\int_\Omega p(\omega)^2 d\sigma - \int \lambda'(\omega)q(\omega)^2 d\sigma = 0 \pmod{\mathfrak{M}_1}$$

for $$\lambda'(\omega) = \lambda(\omega) + 2H_1.$$

Substituting (7.24) into $(7.20)_\lambda$ we obtain

$$f(\omega;x)_{xx} = -2\int_{\Omega_1} \lambda(\omega_1)f(\omega_1;x)^2 d\sigma_1 \cdot f(\omega;x) + 2H_1 f(\omega;x) + \lambda(\omega)f(\omega;x),$$

which is equivalent to the system $(7.23)_{\lambda'}$.

Remark 7.2. It immediately follows from (7.24) that the system $(7.20)_\lambda$ on $\mathfrak{M}_1$ for a given energy H_1 after imbedding into the Russian chain $(7.23)_{\lambda'}$ for $\lambda'(\omega) = \lambda(\omega) + 2H_1$, has zero energy in terms of the Hamiltonian $(7.23)_{\lambda'}$.

After statements 7.1, 7.2 we find it necessary to consider the system (7.23) together with the following invariant

$$(7.25)\qquad \begin{aligned} &\int_\Omega f(\omega;x)^2 d\sigma = 1, \quad \int_\Omega f(\omega;x) f(\omega;x)_x d\sigma = 0, \\ &\int_\Omega f(\omega;x)_x^2 d\sigma - \int_\Omega \lambda(\omega) f(\omega;x)^2 d\sigma = 0, \end{aligned}$$

The infinite dimensional manifold defined by (7.25) is denoted as $\mathfrak{M}_2$ or $(\mathfrak{M}_2)_\lambda$.

Corollary 7.3. The Russian chain $(7.23)_\lambda$ considered on the manifold $\mathfrak{M}_2$ is equivalent to the system (7.15) or $(7.20)_\lambda$, if the latter one is restricted to the manifold $\mathfrak{M}_1 \subset \mathfrak{M}_2$ and the condition $H_1 = 0$ is added. Conversely, any system (7.15) on $\mathfrak{M}_1$ for a given λ is equivalent to the Russian chain $(7.23)_{\lambda'}$ on $(\mathfrak{M}_2)_{\lambda'}$, where $\lambda'(\omega) = \lambda(\omega) + 2H_1$. Of course, it is clear that for given λ the manifold $(\mathfrak{M}_2)_\lambda$ is a submanifold of $\mathfrak{M}_1$ satisfying the equation $H_1 = 0$.

Our simple results show that any system (7.15) on the manifold $\mathfrak{M}_1$ is equivalent to some Russian chain on $\mathfrak{M}_2$. Certainly, the Russian chain is much more interesting itself than $(7.20)_\lambda$. The system $(7.20)_\lambda$, as it is seen from the Hamiltonian H_1 in (7.21) describes harmonic oscillators with frequencies $\lambda(\omega)$ on the unit sphere of $L^2(\Omega,d\sigma)$. This interpretation of the system (7.20) as a sequence of harmonic oscillators on an n-dimensional sphere corresponding to the case $|\Omega| = n$ goes back to Jacobi and was explicitly proposed by Neumann [51]. In the infinite dimensional case the system (7.20) with constraints $\mathfrak{M}_1$ appeared for the first time in the paper [26].

Of course, the system (7.1) on $\mathfrak{M}_2$ corresponds to the trivial zero value of the corresponding Hamiltonian on $\mathfrak{M}_2$. In such a situation it is not very natural to speak about harmonic oscillators restricted to some manifold. It is a free motion restricted to some manifold.

§8. Quantization of the Russian chain.

8.1. The most important feature of the reduction of two-dimensional completely integrable systems to operator one-dimensional is the possibility of simple quantization. First of all, instead of field equations we quantize the usual one-dimensional Hamiltonian systems, though in operator form. Secondly, the simple form of Hamiltonian and conservation laws gives simple rules for quantization: there always is a typical kinetic part and we can always replace p_x by $\partial/\partial q_k$, for example. The rules for quantization are "classical": one replaces p_k and q_k in all the formulae by operators P_k and Q_k satisfying ordinary commutation relations.

$$[P_k,P_{k'}] = [Q_k,Q_{k'}] = 0, \quad [P_k,Q_{k'}] = \delta(k-k').$$

As it was explained before one is allowed to take the representation $p_k = \partial/\partial q_k$.

Now the main result obtained from the existence of infinite numbers of conserved quantities for the Hamiltonian is the existence of infinite numbers of commuting differential operators in P_k, Q_k that commute also with H considered as an operator. As it is supposed to be, these operators are quadratic in P_k and quartic in Q_k. One of the interesting observations here is the following: Bäcklund transformation of H is the canonical transformation on $\mathfrak{M}$ and can be represented as a differential operator anti-commuting with H and other Hamiltonians. This way we come to certain forms of Bethe ansatz.

Now there is no problem to translate this quantization together with commutativity and anticommutativity relations for operator from one-dimensional multicomponent systems to two-dimensional one component (here one component may be one operator component). At the same time we have direct connection with reflection coefficients because our first integrals are immediately expressed in terms of reflection coefficients. This way we immediately reconstruct all the commutativity conditions between different reflection coefficients that were considered by Honerkamp, Faddeev, Sklynin, Bergknoff, Thacker, Takhtadjan, and Kulish.

However, in the same way we obtain similar formulae for arbitrary operator (or multicomponent) nonlinear Schrödinger, multicomponent modified KdV, matrix sin-Gordon or any equation of "second order" in two dimensions. Our approach also gives the possibility to quantize the systems in three dimensions. The best candidate in three dimensions is so-called modified two-dimensional KdV or two-dimensional nonlinear Schrödinger equation , but, these equations are very complicated and probably do not have any physical sense.

On the other hand, we want to mention certain ambiguities in quantization of nonlinear Schrödinger. While eigenfunctions for the second quantization of nonlinear Schrödinger are determined by Bethe ansatz, the eigenfunctions of the quantized stationary nonlinear Schrödinger are eigenfunctions of the quartic oscillators. We can guess that in order to get consistent quantization of two-dimensional completely integrable systems and their stationary counterparts additional "non-elementary" functions (e.g., eigenfunctions of the quartic oscillator) should be introduced.

Remark: In connection with future complete integrability of full quantum four-dimensional Yang-Mills it should be noted immediate appearance of multi-dimensional quartic oscillator. As it had been noticed by Zakharov, four-dimensional classical Yang-Mills implies the following two-dimensional system:

$$U_{xx} - U_{tt} = UU^{+}U.$$

If you change U_{tt} by iU_t you get operator nonlinear Schrodinger that is completely integrable as we had shown. If you add to the right hand side mU for $m \neq U$, then the systems will not be completely integrable, but what to do with this system itself, we don't know. Our guess is complete integrability for both classical and quantum systems.

8.2. We start first from the classical Hamiltonian

$$H = \frac{1}{2}\{\int p_k^2 d\mu_k + \int \lambda(k) q_k^2 + \frac{1}{2}(\int q_k^2 d\mu_k)^2\},$$

where $q(k)$, $p(k)$ are considered as elements of $L^2(\Omega, d\mu_k)$, $\lambda(k)$ is a measurable function (from $L^2(\Omega, d\mu_k)$).

Now we naturally quantize this system. We replace p_k and q_k by operators P_k, Q_k, assuming the following basic commutation relations take place

$$[P_k, P_{k'}] = [Q_k, Q_{k'}] = 0, \quad [P_k, Q_{k'}] = \delta(k-k').$$

The main result concerning operators commuting with H is the following observation, which is based on the corresponding classical fact.

Theorem 8.1. Let us define for any $k \in \Omega$ the following operator

$$G_k = P_k^2 + \lambda(k) Q_k^2 + Q_k^2 \int_\Omega Q_\lambda^2 d\mu_\lambda - \int_\Omega \frac{1}{\lambda(\ell)-\lambda(k)} \{Q_\ell P_k - Q_k P_\ell\}^2 d\mu .$$

Then G_k commute: $[G_k, G_{k'}] = 0$ for all k, k'.

Using linear transformation

$$G \to \int_\Omega h(k) G_k d\mu_k$$

we can represent these operations in a more isotropic form, e.g.,

$$H = \frac{1}{2}\int_\Omega G_k d\mu_k$$

is our Hamiltonian

$$H = \frac{1}{2}\{\int P_k^2 d\mu_k + \int \lambda(k) Q_k^2 d\mu_k + \frac{1}{2}(\int Q_k^2 d\mu_k)^2\}.$$

Now replacing P_k, Q_k by p_k, q_k we obtain from G_k first integral $C(k)$ of the Hamiltonian H. These $C(k)$ are the motivation for the introduction of G_k.

This theorem is proved just by taking commutators of the corresponding operators. However, one can use an analogy with the inverse scattering as $C(k)$ is expressed in terms of $2k/\pi \ \ln|R(k)|^2$.

The same quantization is applied to the Russian chain (6.4), when one replaces f_k by Q_{1k}, g_k by Q_{2k}, and g_{kx} by P_{1k}, f_{kx} by P_{2k}. This way there appears new commutation relations between operators $C[k]$, $K[k]$, $C_2[k]$, $C_3[k]$, when the canonical variables are substituted for operators. In particular, we have

$$[K[k'], C_j[k]] = 0 \quad \text{for } k' \neq k,$$

$$[K[k], K[k']] = 0; \quad [K[k], C_j[k]] = 2C_j[k]: \ j = 2,3.$$

One can combine this quantization of the Russian chain with our reduction of KdV-type equation to a combination of the Russian chain and commuting Hamiltonian flows. These new relations between $K[k]$, $C[k]$, $C_2[k]$, $C_3[k]$ after quantization give rise to a nice commutation relations between quantum Hamiltonians and different quantum scattering coefficients $b(\xi)$, $b^*(\xi)$ for quantum modified KdV, nonlinear Schrödinger, sin-Gordon, massive Thirring models obtained by Honerkamp [40], Thacker [57], Faddeev [56], [39].

§9. Completely integrable two-dimensional systems associated with linear differential operators of the first order.

9.0. This chapter is devoted to the description of the class of equations solvable by the inverse scattering method and associated with linear differential equations of the first order with matrix operator coefficients. In 9.1, this class of equations is described in terms of certain integrodifferential operators following the method of AKNS [1] and Newell [52]. In 9.2, the same class of equations is characterized in terms of coefficients of expansions of eigenfunctions in powers of spectral parameter, see [33], [34]. This enables us to associate with the coefficients of the expansion certain spectral measures $d\Sigma_E$. Then the coefficients of the linear differential operator of the first order are represented as integrals over the spectral measure $d\Sigma_E$ of the product of eigenfunctions of the given and conjugate spectral problem. Such representation immediately leads us to a certain version of the operator Russian chain.

9.1. Let us following [52] present nonlinear (x,t)-dimensional partial differential equations associated with inverse scattering method for matrix (operator) linear differential equations of the first order with matrix (operator) coefficients.

We start with the first order matrix (operator) equation on a Hilbert space H:

$$(9.1)\qquad f_x = -P(x)f + EAf$$

for f from H and diagonal A. We assume, as usual, that

$$(9.2)\qquad P = [A,V].$$

We assume in 9.1 the knowledge of asymptotics of P as $x \to \mp \infty$. Actually all that is needed is the existence of a "monodromy" operator (cf. [52], [33]). In the case of $\|P\| \to 0$ as $x \to \mp \infty$ we have a "monodromy operator" as a scattering operator. One considers a fundamental matrix (operator) solution $\phi(x;E)$ of (9.1). We can define the scattering matrix R(E) of (9.1) as

$$(9.3)\qquad \phi(+\infty;E) = \phi(-\infty;E)R(E).$$

The inverse scattering transform is a mapping between P(x) and R(E) including (if necessary) information on the analytic extension to Im E > 0 or Im E < 0 of the principal corner minors of R(E).

For the corresponding nonlinear (x,t) partial differential equation associated with (9.1) the dependence of R(E;t) on t is described by a linear partial differential equation as in Appendix to §4. We have:

$$\begin{array}{ccccccc} P(x;0) & \rightarrow & R(E;0) & \rightarrow & R(E;t) & \rightarrow & P(x;t). \\ & \text{direct spectral transform} & & \text{linear evolution} & & \text{inverse spectral transform} & \end{array}$$

This class of (x;t) equations is defined as in §5 using an integro-differential operator.

To make our notations clearer we adopt matrix ones. We present operators W on H sometimes in a matrix form $W = (W_{i,j})$ corresponding to some choice of basis of H. Let A be a diagonal matrix:

$$(9.4) \qquad A = (a_i \delta_{ij})$$

If $B = (b_{ij})$, then we put $B_A = [A,B] = ((a_i - a_j)b_{ij})$. We also set $B_D = (b_{D,ij})$ where $b_{D,ij} = 0$ if $a_i = a_j$ and $b_{D,ij} = b_{ij}$ otherwise; $B_K = (b_{K,ij})$, where $b_{K,ij} = b_{ij}$ if $a_i = a_j$ and $b_{K,ij} = 0$ otherwise. So $B_D + B_K = B$.

Now let B_D be known (e.g., B_A is known). We determine in a canonical way some B_K. We put

$$K(B) = \int_x^\infty [P,B_D]_K dx_1.$$

If $B = B_D + B_K$ for $B_K = K(B)$ and $[B_K,P]_K = 0$, then we define

$$\mathcal{D}\cdot B_D = \frac{\partial}{\partial x}\cdot B + [P,B].$$

Now let B_D be known and $B = B_D + K(B_D)$. Let us set

$$\mathcal{D}_R\cdot B_A = \mathcal{D}\cdot(B_D).$$

Then the following assertion is true:

<u>Theorem 9.1</u> ([52]). For entire functions G, Ω, F of E the evolution equation for P(x,t):

$$(9.5) \qquad G(\mathcal{D}_R,t)\cdot P_t = \Omega(\mathcal{D}_R,t)\cdot[C,P] + F(\mathcal{D}_R,t)\cdot x\,[A,P]$$

is equivalent to:

$$G(E,t)\cdot R_t = \Omega(E,t)\cdot[C,R] - F(E,t)\cdot R_E.$$

We consider below <u>only</u> equations with $F \equiv 0$, in other words, only those equations that appear in a traditional AKNS [1] or Zakharov-Shabat scheme [59]. Equations with $F \not\equiv 0$ correspond to equations of Painlevé type (cf. [18]). As we see in §10, these equations can be reduced to a joint action of two commuting Hamiltonian flows. These Hamiltonians (analogical to the Russian chain) arise from the representation of the potential in (9.1) in terms of the integral over the spectral measure of product of eigenfunctions of (9.1) and the conjugate problem.

9.2. Now the structure of equations of evolution associated with (9.1) is analyzed in terms of certain expansions in powers of spectral parameter E at $E = \infty$. As before we consider operators on a Hilbert space H as (infinite) matrices. Potentials V are considered as operators on H. For a diagonal matrix A (9.4) on H we define the potential P in (9.1) the same as in (9.2), denoting P as V_A:

(9.6) $$V_A = [A,V].$$

Lemma 9.2. The following equation

(9.7) $$\frac{\partial\lambda}{\partial x} = [\lambda, V_A - EA]$$

has a unique solution $\lambda = \lambda_A(E,x)$ as a formal power series from $C[[1/E]]$ such that

(9.8) $$\lambda_A = A + \lambda_{1,A}/E + \cdots = \sum_{n=o}^{\infty} \frac{\lambda_{n,A}}{E^n}$$

with $\lambda_{0,A} = A$, $\lambda_{1,A} = - V_A$ and for which

(9.9) $$\lambda_A \equiv A \quad \text{if} \quad V_A = 0.$$

Proof. We use the equation (9.7) and get the following system of equations determining $\lambda_{n,A} = \lambda_{n,A}(x)$:

(9.10) $$\frac{\partial\lambda_{n,A}}{\partial x} = [\lambda_n, V_A] + [A, \lambda_{n+1}]:$$

$n = 0,1,2,\ldots$. Then the conditions (9.8)-(9.9) together with $\lambda_{o,A} = A$ determine all the $\lambda_{n,A}$ from (9.10) consecutively.

The quantities $\lambda_{n,A}$ are functionals in $V_A, (V_A)_x, (V_A)_{xx}, \ldots$. Nonlinear (x,t)-equations of evolution associated with the linear differential matrix (operator) equation of the first order (see 9.1 and Theorem 9.1) can be expressed in terms of the functionals $\lambda_{n,A}$. According to results of Dubrovin [34] or the results of [33], [36] these equations can be represented as

(9.11) $$[A, V_t + \lambda_{n,A}] = 0; \text{ or}$$

$$-[A, V_t] = [A, \lambda_{n,A}] : n = 0, 1, 2, \ldots .$$

The general (x,t)-equations have the form [33], [34], [36]:

(9.12) $$[A, V_t + \sum_{n=1}^{N} \lambda_{n,A} c_n] = 0$$

for constants $c_n (n = 1,\ldots,N)$. Equations (9.11)-(9.12) are Hamiltonian systems in V_A and they admit Lax representations [36]:

$$\frac{dL}{dt} = [M,L]$$

for $L = B\cdot\frac{d}{dx} + BV_A$ and $AB = I$. Here M is a linear differential operator in $\frac{d}{dx}$ of order N.

We relate $\lambda_{n,A}$ with the expansion of eigenfunctions of a linear operator naturally.

We consider now the linear operator associated with the potential V_A:

$$(9.13) \qquad L_A = \frac{d}{dx} + V_A - EA$$

and E is considered as (complex) spectral parameter. In order to specify the dependence L_A on E we put

$$L_{A,E} = \frac{d}{dx} + V_A - EA.$$

The operator conjugate to $L_{A,E}$ is

$$(9.14) \qquad L^t_{A,E} = \frac{d}{dx} - V^t_A + EA^t.$$

Here B^t is a matrix transposed to B. We consider eigenoperators of $L_{A,E}$ and $L^t_{A,E}$:

$$(9.15) \qquad L_{A,E}F = 0 \quad \text{and}$$

$$(9.16) \qquad L^t_{A,E}F' = 0.$$

If we define

$$(9.17) \qquad \check{F} = F,\ \hat{F} = F'^t$$

then conditions (9.15)-(9.16) can be written as equations for right and left eigenfunctions of $L_{A,E}$:

$$(9.18) \quad \begin{aligned} &\text{a)} \quad \frac{d}{dx}\check{F}_E = -V_A\check{F}_E + EA\check{F}_E; \\ &\text{b)} \quad \frac{d}{dx}\hat{F}_E = \hat{F}_E V_A - \hat{F}_E\cdot EA. \end{aligned}$$

Remark 9.3. If $\check{F}_E$ is a solution of (9.15) or (9.18),a), then $\check{F}_E\cdot C$ is also a solution of (9.18),a) for an x-independent C. In this case $C\cdot(\check{F}_E)^{-1}$ is a solution of (9.16) or (9.18),b) whenever $\check{F}_E$ is non-singular.

For example, if one takes $\check{F}^\circ_E$ as any non-singular solution of (9.18),a), then any other solution of (9.18) can be written in the form

$$(9.19) \quad \begin{aligned} \check{F}_E &= \check{F}^\circ_E\cdot C \\ \hat{F}_E &= D\cdot(\check{F}^\circ_E)^{-1}\ . \end{aligned}$$

From the class of solutions of (9.19) we want to single out some special ones that determine spectral measure. We want to consider spectral problems algebraically but not analytically because we want to consider the most general class of potentials

V_A. In order to determine spectral measures the method of the matrix moment problem is used.

First of all a few auxiliary results are presented:

Lemma 9.4. Let $\check{F}_E, \hat{F}_E$ be solutions of (9.18). If

$$k_E = \check{F}_E \cdot \hat{F}_E, \tag{9.20}$$

then k_E satisfies the equation (9.7).

Precisely,

$$\frac{dk_E}{dx} = [k_E, V_A - EA].$$

Proof. We have by (9.18),

$$k_{E,x} = (-V_A\check{F}_E + EA\check{F}_E)\hat{F}_E + \check{F}_E(\hat{F}_E V_A - \hat{F}_E EA) = [k_E, V_A] + E\cdot[A, k_E].$$

The converse assertion is also true:

Lemma 9.5. If k_E satisfies the equation (9.7):

$$\frac{dk_E}{dx} = [k_E, V_A - EA],$$

then there are solutions $\check{F}_E, \hat{F}_E$ of (9.18) such that the representation (9.20) is valid:

$$k_E = \check{F}_E \hat{F}_E.$$

Proof. We choose the following system of fundamental solutions of (9.18):

$$\check{F}^\circ_E \quad \text{and} \quad \hat{F}^\circ_E = (\check{F}^\circ_E)^{-1}.$$

Let us put

$$\ell_E = (\check{F}^\circ_E)^{-1}\cdot k_E \cdot \check{F}^\circ_E = (\check{F}^\circ_E)^{-1} k_E (\hat{F}^\circ_E)^{-1}.$$

Then

$$(\ell_E)_x = -(\check{F}^\circ_E)^{-1}\{-V_A + EA\}k_E\check{F}^\circ_E + (\check{F}^\circ_E)^{-1}[k_E, V_A - EA]\check{F}^\circ_E + (\check{F}^\circ_E)^{-1}k_E\cdot\{-V_A + EA\}\check{F}^\circ_E = 0.$$

Thus ℓ_E = const and $k_E = \check{F}^\circ_E \cdot \ell_E \cdot \hat{F}^\circ_E$. According to (9.19) this means that k_E admits the representation (9.20), and the lemma is proved.

Now we want to find the solutions $\check{F}_E$, F_E of (9.18) such that for $k_E = \check{F}_E \cdot \hat{F}_E$ one has the asymptotic expansion (9.8) exactly for $E \to \infty$ with conditions (9.9) satisfied. In order to explain more precisely what we mean we should recall the Borel definition of asymptotic expansions [61]. The series (9.8) is definitely an asymptotic expansion for an unknown function that may be non-analytic. Nevertheless if one assumes that λ_A is a Hilbert transform of some, maybe complex measure, then the coefficients in the asymptotic expansion (9.8) are nothing but momentae with respect to the spectral measure [2].

First of all in order to get spectral measure we have to show that asymptotic expansion (9.8) can be really realized as a function k_E from (9.20) for some solutions

$\check{F}_E$, $\hat{F}_E$ of (9.18). For this, one needs canonical expansion of solutions of (9.18) in descending powevers of E as $E \to \infty$ [34].

The lemma below is a classical statement on asymptotic expansion of solutions of linear differential equations in the neighborhood of a regular singularity.

Lemma 9.6. (cf. [33], [34]) There are fundamental solutions of the system (9.18) $\check{\phi}_E$ and $\hat{\phi}_E$ tnat can be represented as a formal power series in E when $E \to \infty$ in the following way:

$$(9.21) \qquad \check{\phi}_E = \left\{\sum_{m=o}^{\infty} \check{T}_m E^{-m}\right\} \exp(EAx)$$

$$(9.22) \qquad \hat{\phi}_E = \exp(-EAx) \cdot \left\{\sum_{m=o}^{\infty} \hat{T}_m E^{-m}\right\}$$

with

$$(9.23) \qquad \check{T}_o = I, \quad \hat{T}_o = I$$

and elements of $\check{T}_m$, $\hat{T}_m$ can be reconstructed from elements of A, V, V_x, V_{xx},... and they are such that

$$(9.24) \qquad \check{T}_m = \hat{T}_m = 0: m \geq 1 \quad \text{if } V = 0.$$

Moreover,

$$(9.25) \qquad \hat{\phi}_E = (\check{\phi}_E)^{-1} \quad \text{or}$$

$$\left\{\sum_{m=o}^{\infty} \check{T}_m E^{-m}\right\} \cdot \left\{\sum_{m=o}^{\infty} \hat{T}_m E^{-m}\right\} = I.$$

Proof. Let us consider an asymptotic expansion (9.21) together with the initial condition (9.24). Substituting (9.21) into (9.18) we obtain the following equations for the matrix coefficients $\check{T}_m$: $m \geq 0$:

$$\check{T}_o = I \quad \text{and}$$

$$(9.26) \qquad \check{T}_{n,x} = [A, \check{T}_{n+1}] - V_A \check{T}_n$$

for $n = 0, 1, 2, \ldots$. The equations (9.26) follow from (9.18) because

$$\check{\phi}_{E,x} \overset{\text{def}}{=} \left\{\sum_{m=o}^{\infty} \check{T}_{m,x} E^{-m} + \check{T}_m E^{-m+1} A\right\} \times \exp(EAx) = \left\{\sum_{m=o}^{\infty} (EA - V_A) \check{T}_m E^{-m}\right\} \exp(EAx).$$

If we denote the matrix $\check{T}_n$ by

$$\check{T}_n = (t_{n,ij})$$

then the equations (9.26) are equivalent to:

$$(9.27)\qquad (t_{n,ij})_x = (a_i-a_j)t_{n+1,ij} - \sum_k (a_i-a_k)V_{ik}t_{n,kj}$$

for

$$(9.28)\qquad A = (a_i\delta_{ij}),\quad V = (V_{ij}).$$

Equations (9.2) determine $\check{T}_n$ completely by induction in n.

If $a_i \neq a_j$, then

$$t_{n+1,ij} = (a_i-a_j)^{-1}t_{n,ij,x} + \sum_k (a_i-a_j)^{-1}(a_i-a_k)V_{ik}t_{n,kj}\,.$$

On the other hand, let $a_i = a_j$. Then

$$t_{n+1,ijx} = -\sum_k (a_i-a_k)V_{ik}t_{n+1,kj} = -\sum_{k,a_j\neq a_k} (a_j-a_k)V_{ij}t_{n+1,kj}\,.$$

So $t_{n+1,ij}$ is determined in terms of $t_{n+1,rs}$ for $a_r \neq a_s$. Now $t_{n+1,ij}$ can be determined in such a way that $t_{n+1,ij} = 0$ for $V = 0$.

In order to get (9.25) one gets (9.22) with (9.23)-(9.24) and then use the uniqueness of the $\check{T}_m$. One simply notices that $(\check{\phi}_E)^{-1}$ satisfies all the conditions (9.22)-(9.24) so it is identical with $\hat{\phi}_E$.

One of the most interesting problems is the locality of $\check{T}_m$ in terms of V and A only. We can get from lemma 9.6 immediately representation of λ_A as k_E for some solutions $\check{F}_E$, $\hat{F}_E$ of (9.18).

<u>Lemma 9.7.</u> Let $\check{\phi}_E$ and $\hat{\phi}_E$ satisfy all the conditions (9.21)-(9.25) of the lemma 9.6. Let $\check{F}_E$, $\hat{F}_E$ be solutions of (9.18) such that

$$\check{F}_E = \check{\phi}_E C_1,\quad \hat{F}_E = C_2\hat{\phi}_E$$

and $C_1\cdot C_2 = A$. Then for

$$(9.29)\qquad \eta_A \overset{\text{def}}{=} \check{F}_E\cdot\hat{F}_E = \check{\phi}_E A\hat{\phi}_E$$

we have the same asymptotic expansion for $E\to\infty$ as λ_A:

$$(9.30)\qquad \eta_A \doteq A + \frac{\lambda_{1,A}}{E} + \cdots = \sum_{n=o}^{\infty} \frac{\lambda_{n,A}}{E^n}\,.$$

Here $\lambda_{n,A}$ are differential polynomials in $V, V_x, V_{xx}, \ldots$ and such that

$$\lambda_{o,A} = A \quad\text{and}\quad \lambda_{n,A} = 0 \quad \text{for } n \geq 1 \text{ if } V = 0.$$

<u>Proof.</u> We have

$$\eta_A = \left\{\sum_{m=o}^{\infty} \check{T}_m E^{-m}\right\}\exp(EAx)A \times \exp(-EAx)\left\{\sum_{m=o}^{\infty} \hat{T}_m E^{-m}\right\} =$$

$$= \left\{\sum_{m=o}^{\infty} \check{T}_m E^{-m}\right\} \times A \times \left\{\sum_{m=o}^{\infty} \hat{T}_m E^{-m}\right\} = A - [A,V]E^{-1} + \cdots = \sum_{m=o}^{\infty} \eta_{m,A} E^{-m}.$$

Here $\eta_{m,A}$ are differential polynomials in A, V, V_x, V_{xx},... and by (9.23)-(9.25) $\eta_{m,A} = 0$ for $m \geq 1$ if $V = 0$. Because η_A satisfies (9.7) and λ_A is a unique asymptotic expansion, we have

$$\eta_A \hat{=} \sum_{n=o}^{\infty} \frac{\lambda_{n,A}}{E^n}.$$

§10. Matrix spectral measure and representation of potential as a quadratic form in eigenfunctions. Commuting Hamiltonians associated with linear differential operators of the first order.

10.0. We consider in this chapter the spectral measure $d\Sigma_E$ associated with the linear problem (9.18) in such a way that for eigenfunctions $\check{\phi}_E$, $\hat{\phi}_E$ we have the solution of the moment problem $\int E^n \check{\phi}_E d\Sigma_E \hat{\phi}_E = \lambda_{n,A}$ (n = 0, 1, 2, ...), cf. (9.8). In particular, $\int \check{\phi}_E d\Sigma_E \hat{\phi}_E = A$, $\int E\check{\phi}_E d\Sigma_E \hat{\phi}_E = -V_A$. This way the linear problem (9.18) is replaced by a nonlinear Hamiltonian system, which is completely integrable. The corresponding (operator) Hamiltonian system can be considered as an analogue of the Russian chain (see §11 for the operator generalization of the Russian chain). Analyzing this Hamiltonian system and its first integrals we come to the following conclusion. All the (x,t)-dimensional "completely" integrable systems solvable via the inverse scattering method for the linear operator of the first order can be decomposed into two commuting one-dimensional Hamiltonians. These Hamiltonians are naturally associated with spectral measure $d\Sigma_E$ and moment problem 10.1.

10.1. Let us define now the spectral measure $d\Sigma_E$ associated with the eigenvalue problem for the linear differential equation of the first order (9.1) or (9.18) with operator (matrix) coefficients. We use for this the expansions of 9.2 in powers of spectral parameter E. We define the spectral measure as the measure realizing the momentae $\lambda_{n,A}$ from Lemma 9.2. Of course, in this matrix case the spectral measure is the spectral matrix. We formulate the definition of the spectral matrix measure $d\Sigma_E = (d\sigma_{E,ij})$ in the following way:

Spectral Problem 10.1. Let $\check{\phi}_E$, $\hat{\phi}_E$ be solutions of (9.18) defined in Lemma 9.6 and satisfying all the conditions (9.21)-(9.25) of the Lemma 9.6 The measure $d\Sigma_E$ is called a spectral measure $d\Sigma_E = (d\sigma_{i,j})$ if it is a solution of the following moment problem:

$$(10.1) \qquad \int_E \check{\phi}_E \cdot E^n d\Sigma_E \hat{\phi}_E = \lambda_{n,A}: \quad n = 0,1,2,\ldots .$$

Of course, in (10.1) $\check{\phi}_E$, $\hat{\phi}_E$ can be chosen as arbitrary solutions of (9.18). In this case $d\Sigma_E$ is multiplied by a constant matrix.

The problem (10.1) is looking like a classical matrix moment problem [2], [6], [7], [45]. First of all, an $n \times n$ moment problem is always reduced to n^2 scalar moment problems. The peculiarity of the moment problem (10.1) lies in the following. Following Stieltjes-Borel studies we can always solve the moment problem with momentae $\{\lambda_{n,A}: n = 0,1,\ldots\}$ in the following way:

$$\int_E E^n d\tau_E = \lambda_{n,A}: \quad n = 0,1,2,\ldots . \tag{10.2}$$

The interesting feature of the moment problem (10.1) is the representation of the weight function $d\tau$ in (10.2) in the form $d\tau = \check{\phi}_E d\Sigma_E \hat{\phi}_E$, i.e., with separation of E-part and x-part. This representation of $d\tau$ is achieved because one can treat (10.1) or (10.2) as a moment problem in E depending on an additional parameter x.

We ignore for a moment the problem of uniqueness of the measure $d\tau$ in (10.2) referring the reader to the discussion of exceptional cases in [6], [2]. Another problem treated in detail in [2], [6], [45], [61] is the positivity of $d\tau$ (or $d\Sigma_E$) in self-adjoint cases.

<u>Proof of the existence of a spectral measure</u> $d\Sigma_E$ <u>in 10.1</u>. Let us consider an arbitrary solution (weight matrix or matrix measure) $d\tau = d\tau(E,x)$ of the moment problem (10.2). We want to show that $d\tau(E,x)$ can be represented in the form $\check{\phi}_E d\Sigma_E \hat{\phi}_E$ for $\check{\phi}_E$, $\hat{\phi}_E$ being solutions of (9.18). For this we use equations (9.10) satisfied by $\lambda_{n,A}$:

$$\int_E E^n (d\tau)_x = [\int_E E^n d\tau, V_A] + [A, \int_E E^{n+1} d\tau] = \int_E E^n \cdot [d\tau, V_A - EA] \quad \text{or}$$

$$\int_E E^n \{d\tau_x - [d\tau, V_A - EA]\} = 0: \quad n = 0,1,2,\ldots .$$

In the uniqueness case (cf. [2]) we have $d\tau_x = [d\tau, V_A - EA]$. By lemma 9.5 one obtains

$$d\tau = \check{\phi}_E \cdot d\Sigma_E \cdot \hat{\phi}_E$$

for measure $d\Sigma_E$ independent of x and any solutions $\check{\phi}_E$, $\hat{\phi}_E$ of (9.18). This way the measure $d\Sigma_E$ corresponding to solutions of (9.18) satisfying conditions of lemma 9.6.

Now the solution of $d\Sigma_E$ of the spectral problem 10.1 can be combined with (9.18) in order to separate variables in an (x,t)-dimensional "completely" integrable systems (9.11)-(9.12).

<u>10.2.</u> Let us consider the system of linear differential equations (9.18):

$$\begin{aligned} \check{\phi}_{Ex} &= -V_A \check{\phi}_E + EA\check{\phi}_E; \\ \hat{\phi}_{Ex} &= \hat{\phi}_E V_A - \hat{\phi}_E EA. \end{aligned} \tag{10.3}$$

Then one can consider according to <u>10.1</u> the eigenfunctions $\check{\phi}_E$, $\hat{\phi}_E$ such that for some spectral matrix measure $d\Sigma_E = (d\sigma_{ijE})$ depending on V and A we have the solution of the moment problem:

$$\int E^n \check{\phi}_E \cdot d\Sigma_E \cdot \hat{\phi}_E = \lambda_{n,A}: \quad n = 0,1,2,\ldots . \tag{10.4}$$

The most important are the first two momentae

$$(10.5)\qquad \int\check{\phi}_E d\Sigma_E\cdot\hat{\phi}_E = A; \qquad \int E\check{\phi}_E d\Sigma_E\cdot\hat{\phi}_E = -V_A.$$

The representation for the potential V_A in (10.5) allows us to treat linear differential equation (10.3) as a Hamiltonian system with constraints (10.4) (or only (10.5)). In other words we have, e.g., the following system of equations:

$$(10.6)\qquad \begin{aligned}\check{\phi}_{E,x} &= \int_{E_1} E_1\check{\phi}_{E_1} d\Sigma_{E_1}\cdot\hat{\phi}_{E_1}\cdot\check{\phi}_E + EA\check{\phi}_E\,;\\ \hat{\phi}_{E,x} &= -\hat{\phi}_E\int_{E_1} E_1\check{\phi}_{E_1} d\Sigma_{E_1}\hat{\phi}_{E_1} - \hat{\phi}_E EA\,.\end{aligned}$$

On the other hand one can use the first of the formulae in (10.5) and replace A by a quadratic form in $\check{\phi}_E$, $\hat{\phi}_E$. In this case instead of (10.6) we obtain

$$(10.7)\qquad \begin{aligned}\check{\phi}_{E,x} &= \int_{E_1} (E_1+E)\check{\phi}_{E_1}\cdot d\Sigma_{E_1}\cdot\hat{\phi}_{E_1}\cdot\check{\phi}_E\,;\\ \hat{\phi}_{E,x} &= -\hat{\phi}_E\cdot\int_{E_1} (E_1+E)\check{\phi}_{E_1}\cdot d\Sigma_{E_1}\cdot\hat{\phi}_{E_1}\,.\end{aligned}$$

Both Hamiltonian systems (10.6) and (10.7) are completely integrable systems and both systems have first integrals quartic in $\check{\phi}$, $\hat{\phi}$. However, we consider here only the system (10.7) because for this system the description of Hamiltonian structure and first integrals are the simplest.

<u>10.3</u>. We now study the following Hamiltonian system that is derived in <u>10.1</u> - <u>10.2</u> via a matrix linear differential operator of the first order:

$$(10.8)\qquad \begin{aligned}\check{\phi}_{\lambda,x} &= \int_\mu (\lambda+\mu)\check{\phi}_\mu d\Sigma_\mu\hat{\phi}_\mu\cdot\check{\phi}_\lambda\,;\\ \hat{\phi}_{\lambda,x} &= -\hat{\phi}_\lambda\cdot\int_\mu (\lambda+\mu)\check{\phi}_\mu d\Sigma_\mu\hat{\phi}_\mu:\end{aligned}$$

$\lambda\in\Omega$, where operators (matrices) $\check{\phi}_\lambda$, $\hat{\phi}_\lambda$ are defined on H. Here $d\Sigma_\mu$ $(\mu\in\Omega)$ is a certain matrix measure and the triplet $(\check{\phi}_\lambda,\hat{\phi}_\eta,d\Sigma_\lambda)$ is considered on the space $\bigoplus_{L^2(\Omega,d\Sigma_\mu)} H$.

The system (10.8) possesses a lot of symmetries and many first integrals. At once we present the corresponding results.

<u>Theorem 10.2</u>. For any $\lambda\in\Omega$ the following quantities are the first integrals of the system (10.8):

(10.9) $$K[\lambda] = \hat{\phi}_\lambda \check{\phi}_\lambda \quad \text{and}$$

(10.10) $$C[\lambda[= \hat{\phi}_\lambda \cdot \int_\mu \frac{1}{\lambda-\mu} \check{\phi}_\mu \cdot d\Sigma_\mu \cdot \hat{\phi}_\mu \cdot \check{\phi}_\lambda .$$

Proof. The proof is actually extremely simple: one just differentiates with respect to x and uses (10.8).

This way one gets:

$$\frac{d}{dx} K[\lambda] = \hat{\phi}_\lambda \int (\lambda+\mu) \check{\phi}_\mu d\Sigma_\mu \hat{\phi}_\mu \check{\phi}_\lambda + \hat{\phi}_\lambda \int (-\lambda-\mu) \check{\phi}_\mu d\Sigma_\mu \hat{\phi}_\mu \check{\phi}_\lambda = 0.$$

In order to prove the same assertion for (10.10) we use the following auxiliary statement:

$$\frac{d}{dx} \{\hat{\phi}_\lambda \check{\phi}_\eta d\Sigma_\eta \hat{\phi}_\eta \check{\phi}_\lambda\} = \hat{\phi}_\lambda \int_\mu (\eta-\lambda) \check{\phi}_\mu d\Sigma_\mu \hat{\phi}_\mu \cdot \check{\phi}_\eta d\Sigma_\eta \hat{\phi}_\eta \check{\phi}_\lambda + \hat{\phi}_\lambda \check{\phi}_\eta d\Sigma_\eta \hat{\phi}_\eta \int_\mu (\lambda-\eta) \check{\phi}_\mu d\Sigma_\mu \hat{\phi}_\mu \cdot \check{\phi}_\lambda .$$

In other words we have

$$\frac{d}{dx} C[\lambda] = 0 \text{ for any } \lambda \in \Omega.$$

We denote

$$\hat{\phi}_\lambda = (\hat{\phi}_{\lambda,ij}), \quad \check{\phi}_\lambda = (\check{\phi}_{\lambda,ij})$$

where one can consider $\hat{\phi}_{\lambda,ij}$ and $\check{\phi}_{\lambda,ij}$ as canonical variables and one assumes that $\hat{\phi}_{\lambda,ji}$ is the canonical coordinate-impulse conjugate to canonical coordinate variable $\check{\phi}_{\lambda,ij}$.

In conjugate variables $\check{\phi}$ and $(\hat{\phi})^t$ the general Hamiltonian equations have the form

$$\check{\phi}_x = \frac{\delta H}{\delta(\hat{\phi})^t}, \quad (\hat{\phi})^t_x = -\frac{\delta H}{\delta \check{\phi}} .$$

Here for $F = (F_{ij})_{i,j}$ one denotes for simplicity:

(10.11) $$\frac{\delta H}{\delta F} = \left(\frac{\delta H}{\delta F_{ij}}\right)_{i,j} .$$

Then we consider the following Hamiltonian:

(10.12) $$H_{II} = \operatorname{tr}\{(\int_\mu \check{\phi}_\mu d\Sigma_\mu \hat{\phi}_\mu) \cdot (\int_\mu \mu \check{\phi}_\mu d\Sigma_\mu \hat{\phi}_\mu)\}$$

This Hamiltonian H_{II} is exactly the Hamiltonian of the system (10.8). Indeed, let

$$d\Sigma_\mu = (d\sigma_{ij}), \text{ then}$$

$$H_{II} = \sum_{i_1 i_2 i_3 i_4 i_5, i_6} \int_{\mu_1} \check{\phi}_{\mu_1 i_1 i_2}\, d\sigma_{\mu_1 i_2 i_3} \hat{\phi}_{\mu_1 i_3 i_4} \times$$

$$\times \int_{\mu_2} \mu_2 \check{\phi}_{\mu_2 i_4 i_4}\, d\sigma_{\mu_2 i_5 i_6} \phi_{\mu_2 i_6 i_1} .$$

Then, e.g., in the case of $d\Sigma_\mu = Id\mu$ we have

$$\frac{\delta H_{II}}{\delta \check{\phi}_{\lambda,ij}} = \Sigma_k \check{\phi}_{\lambda jk} (\textstyle\int_\mu \mu \check{\phi}_\mu d\Sigma_\mu \hat{\phi}_\mu) + \lambda \Sigma_\kappa \hat{\phi}_{\lambda jk} (\textstyle\int_\mu \check{\phi}_\mu d\Sigma_\mu \hat{\phi}_\mu)_{ki} ;$$

$$\frac{\delta H_{II}}{\delta \hat{\phi}_{\lambda,ij}} = \Sigma_k (\textstyle\int_\mu \mu \check{\phi}_\mu d\Sigma_\mu \hat{\phi}_\mu)_{jk} \check{\phi}_{\lambda ki} + \lambda \Sigma_k (\textstyle\int_\mu \check{\phi}_\mu d\Sigma_\mu \hat{\phi}_\mu)_{jk} \check{\phi}_{ki} .$$

Now by the definition of symplectic structure we have

$$(\hat{\phi}_{\lambda,ij})_x = - \frac{\delta H_{II}}{\delta \check{\phi}_{\lambda,ji}} \quad ; \quad (\check{\phi}_{\lambda,ij})_x = \frac{\delta H_{II}}{\delta \hat{\phi}_{\lambda,ij}} .$$

These are exactly the equations (10.8).

Let us consider the following auxiliary Hamiltonian

$$H_\lambda = \mathrm{tr}\, C[\lambda] = \mathrm{tr}\{\check{\phi}_\lambda \hat{\phi}_\lambda \cdot \textstyle\int_\mu (\lambda-\mu)^{-1} \check{\phi}_\mu d\Sigma_\mu \hat{\phi}_\mu\}.$$

Let t_λ be the time corresponding to the Hamiltonian H_λ. Then we have:

$$\frac{\partial}{\partial t_\lambda} \cdot \check{\phi}_\eta = (\lambda-\eta)^{-1} \check{\phi}_\lambda \hat{\phi}_\lambda \check{\phi}_\eta \quad : \eta \neq \lambda, \text{ and}$$

$$\frac{\partial}{\partial t_\lambda} \cdot \check{\phi}_\lambda = \textstyle\int_\mu (\lambda-\mu)^{-1} \check{\phi}_\mu d\Sigma_\mu \hat{\phi}_\mu \cdot \check{\phi}_\lambda ;$$

$$\frac{\partial}{\partial t_\lambda} \cdot \hat{\phi}_\eta = -(\lambda-\eta)^{-1} \hat{\phi}_\eta \check{\phi}_\lambda \hat{\phi}_\lambda \quad : \eta \neq \lambda,$$

$$\frac{\partial}{\partial t_\lambda} \cdot \hat{\phi}_\lambda = -\hat{\phi}_\lambda \cdot \textstyle\int_\mu (\lambda-\mu)^{-1} \check{\phi}_\mu d\Sigma_\mu \hat{\phi}_\mu .$$

Now for

$$I_1 = \textstyle\int_\mu \mu \check{\phi}_\mu d\Sigma_\mu \hat{\phi}_\mu$$

we have

$$\textstyle\int \lambda^n \frac{\partial I_1}{\partial t_\lambda} = [\textstyle\int_\mu \check{\phi}_\mu d\Sigma_\mu \hat{\phi}_\mu, \textstyle\int_\lambda \lambda^n \check{\phi}_\lambda d\Sigma_\lambda \hat{\phi}_\lambda] .$$

The most important objects are the following commuting Hamiltonians arising from $C[\lambda]$:

(10.13) $$H_m = \mathrm{tr}\,\{\int_\lambda \lambda^m C[\lambda]d\Sigma_\lambda\} : m = 0, 1, 2, 3, \ldots .$$

These Hamiltonians can be rewritten in the explicit form:

(10.14) $$H_m = \mathrm{tr}\,\{\textstyle\int_\lambda \cdot \int_\mu \lambda^m(\lambda-\mu)^{-1}\check{\phi}_\lambda d\Sigma_\lambda\hat{\phi}_\lambda \times \check{\phi}_\mu d\Sigma_\mu\hat{\phi}_\mu\}: \quad m \geq 0, \quad \text{e.g.,}$$

$$H_o = 0,$$

$$H_1 = \mathrm{tr}\,\{(\textstyle\int_\lambda \check{\phi}_\lambda d\Sigma_\lambda\hat{\phi}_\lambda)^2\},$$

$$H_2 = \mathrm{tr}\,\{(\textstyle\int_\lambda \check{\phi}_\lambda d\Sigma_\lambda\hat{\phi}_\lambda)\cdot(\int_\lambda \lambda\check{\phi}_\lambda d\Sigma_\lambda\hat{\phi}_\lambda)\}$$

$$\ldots$$

Now for

$$I_1 = \textstyle\int_\mu \mu\check{\phi}_\mu d\Sigma_\mu\hat{\phi}_\mu$$

and evolution according to Hamiltonian H_m we have

$$\frac{\partial}{\partial t_{H_m}} I_1 = [\textstyle\int_\mu \check{\phi}_\mu d\Sigma_\mu\hat{\phi}_\mu, \int_\mu \mu^m\check{\phi}_\mu d\Sigma_\mu\hat{\phi}_\mu]: m \geq 0.$$

If as before the spectral (matrix) measure $d\Sigma_\lambda$ is chosen according to 10.1 in such a way that

$$\textstyle\int_\mu \mu^m\check{\phi}_\mu d\Sigma_\mu\hat{\phi}_\mu = \lambda_{m,A}: m = 0, 1, 2, \ldots , \text{ then}$$

$$A = \textstyle\int_\mu \check{\phi}_\mu d\Sigma_\mu\hat{\phi}_\mu$$

$$V_A = -\textstyle\int_\mu \mu\check{\phi}_\mu d\Sigma_\mu\hat{\phi}_\mu$$

Now

$$-\frac{\partial}{\partial t_{H_m}} V_A = [A, \lambda_{m,A}]$$

which is the nonlinear (x, t_{H_m}) evolution equation (9.11) associated with matrix linear differential equation:

$$\frac{d}{dx} F = -V_A F + \lambda A F$$

for a diagonal A.

§11. The operator Russian chain (stationary nonlinear Schrödinger equation).

11.0. In this chapter we consider an operator generalization of the Russian chain of §6. This system is dealt with in 11.1 and 11.2 where the relationship between the operator Russian chain and operator equations of order two introduced in §5 are considered. This relationship is similar to the results of §10, where the systems of order two associated with linear differential equations of order one were decomposed into common action of two commuting one-dimensional Hamiltonian flows.

In [14], [15] we proved that (x,t)-dimensional equations of order two connected with the Schrödinger operator can be represented as the common action of the Russian chain in x-direction and of one of the Hamiltonian flows commuting with it in the t-direction.

11.1. We derive the operator Russian chain generalizing (R) or (7.1) as a stationary case of the operator nonlinear Schrödinger equation (cf. §5).

Let us consider the coupled Operator Nonlinear Schrödinger of §5, (5.6):

$$\begin{cases} i\phi_t = \phi\psi\phi - \phi_{xx} \\ -i\psi_t = \psi\phi\psi - \psi_{xx} \end{cases}$$

and look at its stationary solutions:

$$\phi(x,t) = \phi(x)e^{i\Lambda t};$$

$$\psi(x,t) = e^{-i\Lambda t}\psi(x)$$

for a constant Λ. Then we have the coupled stationary nonlinear Schrödinger:

$$\begin{cases} -\phi\cdot\Lambda = \phi\psi\phi - \phi_{xx}; \\ -\Lambda\cdot\psi = \psi\phi\psi - \psi_{xx} \ . \end{cases}$$

Now we consider H and $L^2(\Omega,d\Sigma_\omega)$-extension of H_o: $H = \oplus_{L^2(\Omega,d\Sigma_\omega)} H_o$ and $\alpha(\omega)$, $\beta(\omega)$ on $L^2(\Omega,d\Sigma_\omega)$ such that $\int_\Omega \alpha(\omega)d\Sigma_\omega\beta(\omega) = 1$. We take $\lambda(\omega)$ as a measurable function $\Omega \to \mathbb{C}$ and define operators $\phi(\omega)$, $\psi(\omega)$ over H_o. We choose operators ϕ and ψ over H, in the following form: for $f \in H$,

$$(\phi\cdot f)(\omega) = \alpha(\omega)\cdot \int_\Omega \phi(\omega_1)d\Sigma_{\omega_1}\cdot f(\omega_1)$$

$$(\psi\cdot f)(\omega) = \psi(\omega)\cdot \int_\Omega \beta(\omega_1)d\Sigma_{\omega_1}\cdot f(\omega_1)$$

$$(\Lambda f)(\omega) = \lambda(\omega)f(\omega).$$

I.e., ϕ and ψ on H are of rank 1 over H_o. Then the stationary nonlinear Schrödinger equation can be reduced from H to H_o in the following multicomponent form:

$$\phi(\omega)_{xx} = \int_\Omega \phi(\omega_1)d\Sigma_{\omega_1}\cdot\psi(\omega_1)\phi(\omega) + \lambda(\omega)\phi(\omega);$$

$$\psi(\omega)_{xx} = \psi(\omega)\cdot\int_\Omega \phi(\omega_1)d\Sigma_{\omega_1}\cdot\psi(\omega_1) + \lambda(\omega)\psi(\omega).$$

In the particular case $\Omega \subseteq \mathbb{C}$, $\lambda(\omega) = -\omega^2$ we get just an Operator Russian Chain.

11.2. We now consider such an aggregate: infinite component operator Stationary Nonlinear Schrödinger equation

(OR) or (11.1)

$$\begin{cases} \phi_{\lambda,xx} = \int\phi_\mu d\Sigma_\mu\tilde{\phi}_\mu\cdot\phi_\lambda - \lambda^2\phi_\lambda; \\ \tilde{\phi}_{\lambda,xx} = \tilde{\phi}_\lambda\cdot\int\phi_\mu d\Sigma_\mu\tilde{\phi}_\mu - \lambda^2\tilde{\phi}_\lambda \end{cases}$$

In other words, ϕ_λ and $\tilde{\phi}_\lambda$ are (left and right) eigenfunctions corresponding to the eigenvalue λ^2 of the operator Schrödinger equation with the potential:

$$U(x) = \int\phi_\mu(x)d\Sigma_\mu\tilde{\phi}_\mu(x).$$

However, from now on we consider (OR) as (an infinite-dimensional) Hamiltonian system. Here we assume of course that ϕ_λ, $\tilde{\phi}_\lambda$ belong to (a Banach) algebra of operators, $d\Sigma_\mu$ is an operator (e.g., matrix) measure on $\mathbb{R}$ or $\mathbb{C}$ and we may impose rather restrictive conditions on U and $\phi,\tilde{\phi}_\lambda$. E.g., we demand

$$U = \int\phi_\mu d\Sigma_\mu\tilde{\phi}_\mu$$

to be a bounded operator (for example, in analogy with the classical inverse scattering $U(x)$ can vanish exponentially on infinity). Also we demand all ϕ_μ, $\tilde{\phi}_\mu$ and $\int\phi_\mu d\Sigma_\mu\phi_\mu$, $\int\phi_\mu d\Sigma_\mu\tilde{\phi}_\mu$, $\int\tilde{\phi}_\mu d\Sigma_\mu\tilde{\phi}_\mu$ to be bounded operators.

All such restrictions will be important if we want to write down all the solutions of (OR). But from the point of view of the proof of the complete integrability and the correct choice of action-angle variables it is necessary to have only a formal scheme. We have conservation laws for (OR) of the type of §6:

<u>Theorem 11.1</u> [14]: For the system (OR) we have the following first integrals

$$K[\lambda] = \tilde{\phi}_{\lambda x}\phi_\lambda - \tilde{\phi}_\lambda\phi_{\lambda x}; \tag{11.2}$$

$$C[\lambda] = \tilde{\phi}_{\lambda x}\phi_{\lambda x} + \lambda^2\tilde{\phi}_\lambda\phi_\lambda - \frac{1}{2}\tilde{\phi}_\lambda\cdot\int\phi_\mu d\Sigma_\mu\tilde{\phi}_\mu\cdot\phi_\lambda + \frac{1}{2}\int\frac{1}{\lambda^2-\eta^2}[\tilde{\phi}_{\lambda x}\phi_\eta - \tilde{\phi}_\lambda\phi_{\eta x}]d\Sigma_\eta[\tilde{\phi}_{\eta x}\phi_\lambda - \tilde{\phi}_\eta\phi_{\lambda x}]. \tag{11.3}$$

Moreover, all the first integrals $C[\lambda]$ are in involution. Here we consider elements of $\phi_{\lambda x}$ $(\tilde{\phi}_{\lambda x})$ as the conjugate variables to the corresponding elements of $\tilde{\phi}_\lambda$ (ϕ_λ) (if one views ϕ, $\tilde{\phi}$ as infinite matrices). In particular we have first integrals of the system (OR) in a traditional form: for an arbitrary constant operator $S(\lambda)$ the following Hamiltonian is in involution with (OR):

$$H_S = \int \mathrm{tr}_H\{d\Sigma_\lambda C[\lambda]S[\lambda]\}$$

In particular, on the solutions of (OR),

$$H_S = \text{const.}$$

Also the Hamiltonian of (OR) has the form

$$H = \frac{1}{2}\int \mathrm{tr}_H\{d\Sigma_\lambda C[\lambda]\}.$$

The most important class of Hamiltonians H_S is the class of "momentae" Hamiltonians. We define,

$$H_n = \int \mathrm{tr}_H\{\lambda^n d\Sigma_\lambda C[\lambda]\}$$

for an integer n.

In general Hamiltonians H_S define a rather complicated evolution for

$$(\phi_\lambda, \phi_{\lambda x}; \tilde{\phi}_\lambda, \tilde{\phi}_{\lambda x}),$$

and of course the evolution of

$$U = \int \phi_\mu d\Sigma_\mu \tilde{\phi}_\mu$$

under the Hamiltonian flow H_S may be nonlocal in U. However, the evolution of U under the Hamiltonian H_n for even $n = 2m \geq 0$ is always <u>local</u> in the sense that the equation of the evolution of U under H_n takes the form of the m-th operator KdV equation (§5): $U_{t_{2m}} = P_{2m}[U, U_x, \ldots, \underbrace{U_{x\ldots x}}_{2m+1}]$. E.g., $U_{t_2} = 6(UU_x + U_xU) - U_{xxx}$ etc.

Now we come to the main result concerning the (OR) [14]. Any equation of order two presented in §5, theorem 5.1, and corresponding to $\alpha(L)$ and $\beta(L)$ for the potential

$$U(x,t) = \int \phi_\mu(x,t) d\Sigma_\mu \tilde{\phi}_\mu(x,t)$$

is equivalent to the evolution of $(\phi_\mu(x,t), \tilde{\phi}_\mu(x,t))$ according to

$$H = \frac{1}{2}\int \mathrm{tr}_H\{d\Sigma_\lambda C[\lambda]\}$$

in x-direction and according to the Hamiltonian of the form H_S in t-direction. E.g., in the case $\alpha(L) = 0$, $\beta(L)$-arbitrary, $M = I$, we have

$$H_S = \int \mathrm{tr}\{\beta(-\lambda^2) d\Sigma_\lambda C[\lambda]\}$$

and

$$U_t = \frac{\partial}{\partial x}\{\int \beta(-\lambda^2)\phi_\lambda \tilde{\phi}_\lambda d\sigma_\lambda\}$$

together with $U_t = R_n[U, U_x, \ldots, \underbrace{U_{x\ldots x}}_{2n+1}]$ for $\beta(\lambda) = \lambda^n$.

11.3. One of the questions that arises in connection with the Russian chain (OR) is the possibility of the representation

$$(11.4) \qquad U(x) = \int \phi_\mu d\Sigma_\mu \tilde{\phi}_\mu$$

for the large class of potentials $U(x)$ in terms of (left and right) eigenfunctions ϕ_μ, $\tilde{\phi}_\mu$ corresponding to the eigenvalue $-\mu^2$ of the operator Schrodinger equation with the potential $U(x)$. We refer to §10 where such a problem is reduced to the matrix moment problem.

Another important problem: how is the measure $d\Sigma_\mu$ connected with the spectral measure naturally associated to $U(x)$? We know already answers to these questions in the scalar case, so we can conjecture the answer in general. In certain cases this conjecture can be proved (see [14], [15], [26], [30]).

In [15], [23]-[25] we considered the representation of the C^∞-periodic (say, finite band) potential $u(x)$ through the sum of squares of its eigenfunctions. For the C^∞-periodic potentials $u(x)$ there are representations (cf. [49], [23]):

$$u(x) = -\sum_{i=o}^{\infty} f^2_{2i+1}\cdot\varepsilon_{2i} + C_1 \quad \text{and}$$

$$u(x) = -\sum_{i=o}^{\infty} f^2_{2i+1}\cdot\varepsilon_{2i+1} + C_2$$

where λ_{2i} are periodic and λ_{2i+1} are antiperiodic eigenvalues - end of lacunae - of $-y'' + (u+\lambda)y = 0$ and for $f_i(x)$ being eigenfunctions corresponding to λ_i ($\varepsilon_j = 0$ if λ_j is a degenerate eigenvalue). As we know (see §10 and §6) there is, however, the one canonical way to present $u(x)$ through f_i^2 when <u>all</u> eigenfunctions f_i^2 are present. Here

$$\lambda_{2n+1} \sim \lambda_{2n} \sim n^2 \frac{\pi^2}{T^2} + O(1) + O(\frac{1}{n^2}) \dots .$$

Naturally singular solutions of the form

$$d\sigma_k = \sum_{i=1}^{\infty} \delta(k-\lambda_i)$$

with arbitrary λ_i may correspond to an arbitrary quasiperiodic potential $u(x)$. However, for the general quasiperiodic potential we have no idea what the spectrum is. Probably we need some diophantine conditions for quasi-periods of $u(x)$.

We have a complete reduction of the inverse scattering problem to the infinite-component Russian chain in the case of $u(x)$ rapidly decreasing on infinity (cf.§§6,7).

Let us define all spectral data in this case. Let $\int_{-\infty}^{\infty} |u(x)|(1+x^2)dx < \infty$ and $f(x,k)$ be the solution of $-f'' + uf = k^2 f$ with

$$f(x,k) \sim e^{ikx}; \ x \to +\infty$$

and

$$f(x,k) = a(k)e^{ikx} + b(k)e^{-ikx}: x \to -\infty.$$

The reflection coefficient r(k) can be defined as

$$r(k) = -\frac{b^*(k)}{a(k)} .$$

It is known that u(x) may have only a finite number of bound states $-\beta_n^2 <\cdots< -\beta_1^2 < 0$. Then the potential is uniquely determined by the reflection coefficient r(k), n bound states $\beta_1^2,\ldots,\beta_n^2$ and n norming constants

$$c_j = \left\{\int_{\infty}^{\infty} f^2(x;i\beta_j)dx\right\}^{-1}: j = 1,\ldots,n.$$

It is possible to express u(x) in terms of $f(x;k)^2$ in the form of (7.23). This was done for the first time by M. Kruscal, C. Gardner, J. Greene, and R. Miura for purely soliton solutions, for which $r(k) \equiv 0$:

$$u(x) = -4\sum_{j=1}^{n} c_j\beta_j f^2(x;i\beta_j).$$

However, for an arbitrary potential with $r(k) \not\equiv 0$ we have in the representation (11.4), both singular and smooth parts.

There is, however, one case when we have a smooth representation only. This is the case of u(x) without bound states treated by Deift & Trubowitz [30]. In this case, a very simple formula exists:

$$u(x) = \frac{2i}{\pi}\int_{-\infty}^{\infty} kr(k)f^2(x,k)dk$$

However, we found that a formula containing both reflectionless and bound stateless potentials exists. This corresponds to (11.4) with both singular and smooth parts:

$$u(x) = \frac{2i}{\pi}\int_{-\infty}^{\infty} kr(k)f^2(x,k)dk - \sum_{j=1}^{n} 4c_j\beta_j f^2(x;i\beta_j).$$

Moreover, according to our previous results we have a more general representation for $R_K[u]$.

<u>Theorem 11.2</u>. In the most general case we have

$$R_K[u] = \frac{i}{2\pi}\int_{-\infty}^{\infty} (-1)^K k^{2K-1} r(k)f^2(x,k)dk$$
$$+ \sum_{j=1}^{n} (-1)^{K+1}\beta_j^{2K-1} c_j f(x;i\beta_j)^2: K = 0,1,2,\ldots .$$

We would like to mention that such a representation of u(x) as (11.4) is not unique! E.g., a finite band potential u(x) can be represented as a linear combination of any n from the 2n+1 squares of eigenfunctions which correspond to the end of the bands. An analogical assertion is true for arbitrary C^{∞}-potential, which is periodic. The proper normalization of the type 11.2 is proposed in [14], [15], [26], and in §6.

Let us consider an operator case: U(x) is hermitian, $U = U^{+}$ and U(x) vanishes asymptotically exponentially or faster, i.e., for some $\varepsilon > 0$,

$$\lim_{|x|\to\infty} [\exp(\varepsilon|x|)U(x)] = 0.$$

We assume also that U(x) doesn't have negative eigenvalues (this restriction may be, however, removed).

Theorem 11.3. Under the assumptions above on U(x) we have the representation

$$U(x) = \frac{2i}{\pi}\int_{\infty}^{\infty} k\phi_1(x,k)\tilde{\phi}_1(x,k)\,dk,$$

where $\phi_1(x,k)$, $\tilde{\phi}_1(x,k)$ are eigenfunctions corresponding to U(x):

$$\phi_{1,xx}(x,k) = U\cdot\phi_1(x,k) - k^2\phi_1(x,k):$$

$$\tilde{\phi}_{1,xx}(x,k) = \tilde{\phi}_1(x,k)\cdot U - k^2\tilde{\phi}_1(x,k).$$

These functions $\phi_1(x,k)$, $\tilde{\phi}_1(x,k)$ are closely connected with Jost eigenfunctions:

$$\phi_1(x,k) \to A(k)e^{ikx}: \quad x \to +\infty$$

$$\tilde{\phi}_1(x,k) \to \tilde{A}(k)e^{ikx}; \quad x \to +\infty$$

and $A(k)\tilde{A}(k) = R(k)$, where R(k) is the reflection coefficient.

The same type of result is true for periodic potential U(x),

$$U(x+T) \equiv U(x): \quad T \neq 0.$$

However, in the last case $d\Sigma_{\mu}$ should be a singular measure with the support at the set of ends of the lacunae of the Bloch spectrum. E.g., let for

$$\left(-\frac{d^2}{dx^2} + U - k^2\right)\phi = 0$$

we have nontrivial forbidden zonae, and U is periodic. Then for eigenvalues $\lambda_1^2,\dots,\lambda_n^2,\dots$: $n < \infty$ being the ends of the forbidden zonae, we have the representation

$$U(x) = \sum_{i=1}^{\infty} \phi_i\tilde{\phi}_i + C$$

with a constant operator C and eigenfunctions ϕ_i, $\tilde{\phi}_i$ corresponding to eigenvalues λ_i^2:

$$\phi_{i,xx} = U\phi_i - \lambda_i^2\phi_i,$$

$$\tilde{\phi}_{i,xx} = \tilde{\phi}_i U - \lambda_i^2\tilde{\phi}_i: \quad i = 1,\dots .$$

Of course, for real $U = U^+$, λ_i are real and $\tilde{\phi}_k = \phi_k^+$, i.e., $U = \sum_{i=1} \phi_i\phi_i^+ + C$.

§12. Three-dimensional completely integrable systems of the Kadomtsev-Petviashvili type.

12.0. There have already been for ten years several examples [59] of (x,y,t) systems that arise from conditions of commutativity of two linear differential operators in $\partial/\partial x$, $\partial/\partial y$, $\partial/\partial t$ displayed in §1, 1.4, (1.15). These equations were called "completely integrable" because a large class of solutions was known. However, not too much was known about these systems in general.

One of the reasons why three and more dimensional systems hardly can be called completely integrable is the absence of the consistent Hamiltonian structure on them. On has here the following problem: we consider certain systems in variables $x_1,\dots,x_n$. Then one of these variables, say x_1, is singled out as time and we try to build a Hamiltonian structure using conservation laws. We find immediately that if $n > 2$ these conservation laws are nonlocal and it is even unclear what canonical shape they have. On the other hand, if we choose another x_i as time, do we have a guarnatee that conservation laws will change into conservation laws? We also want to project Hamiltonian structure to the stationary case, when a solution does not depend on $x_1, x_2,\dots$ etc. If conservation laws in the whole system were independent, involutive and complete, will it be preserved after projection? We find one way to answer these questions. This way is to separate variables $x_1,\dots,x_n$ in a given completely integrable system and present it as a joint action of n commuting Hamiltonian flows (in the usual sense) on infinite dimensional simplectic manifold.

Now, for the first time, we can call some three-dimensional system (1.15) completely integrable because these systems can be: a) represented as Hamiltonian systems; b) can be indeed reduced to action-angle variables.

These phenomena will be discussed below and we shall see how several three-dimensional completely integrable systems of the type (1.15) or [59] can be reduced to one-dimensional operator systems as a result of motion via three Hamiltonian flows that commute. Of course, as you may already guess, this one-dimensional operator system is nothing but the operator Russian chain:

$$\text{(OR)}\qquad \frac{d^2\psi(\zeta)}{dx^2} = -2\int \eta^2\psi(\eta)d\Sigma_\eta\phi(\eta)\cdot\psi(\zeta) + \zeta^2\psi(\zeta);$$

$$\frac{d^2\phi(\zeta)}{dx^2} = -2\phi(\zeta)\cdot\int \eta^2\psi(\eta)d\Sigma_\eta\phi(\eta) + \zeta^2\phi(\zeta).$$

Our way of thinking can be explained in a very simple way. We know (by §7 or §11) that all the higher KdV equations and, generally speaking, all two-dimensional completely integrable systems of order two, can be reduced to an infinite component Russian chain (an infinite component nonlinear stationary Schrödinger). In other words, the introduction of a new spectral variable k allows us to get rid of variable t. Now it is quite natural, that if we add in the (x,y,t) system one variable k we can get rid of variable t, and if we introduce variable λ we can get rid of variable y. This way we pass first from a two-dimensional KdV equation to a multicomponent nonlinear Schrödinger and then from a multicomponent nonlinear Schrödinger to an operator stationary nonlinear Schrödinger (an operator Russian chain). Basically the rule here is the following: If you want to construct the solution u(x,t) of a two-dimensional system of KdV type, then you write it in the form

$$u(x,t) = -2\int k^2 f(x,t,k)^2 d\mu_k$$

where f(k) are eigenfunctions of

$$f_{xx}(k) = u(x,t)f(k) + k^2 f(k),$$

with the evolution in x determined by this Russian chain and the evolution in t governed by any Hamiltonian flow commuting with the Russian chain (or the Hamiltonian H_I). Now if one wants to find the solution u(x,t,y) of an (x,y,t)-dimensional system of KdV type, we present this solution in the form

$$u(x,t,y) = -2\int tr[\zeta^2\psi_\zeta^+ d\Sigma_\zeta\psi_\zeta]$$

where ψ_ζ^+, ψ_ζ arises from a stationary Russian chain

$$\frac{d^2\psi(\zeta)}{dx^2} = -2\psi(\zeta)\int\eta^2\psi^+(\eta)d\Sigma_\eta\psi(\eta) + \zeta^2\psi(\zeta),$$

and two Hamiltonian flows in t and y commuting with the given one in x.

12.1. In order to consider three-dimensional systems we take an example of the Kadomtsev-Petviashvili equation (so-called two-dimensional KdV) in (x,y,t)-dimensions on a Hilbert space H_o.

For an operator $\mathcal{A}$ on H_o, $\mathcal{A} = \mathcal{A}(x,y,t)$ this equation has the following form:

$$(12.1)\qquad 3\mathcal{A}_{yy} = \frac{\partial}{\partial x}(4\mathcal{A}_t + \mathcal{A}_{xxx} - 6(\mathcal{A}_x\mathcal{A} + \mathcal{A}\mathcal{A}_x) + 6i[\int\mathcal{A}_y dx', \mathcal{A}]).$$

In the scalar case (H_o - one-dimensional) one gets the usual Kadomtsev-Petviashvili (see §1):

$$3u_{yy} = \frac{\partial}{\partial x}(4u_t + u_{xxx} - 12uu_x).$$

Now this equation can be represented as a result of common action of three commuting Hamiltonian flows arising from the operator Russian chain.

12.2. In order to reduce (12.1) to three commuting operator Hamiltonians we first reduce (12.1) to two two-dimensional operator systems. They are: the operator non-linear Schrödinger (5.7) and the operator modified KdV (5.5) equations. Both of these equations are particular cases of equations of order two (§5) or the equations discussed in §9. We decided to deduce these equations from the Dirac equation of the first order (see [14], [17], or [52]). If we consider the Kadomtsev-Petviashvili equation over H, then the Dirac equation is considered over $H_o \times H_o$, where H_o is an infinite-dimensional $L^2(\Omega, dM_\mu)$ extension of H.

We proceed from the Dirac equation on $H_o \times H_o$. For the better representation of the Dirac equation we adopt notations of §9 for $A \overset{def}{=} \sigma_3 = \begin{pmatrix} I & 0 \\ 0 & -I \end{pmatrix}$. Then we consider the fundamental (operator) left and right eigenfunctions $\check{\phi}(\zeta)$, $\hat{\phi}(\zeta)$ of the Dirac equation on $H_o \times H_o$:

$$\text{(12.2)} \qquad \begin{aligned} \frac{d}{dx}\check{\phi}(\zeta) &= -V\cdot\check{\phi}(\zeta) + \zeta\sigma_3\cdot\check{\phi}(\zeta) , \\ \frac{d}{dx}\hat{\phi}(\zeta) &= \hat{\phi}(\zeta)\cdot V - \zeta\hat{\phi}(\zeta)\cdot\sigma_3, \end{aligned}$$

for

$$\text{(12.3)} \qquad V = \begin{pmatrix} 0 & Q \\ R & 0 \end{pmatrix}, \quad \sigma_3 = \begin{pmatrix} I & 0 \\ 0 & -I \end{pmatrix},$$

and the operators Q, R on H_o. We can pass from the Dirac equation (12.2) to the operator Schrödinger equation on $H_o \times H_o$. If we put

$$\text{(12.4)} \qquad \check{\Phi}(\zeta) = \check{\phi}(\zeta), \quad \hat{\Phi}(\zeta) = \hat{\phi}(\zeta)\cdot\sigma_3,$$

then iterating equation (12.2) we obtain the Schrodinger equations

$$\text{(12.5)} \qquad \begin{aligned} \frac{d^2}{dx^2}\check{\Phi}(\zeta) &= U\check{\Phi}(\zeta) + \zeta^2\check{\Phi}(\zeta) , \\ \frac{d^2}{dx^2}\hat{\Phi}(\zeta) &= \hat{\Phi}(\zeta)U + \zeta^2\hat{\Phi}(\zeta) , \end{aligned}$$

for the potential

$$\text{(12.6)} \qquad U = V^2 - V_x.$$

According to §§9-10, we can find a certain spectral measure dM_η such that V and U and some canonical system of polynomials in V, V_x, V_{xx}, ... is represented as a system of moments of $\phi(\zeta)dM_\zeta\hat{\phi}(\zeta)$. We can use for this 10.1, the formula (10.1) for the Dirac equation and the eigenfunctions $\check{\phi}$, $\hat{\phi}$ of (12.2) or §11 for the operator Schrödinger (12.5) and eigenfunctions $\check{\Phi}$, $\hat{\Phi}$ in (12.4). We present in Theorem 12.1 the formula for the lowest moments that are obtained from the expressions (10.1) for Hamiltonian densities $\lambda_{n,A}$ of Lemma 9.2 for $A = \sigma_3$. The corresponding explicit formulae were presented in [42], [52] for a general class of potentials V.

Theorem 12.1. For the operator solutions $\check{\phi}(\zeta)$, $\hat{\phi}(\zeta)$ of (12.2) and the operator solutions $\check{\Phi}(\zeta)$, $\hat{\Phi}(\zeta)$ (12.4) of (12.5) there exists a spectral measure $d\Sigma_\zeta$ such that

$$\sigma_3 = \int\check{\phi}(\zeta)d\Sigma_\zeta\hat{\phi}(\zeta); \quad \begin{pmatrix} I & 0 \\ 0 & I\end{pmatrix} = \int\check{\Phi}(\zeta)d\Sigma_\zeta\hat{\Phi}(\zeta);$$

$$-V = \int\zeta\check{\phi}(\zeta)d\Sigma_\zeta\hat{\phi}(\zeta); \quad -V\sigma_3 = \int\zeta\check{\Phi}(\zeta)d\Sigma_\zeta\hat{\Phi}(\zeta); \tag{12.7}$$

$$-(V^2-V_x)\sigma_3 = \int 2\zeta\check{\phi}(\zeta)d\Sigma_\zeta\hat{\phi}(\zeta); \quad -U = \int 2\zeta\check{\Phi}(\zeta)d\Sigma_\zeta\hat{\Phi}(\zeta);$$

and

$$2V^3 - V_{xx} + VV_x - V_xV = \int\check{\phi}(\zeta)(2\zeta)^3dM_\zeta\hat{\phi}(\zeta). \tag{12.8}$$

Here

$$V\sigma_3 = \begin{pmatrix} 0 & -Q \\ R & 0\end{pmatrix}, \quad U = V^2 - V_x = \begin{pmatrix} QR & -Q_x \\ -R_x & RQ\end{pmatrix},$$

$$2V^3 - V_{xx} + VV_x - V_xV = \begin{pmatrix} QR_x-Q_xR & ; \ 2QRQ-Q_{xx} \\ -R_{xx}+2RQR & ; \ RQ_x-R_xQ\end{pmatrix}.$$

12.3. Now we can consider nonlinear two-dimensional evolutionary equations corresponding to the Dirac equation (12.2) described in (9.11)-(9.12) for $A = \sigma_3$. By §10 these equations are equivalent to the common action of two one-dimensional Hamiltonians (10.13)-(10.14) commuting with H_{II} in (10.12). The simplest Hamiltonians are H_m: m = 2,3,4 corresponding to the equations (9.11) for n = 2,3,4. Let us present these three equations as evolutionary Hamiltonian equations on Q and R with three time variables $t_0 = x$, $t_1 = y$, $t_2 = t$:

$$\begin{gathered} R_{t_0} = R_x, \quad Q_{t_0} = Q_x \quad (\text{i.e., } t_0 = x), \\ iR_y = R_{xx} - 2RQR; \quad -iQ_y = Q_{xx} - 2QRQ \quad (\text{i.e., } t_1 = y) \\ R_t = 3R_xQR + 3RQR_x - R_{xxx}, \\ Q_t = 3Q_xRQ + 3QRQ_x - Q_{xxx} \quad (\text{i.e., } t_2 = t). \end{gathered} \tag{12.9}$$

According to §10, these three flows arise naturally from commuting one-dimensional Hamiltonian flows on the infinite-dimensional symplectic manifold corresponding to the Hamiltonian H_{II}:

$$(12.10)\qquad \begin{aligned} \frac{d}{dx}\check{\phi}(\zeta) &= \int_{\mu}(\zeta+\mu)\check{\phi}(\mu)\,d\Sigma_{\mu}\hat{\phi}(\mu)\cdot\check{\phi}(\zeta);\\ \frac{d}{dx}\hat{\phi}(\zeta) &= -\hat{\phi}(\zeta)\cdot\int_{\mu}(\zeta+\mu)\check{\phi}(\mu)\,d\Sigma_{\mu}\hat{\phi}(\zeta). \end{aligned}$$

Of course, Theorem 12.1 and the results of §11 allow us to represent all equations (12.9) in terms of Hamiltonian flows commuting with the Operator Russian Chain system (11.1) or (OR). Let us deduce the Kadomtsev-Petviashvili equation from the equations (12.9). We do know that the equations (12.9) are consistent because they correspond to commuting Hamiltonian flows.

Thus we can write the conditions of consistency

$$(12.11)\qquad R_{yt} = R_{ty},\quad Q_{yt} = Q_{ty}.$$

It is impossible to verify immediately that the conditions (12.11) can be reduced to the following ones. Let

$$Y = QR,\ W = QR_x,\ Z = Q_xR,$$

so $Y_x = W + Z$. Then (12.9) takes the form:

$$\begin{aligned} iR_y &= R_{xx} - 2RY; \quad -iQ_y = Q_{xx} - 2YQ;\\ R_t &= 3R_xY + 3RW - R_{xxx};\\ Q_t &= 3YQ_x + 3ZQ - Q_{xxx}. \end{aligned}$$

Then the conditions (12.11) take the form:

$$(12.12)\qquad R_x(3Y_y+3iY_{xx}+6iW_x) + R(6iY_xY+6i[Y,W]+3W_y-2iY_{xxx}+3iW_{xx}-2iY_t) = 0$$

and

$$(12.12)'\qquad (6iZ_x-3Y_y-3iY_{xx})Q_x + (3iZ_{xx}-2iY_t-6i[Y,Z]-3Z_y+6iYY_x-2iY_{xxx})Q = 0.$$

One makes the following important

Corollary 12.2: If rank (R) and rank (Q) are larger than rank (Y) + rank (W), then system (12.12) or (12.12)' is equivalent to the following:

$$3Y_y - 3iY_y + 6iW_x = 0;$$

$$6iY_xY + 6i[Y,W] + 3W_y - 2iY_{xxx} + 3iW_{xx} - 2iY_t = 0.$$

These last equations can be written in a more convenient form. We have

$$Y_x = W + Z,\ Y_y = (W-Z)_x$$

or

$$(12.13)\qquad W = \frac{1}{2}Y_x + \frac{i}{2}\int Y_y dx;\quad Z = \frac{1}{2}Y_x - \frac{i}{2}\int Y_y dx.$$

The second equation in the Corollary 12.2 can be rewritten taking into account (12.13) as a single (x,y,t)-operator Kadomtsev-Petviashvili equation (12.1):

$$3Y_{yy} = \frac{\partial}{\partial x}\left(4Y_t+Y_{xxx}-6(Y_xY+YY_x)+6i\left[\int Y_y dx, Y\right]\right) \tag{12.14}$$

We must remind the reader that in order to get the equation (12.14) we demanded rank (R) > rank (Y). For this it is natural to take R and Q as operators with Range (R) = Range (Q) = H with codim H in H_o to be infinite. As a model we can look at H_o as a L_2-type space over H: $H_o = \bigoplus_{L^2(\Omega,dM_\mu)} H$, for a certain space $L^2(\Omega,dM_\mu)$

Now we choose an operator R,Q in the following matrix form over H:

$$\begin{aligned}(R)_{ij} &= R_i\alpha_j;\\ (Q)_{ij} &= \beta_iQ_j; \quad i,j\in\Omega\end{aligned} \tag{12.15}$$

for operators R_i, Q_j and $\alpha_j = \alpha(j)$, $\beta_i = \beta(j)$ on H_o over $L^2(\Omega,dM_\mu)$ such that

$$\int_\Omega \alpha(j)dM_j\,\beta(j) = 1. \tag{12.16}$$

Here (12.15) means that R acts on element $\bar\psi = (\psi(j): j\in\Omega)$ of H_o as

$$R\bar\psi = \bar\phi \in H_o,\quad \bar\phi_j = R_j\int_\Omega \alpha(i)dM_j\psi(j): j\in\Omega.$$

Analogically $Q\bar\psi_1 = \bar\phi_1$ if

$$(\bar\phi_1)_j = \beta_j\int_\Omega Q_i dM_i\psi(i): j\in\Omega.$$

Now

$$(Q\cdot R)_{ij} = \beta_i\alpha_j\int_\Omega Q_\lambda dM_\lambda R_\lambda. \tag{12.17}$$

We define c-matrix (an operator on $L^2(\Omega,dM_\lambda)$) as

$$C = (\beta_i\alpha_j),$$

and potential (an operator on H)

$$\mathcal{U} = \int_\Omega Q_\lambda DM_\lambda R_\lambda. \tag{12.18}$$

Thus $QR = C\cdot\mathcal{U} = \mathcal{U}\cdot C$. Taking into account (12.16) we get $C^2 = C$ and C is a constant operator. Substituting

$$Y = QR = C\cdot\mathcal{U}$$

into the equation (12.13) we obtain a single equation for $\mathcal{U}$ only:

$$3\mathcal{U}_{yy} = \frac{d}{dx}\left(4\mathcal{U}_t+\mathcal{U}_{xxx}-6(\mathcal{U}_x\mathcal{U}+\mathcal{U}\mathcal{U}_x) + 6i\left[\int\mathcal{U}_y dx,\mathcal{U}\right]\right). \tag{12.19}$$

12.4. We now consider operator solutions $\check{\phi}_o(\zeta)$, $\hat{\phi}_o(\zeta)$ of the operator Dirac equation (12.2) or solutions $\check{\Phi}_o(\zeta) = \check{\phi}_o(\zeta)$, $\hat{\Phi}_o(\zeta) = \hat{\phi}_o(\zeta)\sigma_3$ of the Schrödinger equations (12.5) for the choice of operators R, Q on H_o as in (12.15)-(12.16). In this case the spectral measure $d\Sigma_\zeta$ depends on $\mathcal{O}\!\mathcal{L}$ and on (Ω, dM_μ) as well. According to Theorem 12.1 or §11 we have (12.5):

$$\frac{d^2}{dx^2}\check{\Phi}_o(\zeta) = U\check{\Phi}_o(\zeta) + \zeta^2\check{\Phi}_o(\zeta);$$

$$\frac{d^2}{dx^2}\hat{\Phi}_o(\zeta) = \hat{\Phi}_o(\zeta)\cdot U + \zeta^2\cdot\hat{\Phi}_o(\zeta)$$

for

$$U = V^2 - V_x = \begin{pmatrix} QR & -Q_x \\ -R_x & RQ \end{pmatrix}. \tag{12.20}$$

Then by Theorem 12.1 (cf. §11) we have for a spectral measure $d\Sigma_\zeta$:

$$U = -2\int \zeta^2\check{\Phi}_o(\zeta)\,d\Sigma_\zeta\,\hat{\Phi}_o(\zeta). \tag{12.21}$$

This is an operator identity on $H_o \times H_o$. We take trace with respect to $H_o^2 = H_o \times H_o/H$ and obtain

$$\mathrm{tr}_{H_o^2/H}(U) = -2\int \mathrm{tr}_{H_o^2/H}(\zeta^2\check{\Phi}_o(\zeta)\,d\Sigma_\zeta\,\hat{\Phi}_o(\zeta)). \tag{12.22}$$

Now

$$\mathrm{tr}_{H_o^2/H}(U) = \mathrm{tr}_{H_o/H}(QR) + \mathrm{tr}_{H_o/H}(RQ) = 2\,\mathrm{tr}_{H_o/H}(QR).$$

But by (12.16)

$$\mathrm{tr}_{H_o/H}(QR) = \mathcal{O}\!\mathcal{L}\ .$$

Thus from (12.22) it follows

$$\mathcal{O}\!\mathcal{L} = -\int \mathrm{tr}_{H_o^2/H}(\zeta^2\check{\Phi}_o(\zeta)\,d\Sigma_\zeta\,\hat{\Phi}_o(\zeta))\ . \tag{12.23}$$

This formula gives the most general expression for the solution of the Operator (x,y,t)-Dimensional Completely Integrable System of the type (1.15) or (12.1):

Corollary 12.3. In order to get the solution $\mathcal{O}\!\mathcal{L}(x,t,y)$ of a matrix (operator) three-dimensional systems (12.1) (or commuting with it) over the Hilbert space H, it is necessary to consider the Hilbert space $H_2 = H_o \times H_o$ for H_o infinite-dimensional over H (being L^2-extension of H) and the corresponding operator Schrodinger equations over H_2:

$$\frac{d^2\check{\Phi}(\eta)}{dx^2} = U\check{\Phi}(\eta) + \eta^2\check{\Phi}(\eta);$$

$$\frac{d^2\hat{\Phi}(\eta)}{dx^2} = \hat{\Phi}(\eta)U + \eta^2\hat{\Phi}(\eta) \tag{OR}$$

being also the operator Russian chain (11.1) for

(12.24) $$U = -2\int \zeta^2 \check{\Phi}(\zeta)\, d\Sigma_\sigma \hat{\Phi}(\zeta).$$

Now we take the solutions $\check{\Phi}(\zeta;x,t,y)$, $\hat{\Phi}(\zeta;x,t,y)$ of the system (OR), (12.24) satisfying simultaneously two Hamiltonian systems in t and y commuting with (OR). These two Hamiltonians can be chosen in a standard form such that, e.g.,

$$U_t = -2\,\frac{\partial}{\partial x}\,\{\int \zeta^{2n}\check{\Phi}(\zeta)\,d\Sigma_\zeta \hat{\Phi}(\zeta)\} \quad \text{for} \quad t = t_{2n}.$$

Then one can represent the solution $\mathcal{Q}$ as:

(12.25) $$\mathcal{Q}(x,t,y) = -2\int \mathrm{tr}_{H_2/H_o}\{\check{\Phi}(\zeta)\zeta^2 d\Sigma_\zeta \hat{\Phi}(\zeta)\}$$

for $\check{\Phi}(\zeta) = \check{\Phi}(\zeta;x,t,y)$; $\hat{\Phi}(\zeta) = \hat{\Phi}(\zeta;x,t,y)$. Moreover, in the case of real and self-adjoint $\mathcal{Q}$, $\mathcal{Q}^+ = \mathcal{Q}$ and self-adjoint problem (OR) we have

$$\hat{\Phi}(\zeta) = \check{\Phi}(\zeta)^+,\ U^+ = U.$$

Then

$$\mathcal{Q}(x,t,y) = -2\int \mathrm{tr}_{H_2/H_o}\{\Phi^+(\zeta;x,t,y)\,d\Sigma_\zeta \Phi(\zeta;x,t,y)\}$$

for

$$\frac{d^2}{dx^2}\,\Phi(\zeta) = -2\Phi(\zeta)\cdot\int \Phi^+(\mu)\,d\Sigma_\mu \Phi(\mu) + \zeta^2\Phi(\zeta).$$

In general, coefficients of the linear differential operator $L = \sum_{j=o}^{n} U_j \frac{d^j}{dx^j}$ can be expressed in terms of products of eigenfunctions of L: $L\psi = \lambda\psi$ and eigenfunctions of the conjugate problem $L^*\psi^* = \lambda\psi^*$. This representation for n = 1 is described in §§9, 10, and for n = 2 in §11. Such a representation and its connection with spectral measures, allows us to reduce <u>any</u> two- (or three-) dimensional completely integrable system of the Zakharov-Shabat type (see §1 for definition) to a common action of two (or three) commuting one-dimensional completely integrable Hamiltonians. We postpone the general discussion for further publications and now briefly consider only the case n = 3. As the case n = 2 is connected with the nonlinear Schrödinger equation (5.7), the case n = 3 is connected with the coupled modified KdV equation (5.5) (see [13], [14]):

$$\begin{cases} \phi_t = -\phi_{xxx} + \phi_x\psi\phi + \phi\psi\phi_x; \\ \psi_t = -\psi_{xxx} + \psi_x\phi\psi + \psi\phi\psi_x . \end{cases}$$

Again we can consider stationary solutions $\phi = e^{\Lambda t}\phi_o$; $\psi = \psi_o e^{-\Lambda t}$ for ϕ_o, ψ_o depending on x only and for a constant Λ. We now consider operators over the $L^2(\Omega,d\Sigma_\omega)$-extension of a Hilbert space H. For the sake of simplicity, for the $L^2(\Omega,d\Sigma_\omega)$ sequence of operators $\vec{A} = (A_\omega)$, $\vec{B} = (B_\omega)$ $(\omega \in \Omega)$ on H, we denote

$$(\vec{A},\vec{B}) = \int_{\Omega} A_\omega d\Sigma_\omega B_\omega .$$

Then the stationary operator modified KdV equation can be represented as the following coupled system of equations of the third order in ϕ_k, ψ_k:

$$(H_{III}) \begin{cases} -\phi_{k,xxx} + (\vec{\phi},\vec{\psi})\phi_{k,x} + (\vec{\phi}_x,\vec{\psi})\phi_k = k^3\phi_k ; \\ -\psi_{k,xxx} + \psi_{k,x}(\vec{\phi},\vec{\psi}) + \psi_k(\vec{\phi},\vec{\psi}_x) = -k^3\psi_k . \end{cases}$$

Now we understand that the moment problem of the type of Problem 10.1, leads to the representation of the coefficients U and V of the eigenvalue problem of the third order

$$(-\frac{d^3}{dx^3} + U\frac{d}{dx} + V)\phi_k = k^3\phi_k ,$$

in terms of the eigenfunctions ϕ_k and eigenfunctions ψ_k of the conjugate eigenvalue problems. Namely, we have the following representation in the case of U and V being operators on H:

$$U = \int_{\Omega} \phi_k d\Sigma_k \psi_k ; \quad V = \int_{\Omega} \phi_{kx} d\Sigma_k \psi_k \quad ,$$

for a certain spectral measure $d\Sigma_k$. E.g., as in §10-11 (see [15]), such a representation leads to the decomposition of the Boussinesq equation $u_{yy} = (6uu_x - u_{xxx})_x$ into the common action of the Hamiltonian H_{III} (scalar case) and one of the Hamiltonians commuting with H_{III}. We mention here only that the Boussinesq equation corresponds to the $\frac{\partial u}{\partial t} = 0$ case in the Kadomtsev-Petviashvili equation; thus there are interesting relations between H_{III} and the Russian chain.

Bibliography.

1. M.J. Ablowitz, D.J. Kaup, A.C. Newell, H. Segur, Stud. Appl. Math. 53(1974), pp. 249-315.
2. N.I. Akhiezer, The classical moment problem, Hafner Publishing Company, 1965.
3. G.G. Appelrot, Mat. Sbornik v. 16 (1892), Proc. Moscow Univ. Sect. Phys.-Math., No. 11 (1894); N.B. Delone, Algebraic integrals of motion of a rigid body about a fixed point, Petersburg, 1892; V.V. Golubev, Lectures on integration of the equations of motion of a rigid body about a fixed point, Published for NSF, Jerusalem, 1960; D.N. Goryachoff, Math. Sbornic., 21 (1900), No. 3; P. Nekrasov, Math. Sbornik., 16 (1892); V.A. Steklov, Transactions of the physical sciences of the society of amateurs of science, v. 8 (1896).
4. H.F. Baker, Proc. Royal Soc. London, Ser. A118 (1928), pp. 584-600.
5. A.A. Belavin, V.E. Zakharov, Pisma JETP 25(1977), No. 12, pp. 603-607.
6. Yu. M. Berezansky, Trudy Moscow Math. Soc., 21 (1970), pp. 47-102; Yu. M. Berezansky, Self-adjoint eigenfunction expansions, Providence, Rhode Island, 1970.
7. Yu. M. Berezansky, Self-adjoint operators in spaces of functions with infinite number of variables, Naukova Dumka, Kiev, 1978.
8. A.S. Budagov, L.A. Takhtajan, Dokl. Acad. Sci., USSR, 235 (1977), No. 4, pp. 805-808.
9. A.S. Budagov, Proceedings of LOMI, 1970, pp. 24-56.
10. J.L. Burchnall, T.W. Chaundy, Proc. London Math. Soc., 21 (1922), pp. 420-440; J.L. Burchnall, T.W. Chaundy, Proc. Royal Soc. London, Ser. A118 (1928), 557-573.

11. F. Calogero, A. Degasperis, Nuovo Cimento 32B (1976), p. 201; F. Calogero, A. Degasperis, Nuovo Cimento 39B (1977), 1.
12. I.V. Cherednik, Funct. and Appl. 12 (1978), 45.
13. D.V. Chudnovsky, C.R. Acad. Sci. Paris, A289(1979), pp. A-429-A-432.
14. D.V. Chudnovsky, Phys. Lett. A, 74A(1979), pp. 185-188.
15. D.V. Chudnovsky, Saclay preprint DPh-T/79/80, J. Phys. (to appear).
16. D.V. Chudnovsky, Lett. Nuovo Cimento, 26(1979), pp. 423-427.
17. D.V. Chudnovsky, C.R. Acad. Sci. Paris A289(1979), pp. A-438-A442.
18. D.V. Chudnovsky, Riemann monodromy problem, isomonodromy deformation equations and completely integrable systems. Lectures. Cargese Summer School on Mathematical Physics, June 1979, D. Reidel Publishing Company, Dordrecht, Holland (to appear).
19. D.V. Chudnovsky, The generalized Riemann-Hilbert problem and the spectral interpretation, Proc. Intl. Meeting "Nonlinear evolution equations and dynamical systems", Lecce, Italy, July 1979, Lect. Notes Physics, Springer-Verlag (to be published).
20. D.V. Chudnovsky, Saclay preprint DPh-T/79/38, Nuovo Cimento B (to appear).
21. D.V. Chudnovsky, International Congress of Mathematicians, Helsinki, Finland 1978, Abstract of short communications, p. 183; Notices of the AMS, 26(1979), No. 1, Abstract 763-35-10, p. A-83-A-84.
22. D.V. Chudnovsky and G.V. Chudnovsky, Nuovo Cimento, 40B (1977), pp. 339-353.
23. D.V. Chudnovsky and G.V. Chudnovsky, Seminaire sur les equations non-lineaires I (1977-1978), Centre de Mathematiques, Ecole Polytechnique, pp. 300.
24. D.V. Chudnovsky and G.V. Chudnovsky, Lett. Nuovo Cimento 22(1978), pp. 31-36; pp. 47-51.
25. D.V. Chudnovsky and G.V. Chudnovsky, C.R. Acad. Sci. Paris A286(1978), pp. A-1075-A-1078.
26. D.V. Chudnovsky and G.V. Chudnovsky, Phys. Lett. A., 72A(1979), pp. 291-293.
27. D.V. Chudnovsky and G.V. Chudnovsky, Saclay preprint, Dph-T/79/110. Z. Phys. D (to appear).
28. D.V. Chudnovsky and G.V. Chudnovsky, Seminar on Spectral Theory, Riemann Problem and Complete Integrability, IHES 1979, Paris (to be published); G.V. Chudnovsky, Padé approximation and the Riemann monodromy problem. Lectures. Cargese Summer School on Mathematical Physics, June 1979, D. Reidel Publishing Company, Dordrecht, Holland (to appear).
29. G. Darboux, C.R. Acad. Sci. Paris, 94(1882), p. 1456.
30. P. Deift, E. Trubowitz, Comm. Pure Appl. Math. 32(1979), 121-200.
31. A. Degasperis, Lecture Notes in Physics, v. 80, Springer, 1978.
32. J. Drach, C.R. Acad. Sci. Paris 167(1918), pp. 743-746; C.R. Acad. Sci. Paris 168(1919), pp. 47-50; pp. 337-340.
33. B.A. Dubrovin, V.B. Matveev, S.P. Novikov, Russian Math. Survey 31(1976), 59.
34. B.A. Dubrovin, Funct. Anal. 11(1977), No. 4, 28-41.
35. R. Garnier, Circolo Math. Palermo, 43(1919), pp. 155-191.
36. I.M. Gelfand, L.A. Dikij, Funct. Anal. 10(1976), No. 1, pp. 17-25.
37. I.M. Gelfand, L.A. Dikij, Russian Math. Survey 20(1975), No. 5, 77.
38. I.M. Gelfand, L.A. Dikij, Funct. Anal. Appl. 13(1979), No. 1, pp. 7-20.
39. H. Grosse, Institut of Theor. Phys., Univ. Wien, UWThPh-1979-19 (to appear).
40. J. Honerkamp, Freiburg, 12p. Preprint THEP 79/8, (see this Les Houches volume).
41. C. Jacobi, Vorlesunger über Dynamic, Berlin, 1843.
42. D. Kaup, J. Math. Anal. Appl. 59(1976), 849-864.
43. W. Koppelman, J. Math. and Mech. (1961), pp. 247-277.
44. S. Kowalevski, Letters to Mittag-Leffler, Letter from 21 November, 1881. Correspondence, Published by the Academy of Sciences, USSR, 1948; S. Kowalevski, Acta Math., 12(1889), pp. 177-232; Acta Math. 14(1890).
45. M.G. Krein, M.A. Krasnoselsky, Uspekhi Math. Nauk 2(1947), No. 3, 60-106.
46. J.M. Kritchiver, S.P. Novikov, Funct. Anal. Appl. 12(1978), pp. 41-52.
47. P.D. Lax, Commun. Pure Appl. Math. 21(1968), pp. 467-490; P.D. Lax, Comm. Pure Appl. Math. 28(1975), pp. 141-200.
48. B.M. Levitan, Math. Sbornik 76(1968), pp. 239-270.
59. H.P. McKean, P. van Moerbeke, Invent. Math., 30(1975), pp. 217-300.
50. H.P. McKean, E. Trubowitz, Comm. Pure Appl. Math. 29(1976), pp. 143-226.
51. O. Neumann, J. Reine Augew. Math. 56(1859), pp. 46-63.
52. A.C. Newell, Proc. Royal Soc. London, A365(1979), pp. 283-311.

53. L.P. Niznik, Nonstationary inverse scattering problem, Naukova Dumka, Kiev, 1973.
54. K. Pohlmeyer, K.-H. Rehren, Freiburg Preprint 1979, J. Math. Phys. (to appear); Comm. Math. Phys. 46(1976), 207-221; Freiburg Priprint THEP 79|16.
55. L. Schlesinger, J. für Reine Angew. Math. 141(1912), pp. 96-145.
56. E.K. Sklyanin, L.A. Takhtadjan, L.D. Fadeev, LOMI preprint P-I-79, Leningrad, 1979, pp. 35.
57. H.B. Thacker, D.Wilkinson, Fermilab, pub. 79/19-THY, February 1979 (to appear); H.B. Thacker, Phys. Rev. D 17(1978), 1031.
58. V.E. Zakharov, A.V. Mikhailov, Zr. Eksp. Teor. Fis. 74(1978), No. 6, pp. 1053-1973.
59. V.E. Zakharov, A.B. Shabat, Funct. Anal. Appl. 8 (1974), No. 3, pp. 54-56; Funct. Anal. Appl. 13(1979), No. 3, pp. 13-22.
60. J.L. Gervais, A. Neveu, E.N.S. Preprint (to be published).
61. E. Borel, Lectures on divergent series, Paris 1928 (English translation ed.by J. Gammel, Los Alamos, LA-6140-TR, 1975).
62. D.V. Chudnovsky, Proc. Natl. Acad. Sci., USA, 75(1978), pp. 4082-4084; Lett. Nuovo Cimento 25(1979), pp. 263-265.

Quantization of Exactly Integrable Field Theoretic Models

- The Operator Transform Method -

J. Honerkamp

Fakultät für Physik
der Universität Freiburg
Hermann-Herder-Str. 3
7800 Freiburg / Germany

Whereas the construction of an exactly integrable model or the discussion of whether a given model is exactly integrable is a more or less mathematical problem, a very interesting question for a physicist is immediately:

- What are the consequences of the exact integrability for the quantization of the theory -
- which relations of the classical canonical structure survive in the quantization procedure -
- which benefit can be derived for the quantization of the theory from the knowledge of the exact integrability?

I would like to discuss these questions in the context of the nonlinear Schrödinger model, where we have already a complete answer. But of great interest are also other models like:

a) The massive Thirring model
b) the Heisenberg model
c) the Toda Lattice model

because pursuing the same strategy one may uncover new aspects which become trivial in the nonlinear Schrödinger model because of its simplicity.
In the following I am going to recapitulate the classical canonical structure of the nonlinear Schrödinger model.
Then I translate the classically defined quantities into quantum operators and I demonstrate

a) that some of these operators can be used as generators of the quantum eigenstates

b) that these eigenstates can be characterized by a wave function which is of a very special structure known as "Bethe-wave function".

c) finally that some other quantities of the classical canonical structure, translated into quantum mechanics, can be regarded as generators of the conserved quantities.
   One obtains in this way an infinite set of commuting observables.

1. The classical canonical structure of the nonlinear Schrödinger equation.

The nonlinear Schrödinger equation reads

$$i\partial_t \Phi = -\partial_x^2 \Phi + 2c\,\Phi^*\Phi\,\Phi \; ; \qquad \partial_x = \frac{\partial}{\partial x} \tag{1}$$

From Zakharov and Shabat [(1)] we know that this nonlinear equation for the field $\Phi(x,t)$ **guarantees** the compatibility of the two linear differential equations

$$L\begin{pmatrix} v_1 \\ v_2 \end{pmatrix} = \zeta\begin{pmatrix} v_1 \\ v_2 \end{pmatrix} \tag{2}$$

$$i\,\partial_t\begin{pmatrix} v_1 \\ v_2 \end{pmatrix} = M\begin{pmatrix} v_1 \\ v_2 \end{pmatrix} \tag{2b}$$

where

$$L = \begin{pmatrix} i\partial_x & \sqrt{c}\,\Phi \\ \sqrt{c}\,\Phi^* & -i\partial_x \end{pmatrix} \; ; \qquad M = \begin{pmatrix} 2\zeta^2 + c\,\Phi^*\Phi & -\sqrt{c}(i\zeta\Phi - \frac{1}{2}\Phi_x) \\ \sqrt{c}(i\zeta\Phi^* + \frac{1}{2}\Phi_x) & -2\zeta^2 + c\,\Phi^*\Phi \end{pmatrix} \tag{3}$$

If $\begin{pmatrix} v_1 \\ v_2 \end{pmatrix}$ is a solution to (2a) with eigenvalue $\zeta = \xi + i\eta$, then $\begin{pmatrix} v_2^* \\ -v_1^* \end{pmatrix}$ is a solution with eigenvalue $\zeta = \xi - i\eta$.

We may define special solutions to the equation (2a) by requiring a special asymptotic behaviour. We assume $\Phi \to 0$ for $|x| \to \infty$ and we define the Jost-solutions $u(\xi,x)$ and $w(\xi,x)$ by $(\zeta = \xi)$.

$$u(\xi,x) \xrightarrow[x\to-\infty]{} \begin{pmatrix} 1 \\ 0 \end{pmatrix} e^{-i\xi x} \tag{4}$$

$$w(\xi,x) \xrightarrow[x\to+\infty]{} \begin{pmatrix} 0 \\ 1 \end{pmatrix} e^{i\xi x} \tag{4b}$$

Then $u = \begin{pmatrix} u_1 \\ u_2 \end{pmatrix}$ and $\bar{u} := \begin{pmatrix} u_2^* \\ -u_1^* \end{pmatrix}$ represent a complete set of solutions, like w and $\bar{w}$, and expanding $u(\xi,x)$ in terms of the w, $\bar{w}$ we obtain

$$\begin{aligned} u(\xi,x,t) &= a(\xi,t)\,\bar{w}(\xi,x,t) + b(\xi,t)\,w(\xi,x,t) \\ &\xrightarrow[x\to+\infty]{} \begin{pmatrix} a(\xi,t)\; e^{-i\xi x} \\ b(\xi,t)\; e^{i\xi x} \end{pmatrix} \end{aligned} \tag{5}$$

The coefficients of this expansion, the Jost-functions $a(\xi)$ and $b(\xi)$, can be obtained by looking at the asymptotic behaviour of the solution $u(\xi,x)$, which may be constructed by converting the differential equation (2a) into an integral equation and by solving this Volterra equation by iteration. One obtains in this way

$$b(\xi;\Phi(x,t),\overset{*}{\Phi}(x,t)) = \sqrt{c}\Big(\int dx\, \overset{*}{\Phi}(x,t)\, e^{-2i\xi x} + \tag{6}$$

$$+ c \int dx_1 dx_2 dx_3\, \Theta(x_1 > x_2 > x_3) e^{-2i\xi(x_1 + x_2 - x_3)} \overset{*}{\Phi}(x_1) \overset{*}{\Phi}(x_2) \Phi(x_3)$$

$$+ \ldots \Big)$$

and

$$a(\xi;\ \Phi(x,t),\ \overset{*}{\Phi}(x,t)) = 1 + \tag{7}$$

$$+ c \int dx_1 dx_2\, \Theta(x_2 > x_1) e^{2i\xi(x_1 - x_2)} \overset{*}{\Phi}(x_1)\, \Phi(x_2) + \ldots$$

The structure of the higher terms of the series (6) and (7) is obvious. The remarkable fact now is, that if $\Phi(x,t)$ and $\overset{*}{\Phi}(x,t)$ evolve in time in accordance with the nonlinear Schrödinger equation (1), then the time evolution of the Jost-functions $a(\xi,t)$ and $b(\xi,t)$ is very simple:

$$a(\xi,t) = a(\xi,0) \tag{8a}$$

$$b(\xi,t) = b(\xi,0) e^{4i\xi^2 t} \tag{8b}$$

This means: $a(\xi)$ and $b(\xi)$ represent another set of coordinates, instead of $\{\Phi(x),\ \overset{*}{\Phi}(x)\}$, with a very simple time behaviour. This transformation can be used to solve the nonlinear equation (1), the time evolution will be calculated in terms of the coordinates $a(\xi)$ and $b(\xi)$; the step which remains is the transformation from the $a(\xi,t)$ and $b(\xi,t)$ back to $\Phi(x,t)$, $\overset{*}{\Phi}(x,t)$, which can be done by an inverse scattering method, which gives this whole procedure its name: the inverse scattering transform method.

By the Poisson brackets

$$\{\Phi(x),\ \overset{*}{\Phi}(y)\} = i\delta(x-y) \tag{9}$$

and the Hamiltonian

$$H = \int dx(-\overset{*}{\Phi}(x)\ \partial_x^2\,\Phi(x) + c(\Phi^*(x)\Phi(x))^2) \tag{10}$$

the field equation reads

$$\partial_t\Phi = \{H,\Phi\} \tag{11}$$

and the equation (8) can be written as

$$\{H,a\} = 0 \quad ; \quad \{H,b(\xi)\} = 4i\xi^2 b(\xi) \tag{12}$$

One may calculate also some other Poisson brackets

$$\{a(\xi),\ a(\xi')\} = \{b(\xi),\ b(\xi')\} = 0 \tag{13}$$

$$\{a(\xi),\ b(\xi')\} = \frac{c}{\xi-\xi'-i\varepsilon}\ a(\xi)\ b(\xi') \tag{14}$$

$$\{b(\xi),\ \overset{*}{b}(\xi')\} = 2\pi ic\overset{*}{a}(\xi)a(\xi)\delta(\xi-\xi') \tag{15}$$

2. The quantized version of the nonlinear Schrödinger model

In quantizing we interpret the field $\Phi(x,t)$ as an operator field, its complex conjugate as the hermitean conjugate field $\Phi^+(x,t)$. As a quantization rule we take

$$[\Phi(x), \Phi^+(y)] = \delta(x-y) \tag{16}$$

The Hamiltonian (10) then goes over to a Hamiltonian operator

$$\underline{H} = \int dx\left(-\Phi^+\partial_x^2\Phi + c\,\Phi^+(x)\Phi^+(x)\Phi(x)\Phi(x)\right) \tag{17}$$

which we have taken as normal ordered. Now the question is: What are the eigenstates and the eigenvalues of this Hamiltonian operator?
It will turn out in a moment that the Jost-functions defined in (5), (6) and (7), when translated into quantum operators, play a central role in the quantized theory namely:

- The odd Jost function $b(\xi)$ generates, as an operator, the quantum eigenstates

- The even Jost function $a(\xi)$ generates, as an operator, the conserved quantities.

Let us make precise these statements.

a) The eigenstates of the Hamiltonian

We define an operator

$$\begin{aligned} B^+(q) &= \frac{1}{\sqrt{2\pi c}}\, b(\xi = q/2\,,\ \Phi \to \Phi_{op}\,,\ \Phi^* \to \Phi^+_{op}) \\ &= \frac{1}{\sqrt{2\pi}}\int dx\, \Phi^+(x)\, e^{-iqx} \\ &+ c\int dx_1 dx_2 dx_3\, \Theta(x_1>x_2>x_3)\, e^{-iq(x_1+x_2-x_3)}\, \Phi^+(x_1)\Phi^+(x_2)\Phi(x_3) \\ &+ \ldots \end{aligned} \tag{18}$$

$B^+(q)$ contains an odd number of Φ or Φ^+'s. For $c = 0$, $B^+(q)$ is a creation operator of a free particle with momentum q and energy q^2, where the ground state or reference state $|\Omega\rangle$ is defined by

$$\Phi(x)|\Omega\rangle = 0,$$

hence

$$B|\Omega\rangle = 0$$

and

$$B^+(q)|\Omega\rangle = |q\rangle = \frac{1}{\sqrt{2\pi}} \int dx\Phi^+(x)e^{-iqx}|\Omega\rangle \tag{19}$$

<u>Proposition:</u>

$B^+(q_1) \ldots B^+(q_N)|\Omega\rangle$ is an exact eigenstate of $\underline{H}$ with eigenvalue $E = \sum_{i=1}^{N} q_i^2$. $(-\infty < q_i < +\infty,\ N = 1,2\ldots)$.

In order to prove this it is only necessary to show that

$$[H, B^+(q)] = q^2 B^+(q) \tag{20}$$

i.e. that the classical Poisson bracket (12)

$$\{H, b(q)\} = iq^2 b(q)$$

may be taken over to the quantum level by omitting a factor i as in going from (9) to (16).

The proof of (20) was given by Faddeev and Sklyanin [(3)] and by Thacker and Wilkinson.[(4)] Sklyanin pointed out that the linear differential equations (2a), (2b) can be translated into differential equations for operator valued components G(q,x) and F(q,x) of the Jost-solution $u = \begin{pmatrix} u_1 \\ u_2 \end{pmatrix}$, interpreted as an operator:

$$\begin{aligned} \partial_x G(q,x) &= i\sqrt{c}\, F(q,x)\, \Phi(x) - i\tfrac{q}{2}\, G(q,x) \\ \partial_x F(q,x) &= -i\sqrt{c}\, \Phi^+(x)\, G(q,x) + i\tfrac{q}{2}\, F(q,x) \end{aligned} \tag{21}$$

Note that this equation is normal ordered and F and G come out to be normal ordered. In a similar way one may translate the equation (2b) and the compatibility of these two linear operator equation requires

$$i\partial_t \Phi = [\Phi, H] \tag{22}$$

We obtain the operator analogue to $u(\xi,x,t)$ if we require

$$\begin{pmatrix} G(q,x) \\ F(q,x) \end{pmatrix} \xrightarrow[x \to -\infty]{} \begin{pmatrix} e^{-iqx/2} \\ 0 \end{pmatrix} \tag{23}$$

Then we obtain exactly

$$\begin{pmatrix} G(q,x) \\ F(q,x) \end{pmatrix} \xrightarrow[x \to +\infty]{} \begin{pmatrix} A(q)e^{-iqx/2} \\ \sqrt{2\pi c}\, B^{+}(q)e^{iqx/2} \end{pmatrix} \tag{24}$$

The time behaviour of the Jost operators can be calculated in the same way as in the classical case [2], but now with help of the equation (21) and the operator version of (2b). Hence one obtains also ($q = \xi/2$)

$$\begin{aligned} A(q,t) &= A(q,0) \\ B^{+}(q,t) &= B^{+}(q,0)e^{iq^2 t} \end{aligned} \tag{25}$$

On the other hand we have from (22)

$$i\partial_t B^{+}(q,t) = [B^{+},H] \tag{26}$$

Therefore by (25)

$$[H,B^{+}(q)] = q^2 B^{+}(q) \tag{27}$$

and likewise

$$[H,A(q)] = 0 \tag{28}$$

(b) The structure of the eigenstates.

The operator $B^{+}(q)$ is represented as an odd series in the elementary fields $\Phi(x),\Phi^{+}(x)$, but the eigenstates $B^{+}(q_1) \ldots B^{+}(q_N)|\Omega\rangle$ when expressed in terms of the elementary fields are monomials:

$$B^{+}(q_1) \ldots B^{+}(q_N)|\Omega\rangle = \int dx_1 \ldots dx_N \chi(x_1 \ldots x_N \,;\, q_1 \ldots q_N)\Phi^{+}(x_1)\ldots\Phi^{+}(x_N)|\Omega\rangle \tag{29}$$

The reason for this simple structure is that there exists an operator - called "particle number operator" -

$$\underline{N} = \int dx\, \Phi^{+}(x)\Phi(x) \tag{30}$$

with

$$[\underline{N},\underline{H}] = 0 \tag{31}$$

and

$$[\underline{N},\Phi^+(x)] = \Phi^+(x) \; ; \; [\underline{N},B^+(q)] = B^+(q) \tag{32}$$

Hence the eigenstate (29) is also eigenstate of $\underline{N}$ with eigenvalue N and is written as a superposition of such other eigenstates of $\underline{N}$, namely of

$$\Phi^+(x_1) \ldots \Phi^+(x_N)|\Omega> \tag{33}$$

The superposition-coefficients are given by the wave-function $\chi(x_1 \ldots x_N, q_1 \ldots q_N)$ which has to be chosen so that the N-particle state (29) is just an eigenstate of $\underline{H}$. This wave function can be calculated (5,6) to be

$$\chi(x_1 \ldots x_N, q_1 \ldots q_N) = e^{i\sum_{K=1}^{N} q_K x_K} \prod_{1 \le K < j \le N} \left(1 + i\lambda(q_K, q_j)\varepsilon(x_K - x_j)\right) \tag{34}$$

where

$$\lambda(q_K, q_j) = \frac{-c}{|q_K - q_j|} \tag{35}$$

and

$$\varepsilon(x_K - x_j) = \begin{cases} +1 & \text{for } x_K > x_j \\ -1 & \text{for } x_K < x_j \end{cases}$$

Formula (34) means that for $x_K \neq x_j$; $K,j = 1,\ldots N$, the wave function is a free one. At $x_K = x_j$ the wave function makes a jump. The height of it is dictated by the dynamics and this jump is just the whole consequence of the dynamics. It is well known that these wave functions are the quantum mechanical eigenfunctions of the N particle Hamiltonian

$$H = -\sum_{i=1}^{N} \frac{\partial^2}{\partial x_i^2} + 2c \sum_{\substack{i,j=1 \\ i \neq j}}^{N} \delta(x_i - x_j) \tag{36}$$

These eigenfunctions were found in 1963 by Lieb and Liniger (7) and already in 1931, Bethe (8) had shown that for the isotropic Heisenberg model on a lattice the eigenfunctions have such a quasi-free structure as in (34). Here one is able to derive this Bethe wave function by looking at the states

$$B^+(q_1)\ldots B^+(q_N)\,|\Omega\rangle$$

The Bethe structure of the wave function has immediate consequences for the S matrix. Choosing $q_1 < q_2 \ldots < q_N$ and $x_1 > x_2 \ldots > x_N$ we get

$$\chi\,(x_1\ldots x_N\,;\,q_1\ldots q_N) = \prod_{1\le K\le j\le N}\big(1 + i\lambda(q_K,q_j)\big)\,e^{i\sum q_K x_K} \tag{37}$$

which shows that an incoming state χ^+ can be defined by

$$\chi^+(x_1\ldots x_N, q_1\ldots q_N) = \prod_{k<j}\big(1 + i\lambda(q_K q_j)\big)^{-1}\chi \tag{38}$$

By choosing $x_1 < x_2 < \ldots < x_N$ we see that an outgoing state can be defined by

$$\chi^-(x_1\ldots x_N\,,\,q_1\ldots q_N) = \prod_{k<j}\big(1 - i\lambda(q_K,q_j)\big)^{-1}\chi \tag{39}$$

Hence

$$\chi^-(x_1,x_2,\ldots x_N\,,\,q_1\ldots q_N) = \prod_{k<j}\frac{1 + i\lambda(q_K,q_j)}{1 - i\lambda(q_K,q_j)}\,\chi^+(x_1\ldots x_N\,,\,q_1\ldots q_N)$$

and

$$S\,(q'_1\ldots q'_N\,;\,q_1\ldots q_N) = \langle -\,;\,q'_1\ldots q'_N\,|\,+\,,\,q_1\ldots q_N\rangle$$

$$= \prod_{k<j}\frac{1 + i\lambda(q_x,q_j)}{1 - i\lambda(q_x,q_j)}\;\delta^{q'_1\ldots q'_N}_{q_1\ldots q_N}\;\delta_{NN'}$$

where $\delta^{q'_1\ldots q'_N}_{q_1\ldots q_N}$ is the symmetrized product of Kronecker-symbols.

Hence we have as consequences of the Bethe structure of the wave function:

(i) Particle number conservation, where particles are created by those fields, in terms of which the eigenstates are monomials so that a wave function can be defined.

(ii) The set of incoming momenta is equal to the set of outgoing momenta:

$$\{q_1 \ldots q_N\} = \{q_1' \ldots q_N'\}$$

(iii) Factorization of the N particle S-matrix in terms of the two particle S-matrix

$$S\,(q_1,q_2) = \frac{1 + i\lambda(q_1,q_2)}{1 - i\lambda(q_1,q_2)} = \frac{1 - \frac{ic}{|q_1-q_2|}}{1 + \frac{ic}{|q_1-q_2|}} \tag{42}$$

The question arises whether also in the other models mentioned at the beginning the eigenstates can be constructed by this strategy and whether they have a Bethe-structure with all these consequences.

Up to now we know that this is true in the massive Thirring model [5] and in the anisotropic Heisenberg model [9]. In the Toda-lattice model [10] one is able to construct in this way the eigenstates of the Hamiltonian. But the field coordinates in terms of which the eigenstates can be written as a superposition of monomials, thereby defining a wave function, are not yet known to me. It may even be the case, that such coordinates do not exist and that the S-matrix for the Toda-lattice model is of a more general structure. [11]

It should be noted that the S-matrix which is calculated in this way within the massive Thirring model is not yet the physical S-matrix. Because in this case the reference state $|o\rangle$ is not the ground state, one has to construct the ground state as a superposition of the eigenstates. The excited states are then the physical states and by this construction one is also able to calculate the fermion-antifermion bound state spectrum. [12]

c) The conserved quantities

By (25) and (28) we know that A(q) is a conserved quantity, therefore so is each coefficient in an expansion like

$$A(q) = \sum_{n=o}^{\infty} q^{-n}\, I_n \quad \text{or} \quad A(q) = \sum_{n=o}^{\infty} q^{n}\, J_n \tag{43}$$

The J_n are nonlocal quantities, we do not know whether they are useful.
For the I_n (n = 0,1,...) we obtain:

$$\begin{aligned} I_o &= 1 \;;\; I_1 = -ic\underline{N} \\ I_2 &= -ic\underline{P} - (ic)^2/2\; (\underline{N}^2 - \underline{N}) \end{aligned} \tag{44}$$

$$I_3 = ic\underline{H} - (ic)^2 (\underline{PN} - \underline{P}) + (ic)^3/6 \; (\underline{N}^3 - 3\underline{N}^2 + 2\underline{N})$$

etc.

where $\underline{N}$ and $\underline{H}$ are given by (30) and (17) and

$$\underline{P} = \tfrac{1}{2}\int dx(\phi^+ \partial_x \phi - \partial_x \phi^+ \phi) \qquad (45)$$

Hence: The even Jost-function in its quantized version generates an infinite set of conserved quantities, to which the Hamiltonian belongs. In order to discuss the eigenfunctions and the eigenvalues of these conserved quantities one takes advantage of the commutator

$$\left[A(q') \, , \, \overset{+}{B}(q)\right] = \frac{-\, ic}{q' - q - i\varepsilon} \overset{+}{B}(q) \; A(q') \qquad (46)$$

From this one deduces that $\overset{+}{B}(q_1)\ldots\overset{+}{B}(q_N)|0\rangle$ is also an eigenstate of $A(q')$ and therefore of all I_n's:

$$A(q')\overset{+}{B}(q_1)\ldots\overset{+}{B}(q_N)|0\rangle = \prod_{i=1}^{N}\left(1 - \frac{ic}{q_i - q' - i\varepsilon}\right)\overset{+}{B}(q_1)\ldots\overset{+}{B}(q_N)|0\rangle$$

By expanding both sides in q' one obtains the eigenvalues of the I_n's .

Note that the commutator (46) is the translation of the Poisson bracket (14), but on the right hand side of (46) one has to take care with the ordering of the operators.

However, the classical Poisson bracket (15) does not have such a simple translation. For the quantum operators one obtains

$$\left[B(q_1) \, , \, \overset{+}{B}(q_2)\right] = \delta(q_1 - q_2)\overset{+}{A}(q_1)A(q_2) + \overset{+}{B}(q_2)B(q_1) \frac{c^2}{(q_1 - q_2 + i\varepsilon)^2}$$

Recently Faddeev(13) developed a method of calculating the commutator

$$\left[A(q) \, , \, B(q')\right]$$

in every integrable model with boson fields. Here A(q) stands for an even Jost-operator, B(q) for an odd operator. For the massive Thirring model however this method seems not to work.(15)

In conclusion one can ascertain that the canonical classical structure provides us with a powerful tool for the construction of eigenstates and conserved quantities in the associated quantum field theory. The remaining problems in the various models may be clarified certainly very soon.

For relativistic theories as for the massive Thirring model, however, these eigenstates and the corresponding S-matrix are not yet the physical quantities, as I mentioned above. Whereas there exists a well established procedure to calculate the physical energy spectrum,[12] the calculation of the physical S-matrix from these eigenstates remains obscure, though there exist some intuitive arguments which, however, have not yet been made precise. It is a very challenging problem, whether the Green's functions of these integrable field theoretic models can be calculated by exploiting the methods developed by Sato and his coworkers.[16]

References:

1) V. E. Zakharov and A. B. Shabat Zk. Eksp. Teor. Fiz. 61, 118 (1971) (Sov. Phys. JETP 34, 62 (1972)

2) V. E. Zakharov and S. V. Manakov Teor. Mat. Fiz. 19, 332 (1974) (Theor. Math. Phys. 19, 551 (1975)

3) L. D. Faddeev and E. K. Sklyanin Dokl. Akad. Nauk. USSR 243, 1430 (1978)
E. K. Sklyanin, Dokl. Akad. Nauk USSR 244, 1337 (1979)

4) H. B. Thacker and D. W. Wilkinson, Phys. Rev. D 19, 3660 (1979)

5) J. Honerkamp, A. Wiesler and P. Weber, Nucl. Phys. B 152, 266 (1979)

6) A. Wiesler, Freiburg preprint Okt. 79

7) E. H. Lieb and W. Liniger, Phys. Rev. 130, 1605 (1963)

8) H. Bethe, Z. Physik 71, 205 (1931)

9) see e.g.
P. P. Kulish and E. K. Sklyanin, Phys. Lett. 70 A,461 (1979)
J. des Cloiseaux and M. Gaudin, Journ. Math. Phys. 7, 1384 (1966)

10) see e.g.
M. Toda, Phys. Reports 18 C, 1 (1975)
H. Flaschka, Phys. Rev. B 9, 1924 (1975) and
Progr. Theor. Phys. 51, 703 (1974)
K. M. Case and S. C. Chiu, J. Math. Phys. 14, 1643 (1973)

11) see e.g.
the lecture by D. Iagolnitzer, this colloquium.

12) see e.g.
H. Bergknoff and H. B. Thacker Phys. Rev. Lett. 42, 135 (1979)
preprint Fermi-Lab Pub-78/84-Thy

13) L. D. Faddeev, preprint Akad. Nauk USSR Lomi-P-2-79

14) H. Grosse, preprint Wien 1979/UWthPh-1979-19

15) P. Weber, private communication

16) M. Sato, T. Miwa, M. Jimbo, a series of papers entitled " Holonimic Quantum Fields ": I, Publ. RIMS, Kyoto Univ. 14, (1977) 223-267, II ibid. 15 (1979), 201-278
RIMS-preprints 260 (1978), 263 (1978) and 267 (1978)
See also a series of short notes
"Studies on Holonimic Quantum Fields" I-XVI in Proc. Japan Acad.

Aspects of Holonomic Quantum Fields

— Isomonodromic deformation and Ising model —

By

Michio Sato, Tetsuji Miwa, Michio Jimbo

Research Institute for Mathematical Sciences

Kyoto University, Kyoto 606, JAPAN

- Text written by M. Jimbo and T. Miwa -

The present article is based on a series of lectures delivered at the international colloquium at Les Houches, September 5-13, 1979. We intend to give here an expository survey on the works done in the past three years 1977-1979, in collaboration with Professor M. Sato and partly with Dr. K. Ueno and Dr. Y. Môri, concerning deformation of linear differential equations and quantum field theory.

We did not try to put everything in this paper. Rather, we restricted ourselves to the following four main topics, divided into four sections one for each:

§1. Field theoretical approach to the Riemann-Hilbert problem on monodromy

§2. General theory of isomonodromic deformation of linear ordinary differential equations

§3. Scaling limit of the two dimensional Ising model

§4. The one dimensional XY model and its scaling limit.

Each section is written rather independently, so the reader can go through with it without much referring to other sections. Since the second topic is mathematically a bit involved, we suggest those who are mathematically less oriented to skip it for the first reading.

Among the topics not included here are the density matrix of impenetrable bose gas and the variational formulas in higher dimensions. For these we refer the reader to the following review paper along with the bibliography quoted therein:

M. Jimbo, T. Miwa, M. Sato and Y. Môri, Holonomic quantum fields — The unanticipated link between deformation theory of differential equations and quantum fields — preprint RIMS 305 (1979), to appear in the Proceedings of international conference on mathematical physics (M∩Φ), Lausanne 1979.

We would like to acknowledge our indebtedness to Professor D. Iagolnitzer for inviting us to the Les Houches Colloquium and providing us with an opportunity to give lectures.

§1. Field theoretical approach to the Riemann-Hilbert problem

1.1 The classical Riemann's problem

We begin our story of monodromy preserving deformation with a prototype in this branch: the classical Riemann's problem. A particular simplicity consists in that it is concerned with linear ordinary differential equations admitting only regular singularities.

Let us write them in the form of a first order system with $m\times m$ matrix coefficients

$$\frac{dY}{dx} = \sum_{\nu=1}^{n} \frac{A_\nu}{x-a_\nu} Y, \quad A_\nu: m\times m \text{ constant matrices,} \tag{1.1.1}$$

having $x = a_1,\cdots,a_n,\infty$ as its regular singularities. Let $Y = Y(x)$ denote the fundamental matrix solution of (1.1.1) normalized as $Y(x_0) = 1$, where $x_0 \in \mathbb{C}-\{a_1,\cdots,a_n\}$ is an arbitrarily chosen base point. For any closed path γ in $\mathbb{C}-\{a_1,\cdots,a_n\}$ with its endpoint x_0, the analytic continuation $\gamma Y(x)$ of $Y(x)$ induces a linear transformation

$$Y(x) \longmapsto \gamma Y(x) = Y(x)M_\gamma \,. \tag{1.1.2}$$

Here $M_\gamma = \gamma Y(x_0)$ is a constant matrix depending on γ only through its homotopy class $[\gamma]$. It is clear that $M_{\gamma_1\gamma_2} = M_{\gamma_1}M_{\gamma_2}$. Thus to each system (1.1.1) (+ its solution matrix) there is associated a representation of the fundamental group, the monodromy representation:

$$\begin{array}{ccc} \pi_1(\mathbb{C}-\{a_1,\cdots,a_n\};x_0) & \xrightarrow{\rho} & GL(m,\mathbb{C}) \\ [\gamma] & \longmapsto & M_\gamma \,. \end{array} \tag{1.1.3}$$

Since $\pi_1(\mathbb{C}-\{a_1,\cdots,a_n\};x_0)$ is a free group with n generators $\gamma_1,\cdots,\gamma_n$ (see Fig. 1.1)

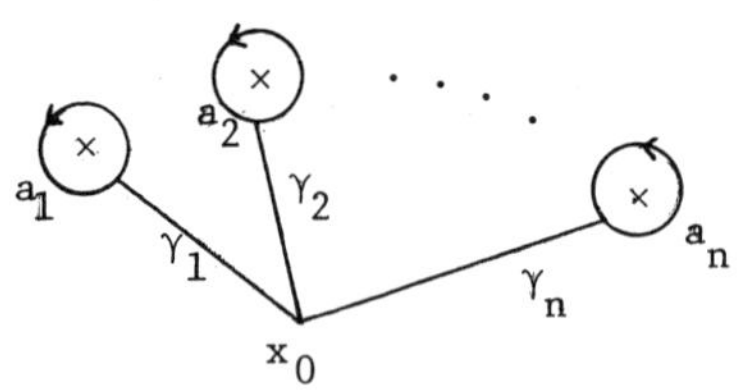

Fig. 1.1

the monodromy group $\rho(\pi_1(\mathbb{C}-\{a_1,\cdots,a_n\};x_0))$ is generated by the n matrices $M_\nu = M_{\gamma_\nu}$ ($\nu=1,\cdots,n$). Viewed as functions of $A_1,\cdots,A_n$ these monodromy matrices M_ν are known to be entire (holomorphic). The Riemann's problem amounts to the inversion of these entire functions. Namely it states as follows:

(1.1.4) Given $a_1,\cdots,a_n$ and $M_1,\cdots,M_n \in GL(m,\mathbb{C})$, find a system of differential equations (1.1.1) whose monodromy representation ρ is the prescribed one: $\rho([\gamma_\nu]) = M_\nu$ $(\nu=1,\cdots,n)$.

Actually in order to specify the solution uniquely it is necessary to refine the data $\{a_1,\cdots,a_n,M_1,\cdots,M_n\}$ slightly.

With an additional condition

(1.1.5) the eigenvalues of $A_1,\cdots,A_n$, $A_\infty = -\sum_{\nu=1}^{n} A_\nu$ do not differ by integers,

the normalized solution $Y(x)$ is shown to have the following properties:

(1.1.6) (0) $Y(x_0) = 1$,

(i) $Y(x)$ is (multivalued) analytic and invertible for $x \neq a_1,\cdots,a_n$, $a_\infty(=\infty)$,

(ii) at $x = a_\nu$ it has the form(*)

$$Y(x) = \hat{Y}^{(\nu)}(x)(x-a_\nu)^{L_\nu},$$

where $\hat{Y}^{(\nu)}(x)$ is holomorphic and invertible at $x = a_\nu$ $(\nu=1,\cdots,n,\infty)$.

Here the exponent matrices L_ν are related to the monodromy through $M_\nu = e^{2\pi i L_\nu}$, so that they constitute a refined notion of the latter. We admit x_0 to be one of the branch points a_ν as well. In this case condition (1.1.6)-(0) should be replaced by

(1.1.6)-(0)' $\hat{Y}^{(\nu)}(x_0) = 1$.

Conversely, as has already been noted by Riemann, such a matrix $Y(x)$ (if ever exists) should necessarily be a solution of a differential equation of the form (1.1.1). The argument is standard. Consider $\frac{dY}{dx} Y^{-1}$. It is single-valued (the monodromy cancels out), holomorphic except for $x = a_1,\cdots,a_n,\infty$ by (i), has a local expression $\hat{Y}^{(\nu)}(x)\frac{L_\nu}{x-a_\nu}\hat{Y}^{(\nu)}(x)^{-1}$ + holomorphic at $x = a_\nu$ (resp. holomorphic at ∞), and hence must be a rational function $\sum_{\nu=1}^{n} \frac{A_\nu}{x-a_\nu}$ with $A_\nu = \operatorname{Res}_{x=a_\nu} \frac{dY}{dx} Y^{-1} = \hat{Y}^{(\nu)}(a_\nu)L_\nu\hat{Y}^{(\nu)}(a_\nu)^{-1}$. It is also unique, for the ratio $Y_1(x)Y_2(x)^{-1}$ of two such matrices $Y_1(x)$, $Y_2(x)$ is seen to be constant $= Y_1(x_0)Y_2(x_0)^{-1} = 1$. Thus the following equivalence holds:

Differential equations $\frac{dY}{dx} = \sum_{\nu=1}^{n} \frac{A_\nu}{x-a_\nu} Y$ with singularity data $a_1,\cdots,a_n$, $A_1,\cdots,A_n$.	$\Longleftrightarrow$	Multivalued matrix $Y = Y(x)$ with monodromy data $a_1,\cdots,a_n$, $L_1,\cdots,L_n$.

(*) For $x = \infty$ replace $x-a_\nu$ by $\frac{1}{x}$.

From now on we pose the Riemann's problem in the following strict sense:

(1.1.7) Given $a_1,\cdots,a_n$ and $L_1,\cdots,L_n$, find $Y(x)$ satisfying (1.1.6).

Its unique solution is denoted by $Y(x) = Y(x_0,x) = Y\begin{pmatrix} & & a_1,\cdots,a_n \\ x_0, & x; & \\ & & L_1,\cdots,L_n \end{pmatrix}$. We remark that a different choice of normalization point x_0 gives rise to a similarity transformation for $A_\nu = A_\nu(x_0)$:

$$(1.1.8) \quad \begin{aligned} Y(x_0',x) &= Y(x_0,x_0')^{-1}Y(x_0,x) \\ A_\nu(x_0') &= Y(x_0,x_0')^{-1}A_\nu(x_0)Y(x_0,x_0') . \end{aligned}$$

1.2 Schlesinger equations

Riemann proposed to study $Y\begin{pmatrix} & & a_1,\cdots,a_n \\ x_0, & x; & \\ & & L_1,\cdots,L_n \end{pmatrix}$ as a function of $n+2$ variables $x_0,x,a_1,\cdots,a_n$: how does it depend on $x_0,a_1,\cdots,a_n$ when the monodromy is kept invariant? This has been worked out after his death by L. Schlesinger. His answer is the following. In terms of $Y(x)$ the necessary and sufficient condition(*) for the monodromy to be preserved is that $Y(x)$ should satisfy a system of linear partial differential equations

$$(1.2.1) \quad \frac{\partial Y}{\partial x_0} = -\sum_{\nu=1}^{n} \frac{A_\nu}{x_0-a_\nu}\, Y, \qquad \frac{\partial Y}{\partial a_\nu} = \left(-\frac{A_\nu}{x-a_\nu} + \frac{A_\nu}{x_0-a_\nu}\right)Y \qquad (\nu=1,\cdots,n).$$

In terms of the coefficients $A_\nu = A_\nu(x_0;a_1,\cdots,a_n)$ the condition is expressed as the non-linear partial differential equations

$$(1.2.2) \quad \begin{aligned} \frac{\partial A_\nu}{\partial x_0} &= \sum_{\mu(\neq\nu)} [A_\mu,A_\nu]\, \frac{-1}{x_0-a_\mu} \\ \frac{\partial A_\nu}{\partial a_\mu} &= [A_\mu,A_\nu]\left(\frac{-1}{a_\nu-a_\mu} + \frac{1}{x_0-a_\mu}\right) \qquad (\mu\neq\nu) \\ &= \sum_{\mu(\neq\nu)} [A_\mu,A_\nu]\left(\frac{1}{a_\nu-a_\mu} - \frac{1}{x_0-a_\mu}\right) \qquad (\mu=\nu). \end{aligned}$$

The Schlesinger's equations (1.2.2) are the integrability condition for (1.2.1) and (1.1.1). Equations (1.2.1) are derived by a similar argument as in §1 applied to $\frac{\partial Y}{\partial a_\nu}\cdot Y^{-1}$. If $x_0 \equiv a_\nu$, (1.2.1) and (1.2.2) are modified. For instance the choice $x_0 = \infty$ simplifies them into

$$(1.2.1)' \quad \frac{\partial Y}{\partial a_\nu} = -\frac{A_\nu}{x-a_\nu}\, Y$$

(*) We assume (1.1.5).

$$(1.2.2)' \qquad \frac{\partial A_\nu}{\partial a_\mu} = \frac{[A_\mu, A_\nu]}{a_\mu - a_\nu} \qquad (\mu \neq \nu)$$

$$= - \sum_{\mu(\neq\nu)} \frac{[A_\mu, A_\nu]}{a_\mu - a_\nu} \qquad (\mu = \nu).$$

1.3 Field theoretical formulation

The Riemann's problem is put into various formulations, among which the Hilbert's approach by boundary value problem is of relevance here. Choose any simple contour Γ passing through the branch points $a_1, \cdots, a_n, \infty$ (Fig. 1.2).

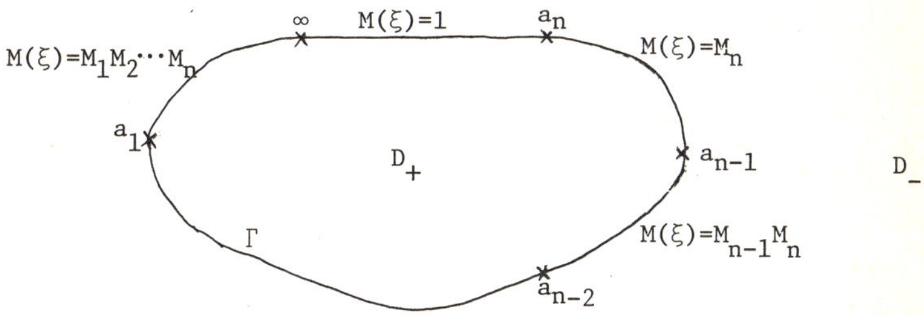

Fig. 1.2

Let $Y_+(x)$, $Y_-(x)$ be the branches of $Y(x)$ inside (D_+) or outside (D_-) this contour Γ respectively, which are analytic continuations of each other through a portion of Γ, say between a_n and ∞. Then the monodromy property of $Y(x)$ is equivalent to the following relation between the boundary values of $Y_\pm(x)$:

$$(1.3.1) \qquad Y_+(\xi^+) = Y_-(\xi^-)M(\xi), \qquad \xi \in \Gamma \; \{a_1, \cdots, a_n, \infty\},$$

where $Y_\pm(\xi^\pm) = \lim_{D_\pm \ni x \to \xi} Y_\pm(x)$ and $M(\xi)$ is a step function whose values are tabulated in Fig. 1.2. Thus the Riemann's problem amounts to finding two matrices $Y_\pm(x)$, holomorphic and invertible in $D_\pm$, growing at most polynomially at $x = a_1, \cdots, a_n, \infty$, and that are related on Γ through (1.3.1). This boundary value problem of Hilbert admits a nice description in the language of quantum field theory. Let us show how it works, assuming that the branch points $a_1 < \cdots < a_n$ all lie on the real line $\Gamma = \mathbb{R}^1$.

Let $\psi^{*(j)}(x)$, $\psi^{(j)}(x)$ $(j=1, \cdots, m,\ x \in \mathbb{R})$ denote free fermion operators in one dimension, satisfying the following anti-commutation relations:

$$(1.3.2) \qquad [\psi^{*(j)}(x), \psi^{(j')}(x')]_+ = \delta_{jj'}\delta(x-x') \;{}^{(*)}$$

$$[\psi^{*(j)}(x), \psi^{*(j')}(x')]_+ = 0, \qquad [\psi^{(j)}(x), \psi^{(j')}(x')]_+ = 0.$$

(*) $[X,Y]_+ = XY+YX$.

What is meant by the term "field operator" will be made clearer in later paragraphs. At the moment it is sufficient to know that they are operators parametrized by $x \in \mathbb{R}$ in a suitable sense, acting on some Hilbert space $\mathcal{F}$. The latter (resp. its dual space $\mathcal{F}^*$) has a distinguished vector, which we denote by $|vac\rangle$ (resp. $\langle vac|$) following the bra-ket notation of Dirac. For an operator $\mathcal{O}$, the inner product of the vector $\mathcal{O}|vac\rangle$ with $\langle vac|$ is called the vacuum expectation value of $\mathcal{O}$, and is denoted by $\langle \mathcal{O} \rangle = \langle vac|\mathcal{O}|vac\rangle$. In our case of free fermion operators we assume the following:

$$(1.3.3) \qquad \begin{aligned} &\langle \psi^{*(j)}(x)\psi^{(j')}(x')\rangle = \langle \psi^{(j)}(x)\psi^{*(j')}(x')\rangle = \delta_{jj'}\frac{1}{2\pi}\frac{i}{x-x'+i0} \\ &\langle \psi^{*(j)}(x)\psi^{*(j')}(x') \quad = 0, \quad \langle \psi^{(j)}(x)\psi^{(j')}(x')\rangle = 0. \end{aligned}$$

The main point is the $+i0$ prescription in (1.3.3). This guarantees that, for any $\mathcal{O}$, the expectation values $\langle \mathcal{O}\psi^{(i)}(x)\rangle$, $\langle \mathcal{O}\psi^{*(i)}(x)\rangle$ (resp. $\langle \psi^{(i)}(x)\mathcal{O}\rangle$, $\langle \psi^{*(i)}(x)\mathcal{O}\rangle$) as functions of x are analytically prolongable to the lower (resp. upper) half complex plane $\mathrm{Im}\, x < 0$ (resp. $\mathrm{Im}\, x > 0$). We shall come to this later.

Now let φ be a field operator satisfying the following commutation relation with the free fields:

$$(1.3.4) \qquad \begin{aligned} \varphi\cdot\psi^{(k)}(x) &= \sum_{j=1}^{m} \psi^{(j)}(x)\cdot\varphi\cdot M(x)_{jk} \\ \varphi\cdot\psi^{*(k)}(x) &= \sum_{j=1}^{m} \psi^{*(j)}(x)\cdot\varphi\cdot({}^{t}M(x)^{-1})_{jk} \end{aligned}$$

where

$$(1.3.5) \qquad \begin{aligned} M(x) &= M_1(x) \cdots M_n(x) \\ M_\nu(x) &= e^{2\pi i L_\nu \theta(-x+a_\nu)} = \begin{cases} 1 & (x > a_\nu) \\ e^{2\pi i L_\nu} & (x < a_\nu). \end{cases} \end{aligned}$$

Let $Y_\pm(x)$ be $m\times m$ matrices whose (j,k)-th elements are given by

$$(1.3.6) \qquad \begin{aligned} Y_+(x)_{jk} &= 2\pi i(x-x_0)\langle \psi^{*(j)}(x_0)\psi^{(k)}(x)\varphi\rangle/\langle\varphi\rangle, \\ Y_-(x)_{jk} &= 2\pi i(x-x_0)\langle \psi^{*(j)}(x_0)\varphi\,\psi^{(k)}(x)\rangle/\langle\varphi\rangle. \end{aligned}$$

Here $x_0 \in \mathbb{R}$ is a normalization parameter. <u>Then these are solutions to the Riemann-Hilbert problem satisfying</u> $Y_\pm(x_0) = 1$.

The reason is simple. As has been mentioned above, each matrix element of $Y_-(x)$ is prolonged analytically to $\mathrm{Im}\, x < 0$. Similarly $Y_+(x)$ has an analytic continuation to $\mathrm{Im}\, x > 0$. This latter statement is seen by rewriting (1.3.6) as $Y_+(x)_{jk} = -2\pi i(x-x_0)\langle \psi^{(k)}(x)\psi^{*(j)}(x_0)\varphi\rangle/\langle\varphi\rangle$ (the remainder term, being proportional to

$\delta(x-x_0)$ by (1.3.2), is cancelled by the factor $2\pi i(x-x_0)$ in front). Finally the commutation relations (1.3.4) ensure the relation $Y_-(x) = Y_+(x)M(x)$. The normalization condition is seen from $\langle\psi^{*(j)}(x_0)\psi^{(k)}(x)\varphi\rangle = \delta_{jk}\frac{1}{2\pi}\frac{i}{x_0-x+i0}\langle\varphi\rangle$ + holomorphic term at $x = x_0$). Our problem is then: how to find such an operator φ ?

It should be mentioned that the above heuristic argument is not quite sufficient. One should check that the matrix $Y_\pm(x)$ constructed through (1.3.6) satisfy stronger requirement (1.1.6), namely the local behavior (ii) including $x = \infty$ and its invertibility (i). For this we need a closer look at the structure of the operator φ.

Before closing this paragraph we give one useful observation. If φ, φ' satisfy the commutation relation of the type (1.3.4) corresponding to matrices $M(x)$ and $M'(x)$ respectively, then the product $\varphi\varphi'$ has the same property with the corresponding matrix $M(x)M'(x)$. In fact the totality of such φ's form a multiplicative group G (the Clifford group), and the correspondence $\varphi \longmapsto M(x)$ gives a representation of G. In particular it is sufficient to construct φ for $n = 1$, since the general case of (1.3.5) is obtained as a product $\varphi_1\cdots\varphi_n$. But the Riemann's problem with two branch points $x = a,\infty$ admits an elementary solution $(x-a)^L$. As we shall see this enables us to construct explicitly the operator φ for the Riemann's problem.

1.4 Theory of Clifford group

The operator φ introduced in 1.3 has the following remarkable property. If we write as

$$(1.4.1) \qquad \varphi\psi^{(j)}(x) = T(\psi^{(j)}(x))\varphi, \qquad \varphi\psi^{*(j)}(x) = T(\psi^{*(j)}(x))\varphi$$

we see from (1.3.4) that $T(\psi^{(j)}(x))$, $T(\psi^{*(j)}(x))$ are again linear combinations of free fields. To get insight into the algebraic structure of the problem, let us study this situation in the case of finite degrees of freedom — the theory of Clifford group.

Let W be a complex vector space of even dimensions N, and let $\langle\ ,\ \rangle$ be a non-degenerate symmetric bilinear form on W. Let $A(W)$ be the Clifford algebra, an algebra generated by W with the defining relation

$$(1.4.2) \qquad ww'+w'w = \langle w,w'\rangle \in \mathbb{C} \quad \text{for } w, w' \in W.$$

The Clifford group $G(W)$ is by definition the set of invertible elements $g \in A(W)$ such that

$$(1.4.3) \qquad T_g(w) = gwg^{-1} \in W \quad \text{for any } w \in W.$$

Clearly the linear transformation T_g belongs to the orthogonal group $O(W) = \{T\in GL(W)\,|\,\langle T(w),T(w')\rangle = \langle w,w'\rangle \text{ for } w, w' \in W\}$, and we have the exact sequence of group homomorphisms

$$(1.4.4)\qquad 1 \longrightarrow GL(1) \longrightarrow G(W) \longrightarrow O(W) \longrightarrow 1 . \qquad g \longmapsto T_g$$

In other words g is uniquely determined from T_g up to a multiplicative constant factor. To make explicit this correspondence we need the notion of normal ordering.

Let $W = V^* \oplus V$ be a decomposition of W into a direct sum of ordered pair of subspaces (V^*,V) such that

$$(1.4.5)\qquad \begin{aligned} &\langle v_1^*, v_2^* \rangle = 0 \qquad \text{for any} \quad v_1^*, v_2^* \in V^* \\ &\langle v_1, v_2 \rangle = 0 \qquad \text{for any} \quad v_1, v_2 \in V. \end{aligned}$$

Thus V^* and V generate in $A(W)$ the Grassmann algebras $\Lambda(V^*)$ and $\Lambda(V)$, respectively. There exists a unique isomorphism of left $\Lambda(V^*)$- and right $\Lambda(V)$- module, called the normal ordering

$$(1.4.6)\qquad \Lambda(W) = \Lambda(V^*)\wedge\Lambda(V) \xrightarrow{\sim} A(W) = \Lambda(V^*)\cdot\Lambda(V) , \qquad \lambda \longmapsto :\lambda:$$

such that $:1: = 1$. For $a \in A(W)$, the term of degree 0 of the corresponding element in $\Lambda(W)$, with respect to the grading $\Lambda(W) = \mathbb{C} \oplus W \oplus \Lambda^2(W) \oplus \cdots$, is denoted by $\langle a\rangle \in \mathbb{C}$ and is called the (vacuum) expectation value of a.

Example. For $w_1, w_2, \cdots \in W$ we have $w_1 = :w_1:$, $w_1w_2 = \langle w_1w_2\rangle + :w_1w_2:$, $w_1w_2w_3 = \langle w_1w_2\rangle w_3 - \langle w_1w_3\rangle w_2 + \langle w_2w_3\rangle w_1 + :w_1w_2w_3:$. In general (Wick's theorem)

$$(1.4.7)\qquad w_1\cdots w_k = \sum \operatorname{sgn}\begin{pmatrix} 1 & 2 & \cdots & & \cdots & k \\ \mu_1 & \cdots & \mu_r & \nu_1 & \cdots & \nu_s \end{pmatrix} \langle w_{\mu_1} w_{\mu_2}\rangle \cdots \langle w_{\mu_{r-1}} w_{\mu_r}\rangle \times :w_{\nu_1} \cdots w_{\nu_s}:$$

(summed over partitions $\{1,\cdots,k\} = \{\mu_1,\cdots,\mu_r\}\cup\{\nu_1,\cdots,\nu_s\}$, $\mu_1 < \cdots < \mu_r$, $\nu_1 < \cdots < \nu_s$, $r = k-s$ even).

Each decomposition $W = V^* \oplus V$ allows us to realize $A(W)$ as an "operator algebra" of free fermion. For this purpose we introduce two vector spaces, on which $A(W)$ acts from the left or from the right:

$$(1.4.8)\qquad \begin{aligned} \mathcal{F} &= A(W) \bmod A(W)V \\ \mathcal{F}^* &= A(W) \bmod V^*A(W). \end{aligned}$$

If we denote by $|\mathrm{vac}\rangle$ (resp. $\langle \mathrm{vac}|$) the residue class of $1 \in A(W)$ in $\mathcal{F}$ (resp. $\mathcal{F}^*$), we have

$$(1.4.9)\qquad \begin{aligned} \mathcal{F} &= A(W)|vac\rangle = \Lambda(V^*)|vac\rangle \\ \mathcal{F}^* &= \langle vac|A(W) = \langle vac|\Lambda(V) \\ A(W) &\cong End_{\mathbb{C}}(\mathcal{F}) \cong End_{\mathbb{C}}(\mathcal{F}^*). \end{aligned}$$

These vector spaces are dual to each other through the (well defined) map $\mathcal{F}^*\times\mathcal{F}\longrightarrow\mathbb{C}$, $(\langle vac|a_1,\ a_2|vac\rangle)\longmapsto\langle a_1a_2\rangle$. Furthermore, there exists a unique "number operator" $\mathbf{N}\in A(W)$ with the properties $[\mathbf{N},v]=-v\ (v\in V)$, $[\mathbf{N},v^*]=v^*\ (v^*\in V^*)$, $\langle\mathbf{N}\rangle=0$. $\mathcal{F}$ (resp. $\mathcal{F}^*$) is decomposed into a direct sum of eigenspaces $\mathcal{F}_k$ (resp. $\mathcal{F}_k^*$) corresponding to the eigenvalues $k=0,1,\cdots,N/2$. By choosing a basis $v_j^*\in V^*$ (resp. $v_j\in V$) these "k-particle sectors" $\mathcal{F}_k$ (resp. $\mathcal{F}_k^*$) are spanned by the "state vectors" of the form $\sum_{j_1<\cdots<j_k} c_{j_1\cdots j_k}v_{j_1}^*\cdots v_{j_k}^*|vac\rangle$ (resp. $\langle vac|\sum_{j_1<\cdots<j_k} c'_{j_1\cdots j_k}v_{j_1}\cdots v_{j_k}$), and $V^*\mathcal{F}_k=\mathcal{F}_{k+1}$, $V\mathcal{F}_k=\mathcal{F}_{k-1}$.(*) In this sense elements of V^* (resp. V) are called the creation (resp. annihilation) operators of free fermion.(**) What we get here is a miniature of quantum field theory.

Now we proceed to a formula which recovers $g\in G(W)$ from $T_g\in O(W)$. Choose any basis $w_1,\cdots,w_N$ of W, and set

$$(1.4.10)\qquad J=(\langle w_j,w_k\rangle)_{j,k=1,\cdots,N}=K+{}^tK,\quad K=(\langle w_jw_k\rangle)_{j,k=1,\cdots,N}{}^{(***)}$$

$$E_+=J^{-1}K,\quad E_-=J^{-1}\,{}^tK.$$

Formula 1. If $E_++E_-T_g$ is invertible, we have

$$(1.4.11)\qquad g=\langle g\rangle:e^{\rho/2}:,\quad \rho=\sum_{j,k=1}^{N}R_{jk}w_jw_k\in\Lambda^2(W)$$

where

$$(1.4.12)\qquad \langle g\rangle^2=nr(g)\det(E_++E_-T_g)$$

$$(1.4.13)\qquad R=(R_{jk})_{j,k=1,\cdots,N}=(T_g-1)(E_++E_-T_g)^{-1}J^{-1}.$$

Here we have set $nr(g)=gg^*=g^*g\in\mathbb{C}$ for $g\in G(W)$, with $*$ denoting the unique antiautomorphism of $A(W)$ such that $w^*=-w\ (w\in W)$. Note that $e^{\rho/2}=1+\frac{1}{1!}\frac{\rho}{2}+\frac{1}{2!}(\frac{\rho}{2})^2+\cdots$ is always a finite series in $\Lambda(W)$.

(*) $\mathcal{F}_{-1}=0$, $\mathcal{F}_{\frac{N}{2}+1}=0$.

(**) The term "fermion" refers to the anticommutation relation (1.4.2).

(***) The notation is a little bit confusing. $\langle w_j,w_k\rangle$ stands for the inner product (1.4.2), and $\langle w_jw_k\rangle$ for the expectation value of a product w_jw_k.

More generally the closure of $G(W)$ in $A(W)$ is characterized as

$$\overline{G(W)} = \{c{:}w^{(1)}\cdots w^{(\ell)}e^{\rho/2}{:} \mid c\in\mathbf{C},\ w^{(1)},\cdots,w^{(\ell)}\in W,\ \rho\in\Lambda^2(W)\}. \tag{1.4.14}$$

We briefly indicate here how (1.4.13) is derived. Let $w = \sum_{j=1}^{N} c_j w_j$ be an arbitrary element of W. Setting $\tilde{g} = {:}e^{\rho/2}{:}$ we calculate the normal product representation of $w\tilde{g}$ and $\tilde{g}w$ by using the rule $w{:}w^{(1)}\cdots w^{(\ell)}{:} = {:}ww^{(1)}\cdots w^{(\ell)}{:} + \sum_{k=1}^{\ell}(-)^{k-1}\langle ww^{(k)}\rangle \times {:}w^{(1)}\cdots \overset{k}{\wedge}\cdots w^{(\ell)}{:}$. The result is of the form

$$\begin{aligned} w\tilde{g} &= {:}w'e^{\rho/2}{:}, & w' &= \sum_{j=1}^{N} c'_j w_j \\ \tilde{g}w &= {:}w''e^{\rho/2}{:}, & w'' &= \sum_{j=1}^{N} c''_j w_j \end{aligned} \tag{1.4.15}$$

where the coefficients $\vec{c}\,' = {}^t(c'_1,\cdots,c'_N)$, $\vec{c}\,'' = {}^t(c''_1,\cdots,c''_N)$ are determined from R and $\vec{c} = {}^t(c_1,\cdots,c_N)$ as

$$\begin{aligned} \vec{c}\,' &= (1-R\,{}^tK)\vec{c} \\ \vec{c}\,'' &= (1+RK)\vec{c}\ . \end{aligned} \tag{1.4.16}$$

To attain $g = \text{const}\cdot\tilde{g}$ it is sufficient to choose R so that $T_g = T_{\tilde{g}}$. This amounts to setting $\vec{c}\,'' = T_g\vec{c}\,'$, i.e. $1+RK = (1-R\,{}^tK)T_g$. Solving this for R we get (1.4.13).

Products of elements of the form (1.4.11) are calculated as follows. For convenience we employ the vector notation $\vec{w} = (w_1,\cdots,w_N)$.

Formula 2. Let $g_\nu = \langle g_\nu\rangle{:}e^{\rho_\nu/2}{:}$, $\rho_\nu = \vec{w}R_\nu{}^t\vec{w}$ $(\nu=1,\cdots,n)$, and set

$$R = \begin{pmatrix} R_1 & & \\ & \ddots & \\ & & R_n \end{pmatrix}, \quad A = \begin{pmatrix} 0 & K & \cdots & K \\ -{}^tK & 0 & \ddots & \vdots \\ \vdots & \ddots & \ddots & K \\ -{}^tK & \cdots & -{}^tK & 0 \end{pmatrix} \tag{1.4.17}$$

Assuming that $1-RA = {}^t(1-AR)$ is invertible, we have

$$g_1\cdots g_n = \langle g_1\cdots g_n\rangle{:}\exp\frac{1}{2}(\vec{w}R_{12\cdots n}{}^t\vec{w}){:} \tag{1.4.18}$$

with

$$\begin{aligned} \langle g_1\cdots g_n\rangle &= \langle g_1\rangle\cdots\langle g_n\rangle\det(1-RA)^{1/2} \\ R_{12\cdots n} &= \sum_{\mu,\nu=1}^{n}[(1-RA)^{-1}R]_{\mu\nu} \end{aligned} \tag{1.4.19}$$

where $[\ \ ]_{\mu\nu}$ denotes the (μ,ν)-th block according to the partition in (1.4.17).

For physicists it is perhaps clear that formulas (1.4.11) ∿ (1.4.19) are no more than a Wick's theorem applied to the Clifford group.

1.5 Construction of operators

We now go back to the continuum and construct an operator φ in 1.3.

To go from discrete to continuum we first proceed most naïvely; replace sums by integrals and "apply" the theory in 1.4. As a vector space W of free fermions we choose the space of 2m tuple of functions $w = {}^t(w_1^*(x),\cdots,w_m^*(x),\ w_1(x),\cdots,w_m(x))$ equipped with the inner product

$$\langle w,w'\rangle = \sum_{j=1}^{m}\int_{-\infty}^{+\infty} dx(w_j^*(x)w_j'(x) + w_j(x)w_j^{*\prime}(x)). \tag{1.5.1}$$

By the identification $w = \sum_{j=1}^{m}\int_{-\infty}^{+\infty} dx(w_j^*(x)\psi^{*(j)}(x) + w_j(x)\psi^{(j)}(x))$ (1.5.1) is equivalent to (1.3.2). We set

$$\hat{\psi}^{*(j)}(p) = \int_{-\infty}^{+\infty} dx\, e^{ixp}\psi^{*(j)}(x), \qquad \hat{\psi}^{(j)}(p) = \int_{-\infty}^{+\infty} dx\, e^{-ixp}\psi^{(j)}(x) \tag{1.5.2}$$

where $\hat{\psi}^{*(j)}(-p)$, $\hat{\psi}^{(j)}(p)$ are creation or annihilation operators according as $p<0$ or $p>0$, respectively, satisfying

$$[\hat{\psi}^{*(j)}(p),\hat{\psi}^{*(j')}(p')]_+ = 0, \qquad [\hat{\psi}^{(j)}(p),\hat{\psi}^{(j')}(p')]_+ = 0 \tag{1.5.3}$$

$$[\hat{\psi}^{*(j)}(p),\hat{\psi}^{(j')}(p')]_+ = \delta_{jj'}\delta(p-p')$$

$$\hat{\psi}^{*(j)}(-p)|vac\rangle = 0, \qquad \hat{\psi}^{(j)}(p)|vac\rangle = 0 \qquad (p>0) \tag{1.5.4}$$

$$\langle vac|\hat{\psi}^{*(j)}(-p) = 0, \qquad \langle vac|\hat{\psi}^{(j)}(p) = 0 \qquad (p<0).$$

In the basis $\psi^{*(j)}(x)$, $\psi^{(j)}(x)$ the matrices (1.4.10) are now integral operators with the matrix kernels

$$E_\pm(x,x') = \frac{1}{2\pi}\frac{\pm i}{x-x'\pm i0}\cdot I_{2m} \quad . \tag{1.5.5}$$

To put it differently we have chosen a decomposition $W = V^*\oplus V$ such that V^* (resp. V) consists of 2m tuples, whose components $w_i^*(x)$, $w_i(x)$ are boundary values of holomorphic functions in the upper (resp. lower) half plane $\mathrm{Im}\,x>0$ (resp. $\mathrm{Im}\,x<0$). We are to find an element φ of the "Clifford group" $G(W)$ corresponding to the "orthogonal transformation"

$$T_\varphi :\ (\vec{\psi}^*(x),\vec{\psi}(x)) \longmapsto (\vec{\psi}^*(x),\vec{\psi}(x))\begin{pmatrix} {}^tM(x)^{-1} & 0 \\ 0 & M(x)\end{pmatrix} \tag{1.5.6}$$

$$\vec{\psi}^*(x) = (\psi^{*(1)}(x), \cdots, \psi^{*(m)}(x))$$

$$\vec{\psi}(x) = (\psi^{(1)}(x), \cdots, \psi^{(m)}(x)).$$

Lemma. Suppose there exist invertible matrices $Y_\pm$ satisfying

$$(1.5.7). \qquad E_-Y_+^{\pm 1}E_+ = 0, \quad E_+Y_-^{\pm 1}E_- = 0, \quad Y_- = Y_+T.$$

Then R in (1.4.13) is given by

$$(1.5.8) \qquad R = (Y_+^{-1}-Y_-^{-1})(E_+Y_-+E_-Y_+)J^{-1}.$$

In our context we choose $Y_\pm$ to be multiplication by matrix functions $\begin{pmatrix} {}^tY_\pm(x)^{-1} & 0 \\ 0 & Y_\pm(x) \end{pmatrix}$. Conditions (1.5.7) then state that $Y_\pm(x)$, $Y_\pm(x)^{-1}$ should be boundary values of holomorphic functions on $\mathrm{Im}\, x \gtrless 0$, and that $Y_-(x) = Y_+(x)M(x)$ $(x \in \mathbb{R})$. We have thus shown a converse to a statement in 1.3; once the Riemann-Hilbert problem is solved, an operator φ satisfying (1.3.4) is found explicitly as

$$(1.5.9) \qquad \varphi = \langle\varphi\rangle :\exp\left(\int_{-\infty}^{+\infty}\int_{-\infty}^{+\infty} dxdx'\, \vec{\psi}(x)R(x,x')\,{}^t\vec{\psi}^*(x')\right):$$

$$R(x,x') = (Y_+(x)^{-1}-Y_-(x)^{-1})\frac{1}{2\pi}\left(\frac{i}{x-x'+i0}\, Y_-(x') + \frac{-i}{x-x'-i0}\, Y_+(x')\right).$$

As remarked in 1.3, the case of only two branch points a, ∞ admits of an elementary solution $Y_\pm(x) = (x-a\pm i0)^L$. Correspondingly we have the following formula for $\varphi = \varphi(a;L)$ (with the normalization $\langle\varphi\rangle = 1$)

$$(1.5.10) \qquad \varphi(a;L) = :\exp\left(\int_{-\infty}^{+\infty}\int_{-\infty}^{+\infty} dxdx'\, \vec{\psi}(x)R(x-a,x'-a;L)\,{}^t\vec{\psi}^*(x')\right):$$

$$R(x,x';L) = -2i\sin\pi L\cdot x_-^{-L}x_-'^{L}\, \frac{1}{2\pi}\left(\frac{i}{x-x'+i0}\, e^{-\pi i L} + \frac{-i}{x-x'-i0}\, e^{\pi i L}\right).^{(*)}$$

By applying the product formula (1.4.18) we may derive an infinite series expression for Y in the general case. Using (1.4.15)-(1.4.16), $\langle w^{(1)}:w'e^{\rho/2}:\rangle = \langle w^{(1)}w'\rangle$ and the Neumann series expansion $(1-RA)^{-1}R = R+RAR+\cdots$ we find

$$(1.5.11) \qquad Y(x_0,x) = 1 + 2\pi i(x-x_0)\sum_{\mu,\nu=1}^{n} Z_{\mu\nu}(x_0,x),$$

where

(*) $x_-^L = 0 \quad (x > 0), \; = |x|^L \quad (x < 0).$

$$(1.5.12)\qquad Z_{\mu\nu}(x_0,x) = \delta_{\mu\nu}\int_{-\infty}^{0}\int_{-\infty}^{0} dx_1 dx_2 \frac{1}{2\pi}\frac{i}{x_0-x_1-a_\mu} R(x_1,x_2;L_\nu)\frac{1}{2\pi}\frac{i}{x_2+a_\nu-x}$$
$$+ \sum_{\ell=1}^{\infty}\sum_{\nu_1,\cdots,\nu_{\ell-1}=1}^{n}\int_{-\infty}^{0}\cdots\int_{-\infty}^{0} dx_1\cdots dx_{2\ell+2}\frac{1}{2\pi}\frac{i}{x_0-x_1-a_\mu} R(x_1,x_2;L_\mu)A_{\mu\nu_1}(x_2,x_3)\cdots$$
$$R(x_{2\ell-1},x_{2\ell};L_{\nu_{\ell-1}})A_{\nu_{\ell-1}\nu}(x_{2\ell},x_{2\ell+1})R(x_{2\ell+1},x_{2\ell+2};L_\nu)\frac{1}{2\pi}\frac{i}{x_{2\ell+2}+a_\nu-x}\ .$$

We have set $A_{\mu\nu}(x,x') = \frac{1}{2\pi}(1-\delta_{\mu\nu})\frac{i}{x+a_\mu-x'-a_\nu-i\varepsilon_{\mu\nu}0}$, $\varepsilon_{\mu\nu} = \mathrm{sgn}(\mu-\nu)$.

At this stage our formal argument is justified with mathematical rigor. Making use of the lemma below one can show that (1.5.12) is convergent and analytic for complex $(x_0,x) \in (\mathbb{C}-\Gamma_\mu)\times(\mathbb{C}-\Gamma_\nu)$ (*), provided that L_ν s are close to 0 (Fig. 1.3).

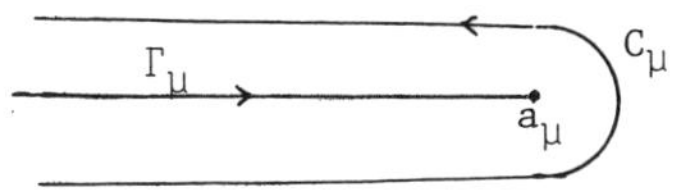

Fig. 1.3

Lemma. If all the eigenvalues of L lie in the strip $|\mathrm{Re}\lambda| < 1/2$, the integral operator

$$f(x) \longmapsto \int_{-\infty}^{0}\frac{dx'}{2\pi}|x|^{L}\frac{i}{x-x'+i\varepsilon}|x'|^{-L}f(x') \qquad (\varepsilon \geqq 0)$$

is bounded in $L^2(-\infty,0;dx)^m$.

Local behavior of Y at $x = a_\nu$ is verified by rewriting (1.5.11) as

$$(1.5.13)\qquad Y(x_0,x) = (x_0-a_\mu)^{-L_\mu}\hat{Y}_{\mu\nu}(x_0,x)(x-a_\nu)^{L_\nu}$$

$$(1.5.14)\qquad \hat{Y}_{\mu\nu}(x_0,x) = \delta_{\mu\nu}-2\pi i(x-x_0)\int_{C_\mu}dx_1\int_{C_\nu}dx_2\frac{1}{2\pi}\frac{i}{x_0-x_1}(x_1-a_\mu)^{L_\mu}\Big(\frac{1}{2\pi}\frac{i}{x_1-x_2}(1-\delta_{\mu\nu})$$
$$+\sum_{\rho(\neq\mu)}\sum_{\sigma(\neq\nu)}Z_{\rho\sigma}(x_1,x_2)\Big)(x_2-a_\nu)^{-L_\nu}\frac{1}{2\pi}\frac{i}{x_2-x}\ .$$

Here x_0 and x are supposed to lie inside the contours C_μ and C_ν, respectively, of Fig. 1.3. Local behavior at $x = \infty$ is checked through a little argument using the estimate $|Z_{\mu\nu}(x_0,x)| = O(\frac{1}{\sqrt{|x|}})$ $(|x| \to \infty)$, which follows from (1.5.12).

In this way, for small L_ν, quantum field theory provides a new method of solving the Riemann's problem.

(*) To be precise we argue first for the case $\mathrm{Im}\,a_1 > \cdots > \mathrm{Im}\,a_n$. The original case is attained as a limiting case (L^2-convergence).

1.6 τ function

Let us turn to the vacuum expectation value $\tau = \langle \varphi(a_1;L_1)\cdots\varphi(a_n;L_n)\rangle$ itself. It is shown that its logarithmic derivative is expressible in terms of a solution to the Schlesinger's equations (1.2.2) as follows:

$$(1.6.1)\qquad d\log\tau = \frac{1}{2}\sum_{\mu\neq\nu}\operatorname{trace} A_\mu A_\nu \frac{da_\mu - da_\nu}{a_\mu - a_\nu}\ .$$

The right hand side represents a closed 1-form for any solution of (1.2.2). This fact will be generalized in §2 to the case admitting irregular singularities of arbitrary rank.

We give below several arguments to derive (1.6.1).

(1) The formula (1.4.19) enables us to express $\log\tau$ as a Neumann series $\frac{1}{2}\log\det(1-RA) = \frac{-1}{2}\sum_{\ell=1}^{\infty}\operatorname{trace}\frac{1}{\ell}(RA)^\ell$. Unfortunately the kernel $(RA)^\ell(x,x')$ is singular on the diagonal $x=x'$, and its trace is meaningless by itself. Nevertheless its derivative does make sense as a convergent series:

$$(1.6.2)\qquad d\log\tau = -\frac{2i}{\pi}\sum_{\ell=2}^{\infty}\sum_{\nu_1,\cdots,\nu_\ell=1}^{n}\int_{-\infty}^{0}\cdots\int_{-\infty}^{0}dx_1\cdots dx_{2\ell}$$
$$\times\operatorname{trace}(L_{\nu_1}\sin^2\pi L_{\nu_1}da_{\nu_1}|x_1|^{L_{\nu_1}-1}A_{\nu_1\nu_2}(x_1,x_2)R(x_2,x_3;L_{\nu_2})\cdots$$
$$\cdots R(x_{2\ell-2},x_{2\ell-1};L_{\nu_\ell})A_{\nu_\ell\nu_1}(x_{2\ell-1},x_{2\ell})|x_{2\ell}|^{-L_{\nu_1}-1}).$$

Comparing this with

$$(1.6.3)\qquad \frac{\partial}{\partial x}\hat{Y}_{\nu\nu}(a_\nu,x)\big|_{x=a_\nu}$$
$$= -\frac{2i}{\pi}\int_{-\infty}^{0}\int_{-\infty}^{0}dx_1dx_2\sin^2\pi L_\nu|x_1|^{L_\nu-1}\sum_{\rho(\neq\nu)}\sum_{\sigma(\neq\nu)}Z_{\rho\sigma}(x_1+a_\nu,x_2+a_\nu)|x_2|^{-L_\nu-1}$$

we find that

$$(1.6.4)\qquad d\log\tau = \sum_{\nu=1}^{n}\operatorname{trace}(L_\nu Y_1^{(\nu)}da_\nu),\qquad Y_1^{(\nu)} = \frac{\partial}{\partial x}\hat{Y}_{\nu\nu}(a_\nu,x)\big|_{x=a_\nu}\ .$$

On the other hand,

$$(1.6.5)\qquad Y_\nu(x) = \lim_{x_0\to a_\nu}(x_0-a_\nu)^{L_\nu}Y(x_0,x)$$
$$= \hat{Y}_{\nu\nu}(a_\nu,x)(x-a_\nu)^{L_\nu}$$

is the solution of Riemann's problem with the normalization (1.1.6)-(0)'. If we set

$$(1.6.6)\qquad \frac{dY_\nu}{dx}Y_\nu^{-1} = \sum_{\mu=1}^{n}\frac{A_\mu^{(\nu)}}{x-a_\mu}$$

we have

$$(1.6.7)\qquad A_\nu^{(\nu)} = L_\nu,\quad Y_1^{(\nu)} + [Y_1^{(\nu)}, L_\nu] = \sum_{\mu(\neq\nu)}\frac{A_\mu^{(\nu)}}{a_\nu-a_\mu}$$

so that

$$(1.6.8)\qquad \mathrm{trace}(L_\nu Y_1^{(\nu)}da_\nu) = \sum_{\mu(\neq\nu)}\mathrm{trace}(A_\nu^{(\nu)}A_\mu^{(\nu)})\frac{da_\nu}{a_\nu-a_\mu}.$$

Since A_ν's with different normalizations are similar to each other (see (1.1.8)), trace $A_\nu^{(\nu)}A_\mu^{(\nu)}$ = trace $A_\nu A_\mu$ is independent of x_0, and we have (1.6.1).

(2) The above procedure is carried out at the level of field operators. Differentiation of (1.5.10) yields

$$(1.6.9)\qquad \frac{d}{da}\varphi(a;L) = :\frac{d}{da}\left(\frac{\rho(a;L)}{2}\right)e^{\rho(a;L)/2}:$$

$$= 2\pi i:\vec{\psi}_{-L}(a)L\,{}^t\vec{\psi}^*_{{}^tL}(a)e^{\rho(a;L)/2}:.$$

Here $\varphi(a;L) = :e^{\rho(a;L)/2}:$, and the row vectors $\vec{\psi}_{-L}(a) = (\psi_{-L}^{(1)}(a),\cdots,\psi_{-L}^{(m)}(a))$, $\vec{\psi}^*_{{}^tL}(a) = (\psi^{*(1)}_{{}^tL}(a),\cdots,\psi^{*(m)}_{{}^tL}(a))$ are given by

$$(1.6.10)\qquad \vec{\psi}_{-L}(a) = \frac{-1}{\pi}\int_{-\infty}^{a}dx\vec{\psi}(x)\sin\pi L\cdot|x-a|^{-L-1}$$

$$\vec{\psi}^*_{{}^tL}(a) = \frac{1}{\pi}\int_{-\infty}^{a}dx\vec{\psi}^*(x)\sin\pi{}^tL\cdot|x-a|^{{}^tL-1}.$$

Moreover we have the formal operator expansion

$$(1.6.11)\qquad \vec{\psi}(x)\varphi(a;L) = (\vec{\varphi}_{-L}(a;L) + \vec{\varphi}_{-L-1}(a;L)\cdot(x-a) + \cdots)(x-a)^L$$

$$\vec{\varphi}_{-L-k}(a;L) = :\vec{\psi}_{-L-k}(a)e^{\rho(a;L)/2}:\quad (k=0,1,2,\cdots).$$

This implies

$$(1.6.12)\qquad \hat{Y}_{\nu\nu}(a_\nu,x) = 1 + Y_1^{(\nu)}(x-a_\nu) + \cdots$$

$$(Y_1^{(\nu)})_{jk} = 2\pi i\tau^{-1}\langle\varphi(a_1;L_1)\cdots:\psi^{*(j)}_{{}^tL_\nu}(a_\nu)\psi^{(k)}_{-L_\nu}(a_\nu)e^{\rho(a_\nu;L_\nu)/2}:\cdots\varphi(a_n;L_n)\rangle.$$

Hence (1.6.4) follows from (1.6.9) and (1.6.12).

(3) It is also possible to start from the formula (1.4.12) in 1.4. Since

$nr(gg') = nr(g)nr(g')$ we have

$$\frac{\langle g_1 \cdots g_n \rangle^2}{\langle g_1 \rangle^2 \cdots \langle g_n \rangle^2} = \frac{\det(E_+ + E_- T)}{\prod_{\nu=1}^{n} \det(E_+ + E_- T_\nu)}, \quad T_\nu = T_{g_\nu}, \quad T = T_1 \cdots T_n. \tag{1.6.13}$$

The effect of a variation δT_ν is calculated as

$$\delta \log \frac{\langle g_1 \cdots g_n \rangle}{\langle g_1 \rangle \cdots \langle g_n \rangle} = \frac{1}{2} \operatorname{trace}((E_+ + E_- T)^{-1} E_- \delta T - \sum_{\nu=1}^{n} (E_+ + E_- T_\nu)^{-1} E_- \delta T_\nu). \tag{1.6.14}$$

Choosing decompositions (1.5.7) of $T = Y_+^{-1} Y_-$, $T_\nu = Y_{\nu+}^{-1} Y_{\nu-}$ we get $(E_+ + E_- T)^{-1} E_- = Y_-^{-1}(E_+ Y_- + E_- Y_+) E_- = T^{-1} Y_+^{-1} E_- Y_+$ and likewise for T_ν. In our case, T, T_ν, $Y_\pm$ and $Y_{\nu\pm}$ are multiplication operators by $M(x)$, $M_\nu(x)$, $Y(x \pm i0)$ and $(x-a_\nu \pm i0)^{L_\nu}$, respectively.(*) Assuming $a_1 < \cdots < a_n$ we find from (1.3.5) that

$$dM(x) \cdot M(x)^{-1} = 2\pi i \sum_{\nu=1}^{n} L_\nu \delta(x-a_\nu) da_\nu . \tag{1.6.15}$$

Hence the right hand side of (1.6.14) is interpreted as

$$\begin{aligned} &\int dx\, 2\pi i \sum_{\nu=1}^{n} \delta(x-a_\nu) da_\nu \operatorname{trace}((Y(x)^{-1} \frac{1}{2\pi} \frac{-i}{x-x'-i0} Y(x') \\ &\qquad - (x-a_\nu)^{-L_\nu} \frac{1}{2\pi} \frac{-i}{x-x'-i0} (x'-a_\nu)^{L_\nu}) L_\nu) \Big|_{x=x'} \\ &= \sum_{\nu=1}^{n} da_\nu \operatorname{trace}(Y_1^{(\nu)} L_\nu + [Y_1^{(\nu)}, L_\nu] L_\nu + O(x-a_\nu)) \Big|_{x \mapsto a_\nu} , \end{aligned} \tag{1.6.16}$$

which is (1.6.4).

References

Field theoretical approach to the Riemann-Hilbert problem was initiated by

[1] M. Sato, T. Miwa and M. Jimbo, Holonomic quantum fields. II, Publ. RIMS, Kyoto Univ. 15 (1979), 201-278.

Classical works in this branch including Schlesinger equations are cited in the references of [1].

The theory of Clifford group is expounded in

[2] M. Sato, T. Miwa and M. Jimbo, Holonomic quantum fields. I, ibid. 14 (1978), 223-267.

Main results of [2] are summarized in the Appendix 1 of

[3] M. Jimbo, T. Miwa, Y. Môri and M. Sato, Density matrix of impenetrable bose gas and the fifth Painlevé transcendent, preprint RIMS-303, Kyoto Univ. (1979).

(*) More precisely those of double size $\begin{pmatrix} {}^tM(x)^{-1} & \\ & M(x) \end{pmatrix}$, etc.

§2. Monodromy preserving deformation and τ function

2.1 Singularity data

Let us consider the following system of linear ordinary differential equations on $\mathbb{P}^1$, for an $m\times m$ matrix $Y(x)$

$$\frac{d}{dx}Y(x) = A(x)Y(x). \tag{2.1.1}$$

Here $A(x)$ is a rational $m\times m$ matrix given by the following partial fractions at $n+1$ distinct points $a_1,\cdots,a_n$ and $a_\infty=\infty$.

$$A(x) = A_\infty(x) + A_1(x) + \cdots + A_n(x), \tag{2.1.2}$$

$$A_\nu(x) = \begin{cases} -\sum_{j=1}^{r_\infty} A_{\infty,-j}x^{j-1} & (\nu=\infty) \\ \sum_{j=0}^{r_\nu} A_{\nu,-j}\dfrac{1}{(x-a_\nu)^{j+1}} & (\nu\neq\infty). \end{cases} \tag{2.1.3}$$

If $r_\nu=0$, a_ν is called a regular singularity for the system (2.1.1). If $r_\nu>0$, it is called an irregular singularity of rank r_ν.

We assume that $A_{\nu,-r_\nu}$ $(\nu=1,\cdots,n,\infty)$ is diagonalizable by an invertible matrix $G^{(\nu)}$.

$$A_{\nu,-r_\nu} = G^{(\nu)}T^{(\nu)}_{-r_\nu}G^{(\nu)-1} \tag{2.1.4}$$

$$T^{(\nu)}_{-r} = (\delta_{\alpha\beta}t^{(\nu)}_{-r_\nu,\beta})_{\alpha,\beta=1,\cdots,m} \quad : \text{diagonal.} \tag{2.1.5}$$

We normalize (2.1.1) so that $A_{\infty-r_\infty}$ is diagonal and choose $G^{(\infty)}=1$. We call $G^{(\nu)}$ the gauge transformation from a_ν to ∞. It represents the difference between two diagonalizations of highest poles at a_ν and ∞.

Along with (2.1.1) we consider the following transformed system

$$\frac{d}{dx}Y^{(\nu)}(x) = A^{(\nu)}(x)Y^{(\nu)}(x), \tag{2.1.6}$$

$$A(x) = G^{(\nu)}A^{(\nu)}(x)G^{(\nu)-1}. \tag{2.1.7}$$

We denote by $A^{(\nu)}_j$ $(\nu=1,\cdots,n,\infty;\ j=-r_\nu,-r_\nu+1,\cdots)$ the coefficients of the Laurent expansion of $A^{(\nu)}(x)$ at $x=a_\nu$.

$$(2.1.8)\qquad A^{(\nu)}(x) = \begin{cases} -\sum\limits_{j=-r_\infty}^{\infty} A_j^{(\infty)} x^{-j-1} & (\nu=\infty) \\ \sum\limits_{j=-r_\nu}^{\infty} A_j^{(\nu)} (x-a_\nu)^{j-1} & (\nu\neq\infty). \end{cases}$$

The systems (2.1.1) and (2.1.6) are specified by the following table, which we call <u>the singularity data</u>.

$$(2.1.9)\qquad \begin{aligned} &\infty\ ;\ A_{-r_\infty}^{(\infty)} = T_{-r_\infty}^{(\infty)},\ A_{-r_\infty+1}^{(\infty)},\ \cdots,\ A_0^{(\infty)},\ G^{(\infty)} = 1 \\ &a_1;\ A_{-r_1}^{(1)} = T_{-r_1}^{(1)},\ A_{-r_1+1}^{(1)},\ \cdots,\ A_0^{(1)},\ G^{(1)} \\ &a_n;\ A_{-r_n}^{(n)} = T_{-r_n}^{(n)},\ A_{-r_n+1}^{(n)},\ \cdots,\ A_0^{(n)},\ G^{(n)} \end{aligned}$$

The above construction implies the following constraint.

$$(2.1.10)\qquad A_0^{(\infty)} + G^{(1)}A_0^{(1)}G^{(1)-1} + \cdots + G^{(n)}A_0^{(n)}G^{(n)-1} = 0.$$

Further we assume the following.

$$(2.1.11)\qquad t_{-r_\nu,\alpha}^{(\nu)} \neq t_{-r_\nu,\beta}^{(\nu)} \qquad (\alpha\neq\beta) \qquad \text{if } r_\nu > 0,$$

$$t_{0,\alpha}^{(\nu)} - t_{0,\beta}^{(\nu)} \notin \mathbb{Z} - \{0\} \qquad \text{if } r_\nu = 0.$$

We denote the space of singularity data (2.1.9) with constraints (2.1.10) and (2.1.11) by $\mathcal{M}$.

<u>2.2 Monodromy data</u>

Under the assumption (2.1.11) we can find uniquely <u>a formal solution</u> of (2.1.6) of the following form.

$$(2.2.1)\qquad Y^{(\nu)}(x) = \hat{Y}^{(\nu)}(x)e^{T^{(\nu)}(x)},$$

$$(2.2.2)\qquad \hat{Y}^{(\nu)}(x) = \begin{cases} \sum\limits_{j=0}^{\infty} Y_j^{(\infty)} x^{-j}, & Y_0^{(\infty)} = 1 \qquad (\nu=\infty) \\ \sum\limits_{j=0}^{\infty} Y_j^{(\nu)} (x-a_\nu)^{j}, & Y_0^{(\nu)} = 1 \qquad (\nu\neq\infty), \end{cases}$$

$$(2.2.3)\qquad T^{(\nu)}(x) = \begin{cases} \sum_{j=1}^{r_\infty} T_{-j}^{(\infty)} \frac{x^j}{(-j)} + T_0^{(\infty)} \log(\frac{1}{x}) & (\nu=\infty) \\ \sum_{j=1}^{r_\nu} T_{-j}^{(\nu)} \frac{(x-a_\nu)^{-j}}{(-j)} + T_0^{(\nu)} \log(x-a_\nu) & (\nu\neq\infty). \end{cases}$$

Here $T_{-j}^{(\nu)}$ $(\nu=1,\cdots,n,\infty;\ j=0,\cdots,r_\nu)$ are diagonal matrices. $T_0^{(\nu)}$ is called the exponent matrix of formal monodromy at $x=a_\nu$. The constraint (2.1.10) implies

$$(2.2.4)\qquad \mathrm{trace}(T_0^{(\infty)}+T_0^{(1)}+\cdots+T_0^{(n)}) = 0,$$

which is called the Fuchs' relation.

We note that $T_j^{(\nu)}$ and $Y_j^{(\nu)}$ are rational functions of the singularity data (2.1.9).

In order to describe the global properties of solutions to (2.1.1), we prepare some notations on topology. We set

$$(2.2.5)\qquad \mathcal{R} = \mathbb{P}^1 - \{a_1, \cdots, a_n, \infty\}$$

and denote by $\tilde{\mathcal{R}}$ and $\pi: \tilde{\mathcal{R}} \to \mathcal{R}$ the universal covering of $\mathcal{R}$ and the covering map, respectively. We choose a small neighborhood V_ν of a_ν in $\mathbb{P}^1$ and a set of sectors $\mathcal{S}_1^{(\nu)}, \cdots, \mathcal{S}_{k_\nu+1}^{(\nu)}$ in $\pi^{-1}(V_\nu-\{a_\nu\})$ such that $\mathcal{S}_\ell^{(\nu)} \cap \mathcal{S}_{\ell+1}^{(\nu)} \neq \phi$, $\pi(\mathcal{S}_1^{(\nu)})=\pi(\mathcal{S}_{k_\nu+1}^{(\nu)})$ and $\pi(\bigcup_{\ell=1}^{k_\nu}\mathcal{S}_\ell^{(\nu)})=V_\nu-\{a_\nu\}$. As a special choice, we choose, for small $\delta > 0$, a set of sectors $\mathcal{S}_{\ell,\delta}^{(\nu)}$ $(\ell=1,\cdots,2r_\nu+1)$:

$$(2.2.6)\qquad \pi(\mathcal{S}_{\ell,\delta}^{(\nu)}) = \begin{cases} \{x\in V_\nu \,|\, \frac{\pi(\ell-1)}{r_\infty}-\delta < \arg(\frac{1}{x}) < \frac{\pi\ell}{r_\infty}\} & (\nu=\infty) \\ \{x\in V_\nu \,|\, \frac{\pi(\ell-1)}{r_\nu}-\delta < \arg(x-a_\nu) < \frac{\pi\ell}{r_\nu}\} & (\nu\neq\infty). \end{cases}$$

For $r_\nu=0$ we set $\mathcal{S}_1^{(\nu)}=V_\nu-\{a_\nu\}$.

We choose a point $x_\nu\in\pi(\mathcal{S}_1^{(\nu)})$ and a closed path γ_ν in V_ν, with its endpoint x_ν, which encircles a_ν anticlockwise. We choose a path $\gamma_{\nu\infty}$ in $\mathcal{R}$ starting from x_∞ to x_ν such that $(\gamma_{n\infty}^{-1}\gamma_n\gamma_{n\infty})\cdots(\gamma_{1\infty}^{-1}\gamma_1\gamma_{1\infty})$ is homotopic to γ_∞^{-1}.

If $r_\nu=0$, the infinite series (2.2.2) or equivalently

$$(2.2.7)\qquad Y^{(\nu)}(x) = \begin{cases} \hat{Y}^{(\infty)}(x)x^{-T_0^{(\infty)}} & (\nu=\infty) \\ \hat{Y}^{(\nu)}(x)(x-a_\nu)^{T_0^{(\nu)}} & (\nu\neq\infty) \end{cases}$$

represents a local solution to (2.1.6).

If $r_\nu > 0$, the infinite series (2.2.2) is, in general, divergent. Nevertheless, for a sufficiently small $\delta > 0$, there exists a unique holomorphic and invertible solution $Y_\ell^{(\nu)}(x)$ to (1.1.6) in $\mathcal{S}_{\ell,\delta}^{(\nu)}$ having the asymptotic expansion

$$(2.2.8) \qquad Y_\ell^{(\nu)}(x) \sim Y^{(\nu)}(x) \quad \text{in} \quad \mathcal{S}_{\ell,\delta}^{(\nu)}.$$

Since $Y_\ell^{(\nu)}(x)$ and $Y_{\ell+1}^{(\nu)}(x)$ satisfy the same equation (2.1.6), there exists a constant matrix $S_\ell^{(\nu)}$ such that

$$(2.2.9) \qquad Y_{\ell+1}^{(\nu)}(x) = Y_\ell^{(\nu)}(x)S_\ell^{(\nu)}.$$

$S_\ell^{(\nu)}$ $(\ell=1,\cdots,2r_\nu)$ are called the Stokes multipliers at the irregular singularity a_ν.

We also note that

$$(2.2.10) \qquad Y_{2r_\nu+1}^{(\nu)}(x^+) = Y_1^{(\nu)}(x)e^{2\pi i T_0^{(\nu)}}.$$

Here, for a point $x \in \mathcal{S}_1^{(\nu)}$, we denote by x^+ the point in $\mathcal{S}_{2r_\nu+1}^{(\nu)}$ satisfying $\pi(x^+)=\pi(x)$.

From (2.2.9) and (2.2.10) we see that the analytic continuation $Y_1^{(\nu)}(x^+)$ of $Y_1^{(\nu)}(x)$ is given by

$$(2.2.11) \qquad Y_1^{(\nu)}(x^+) = Y_1^{(\nu)}(x)e^{2\pi i T_0^{(\nu)}} S_{2r_\nu}^{(\nu)-1} \cdots S_1^{(\nu)-1}$$

Now, let us compare the analytic continuation of $Y_1^{(\infty)}(x)$ along $\gamma_{\nu\infty}$ and $Y_1^{(\nu)}(x)$ in $\mathcal{S}_{1,\delta}^{(\nu)}$. Since $Y_1^{(\infty)}(x)$ and $G^{(\nu)}Y_1^{(\nu)}(x)$ satisfy the same equation (2.1.1), there exists a constant matrix $C^{(\nu)}$ such that

$$(2.2.12) \qquad Y_1^{(\infty)}(x) = G^{(\nu)}Y_1^{(\nu)}(x)C^{(\nu)}.$$

We call $C^{(\nu)}$ the connection matrix from a_ν to ∞.

From (2.2.11) and (2.2.12) we have

$$(2.2.13) \qquad Y_1^{(\infty)}(x^+) = Y_1^{(\infty)}(x)M^{(\nu)} \qquad (x \in \mathcal{S}_1^{(\nu)},\ x^+ \in \mathcal{S}_{2r_\nu+1}^{(\nu)},\ \pi(x)=\pi(x^+)),$$

$$(2.2.14) \qquad M^{(\nu)} = C^{(\nu)-1}e^{2\pi i T_0^{(\nu)}} S_{2r_\nu}^{(\nu)-1} \cdots S_1^{(\nu)-1}C^{(\nu)}.$$

$M^{(\nu)}$ is called the monodromy matrix of $Y_1^{(\infty)}(x)$ corresponding to the path $\gamma_{\nu\infty}^{-1}\gamma_\nu\gamma_{\nu\infty}$

The homotopy equivalence

$$(2.2.15)\qquad (\gamma_{n\infty}^{-1}\gamma_n\gamma_{n\infty}) \cdots (\gamma_{1\infty}^{-1}\gamma_1\gamma_{1\infty}) \sim \gamma_\infty^{-1}$$

implies

$$(2.2.16)\qquad e^{2\pi i T_0^{(\infty)}} S_{2r_\infty}^{(\infty)-1} \cdots S_1^{(\infty)-1} (C^{(n)-1} e^{2\pi i T_0^{(n)}} S_{2r_n}^{(n)-1} \cdots S_1^{(n)-1} C^{(n)})$$

$$\cdots \times (C^{(1)-1} e^{2\pi i T_0^{(1)}} S_{2r_1}^{(1)-1} \cdots S_1^{(1)-1} C^{(1)}) = 1.$$

Thus for every singularity data we obtained solution matrices $Y_\ell^{(\nu)}$ $(\nu=1,\cdots,n,\infty;$ $\ell=1,\cdots,2r_\nu+1)$ with the following monodromy data:

$$(2.2.17)\qquad \begin{array}{l} \infty\ ;\ T_{-r_\infty}^{(\infty)}, \cdots, T_0^{(\infty)}, S_1^{(\infty)}, \cdots, S_{k_\infty}^{(\infty)}, C^{(\infty)}=1 \\ a_1;\ T_{-r_1}^{(1)}, \cdots, T_0^{(1)}, S_1^{(1)}, \cdots, S_{k_1}^{(1)}, C^{(1)} \\ \vdots \\ a_n:\ T_{-r_n}^{(n)}, \cdots, T_0^{(n)}, S_1^{(n)}, \cdots, S_{k_n}^{(n)}, C^{(n)}. \end{array}$$

In the present case $k_\nu=2r_\nu$.

2.3 From Y(x) to A(x)

Now we consider the problem in a converse way; namely we start from an $m\times m$ matrix $Y(x)$ with the following monodromy properties and show that it solves a system of linear ordinary differential equations of the form (2.1.1) with a rational coefficient $A(x)$.

Monodromy properties.

i) $Y(x)$ is holomorphic in $\tilde{\mathcal{R}}$, and there exist $m\times m$ constant matrices $M^{(\nu)}$ $(\nu=1,\cdots,n,\infty)$ such that

$$(2.3.1)\qquad Y(x^+) = Y(x)M^{(\nu)} \qquad (x\in\mathcal{S}_1^{(\nu)},\ x^+\in\mathcal{S}_{k_\nu+1}^{(\nu)},\ \pi(x)=\pi(x^+)).$$

ii) There exist $m\times m$ constant matrices $C^{(\nu)}$ $(\nu=1,\cdots,n,\infty;\ C^{(\infty)}=1)$ and $S_1^{(\nu)},\cdots,$ $S_{k_\nu}^{(\nu)}$ $(\nu=1,\cdots,n,\infty)$ such that $Y(x)C^{(\nu)-1}, Y(x)C^{(\nu)-1}S_1^{(\nu)}, \cdots, Y(x)C^{(\nu)-1}S_1^{(\nu)}\cdots S_{k_\nu}^{(\nu)}$ have the same asymptotic expansion in $\mathcal{S}_1^{(\nu)},\cdots,\mathcal{S}_{k_\nu+1}^{(\nu)}$, respectively:

$$(2.3.2)\qquad Y(x)C^{(\nu)-1}S_1^{(\nu)}\cdots S_\ell^{(\nu)} \sim G^{(\nu)}\hat{Y}^{(\nu)}(x)e^{T^{(\nu)}(x)},$$

for some $G^{(\nu)}$ $(\nu=1,\cdots,n,\infty;\ G^{(\infty)}=1)$ and for $\hat{Y}^{(\nu)}(x)$ and $T^{(\nu)}(x)$ given by (2.2.2) and (2.2.3), respectively. Here we assume that the Fuchs' relation (2.2.4) holds.

The above monodromy properties imply the following.

iii) $Y(x)$ is invertible in $\tilde{\mathcal{R}}$.

iv) $A(x)=\frac{d}{dx}Y(x)\cdot Y(x)^{-1}$ is rational on $\mathbb{P}^1$.

We also note that

v) $Y(x)$ is uniquely determined by the monodromy data (2.2.17).

We shall give a sketch of proof.

First (2.3.2) implies that

$$\det S_\ell^{(\nu)} = 1. \tag{2.3.3}$$

We set

$$d(x) = \det Y(x)\cdot e^{-\mathrm{trace}\left(\sum_{\nu=1,\cdots,n,\infty} T^{(\nu)}(x)-T_0^{(\infty)}\log(\frac{1}{x})\right)}. \tag{2.3.4}$$

(2.2.14) with $2r_\nu$ replaced by k_ν and (2.3.3) imply that $d(x)$ is single-valued holomorphic in $\mathcal{R}$. Moreover (2.3.2) implies $d(x)$ is also holomorphic at $x=a_\nu$ $(\nu=1,\cdots,n)$. Finally the Fuchs' relation (1.2.4) implies that $d(x)$ is holomorphic at $x=\infty$. Thus $d(x)$ is a non zero constant, hence $Y(x)$ is invertible in $\mathcal{R}$.

The differentiation with respect to x does not change the monodromy property (2.3.1). Hence $\frac{d}{dx}Y(x)\cdot Y(x)^{-1}$ is single-valued. Since the exponential factors in (2.3.2) cancel out, it is rational on $\mathbb{P}^1$.

To prove v), let us assume $Y_1(x)$ and $Y_2(x)$ satisfy monodromy properties i) and ii) with the same monodromy data (2.2.17). Since $M^{(\nu)}$ is determined by the monodromy data, $Y_1(x)Y_2(x)^{-1}$ is single-valued in $\mathcal{R}$. Then (2.3.2) and the normalization $G^{(\infty)}=1$ imply $Y_1(x)Y_2(x)^{-1}=1$.

We note that the partial fractions (2.1.3) for $A(x)$ is given by

$$A^{(\nu)}(x) \equiv \begin{cases} \hat{Y}^{(\infty)}(x)\frac{d}{dx}T^{(\infty)}(x)\hat{Y}^{(\infty)}(x)^{-1} \mod x^{-2} \\ \hat{Y}^{(\nu)}(x)\frac{d}{dx}T^{(\nu)}(x)\hat{Y}^{(\nu)}(x)^{-1} \mod (x-a_\nu)^0. \end{cases} \tag{2.3.5}$$

This fact follows from the identity

$$A^{(\nu)}(x) = \frac{d}{dx}\hat{Y}^{(\nu)}(x)\cdot\hat{Y}^{(\nu)}(x)^{-1} + \hat{Y}^{(\nu)}(x)\frac{d}{dx}T^{(\nu)}(x)\hat{Y}^{(\nu)}(x)^{-1}. \tag{2.3.6}$$

2.4 Deformation equations for $Y(x)$

Now we consider a family $\{Y(x,t)\}_{t\in\mathfrak{T}}$ of $m\times m$ matrices $Y(x,t)$ having monodromy properties in the sense of 2.3. We assume that $Y(x,t)$ is holomorphic in $t=(t_1,\cdots,t_N)$. Thus the monodromy data are holomorphic in t. Moreover we impose the following isomonodromy properties on partial monodromy data.

$$(2.4.1)\qquad \begin{aligned} &dT_0^{(\infty)} = 0,\ dS_1^{(\infty)} = 0,\ \cdots,\ dS_{k_\infty}^{(\infty)} = 0 \\ &dT_0^{(1)} = 0,\ dS_1^{(1)} = 0,\ \cdots,\ dS_{k_1}^{(1)} = 0,\ dC^{(1)} = 0 \\ &\quad\vdots \\ &dT_0^{(n)} = 0,\ dS_1^{(n)} = 0,\ \cdots,\ dS_{k_n}^{(n)} = 0,\ dC^{(n)} = 0 \end{aligned}$$

Namely, we assume that formal monodromies, Stokes multipliers and connection matrices are independent of t. Then what is the consequence on $Y(x,t)$? The answer is given by a system of linear total differential equations for $Y(x,t)$,

$$(2.4.2)\qquad dY(x,t) = \Omega(x,t)Y(x,t).$$

Here $\Omega(x,t)=\sum_{j=1}^{N} R_j(x,t)dt_j$ is a 1 form of t with $m\times m$ coefficient matrices $R_j(x,t)$ $(j=1,\cdots,N)$. (2.4.2) means the following simultaneous system.

$$(2.4.3)\qquad \frac{\partial Y(x,t)}{\partial t_j} = R_j(x,t)Y(x,t).$$

We shall give an explicit form of $\Omega(x,t)$. Since x dependence is essential in the following argument, we abbreviate $Y(x,t)$, $\Omega(x,t)$, etc. to $Y(x)$, $\Omega(x)$ etc.

First (2.4.1) implies that

$$(2.4.4)\qquad dM^{(\nu)} = 0 \qquad (\nu=1,\cdots,n,\infty).$$

Hence $\Omega(x)=dY(x)\cdot Y(x)^{-1}$ is single-valued in $\mathcal{R}$, at $x=a_\nu$ using (2.3.2) and (2.4.1) we have

$$(2.4.5)\qquad \Omega(x) = \begin{cases} d\hat{Y}^{(\infty)}(x)\cdot\hat{Y}^{(\infty)}(x)^{-1} + \hat{Y}^{(\infty)}(x)d'T^{(\infty)}(x)\hat{Y}^{(\infty)}(x)^{-1} & (\nu=\infty) \\ dG^{(\nu)}\cdot G^{(\nu)-1}+G^{(\nu)}(d\hat{Y}^{(\nu)}(x)\cdot\hat{Y}^{(\nu)}(x)^{-1}+\hat{Y}^{(\nu)}(x)d'T^{(\nu)}(x)\hat{Y}^{(\nu)}(x)^{-1})G^{(\nu)-1} & (\nu\neq\infty), \end{cases}$$

where

$$(2.4.6)\qquad d'T^{(\nu)}(x) = \begin{cases} \sum_{j=1}^{r_\infty} dT_{-j}^{(\infty)}\cdot\frac{x^j}{(-j)} & (\nu=\infty) \\ \sum_{j=1}^{r_\nu} dT_{-j}^{(\nu)}\cdot\frac{(x-a_\nu)^{-j}}{(-j)} - \sum_{j=0}^{r_\nu} T_{-j}^{(\nu)}(x-a_\nu)^{-j-1}da_\nu & (\nu\neq\infty). \end{cases}$$

Thus $\Omega(x)$ is rational in x and its partial fractions $\Omega(x)=\Omega_\infty(x)+\Omega_1(x)+\cdots+\Omega_n(x)$ are determined by

$$(2.4.7)\qquad \Omega_\nu(x) \equiv \begin{cases} \hat{Y}^{(\infty)}(x)d'T^{(\infty)}(x)\hat{Y}^{(\infty)}(x)^{-1} \mod x^{-1} & (\nu=\infty) \\ G^{(\nu)}\hat{Y}^{(\nu)}(x)d'T^{(\nu)}(x)\hat{Y}^{(\nu)}(x)^{-1}G^{(\nu)-1} \mod (x-a_\nu)^0 & (\nu\neq\infty) \end{cases}$$

Namely, if we set

$$(2.4.8)\qquad \hat{Y}^{(\nu)}(x)d'T^{(\nu)}(x)\hat{Y}^{(\nu)}(x)^{-1} = \begin{cases} \sum_{j=-r_\infty}^{\infty} \Phi_j^{(\infty)}(\frac{1}{x})^j & (\nu=\infty) \\ \sum_{j=-r_\nu-1}^{\infty} \Phi_j^{(\nu)}(x-a_\nu)^j & (\nu\neq\infty), \end{cases}$$

we have

$$(2.4.9)\qquad \Omega_\nu(x) = \begin{cases} \sum_{j=0}^{r_\infty} \Phi_{-j}^{(\infty)}x^j & (\nu=\infty) \\ \sum_{j=1}^{r_\nu+1} G^{(\nu)}\Phi_{-j}^{(\nu)}G^{(\nu)-1}(x-a_\nu)^{-j} & (\nu\neq\infty). \end{cases}$$

2.5 Deformation equations for $G^{(\nu)}$

If we subtract from $\Omega(x)$ its singular part $\Omega_\nu(x)$ at $x=a_\nu$, and restrict x to a_ν, we obtain a linear system of total differential equation for $G^{(\nu)}$

$$(2.5.1)\qquad dG^{(\nu)} = \Theta^{(\nu)}G^{(\nu)},$$

$$(2.5.2)\qquad \Theta^{(\nu)} = (\Omega(x)-\Omega_\nu(x))\Big|_{x=a_\nu} + G^{(\nu)}(Y_1^{(\nu)}da_\nu-\Phi_0^{(\nu)})G^{(\nu)-1}.$$

Since $Y(x)=G^{(\nu)}Y^{(\nu)}(x)S_\ell^{(\nu)-1}\cdots S_1^{(\nu)-1}C^{(\nu)}$, we obtain the linear system for $Y^{(\nu)}(x)$ from (2.4.1), (2.4.2) and (2.5.1).

$$(2.5.3)\qquad dY^{(\nu)}(x) = \Omega^{(\nu)}(x)Y^{(\nu)}(x),$$

$$(2.5.4)\qquad \Omega^{(\nu)}(x) = G^{(\nu)-1}(\Omega^{(\nu)}(x)-\Theta^{(\nu)})G^{(\nu)}.$$

In the following we shall denote $Y(x)$ and $\Omega(x)$ by $Y^{(\infty)}(x)$ and $\Omega^{(\infty)}(x)$, respectively.

2.6 Deformation parameters

We have derived linear systems for $Y^{(\nu)}(x)$ and $G^{(\nu)}$, assuming that the partial monodromy data are independent of deformation parameters t : (2.4.1). The next question is, what is the universal family of isomonodromy deformation. The most optimistic answer is that the rest of the monodromy data

$$(2.6.1)\qquad \begin{array}{l} a_1, \cdots , a_n, \\ T^{(\infty)}_{-r_\infty}, \cdots , T^{(\infty)}_{-1}, \\ T^{(1)}_{-r_1}, \cdots , T^{(1)}_{-1}, \\ \vdots \qquad\quad \vdots \\ T^{(n)}_{-r_n}, \cdots , T^{(n)}_{-1} \end{array}$$

can be chosen as independent parameters of deformation. In fact this turns out to be true.

Since $d'T^{(\nu)}(x)$ in (2.4.6) is written as a linear combination of da_ν $(\nu=1,\cdots,n)$, $dt^{(\nu)}_{j\alpha}$ $(j=-r_\nu,\cdots,-1,\ \alpha=1,\cdots,m;\ \nu=1,\cdots,n,\infty)$, the linear systems (2.5.1) and (2.5.3) can be regarded as linear systems of partial differential equations with independent variables (2.6.1). For the existence of solutions to these systems, the following integrability conditions should be satisfied:

$$(2.6.2)\qquad d\Theta^{(\nu)} = \Theta^{(\nu)} \wedge \Theta^{(\nu)}$$

$$(2.6.3)\qquad d\Omega^{(\nu)} = \Omega^{(\nu)} \wedge \Omega^{(\nu)}.$$

These are equations of 2-forms. In the next section, we reduce (2.6.2) and (2.6.3) to a completely integrable system of non linear total differential equations. This implies that given an $m\times m$ matrix $Y(x)$ satisfying monodromy data (2.2.7), for any sufficiently small deviation in deformation parameters (2.6.1) there exists a matrix with the same partial monodromy data, i.e. formal monodromies, Stokes multipliers and connection matrices.

2.7 Deformation equations for $A^{(\nu)}(x)$

In 2.3 we have seen that the monodromy properties for $Y^{(\nu)}(x)$ $(\nu=1,\cdots,n,\infty)$ imply linear systems

$$(2.7.1)\qquad \frac{\partial}{\partial x}Y^{(\nu)}(x) = A^{(\nu)}(x)Y^{(\nu)}(x) \qquad\qquad (\nu=1,\cdots,n,\infty).$$

Then in 2.4 we have seen that the isomonodromy properties imply further linear systems

(2.7.2) $\quad dY^{(\nu)}(x) = \Omega^{(\nu)}(x)Y^{(\nu)}(x) \qquad (\nu=1,\cdots,n,\infty).$

Differentiating (2.7.1) and (2.7.2) we obtain

(2.7.3) $\quad d\frac{\partial}{\partial x}Y^{(\nu)}(x) = dA^{(\nu)}(x)\cdot Y^{(\nu)}(x) + A^{(\nu)}(x)\Omega^{(\nu)}(x)Y^{(\nu)}(x)$

(2.7.4) $\quad \frac{\partial}{\partial x}dY^{(\nu)}(x) = \frac{\partial}{\partial x}\Omega^{(\nu)}(x)\cdot Y^{(\nu)}(x) + \Omega^{(\nu)}(x)A^{(\nu)}(x)Y^{(\nu)}(x)$

Subtracting (2.7.4) from (2.4.3) we have $0=(dA^{(\nu)}(x)-\frac{\partial}{\partial x}\Omega^{(\nu)}(x)-[\Omega^{(\nu)}(x),A^{(\nu)}(x)])Y^{(\nu)}(x)$, or equivalently

(2.7.5) $\quad dA^{(\nu)}(x) = \frac{\partial}{\partial x}\Omega^{(\nu)}(x) + [\Omega^{(\nu)}(x),A^{(\nu)}(x)].$

Note that $\Omega^{(\nu)}(x)$ is rational in x and its coefficient matrices are rational functions of singularity data (see Appendix 1 and (2.4.6), (2.4.7)). Thus along with (2.5.1) we have obtained a non linear system of total differential equations for singlarity data (2.1.9).

Comparing formal expansions at $x=a_\nu$ of both sides of (2.7.5) we obtain infinite number of systems. Of course only finite number of them are independent. Namely, if we set

(2.7.6) $\quad dA^{(\nu)}(x) - \frac{\partial}{\partial x}\Omega^{(\nu)}(x) - [\Omega^{(\nu)}(x),A^{(\nu)}(x)]$

$$= \begin{cases} \sum_j \Xi_j^{(\infty)}x^{-j-1} & (\nu=\infty) \\ \sum_j \Xi_j^{(\nu)}(x-a_\nu)^{j+1} & (\nu\neq\infty), \end{cases}$$

then we have

(2.7.7) $\quad \Xi_j^{(\nu)} = 0 \qquad \text{if} \quad j \leq -r .$

Moreover all the entries of $\Xi_j^{(\nu)}$ $(\nu=1,\cdots,n,\infty;\ j=-r_\nu+1,\ -r_\nu+2,\cdots)$ belong to the ideal $\mathcal{J}$ of differential forms on $\mathcal{M}$ generated by

(2.7.8) $\quad (\Xi_j^{(\nu)})_{\alpha\beta} \qquad (\nu=1,\cdots,n,\ ;\ j=-r_\nu+1,\cdots,-1;\ \alpha\neq\beta),$

$(\Xi_0^{(\nu)})_{\alpha\beta} \qquad (\nu=1,\cdots,n),$

$(dG^{(\nu)}-\Theta^{(\nu)}G^{(\nu)})_{\alpha\beta} \qquad (\nu=1,\cdots,n).$

Hence the unknown functions of our deformation equations are

$$(2.7.9)\qquad (A_j^{(\nu)})_{\alpha\beta}\qquad (\nu=1,\cdots,n,\infty;\ j=-r_\nu+1,\cdots,-1;\ \alpha\neq\beta),$$

$$(A_0^{(\nu)})_{\alpha\beta}\qquad (\nu=1,\ldots,n),$$

$$G^{(\nu)}_{\alpha\beta}\qquad (\nu=1,\ldots,n),$$

while the independent variables are deformation parameters (2.6.1).

Finally we can prove that the ideal $\mathcal{I}$ is d-closed:

$$(2.7.10)\qquad d\mathcal{I}\subset\mathcal{I}.$$

Thus by Frobenius' theorem our system

$$(2.7.11)\qquad \mathcal{I}=0$$

of deformation equations is completely integrable.

The complete integrability assures the existence of local solutions for an arbitrary monodromy data. As for global properties of solutions of the non-linear differential equations (2.7.11) it is conjectured that their only singularities are poles except for the following fixed singularities of the system itself:

$$(2.7.12)\qquad a_\mu=a_\nu\qquad \text{for some}\quad \mu\neq\nu,$$

$$(2.7.13)\qquad t^{(\nu)}_{-r_\nu\alpha}=t^{(\nu)}_{-r_\nu\beta}\qquad \text{for some}\ \alpha\neq\beta\ \text{and some}\ \nu\ \text{with}\ r_\nu\geq 1.$$

2.8 τ function

We denote by ω the following 1-form on $\mathcal{M}$:

$$(2.8.1)\qquad \omega=\sum_{\nu=1,\cdots,n}\omega_\nu,$$

$$(2.8.2)\qquad \omega_\nu=-\operatorname*{Res}_{x=a_\nu}\operatorname{trace}\hat{Y}^{(\nu)}(x)^{-1}\frac{\partial}{\partial x}\hat{Y}^{(\nu)}(x)d'T^{(\nu)}(x).$$

In terms of the coefficients $Y_j^{(\nu)}$ in (2.2.2) ω_ν is written as

$$(2.8.3)\qquad \omega_\nu=\begin{cases}\operatorname{trace}\displaystyle\sum_{j=1}^{r_\infty}Z_j^{(\infty)}dT_{-j}^{(\infty)} & (\nu=\infty)\\[2ex] \operatorname{trace}\displaystyle\sum_{j=1}^{r_\nu}Z_j^{(\nu)}dT_{-j}^{(\nu)}+\sum_{j=1}^{r_\nu+1}jZ_j^{(\nu)}T_{-j+1}^{(\nu)}da_\nu & (\nu\neq\infty),\end{cases}$$

$$(2.8.4)\qquad Z_1^{(\nu)} = Y_1^{(\nu)}, \qquad Z_2^{(\nu)} = Y_2^{(\nu)} - \frac{1}{2} Y_1^{(\nu)2},$$

$$Z_3^{(\nu)} = Y_3^{(\nu)} - \frac{2}{3} Y_1^{(\nu)} Y_2^{(\nu)} - \frac{1}{3} Y_2^{(\nu)} Y_1^{(\nu)} + \frac{1}{3} Y_1^{(\nu)3}$$

$$\cdots$$

$$Z_j^{(\nu)} = -\frac{1}{j} \sum_{p=1}^{j} \sum_{\substack{\ell_1+\cdots+\ell_p=j \\ \ell_1,\cdots,\ell_p \geq 1}} \ell_p (-Y_{\ell_1}^{(\nu)}) \cdots (-Y_{\ell_p}^{(\nu)}).$$

We note that the coefficients $Z_j^{(\nu)}$ are polynomials of unknown function (2.7.9) with rational coefficients in the independent variables (2.6.1).

The fundamental property of ω is the closedness:

$$(2.8.5)\qquad d\omega \equiv 0 \qquad \text{mod } \mathcal{J}.$$

Corresponding to the conjecture in 2.7, it is conjectured that on each solution leaf the only singularities of ω other than the fixed singularities (2.7.12), (2.7.13) are simple poles whose residues are positive integers.

The closedness of ω implies the existence of a function τ of the deformation parameters satisfying

$$(2.8.6)\qquad \omega = d \log \tau.$$

We call this function (with a constant multiple undetermined) the <u>τ-function</u>. The above conjecture implies that the τ-function is multi-valued holomorphic except for the fixed singularities.

2.9 Schlesinger transformation and τ quotient

Given an $m\times m$ matrix $Y(x)$ with monodromy data (2.2.17) we can construct another matrix $Y'(x)$ with the same monodromy data except for <u>integer differences</u> in expontnts $t_{0\alpha}^{(\nu)}$ $(\nu=1,\cdots,n,\infty;\ \alpha=1,\cdots,m)$. Since it was L. Schlesinger who considered such a transformation systematically in the case of regular singularities, we call it a <u>Schlesinger transformation</u>.

Choose diagonal matrices $L^{(\nu)} = (\delta_{\alpha\beta}\ell_\beta^{(\nu)})_{\alpha,\beta=1,\cdots,m}$ $(\nu=1,\cdots,n,\infty)$ with integer entries subject to the constraint

$$(2.9.1)\qquad \sum_{\nu=1,\cdots,n,\infty} \sum_{\alpha=1}^{m} \ell_\alpha^{(\nu)} = 0.$$

A Schlesinger transformation is called of type

$$(2.9.2)\qquad \begin{Bmatrix} \infty & a_1 & \cdots & a_n \\ L^{(\infty)} & L^{(1)} & \cdots & L^{(n)} \end{Bmatrix}$$

if it induces the following change of exponents:

$$(2.9.3)\qquad T_o^{(\nu)} \longmapsto T_0^{(\nu)} + L^{(\nu)} \qquad (\nu=1,\cdots,n,\infty).$$

In general, a Schlesinger transformation is achieved by a multiplication by a rational matrix $R(x)$:

$$(2.9.4)\qquad Y'(x) = R(x)Y(x).$$

The condition so that (2.9.4) gives the desired Schlesinger transformation is the following.

$$(2.9.5)\qquad R(x)\hat{Y}^{(\infty)}(x)x^{L^{(\infty)}} = \hat{Y}^{(\infty)\prime}(x),$$

$$(2.9.6)\qquad R(x)G^{(\nu)}\hat{Y}^{(\nu)}(x)(x-a_\nu)^{-L^{(\nu)}} = G^{(\nu)\prime}\hat{Y}^{(\nu)\prime}(x) \qquad (\nu\neq\infty),$$

$$(2.9.7)\qquad \hat{Y}^{(\nu)\prime}(x) = \begin{cases} \sum\limits_{j=0}^{\infty} Y_j^{(\infty)\prime}x^{-j}, & Y_0^{(\infty)\prime} = 1 \qquad (\nu=\infty) \\ \sum\limits_{j=0}^{\infty} Y_j^{(\nu)\prime}(x-a_\nu)^j, & Y_0^{(\nu)\prime} = 1 \qquad (\nu\neq\infty) \end{cases}$$

We define the length $N\begin{Bmatrix} \infty & a_1 & \cdots & a_n \\ L^{(\infty)} & L^{(1)} & \cdots & L^{(n)} \end{Bmatrix}$ of the Schlesinger transformation of type $\begin{Bmatrix} \infty & a_1 & \cdots & a_2 \\ L^{(\infty)} & L^{(1)} & \cdots & L^{(n)} \end{Bmatrix}$ by

$$(2.9.8)\qquad N\begin{Bmatrix} \infty & a_1 & \cdots & a_n \\ L^{(\infty)} & L^{(1)} & \cdots & L^{(n)} \end{Bmatrix} = \sum_{\nu=1,\cdots,n,\infty} \quad \sum_{\ell_\alpha^{(\nu)}>0} \ell_\alpha^{(\nu)}.$$

We shall give two examples of Schlesinger transformations of length 1. We set $E_{\alpha_0} = (\delta_{\alpha\alpha_0}\delta_{\beta\alpha_0})_{\alpha,\beta=1,\cdots,m}$

Type $\begin{Bmatrix} \infty & a_1 & \cdots & a_n \\ -E_{\alpha_0}+E_{\beta_0} & 0 & \cdots & 0 \end{Bmatrix}$:

$$(2.9.9)\qquad R(x) = E_{\alpha_0}x + R_0$$

where $R_{0,\alpha\beta}$ is given by

(2.9.10)

	$\beta=\alpha_0$	$\beta=\beta_0$	$\beta\neq\alpha_0,\beta_0$
$\alpha=\alpha_0$	$\dfrac{-Y^{(\infty)}_{2,\alpha_0\beta_0}+\sum_{\gamma(\neq\alpha_0)} Y^{(\infty)}_{1,\alpha_0\gamma}Y^{(\infty)}_{1,\gamma\beta_0}}{Y^{(\infty)}_{1,\alpha_0\beta_0}}$	$-Y^{(\infty)}_{1,\alpha_0\beta_0}$	$-Y^{(\infty)}_{\alpha_0\beta}$
$\alpha=\beta_0$	$1/Y^{(\infty)}_{1,\alpha_0\beta_0}$	0	0
$\alpha\neq\alpha_0,\beta_0$	$-Y^{(\infty)}_{1,\alpha\beta_0}/Y^{(\infty)}_{1,\alpha_0\beta_0}$	0	$\delta_{\alpha\beta}$.

Type $\begin{Bmatrix} \infty & a_1 & \cdots & a_{\mu_0} & \cdots & a_n \\ -E_{\alpha_0} & 0 & \cdots & E_{\beta_0} & \cdots & 0 \end{Bmatrix}$

(2.9.11) $\quad R(x) = E_{\alpha_0}(x-a_{\mu_0}) + R_0$

where $R_{0,\alpha\beta}$ is given by

(2.9.12)

	$\beta=\alpha_0$	$\beta\neq\alpha_0$
$\alpha=\alpha_0$	$\dfrac{\sum_{\gamma(\neq\alpha_0)} Y^{(\infty)}_{1,\alpha_0\gamma}G^{(\mu_0)}_{\gamma\beta_0}}{G^{(\mu_0)}_{\alpha_0\beta_0}}$	$-Y^{(\infty)}_{1,\alpha_0\beta}$
$\alpha\neq\alpha$	$-G^{(\mu_0)}_{\alpha\beta_0} / G^{(\mu_0)}_{\alpha_0\beta_0}$	$\delta_{\alpha\beta}$.

We denote by $q\begin{Bmatrix} \infty & a_1 & \cdots & a_n \\ L^{(\infty)} & L^{(1)} & \cdots & L^{(n)} \end{Bmatrix}$ the quotient of the τ function corresponding to $Y'(x)$ by the τ function corresponding to $Y(x)$, and we call it the <u>τ quotient</u>. It is remarkable that contrary to the τ function itself the τ quotient can be expressed in terms of $G^{(\nu)}$ and $Y^{(\nu)}_j$.

For example, we have

$$(2.9.13)\qquad q\left\{\begin{matrix} \infty & a_1 \ \dots \ a_n \\ -E_{\alpha_0}+E_{\beta_0} & 0 \ \dots \ 0 \end{matrix}\right\} = Y^{(\infty)}_{1,\alpha_0\beta_0},$$

$$(2.9.14)\qquad q\left\{\begin{matrix} \infty & a_1 & \dots & a_{\mu_0} & \dots & a_n \\ -E_{\alpha_0} & 0 & \dots & E_{\beta_0} & \dots & 0 \end{matrix}\right\} = C^{(\mu_0)}_{\alpha_0\beta_0}$$

In general, $q\left\{\begin{matrix} \infty & a_1 & \dots & a_n \\ L^{(\infty)} & L^{(1)} & \dots & L^{(n)} \end{matrix}\right\}$ is given as the determinant of a matrix of size $N\left\{\begin{matrix} \infty & a_1 & \dots & a_n \\ L^{(\infty)} & L^{(1)} & \dots & L^{(n)} \end{matrix}\right\}$ whose entries are polynomial in $G^{(\nu)}_{\alpha\beta}$, $(G^{(\nu)-1})_{\alpha\beta}$, $Y^{(\nu)}_{j,\alpha\beta}$ with rational coefficients in $a_1,\dots,a_n$.

References

The general framework of monodromy preserving deformation with irregular singularities was given by

[1] K. Ueno, Master's thesis, Kyoto University (1979).

See also

[2] B. Klares, Sur une classe de connexions relatives, C. R. Acad. Sc. Paris, 288 (1979), 205-208.

[3] H. Flaschka and A. C. Newell, Monodromy and spectral preserving deformatoins I, preprint Clarkson college of tech. (1979).

The proof of complete integrability and the general definition of τ function was given by

[4] M. Jimbo, T. Miwa and K. Ueno, Monodromy preserving deformation of linear ordinary differential equations with rational coefficients I, preprint RIMS 315, Kyoto Univ. (1980).

As for Schlesinger transformatoins and τ-quotients, see

[5] M. Jimbo and T. Miwa, Monodromy preserving deformation of linear ordinary differential euqations with rational coefficients II, preprint RIMS 316, Kyoto Univ. (1980).

In this chapter, we have not mentioned special examples such as Schlesinger equation and Painlevé equations. Schlesinger equation is discussed in chapter 1. As for Painlevé equations, see

[6] K. Okamoto, Polynomial Hamiltonians associated to the Painlevé equations, preprint Tokyo Univ. (1980).

§3. Scaling limit of the 2 dimensional Ising model

3.1 Formulation of the problem

The 2 dimensional Ising model with nearest neighbor interaction is a celebrated non-trivial but exactly solvable model in statistical mechanics. Consider a rectangular lattice of size $M\times N$. To each site (j,k) there is attached a random variable σ_{jk} ("spin") taking the values of ± 1. The probability that a specific configuration $\sigma = \{\sigma_{jk}\}_{\substack{j=1,\cdots,M\\ k=1,\cdots,N}}$ takes place is given by $Z_{MN}^{-1}e^{-\beta E(\sigma)}$ (*), where

$$(3.1.1)\qquad E(\sigma) = -E_1\sum_{j=1}^{M}\sum_{k=1}^{N}\sigma_{jk}\sigma_{j+1\,k} - E_2\sum_{j=1}^{M}\sum_{k=1}^{N}\sigma_{jk}\sigma_{j\,k+1}$$

(cyclic boundary condition)

with positive constants E_1, E_2. The normalization constant

$$(3.1.2)\qquad Z_{MN} = \sum_{\sigma} e^{-\beta E(\sigma)}$$

(summed over all $\sigma_{jk} = \pm 1$)

has been calculated for the first time by Onsager. Here we are concerned with the n-point spin correlation functions

$$(3.1.3)\qquad \langle\sigma_{j_1k_1}\cdots\sigma_{j_nk_n}\rangle = Z_{MN}^{-1}\sum_{\sigma}\sigma_{j_1k_1}\cdots\sigma_{j_nk_n}e^{-\beta E(\sigma)}.$$

The Ising model admits of a variety of approaches, as is always the case for a "good" problem. One of the standard ways is to introduce the "transfer matrix" V and the "spin operator" $s_{jk} = V^k s_j V^{-k}$, both acting on a linear space of dimensions 2^M, and to rewrite (3.1.3) as

$$(3.1.4)\qquad \langle\sigma_{j_1k_1}\cdots\sigma_{j_nk_n}\rangle = \mathrm{Trace}(s_{j_1k_1}\cdots s_{j_nk_n}V^N)/\mathrm{Trace}(V^N)\qquad (k_1\leq\cdots\leq k_n).$$

Denoting by $|vac\rangle$ the unique normalized eigenvector of V corresponding to its largest eigenvalue, we have for large M, N

$$(3.1.4)'\qquad \langle\sigma_{j_1k_1}\cdots\sigma_{j_nk_n}\rangle \sim \langle vac|\, s_{j_1k_1}\cdots s_{j_nk_n}\,|vac\rangle.$$

Onsager observed in effect the following: it is possible to associate free fermion operators p_j, q_j in such a way that s_{jk}, V belong to the Clifford group. This is the key to the solvability of the problem.

In the limit of infinite lattice $M, N\to\infty$ there appears a critical temperature

(*) $\beta^{-1} = kT$, k: Boltzmann constant, T: temperature.

T_c which sharply seperates the ordered $(T<T_c)$ and disordered $(T>T_c)$ phases. The behaviors of (3.1.2), (3.1.3) in the critical region $T\sim T_c$ are of particular interest. In the sequel we shall concentrate upon the scaling limit

$$(3.1.5)\qquad T\to T_c,\quad j_\nu^2+k_\nu^2\to\infty$$

$$\mathrm{const.}\,|T-T_c|\cdot(j_\nu,k_\nu)=(a_\nu^1,a_\nu^2)\qquad \text{fixed}\quad(\nu=1,\cdots,n).$$

It is possible to take this limit at the level of operators. Apart from a scaling factor proportional to $|T-T_c|^{1/8}$, the spin operator s_{jk} tends to the following two dimensional fields with relativistic covariance:(*)

$$(3.1.6)\qquad T\lesssim T_c\qquad \varphi_F(x)=:e^{\rho_F(x)/2}:$$

$$T\gtrsim T_c\qquad \varphi^F(x)=:\psi_0(x)e^{\rho_F(x)/2}:$$

$$(3.1.7)\qquad \rho_F(x)=\int_{-\infty}^{+\infty}\int_{-\infty}^{+\infty}\frac{du}{2\pi|u|}\frac{du'}{2\pi|u'|}\frac{-i(u-u')}{u+u'-i0}e^{-im(x^-(u+u')+x^+(u^{-1}+u'^{-1}))}\psi(u)\psi(u')$$

$$(3.1.8)\qquad \psi_0(x)=\int_{-\infty}^{+\infty}\frac{du}{2\pi|u|}e^{-im(x^-u+x^+u^{-1})}\psi(u)\qquad (x^\pm=(x^0\pm x^1)/2)$$

In the above $\psi(u)$ signifies the creation $(u<0)$ — annihilation $(u>0)$ operator of free fermion carrying the energy-momentum $(p^0,p^1)=(m\frac{u+u^{-1}}{2},m\frac{u-u^{-1}}{2})$. It obeys the canonical anti-commutation relations

$$(3.1.9)\qquad [\psi(u),\psi(u')]_+=2\pi|u|\delta(u+u').$$

Accordingly the n-point correlation functions (3.1.4)' (with a scale factor removed) tend respectively to $\langle\varphi_F(a_1)\cdots\varphi_F(a_n)\rangle$ or $\langle\varphi^F(a_1)\cdots\varphi^F(a_n)\rangle$ as $T\to T_c\mp 0$. The free fermion $p_j\pm q_j$ are scaled to give the massive free relativistic Majorana field

$$(3.1.10)\qquad \psi_\pm(x)=\sqrt{\frac{m}{2}}\int_{-\infty}^{+\infty}\frac{du}{2\pi|u|}\sqrt{0+iu}^{\,\pm1}\,e^{-im(x^-u+x^+u^{-1})}\psi(u).$$

(*) Here we make the "Wick rotation" and consider the region where $x=(x^0,x^1)=\mathrm{const.}\,|T-T_c|(-\sqrt{-1}\,j,k)$ is real.

From the explicit formula (3.1.6) it is apparent that $\varphi_F(x)$, $\varphi^F(x)$ "belong to the Clifford group". In fact they satisfy with $\psi_\pm(x)$ the peculiar commutation relations

$$\varphi_F(a)\psi_\pm(x) = \varepsilon(x^1 - a^1)\psi_\pm(x)\varphi_F(a) \tag{3.1.11}$$

$$\varphi^F(a)\psi_\pm(x) = \varepsilon(a^1 - x^1)\psi_\pm(x)\varphi^F(a)$$

for $x-a$ sapcelike.

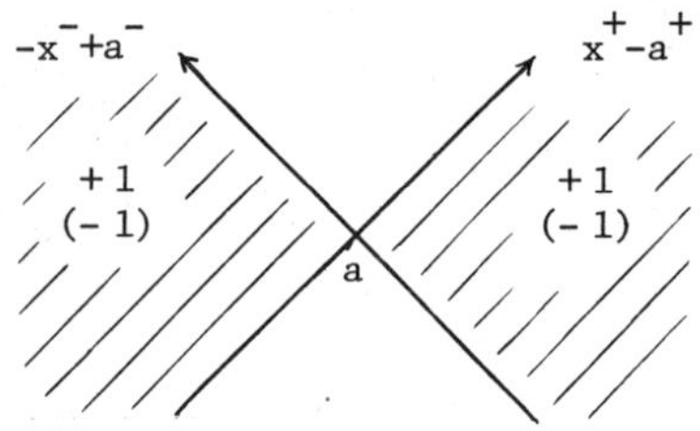

Fig. 3.1.

In what follows we shall show how the scaled n-point functions $\langle\varphi(a_1)\cdots\varphi(a_n)\rangle$ $(\varphi=\varphi_F$ or $\varphi^F)$ are calculated by appealing to the monodromy preserving deformation theory.

3.2 Wave functions

The monodromy problem comes in when we consider instead of $\langle\varphi(a_1)\cdots\varphi(a_n)\rangle$ the following objects as functions of x.

$$w_{F\nu}(x) = {}^t(w_{F\nu+}(x),\ w_{F\nu-}(x)) \tag{3.2.1}$$

$$w_{F\nu\pm}(x) = \frac{\langle\psi_\pm(x)\,\varphi_F(a_1)\cdots\varphi^F(a_\nu)\cdots\varphi_F(a_n)\rangle}{\langle\varphi_F(a_1)\cdots\varphi_F(a_n)\rangle}, \qquad \nu = 1,\cdots,n.$$

As an illustration first let $n=1$. In this case (3.2.1.) is explicitly calculated as

$$\sqrt{\frac{m}{2}}^{\,-1} w_F(x) = \sqrt{\frac{m}{2}}^{\,-1}\begin{pmatrix}\langle\psi_+(x)\psi_0(a)\rangle\\ \langle\psi_-(x)\psi_0(a)\rangle\end{pmatrix} \tag{3.2.2}$$

$$= \int_0^\infty \frac{du}{2\pi u}\begin{pmatrix} e^{\frac{\pi i}{4}}\sqrt{u} \\ e^{-\frac{\pi i}{4}}\sqrt{u}^{\,-1}\end{pmatrix} e^{-im((x^- - a^-)u + (x^+ - a^+)u^{-1})}$$

$$= \frac{i}{2\sqrt{\pi}} \begin{pmatrix} \dfrac{1}{\sqrt{m(-x^-+a^-)}} \\ \dfrac{1}{\sqrt{m(x^+-a^+)}} \end{pmatrix} e^{-2m\sqrt{(-x^-+a^-)(x^+-a^+)}}$$

Hereafter we allow the variables $x, a, \cdots$ to be complex, in particular to run in the Euclidean space $\{x^0(=-ix^2) \in i\mathbb{R}, x^1 \in \mathbb{R}\}$. We often use the notation $z = -x^-$, $z^* = x^+$, $a_\nu = -a_\nu^-$, $a_\nu^* = a_\nu^+$. The point is that, due to the square-root in (3.2.2), (the Euclidean continuation of) $w_F(x)$ is a double-valued function, changing its sign each time when prolonged around the branch point $x = a$ in the Euclidean region. This was to be expected from the commutation relation (3.1.11) by the same argument as in 1.3. Since $\langle vac|\psi_\pm(x)$ (resp. $\psi_\pm(x)|vac\rangle$) contains only the positive (resp. negative) frequency part, the expectation value $\langle\psi_\pm(x)\,\varphi^F(a)\rangle$ (resp. $\langle\varphi^F(a)\,\psi_\pm(x)\rangle$) is analytically prolongable to the upper (resp. lower) half Euclidean plane $\mathrm{Im}(x^\mp - a^\mp) < 0$ (resp. $\mathrm{Im}\,(x^\mp - a^\mp) > 0$). For $x^0 = a^0$, $x^1 > a^1$ they are analytic continuations of each other, while for $x^1 < a^1$ they differ by sign. This shows that $w_F(x)$ has a monodromy -1 around $x = a$.

For general n, (3.2.1) is expressible as an infinite series (cf. Formula 2, 1.4)

$$(3.2.3)\qquad \sqrt{\frac{m}{2}}^{\,-1} w_{F\nu\pm}(x) = \int_0^\infty \frac{du}{2\pi u}\sqrt{0+iu}^{\,\pm 1}\, e^{-im((x^- - a_\nu^-)u + (x^+ - a_\nu^+)u^{-1})}$$

$$+ \sum_{\ell=0}^{\infty} \sum_{\nu_0,\cdots,\nu_\ell=1}^{n} \varepsilon_{\nu_0\nu_1}\varepsilon_{\nu_1\nu_2}\cdots\varepsilon_{\nu_\ell\nu} \int_{-\infty}^{+\infty}\cdots\int_{-\infty}^{+\infty} \frac{du_0}{2\pi|u_0|}\cdots\frac{du_\ell}{2\pi|u_\ell|}\frac{du}{2\pi|u|}$$

$$\times\, \theta(u_0)\theta(-\varepsilon_{\nu_0\nu_1}u_1)\cdots\theta(-\varepsilon_{\nu_\ell\nu}u)\sqrt{0+iu_0}^{\,\pm 1}\frac{i(u_0+u_1)}{u_0-u_1+i0}\cdots\frac{i(u_\ell+u)}{u_\ell-u+i0}$$

$$\times \exp(-im((x^- - a_{\nu_0}^-)u_0 + (a_{\nu_0}^- - a_{\nu_1}^-)u_1 + \cdots + (a_{\nu_\ell}^- - a_\nu^-)u$$

$$+ (x^+ - a_{\nu_0}^+)u_0^{-1} + (a_{\nu_0}^+ - a_{\nu_1}^+)u_1^{-1} + \cdots + (a_{\nu_\ell}^+ - a_\nu^+)u^{-1}))$$

where $\varepsilon_{\mu\nu} = -1\ (\mu<\nu),\ =0\ (\mu=\nu),\ =1\ (\mu>\nu)$ and $\theta(u) = 1\ (u>0),\ =0\ (u<0)$. Likewise the τ-function $\tau_F = \langle\varphi_F(a_1)\cdots\varphi_F(a_n)\rangle$ itself has a representation

$$(3.2.4)\qquad \log\tau_F = -\sum_{\ell=2}^{\infty}\frac{1}{2\ell}\sum_{\nu_1,\cdots,\nu_\ell=1}^{n}\varepsilon_{\nu_1\nu_2}\cdots\varepsilon_{\nu_\ell\nu_1}\int_{-\infty}^{+\infty}\cdots\int_{-\infty}^{+\infty}\frac{du_1}{2\pi|u_1|}\cdots\frac{du_\ell}{2\pi|u_\ell|}$$

$$\times\,\theta(-\varepsilon_{\nu_\ell\nu_1}u_1)\cdots\theta(-\varepsilon_{\nu_{\ell-1}\nu_\ell}u_\ell)\frac{i(u_1+u_2)}{u_1-u_2+i0}\cdots\frac{i(u_\ell+u_1)}{u_\ell-u_1+i0}$$

$$\times \exp(-im((a^-_{\nu_\ell}-a^-_{\nu_1})u_1+(a^-_{\nu_1}-a^-_{\nu_2})u_2+\cdots+(a^-_{\nu_{\ell-1}}-a^-_{\nu_\ell})u_\ell$$

$$+(a^+_{\nu_\ell}-a^+_{\nu_1})u_1^{-1}+(a^+_{\nu_1}-a^+_{\nu_2})u_2^{-1}+\cdots+(a^+_{\nu_{\ell-1}}-a^+_{\nu_\ell})u_\ell^{-1})).$$

The series (3.2.3), (3.2.4) are convergent for $\mathrm{Im}(-x^{\mp}+a^{\mp}_{\mu})>0$ $(\mu=1,\cdots,n)$ and sufficiently large $\mathrm{Im}(-a^{\mp}_1+a^{\mp}_2),\cdots,\ \mathrm{Im}(-a^{\mp}_{n-1}+a^{\mp}_n)>0.$(*) The above argument on the double valuedness of $w_{F\nu}(x)$ applies to the general case as well. Let us take a closer look at the properties of $w_{F\nu}(x)=w$ restricted to the Euclidean region.

Firstly it satisfies the Euclidean Dirac equation

$$\begin{pmatrix} m & -\frac{\partial}{\partial z} \\ -\frac{\partial}{\partial z^*} & m \end{pmatrix} w = 0 \quad \text{outside} \quad a_1,\cdots,a_n. \tag{3.2.5}$$

This is a direct consequence of the Dirac equation $\frac{\partial}{\partial x^+}\psi_+(x) = m\psi_-(x)$, $\frac{\partial}{\partial x^-}\psi_-(x) = -m\psi_+(x)$ for the Majorana fields $\psi_\pm(x)$. Secondly it has the monodromy property

(3.2.6) $\quad w$ changes sign when prolonged around $a_1,\cdots,a_n$.

Finally we have the growth conditions

$$|w| = O\left(\frac{1}{\sqrt{|z-a_\nu|}}\right) \qquad z\to a, \quad \nu=1,\cdots,n, \tag{3.2.7$_\nu$}$$

$$|w| = O(e^{-2m|z|}) \qquad |z|\to\infty. \tag{3.2.7$_\infty$}$$

Properties (3.2.5) ∿ (3.2.7) follow either from (3.2.3), or directly from (3.2.1). To see (3.2.7) at the operator level, we need the following short distance operator expansion abtained by applying (1.4.15) ∿ (1.4.16)

$$\sqrt{\frac{m}{2}}^{-1}\psi(x)\,\varphi_F(a) = \frac{1}{2}\varphi^F(a)\cdot w_0 + \frac{-1}{m}\frac{\partial}{\partial a^-}\varphi^F(a)\cdot w_1+\cdots$$
$$+\frac{1}{2}\varphi^F(a)\cdot w_0^* + \frac{1}{m}\frac{\partial}{\partial a^+}\varphi^F(a)\cdot w_1^*+\cdots \tag{3.2.8}$$

$$\sqrt{\frac{m}{2}}^{-1}\psi(x)\,\varphi^F(a) = \frac{i}{2}\varphi_F(a)\cdot w_0 + \frac{-i}{m}\frac{\partial}{\partial a^-}\varphi_F(a)\cdot w_1+\cdots$$
$$-\frac{i}{2}\varphi_F(a)\cdot w_0^* + \frac{-i}{m}\frac{\partial}{\partial a^+}\varphi_F(a)\cdot w_1^*+\cdots .$$

Here $w_\ell = w_\ell(-x^-+a^-,\ x^+-a^+)$, $w^*_\ell = w^*_\ell(-x^-+a^-,\ x^+-a^+)$ are solutions of the Dirac equation with the local behaviors

---

(*) Using the covariance under $a^{\mp}_\nu \mapsto e^{\pm i\theta}a^{\mp}_\nu$ we may also cover the region where a_ν's are real.

$$(3.2.9)\qquad w_\ell(z,z^*) = \begin{pmatrix} (mz)^{\ell-\frac{1}{2}}\left(\dfrac{1}{(\ell-\frac{1}{2})!} + \dfrac{m^2 zz^*}{(\ell+\frac{1}{2})!} + \cdots\right) \\[2ex] (mz)^{\ell+\frac{1}{2}}\left(\dfrac{1}{(\ell+\frac{1}{2})!} + \dfrac{m^2 zz^*}{(\ell+\frac{3}{2})!} + \cdots\right) \end{pmatrix}$$

$$w^*_\ell(z,z^*) = \begin{pmatrix} (mz^*)^{\ell+\frac{1}{2}}\left(\dfrac{1}{(\ell+\frac{1}{2})!} + \dfrac{m^2 zz^*}{(\ell+\frac{3}{2})!} + \cdots\right) \\[2ex] (mz^*)^{\ell-\frac{1}{2}}\left(\dfrac{1}{(\ell-\frac{1}{2})!} + \dfrac{m^2 zz^*}{(\ell+\frac{1}{2})!} + \cdots\right) \end{pmatrix}.$$

Explicitly they are given by the modified Bessel functions

$$(3.2.10)\qquad w_\ell(z,z^*) = \begin{pmatrix} e^{i(\ell-\frac{1}{2})\theta} I_{\ell-\frac{1}{2}}(mr) \\[1ex] e^{i(\ell+\frac{1}{2})\theta} I_{\ell+\frac{1}{2}}(mr) \end{pmatrix}, \qquad w^*_\ell(z,z^*) = \begin{pmatrix} e^{-i(\ell+\frac{1}{2})\theta} I_{\ell+\frac{1}{2}}(mr) \\[1ex] e^{-i(\ell-\frac{1}{2})\theta} I_{\ell-\frac{1}{2}}(mr) \end{pmatrix}$$

where $z = re^{i\theta}/2$, $z^* = re^{-i\theta}/2$. It is shown that any local solution w of (3.2.5) having the monodromy -1 and growing at most as $O\left(\dfrac{1}{|z-a|^{N+\frac{1}{2}}}\right)$ is expanded uniquely as $w = \sum_{\ell=-N}^{\infty} c_\ell w_\ell(z-a, z^*-a^*) + \sum_{\ell=-N}^{\infty} c^*_\ell w^*_\ell(z-a, z^*-a^*)$. The following formula is sometimes useful.

$$(3.2.11)\qquad \frac{\partial}{\partial z} w_\ell = m w_{\ell-1}, \quad \frac{\partial}{\partial z} w^*_\ell = m w^*_{\ell+1}, \quad \frac{\partial}{\partial z^*} w_\ell = m w_{\ell+1}, \quad \frac{\partial}{\partial z^*} w^*_\ell = m w^*_{\ell-1},$$

$$M_F w_\ell = \ell w_\ell, \qquad M_F w^*_\ell = -\ell w^*_\ell$$

where M_F denotes the infinitesimal generator with spin $\frac{1}{2}$ of the Euclidean rotation around the origin

$$(3.2.12)\qquad M_F = z\frac{\partial}{\partial z} - z^*\frac{\partial}{\partial z^*} + \frac{1}{2}\begin{pmatrix} 1 & \\ & -1 \end{pmatrix}.$$

3.3 Linear differential equations for wave functions

Now we show that the properties (3.2.5) $\sim$ (3.2.7) are sufficient to characterize the wave functions in terms of a system of linear partial differential equations.

First let us show that the vector space $W_{a_1,\ldots,a_n}$ of functions satisfying

(3.2.5) ∿ (3.2.7) is finite dimensional. Set

$$(3.3.1)\qquad I_F(w,w') = \frac{m^2}{2}\iint i\,dz\wedge dz^*(w_+w'^*_+ + w_-w'^*_-)$$

$$= I_F(w',w)^*, \qquad\qquad w, w' \in W_{a_1,\cdots,a_n}.$$

Property (3.2.6) guarantees that the integrand is single-valued, while (3.2.7) ensures the convergence. Hence (3.3.1) defines a positive definite hermitian inner product on $W_{a_1,\cdots,a_n}$. Using the Euclidean Dirac equation (3.2.5) we have $m\,i\,dz\wedge dz^*(w_+w'^*_+ + w_-w'^*_-) = i\,d(w_+w'^*_-dz)$, so that only the boundary term contributes to (3.3.1):

$$(3.3.2)\qquad I_F(w,w') = \frac{im}{2}\sum_{\nu=1}^{n}\lim_{\varepsilon\to 0}\oint_{|z-a_\nu|=\varepsilon} w_+w'^*_-dz$$

$$= -\sum_{\nu=1}^{n} c_0^{(\nu)}(w)\bar{c}_0^{(\nu)}(w')^*$$

where we have set

$$(3.3.3)\qquad w = \sum_{\ell=0}^{\infty} c_\ell^{(\nu)}(w)\cdot w_\ell(z-a_\nu, z^*-a^*_\nu) + \sum_{\ell=0}^{\infty}\bar{c}_\ell^{(\nu)}(w)\cdot w^*_\ell(z-a_\nu, z^*-a_\nu)$$

for $w\in W_{a_1,\cdots,a_n}$. In particular, if $c_0^{(\nu)}(w) = 0$ for $\nu = 1,\cdots,n$ then $w = 0$. This implies $\dim W_{a_1,\cdots,a_n} \leqq n$ and that there are no bounded function (other than 0) in $W_{a_1,\cdots,a_n}$.

Actually $W_{a_1,\cdots,a_n}$ is exactly n dimensional. Making use of some functional anslysis one proves the existence of n elements $w_{C1,\cdots,Cn}$ of $W_{a_1,\cdots,a_n}$ with the following normalized local behavior:

$$(3.3.4)\qquad w_{C\mu} = \delta_{\mu\nu}w_0(z-a_\nu,z^*-a^*_\nu) + \alpha_{\mu\nu}w_1(z-a_\nu,z^*-a^*_\nu) + \cdots$$

$$+ \beta_{\mu\nu}w^*_0(z-a_\nu,z^*-a^*_\nu) + \Big(\sum_{\sigma=1}^{n}\beta_{\mu\sigma}\alpha^*_{\sigma\nu}\Big)w^*_1(z-a_\nu,z^*-a^*_\nu) + \cdots$$

$$(z\to a_\nu,\quad \mu,\nu = 1,\cdots,n).$$

(The argument employed here is analogous to the existence proof of Abelian differentials on a compact Riemann surface.) In terms of the coefficients $\beta_{\mu\nu}$, the positive definiteness of (3.3.1) is equivalent to

(3.3.5) $\beta = (\beta_{\mu\nu})$ is negative definite and hermitian, and hence $\beta = -e^{2H}$ for a unique $H = {}^tH^*$.

Moreover, by calculating $-\iint id(w_{C\mu+}\frac{\partial w_{C\nu+}}{\partial z}dz) = \iint id(w_{C\mu-}\frac{\partial w_{C\nu-}}{\partial z}dz^*)$ for $\mu \neq \nu$ we obtain

(3.3.6) $\qquad \alpha = (\alpha_{\mu\nu})$ is symmetric.

Likewise $-\iint id(w_{C\mu+}w_{C\nu+}dz) = \iint id(w_{C\mu-}w_{C\nu-}dz^*)$ gives

(3.3.7) $\qquad \beta^t\beta = 1, \quad$ i.e. $\quad H = -{}^tH.$

Let us derive the differential equations for $w_{C\nu}$. Note that the Euclidean Dirac operator (3.2.5) commutes with M_F in (3.2.12). Hence $M_F w_{C\nu}$ satisfies (3.2.5), as well as (3.2.6), (3.2.7) except for the growth conditions at $z = a_1, \cdots, a_n$. On the other hand, $\frac{\partial}{\partial z}w_{C\nu}$ and $\frac{\partial}{\partial z^*}w_{C\nu}$ share the same properties including the growth condition $O(\frac{1}{\sqrt{|z-a_\nu|}^3})$. By matching the singularities of $M_F w_{C\nu}$ by linear combinations of $\frac{\partial}{\partial z}w_{C\mu}$, $\frac{\partial}{\partial z^*}w_{C\mu}$ and $w_{C\mu}$ and applying the finite dimensionality argument above, we obtain a system of linear differential equations of the form

$$\begin{pmatrix} M_F w_{C1} \\ \vdots \\ M_F w_{Cn} \end{pmatrix} = (A\frac{\partial}{\partial z} + B\frac{\partial}{\partial z^*} + F)\begin{pmatrix} w_{C1} \\ \vdots \\ w_{Cn} \end{pmatrix}, \qquad A, B, F : n \times n \text{ constant matrices.} \tag{3.3.8}$$

Using the recursion formula (3.2.11) the coefficient matrices A, B, F are identified as

$$A = \begin{pmatrix} a_1 & & \\ & \ddots & \\ & & a_n \end{pmatrix}, \qquad B = -G^{-1}A^*G, \qquad F = [\alpha, mA] \tag{3.3.9}$$

$$A^* = \begin{pmatrix} a_1^* & & \\ & \ddots & \\ & & a_n^* \end{pmatrix}, \qquad G = -\beta^{-1} = e^{-2H}.$$

The coefficients of w_0^* and w_1 in the local expansion of (3.3.8) give respectively

$$GFG^{-1} = -[\alpha^*, mA^*] \tag{3.3.10}$$

$$[C_2, mA] + \alpha - F\alpha - mA^* + G^{-1}mA^*G = 0 \tag{3.3.11}$$

with $C_2 = (c_2^{(\nu)}(w_{F\mu}))_{\mu,\nu=1,\cdots,n}$. In particular the diagonal of (3.3.11) gives

$$\alpha_{\mu\mu} = \sum_{\nu(\neq\mu)} \frac{f_{\mu\nu}f_{\nu\mu}}{m(a_\mu - a_\nu)} + ma_\mu^* - (G^{-1}mA^*G)_{\mu\mu}. \tag{3.3.11'}$$

Since the (Euclidean) wave functions $w_{F\nu}$ of 3.2 belongs to $W_{a_1,\cdots,a_n}$, they are linear combinations of the basis $w_{C\nu}$:

$$(3.3.12)\qquad \sqrt{\frac{m}{2}}^{-1} w_{F\mu} = \frac{i}{2}\sum_{\nu=1}^{n} c_{\mu\nu} w_{C\nu}.$$

The short distance expansion (3.2.8) enables us to calculate the local behavior of $w_{F\mu}$. We find that

$$(3.3.13)\qquad \sqrt{\frac{m}{2}}^{-1} c_0^{(\nu)}(w_{F\mu}) = \begin{cases} \frac{i}{2} & (\mu = \nu) \\ \mp\frac{1}{2}\hat{\tau}_F^{\mu\nu} & (\mu \lessgtr \nu) \end{cases}$$

$$\sqrt{\frac{m}{2}}^{-1} \bar{c}_0^{(\nu)}(w_{F\mu}) = \begin{cases} -\frac{i}{2} & (\mu = \nu) \\ \mp\frac{1}{2}\hat{\tau}_F^{\mu\nu} & (\mu \lessgtr \nu) \end{cases}$$

where $\hat{\tau}_F^{\mu\nu}$ denotes the Euclidean continuation of $<\varphi_F(a_1)\cdots\varphi^F(a_\mu)\cdots\varphi^F(a_\nu)\cdots\varphi_F(a_n)>/<\varphi_F(a_1)\cdots\varphi_F(a_n)>$. Setting $T = (i\varepsilon_{\mu\nu}\hat{\tau}_F^{\mu\nu})$, $C = (c_{\mu\nu})$ and comparing the local behaviors of (3.3.12) we have $1 - T = C$, $-1 - T = C\beta$. Since $1-\beta = 1+G^{-1}$ is invertible we obtain

$$(3.3.14)\qquad \begin{aligned} T &= (1-G)(1+G)^{-1} = \tanh H, \\ C &= 2G(1+G)^{-1} = e^{-H}(\cosh H)^{-1}. \end{aligned}$$

In particular C is invertible. Hence $w_{F\nu}$'s also span the vector space $W_{a_1,\cdots,a_n}$.

3.4. Deformation equations

So far we have fixed the position of branch points a_1, a_1^*, $\cdots$, a_n, a_n^*, and regarded $w_{C\nu}$'s as functions of z, z^*. Next let us consider their dependence on a_μ, a_μ^*. It is easy to see that their derivatives $\frac{\partial w_{C\nu}}{\partial a_\mu}$ $\frac{\partial w_{C\nu}}{\partial a_\mu^*}$ share the properties (3.2.5) $\sim$ (3.2.7) with the growth condition at most $O(\frac{1}{\sqrt{|z-a_\lambda|}^3})$ $(z \to a_\lambda)$. Thus the same argument as in 3.3 (matching of the singularities) yields the following system of linear total differential equations for $w_{C\nu}$'s :

$$(3.4.1)\qquad \begin{pmatrix} dw_{C1} \\ \vdots \\ dw_{Cn} \end{pmatrix} = (-\,dA\cdot\frac{\partial}{\partial z} - G^{-1}dA^*\cdot G\frac{\partial}{\partial z^*} + \Theta)\begin{pmatrix} w_{C1} \\ \vdots \\ w_{Cn} \end{pmatrix}.$$

Here d denotes the exterior differentiation with respect to a_ν, a_ν^* :

$dw = \sum_{\nu=1}^{n} (\frac{\partial w}{\partial a_\nu} da_\nu + \frac{\partial w}{\partial a^*_\nu} da^*_\nu)$, and $\Theta = (\theta_{\mu\nu})$ is a matrix of 1-forms related to $F = (f_{\mu\nu})$ in (3.3.9) through

$$\text{(3.4.2)} \qquad \theta_{\mu\nu} = \begin{cases} - f_{\mu\nu} \dfrac{da_\mu - da_\nu}{a_\mu - a_\nu} & (\mu \neq \nu) \\ 0 & (\mu = \nu). \end{cases}$$

The Euclidean Dirac equation (3.2.5), equations (3.3.8) and (3.4.1) characterize (to within a finite number of integration constants) the wave functions $w_{C1},\cdots,w_{Cn}$ as functions of total set of variables $z,z^*,a_1,a_1^*,\cdots,a_n,a_n^*$. As the integrability condition for them (i.e. cross differentiation $\frac{\partial}{\partial a_\nu} M_F = M_F \frac{\partial}{\partial a_\nu}$, $\frac{\partial}{\partial a_\mu}\frac{\partial}{\partial a_\nu} = \frac{\partial}{\partial a_\nu}\frac{\partial}{\partial a_\mu}$, etc.) we obtain a system of non-linear total differential equations, the "deformation equations" for the coefficients F, G:

$$\text{(3.4.3)} \qquad \begin{aligned} dF &= [\Theta,F] + m^2([dA,G^{-1}A^*G] + [A,G^{-1}dA^*\cdot G]) \\ dG &= -G\Theta + \Theta^*G. \end{aligned}$$

Here $\Theta^* = (\theta^*_{\mu\nu})$ is the complex conjugate of (3.4.2),(*) and F, G are subject to the symmetry

$$\text{(3.4.4)} \qquad {}^tF = -F, \quad {}^tF^* = GFG^{-1}(= -F^*), \quad {}^tG = G^{-1} = G^*.$$

In deriving (3.4.3) we have used the following fact. If $p_\nu(\frac{\partial}{\partial z},\frac{\partial}{\partial z^*})$ is a differential operator with constant coefficients and if $\sum_{\nu=1}^{n} p_\nu(\frac{\partial}{\partial z},\frac{\partial}{\partial z^*}) w_{C\nu} = 0$, then $p_\nu(\frac{\partial}{\partial z},\frac{\partial}{\partial z^*}) \equiv 0$ modulo $\frac{\partial^2}{\partial z \partial z^*} - m^2$. This is proved by examining the singularities.

As an example let us write (3.4.3) down in the simplest nontrivial case $n = 2$. Regarding the Euclidean covariance we may set $mA = \frac{\theta}{2}\begin{pmatrix}1 & \\ & -1\end{pmatrix} = mA^*$ $(\theta = m|a_1 - a_2|)$ and

$$\text{(3.4.5)} \qquad F = f\begin{pmatrix} & -i \\ i & \end{pmatrix}, \qquad H = -\frac{\psi}{2}\begin{pmatrix} & -i \\ i & \end{pmatrix}$$

$$G = \begin{pmatrix} \cosh\psi & -i\sinh\psi \\ i\sinh\psi & \cosh\psi \end{pmatrix}, \qquad \Theta = -f\frac{d\theta}{\theta}\begin{pmatrix} & -i \\ i & \end{pmatrix}$$

$$f^* = f, \qquad \psi^* = \psi.$$

(*) Our definition of Θ^* here differes from [5] by sign.

Then (3.4.3) reduces to

$$(3.4.6)\qquad f = \frac{1}{2}\theta\frac{d\psi}{d\theta}, \qquad \frac{d^2\psi}{d\theta^2} + \frac{1}{\theta}\frac{d\psi}{d\theta} = 2\sinh 2\psi,$$

or equivalently to a Painleve equation of the third kind for $\eta = e^{-\psi}$

$$(3.4.7)\qquad \frac{d^2\eta}{d\theta^2} = \frac{1}{\eta}\left(\frac{d\eta}{d\theta}\right)^2 - \frac{1}{\theta}\frac{d\eta}{d\theta} + \eta^3 - \frac{1}{\eta}.$$

It is instructive to note that the deformation theory developed here is included in the scheme of §2 for ordinary linear differential equations. Let

$$(3.4.8)\qquad \begin{pmatrix} w_+ \\ w_- \end{pmatrix} = \int \frac{du}{2\pi u}\begin{pmatrix} \sqrt{u} \\ \sqrt{u}^{-1} \end{pmatrix} e^{m(zu+z^*u^{-1})}\hat{w}(u)$$

be a formal Laplace transformation to solve the Euclidean Dirac equation (3.2.5). Then the systems (3.3.8) and (3.4.1) are rewritten in terms of an n component column vector $\hat{w}_C = {}^t(\hat{w}_{C1},\cdots,\hat{w}_{Cn})$ as

$$(3.4.9)\qquad \frac{d}{du}\hat{w}_C = \left(\frac{G^{-1}mA^*G}{u^2} - \frac{F}{u} - mA\right)\hat{w}_C$$

$$(3.4.10)\qquad d\hat{w}_C = \left(-\frac{G^{-1}mdA^*\cdot G}{u} + \Theta - umdA\right)\hat{w}_C.$$

In other words we are dealing with the deformation theory of linear ordinary differential equations (3.4.9) having irregular singularities of rank one at $u=0$ and $u=\infty$. The system (3.4.3) are nothing but the deformation equations for (3.4.9) $\sim$ (3.4.10).

3.5 Correlation functions

By now it is relatively an easy task to obtain a closed expression for $d\log\tau_F$. The key is the short distance expansion (3.2.8). From (3.2.1) and (3.2.8) it follows that the second coefficients of the local expansions of $w_{F\nu\pm}(x)$ are given by

$$(3.5.1)\qquad \sqrt{\frac{m}{2}}^{-1} c_1^{(\nu)}(w_{F\nu}) = \frac{i}{m}\frac{\partial}{\partial a_\nu}\tau_F/\tau_F$$

$$\sqrt{\frac{m}{2}}^{-1} \bar{c}_1^{(\nu)}(w_{F\nu}) = \frac{-i}{m}\frac{\partial}{\partial a_\nu^*}\tau_F/\tau_F.$$

This shows that $d\log\tau_F = \sum_{\nu=1}^{n}\sqrt{\frac{m}{2}}^{-1}(-imc_1^{(\nu)}(w_{F\nu})da_\nu + im\bar{c}_1^{(\nu)}(w_{F\nu})da_\nu^*)$, or by (3.3.12) and (3.3.14) that

$$(3.5.2)\qquad d\log\tau_F = \frac{1}{2}\sum_{\mu,\nu=1}^{n} c_{\nu\mu}(\alpha_{\mu\nu}mda_\nu - (\beta\alpha^*)_{\mu\nu}mda_\nu^*)$$

$$= \frac{1}{2} \text{ trace } (1 - T)\alpha m dA \quad + \quad \text{complex conjugate.}$$

Using (3.3.11)' and noting trace $T\alpha m dA = \text{trace } m dA^t\alpha^t T = -\frac{1}{2}$ trace $T\Theta$, we have

(3.5.3) $$d \log \tau_F = \frac{1}{4} \text{ trace } (T\Theta - \Theta^*T) + \frac{1}{2}\omega,$$

(3.5.4) $$\omega = -\frac{1}{2} \text{ trace } (F\Theta - \Theta^*GFG^{-1}) + m^2 \text{ trace } (d(AA^*) - G^{-1}A^*GdA - GAG^{-1}dA^*).$$

Finally the first term of (3.5.3) is integrated as $\frac{1}{2} d \log \det \cosh H$, by using (3.4.3). The 1-form (3.5.4) is shown to be closed for any solution of the deformation equations (3.4.3). Summing up we obtain the formula

(3.5.5) $$\tau_F = \text{const}\cdot\sqrt{\det \cosh H} \, \exp(\frac{1}{2}\int\omega)$$

where $\int\omega$ denotes a primitive of (3.5.4). We remark that the closed 1-form associated with (3.4.9) $\sim$(3.4.10) in the sense of §2 coincides with (3.5.4) except for the exact differential term $m^2 d \,\text{trace}\, AA^*$.

The ratios $\hat{\tau}_F^{\mu\nu}$ of (Euclidean) $\langle \varphi_F(a_1)\cdots\varphi^F(a_\mu)\cdots\varphi^F(a_\nu)\cdots\varphi_F(a_n)\rangle$ to τ_F are obtained directly from (3.3.14) as

(3.5.6) $$\hat{\tau}_F^{\mu\nu} = \pm(i \tanh H)_{\mu\nu} \qquad (\mu \gtrless \nu).$$

More generally we have

(3.5.7) $$\langle \varphi_F(a_1)\cdots\varphi^F(a_{\nu_1})\cdots\varphi^F(a_{\nu_2})\cdots\varphi^F(a_{\nu_m})\cdots\varphi_F(a_n)\rangle$$
$$= \tau_F\cdot\text{Pfaffian } (i(\tanh H)_{\nu\nu'})_{\nu,\nu'=\nu_1,\cdots,\nu_m}.$$

In particular the (Euclidean) correlation function $\tau^F = \langle\varphi^F(a_1)\cdots\varphi^F(a_n)\rangle$ has an expression

(3.5.8) $$\tau^F = \text{const}\cdot\sqrt{\det i \sinh H} \, \exp(\frac{1}{2}\int\omega).$$

For $n = 2$ (3.5.5) and (3.5.8) reduce to

(3.5.9) $$\left.\begin{matrix}\tau_F \\ \tau^F\end{matrix}\right\} = \text{const}\cdot\left\{\begin{matrix}\cosh\frac{\psi}{2} \\ \sinh\frac{\psi}{2}\end{matrix}\right\} \exp\left[\frac{1}{2}\int_\infty^\theta dt\, \frac{t}{2}\left(\left(\frac{d\psi}{dt}(t)\right)^2 - \sinh^2\psi(t)\right)\right].$$

3.6 Field theory of Ising model in the scaling limit

The continuum model $\varphi^F(x)$ so far discussed is of equal interest from the field theoretical point of view. It is at present the only nontrivial massive model of relativistic field theory whose n point functions are known exactly in a closed form. It is also the simplest model whose S-matrix has the factorization property; there are no particle production, and the S-matrix in the N-particle sector is the product of $N(N-1)/2$ pairwise 2-body S-matrices

$$(3.6.1) \qquad S_{2,2} = -1.$$

Moreover the asymptotic fields $\phi_{\substack{in\\out}}(x)$ of $\varphi^F(x)$ are known explicitly. They are neutral free bosons (*) whose creation - annihilation operators $\phi^\dagger_{\substack{in\\out}}(u)$, $\phi_{\substack{in\\out}}(u)$ are expressed in terms of the auxiliary fermion $\psi^\dagger(u)$ $(=\psi(-u))$, $\psi(u)$ as

$$(3.6.2) \qquad \phi^\dagger_{\substack{in\\out}}(u) = \psi^\dagger(u)\cdot(-)^{N^{\mp}(u)}$$

$$\phi_{\substack{in\\out}}(u) = (-)^{N^{\mp}(u)}\cdot\psi(u)$$

$$N^{\mp}(u) = \int_0^\infty \frac{du'}{2\pi u'}\,\theta(\pm(u-u'))\psi^\dagger(u')\psi(u')$$

$$(u > 0).$$

Since $\phi^\dagger_{\substack{in\\out}}(u)\phi_{\substack{in\\out}}(u) = \psi^\dagger(u)\psi(u)$, the above formula is reciprocal, and it is possible to construct free fermion out of free boson. Actually we may mimic all the constructions by starting from neutral free boson $\phi(x)$ to obtain analogous fields $\varphi_B(x)$, $\varphi^B(x) = {}^t(\varphi^B_+(x),\varphi^B_-(x))$. The latter is a 2-component spinor obeying the fermi statistics. By interchanging ϕ and ψ in (3.6.2) we obtain the relation between the asymptotic fields $\psi^\dagger_{\substack{in\\out}}(u)$, $\psi_{\substack{in\\out}}(u)$ of $\varphi^B(x)$ and the auxiliary boson $\phi^\dagger(u)$, $\phi(u)$. The n point functions are related to τ_F, τ^F through the simple formula

$$(3.6.3) \qquad \langle\varphi_F(a_1)\cdots\varphi_F(a_n)\rangle\langle\varphi_B(a_1)\cdots\varphi_B(a_n)\rangle = \sqrt{\det\cosh H}\ .$$

It is straightforward to incorporate a general phase factor $e^{2\pi i\ell}$ ($\ell\in\mathbb{C}$, or it can be a matrix) in place of the simplest monodromy -1. To achieve this one starts from the free Dirac (i.e. charged) fields instead of Majorana fields. The deformation theory of 3.4, 3.5 holds without any change (except for the symmetry conditions for F, G). The relevant fields are now related to the solution of the Federbush model

$$(3.6.4) \qquad \mathcal{L}_{int} = -g\,\varepsilon_{\mu\nu}J^\mu_I(x)J^\nu_{II}(x)$$

(*) $\varphi_F(x),\varphi^F(x)$ are bosons satisfying the microcausality $[\varphi(x),\varphi(x')]=0$ for $(x-x')^2<0$.

(current - pseudocurrent interaction between two species of fermions). The coupling constant is related to the exponent of the monodromy

$$g = 2\pi\ell. \tag{3.6.5}$$

References

There are an enormous number of literatures concerning the Ising model. See e.g.

[1] B.M. McCoy and T.T. Wu, The two-dimensional Ising model, Harvard University Press, 1973.

Explicit formulas of the spin operators and correlation functions on the lattice are given from the viewpoint of the Clifford group in

[2] M. Sato, T. Miwa and M. Jimbo, Holonomic quantum fields V, preprint RIMS 267, Kyoto Univ. (1978).

Some recent developments in this branch are quoted in the references of [2].

Deformation theory and closed expression of correlation functions were introduced by

[3] M. Sato, T. Miwa and M. Jimbo, Proc. Japan Acad. 53A (1977), 147-152, 153-158, 183-185.

This work is a generalization of the monumental papers

[4] E. Barouch, B.M. McCoy and T.T. Wu, Phys. Rev. Lett. 31 (1973) 1409-1411,
T.T. Wu, B.M. McCoy, C.A. Tracy and E. Barouch, Phys. Rev. B13 (1976) 316-374.

in which the 2 point functions are obtained in terms of a Painlevé transcendent of the third kind. Details along with some generalizations of [3] are found in

[5] M. Sato, T. Miwa and M. Jimbo, Publ. RIMS, Kyoto Univ. 15 (1979) 577-629.
________, Holonomic quantum fields IV, to appear in Publ. RIMS, Kyoto Univ. 15 (1980).
________, Supplement to Holonomic quantum fields IV, preprint RIMS 304, Kyoto Univ. (1979).

Extension of the above sheme to higher dimensions is briefly touched in

[6] M. Jimbo, T. Miwa, M. Sato and Y. Môri, Holonomic quantum fields — The unanticipated link between deformation theory of differential equations and quantum fields —, preprint RIMS 305, Kyoto Univ. (1979)

and the references cited therein.

§4. 1 dimensional XY model and its scaling limit

4.1 Free fermion operator

1 dimensional XY model is defined by the Hamiltonian

$$(4.1.1)\qquad H = -\frac{1}{4}\sum_{m\in\mathbf{Z}}((1+\gamma)\sigma^x_m\sigma^x_{m+1}+(1-\gamma)\sigma^y_m\sigma^y_{m+1}+2h\sigma^z_m).$$

Here γ and h are real parameters and σ^x_m, σ^y_m and σ^z_m are the following spin matrices.

$$(4.1.2)\qquad \sigma^x_m = \cdots\otimes\begin{pmatrix}1&\\&1\end{pmatrix}\otimes\overset{m}{\begin{pmatrix}&1\\1&\end{pmatrix}}\otimes\begin{pmatrix}1&\\&1\end{pmatrix}\otimes\cdots,$$

$$\sigma^y_m = \cdots\otimes\begin{pmatrix}1&\\&1\end{pmatrix}\otimes\begin{pmatrix}&-i\\i&\end{pmatrix}\otimes\begin{pmatrix}1&\\&1\end{pmatrix}\otimes\cdots,$$

$$\sigma^z_m = \cdots\otimes\begin{pmatrix}1&\\&1\end{pmatrix}\otimes\begin{pmatrix}1&\\&-1\end{pmatrix}\otimes\begin{pmatrix}1&\\&1\end{pmatrix}\otimes\cdots.$$

These spin matrices $\sigma^x_m = \sigma^1_m$, $\sigma^y_m = \sigma^2_m$ and $\sigma^z_m = \sigma^3_m$ satisfy the commutation relations

$$(4.1.3)\qquad [\sigma^j_m,\sigma^k_{m'}] = 2i\sum_{\ell=1}^{3}\varepsilon_{jk\ell}\delta_{mm'}\sigma^\ell_{m'},$$

$$[\sigma^j_m,\sigma^k_m]_+ = 2\delta_{jk},$$

where $\varepsilon_{jk\ell}$ is the anti-symmetric tensor normalized as $\varepsilon_{123} = 1$. The Heisenberg equation reads as

$$(4.1.4)\qquad \dot\sigma^x_m = \frac{1}{2}((1-\gamma)(-\sigma^z_m\sigma^y_{m+1}-\sigma^y_{m-1}\sigma^z_m)+2h\sigma^y_m),$$

$$\dot\sigma^y_m = \frac{1}{2}((1+\gamma)(\sigma^z_m\sigma^x_{m+1}+\sigma^x_{m-1}\sigma^z_m)-2h\sigma^x_m),$$

$$\dot\sigma^z_m = \frac{1}{2}(-(1+\gamma)(\sigma^y_m\sigma^x_{m+1}+\sigma^x_{m-1}\sigma^y_m)+(1-\gamma)(\sigma^x_m\sigma^y_{m+1}+\sigma^y_m\sigma^x_{m+1}))$$

where $\cdot$ denotes the time derivative.

The main problems we are concerned with are the following:

i) Diagonalization of the Hamiltonian.

ii) Calculation of matrix elements of spin matrices.

iii) Characterization of correlation functions.

The key to the problem i) is to introduce free fermion operators.

$$(4.1.5)\qquad p_m = \sigma^x_m\sigma^z_{m-1}\sigma^z_{m-2}\cdots,$$

$$q_m = i\sigma^y_m\sigma^z_{m-1}\sigma^z_{m-2}\cdots.$$

They are fermions because they satisfy the anti-commutation relations

$$(4.1.6)\qquad [p_m,p_{m'}]_+ = 2\delta_{mm'},\quad [q_m,q_{m'}]_+ = -2\delta_{mm'},\quad [p_m,q_{m'}]_+ = 0.$$

The key point is that the anti-commutators are not operators but they are constant. By (4.1.6) we can define a symmetric inner product in the vector space generated by free fermion. We call this vector space the <u>orthogonal space of free fermions</u>.

p_m and q_m are free because the Hamiltonian is <u>quadratic</u> in them:

$$(4.1.7)\qquad H = \frac{1}{4}\sum_{m\in\mathbf{Z}}((1+\gamma)q_mp_{m+1}-(1-\gamma)p_mq_{m+1}-2hq_mp_m),$$

or equivalently the Heisenberg equation for them is <u>linear</u>:

$$(4.1.8)\qquad \dot{p}_m = \frac{1}{2}((1+\gamma)q_{m-1}+(1-\gamma)q_{m+1}-2hq_m),$$

$$\dot{q}_m = \frac{1}{2}(-(1+\gamma)p_{m+1}-(1-\gamma)p_{m-1}+2hp_m).$$

In order to diagonalize (4.1.7) or (4.1.8) we should take account of parameters γ and h. There exist the following three different phases:

$$(4.1.9)\qquad \mathcal{R}_1 : h < -1,\qquad \gamma > 0,$$

$$\mathcal{R}_2 : -1 < h < 1,\quad \gamma > 0,$$

$$\mathcal{R}_3 : 1 < h,\qquad \gamma > 0.$$

We shall be concerned with the phase $\mathcal{R}_2$. The diagonalization is achieved by the following Fourier transformation. Namely for $\theta \in \mathbb{R}/2\pi\mathbf{Z}$ we set

$$(4.1.10)\qquad \psi^\dagger(-\theta) = \frac{1}{2}\left(\sqrt{a(\theta)}\sum_{m\in\mathbf{Z}} e^{-i(m+1)\theta}p_m + \sqrt{a(\theta)}^{-1}\sum_{m\in\mathbf{Z}} e^{-im\theta}q_m\right),$$

$$\psi(\theta) = \frac{1}{2}\left(\sqrt{a(\theta)}\sum_{m\in\mathbf{Z}} e^{-i(m+1)\theta}p_m - \sqrt{a(\theta)}^{-1}\sum_{m\in\mathbf{Z}} e^{-im\theta}q_m\right).$$

Here we introduce $a(\theta)$, $\omega(\theta)$ and $b(\theta)$ by

$$(4.1.11)\qquad a(\theta) = b(\theta)/b(-\theta),$$

$$(4.1.12)\qquad \omega(\theta) = 1/b(\theta)b(-\theta),$$

$$(4.1.13)\qquad b(\theta) = 1/\sqrt{(1-\alpha_+^{-1}e^{i\theta})(1-\alpha_-^{-1}e^{i\theta})},$$

$$(4.1.14)\qquad \alpha_\pm = \begin{cases} \dfrac{h\pm i\sqrt{1-h^2-\gamma^2}}{1-\gamma} & \text{if} \quad h^2+\gamma^2 \le 1 \\ \dfrac{h\pm\sqrt{h^2+\gamma^2-1}}{1-\gamma} & \text{if} \quad h^2+\gamma^2 \ge 1. \end{cases}$$

Note that in $\mathcal{R}_2$ $b(\theta)$ is holomorphic in the unit disk $|e^{i\theta}| < 1$.

The commutation relations, the Hamiltonian and the Heisenberg equation read as

$$(4.1.15)\qquad [\psi(\theta),\psi(\theta')]_+ = 0, \quad [\psi^\dagger(\theta),\psi^\dagger(\theta')]_+ = 0, \quad [\psi^\dagger(\theta),\psi(\theta')]_+ = 2\pi\delta(\theta-\theta'),$$

$$(4.1.16)\qquad H = \int \frac{d\theta}{2\pi}\,\omega(\theta)\psi^\dagger(\theta)\psi(\theta),$$

$$(4.1.17)\qquad \dot{\psi}^\dagger(\theta) = \omega(\theta)\psi^\dagger(\theta),$$

$$\dot{\psi}(\theta) = -\omega(\theta)\psi(\theta).$$

4.2 Vacuum

The vacuum state $|vac>$ is defined to be annihilated by the annihilation operator $\psi(\theta)$.

$$(4.2.1)\qquad \psi(\theta)\,|vac> = 0.$$

This is an eigenstate of the Hamiltonian with the zero eigenvalue.

The n-particle state $|\theta_1,\cdots,\theta_n>$ is defined to be created from the vacuum by the creation operator $\psi^\dagger(\theta)$.

$$(4.2.2)\qquad |\theta_1,\cdots,\theta_n> = \psi^\dagger(\theta_1)\cdots\psi^\dagger(\theta_n)\,|vac>.$$

From (4.1.15) and (4.1.16) we have

$$(4.2.3)\qquad H|\theta_1,\cdots,\theta_n> = \sum_{j=1}^{n}\omega(\theta_j)\,|\theta_1,\cdots,\theta_n>.$$

A general state is a superposition of the above eigenstates.

$$(4.2.4)\qquad \sum_{\ell=0}^{\infty}\int\cdots\int\frac{d\theta_1}{2\pi}\cdots\frac{d\theta_\ell}{2\pi}\,f_\ell(\theta_1,\cdots,\theta_\ell)\,|\theta_1,\cdots,\theta_n> .$$

Here f_0 is a constant and $f_\ell(\theta_1,\cdots,\theta_\ell)$ is a function in ℓ variables. The totality of such state is called the Fock space.

The operation of $\psi^\dagger(\theta)$ and $\psi(\theta)$ on the Fock space is given by

$$(4.2.5)\qquad \psi^\dagger(\theta)\,|\theta_1,\cdots,\theta_n> = |\theta,\theta_1,\cdots,\theta_n>,$$

(4.2.6) $$\psi(\theta)|\theta_1,\cdots,\theta_n\rangle = \sum_{j=1}^{n} (-)^{j-1}\delta(\theta-\theta_j)|\theta_1,\cdots,\theta_{j-1},\theta_{j+1},\cdots,\theta_n\rangle.$$

A general operator $\mathcal{O}$ is of the form

(4.2.7) $$\mathcal{O} = \sum_{\ell,\ell'=0}^{\infty} \int\cdots\int \frac{d\theta_1}{2\pi}\cdots\frac{d\theta_\ell}{2\pi}\int\cdots\int\frac{d\theta'_1}{2\pi}\cdots\frac{d\theta'_{\ell'}}{2\pi}$$
$$\times f_{\ell,\ell'}(\theta_1,\cdots,\theta_\ell;\theta'_1,\cdots,\theta'_{\ell'})\psi^\dagger(\theta_1)\cdots\psi^\dagger(\theta_\ell)\psi(\theta'_1)\cdots\psi(\theta'_{\ell'}).$$

Note that in the expression (4.2.7) annihilation operators are in the right while creation operator are in the left. This is called the normal ordering. The totality of such operators is called the operator algebra.

The dual Fock space is generated by the (dual) vacuum $\langle vac|$ and (dual) n-particle state $\langle\theta_1,\cdots,\theta_n|$ defined by

(4.2.8) $$\langle vac|\psi^\dagger(\theta) = 0,$$

(4.2.9) $$\langle\theta_1,\cdots,\theta_n| = \langle vac|\psi(\theta_1)\cdots\psi(\theta_n).$$

We normalize the constant multiple so that the inner product between $\langle vac|$ and $|vac\rangle$ is 1.

(4.2.10) $$\langle vac|vac\rangle = 1.$$

The vacuum-to-vacuum matrix element of an operator $\mathcal{O}$ is called the vacuum expectation value of $\mathcal{O}$ and is denoted by $\langle\mathcal{O}\rangle$.

(4.2.11) $$\langle\mathcal{O}\rangle = \langle vac|\mathcal{O}|vac\rangle.$$

If $\mathcal{O}$ is normal ordered as (4.2.7) $\langle\mathcal{O}\rangle$ is nothing but the constant $f_{0,0}$. The correlation function is by definition the vacuum expectation value of a product of spin operators. For example the 2-point correlation function of σ^x_m is $\langle\sigma^x_{m_1}\sigma^x_{m_2}\rangle$.

The calculation of correlation functions is carried out in the following two steps. The first step is to give a normal ordered expression for spin operators. The second step is to apply the Wick's theorem to the product of normal ordered operators. As we shall see in the following sections, the first step reduces to a Wiener-Hopf factorization problem and the second step reduces to a Fredholm integral equation.

4.3 Spin operator

Let us consider σ^x_m with the boundary condition $\sigma^x_m \longrightarrow 1$ for $m \longrightarrow -\infty$ in the phase $\mathcal{R}_2$. This amounts to considering the following product.

(4.3.1) $\quad \bar{\sigma}^x_m = (p_m q_{m-1})(p_{m-1}q_{m-2}) \cdots$

$\bar{\sigma}^x_m$ is characterized up to constant multiple by the following commutation relations.

$$(4.3.2) \qquad \bar{\sigma}^x_m p_{m'} = \begin{cases} p_{m'}\bar{\sigma}^x_m & m' \geq m+1 \\ -p_{m'}\bar{\sigma}^x_m & m' \leq m \end{cases},$$

$$\bar{\sigma}^x_m q_{m'} = \begin{cases} q_{m'}\bar{\sigma}^x_m & m' \geq m \\ -q_{m'}\bar{\sigma}^x_m & m' \leq m-1 \end{cases}.$$

In other words, the inner automorphism of the operator algebra induced by $\bar{\sigma}^x_m$ <u>preserves</u> the orthogonal space of free fermion.

In general, this characteristic property is equivalent to the following characteristic form of operator

$$(4.3.3) \qquad :cw_1 \cdots w_k e^{\rho/2}:,$$

where c is constant, $w_1, \cdots, w_k$ belong to the orthogonal space and ρ is a quadratic form of free fermions. The symbol $:\ :$ means the normal product, namely, in (4.3.3) one should expand the exponential and rewrite it in the normal product form.

$\bar{\sigma}^x_m$ has the following normal product form.

$$(4.3.4) \qquad \bar{\sigma}^x_m = \langle\bar{\sigma}^x_m\rangle : e^{\rho_m/2}:,$$

$$(4.3.5) \qquad \rho_m = \iint \frac{d\theta}{2\pi}\frac{d\theta'}{2\pi} (\psi^\dagger(-\theta)\psi(\theta)) \frac{1}{\sqrt{\omega(\theta)}\sqrt{\omega(\theta')}}$$

$$\times \begin{pmatrix} \omega(\theta)-\omega(\theta') & -\omega(\theta)-\omega(\theta') \\ \omega(\theta)+\omega(\theta') & -\omega(\theta)+\omega(\theta') \end{pmatrix} \begin{pmatrix} \psi^\dagger(-\theta') \\ \psi(\theta') \end{pmatrix} \cdot \frac{e^{i(m-1)(\theta+\theta')}}{1-e^{-i(\theta+\theta'-i0)}}$$

In the next section we shall explain how to derive (4.3.5) from (4.3.2).

<u>4.4 Wiener-Hopf factorization</u>

As a basis of the orthogonal space of free fermions we use $(p(\theta)\frac{d\theta}{2\pi}, q(\theta)\frac{d\theta}{2\pi})$:

$$(4.4.1) \qquad p(\theta) = \sqrt{a(\theta)}^{-1}(\psi^\dagger(-\theta)+\psi(\theta)),$$

$$(4.4.2) \qquad q(\theta) = \sqrt{a(\theta)}(\psi^\dagger(-\theta)-\psi(\theta)).$$

The kernel $T(\theta,\theta')\frac{d\theta'}{2\pi}$ of the orthogonal transformation T in the orthogonal space of free fermions induced by $\bar{\sigma}^x_0$ (see (4.3.2)) is given by

(4.4.3) $$T(\theta,\theta') = \sum_{m\geq 0} e^{im(\theta-\theta')} - \sum_{m\leq -1} e^{im(\theta-\theta')}.$$

We set

(4.4.4) $$P = \frac{1-T}{2}.$$

The operators P and $1-P$ are projection operators with the following properties: If $X_\pm$ is the multiplication operator with the kernel $X_\pm(\theta)\delta(\theta-\theta')d\theta'$ where $X_\pm(\theta)$ is holomorphic for $|e^{\pm i\theta}| < 1$, then we have

(4.4.5) $$PX_+(1-P) = 0, \quad (1-P)X_-P = 0.$$

The projection operator $E_\pm$ of the orthogonal space of free fermions to the subspace of creation operators (resp. annihilation operators) is the multiplication by the following $E_+(\theta)$ (resp. $E_-(\theta)$) in the above basis:

(4.4.6) $$E_\pm(\theta) = \frac{1}{2}\begin{pmatrix} 1 & \pm a(\theta) \\ \pm a(\theta)^{-1} & 1 \end{pmatrix}.$$

We set

(4.4.7) $$E = E_+ - E_-.$$

Let $\begin{pmatrix} R^{++}(\theta,\theta') & R^{+-}(\theta,\theta') \\ R^{-+}(\theta,\theta') & R^{--}(\theta,\theta') \end{pmatrix}$ be the kernel of the quadratic form ρ_0 with respect to the above basis. Noting that

(4.4.8) $$[p(\theta),p(\theta')]_+ = 2.2\pi\delta(\theta+\theta'), [q(\theta),q(\theta')]_+ = -2.2\pi\delta(\theta+\theta'), [p(\theta),q(\theta')]_+ = 0,$$

the formula ( ) is rewritten as

(4.4.9) $$W(1-P+EP) = -2P,$$

where W is the integral operator with kernel $W(\theta,\theta')\frac{d\theta'}{2\pi}$ given by

(4.4.10) $$W(\theta,\theta') = \begin{pmatrix} 2R^{++}(\theta,-\theta') & -2R^{+-}(\theta,-\theta') \\ 2R^{-+}(\theta,-\theta') & -2R^{--}(\theta,-\theta') \end{pmatrix}.$$

If we find $X_\pm(\theta)$, which is holomorphic and invertible for $|e^{\pm i\theta}|<1$, satisfying

(4.4.11) $$X_- = X_+E,$$

a solution W for the integral equation is given by

(4.4.12) $$W = -2X_-^{-1}PX_+.$$

In the present case of (4.4.6) we can take

(4.4.13) $$X_+(\theta) = \begin{pmatrix} b(\theta)^{-1} & \\ & b(\theta) \end{pmatrix},$$

(4.4.14) $$X_-(\theta) = \begin{pmatrix} & b(-\theta)^{-1} \\ b(-\theta) & \end{pmatrix}.$$

Finally changing the basis from $(p(\theta)\frac{d\theta}{2\pi}, q(\theta)\frac{d\theta}{2\pi})$ to $(\psi^\dagger(-\theta)\frac{d\theta}{2\pi}, \psi(\theta)\frac{d\theta}{2\pi})$ we obtain (4.3.5) with $m=0$. The general m case is obtained by the translation.

4.5 Scaling limit

Thus far we have argued the lattice theory. Now let us consider <u>the continuum limit</u>. By this we mean the following limit.

(4.5.1) $$\varepsilon \longrightarrow 0,\ m \longrightarrow \infty,\ \theta \longrightarrow 0 \quad \text{keeping} \quad m\varepsilon = a \quad \text{and} \quad \theta/\varepsilon = p \quad \text{fixed.}$$

In order to obtain non trivial results we should take simultaneously a certain limit to <u>critical values</u> of parameters γ and h. Here we consider the limit

(4.5.2) $$\gamma = 1,\ h = 1-\mu\varepsilon \quad (\varepsilon \to 0,\ \mu\text{: fixed}).$$

We shall list up the <u>scaling limit</u> of several quantities below.

(4.5.3) $$\sqrt{\varepsilon}\, b(\theta) \longrightarrow b(p) = \frac{1}{\sqrt{\mu-ip}},$$

(4.5.4) $$a(\theta) \longrightarrow a(p) = \sqrt{\mu+ip}/\sqrt{\mu-ip},$$

(4.5.5) $$\varepsilon^{-1}\omega(\theta) \longrightarrow \omega(p) = \sqrt{p^2+\mu^2},$$

(4.5.6) $$\sqrt{\varepsilon}\,\psi^\dagger(\theta) \longrightarrow \psi^\dagger(p),$$

(4.5.7) $$\sqrt{\varepsilon}\,\psi(\theta) \longrightarrow \psi(p).$$

The anti-commutation relation (4.1.15) is scaled to

(4.5.8) $$[\psi^\dagger(p),\psi^\dagger(p')]_+ = 0,\ [\psi(p),\psi(p')]_+ = 0,\ [\psi^\dagger(p),\psi(p')]_+ = 2\pi\delta(p-p').$$

(4.5.9) $$\sqrt{\varepsilon}^{-1}p_m \longrightarrow p(a) = \int \frac{dp}{2\pi}\,\sqrt{a(p)}^{-1}(\psi^\dagger(-p)+\psi(p))e^{iap},$$

$$(4.5.10) \qquad \sqrt{\varepsilon}^{\,-1} q_m \longrightarrow q(a) = \int \frac{dp}{2\pi} \sqrt{a(p)}\, (\psi^\dagger(-p)-\psi(p)) e^{iap},$$

$$(4.5.11) \qquad \varepsilon^{-1} H \longrightarrow H = \int \frac{dp}{2\pi} \omega(p)\psi^\dagger(p)\psi(p),$$

$$(4.5.12) \qquad (2\varepsilon)^{-1/8} \bar{\sigma}^x_m \longrightarrow \varphi(a) = :e^{\rho(a)/2}:,$$

$$(4.5.13) \qquad \rho(a) = \iint \frac{dp}{2\pi}\frac{dp'}{2\pi} (\psi^\dagger(-p)\psi(p)) \frac{1}{\sqrt{\omega(p)}\sqrt{\omega(p')}}$$

$$\times \begin{pmatrix} \omega(p)-\omega(p') & -\omega(p)-\omega(p') \\ \omega(p)+\omega(p') & -\omega(p)+\omega(p') \end{pmatrix} \frac{-ie^{ia(p+p')}}{p+p'-i0} \begin{pmatrix} \psi^\dagger(-p') \\ \psi(p') \end{pmatrix} .$$

The commutation relation (4.3.2) is scaled to

$$(4.5.14) \qquad \varphi(a)p(x) = \begin{cases} p(x)\,\varphi(a) & x > a \\ -p(x)\,\varphi(a) & x < a, \end{cases}$$

$$(4.5.14) \qquad \varphi(a)q(x) = \begin{cases} q(x)\,\varphi(a) & x > a \\ -q(x)\,\varphi(a) & x < a. \end{cases}$$

We note that (4.5.13) follows either from (4.3.5) by scaling or from (4.5.14) and (4.5.15) by a similar integral equation as (4.4.9). We note also that $\varphi(a)$ is identical with the (time zero) scaling limit of Ising spin below the critical temperature.

4.6 Operator product and Fredholm equation

Now our main interest is in the operator product $\varphi(a_1)\cdots\varphi(a_n)$ and the correlation function $<\varphi(a_1)\cdots\varphi(a_n)>$.

The quadratic kernel for $\varphi(a)$ was obtained by solving the integral equation (4.4.9), which is a consequence of the commutation relation (4.5.14), (4.5.15).

Let us assume, for a while, that $n = 2$ and $a_1 < a_2$. Then the product $\varphi(a_1)\varphi(a_2)$ satisfy the commutation relation

$$(4.6.1) \qquad \varphi(a_1)\,\varphi(a_2)p(x) = \begin{cases} p(x)\,\varphi(a_1)\,\varphi(a_2) & x < a_1 \text{ or } x > a_2 \\ -p(x)\,\varphi(a_1)\,\varphi(a_2) & a_1 < x < a_2, \end{cases}$$

$$\varphi(a_1)\,\varphi(a_2)q(x) = \begin{cases} q(x)\,\varphi(a_1)\,\varphi(a_2) & x < a_1 \text{ or } x > a_2 \\ -q(x)\,\varphi(a_1)\,\varphi(a_2) & a_1 < x < a_2. \end{cases}$$

From this we can deduce an integral equation similar to (4.4.9). But this time we cannot expect a simple solution like (4.4.12). Instead the solution is characterized by monodromy properties and by a system of linear differential equations. To see this we use a different integral equation from (4.4.9), which we shall explain in this section.

It is natural to enlarge the space of free fermions and the operator algebra as follows.

We introduce copies $\psi^{\dagger(j)}(p)$ and $\psi^{(j)}(p)$ $(j=1,\cdots,2n)$ of creation and annihilation operators, respectively. We assume the following commutation relation

$$(4.6.2)\qquad [\psi^{\dagger(j)}(p),\psi^{\dagger(k)}(p')]_+ = 0,\quad [\psi^{(j)}(p),\psi^{(k)}(p')]_+ = 0,$$

$$[\psi^{(j)}(p),\psi^{\dagger(k)}(p')]_+ = \begin{cases} 2\pi\delta(p-p') & \text{if } |j-k| = n \\ \lambda_{j-n\,k-n}2\pi\delta(p-p') & \text{if } n+1 \le j < k \le 2n \\ 0 & \text{otherwise.}\end{cases}$$

Here λ_{jk} $(1 \le j < k \le n)$ are arbitrary complex parameters and we set

$$(4.6.3)\qquad \Lambda = \begin{pmatrix} 0 & \lambda_{12} & \lambda_{13} & \cdots & \lambda_{1n} \\ 0 & 0 & \lambda_{23} & \cdots & \lambda_{2n} \\ \vdots & \vdots & \vdots & & \vdots \\ 0 & 0 & 0 & \cdots & 0 \end{pmatrix}.$$

For $a_1 < \ldots < a_n$, we denote by φ_j the operator $:e^{\rho^{(j)}/2}:$ with the following quadratic form $\rho^{(j)}$.

$$(4.6.4)\qquad \rho^{(j)} = \iint\frac{dp}{2\pi}\frac{dp'}{2\pi}\,(\psi^{\dagger(j+n)}(-p)\psi^{(j+n)}(p))R^{(j)}(p,p')\begin{pmatrix}\psi^{\dagger(j+n)}(-p') \\ \psi^{(j+n)}(p')\end{pmatrix}.$$

$$(4.6.5)\qquad R^{(j)}(p,p') = \frac{1}{\sqrt{\omega(p)}\,\sqrt{\omega(p')}}\begin{pmatrix}\omega(p)-\omega(p') & -\omega(p)-\omega(p') \\ \omega(p)+\omega(p') & -\omega(p)+\omega(p')\end{pmatrix}\frac{-ie^{ia_j(p+p')}}{p+p'-i0}.$$

We define $2n \times 2n$ matrix kernel $W(p,p') = (W^{\varepsilon\varepsilon'}_{jj'}(p,p'))_{\varepsilon,\varepsilon'=\pm;j,j'=1,\cdots,n}$ by

$$(4.6.6)\qquad W^{++}_{jj'}(p,p') = \langle\psi^{(j')}(-p')\psi^{(j)}(-p)\varphi_1\cdots\varphi_n\rangle/\tau_n\,,$$

$$W^{-+}_{jj'}(p,p') = -\langle\psi^{(j')}(-p')\varphi_1\cdots\varphi_n\psi^{\dagger(j)}(p)\rangle/\tau_n\,,$$

$$W^{+-}_{jj'}(p,p') = \langle\psi^{(j)}(-p')\varphi_1\cdots\varphi_n\psi^{\dagger(j')}(p')\rangle/\tau_n\,,$$

$$W^{--}_{jj'}(p,p') = -\langle\varphi_1\cdots\varphi_n\psi^{\dagger(j)}(p)\psi^{\dagger(j')}(p')\rangle/\tau_n\,,$$

$$(4.6.7)\qquad \tau_n = \langle\varphi_1\cdots\varphi_n\rangle.$$

The quadratic form ρ_n for $\varphi_1\cdots\varphi_n = \tau_n:e^{\rho_n/2}:$ is determined by the above two particle matrix elements:

$$(4.6.8)\quad \rho_n = \sum_{j,j'=1}^{n} \iint \frac{dp}{2\pi}\frac{dp'}{2\pi}(\psi^{\dagger(j+n)}(-p)\ \psi^{(j+n)}(p)) \begin{pmatrix} W^{++}_{jj'}(pp') & W^{+-}_{jj'}(pp') \\ W^{-+}_{jj'}(pp') & W^{--}_{jj'}(pp') \end{pmatrix} \begin{pmatrix} \psi^{\dagger(j'+n)}(-p') \\ \psi^{(j'+n)}(p') \end{pmatrix}.$$

Now we shall write down the integral equation which characterizes the kernel $W(p,p')$. We denote by $R^{(j)}$ the operator with the kernel $R^{(j)}(p,p')\frac{dp'}{2\pi}$. We also denote by K (resp. ${}^tK$) the operator with the kernel $\begin{pmatrix}0&0\\1&0\end{pmatrix}\delta(p+p')dp'$ (resp. $\begin{pmatrix}0&1\\0&0\end{pmatrix}\delta(p+p')dp'$). We set

$$R = \begin{pmatrix} R_1 & & \\ & \ddots & \\ & & R_n \end{pmatrix} \quad\text{and}\quad A = \begin{pmatrix} 0 & \lambda_{12}K & \cdots\cdots & \lambda_{1n}K \\ -\lambda_{12}{}^tK & 0 & \cdots\cdots & \lambda_{2n}K \\ \vdots & \vdots & & \vdots \\ -\lambda_{1n}{}^tK & -\lambda_{2n}{}^tK & \cdots & 0 \end{pmatrix}.$$

Then the operator W with the kernel $W(p,p')\frac{dp'}{2\pi}$ satisfy the following Fredholm integral equation

$$(4.6.9)\qquad (1-RA)W = R.$$

Moreover the correlation function τ_n is nothing but the square root of the Fredholm determinant.

$$(4.6.10)\qquad \tau_n = (\det(1-RA))^{\frac{1}{2}}.$$

4.7 Operator expansion

Before proceeding further we need the asymptotic formula at $p = \infty$ for the products of $\varphi(a)$ and $\psi^\dagger(p)$, $\psi(p)$. Making use of the explicit form (4.5.13) of the kernel, we obtain the following operator expansion at $p = \infty$.

$$(4.7.1)\qquad [\psi^\dagger(-p)\pm\psi(p),\varphi(a)] = 2ip^{\pm\frac{1}{2}-1}e^{-iap}(\varphi^{(\mp)}(a)-\frac{i}{p}\frac{\partial}{\partial a}\varphi^{(\mp)}(a)+\cdots),$$

where the operators $\varphi^{(\pm)}(a)$ is defined by

$$(4.7.2)\qquad \varphi^{(\pm)}(a) =:(\int\frac{dp}{2\pi}\sqrt{\omega(p)}^{\pm1}(\psi^\dagger(-p)\pm\psi(p))e^{iap})e^{\rho(a)/2}:.$$

We note that $\varphi^{(-)}(a)$ is identical with the (time zero) scaling limit of Ising spin avobe the critical temperature.

We also need the formula for the products of $\varphi^{(\pm)}(a)$ and $\psi^\dagger(p)$, $\psi(p)$:

$$(4.7.3)\qquad [\psi^\dagger(-p)\pm\psi(p),\varphi^{(\pm)}(a)]_+ = \pm2p^{\pm\frac{1}{2}}e^{-iap}(\varphi(a)-\frac{i}{p}\frac{\partial}{\partial a}\varphi(a)+\cdots),$$

$$(4.7.4)\qquad [\psi^{\dagger}(-p)\pm\psi(p),\boldsymbol{\varphi}^{(\mp)}(a)]_{+} = 2ip^{\pm\frac{1}{2}-2}e^{-iap}(\tilde{\boldsymbol{\varphi}}^{(\mp)}(a)-\frac{i}{p}\frac{\partial}{\partial a}\tilde{\boldsymbol{\varphi}}^{(\mp)}(a)+\cdots),$$

where the operators $\tilde{\boldsymbol{\varphi}}^{(\pm)}(a)$ are defined by

$$(4.7.5)\qquad \tilde{\boldsymbol{\varphi}}^{(\pm)}(a)=:\left(\int\frac{dp}{2\pi}p\sqrt{\omega(p)}^{\pm1}(\psi^{\dagger}(-p)\pm\psi(p))e^{iap}\right)\left(\int\frac{dp}{2\pi}\sqrt{\omega(p)}^{\pm1}(\psi^{\dagger}(-p)\pm\psi(p))e^{iap}\right)e^{\rho(a)/2}:.$$

Let us explain how to use the above formulas to obtain the asymptotic expansions of $W^{\varepsilon\varepsilon'}_{jj'}(p,p')$. The anti-commutation relation (4.6.2) implies that $\psi^{\dagger(j)}(p)$, $\psi^{(j)}(p)$ commute with $\boldsymbol{\varphi}^{(k)}$ if $j\neq k$. Hence we have $\langle\psi^{(j')}(-p')\psi^{(j)}(-p)\boldsymbol{\varphi}_1\cdots\boldsymbol{\varphi}_n\rangle = \langle\psi^{(j')}(-p')\boldsymbol{\varphi}_1\cdots\psi^{(j)}(-p)\boldsymbol{\varphi}_j\cdots\boldsymbol{\varphi}_n\rangle$, $0 = \langle\psi^{(j')}(-p')\boldsymbol{\varphi}_1\cdots\boldsymbol{\varphi}_n\psi^{(j)}(-p)\rangle = \langle\psi^{(j')}(-p')\boldsymbol{\varphi}_1\cdots\boldsymbol{\varphi}_j\psi^{(j)}(-p)\cdots\boldsymbol{\varphi}_n\rangle$ and $\langle\psi^{(j')}(-p')\psi^{(j)}(-p)\boldsymbol{\varphi}_1\cdots\boldsymbol{\varphi}_n\rangle = \langle\psi^{(j')}(-p')\boldsymbol{\varphi}_1\cdots[\psi^{(j)}(-p),\boldsymbol{\varphi}_j]\cdots\boldsymbol{\varphi}_n\rangle$.

It is convenient to eleminate the p-dependence of $W^{\varepsilon\varepsilon'}_{jj'}(p,p')$ by taking the limit $p\to\infty$. Thus we set

$$(4.7.6)\qquad W^{(1)\varepsilon\varepsilon'}_{jj'}(p')=\lim_{p\to\infty}\frac{\varepsilon p^{1+\varepsilon}}{2i}\sqrt{\omega(p')}\,e^{-ia_jp}(W^{+\varepsilon'}_{jj'}(p,-p')-\varepsilon W^{-\varepsilon'}_{jj'}(p,-p')),$$

$$=\begin{cases}\sqrt{\omega(p')}\langle\psi^{(j')}(p')\boldsymbol{\varphi}_1\cdots\boldsymbol{\varphi}^{(\varepsilon)}_j\cdots\boldsymbol{\varphi}_n\rangle/\tau_n & (\varepsilon'=+)\\ \sqrt{\omega(p')}\langle\boldsymbol{\varphi}_1\cdots\boldsymbol{\varphi}^{(\varepsilon)}_j\cdots\boldsymbol{\varphi}_n\psi^{\dagger(j')}(-p)\rangle/\tau_n & (\varepsilon'=-).\end{cases}$$

In the next section we study the monodromy property of the matrix $W^{(1)}(p) = (W^{(1)\varepsilon\varepsilon'}_{jj'}(p))_{\varepsilon,\varepsilon'=\pm;j,j'=1,\cdots,n}$.

4.8 Monodromy property

The Fredholm integral equation can be solved by using a Neumann series. Taking the limit of (4.7.6) we obtain the following series for $W^{(1)}(p)$.

$$(4.8.1)\qquad W^{(1)\varepsilon\varepsilon'}_{jj'}(p) = e^{-ia_{j'}p}\{\delta_{jj'}(\delta_{1\varepsilon}\omega+\delta_{-1\varepsilon}\varepsilon')$$
$$+\sum_{\ell=1}^{\infty}\sum_{j_2,\dots,j_\ell=1}^{n}\int\frac{dp_1}{2\pi i\omega_1}\cdots\int\frac{dp_\ell}{2\pi i\omega_\ell}\lambda_{j_1j_2}\cdots\lambda_{j_\ell j_{\ell+1}}\times$$
$$\times e^{i(a_{j_1}-a_{j_2})p_1+\cdots+i(a_{j_\ell}-a_{j_{\ell+1}})p_\ell}$$
$$\times\frac{\omega-\varepsilon'\varepsilon_{j_1j_2}\omega_1}{p-p_1+i0}\cdot\frac{\omega_1+\varepsilon_{j_1j_2j_3}\omega_2}{p_1-p_2+i0}\cdots\frac{\omega_{\ell-1}+\varepsilon_{j_{\ell-1}j_\ell j_{\ell+1}}\omega_\ell}{p_{\ell-1}-p_\ell+i0}(\delta_{1\varepsilon}\omega_\ell-\delta_{-1\varepsilon}\varepsilon_{j_\ell j_{\ell+1}}),$$

where we set $\lambda_{jk} = \lambda_{kj}$ $(j > k)$, $\lambda_{jj} = 0$; $\omega = \omega(p)$, $\omega_j = \omega(p_j)$; $\varepsilon_{jk} = \begin{cases} 1 & a_j > a_k, \\ 0 & a_j = a_k \\ -1 & a_j < a_k \end{cases}$
$\varepsilon_{jk\ell} = \varepsilon_{jk}\varepsilon_{k\ell}$ and $j_1 = j'$, $j_{\ell+1} = j$.

Further we denote by $W^{(2)}(p)$ (resp. $W^{(3)}(p)$) the $2n\times 2n$ matrix whose elements are given by (4.8.1) with $p-p_1+i0$ replaced by $p-p_1-i0$ (resp. $p-p_1-i\varepsilon_{j_1j_2}0$). $W^{(1)}(p)$ (resp. $W^{(2)}(p)$) is well-defined in the upper half plane $\mathcal{D}_1$ (resp. in the lower half plane $\mathcal{D}_2$). Since $\omega(p) = \sqrt{p^2+\mu^2}$ has branch points at $p = \pm i\mu$, $W^{(1)}(p)$ (resp. $W^{(2)}(p)$) has a branch point $p = i\mu$ in $\mathcal{D}_1$ (resp. $p = -i\mu$ in $\mathcal{D}_2$). The contour of p_1 integration for $W^{(3)}(p)$ can be deformed into $C_\pm$ according as $\varepsilon_{j_1j_2} \gtrless 0$. Hence $W^{(3)}(p)$ is well-defined in the region $\mathcal{D}_3$.

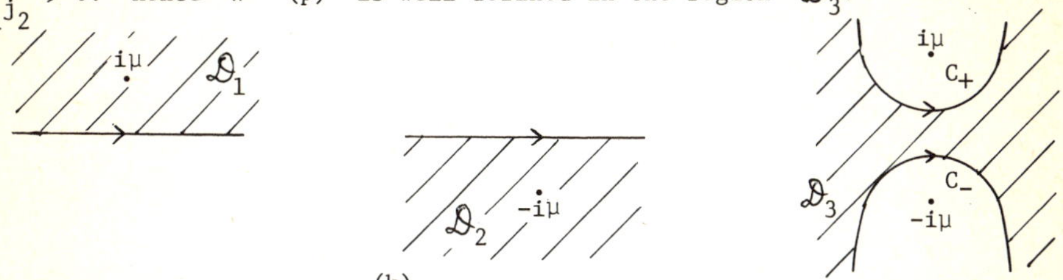

The relations between $W^{(k)}(p)$ $(k=1,2,3)$ is easily seen from (4.8.1) using the identity $\frac{i}{p+i0} + \frac{-i}{p-i0} = 2\pi\delta(p)$.

$$(4.8.2) \qquad W^{(3)}(p) = W^{(1)}(p) \begin{pmatrix} 1 & \\ & 1+2\Lambda \end{pmatrix},$$

$$(4.8.3) \qquad W^{(2)}(p) = W^{(3)}(p) \begin{pmatrix} 1+2\,{}^t\Lambda & \\ & 1 \end{pmatrix}.$$

We set

$$(4.8.4) \qquad Y^{(k)\varepsilon\varepsilon'}_{jj'}(p) = \frac{1}{2}\,(W^{(k)\varepsilon+}_{jj'}(p) + \varepsilon' W^{(k)\varepsilon-}_{jj'}(p)).$$

We note that

$$(4.8.5) \qquad Y^{(1)\varepsilon\pm}_{jj'}(p) = \begin{cases} \pm\frac{\sqrt{\omega(p)}}{2} \langle \varphi_1 \cdots [\psi^{\dagger(j)}(-p) \pm \psi^{(j)}(p), \varphi_{j'}] \cdots \varphi_j^{(\varepsilon)} \cdots \varphi_n \rangle / \tau_n & (j > j') \\ \pm\frac{\sqrt{\omega(p)}}{2} \langle \varphi_1 \cdots [\psi^{\dagger(j)}(-p) \pm \psi^{(j)}(p), \varphi_j^{(\varepsilon)}]_+ \cdots \varphi_n \rangle / \tau_n & (j = j') \\ \mp\frac{\sqrt{\omega(p)}}{2} \langle \varphi_1 \cdots \varphi_j^{(\varepsilon)} \cdots [\psi^{\dagger(j)}(-p) \pm \psi^{(j)}(p), \varphi_{j'}] \cdots \varphi_n \rangle / \tau_n & (j < j'). \end{cases}$$

The relations (4.8.2) and (4.8.3) read as

$$(4.8.6) \qquad Y^{(3)}(p) = Y^{(1)}(p) \begin{pmatrix} 1+\Lambda & -\Lambda \\ -\Lambda & 1+\Lambda \end{pmatrix},$$

$$(4.8.7) \qquad Y^{(2)}(p) = Y^{(3)}(p) \begin{pmatrix} 1+{}^t\Lambda & {}^t\Lambda \\ {}^t\Lambda & 1+{}^t\Lambda \end{pmatrix}.$$

The monodromy property (see 2.1 and 2.2) of $Y^{(k)}$ $(k=1,2,3)$ is explicit if we write them as

(4.8.8) $$Y^{(k)}(p) = \hat{Y}^{(k)}(p)\omega(p)^D e^{A_\infty p}$$

where $D = \begin{pmatrix} 1 & & & & & \\ & \ddots & & & & \\ & & 1 & & & \\ & & & 0 & & \\ & & & & \ddots & \\ & & & & & 0 \end{pmatrix}$ and $A_\infty = \begin{pmatrix} -ia_1 & & & & & \\ & \ddots & & & & \\ & & -ia_n & & & \\ & & & -ia_1 & & \\ & & & & \ddots & \\ & & & & & -ia_n \end{pmatrix}$.

Here $\hat{Y}^{(k)}(p)$ is holomorphic and invertible in $\mathcal{D}_k$ and has the following infinite series expression.

(4.8.9) $$\hat{Y}^{(k)\varepsilon\varepsilon'}_{jj'}(p) = \delta_{jj'}\delta_{\varepsilon\varepsilon'}$$

$$+ \sum_{\ell=1}^{\infty} \sum_{j_2,\cdots j_\ell=1}^{n} \int\frac{dp_1}{2\pi i\omega_1} \cdots \int\frac{dp_\ell}{2\pi i\omega_\ell} \lambda_{j_1j_2}\cdots\lambda_{j_\ell j_{\ell+1}} e^{i(a_{j_1}-a_{j_2})p_1+\cdots+i(a_{j_\ell}-a_{j_{\ell+1}})p_\ell}$$

$$\times \frac{\delta_{1\varepsilon'}-\delta_{-1\varepsilon'}\varepsilon_{j_1j_2}\omega_1}{p-p_1+i\varepsilon^{(k)}_{j_1j_2}0} \cdot \frac{\omega_1+\varepsilon_{j_1j_2j_3}\omega_2}{p_1-p_2+i0} \cdots \frac{\omega_{\ell-1}+\varepsilon_{j_{\ell-1}j_\ell j_{\ell+1}}\omega_\ell}{p_{\ell-1}-p_\ell+i0} (\delta_{1\varepsilon}\omega_\ell-\delta_{-1\varepsilon}\varepsilon_{j_\ell j_{\ell+1}}).$$

where $\varepsilon^{(k)}_{j_1j_2} = \begin{cases} 1 & k=1 \\ -1 & k=2 \\ -\varepsilon_{j_1j_2} & k=3 \end{cases}$.

$Y(p) = Y^{(3)}(p)$ has an irregular singularity at $p = \infty$ and regular singularities at $p = \pm i\mu$.

At $p = \infty$ we take sectors

(4.8.10) $$\mathcal{S}_k : (k-2)\pi < \arg \frac{1}{x} < k\pi \qquad (k=1,2,3)$$

and denote by $Y_k(p)$ the analytic continuation of $Y^{(3)}(p)$ from the real axis. $Y_k(p)$ $(k=1,2,3)$ have the same asymptotic expansion in each sectors. The Stokes multipliers S_k $(k=1,2)$, which is defined so that $Y_{k+1}(p) = Y_k(p)S_k$, is given by

(4.8.11) $$S_1 = \begin{pmatrix} 1+{}^t\Lambda & {}^t\Lambda \\ {}^t\Lambda & 1+{}^t\Lambda \end{pmatrix}\begin{pmatrix} -1 & \\ & 1 \end{pmatrix}\begin{pmatrix} 1+{}^t\Lambda & {}^t\Lambda \\ {}^t\Lambda & 1+{}^t\Lambda \end{pmatrix}^{-1},$$

(4.8.12) $$S_2 = \begin{pmatrix} 1+\Lambda & -\Lambda \\ -\Lambda & 1+\Lambda \end{pmatrix}^{-1}\begin{pmatrix} -1 & \\ & 1 \end{pmatrix}\begin{pmatrix} 1+\Lambda & -\Lambda \\ -\Lambda & 1+\Lambda \end{pmatrix}.$$

The exponents of formal monodromy at $p = \infty$ are $(1,\cdots,1,0,\cdots,0)$.

Let $\gamma_\pm$ the closed path starting from $p = 0$ and encircling $\pm i\mu$ anticlockwise. Let $(\gamma_\pm Y)(p)$ be the analytic continuation of $Y(p)$ along the path $\gamma_\pm$.

Then the monodromy matrices $M_{\pm}$ at $\pm i\mu$, which is defined so that $(\gamma_{\pm}Y)(p)=Y(p)M_{\pm}$, is given by

$$(4.8.13)\qquad M_+ = S_2\ , \quad M_- = S_1.$$

The exponents of formal monodromy at $\pm i\mu$ are $(\frac{1}{2},\cdots,\frac{1}{2},0,\cdots,0)$.

4.9 Deformation theory

In this section we show that the monodromy property of $Y^{(k)}(p)$ leads us to linear and non-linear system of differential equations.

First let us consider $\frac{\partial}{\partial p}Y^{(k)}(p)\cdot Y^{(k)}(p)^{-1}$. Since it does not depend on the species $k=1,2,3$, we shall drop the index k.

The multi-valuedness of $Y(p)$ cancels and $\frac{\partial}{\partial p}Y(p)\cdot Y(p)^{-1}$ is single-valued holomorphic except for $p = \pm i\mu$ and $p = \infty$. The singular part of $\frac{\partial}{\partial p}Y(p)\cdot Y(p)^{-1}$ at $p = \pm i\mu$ is given by

$$(4.9.1)\qquad \frac{\partial}{\partial p}Y(p)\cdot Y(p)^{-1} = \begin{cases} \dfrac{A_+}{p-i\mu}+\cdots & \text{at } p = i\mu \\[2ex] \dfrac{A_-}{p-i\mu}+\cdots & \text{at } p = -i\mu, \end{cases}$$

$$(4.9.2)\qquad A_+ = \hat{Y}^{(1)}(i\mu)\begin{pmatrix} \frac{1}{2} & & & & \\ & \ddots & & & \\ & & \frac{1}{2} & & \\ & & & 0 & \\ & & & & \ddots \\ & & & & & 0 \end{pmatrix}\hat{Y}^{(1)}(i\mu)^{-1},$$

$$(4.9.3)\qquad A_- = \hat{Y}^{(2)}(-i\mu)\begin{pmatrix} \frac{1}{2} & & & & \\ & \ddots & & & \\ & & \frac{1}{2} & & \\ & & & 0 & \\ & & & & \ddots \\ & & & & & 0 \end{pmatrix}\hat{Y}^{(2)}(-i\mu)^{-1}.$$

At $p = \infty$ $\frac{\partial}{\partial p}Y(p)\cdot Y(p)^{-1}$ is of the form

$$(4.9.4)\qquad \frac{\partial}{\partial p}Y(p)\cdot Y(p)^{-1} = A_\infty + O(\frac{1}{p}).$$

From (4.9.1) and (4.9.4) we conclude that $Y(p)$ satisfy the following system of linear ordinary differential equations.

$$(4.9.5)\qquad \frac{\partial}{\partial p}Y(p) = A(p)Y(p),\quad A(x) = \frac{A_+}{p-i\mu} + \frac{A_-}{p+i\mu} + A_\infty.$$

For the later convenience we write down the relations between $A_{\pm}$, A_∞ and the coefficients Y_j $(j=1,2,\cdots)$ of the expansion of $\hat{Y}(p) = 1+\sum_{j=1}^{\infty} Y_j/p^j$ at $p = \infty$.

(4.9.6) $[Y_1, A_\infty] + D = A_+ + A_-,$

(4.9.7) $[Y_2, A_\infty] + [A_\infty, Y_1]Y_1 + [Y_1, D] - Y_1 = i\mu(A_+ - A_-).$

Now let us <u>deform</u> $a_1, \cdots, a_n$ and μ keeping Λ fixed. Namely we deform $Y(p)$ keeping ths Stokes multipliers, connection matrices and the exponents of formal monodromy fixed. This isomonodromy property assures that $dY(p)\cdot Y(p)^{-1}$ is single-valued (d is the exterior differentiation with respect to $a_1, \ldots, a_n$ and μ). Then by a similar argument as above we can derive the following <u>system of linear total differential equations</u>.

(4.9.8) $$dY(p) = \Omega(p)Y(p), \quad \Omega(p) = p\,dA_\infty + [Y_1, dA_\infty] + \left(\frac{-iA_+}{p-i\mu} + \frac{iA_-}{p+i\mu}\right)d\mu.$$

The compatibility condition between (4.9.5) and (4.9.8) is written as

(4.9.9) $$dA(p) = \frac{\partial}{\partial p}\Omega(p) + [\Omega(p), A(p)].$$

Comparing the singular parts at $p = \pm i\mu$ in (4.9.9) we obtain the following <u>system of non linear total differential equations</u> for $A_\pm$.

(4.9.10) $$dA_+ = [d(i\mu A_\infty) + [Y_1, dA_\infty] + A_- d\log\mu, A_+]\ ,$$

(4.9.11) $$dA_- = [d(-i\mu A_\infty) + [Y_1, dA_\infty] + A_+ d\log\mu, A_-]\ .$$

We note that from (4.9.6) $[Y_1, dA_\infty]$ is solved in terms of $A_\pm$.

4.10 Correlation function

In this section we shall give closed expressions for the correlation functions of operators $\varphi(a)$, $\varphi^{(+)}(a)$ and $\varphi^{(-)}(a)$.

The Y_1 term in the expansion of $\hat{Y}(p) = 1 + \sum_{j=1}^{\infty} Y_j/p^j$ at $p = \infty$ has the following operator expression as a consequence of (4.7.1) $\sim$ (4.7.5) and (4.8.5).

(4.10.1) $$Y_1^{++} = \frac{1}{\tau_n}\begin{pmatrix} -i\frac{\partial}{\partial a_1}\langle\varphi_1\varphi_2\cdots\varphi_n\rangle & -i\langle\varphi_1^{(+)}\varphi_2^{(-)}\cdots\varphi_n\rangle & \cdots & -i\langle\varphi_1^{(+)}\varphi_2\cdots\varphi_n^{(-)}\rangle \\ i\langle\varphi_1^{(-)}\varphi_2^{(+)}\cdots\varphi_n\rangle & -i\frac{\partial}{\partial a_2}\langle\varphi_1\varphi_2\cdots\varphi_n\rangle & \cdots & -i\langle\varphi_1\varphi_2^{(+)}\cdots\varphi_n^{(-)}\rangle \\ \vdots & \vdots & & \vdots \\ i\langle\varphi_1^{(-)}\varphi_2\cdots\varphi_n^{(+)}\rangle & i\langle\varphi_1\varphi_2^{(-)}\cdots\varphi_n^{(+)}\rangle & \cdots & -i\frac{\partial}{\partial a_n}\langle\varphi_1\varphi_2\cdots\varphi_n\rangle \end{pmatrix}$$

$$(4.10.2)\qquad Y_1^{-+} = \frac{1}{\tau_n}\begin{pmatrix} 0 & -i\langle\varphi_1^{(-)}\varphi_2^{(-)}\cdots\varphi_n\rangle & \cdots & -i\langle\varphi_1^{(-)}\varphi_2\cdots\varphi_n^{(-)}\rangle \\ i\langle\varphi_1^{(-)}\varphi_2^{(-)}\cdots\varphi_n\rangle & 0 & \cdots & -i\langle\varphi_1\varphi_2^{(-)}\cdots\varphi_n^{(-)}\rangle \\ \vdots & \vdots & & \vdots \\ i\langle\varphi_1^{(-)}\varphi_2\cdots\varphi_n^{(-)}\rangle & i\langle\varphi_1\varphi_2^{(-)}\cdots\varphi_n^{(-)}\rangle & \cdots & 0 \end{pmatrix}$$

$$(4.10.3)\qquad Y_1^{+-} = \frac{1}{\tau_n}\begin{pmatrix} 0 & i\langle\varphi_1^{(+)}\varphi_2^{(+)}\cdots\varphi_n\rangle & \cdots & -i\langle\varphi_1^{(+)}\varphi_2\cdots\varphi_n^{(+)}\rangle \\ -i\langle\varphi_1^{(+)}\varphi_2^{(+)}\cdots\varphi_n\rangle & 0 & \cdots & -i\langle\varphi_1\varphi_2^{(+)}\cdots\varphi_n^{(+)}\rangle \\ \vdots & \vdots & & \vdots \\ -i\langle\varphi_1^{(+)}\varphi_2\cdots\varphi_n^{(+)}\rangle & -i\langle\varphi_1\varphi_2^{(+)}\cdots\varphi_n^{(+)}\rangle & \cdots & 0 \end{pmatrix}$$

$$(4.10.4)\qquad Y_1^{--} = \frac{1}{\tau_n}\begin{pmatrix} -i\frac{\partial}{\partial a_1}\langle\varphi_1\varphi_2\cdots\varphi_n\rangle & i\langle\varphi_1^{(-)}\varphi_2^{(+)}\cdots\varphi_n\rangle & \cdots & i\langle\varphi_1^{(-)}\varphi_2\cdots\varphi_n^{(+)}\rangle \\ -i\langle\varphi_1^{(+)}\varphi_2^{(-)}\cdots\varphi_n\rangle & -i\frac{\partial}{\partial a_2}\langle\varphi_1\varphi_2\cdots\varphi_n\rangle & \cdots & i\langle\varphi_1\varphi_2^{(-)}\cdots\varphi_n^{(+)}\rangle \\ \vdots & \vdots & & \vdots \\ -i\langle\varphi_1^{(+)}\varphi_2\cdots\varphi_n^{(-)}\rangle & -i\langle\varphi_1\varphi_2^{(+)}\cdots\varphi_n^{(-)}\rangle & \cdots & -i\frac{\partial}{\partial a_n}\langle\varphi_1\varphi_2\cdots\varphi_n\rangle \end{pmatrix}$$

First let us consider $\tau_n = \langle\varphi_1\cdots\varphi_n\rangle$. Note that (4.5.8) and (4.5.13) are scale invariant under the scale transformation $a \to c^{-1}a$, $\mu \to c\mu$, $p \to cp$, $\psi^\dagger(p) \to \sqrt{c}\,\psi^\dagger(cp)$ and $\psi(p) \to \sqrt{c}\,\psi(cp)$. Hence τ_n is invariant under the scale transformation $\mu \to c\mu$, $p \to cp$. Thus it is sufficient to consider

$$(4.10.5)\qquad \omega = d\log\tau_n$$

keeping μ fixed. Then from the diagonals of (4.10.1) and (4.10.4) we obtain

$$(4.10.6)\qquad \omega \equiv -\frac{1}{2}\,\mathrm{trace}\; Y_1 dA_\omega \quad \mathrm{mod}\ d\mu.$$

Finally using the diagonal part of (4.9.7) we obtain

$$(4.10.7)\qquad \omega \equiv \frac{1}{2}\,\mathrm{trace}((A_+-A_-)i\mu dA_\infty + \tfrac{1}{2}[Y_1, dA_\infty](A_++A_-)) \quad \mathrm{mod}\ d\mu.$$

Correlation functions containing $\varphi^{(\pm)}$ is calculated as follows: If the number of $\varphi^{(\pm)}$ is two, they are given in (4.10.1) ∿ (4.10.4) divided by τ_n. If the number of $\varphi^{(\pm)}$ is greater than two, they are written as Pfaffians. For example

$$(4.10.8)\quad \langle\varphi_1^{(-)}\varphi_2^{(-)}\varphi_3^{(+)}\varphi_4^{(+)}\rangle/\tau_4 = (\langle\varphi_1^{(-)}\varphi_2^{(-)}\varphi_3\varphi_4\rangle\langle\varphi_1\varphi_2\varphi_3^{(+)}\varphi_4^{(+)}\rangle$$
$$-\langle\varphi_1^{(-)}\varphi_2\varphi_3^{(+)}\varphi_4\rangle\langle\varphi_1\varphi_2^{(-)}\varphi_3\varphi_4^{(+)}\rangle+\langle\varphi_1^{(-)}\varphi_2\varphi_3\varphi_4^{(+)}\rangle\langle\varphi_1\varphi_2^{(-)}\varphi_3^{(+)}\varphi_4\rangle)/\tau_4^2.$$

4.11 Two point function

In this section we show that if $n=2$ the deformation equations (4.9.10) and (4.9.11) reduce to the Painlevé equation of the fifth kind.

First we note that if $n=2$ $S_1=M_-$ and $S_2=M_+$ split into two 2×2 blocks.

$$(4.11.1)\quad S_1 = M_- = \begin{pmatrix} -1 & & & \\ & -1 & 2\lambda & \\ & & 1 & \\ -2\lambda & & & 1 \end{pmatrix},$$

$$(4.11.2)\quad S_2 = M_+ = \begin{pmatrix} -1 & & & 2\lambda \\ & -1 & & \\ & -2\lambda & 1 & \\ & & & 1 \end{pmatrix}.$$

Hence $Y(p)$ itself splits into (1,4) and (2,3) blocks. We set

$$(4.11.3)\quad Z(p) = \begin{pmatrix} Y_{11}(p) & Y_{14}(p) \\ Y_{41}(p) & Y_{44}(p) \end{pmatrix} e^{ia_2 p},$$

and reconstruct the deformation theory in 2×2 matrix form.

It is known (see [3]) that the deformation equations (4.9.10) and (4.9.11) for 2×2 matrices $A_\pm$ reduce to the following fifth Painlevé equation

$$(4.11.4)\quad \frac{d^2y}{dt^2} = \left(\frac{1}{2y}+\frac{1}{y-1}\right)\left(\frac{dy}{dt}\right)^2 - \frac{1}{t}\frac{dy}{dt} + \frac{(y-1)^2}{t^2}\left(\alpha y+\frac{\beta}{y}\right) + \frac{\gamma}{t}y + \frac{\delta y(y+1)}{y-1},$$

by setting

$$(4.11.5)\quad A_+ = \begin{pmatrix} \lambda_1+z & u(\lambda_2-\lambda_1-z) \\ u^{-1}z & \lambda_2-z \end{pmatrix},\quad A_- = \begin{pmatrix} \kappa_1-\lambda_1-z & v(-\kappa_1+\lambda_1+\mu_2+z) \\ v^{-1}(\kappa_1-\lambda_1-\mu_1-z) & \kappa_2-\lambda_2+z \end{pmatrix},$$

$t=2\mu(a_1-a_2)$ and $y=u^{-1}v$. Here (λ_1,λ_2), (μ_1,μ_2) and (κ_1,κ_2) are the exponents of formal monodromy at $p=i\mu$, $-i\mu$ and ∞, respectively. The parameters $\alpha,\beta,\gamma,\delta$

is given by

$$(4.11.6)\qquad \alpha = \frac{1}{2}(\kappa_1-\lambda_1-\mu_2)^2,\quad \beta = -\frac{1}{2}(\kappa_1-\lambda_2-\mu_1)^2,\quad \gamma = -1+\lambda_1-\lambda_2+\mu_1-\mu_2,\quad \delta = -\frac{1}{2}.$$

In the present case $\lambda_1 = \frac{1}{2}$, $\lambda_2 = 0$, $\mu_1 = \frac{1}{2}$, $\mu_2 = 0$, $\kappa_1 = 1$, $\kappa_2 = 0$. Hence we have

$$(4.11.7)\qquad \alpha = \frac{1}{8}\,,\quad \beta = -\frac{1}{8}\,,\quad \gamma = 0\,,\quad \delta = -\frac{1}{2}\,.$$

We note that (4.11.4) is rewritten into the following first order system.

$$(4.11.8)\qquad t\frac{dy}{dt} = -ty-2z(y-1)^2-(y-1)(-\nu_1 y+\nu_2+\nu_3),$$

$$t\frac{dz}{dt} = -yz(\nu_1-z)-y^{-1}(\nu_2-z)(\nu_3-z).$$

where $\nu_1 = \kappa_1-\lambda_1-\mu_2$, $\nu_2 = \lambda_2-\lambda_1$ and $\nu_3 = \kappa_1-\lambda_1-\mu_1$. In the present case $\nu_1 = \frac{1}{2}$, $\nu_2 = -\frac{1}{2}$ and $\nu_3 = 0$.

In terms of y and z $\frac{d}{dt}\log\tau_2$ is given by

$$(4.11.9)\qquad \frac{d}{dt}\log\tau_2 = \frac{z}{2} + \frac{1}{2t}(-2z^2-yz(\frac{1}{2}-z) + y^{-1}z(\frac{1}{2}+z)).$$

If we set $\sigma = 2t\frac{d}{dt}\log\tau_2$, it satisfies the following second order non linear equation.

$$(4.11.10)\qquad (t\frac{d^2\sigma}{dt^2})^2 = (\sigma - t\frac{d\sigma}{dt} + 2(\frac{d\sigma}{dt})^2)^2 + 4(\frac{d\sigma}{dt})^2(\frac{1}{2} - \frac{d\sigma}{dt})(\frac{1}{2} + \frac{d\sigma}{dt}).$$

References

This review is based on

[1] M. Sato, T. Miwa and M. Jimbo, Holonomic quantum fields V, RIMS preprint 267 (1978), to appear in Publ. RIMS, Kyoto Univ.

[2] M. Jimbo, T. Miwa and M. Sato, Studies on holonomic quantum fields XV, Proc. Japan Acad. 55 A (1979) 267-272.

[3] M. Jimbo, T. Miwa, Y. Mori and M. Sato, Density matrix of impenetrable bose gas and the fifth Painlevé transcendent, RIMS preprint 303 (1979), to appear in Physica D.

In [2] the correlation function for the non-linear Schrödinger equation with finite particle density at the infinite coupling is studied in detail. See also

[4] H. G. Vaidya and C. A. Tracy, One particle reduced density matrix of impenetrable bosons in one dimension at zero temperature, J. Math. Phys. 20 (1979), 2291-2312.

RECENT RESULTS FOR THE PLANAR ISING MODEL

by

D. B. ABRAHAM

Theoretical Chemistry Department
University of Oxford
1 South Parks Road
Oxford OX1 3TG

The planar Ising ferromagnet has played a vital role in the theory of phase transitions since the pioneering work of Onsager [1], who showed that the phenomenon of change of phase could be obtained from Gibbsian statistical mechanics without extra ad hoc assumptions. The burgeoning of the study of critical phenomena, largely a result of the work of the King's College group [2] led, amongst other things, to the scaling hypothesis which can be subjected to rigorous tests using extensions of Onsager's original work [3]. It has long been realised that the enhanced fluctuations typical of the critical region occur on a length scale ξ, called the correlation length, which diverges as the temperature $T \to T_c \pm$, T_c being the critical temperature. In the limit as $\xi \to \infty$, the lattice can be replaced by a continuum; the models thus obtained turn out to be Euclidean quantum field theories, one of which, obtained as $T \to T_c -$, has symmetry breaking. In recent work, four separate groups [4,5,6,7] have obtained the n-point functions both for the Ising model and for its associated field theories. Other advances have focused on boundary perturbations in restricted geometries, leading to refinement of scaling theory [8] and of ideas on phase separation and surface tension [9].

In the following, the approach used by the author to investigate the above problems will be given. It relies on the transfer matrix and theory of spinor algebras.

Definition of the planar Ising ferromagnet:

Consider a subset Λ of Z^2 of the form $\Lambda = \{(n,m);\ 1 \leq n \leq N,\ 1 \leq m \leq M\}$. Assign to each point of $\underline{r}$ of Z^2 a classical spin variable $\sigma(\underline{r}) = \pm 1$. A configuration on Λ_{NM} is an assignment of $\sigma(\underline{r})$ for each $\underline{r}$ in Λ, denoted $\{\sigma\}$; the set of all possible configurations is the configuration space denoted $\{-1, 1\}^{NM}$. An energy

$$E(\{\sigma\}) = -\sum_{|i-j|=1} J(i-j)\,\sigma(i)\,\sigma(j) - H\sum_i \sigma(i) - \sum_{i\in\partial\Lambda} H(i)\,\sigma(i) \quad (1)$$

is associated with each configuration; the $J(x)>0$ are called ferromagnetic couplings and the H and $H(i)$ are magnetic fields; $\partial\Lambda$ is the set of perimeter points of Λ. The statistical mechanics of the model is given by the canonical probability measure

$$P\Lambda, B(\Lambda)\ (\{\sigma\}) = Z^{-1} \exp - \beta E(\{\sigma\}) \quad (2)$$

where Z, called the partition function, normalises the measure; the system is supposed to be in equilibrium with a heat bath at temperature T and $\beta = 1/k_B T$ where k_B is Boltzmann's constant. The label $B(\Lambda)$ given specifies the boundary fields $H(i),\ i\varepsilon\partial\Lambda$.

<u>The Transfer Matrix</u>

The essential idea behind the transfer matrix is to factorise the exponential in (2) and to write the sum over spin variables in each column of Λ as a matrix product. The matrix associated with each adjacent pair of columns is called the transfer matrix.

The n-point correlation function for the probability measure (1) is defined in terms of the expectation $<\ >$ by

$$G((\underline{r})_n \mid \Lambda, B(\Lambda)) = <\prod_1^n \sigma(\underline{r}_j)>_{\Lambda,B(\Lambda)} \quad (3)$$

Here $\underline{r}_j = (x_j, y_j)$ in Cartesian form and the notation

$$(\underline{r})_n = (r_1, \ldots . r_n) \quad (4)$$

will be used. It is a standard result that, on a lattice Λ with toroidal boundary conditions denoted $B = t$, (2.1) may be written

$$G((\underline{r})_n \mid \Lambda, t) = \mathrm{Tr}\left\{V^{x_1}\sigma^x(y_1)\,V^{x_2-x_1}\sigma^x(y_2)\ldots \sigma^x(y_{n-1})\,V^{x_n-x_{n-1}}\sigma^x(y_n)\,V^{N-x_n}\right\}\Big/\mathrm{Tr}\,V^N \quad (5)$$

where

$$V = V_2^{\frac{1}{2}}\,V_1\,V_2^{\frac{1}{2}} \quad (6)$$

with

$$V_2 = \exp K_2 \sum_1^M \sigma^x(j)\,\sigma^x(j+1) \quad (7)$$

and

$$V_1 = (2 \sinh 2K_1)^{M/2} \exp - K_1^* \sum_1^M \sigma^z(j) \qquad (8)$$

The parameters in the above are

$$K_j = \beta J_j \ , \ j = 1,2 \qquad (9)$$

and

$$\exp - 2K^* = \tanh K \qquad (10)$$

The * operation in (2.8) is an involution. It is involved in the duality transformation for this model[4]. The spin operators $\sigma^x(j)$ are defined as shown in table 1.

Table 1 : Hilbert space

site in column	1	2 . . .	j	j+1 . . .	M
$\mathfrak{h}_M$	$\mathfrak{h}$	$\otimes\mathfrak{h}$	$\otimes\mathfrak{h}$	. . .	$\otimes\mathfrak{h}$
$\sigma^\alpha(j)$	1	$\otimes 1$ . . .	$\otimes\sigma^\alpha$	$\otimes 1$ . . .	$\otimes 1$
$f^\alpha(j)$	U	$\otimes U$ . . U	$\otimes\sigma^\alpha$	$\otimes 1$ . . .	$\otimes 1$

In the above $\mathfrak{h}$ is a 2-dimensional complex Hilbert space $\alpha = x,y,z$ and

$$\sigma^x = \begin{pmatrix} 0 & 1 \\ 1 & 0 \end{pmatrix} \qquad \sigma^y = \begin{pmatrix} 0 & -i \\ i & 0 \end{pmatrix} \qquad \sigma^z = \begin{pmatrix} 1 & 0 \\ 0 & -1 \end{pmatrix}$$

$$U = -\sigma^z$$

Notice that if we define

$$f^+(j) = (f^x(j) + i\, f^y(j))/2$$

then the $f(j)$ are Fermi operators, that is

$$[\, f(j),\ f(k)\,]_+ = 0, \quad [\, f(j),\ f(k)^\dagger]_+ = \delta_{jk}$$

The basis for each is $\begin{pmatrix} -1 \\ 1 \end{pmatrix}$ and $\begin{pmatrix} 1 \\ 1 \end{pmatrix}$

Table 2 shows the relationship of the steps followed in solving the model.

Table 2 : Organisation of solution

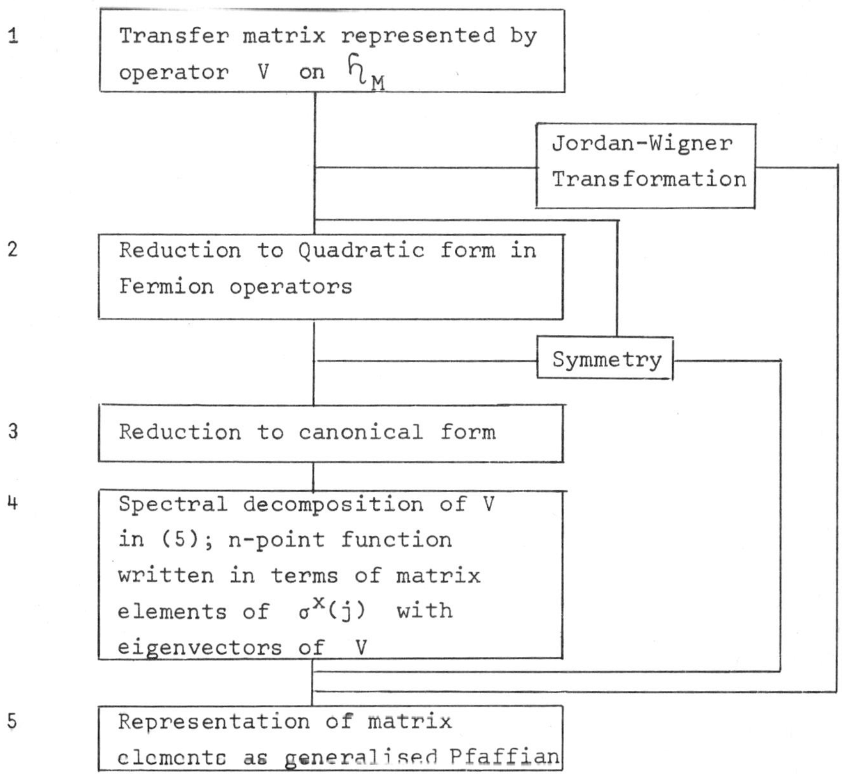

Remarks :

1. The Jordan-Wigner transformation is shown in line 3 of table 1.

2. The symmetry used in step 2 is invariance of V under unitary transformation by

$$P = \exp\left\{ i \frac{\pi}{2} \sum_1^M \sigma^z(j) \right\} \cdot (-1)^M \tag{11}$$

Since $P^2 = 1$ we can project $\mathfrak{h}_M$ on to subspaces $\mathfrak{h}_M(\pm)$ by projectors $P_\pm = (1 \pm P)/2$. Using the usual basis

$$|o\rangle \ , f^+(j)|o\rangle \ , \ . \ . \ . \tag{12}$$

where $f(j)|o\rangle = 0 \ \forall j$, we see that $\mathfrak{h}_M(+)$ $\left(\text{resp. } \mathfrak{h}_M(-)\right)$ contains states with even (resp. odd) number of fermions.

3. <u>2 → 3: Symmetry</u> : here we use the translation defined by

$$\tau\, \sigma^\alpha(j)\, \tau^\dagger = \sigma^\alpha(j-1) \qquad j = 2, \cdots M$$

$$\tau\, \sigma^\alpha(1)\, \tau^\dagger = \sigma^\alpha(M) \tag{13}$$

The non-local character of the Jordan Wigner transformation means that τ may well act on states of (12) in a curious way. Nevertheless, let us define

$$F^\dagger(k) = M^{-\frac{1}{2}} \sum_1^M e^{ikj}\, f^\dagger(j) \tag{14}$$

Then we have the

Lemma:

If

$$\exp ik_j M = (-1)^{r-1} \qquad j = 1, \cdots r \tag{15}$$

then

$$\tau\ F^\dagger(k_1) \quad F^\dagger(k_r)|o\rangle = \exp(ik)\, F^\dagger(k_1) \quad F^\dagger(k_r)\ |o\rangle \tag{16}$$

with

$$k = \sum_1^r k_j \quad (\text{mod } 2\pi) \tag{17}$$

Subremarks:

1. When r is even we use antiperiodic wave numbers k_j.
2. The proof of this is a special case of the implementation of cyclic boundary conditions in the Bethe ansatz.

4. <u>Step 3</u> : Note that from (14) and (15) the $F(k)$ are also Fermi operators in the sense that

$$\left[F(k_1)\ ,\ F(k_2)\right]_+ = 0 \tag{18}$$

$$\left[F(k_1)\ ,\ F^\dagger(k_2)\right]_+ = \delta(k_1, k_2) \tag{19}$$

provided

$$\exp i\, M(k_1 + k_2) = 1$$

But

$$\left[F(k_1)\ ,\ F(k_2)^\dagger\right]_+ = \frac{2}{M(e^{i(k_2 - k_1)} - 1)} \tag{20}$$

if $\exp iM(k_1 + k_2) = -1$. The latter rather subtle effect is central to our development.

5. Step 3 continued :

write

$$V = V_+ P_+ + V_- P_- \tag{21}$$

Then, for instance, with M even

$$V_+ = \overset{<}{>}\Pi_0^\pi V_+(\beta) \tag{22}$$

where $\exp(i\beta M) = -1$ and

$$V_+(\beta) = V_{2+}(\beta)^{\frac{1}{2}} V_{1+}(\beta) V_{2+}(\beta)^{\frac{1}{2}} \tag{23}$$

using (18) and (19). Now

$$V_2(\beta) = \exp 2K_2 \quad \cos\beta \left[F(\beta)^\dagger F(\beta) + F(-\beta)^\dagger F(-\beta) - 1\right]$$
$$- i \sin\beta \left[F(-\beta)^\dagger F(\beta)^\dagger - F(\beta)F(-\beta)\right]$$

$$V_1(\beta) = \exp -2K_1^* \left[F(\beta)^\dagger F(\beta) + F(-\beta)^\dagger F(-\beta) - 1\right] \tag{24}$$

Completing step 3 just requires a canonical transformation of the type

$$G_\pm(k)^\dagger = \cos\theta(k)\, F(k)^\dagger - i \sin\theta(k)\, F(-k) \tag{25}$$

where $\exp(ikM) = \pm 1$ indexes $G(k)^+$ and

$$e^{2i\theta(k)} = \left(\frac{A}{B}\right)^{\frac{1}{2}} e^{i\omega} \left[\frac{(e^{i\omega} - A)(e^{i\omega} - B^{-1})}{(e^{i\omega} - A^{-1})(e^{i\omega} - B)}\right]^{\frac{1}{2}} \tag{26}$$

the correct branch of the square root being such that $2\theta(\pi) = 0$, mod 2π. The parameters A and B are given by

$$A = \coth K_1^* \coth K_2 \qquad B = \coth K_1^* \tanh K_2 \tag{27}$$

Using the notation

$$|(\omega)_n>_\pm = \prod_1^n G_\pm(\omega_j)^\dagger \,|o>_\pm \tag{28}$$

with

$$G_\pm(\omega)\,|o>_\pm = 0 \tag{29}$$

the results in table 3 can be derived

Table 3 : eigenvectors of V

$T > T_c$			$T < T_c$	
$\mathfrak{h}^{(+)}_M$	$\mathfrak{h}^{(-)}_M$		$\mathfrak{h}^{(+)}_M$	$\mathfrak{h}^{(-)}_M$
$\lvert 0\rangle_+$		↓	$\lvert 0\rangle_+$	$\lvert 0\rangle_-$
	$\lvert(\omega)_{2n+1}\rangle_-$		$\lvert(\omega)_{2n}\rangle_+$	$\lvert(\omega)_{2n}\rangle_-$
$\lvert(\omega)_{2n}\rangle_+$		decreasing eigenvalue		

$$\hat{V}\,|(\omega)_n\rangle = \exp\left(-\sum_1^n \gamma(\omega_j)\right)|(\omega)_n\rangle \tag{30}$$

$$\tau\,|(\omega)_n\rangle = \exp\left(i\sum_1^n \omega_j\right)|(\omega)_n\rangle \tag{31}$$

$$\cosh\gamma(\omega) = \cosh 2K_1^* \cosh 2K_2 - \sinh 2K_1^* \sinh 2K_2 \cos\omega \tag{32}$$

Remarks on table 2, continued:

6. Step 4 : Taking the thermodynamic limit $N,M\to\infty$ in (5) gives

$$G((\underline{r})_n\,|t) = \sum_{m_1,\ldots m_{n-1}=0}^{\infty} \int_{-\pi}^{\pi}\!\!\int d(\omega_1)_{m_1}\ldots d(\omega_{n-1})_{m_{n-1}} \prod_1^{n-1} \frac{1}{(2\pi)^{m_j} m_j!}$$

$$F^x((e^{i\omega})_{m_1}) \prod_1^{n-2} F^x((e^{i\omega_j})_{m_j}\ (e^{i\omega_{j+1}})_{m_{j+1}})\ F^x(e^{i\omega_{n-1}})_{m_{n-1}}$$

$$\exp \sum_{\ell=1}^{n-1} \sum_{k=1}^{m_\ell} \left[-\gamma(\omega_{k\ell})(x_{\ell+1}-x_\ell) + i(y_{\ell+1}-y_\ell)\omega_{k\ell}\ \mathrm{sgn}\ \ell\right] \tag{33}$$

where the matrix elements are

$$F^x\left[(\omega)_m|(\omega)_n\right] = {}_\pm\langle(\omega)_m|\sigma^x(1)|(\omega)_n\rangle_\mp \tag{34}$$

Determination of matrix elements

Note that (20) and (25) imply that

$$G_+(\omega) = \frac{-\mathcal{P}}{\pi i} \int_{-\pi}^{\pi} d\omega' \frac{1}{e^{i(\omega-\omega')}-1} \left\{ \left(\frac{\Theta(e^{i\omega'})}{\Theta(e^{i\omega})} + 1 \right) G_-(\omega') + \left(\frac{\Theta(e^{i\omega'})}{\Theta(e^{i\omega})} + 1 \right) G_-(-\omega')^+ \right\} \quad (35)$$

where

$$\Theta(e^{i\omega}) = \exp - 2i\theta(\omega) \quad (36)$$

Using the result that

$$_-\langle(\omega)_n| \; G_+(\omega)|o\rangle_+ = 0 \quad (37)$$

and the notation

$$_-\langle(\omega)_n|o\rangle_+ = F((\omega)_n) \quad (38)$$

we derive the recurrence

$$(YF)(e^{i\omega})_n = \sum_2^n (-1)^j \, h(\omega_1,\omega_j) \, F(\Delta_{1j}(\omega_n)) \quad (39)$$

where

$$h(u,v) = 2 \frac{e^{i(u+v)}}{e^{i(u+v)}-1} \left(\frac{1}{\Theta(e^{iu})\Theta(e^{iv})} - 1 \right) \quad (40)$$

and

$$(YF)(e^{i\omega})_n = \frac{\mathcal{P}}{\pi i} \int_{\pi}^{\pi} d\omega \frac{1}{e^{i(\omega-\omega_1)}-1} \left(1 + \frac{\Theta(e^{i\omega})}{\Theta(e^{i\omega_1})} \right) F(e^{i\omega}, e^{i\omega_2}, \ldots e^{i\omega_n}) \quad (41)$$

Finally, for index sets $I \subset J$,

$$\Delta_I (\omega)_J = (\omega)_{J/I} \quad (42)$$

Thus we have to invert Y ; this can be done using the theory of Wiener-Hopf factorisation and semi-infinite Töplitz forms [4] . To determine $F^x((e^{i\omega})_n|(e^{i\omega})_m)$, use (35) and note the relationship (from table 1)

$$\sigma^x(1) = f(1)^+ + f(1) \tag{43}$$

The results are collected together in table 4.

Table 4 : Matrix elements for V

$T > T_c$:

$$F((e^{i\omega})m|(e^{i\omega})m,2n) = \sum_{2}^{2n} (-1)^j f_>(\omega_1,\omega_j) F(\Delta_{1j}(e^{i\omega})m|(e^{i\omega})m,2n) \tag{44}$$

with

$$f_>(\omega_1,\omega_j) = \frac{e^{i(\omega_1+\omega_j)}}{e^{i(\omega_1+\omega_j)}-1} (g_>(\omega_1)g_>(-\omega_j)+\varepsilon(j)g_>(\omega_j)g_>(-\omega_1)) \tag{45}$$

where

$$\varepsilon(j) = 1 \text{ (resp. } -1) \text{ if } m < j < 2n \text{ (resp. } 1 < j < m) \tag{46}$$

and

$$g_>(\omega) = [(e^{i\omega}-A)(e^{i\omega}-B^{-1})]^{-\frac{1}{2}} \tag{47}$$

boundary condition is

$$F(\phi) = (1 - (\sinh 2K_1 \sinh 2K_2)^2)^{1/8}/\cosh K_1^* \tag{48}$$

$$F^x((e^{i\omega})m|(e^{i\omega})m,2n) = 0 \tag{49}$$

$$F^x((e^{i\omega})m|(e^{i\omega})m,2n+1) = \sum_{j=1}^{2n+1} (-1)^j h_>(\omega_j) F(\Delta_j(e^{i\omega})m|(e^{i\omega})m,2n+1 \tag{50a}$$

$$h_>(\omega) = (B/A)^{\frac{1}{2}} e^{i\omega} / [(e^{i\omega} - A^{-1})(e^{i\omega} - B)]^{\frac{1}{2}} \tag{50b}$$

$T < T_c$:

$$F^x((e^{i\omega})m|(e^{i\omega})m,2n+1) = 0 \tag{51}$$

$$F^x((e^{i\omega})m (e^{i\omega})m,2n) = \sum_{2}^{2n} (-1)^j f_<(\omega_1,\omega_j) F^x(\Delta_{1j}(e^{i\omega})m|(e^{i\omega})m,2n \tag{52}$$

with

$$f_<(\omega_1,\omega_j) = \frac{e^{i(\omega_1+\omega_j)}}{e^{i(\omega_1+\omega_j)}-1} (g_<(\omega_1)g_<(-\omega_j)+\varepsilon(j)g_<(\omega_j)g_<(-\omega_1) \tag{53}$$

$$g_<(\omega) = [(e^{i\omega}-A)/(e^{i\omega}-B)]^{\frac{1}{2}} \tag{54}$$

$$F^x(\phi) = m^* = (1 -(\sinh 2K_1 \sinh 2K_2)^{-2})^{1/8} \tag{55}$$

$\varepsilon(j)$ is given by (46) above.

Graphical Representation

By combining (33) and the results of table 4 a graphical representation of the n-point function can be obtained. Consider (n - 1) disjoint subsets of vertices, denoted V_j, of the vertex set V of a graph $\mathcal{G} = \{V, E\}$. The vertices of V_j are labelled by the integration variables $\omega_{jk} \in [-\pi,\pi]$, $k = 1, \ldots m_j$; notice that $V_j = \phi$ is also allowed. There is an associated vertex weight

$$w_{k\ell} = \frac{1}{2\pi} \exp\left[- \gamma(w_{k\ell})(x_{\ell+1} - x_\ell) + i(y_{\ell+1} - y_\ell)\right] \quad \mathrm{sgn}\, \ell \tag{56}$$

The allowed edge set E can be partitioned in the form

$$E = E_{n-1} \bigcup_{1}^{n-2} (E_j \cup E_{j,j+1}) \tag{57}$$

where the edges of E_j connect vertices in V_j whereas those of $E_{j,j+1}$ connect a vertex in V_j to one in V_{j+1}. Let the terminal vertices of an edge be ω_1 and ω_2. Then the edge weight is given by $e(\omega_1,\omega_2) = f(\omega_1\omega_2)$ from (45) and (53) of table 4, with $\varepsilon(2) = 1$ (resp. -1) if $e \in E_{j,j+1}$ (resp. E_j) for any j .

The rules for drawing the graphs are as follows:

<u>$T < T_c$:</u> $(m_j + m_{j+1})$ for $j = 1,\ldots, n-2$ and m_1 and m_{n-1} must be even; n must be even (for toroidal boundary conditions). Allowed graphs are unions of disjoint cycles such that each and every vertex has valence 2. With + boundary conditions any n is allowed.

<u>$T > T_c$:</u> both $(m_j + m_{j+1})$ for $j = 1,\ldots n-2$ and m_1 and m_{n-1} must be odd; n must be even. Each matrix element produces one termination with weight $h_>(\omega)$. Graphs are unions of disjoint chains and cycles compatible with these rules.

<u>Remarks</u>

1. The integrals over are in general of principal part type ; recalling the Poincaré - Bertrand formula, the order must be specified; given any chain of edges from $E_{j,j+1}$, the principle part integrals are ordered from either end of the chain. This has important consequences for truncation.

2. <u>The scaling limit</u>: As we have seen, each vertex carries a weight $\exp(-x\gamma(\omega) + iy\omega)$ where x and y are relative displacements. Take the formal limit $x,y \to \infty$, $\gamma(o) \to o$ such that $\gamma(o)x = s_1$,

$\gamma(o)y = s_2$ are fixed. Replace the wave number ω by $p = \omega/\gamma(o)$: then $p \ (-\infty,\infty)$ in the limit and the vertex weight is $\exp \ - s_1 (1 + p^2)^{\frac{1}{2}} + i\, p\, s_2$ which is the appropriate Euclidean form for particle mass 1. We can think of the fermions in the theory as particles produced by scattering off the spins.

3. The linked cluster theorem can be applied to obtain an apparently more compact representation of the n-point functions as in [5] , [6] and [7] . A challenging question is to obtain a similar form for truncated n-point functions and thereby to investigate analyticity in the field at zero field for $T > T_c$. The nature of the singularity for $T > T_c$ is still a mystery.

4. Wu at al. [10] obtained a remarkable representation of the two point function in the scaling limit in terms of a Painlevé transcendent of the third kind. This has been generalised notionally by Sato et al. see this volume .

5. The method described here may be applied with trivial modifications to the ground state X - Y chain problem and to the one dimensional Bose gas with infinite δ function repulsion.

References

1. L. Onsager, Phys. Rev. 65, 117 (1944) ; B. Kaufman and L. Onsager, Phys. Rev. 76, 1244 (1949) ; T.D. Schultz, D.C. Mattis and E.H. Lieb, Rev.Mod.Phys. 36, 856 (1964).
2. C. Domb, in Phase Transititions and Critical Phenomena Vol.3, eds. C. Domb and M.S. Green, M.E. Fisher, 1974 : Acedemic Press, London, New York. Lectures in Theoretical Physics VII C, 1964 : Boulder : University of Colorado Press. Rep. Progr. Phys. 30, 615 (1967).
3. C.N. Yang, Phys. Rev. 85, 808, 1952. D.B. Abraham, J. Statist. Phys. 19, 349 (1978) and references therein.
4. D.B. Abraham, Phys. Letts. 61A, 271 (1977) ; Commun. Math. Phys. 59, 17 (1978), 60, 181, 205 (1978).
5. R.Z. Bariev, Phys. Letts. 64A, 169 (1977).
6. B.M. McCoy, C.A. Tracy and T.T. Wu, Phys. Rev. Letts. 38, 793 (1977).
7. See work by Sato, Miwa and Jimbo reported in this volume.
8. M.E. Fisher and H. Au Yang, preprint.
9. D.B. Abraham and P. Reed, Phys. Rev. Letts. 33, 377 (1974) ; Commun. Math. Phys. 49, 35 (1976) ; D.B. Abraham, to be published (1980).
10. E. Barouch, B.M. McCoy, and T.T. Wu, Phys. Rev. Letts. 31, 1409 (1973) ; T.T. Wu, B.M. McCoy, C.A. Tracy and E. Barouch, Phys. Rev. B13, 316 (1976) ; T.T. Wu, C.A. Tracy and B.M. McCoy, J. Math. Phys. 18, 1058 (1977).

Selected Issues from

Lecture Notes in Mathematics

Vol. 662: Akin, The Metric Theory of Banach Manifolds. XIX, 306 pages. 1978.

Vol. 665: Journées d'Analyse Non Linéaire. Proceedings, 1977. Edité par P. Bénilan et J. Robert. VIII, 256 pages. 1978.

Vol. 667: J. Gilewicz, Approximants de Padé. XIV, 511 pages. 1978.

Vol. 668: The Structure of Attractors in Dynamical Systems. Proceedings, 1977. Edited by J. C. Martin, N. G. Markley and W. Perrizo. VI, 264 pages. 1978.

Vol. 675: J. Galambos and S. Kotz, Characterizations of Probability Distributions. VIII, 169 pages. 1978.

Vol. 676: Differential Geometrical Methods in Mathematical Physics II, Proceedings, 1977. Edited by K. Bleuler, H. R. Petry and A. Reetz. VI, 626 pages. 1978.

Vol. 678: D. Dacunha-Castelle, H. Heyer et B. Roynette. Ecole d'Eté de Probabilités de Saint-Flour. VII-1977. Edité par P. L. Hennequin. IX, 379 pages. 1978.

Vol. 679: Numerical Treatment of Differential Equations in Applications, Proceedings, 1977. Edited by R. Ansorge and W. Törnig. IX, 163 pages. 1978.

Vol. 681: Séminaire de Théorie du Potentiel Paris, No. 3, Directeurs: M. Brelot, G. Choquet et J. Deny. Rédacteurs: F. Hirsch et G. Mokobodzki. VII, 294 pages. 1978.

Vol. 682: G. D. James, The Representation Theory of the Symmetric Groups. V, 156 pages. 1978.

Vol. 684: E. E. Rosinger, Distributions and Nonlinear Partial Differential Equations. XI, 146 pages. 1978.

Vol. 690: W. J. J. Rey, Robust Statistical Methods. VI, 128 pages. 1978.

Vol. 691: G. Viennot, Algèbres de Lie Libres et Monoïdes Libres. III, 124 pages. 1978.

Vol. 693: Hilbert Space Operators, Proceedings, 1977. Edited by J. M. Bachar Jr. and D. W. Hadwin. VIII, 184 pages. 1978.

Vol. 696: P. J. Feinsilver, Special Functions, Probability Semigroups, and Hamiltonian Flows. VI, 112 pages. 1978.

Vol. 702: Yuri N. Bibikov, Local Theory of Nonlinear Analytic Ordinary Differential Equations. IX, 147 pages. 1979.

Vol. 704: Computing Methods in Applied Sciences and Engineering, 1977, I. Proceedings, 1977. Edited by R. Glowinski and J. L. Lions. VI, 391 pages. 1979.

Vol. 710: Séminaire Bourbaki vol. 1977/78, Exposés 507–524. IV, 328 pages. 1979.

Vol. 711: Asymptotic Analysis. Edited by F. Verhulst. V, 240 pages. 1979.

Vol. 712: Equations Différentielles et Systèmes de Pfaff dans le Champ Complexe. Edité par R. Gérard et J.-P. Ramis. V, 364 pages. 1979.

Vol. 716: M. A. Scheunert, The Theory of Lie Superalgebras. X, 271 pages. 1979.

Vol. 720: E. Dubinsky, The Structure of Nuclear Fréchet Spaces. V, 187 pages. 1979.

Vol. 724: D. Griffeath, Additive and Cancellative Interacting Particle Systems. V, 108 pages. 1979.

Vol. 725: Algèbres d'Opérateurs. Proceedings, 1978. Edité par P. de la Harpe. VII, 309 pages. 1979.

Vol. 726: Y.-C. Wong, Schwartz Spaces, Nuclear Spaces and Tensor Products. VI, 418 pages. 1979.

Vol. 727: Y. Saito, Spectral Representations for Schrödinger Operators With Long-Range Potentials. V, 149 pages. 1979.

Vol. 728: Non-Commutative Harmonic Analysis. Proceedings, 1978. Edited by J. Carmona and M. Vergne. V, 244 pages. 1979.

Vol. 729: Ergodic Theory. Proceedings 1978. Edited by M. Denker and K. Jacobs. XII, 209 pages. 1979.

Vol. 730: Functional Differential Equations and Approximation of Fixed Points. Proceedings, 1978. Edited by H.-O. Peitgen and H.-O. Walther. XV, 503 pages. 1979.

Vol. 731: Y. Nakagami and M. Takesaki, Duality for Crossed Products of von Neumann Algebras. IX, 139 pages. 1979.

Vol. 733: F. Bloom, Modern Differential Geometric Techniques in the Theory of Continuous Distributions of Dislocations. XII, 206 pages. 1979.

Vol. 735: B. Aupetit, Propriétés Spectrales des Algèbres de Banach. XII, 192 pages. 1979.

Vol. 738: P. E. Conner, Differentiable Periodic Maps. 2nd edition, IV, 181 pages. 1979.

Vol. 742: K. Clancey, Seminormal Operators. VII, 125 pages. 1979.

Vol. 755: Global Analysis. Proceedings, 1978. Edited by M. Grmela and J. E. Marsden. VII, 377 pages. 1979.

Vol. 756: H. O. Cordes, Elliptic Pseudo-Differential Operators – An Abstract Theory. IX, 331 pages. 1979.

Vol. 760: H.-O. Georgii, Canonical Gibbs Measures. VIII, 190 pages. 1979.

Vol. 762: D. H. Sattinger, Group Theoretic Methods in Bifurcation Theory. V, 241 pages. 1979.

Vol. 765: Padé Approximation and its Applications. Proceedings, 1979. Edited by L. Wuytack. VI, 392 pages. 1979.

Vol. 766: T. tom Dieck, Transformation Groups and Representation Theory. VIII, 309 pages. 1979.

Vol. 771: Approximation Methods for Navier-Stokes Problems. Proceedings, 1979. Edited by R. Rautmann. XVI, 581 pages. 1980.

Vol. 773: Numerical Analysis. Proceedings, 1979. Edited by G. A. Watson. X, 184 pages. 1980.

Vol. 775: Geometric Methods in Mathematical Physics. Proceedings, 1979. Edited by G. Kaiser and J. E. Marsden. VII, 257 pages. 1980.